Lecture Notes in Computer Science 5084

Commenced Publication in 1973
Founding and Former Series Editors:
Gerhard Goos, Juris Hartmanis, and Jan van Leeuwen

James F. Peters Andrzej Skowron (Eds.)

Transactions on Rough Sets VIII

Springer

Editors-in-Chief

James F. Peters
University of Manitoba
Department of Electrical and Computer Engineering
Winnipeg, Manitoba, R3T 5V6, Canada
E-mail: jfpeters@ee.umanitoba.ca

Andrzej Skowron
Warsaw University
Institute of Mathematics
Banacha 2, 02-097 Warsaw, Poland
E-mail: skowron@mimuw.edu.pl

Library of Congress Control Number: 2008932437

CR Subject Classification (1998): F.4, F.1, I.2, H.2

LNCS Sublibrary: SL 1 – Theoretical Computer Science and General Issues

ISSN 1861-2059
ISBN-10 3-540-85063-5 Springer Berlin Heidelberg New York
ISBN-13 978-3-540-85063-2 Springer Berlin Heidelberg New York

Springer is a part of Springer Science+Business Media

springer.com

Printed in Germany

Typesetting: Camera-ready by author, data conversion by Scientific Publishing Services, Chennai, India
Printed on acid-free paper SPIN: 12327973 06/3180 5 4 3 2 1 0

Preface

Volume VIII of the *Transactions on Rough Sets (TRS)* contains a wide spectrum of contributions to the theory and applications of rough sets.

The pioneering work by Prof. Zdzisław Pawlak led to the introduction of knowledge representation systems during the early 1970s and the discovery of rough sets during the early 1980s. During his lifetime, he nurtured worldwide interest in approximation, approximate reasoning, and rough set theory and its applications[1]. Evidence of the influence of Prof. Pawlak's work can be seen in the growth in the rough-set literature that now includes over 4000 publications by more than 1900 authors in the rough set database[2] as well as the growth and maturity of the International Rough Set Society[3].

This volume of *TRS* presents papers that introduce a number of new advances in the foundations and applications of artificial intelligence, engineering, logic, mathematics, and science. These advances have significant implications in a number of research areas. In addition, it is evident from the papers included in this volume that rough set theory and its application form a very active research area worldwide. A total of 58 researchers from 11 countries are represented in this volume, namely, Australia, Canada, Chile, Germany, India, Poland, P.R. China, Oman, Spain, Sweden, and the USA. Evidence of the vigor, breadth, and depth of research in the theory and applications rough sets can be found in the articles in this volume.

This volume contains 17 papers that explore a number of research streams. These research streams are represented by papers on rough consequence logics (M.W. Bunder, Mohua Banerjee, and Mihir K. Chakraborty), approximations for incomplete data (Jerzy W. Grzymała-Busse and Wojciech Rzasa), dominance-based rough set approach and multi-criteria decision analysis (Ye Chen, Kevin W. Li, Jason Levy, Keith W. Hipel, and D. Marc Kilgour), partially ordered monads (Patrik Eklund and M. Ángeles Galán), biological control of leafy spurge weed (Mohamed T. Elhadi and Wojciech Ziarko), Pawlak flow graphs and granular computing (Huawen Liu, Jigui Sun, and Huijie Zhang), generalized indiscernibility relations (Wojciech Jaworski), information quanta and approximation operators (Marcin Wolski), reducts and decision rules (Mikhail Ju. Moshkov, Marcin Piliszczuk, and Beata Zielosko), evolutionary rough k-medoid clustering (Georg Peters, Martin Lampart, and Richard Weber), learning complex concepts (Tuan Trung Nguyen), rough set database system representing 3421 publications

[1] See, *e.g.*, Pawlak, Z., Skowron, A.: Rudiments of rough sets. Information Sciences 177, 3–27 (2007); Pawlak, Z., Skowron, A.: Rough sets: Some extensions. Information Sciences 177, 28–40 (2007); Pawlak, Z., Skowron, A.: Rough sets and Boolean reasoning. Information Sciences 177, 41–73 (2007).

[2] `http://rsds.wsiz.rzeszow.pl/rsds.php`

[3] `http://roughsets.home.pl/www/`

and 1913 authors (Zbigniew Suraj and Piotr Grochowalski), attribute reduction and machine learning (Yiyu Yao, Yan Zhao, Jue Wang, and Suqing Han), rough set theory and applications in China (Guoyin Wang, Qingua Zhang, Houkan Huang, Dongyi Ye, Qinghua Hu, Xuegang Hu, Zhongzhi Shi, Yongli Li, Lin Shang, Liping An, Ying Sai, Shanben Chen, Jiye Liang, Keyun Qin, Huanglin Zeng, Keming Xie, Duoqian Miao, Fan Min, Zhaocong Wu, Weizhi Wu, and Jianhua Dai), rough neural computing in classifying power system signals (Liting Han and James F. Peters).

The editors of this volume extend their hearty thanks to the following reviewers: Jan Bazan, Mohua Banerjee, Maciej Borkowski, Cory Butz, Jerzy W. Grzymała-Busse, Mihir K. Chakraborty, Krzysztof Dembczynski, Salvatore Greco, Bożena Kostek, Churn Jung Liau, Pawan Lingras, Dan Lockery, Sushmita Mitra, Mikhail Ju. Moshkov, Son Nguyen, Hoa Nguyen, Piero Pagliani, Uma Shankar, Zbginiew Suraj, Marcin Szczuka, Sheela Ramanna, Dominik Ślęzak, Jarosław Stepaniuk, Piotr Synak, Piotr Wasilewski, Arkadiuz Wojna, Marcin Wolski, Jakub Wróblewski, Wei Zhi Wu, JingTao Yao, Yiyu Yao, and Wojciech Ziarko.

This issue of *TRS* has been made possible thanks to the laudable efforts of a great many generous persons and organizations. The editors and authors of this volume also extend an expression of gratitude to Alfred Hofmann, Ursula Barth, Christine Günther and the LNCS staff at Springer for their support in making this volume of *TRS* possible. In addition, the editors of this volume extend their thanks to Marcin Szczuka for his consummate skill and care in the compilation of this volume.

April 2008

James F. Peters
Andrzej Skowron

LNCS Transactions on Rough Sets

This journal subline has as its principal aim the fostering of professional exchanges between scientists and practitioners who are interested in the foundations and applications of rough sets. Topics include foundations and applications of rough sets as well as foundations and applications of hybrid methods combining rough sets with other approaches important for the development of intelligent systems.

The journal includes high-quality research articles accepted for publication on the basis of thorough peer reviews. Dissertations and monographs up to 250 pages that include new research results can also be considered as regular papers. Extended and revised versions of selected papers from conferences can also be included in regular or special issues of the journal.

Table of Contents

Some Rough Consequence Logics and their Interrelations

Martin W. Bunder[1], Mohua Banerjee[2], and Mihir K. Chakraborty[3]

[1] School of Mathematics and Applied Statistics, University of Wollongong, Wollongong, NSW 2522, Australia
martin_bunder@uow.edu.au

[2] Department of Mathematics and Statistics, Indian Institute of Technology Kanpur 208016, India
mohua@iitk.ac.in

[3] Department of Pure Mathematics, University of Calcutta
35 Ballygunge Circular Road, Kolkata 700019, India
mihirc99@vsnl.com

Abstract. This paper considers a number of alternative rough consequence logics which come in a natural way from the logics L_r and $\mathcal{L}_\mathcal{R}$ previously studied by Chakraborty and Banerjee. The systems have been compared to variants of $S5$, and the logic $Triv$ of Hughes and Cresswell. A comparison has also been made with L_r and $\mathcal{L}_\mathcal{R}$, and therefore with Jaśkowski's discussive logic J, as J is equivalent to $\mathcal{L}_\mathcal{R}$.

1 Introduction

Since the inception of the theory of rough sets (Pawlak [1]), its close link with the system $S5$ of modal logic was apparent. In rough set theory, one has to start with an approximation space which is a set X (of objects) with an equivalence relation R (the indiscernibility of objects by a given collection of attributes). A rough set may be viewed as a triple $\langle X, R, A\rangle$, $A \subseteq X$. The Kripke structure for $S5$ on the other hand, is also a set (of worlds) with an equivalence relation (the accessibility relation). The interpretation of any well-formed formula (wff) in $S5$ is a set in the Kripke structure, viz. the set of worlds where the wff is true. The usual Kripke semantics of a wff in $S5$, thus, turns out to be a rough set. So, $S5$ can be given a rough-set semantics in a natural way.

There is a very significant notion in rough set theory, viz. the notion of *rough equality* - two sets A and B are roughly equal in the approximation space $\langle X, R\rangle$, if they have the same lower and upper approximations relative to $\langle X, R\rangle$. The expression for the corresponding rough equivalence of wffs α and β in the language of $S5$ is $(L\alpha \leftrightarrow L\beta) \wedge (M\alpha \leftrightarrow M\beta)$, L, M, being the necessity and possibility operators respectively. This expression is abbreviated as $\alpha \approx \beta$ (Banerjee and Chakraborty [2]). Not much attention is paid to wffs of the form $\alpha \approx \beta$ in modal logic literature. We know little about properties of such formulae. This fact was noted in Chakraborty and Banerjee [3] and a good deal of development, e.g. the Lindenbaum-like construction on the wffs of $S5$ with $\approx$ instead of the usual

J.F. Peters and A. Skowron (Eds.): Transactions on Rough Sets VIII, LNCS 5084, pp. 1–20, 2008.

bi-implication $\leftrightarrow$ has appeared in subsequent papers (Banerjee and Chakraborty [2], Wasilewska and Vigneron [4]). In $S5$, $\alpha \leftrightarrow \beta$ 'says' that the set of worlds where α holds is the same as the set of worlds where β holds, while $\alpha \approx \beta$ would say that the two sets are roughly equal. The usual modus ponens (MP) rule reads as 'if α and $\alpha \rightarrow \gamma$ are derived then γ may be derived'. In the rough set context (i.e. interpretation), a natural intuition leads to the following.

If $\alpha, \beta \rightarrow \gamma$ and $\alpha \approx \beta$ are all "derivable" from a set Γ of wffs then γ should also be derivable from Γ.

That is, to derive γ, it would be enough to know if α and $\beta \rightarrow \gamma$ are derivable from Γ, where β is a 'close associate' of α, not necessarily exactly α. (So α and β hold in sets of worlds that are *roughly* equal.) One could see that such a rule might be given the name "rough modus ponens" (RMP), which is a generalization of modus ponens. Formally, the consequence relation generated by this kind of modus ponens should be distinct from the standard consequence of $S5$ (henceforth denoted by $\vdash_{S5}$ or by $\vdash$ if there is no confusion). If the former consequence is denoted by the symbol $\mathrel{|\!\!\sim}$ then the above generalised modus ponens may be written as

$$(\mathrm{RMP}) \quad \frac{\Gamma \mathrel{|\!\!\sim} \alpha \qquad \Gamma \mathrel{|\!\!\sim} \beta \rightarrow \gamma \qquad \Gamma \vdash \alpha \approx \beta}{\Gamma \mathrel{|\!\!\sim} \gamma}.$$

This, as well as a seemingly weaker version of it, viz.

$$(MP_{\approx}) \quad \frac{\Gamma \mathrel{|\!\!\sim} \alpha \qquad \Gamma \mathrel{|\!\!\sim} \beta \rightarrow \gamma \qquad \vdash \alpha \approx \beta}{\Gamma \mathrel{|\!\!\sim} \gamma}$$

shall be discussed in this paper.

As $\Gamma \vdash \alpha \approx \alpha$ and $\vdash \alpha \approx \alpha$ hold always, taking β to be the same as α one can immediately see that RMP or $MP_{\approx}$ reduce to the standard MP rule. Some more comments on this aspect are made towards the end of this section.

In fact, thinking of $\alpha \approx \beta$ as a double rough implication $((\alpha \rightsquigarrow \beta) \wedge (\beta \rightsquigarrow \alpha))$, it seems natural that only $\alpha \rightsquigarrow \beta$ should be required in $MP_{\approx}$. It may be observed that $\rightsquigarrow$ would represent the notion of *rough inclusion.* So with

$$\alpha \rightsquigarrow \beta := (M\alpha \rightarrow M\beta) \wedge (L\alpha \rightarrow L\beta),$$

we also consider the rule:

$$(MP_{\rightsquigarrow}) \quad \frac{\Gamma \mathrel{|\!\!\sim} \alpha \quad \Gamma \mathrel{|\!\!\sim} \beta \rightarrow \gamma \quad \vdash \alpha \rightsquigarrow \beta}{\Gamma \mathrel{|\!\!\sim} \gamma}.$$

As a first step towards the study of consequence of the above sort, Banerjee and Chakraborty [3] defined a logic L_r of rough consequence based on $S5$ and the rule

$$(\mathrm{RMP}_1) \quad \frac{\Gamma \mathrel{|\!\!\sim} \alpha \qquad \vdash \beta \rightarrow \gamma \qquad \vdash M\alpha \rightarrow M\beta}{\Gamma \mathrel{|\!\!\sim} \gamma}.$$

It should be stated that in the original version (Chakraborty and Banerjee [3]) two rules were taken, RMP_1 and RMP_2, but subsequently in Bunder [5] and Banerjee [6] it was noticed that the latter follows from the former one.

Note 1. RMP_1 is weaker than $\mathrm{MP}_{\approx}$ and RMP in that $\vdash \beta \rightarrow \alpha$ replaces $\Gamma \mid\!\sim \beta \rightarrow \alpha$, but stronger in that $\vdash M\alpha \rightarrow M\beta$ replaces $\vdash \alpha \approx \beta$.

Note 2. Banerjee and Chakraborty [2] and Bunder [5] have shown that (RMP_1) is equivalent to the simpler rule:

$$(\mathrm{R}_1) \qquad \frac{\Gamma \mid\!\sim \alpha \qquad \vdash M\alpha \rightarrow M\beta}{\Gamma \mid\!\sim \beta}.$$

(RMP_1) gives a semantics (Banerjee and Chakraborty [2]) which is quite different to that intended for RMP.

The logic has been extended recently by Banerjee [6] by adding another rule viz.

$$(\mathrm{R}_2) \qquad \frac{\Gamma \mid\!\sim M\alpha \qquad \Gamma \mid\!\sim M\beta}{\Gamma \mid\!\sim M\alpha \wedge M\beta}.$$

The new logic, which we will call $\mathcal{L}_{\mathcal{R}}$, is capable of capturing the rough logic proposed by Pawlak [7] in 1987. A notion of *rough truth* was introduced in the paper [7]. In modal language, α would be termed roughly true in a domain, provided α 'possibly' holds at every world in it. This was extended to define *rough validity* and *rough semantic consequence* in [3,2], giving rise to a 'rough truth semantics'. In this semantics the rule RMP_1, is accorded a sound interpretation: if α is a rough semantic consequence of Γ, the set of worlds where β holds is always included in the set of worlds where γ holds, and if the set of worlds where α holds is always roughly *upper* included in the set of worlds where β holds then γ is a rough semantic consequence of Γ. RMP_1 being equivalent to R_1, the interpretation becomes simpler: if α is a rough semantic consequence of Γ, and if the set of worlds where α holds is always roughly upper included in the set of worlds where β holds then β is a rough semantic consequence of Γ. In terms of rough truth, R_1 may be equivalently interpreted as: if α is a rough semantic consequence of Γ, and β is roughly true whenever α is, β is also a rough semantic consequence of Γ.

In Banerjee [6], it was observed that the classical conjunction rule $\{\alpha, \beta\} \mid\!\sim \alpha \wedge \beta$ is not sound in the rough truth semantics: even if α and β are both roughly true in a domain, $\alpha \wedge \beta$ may not be so. However, $M\alpha \wedge M\beta$ is roughly true in this situation, i.e. at every world, *independently*, α possibly holds and so does β. So the rule R_2 is sound. R_2, along with R_1, in fact, helps to establish that $\mathcal{L}_{\mathcal{R}}$ is sound and complete with respect to the rough truth semantics. It is also gratifying to find that the logic $\mathcal{L}_{\mathcal{R}}$ turns out to be equivalent to Jaśkowski's discussive logic J [8], proposed long back in 1948 in a different context altogether.

An analysis of the rules RMP and $MP_{\approx}$ gives rise to an interesting class of similar modus ponens rules, that take into account various levels of inclusion between two sets in a rough context. Thus various rough consequences emerge. It may be remarked here that rough logics have been developed from various other angles (e.g. Rasiowa and Skowron [9], or cf. Orłowska [10]), but none have the agendum to reflect 'roughness' within the derivation rule(s). This reflection of roughness may be said to be the essence of the various 'rough' modus ponens rules in the present paper. It will be clear in Section 3, when the rules are specified, that these rules correspond to the natural meaning of the notion of rough equality, rough upper inclusion, rough lower inclusion, rough inclusion and the like.

The aim of this paper is to present a picture of all these consequences (and the corresponding logics, including $S5$ itself and those defined by (R_1) and (R_2)), in a unified manner and to study the interrelations among them.

2 Two Variants of $S5$

We consider two variants of $S5$ (called $S5(1)$ and $S5(2)$) that allow hypotheses.

Definition 1. $S5(1)$:

Axiom Schemes

(Propositional Logic) 1 $\alpha \to \beta \to \alpha$

2 $(\alpha \to \beta \to \gamma) \to (\alpha \to \beta) \to \alpha \to \gamma$

3 $(\sim \alpha \to \sim \beta) \to \beta \to \alpha$

(Modal) 4 $L(\alpha \to \beta) \to L\alpha \to L\beta$

5 $L\alpha \to \alpha$

6 $M\alpha \to LM\alpha$

Rules

Let Γ, Δ be sets of wffs and α a wff.

(Ax) α *is an axiom* $\Rightarrow\ \vdash \alpha$ *(axioms)*

(id) $\alpha \vdash \alpha$ *(identity)*

(wk) $\Gamma \vdash \alpha \ \Rightarrow\ \Gamma, \Delta \vdash \alpha$ *(weakening)*

(N) $\vdash \alpha \ \Rightarrow\ \vdash L\alpha$ *(necessitation)*

(MP) $\Gamma \vdash \alpha,\ \Gamma \vdash \alpha \to \beta \ \Rightarrow\ \Gamma \vdash \beta$ *(modus ponens)*

We say $\Gamma \vdash \alpha$ is derivable in $S5(1)$, if it is obtained by finite applications of the axioms and rules. Sometimes we write $\Gamma \vdash_{S5(1)} \alpha$, if the emphasis is needed.

It may be observed that the *overlap* rule is derivable from the identity (id) and weakening (wk) rules above:

(ov) $\alpha \in \Gamma \ \Rightarrow\ \Gamma \vdash \alpha$ *(overlap)*

This version of $S5$ is equivalent to the one given in Hughes and Cresswell [11], which has, instead of (N):

$$(N1) \qquad \frac{\Gamma, \Delta \vdash \alpha}{\Gamma, L\Delta \vdash L\alpha}$$

where Γ contains only formulas of the form $L\beta$, $\triangle$ none of these, and

$$L\triangle = \{L\beta \mid \beta \in \triangle\}.$$

The formulation is also equivalent to one which has:

$$\delta_1, \ldots, \delta_n \vdash \alpha \quad \Leftrightarrow \quad \vdash \delta_1 \rightarrow \cdots \rightarrow \delta_n \rightarrow \alpha$$

By $\vdash \alpha$ we will mean $\phi \vdash \alpha$.

The other version of $S5$ that we consider, is the following (cf. e.g. Fitting [12]).

Definition 2. $S5(2)$:

This is $S5(1)$ except that (N) is replaced by

$$\text{(N2)} \qquad \Gamma \vdash_{S5(2)} \alpha \quad \Rightarrow \quad \Gamma \vdash_{S5(2)} L\alpha$$

We write $\Gamma \vdash_{S5(2)} \alpha$, if this is derivable by finite applications of the $S5(2)$ axioms and rules.

Note that $S5(1)$, $S5(2)$ and $S5$ are theorem equivalent.

Theorem 1. $\vdash \alpha \Leftrightarrow \vdash_{S5(1)} \alpha \Leftrightarrow \vdash_{S5(2)} \alpha$

Proof. This follows from the fact the (Ax), (ov), (MP) and (N2) can only be used to prove $\vdash_{S5(i)} \alpha$ ($i = 1$ or 2) when Γ is empty. □

We note that $S5$, $S5(1)$ and $S5(2)$ satisfy standard logical properties such as:

Theorem 2. *In $S5(1)$ and $S5(2)$:*

1. *Monotonicity holds.*
   i.e. $\Gamma \vdash \alpha \quad \Rightarrow \quad \Gamma, \triangle \vdash \alpha$
2. *Compactness holds.*
   i.e. $\Gamma \vdash \alpha \quad \Rightarrow \quad (\exists \delta_1, \ldots, \delta_n \in \Gamma)\ \delta_1, \ldots, \delta_n \vdash \alpha$
3. *Cut holds.*
   i.e. $\Gamma, \alpha \vdash \beta\ \&\ \triangle \vdash \alpha \quad \Rightarrow \quad \Gamma, \triangle \vdash \beta$

We can also prove, in the standard way:

Theorem 3. *(Substitution of Equivalence Theorem for $S5$)*
If δ' is the result of replacing zero or more occurrences of α in δ by β then

$$\vdash L(\alpha \leftrightarrow \beta) \rightarrow (\delta \leftrightarrow \delta') \tag{1}$$

Theorem 4. *(The Deduction Theorem for $S5(1)$)*

$$\Gamma, \alpha \vdash_{S5(1)} \beta \quad \Rightarrow \quad \Gamma \vdash_{S5(1)} \alpha \rightarrow \beta$$

The deduction theorem however fails for $S5(2)$. This follows from Theorem 4 and the proof below.

Theorem 5. *S5(2) is strictly stronger than S5(1).*

Proof. Obviously

$$\Gamma \vdash_{S5(1)} \alpha \quad \Rightarrow \quad \Gamma \vdash_{S5(2)} \alpha.$$

$\alpha \vdash_{S5(2)} L\alpha$ follows by (N2). If also $\alpha \vdash_{S5(1)} L\alpha$, by the deduction theorem $\vdash_{S5(1)} \alpha \rightarrow L\alpha$, which is false. □

Note 3. When claiming α is unprovable in $S5(1)$ or $S5(2)$, we use Theorem 1 and the decision procedure for $S5$ from Hughes and Cresswell [11].

Many properties of $S5$ will be used in this paper. The following have proofs in Hughes and Cresswell [11] or can be easily derived. Some are used in the proof of the deduction theorem for $S5(2)$ which follows.

Theorem 6. *The following are valid in* $S5$.

(i) $\vdash M(\alpha \rightarrow \beta) \leftrightarrow (L\alpha \rightarrow M\beta)$
(ii) $\vdash \alpha \rightarrow \beta \Rightarrow \vdash M\alpha \rightarrow M\beta$
(iii) $\vdash \alpha \rightarrow M\alpha$
(iv) $\vdash MM\alpha \leftrightarrow M\alpha$
(v) $\vdash LM\alpha \leftrightarrow M\alpha$
(vi) $\vdash ML\alpha \leftrightarrow L\alpha$
(vii) $\vdash LL\alpha \leftrightarrow L\alpha$
(viii) $\vdash M(\alpha \vee \beta) \leftrightarrow M\alpha \vee M\beta$
(ix) $\vdash L(\alpha \wedge \beta) \leftrightarrow L\alpha \wedge L\beta$
(x) $\vdash M\alpha \rightarrow \beta \Leftrightarrow \vdash \alpha \rightarrow L\beta$
(xi) $\vdash M\alpha \rightarrow M\beta \Leftrightarrow \vdash \alpha \rightarrow M\beta$
(xii) $\vdash L\alpha \rightarrow L\beta \Leftrightarrow \vdash L\alpha \rightarrow \beta$
(xiii) $\vdash L(L\alpha \vee \beta) \leftrightarrow L\alpha \vee L\beta$
(xiv) $\vdash M(M\alpha \wedge \beta) \leftrightarrow M\alpha \wedge M\beta$
(xv) $\vdash L(\alpha \rightarrow L\beta) \leftrightarrow M\alpha \rightarrow L\beta$
(xvi) $\vdash L(M\alpha \rightarrow \beta) \leftrightarrow M\alpha \rightarrow L\beta$

Theorem 7.

$$\triangle, \delta \vdash_{S5(2)} \alpha \tag{2}$$

$$\Leftrightarrow$$

$$\triangle \vdash_{S5(2)} L\delta \rightarrow L\alpha \tag{3}$$

Proof. ($\Rightarrow$): By induction on the derivation of (2).
Case 1. α is an axiom or $\alpha \in \triangle$. By (Ax)/(ov) and (N2), $\vdash_{S5(2)} L\alpha$, so by propositional logic and (wk) we have (3).
Case 2. $\alpha = \delta$. By propositional logic and (wk) we have (3).
Case 3. (2) comes by (N2) from $\triangle, \delta \vdash \beta$, where $\alpha = L\beta$.
By the induction hypothesis we have $\Delta \vdash L\delta \rightarrow L\beta$. By Theorem 6(vii) we obtain (3).
Case 4. (2) comes by (MP) from $\triangle, \delta \vdash_{S5(2)} \beta \rightarrow \alpha$ and $\triangle, \delta \vdash_{S5(2)} \beta$.
By the induction hypothesis:

$$\triangle \vdash_{S5(2)} L\delta \rightarrow \ L(\beta \rightarrow \alpha)$$
$$\text{and} \ \ \triangle \vdash_{S5(2)} L\delta \rightarrow L\beta.$$

As by Axiom 4, $\vdash_{S5(2)} L(\beta \rightarrow \alpha) \rightarrow L\beta \rightarrow L\alpha$, (3) follows.
($\Leftarrow$): By (id), (N2) and (wk),

$$\triangle, \delta \vdash_{S5(2)} L\delta.$$

So if (3) holds, (2) follows by (wk), (MP) and Axiom 5. □

Note 4. (i) Marek and Truszczyński [13] have a similar result on page 200.
(ii) Theorems 1, 7, Axiom 4 and (vii) of Theorem 6 give:

$$\delta_1, \ldots, \delta_n \vdash_{S5(2)} \alpha \ \ \Leftrightarrow \ \ \vdash_{S5} L\delta_1 \rightarrow \ldots \rightarrow L\delta_n \rightarrow L\alpha.$$

So we have a decision procedure for judgements of $S5(2)$ with finite contexts.

3 The "Modus Ponens" (MP) Rules

Looking at the rules $MP_{\approx}$ and $MP_{\rightsquigarrow}$ presented in the Introduction, we check in what possible ways, the $S5$-premisses in these (viz. $\vdash \alpha \approx \beta$, $\vdash \alpha \rightsquigarrow \beta$ respectively) could be changed.

With wffs α, β, one can form the following implications using the connectives L, M:

(i) $L\alpha \rightarrow L\beta$, (ii) $L\alpha \rightarrow \beta$, (iii) $L\alpha \rightarrow M\beta$, (iv) $\alpha \rightarrow L\beta$, (v) $\alpha \rightarrow \beta$, (vi) $\alpha \rightarrow M\beta$, (vii) $M\alpha \rightarrow L\beta$, (viii) $M\alpha \rightarrow \beta$, (ix) $M\alpha \rightarrow M\beta$, (x) $M(\alpha \rightarrow \beta)$, (xi) $L(\alpha \rightarrow \beta)$.

An equivalence relation then emerges:

$\vdash I_1$ if and only if $\vdash I_2$, where I_1, I_2 are any of the above implications.

It is easy to check that we obtain the equivalence classes:

$C_1 : \{(\text{vi}), (\text{ix})\}$, $C_2 : \{(\text{i}), (\text{ii})\}$, $C_3 : \{(\text{v}), (\text{xi})\}$, $C_4 : \{(\text{iv}), (\text{vii}), (\text{viii})\}$, $C_5 : \{(\text{iii}), (\text{x})\}$.

Thus one could consider five representative rules, which we name MP_1, MP_2, MP_3, MP_4 and MP_5, corresponding to the classes $C_1 - C_5$ respectively.

$\underline{MP_1}$
$$\frac{\Gamma \mathrel{|\!\sim} \alpha \quad \Gamma \mathrel{|\!\sim} \beta \rightarrow \gamma \quad \vdash M\alpha \rightarrow M\beta}{\Gamma \mathrel{|\!\sim} \gamma}$$

$\underline{MP_2}$
$$\frac{\Gamma \mathrel{|\!\sim} \alpha \quad \Gamma \mathrel{|\!\sim} \beta \rightarrow \gamma \quad \vdash L\alpha \rightarrow L\beta}{\Gamma \mathrel{|\!\sim} \gamma}$$

$\underline{MP_3}$
$$\frac{\Gamma \mathrel{|\!\sim} \alpha \quad \Gamma \mathrel{|\!\sim} \beta \rightarrow \gamma \quad \vdash \alpha \rightarrow \beta}{\Gamma \mathrel{|\!\sim} \gamma}$$

$\underline{MP_4}$
$$\frac{\Gamma \mathrel{|\!\sim} \alpha \quad \Gamma \mathrel{|\!\sim} \beta \rightarrow \gamma \quad \vdash M\alpha \rightarrow \beta}{\Gamma \mathrel{|\!\sim} \gamma}$$

$$\underline{MP_5} \qquad \frac{\Gamma \mid\!\sim \alpha \quad \Gamma \mid\!\sim \beta \to \gamma \quad \vdash L\alpha \to M\beta}{\Gamma \mid\!\sim \gamma}$$

Observation 1. *Consider the S5-premise of any of the MP_i rules (e.g. $M\alpha \to M\beta$ in MP_1). If α and β are reversed, then taking $\beta = \gamma$ and α as any S5-theorem, we will find that the corresponding MP rule, and Rule (S5) below, lead to triviality: $\Gamma \mid\!\sim \beta$, for any β and any Γ.*

This illustrates that we have all the interesting S5 implications possible in such rules.

Notation. For rules of inference R, R', we write $R \Rightarrow R'$, when R' follows from the assumption that R holds.

Observation 2. $MP_2 \Rightarrow MP_3 \Rightarrow MP_4$, *and* $MP_5 \Rightarrow MP_1$.

Note 5. MP_1, MP_2, MP_3, MP_5, $MP_{\leadsto>}$ and $MP_{\approx}$ all have a special case (when $\alpha = \beta$):

$$\underline{MP_0} \qquad \frac{\Gamma \mid\!\sim \alpha \quad \Gamma \mid\!\sim \alpha \to \gamma}{\Gamma \mid\!\sim \gamma}$$

Observation 3. $MP_2 \Rightarrow MP_{\leadsto>} \Rightarrow MP_{\approx}$.

In the next section, we shall show how logics for the rules $MP_0 - MP_5$, $MP_{\approx}$, $MP_{\leadsto>}$ reduce to four distinct systems, viz. *Triv* (Hughes and Cresswell [11]), $S5(1)$, $S5(2)$ and a seemingly new logic corresponding to MP_4.

Note 6. $M\alpha \to \beta$ is the discursive or discussive implication of Jaśkowski [8] (see also da Costa and Dubikajtis [14]).

The discussive modus ponens rule is the first of two special cases of MP_4:

$$\underline{MP_4(1)} \qquad \frac{\Gamma \mid\!\sim \alpha \quad \Gamma \mid\!\sim M\alpha \to \gamma}{\Gamma \mid\!\sim \gamma}$$

$$\underline{MP_4(2)} \qquad \frac{\Gamma \mid\!\sim \alpha \quad \vdash M\alpha \to \beta}{\Gamma \mid\!\sim \beta}$$

4 The Systems Lr_i

We now define a logic for each rule MP_i, $i = 0, ..., 5, \approx, \leadsto>$.

Definition 3. $\mathbf{Lr_i}$:

Lr_i is the logic based on MP_i and the rules:

(ov) $\qquad \alpha \in \Gamma \;\Rightarrow\; \Gamma \mid\!\sim_i \alpha,$

$(S5)$ $\qquad \vdash \alpha \;\Rightarrow\; \Gamma \mid\!\sim_i \alpha.$

We use $\mid\!\sim_i$ for provability in Lr_i, unless the system being used is clear from the context.

The basic logical properties hold for all Lr_is as they did for $S5(1)$ and $S5(2)$.

Theorem 8. *Monotonicity, Compactness and Cut hold for Lr_i, where $i = 0, ..., 5, \sim>, \approx$.*

Notation. (i) We shall write $L_1 \sim L_2$ for two logics $L_1,\ L_2$, to express that L_1 is equivalent to L_2, i.e. α is derivable from Γ in L_1 if and only if it is derivable from Γ in L_2.
(ii) $L_1 \preceq L_2$ (or, equivalently, $L_2 \succeq L_1$) will denote that L_2 is stronger than L_1, i.e. if α is derivable from Γ in L_1, it is derivable from Γ in L_2, and
(iii) $L_1 \prec L_2$ (or, equivalently, $L_2 \succ L_1$) will denote that L_2 is strictly stronger than L_1, i.e. if α is derivable from Γ in L_1, it is derivable from Γ in L_2, but not conversely.

Note 7. Logics $L_1,\ L_2$ will be termed *independent* of each other, provided there are Γ, α such that α is derivable from Γ in L_1 but not in L_2, and, on the other hand, there are Δ, β such that β is derivable from Δ in L_2 but not in L_1.

Observation 4. *Because of Observation 2, we get $Lr_4 \preceq Lr_3 \preceq Lr_2$, and $Lr_1 \preceq Lr_5$.*

Lr_0, Lr_1, Lr_2, Lr_3, Lr_5, $Lr_{\sim>}$, and $Lr_{\approx}$ all have MP_0, or equivalently, the valid judgement:

$$\alpha,\ \alpha \rightarrow \beta \mid\!\sim_i \beta.$$

L_r and $\mathcal{L}_{\mathcal{R}}$ do not have either, nor, as we will show, does Lr_4.
L_r and $\mathcal{L}_{\mathcal{R}}$ have the deduction theorem

$$\Gamma, \alpha \mid\!\sim \beta \quad \Rightarrow \quad \Gamma \mid\!\sim \alpha \rightarrow \beta.$$

as does $S5(1)$.

We will show that Lr_0, Lr_1, Lr_3 and Lr_5 also have this property. The other systems have a weakened form.

Lemma 1. *For $i = 0, 1, 2, 3, 4, 5, \sim>, \approx$*

$$(i)\quad \frac{\mid\!\sim_i L\alpha}{\mid\!\sim_i \alpha} \qquad\qquad (ii)\quad \frac{\mid\!\sim_i \alpha}{\mid\!\sim_i L\alpha}.$$

Proof. These are straightforward for all systems except Lr_4. For Lr_4:

(i) By Axiom 5 and Theorem 6(vi), $\vdash ML\alpha \rightarrow \alpha$, so if $\mid\!\sim_4 L\alpha$, $\mid\!\sim_4 \alpha$ follows by $MP4(2)$.

(ii) By induction on the derivation of $\mid\!\sim_4 \alpha$.

<u>Case 1.</u> If $\vdash \alpha$ then $\vdash L\alpha$ and so $\mid\!\sim_4 L\alpha$.

Case 2. $\mathrel{|\!\!\sim}_4 \alpha$ comes by MP_4 from $\mathrel{|\!\!\sim}_4 \gamma$, $\mathrel{|\!\!\sim}_4 \beta \rightarrow \alpha$ and $\vdash M\gamma \rightarrow \beta$.

By the induction hypothesis we have $\mathrel{|\!\!\sim}_4 L(\beta \rightarrow \alpha)$. By Axiom 4 and Theorem 6(vi),
$\mathrel{|\!\!\sim}_4 ML(\beta \rightarrow \alpha) \rightarrow L\beta \rightarrow L\alpha$, so by $MP_4(1)$, $\mathrel{|\!\!\sim}_4 L\beta \rightarrow L\alpha$.
From $\vdash M\gamma \rightarrow \beta$ we have $\vdash L(M\gamma \rightarrow \beta)$ by (N) and so, by Axiom 4 and Theorem 6(v), $\vdash M\gamma \rightarrow L\beta$. Now using $\mathrel{|\!\!\sim}_4 \gamma$ and MP_4 we have $\mathrel{|\!\!\sim}_4 L\alpha$. □

One can prove substitution of equivalence theorems for Lr_i.

Theorem 9. *(Substitution of Equivalence for Lr_i with $i = 1, 2, 3, 4, 5, \rightsquigarrow, \approx$)*

If δ' is the result of replacing zero or more occurrences of α in δ by β then

(i) $\mathrel{|\!\!\sim}_i L(\alpha \leftrightarrow \beta) \rightarrow (\delta \leftrightarrow \delta')$

(ii) *If $\mathrel{|\!\!\sim}_i \alpha \leftrightarrow \beta$ then $\mathrel{|\!\!\sim}_i \delta \leftrightarrow \delta'$.*

Proof. (i) By Theorem 3 and $(S5)$.

(ii) If $\mathrel{|\!\!\sim}_i \alpha \leftrightarrow \beta$, by Lemma 1(ii) $\mathrel{|\!\!\sim}_i L(\alpha \leftrightarrow \beta)$ and if MP_0 holds we have

$$\mathrel{|\!\!\sim}_i \delta \leftrightarrow \delta'.$$

In the $i = 4$ case we use Theorem 3 and Theorem 6(vi) to give

$$\vdash ML(\alpha \leftrightarrow \beta) \rightarrow (\delta \leftrightarrow \delta').$$

$\mathrel{|\!\!\sim}_4 L(\alpha \leftrightarrow \beta)$ and $MP_4(2)$ give $\mathrel{|\!\!\sim}_4 \delta \leftrightarrow \delta'$. □

We note two more properties of Lr_i. Let $\alpha \rightarrow_i \beta$ denote the $S5$-implication in MP_i (e.g. $\alpha \rightarrow_1 \beta := M\alpha \rightarrow M\beta$).

Theorem 10. *(i)* $$\frac{\Gamma \mathrel{|\!\!\sim}_i \alpha \qquad \vdash \alpha \rightarrow_i \beta}{\Gamma \mathrel{|\!\!\sim}_i \beta}.$$

(ii) $$\frac{\Gamma,\ \alpha \mathrel{|\!\!\sim}_i \gamma \qquad \vdash \beta \rightarrow_i \alpha}{\Gamma,\ \beta \mathrel{|\!\!\sim}_i \gamma}.$$

Proof. (i) This is MP_i with $\beta = \gamma$.
(ii) By induction on the derivation of

$$\Gamma,\ \alpha \mathrel{|\!\!\sim}_i \gamma. \qquad (4)$$

Case 1. If $\gamma \in \Gamma$ then $\Gamma,\ \beta \mathrel{|\!\!\sim}_i \gamma$.

Case 2. If $\gamma = \alpha$ then $\Gamma,\ \beta \mathrel{|\!\!\sim}_i \beta$, and the result follows by (i), as $\vdash\ \beta \rightarrow_i \alpha$.

<u>Case 3.</u> If $\vdash \gamma$ then $\Gamma,\ \beta \mathrel{|\!\!\sim_i} \gamma$.

<u>Case 4.</u> (4) comes by MP_i from

$$\Gamma,\ \alpha \mathrel{|\!\!\sim_i} \delta$$

$$\Gamma,\ \alpha \mathrel{|\!\!\sim_i} \eta \rightarrow \gamma$$

and $$\vdash\ \delta \rightarrow_i \eta.$$

By the induction hypothesis

$$\Gamma,\ \beta \mathrel{|\!\!\sim_i} \delta$$

and $$\Gamma,\ \beta \mathrel{|\!\!\sim_i} \eta \rightarrow \gamma$$

So the result holds by MP_i. □

Theorem 10 holds for $\approx$ as $\rightarrow_i$.

Let us note a nontrivial rough implication and two rough equivalences.

Lemma 2. *(i)* $\vdash \alpha \leadsto (\alpha \rightarrow L\alpha)$.
(ii) $\vdash (M\alpha \rightarrow \alpha) \approx (\alpha \rightarrow L\alpha)$.

(iii) $\vdash \alpha \approx L\alpha \vee (M\alpha \wedge \sim \alpha)$.

Proof. (i) $\vdash L\alpha \rightarrow (\alpha \rightarrow L\alpha)$, so by (N), Axiom 4 and Theorem 6(vii),

$$\vdash L\alpha \rightarrow L(\alpha \rightarrow L\alpha). \tag{5}$$

By Theorem 6(vi), $\vdash M\alpha \rightarrow (L\alpha \rightarrow ML\alpha)$, and so by Theorem 6(i), $\vdash M\alpha \rightarrow M(\alpha \rightarrow L\alpha)$. Finally by (5), $\vdash \alpha \leadsto \alpha \rightarrow L\alpha$.

(ii) The following equivalences hold in $S5$.
By Theorem 6(xvi) and (xv):
$\vdash L(M\alpha \rightarrow \alpha) \leftrightarrow M\alpha \rightarrow L\alpha$, and $\vdash M\alpha \rightarrow L\alpha \leftrightarrow L(\alpha \rightarrow L\alpha)$.
Also by Theorem 6(i), (v) and (vi):
$\vdash M(M\alpha \rightarrow \alpha) \leftrightarrow M\alpha \rightarrow M\alpha$, $\vdash M\alpha \rightarrow M\alpha \leftrightarrow L\alpha \rightarrow L\alpha$,
$\vdash L\alpha \rightarrow L\alpha \leftrightarrow L\alpha \rightarrow ML\alpha$, and $\vdash L\alpha \rightarrow ML\alpha \leftrightarrow M(\alpha \rightarrow L\alpha)$.
Hence $\vdash (M\alpha \rightarrow \alpha) \approx (\alpha \rightarrow L\alpha)$.

(iii) In $S5$, by Theorem 6(xiii), (ix), (vii) and (v),
$\vdash L(L\alpha \vee (M\alpha \wedge \sim \alpha)) \leftrightarrow L\alpha$.
Also, by Theorem 6(viii), (xiv), (vi), (iv), Axiom 5 and Theorem 6(iii),
$\vdash M(L\alpha \vee (M\alpha \wedge \sim \alpha)) \leftrightarrow M\alpha$.
Hence $\vdash \alpha \approx L\alpha \vee (M\alpha \wedge \sim \alpha)$. □

5 $Lr_1 \sim Lr_5 \sim Triv$

We first prove a few properties of Lr_1.

Theorem 11. *(Deduction Theorem for Lr_1)*

$$\Gamma,\ \delta \vdash_1 \beta \quad \Rightarrow \quad \Gamma \vdash_1 \delta \rightarrow \beta.$$

Proof. By induction on the derivation of

$$\Gamma, \delta \vdash_1 \beta. \tag{6}$$

<u>Case 1.</u> $\beta = \delta$ In this case $\vdash \delta \rightarrow \beta$ and by $(S5)$ $\Gamma \vdash_1 \delta \rightarrow \beta$.

<u>Case 2.</u> $\beta \in \Gamma$ In this case we have

$$MP_0 \dfrac{\Gamma \vdash_1 \beta \qquad (S5)\dfrac{\vdash \beta \rightarrow \delta \rightarrow \beta}{\Gamma \vdash_1 \beta \rightarrow \delta \rightarrow \beta}}{\Gamma \vdash_1 \delta \rightarrow \beta}.$$

<u>Case 3.</u> (6) comes by MP_1 from

$$\Gamma,\ \delta \vdash_1 \alpha$$
$$\Gamma,\ \delta \vdash_1 \gamma \rightarrow \beta$$
and $$\vdash M\alpha \rightarrow M\gamma.$$

The latter gives using Theorem 6 (i), $\vdash M(\delta \rightarrow \alpha) \rightarrow M(\delta \rightarrow \gamma)$.
By the induction hypothesis $\Gamma \vdash_1 \delta \rightarrow \alpha$
and $\Gamma \vdash_1 \delta \rightarrow \gamma \rightarrow \beta$.
Also $\vdash (\delta \rightarrow \gamma \rightarrow \beta) \rightarrow (\delta \rightarrow \gamma) \rightarrow (\delta \rightarrow \beta)$
and so by(S5) andMP_0, $\Gamma \vdash_1 (\delta \rightarrow \gamma) \rightarrow \delta \rightarrow \beta$.
Now byMP_1, $\Gamma \vdash_1 \delta \rightarrow \beta$. □

Lemma 3. *(i)* $\vdash_1 M\alpha \rightarrow \alpha$,
(ii) $\vdash_1 \alpha \rightarrow L\alpha$.

Proof. (i) From $M\alpha \vdash_1 M\alpha$, $M\alpha \vdash_1 \alpha \rightarrow \alpha$, $\vdash MM\alpha \rightarrow M\alpha$ and MP_1 we get $M\alpha \vdash_1 \alpha$. The result follows by Theorem 11.

(ii) By (i) $\vdash_1 M\alpha \rightarrow \alpha$, so as, by Theorem 6(i) and (v), $\vdash M(M\alpha \rightarrow \alpha) \rightarrow M(\alpha \rightarrow L\alpha)$, the result follows by $S5$ and MP_1. □

Theorem 12. *Lr_1 is equivalent to Lr_5.*

Proof. (i) By Observation 4, $Lr_1 \preceq Lr_5$.
(ii) MP_5 is derivable in Lr_1:

Using Lemma 3(ii) and (MP_0),

$$\dfrac{\dfrac{\dfrac{\Gamma \mid\!\sim_1 \alpha \qquad \Gamma \mid\!\sim_1 \alpha \to L\alpha}{\Gamma \mid\!\sim_1 L\alpha} \qquad \Gamma \mid\!\sim_1 \beta \to L\beta \qquad \dfrac{\vdash L\alpha \to M\beta}{\vdash ML\alpha \to M\beta}}{\Gamma \mid\!\sim_1 L\beta}\,(MP_1) \qquad \Gamma \mid\!\sim_1 L\beta \to \beta \qquad \Gamma \mid\!\sim_1 \beta \to \gamma}{\Gamma \mid\!\sim_1 \gamma}$$

□

We recall that the *Trivial* system (Hughes and Cresswell [11]) $Triv$ comprises the classical propositional logic axioms, the modal axiom $L\alpha \leftrightarrow \alpha$, and the rule $\vdash \alpha,\ \vdash \alpha \to \beta \ \Rightarrow\ \vdash \beta$.

Theorem 13. *Lr_1 and Lr_5 are equivalent to $Triv$.*

Proof. By Lemma 3(ii) and Theorem 12, $\mid\!\sim \alpha \leftrightarrow L\alpha$, both in Lr_1 and Lr_5. All the $S5$ axioms and rules are available in these two systems as well, and so the classical propositional logic axioms and rules are inherited too. Thus $Triv \preceq Lr_1 \sim Lr_5$. For the converse, we note that the theorems and valid rules of Lr_1 (Lr_5) can be transformed using $\vdash \alpha \leftrightarrow L\alpha$ to the postulates of $Triv$. Hence the systems are equivalent. □

6 $Lr_2 \sim Lr_{\rightsquigarrow} \sim Lr_{\approx} \sim S5(2) \prec Lr_1$

Theorem 14. *Lr_2, $Lr_{\rightsquigarrow}$, $Lr_{\approx}$ and $S5(2)$ are equivalent.*

Proof. We first show that MP_2, $MP_{\rightsquigarrow}$, $MP_{\approx}$ are admissible in $S5(2)$.

If we have $\Gamma \vdash_{S5(2)} \alpha$ and $\Gamma \vdash_{S5(2)} \beta \to \gamma$, we have, by compactness subsets $\{\delta_1, \ldots, \delta_i\}$ and $\{\delta_j, \ldots, \delta_k\}$ of Γ such that

$$\delta_1, \ldots, \delta_i \vdash_{S5(2)} \alpha$$
$$\text{and} \quad \delta_j, \ldots, \delta_k \vdash_{S5(2)} \beta \to \gamma$$

and so by monotonicity, for $n = max(i, k)$,

$$\delta_1, \ldots, \delta_n \vdash_{S5(2)} \alpha$$

$$\delta_1, \ldots, \delta_n \vdash_{S5(2)} \beta \to \gamma.$$

Then by Theorem 7 and Axiom 4,

$$\vdash_{S5(2)} L\delta_1 \to \ldots \to L\delta_n \to L\alpha$$

and $\quad \vdash_{S5(2)} L\delta_1 \to \ldots \to L\delta_n \to L\beta \to L\gamma.$

If also $\vdash L\alpha \to L\beta$ (which follows from $\vdash \alpha \rightsquigarrow \beta$ as well as $\vdash \alpha \approx \beta$), we have

$$\vdash_{S5(2)} L\delta_1 \to \ldots \to L\delta_n \to L\gamma$$

and by Theorem 7 and monotonicity:

$$\Gamma \vdash_{S5(2)} \gamma.$$

Now we prove ($N2$) in Lr_2, $Lr_{\rightsquigarrow}$ and $Lr_{\approx}$. Let $\mathrel{|\!\sim}$ denote any of $\mathrel{|\!\sim}_2$, $\mathrel{|\!\sim}_{\rightsquigarrow}$, $\mathrel{|\!\sim}_{\approx}$. If $\Gamma \mathrel{|\!\sim} \alpha$ then using $\vdash \alpha \rightarrow (M\alpha \rightarrow \alpha)$ and $\vdash \alpha \rightarrow (\alpha \rightarrow L\alpha) \rightarrow L\alpha$ we can prove: $\Gamma \mathrel{|\!\sim} M\alpha \rightarrow \alpha$ and $\Gamma \mathrel{|\!\sim} (\alpha \rightarrow L\alpha) \rightarrow L\alpha$.

Then by Lemma 2(ii), from which follows $\vdash L(M\alpha \rightarrow \alpha) \rightarrow L(\alpha \rightarrow L\alpha)$ and $\vdash (M\alpha \rightarrow \alpha) \rightsquigarrow (\alpha \rightarrow L\alpha)$, we have $\Gamma \mathrel{|\!\sim} L\alpha$.

As MP_0 holds in Lr_2, $Lr_{\rightsquigarrow}$, $Lr_{\approx}$, the systems are all equivalent. □

Theorem 15. *Lr_1 is strictly stronger than Lr_2.*

Proof. Lr_1 clearly contains all of $S5(2)$ and so Lr_2. We have $\mathrel{|\!\sim}_1 \alpha \rightarrow L\alpha$ by Lemma 3(ii), but $\nvdash \alpha \rightarrow L\alpha$ and so, by Theorem 1, $\not\mathrel{|\!\sim}_2 \alpha \rightarrow L\alpha$. Hence Lr_1 is strictly stronger than Lr_2. □

7 $Lr_4 \prec Lr_3 \sim Lr_0 \sim S5(1) \prec Lr_2$

Theorem 16. *Lr_0 is equivalent to Lr_3.*

Proof. MP_0 holds in Lr_3. We now show that MP_3 holds in Lr_0.

$$MP_0\ \dfrac{\Gamma \mathrel{|\!\sim}_0 \alpha \qquad (S5)\dfrac{\vdash \alpha \rightarrow \beta}{\Gamma \mathrel{|\!\sim}_0 \alpha \rightarrow \beta}}{MP_0\ \dfrac{\Gamma \mathrel{|\!\sim}_0 \beta \qquad\qquad \Gamma \mathrel{|\!\sim}_0 \beta \rightarrow \gamma}{\Gamma \mathrel{|\!\sim}_0 \gamma}}$$

□

Theorem 17. *$S5(1)$ is equivalent to Lr_0.*

Proof. The postulates of $S5(1)$ hold in Lr_0 by definition of the system, and in particular, due to Theorem 8 and Lemma 1(ii). Clearly, the postulates of Lr_0 hold in $S5(1)$. □

Corollary 1. *$S5(1)$ is equivalent to Lr_3.*

In view of Theorems 1, 14 and Corollary 1:

Theorem 18. *Lr_2, Lr_3 and $S5$ have the same theorems.*

The following Lemma allows us to show that Lr_0 is strictly stronger than Lr_4.

Lemma 4. *If*

$$\delta_1, \ldots, \delta_n \mathrel{|\!\sim}_4 \alpha, \tag{7}$$

then for some i, $1 \leq i \leq n$

$$\vdash \delta_i \rightarrow M\delta_1 \rightarrow \ldots \rightarrow M\delta_{i-1} \rightarrow M\delta_{i+1} \ldots \rightarrow M\delta_n \rightarrow \alpha. \tag{8}$$

Proof. By induction on the derivation of (7).
Case 1. $\vdash \alpha$. Then (8) follows.

Case 2. $\alpha = \delta_i$. (8) follows.

Case 3. (7) comes by MP_4 from:

$$\delta_1, \ldots, \delta_n \mathrel{|\!\!\sim_4} \beta,$$

and

$$\delta_1, \ldots, \delta_n \mathrel{|\!\!\sim_4} \gamma \to \alpha$$

$$\vdash M\beta \to \gamma. \tag{9}$$

By the induction hypothesis, for some i and j:

$$\vdash \delta_j \to \ M\delta_1 \to \ldots \to M\delta_{j-1} \to M\delta_{j+1} \to \ldots \to M\delta_n \to \beta$$

$$\vdash \delta_i \to \ M\delta_1 \to \ldots \to M\delta_{i-1} \to M\delta_{i+1} \to \ldots \to M\delta_n \to \gamma \to \alpha \tag{10}$$

Using Theorem 6(ii), (i) and (v):
$\vdash M\delta_j \to \ M\delta_1 \to \ldots \to M\delta_{j-1} \to M\delta_{j+1} \to \ldots \to M\delta_n \to M\beta.$
So by (9)
$\vdash M\delta_j \to \ M\delta_1 \to \ldots \to M\delta_{j-1} \to M\delta_{j+1} \to \ldots \to M\delta_n \to \gamma,$
and by (10) and Theorem 6(iii) we obtain (8). □

Note 8. (i) Writing $\alpha \Rightarrow \beta$ for the discussive implication $M\alpha \to \beta$, this Lemma gives:

$$\delta_1, \ldots, \delta_n \mathrel{|\!\!\sim_4} \alpha \quad \Rightarrow \quad \vdash \delta_i \to \delta_1 \Rightarrow \ldots \delta_{i-1} \Rightarrow \delta_i \Rightarrow \ldots \delta_n \Rightarrow \alpha.$$

(ii) As $\vdash \delta_i \to \ldots$ implies $\mathrel{|\!\!\sim_4} \delta_i \to \ldots$, the Lemma is also a form of deduction theorem for Lr_4.

Theorem 19. *Lr_3 is strictly stronger than Lr_4.*

Proof. MP_4 is clearly valid in Lr_0. We have:

$$p \to q, \ \ p \mathrel{|\!\!\sim_0} q,$$

but not $\vdash (p \to q) \to Mp \to q$
nor $\vdash p \to M(p \to q) \to q$.
So by Lemma 4,

$$p \to q, \ \ p \not\mathrel{|\!\!\sim_4} q.$$

□

Using Theorems 16, 17, 5 and 14, we thus obtain

Theorem 20. $Lr_4 \prec Lr_3 \sim Lr_0 \sim S5(1) \prec Lr_2$.

8 The Banerjee and Chakraborty Systems, Jaśkowski's J and the Systems Lr_i

Jaśkowski's *discussive logic* J is given (see da Costa and Doria [15]) by:

Definition 4. J $\Gamma \vdash_J \alpha \Leftrightarrow M\Gamma \vdash_{S5(1)} M\alpha.$

We have in Banerjee [6]

Theorem 21. *J is equivalent to $\mathcal{L}_{\mathcal{R}}$.*

Further,

Theorem 22. *L_r is strictly weaker than $\mathcal{L}_{\mathcal{R}}$.*

Proof. $\alpha,\ M\alpha \to \beta \vdash_J \beta$, but $p, Mp \to q \not\vdash_r q$. The latter requires (see Bunder [5]) $\vdash Mp \to Mq$ or $\vdash M(Mp \to q) \to Mq$, both of which fail. □

Theorem 23. *Lr_1 is strictly stronger than J.*

Proof. From Theorem 13 it follows that Lr_1 is stronger than or equivalent to J.

$$p,\ p \to q \mathrel{|\!\!\sim}_1 q$$

holds but

$$p,\ p \to q \not\vdash_J q$$

as

$$Mp,\ M(p \to q) \not\vdash_{S5(1)} Mq.$$

□

Theorem 24. *J is strictly stronger than Lr_4.*

Proof. MP_4 is derivable in J:
$\Gamma \vdash_J \alpha$, $\Gamma \vdash_J \beta \to \gamma$ and $\vdash M\alpha \to \beta$ imply, by definition of $\vdash_J$, $M\Gamma \vdash M\alpha$, $M\Gamma \vdash M(\beta \to \gamma)$. But the last gives, by Theorem 6(i), $M\Gamma \vdash L\beta \to M\gamma$. Also, $\vdash M\alpha \to L\beta$, using rule (N) and Axiom 6 of $S5$ on $\vdash M\alpha \to \beta$. Thus, $M\Gamma \vdash M\alpha \to M\gamma$, and as $M\Gamma \vdash M\alpha$, we get $M\Gamma \vdash M\gamma$, i.e. $\Gamma \vdash_J \gamma$.
We note that $M\alpha \vdash_J \alpha$, but $Mp \not\vdash p$. As $Lr_4 \prec S5(1)$ (Theorem 20), $Mp \not\mathrel{|\!\!\sim}_4 p$.

Theorem 25. *L_r and J are both independent of each of Lr_2 and Lr_3.*

Proof. Given Theorems 22, 21 and 20, all we require is to show:

(i) there is α such that $\mathrel{|\!\!\sim}_r \alpha$, but $\not\mathrel{|\!\!\sim}_2 \alpha$, and
(ii) there is α such that $\vdash \alpha$, but $\not\vdash_J \alpha$.

(i) We have by Theorem 6(v) and (i),

$$\vdash M(M\alpha \to \alpha),$$

and so

$$\mid\!\sim_r M\alpha \rightarrow \alpha.$$

But

$$\nvdash Mp \rightarrow p,$$

and so

$$\not\mid\!\sim_{S5(2)} Mp \rightarrow p,$$

i.e.

$$\not\mid\!\sim_2 Mp \rightarrow p.$$

(ii) As demonstrated for Theorem 23,

$$p,\ p \rightarrow q\ \vdash\ q$$

holds but

$$p,\ p \rightarrow q \nvdash_J q.$$

□

Theorem 26. *L_r is independent of Lr_4.*

Proof.

$$p, Mp \rightarrow q \mid\!\sim_4 q$$

($MP_4(1)$), but as noted in the proof of Theorem 22,

$$p, Mp \rightarrow q \not\mid\!\sim_r q.$$

As $Lr_4 \prec Lr_2$, we get the other side through part (i) in the proof of Theorem 25. □

9 Some Extended Systems

9.1 The + Systems

Definition 5. $\mathbf{Lr}_\mathbf{i}^+$:

The system Lr_i^+, where $i = 1, 2, 3, 4, \rightsquigarrow$ or $\approx$, is the system Lr_i with the $S5$-premise in MP_i modified to include Γ on the left of the $\vdash$.

The modified MP_i will be called MP_i^+ and the modified $\mid\!\sim_i$, $\mid\!\sim_i^+$.

As $\dfrac{\vdash \gamma}{\Gamma \vdash \gamma}$ holds each MP_i follows from MP_i^+ and so:

Theorem 27. *Each Lr_i^+ is stronger than or equivalent to Lr_i.*

Theorem 28. *Lr_1^+ and Lr_5^+ are equivalent to Lr_1.*

Proof. $\vdash \alpha \leftrightarrow L\alpha$ and the substitution of equivalence rule will be provable in Lr_1^+, so any valid judgement $\Gamma \mathrel{|\!\!\sim}_1^+ \beta$ can be transformed into a valid judgement of classical propositional logic. Thus by Theorem 13, $\Gamma \mathrel{|\!\!\sim}_1 \beta$ is valid as well. Similarly for Lr_5^+. □

Theorem 29. *$Lr_2^+, Lr_{\approx}^+$ and $Lr_{\rightsquigarrow}^+$ are all equivalent to Lr_2.*

Proof. We show that MP_2^+ is derivable in Lr_2. It then follows that $MP_{\rightsquigarrow}^+$ and $MP_{\approx}^+$ (rule (RMP) of the Introduction) are derivable too.
If $\Gamma \vdash L\alpha \to L\beta$, then for a finite subset $\{\delta_1, \ldots, \delta_n\}$ of Γ, $\delta_1, \ldots, \delta_n \vdash L\alpha \to L\beta$.
By the deduction theorem for $S5(1)$ and propositional calculus:

$$\vdash L\alpha \to \delta_1 \to \ldots \to \delta_n \to L\beta$$

and so

$$\vdash L\alpha \to \delta_1 \to \ldots \to \delta_n \to \beta.$$

By (N), Axiom 4 and Theorem 6(vii),

$$\vdash L\alpha \to L(\delta_1 \to \ldots \to \delta_n \to \beta.) \tag{11}$$

We next have given $\Gamma \mathrel{|\!\!\sim}_2 \beta \to \gamma$:

$$MP_0\ \frac{\Gamma \mathrel{|\!\!\sim}_2 \delta_n \qquad MP_0\ \dfrac{\Gamma \mathrel{|\!\!\sim}_2 \beta \to \gamma \qquad \dfrac{\vdash (\beta \to \gamma) \to \delta_n \to (\delta_n \to \beta) \to \gamma}{\Gamma \mathrel{|\!\!\sim}_2 (\beta \to \gamma) \to \delta_n \to (\delta_n \to \beta) \to \gamma}}{\Gamma \mathrel{|\!\!\sim}_2 \delta_n \to (\delta_n \to \beta) \to \gamma}}{\Gamma \mathrel{|\!\!\sim}_2 (\delta_n \to \beta) \to \gamma}$$

Repeating the process we obtain

$$\Gamma \mathrel{|\!\!\sim}_2 (\delta_1 \to \ldots \to \delta_n \to \beta) \to \gamma.$$

Now using (11) and $\Gamma \mathrel{|\!\!\sim} \alpha$ we have by MP_2,

$$\Gamma \mathrel{|\!\!\sim}_2 \gamma.$$

So MP_2^+ holds. □

Theorem 30. *Lr_3^+ is equivalent to Lr_3.*

Proof. Let $\Gamma \vdash \alpha \to \beta$. Then by compactness, for some finite subset $\{\delta_1, ..., \delta_n\}$ of Γ, $\delta_1, ..., \delta_n \vdash \alpha \to \beta$. As the deduction theorem holds for $S5(1)$,

$$\vdash \delta_1 \to \ldots \to \delta_n \to \alpha \to \beta$$
$$\mathrel{|\!\!\sim}_3 \delta_1 \to \ldots \to \delta_n \to \alpha \to \beta$$

and by MP_0,

$$\Gamma \mathrel{|\!\!\sim}_3 \alpha \to \beta.$$

Now if $\Gamma \mathrel{|\!\!\sim}_3 \alpha$ and $\Gamma \mathrel{|\!\!\sim}_3 \beta \to \gamma$, MP_0 gives $\Gamma \mathrel{|\!\!\sim}_3 \gamma$.

So MP_3^+ holds in Lr_3. □

Theorem 31. *Lr_4^+ is equivalent to Lr_0.*

Proof. We show that MP_0 is derivable in Lr_4^+.
We assume $\Gamma \mathrel{|\!\!\sim}_4^+ \alpha$. As, clearly, Lr_4^+ is weaker than or equivalent to Lr_3^+, which is equivalent to $S5(1)$, we have $\Gamma \vdash \alpha$, and so $\Gamma \vdash M\alpha \to \alpha$.
If also $\Gamma \mathrel{|\!\!\sim}_4^+ \alpha \to \beta$, we have, by MP_4^+, $\Gamma \mathrel{|\!\!\sim}_4^+ \beta$. □

L_r and $\mathcal{L}_\mathcal{R}$ can be extended to L_r^+ and $\mathcal{L}_\mathcal{R}{}^+$ (just as for Lr_i), by adding Γ before the $\vdash$ in the modus ponens rules. However, it seems that, when this is done, the two variants of L_r in Bunder [5] and Banerjee [6] are no longer equivalent. In fact it seems there are at least three variants of L_r^+ and three of $\mathcal{L}_\mathcal{R}{}^+$. We have some results on these but most of this work is left for later.

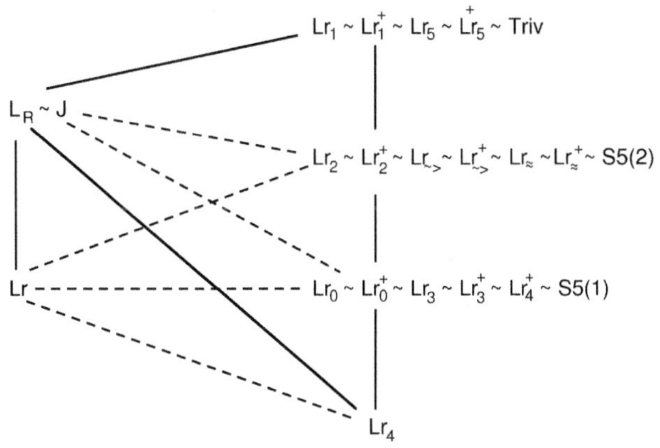

Fig. 1. The consequences

10 Conclusions

We have studied a number of rough consequence logics which come in a natural way from (RMP) and have studied their relative strengths. These logics are represented in the following figure. Here a downward line leads from a system to a system that is strictly weaker. A dotted line joining two systems indicates that the systems are independent.

It is interesting to note that the logics L_r and $\mathcal{L}_\mathcal{R}(\sim J)$ are independent of almost all the other systems. Neither L_r, nor $\mathcal{L}_\mathcal{R}$ has standard modus ponens, while all the Lr_is, except Lr_4, do. All systems except Lr_2 (and those equivalent to it) and Lr_4 have a standard deduction theorem.

L_r, $\mathcal{L}_\mathcal{R}$ and also Lr_4 are *paraconsistent*, i.e. $\alpha, \sim \alpha \mathrel{|\!\!\sim} \beta$ fails. None of the other systems are paraconsistent.

As for semantics, one can see that little is missing – considering the equivalences established between the various systems. $S5(1)$ has a well-known semantics, Fitting [12] gives a semantics for a system equivalent to $S5(2)$, while those for L_r and $\mathcal{L}_\mathcal{R}$ (as mentioned in the Introduction) can be found in [3,6]. Only Lr_4 remains. However, the soundness of the rule MP_4 is not a problem, as (again from the established relationships) it would be sound in each of the logics $\mathcal{L}_{\mathcal{R}}$, $S5(1)$, $S5(2)$.

The observation that all the systems cluster into a few equivalence classes, should be of help in possible applications of these systems. One can foresee situations where approximate reasoning would be required, in particular, where the antecedent of the major premise of MP does not fully match with the minor premise, but only does so roughly. Looking into concrete applications of the logics, is however, beyond the scope of the present work.

As yet unsolved problems include:

1. What are the semantics for the new system Lr_4?
2. What are the relations between the various L_r^+ and $\mathcal{L}_{\mathcal{R}}{}^+$ systems?
3. How do these results change if we replace our use of $S5$ by that of a weaker modal logic?

References

1. Pawlak, Z.: Rough sets. Int. J. Computer and Information Sciences 11(5), 341–356 (1982)
2. Banerjee, M., Chakraborty, M.K.: Rough consequence and rough algebra. In: Ziarko, W.P. (ed.) Rough Sets, Fuzzy Sets and Knowledge Discovery, Proc. Int. Workshop on Rough Sets and Knowledge Discovery (RSKD 1993), Canada, pp. 196–207. Springer, London (1994)
3. Chakraborty, M.K., Banerjee, M.: Rough consequence. Bull. Polish Acad. Sc(Math.) 41(4), 299–304 (1993)
4. Wasilewska, A., Vigneron, L.: Rough equality algebras. In: Wang, P. (ed.) Proc. Int. Workshop on Rough Sets and Soft Computing at 2nd Annual Joint Conf. on Inf. Sc (JCIS 1995), pp. 26–30 (1995)
5. Bunder, M.W.: Rough consequence and Jaśkowski's D2 logics. Technical report 2/04, School of Mathematics and Applied Statistics, University of Wollongong, Australia (2004)
6. Banerjee, M.: Logic for rough truth. Fund. Informaticae 71(2-3), 139–151 (2006)
7. Pawlak, Z.: Rough logic. Bull. Polish Acad. Sc (Tech. Sc.) 35, 253–258 (1987)
8. Jaśkowski, S.: Propositional calculus for contradictory deductive systems. Studia Logica 24, 143–157 (1969); First published in Polish in Studia Scientarium Torunensis Sec A II 55–77 (1948)
9. Rasiowa, H., Skowron, A.: Rough concepts logic. In: Skowron, A. (ed.) SCT 1984. LNCS, vol. 208, pp. 288–297. Springer, Heidelberg (1985)
10. Orłowska, E.: Incomplete Information: Rough Set Analysis. Springer, Berlin (1998)
11. Hughes, G.E., Cresswell, M.J.: A New Introduction to Modal Logic. Routledge, London (1996)
12. Fitting, M.: Proof Methods for Modal and Intuitionistic Logics. D. Reidel, Dordrecht (1983)
13. Marek, V.W., Truszczyński, M.: Nonmonotonic Logic. Springer, Berlin (1993)
14. da Costa, N.C.A., Dubikjatis, L.: On Jaśkowski's discussive logic. In: Arruda, A.I., da Costa, N.C.A., Chaqui, R. (eds.) Non-Classical Logics, Model Theory and Computability, pp. 37–56. North Holland, Amsterdam (1977)
15. da Costa, N.C.A., Doria, F.A.: On Jaśkowski's discussive logics. Studia Logica 54, 33–60 (1995)

Local and Global Approximations for Incomplete Data

Jerzy W. Grzymała-Busse[1,2] and Wojciech Rząsa[3]

[1] Department of Electrical Engineering and Computer Science, University of Kansas, Lawrence, KS 66045, USA

[2] Institute of Computer Science, Polish Academy of Sciences, 01–237 Warsaw, Poland

[3] Department of Computer Science, University of Rzeszow, 35–310 Rzeszow, Poland

Abstract. For completely specified decision tables lower and upper approximations are unique, the lower approximation is the largest definable set contained in the approximated set X and the upper approximation of X is the smallest definable set containing X. For incomplete decision tables the existing definitions of upper approximations provide sets that, in general, are not minimal definable sets. The same is true for generalizations of approximations based on relations that are not equivalence relations. In this paper we introduce two definitions of approximations, local and global, such that the corresponding upper approximations are minimal. Local approximations are more precise than global approximations. Global lower approximations may be determined by a polynomial algorithm. However, algorithms to find both local approximations and global upper approximations are NP-hard. Additionally, we show that for decision tables with all missing attribute values being lost, local and global approximations are equal to one another and that they are unique.

1 Introduction

Development of appropriate methodology to incomplete data sets is crucial since many real-life data sets have missing attribute values. Mining incomplete data requires either a preprocessing (filling in missing attribute values before the main process of rule set induction, decision tree generation, etc.) or mining the data set taking into account that it is incomplete. In this paper we will use the latter approach.

Initially rough set theory was applied to complete data sets (with all attribute values specified). Recently rough set theory was extended to handle incomplete data sets (with missing attribute values) [1,2,3,4,5,6,8,9,10,11,20,21,22,23]. We observe intensive research activity in two areas: rough set approaches to handle incomplete data, mostly in the form of decision tables with missing attribute values, and, in many attempts to study generalizations of the standard indiscernibility relation used to describe decision tables. In the latter area concerned relations are not equivalence relations. Our paper contributes to both research areas.

J.F. Peters and A. Skowron (Eds.): Transactions on Rough Sets VIII, LNCS 5084, pp. 21–34, 2008.

In general, incomplete decision tables are described by characteristic relations, in a similar way as complete decision tables are described by indiscernibility relations [3,4,5,6].

In spite of the fact that input data are presented as decision tables in applications of rough set theory, in theory oriented research such information is frequently expressed as approximation spaces and neighborhood systems [12,17].

Our main objective is to study two novel kinds of approximations: local and global. Both of the two kinds of approximations are optimal in some sense. It means that lower approximations, local and global, are the largest sets that are locally and globally definable, respectively, and contained in the approximated set X. Similarly, upper approximations, local and global, are the smallest sets that are locally and globally definable, respectively, containing the approximated set X. As it will be shown the two kinds of approximations coincide for complete data, and they may differ for incomplete data sets.

A preliminary version of this paper was presented at the Fifth International Conference on Rough Sets and Current Trends in Computing, Kobe, Japan, November 6–8, 2006 [7].

2 Basic Notions

We assume that the input data sets are presented in the form of a *decision table.* An example of a decision table is shown in Table 1. Rows of the decision table represent *cases*, while columns are labeled by *variables.* The set of all cases will be denoted by U. In Table 1, U = {1, 2, ..., 8}. Some variables are called *attributes* while one selected variable is called a *decision* and is denoted by d. The set of all attributes will be denoted by A. In Table 1, A = {*Temperature*, *Headache*, *Nausea*}. Any decision table defines a function ρ that maps the direct product of U and A into the set of all values. For example, in Table 1, $\rho(1, Temperature) = high$. A decision table with completely specified function ρ will be called *completely specified*, or, for the sake of simplicity, *complete.*

For a complete decision table *indiscernibility* relation ind_B is defined according to the following formula

$$ind_B = \{(x, y) \in U^2 \mid \rho(x, a) = \rho(y, a), a \in B\}.$$

For any $B \subseteq A$, ind_B is an equivalence relation. Let $[x]_B$ denotes the equivalence class containing x with respect to the relation ind_B, let $I_A = \{[x]_A \mid x \in U\}$ and $I = \{[x]_B | x \in U, B \subseteq A\}$. It is known that every set $X \subseteq U$ may be presented as a union of some elements of the family I if and only if it can be presented as a union of some elements of the family I_A. Elements of the family I_A are called *elementary* sets. Every set $X \subseteq U$ that is a union of some elementary sets is called *definable.* We assume that the empty set is definable and we denote the family of all definable sets by D. It was observed in [16,17] that a pair (U, D) is a topological space with a topology of open-closed sets. This topology is equivalent to a topology defined by a base I_A as well as defined by a base I. A family I_A is a base of D with the smallest cardinality. The largest definable set $\underline{X}$ contained

Table 1. An incomplete decision table

	Attributes			Decision
Case	Temperature	Headache	Nausea	Flu
1	high	?	no	yes
2	very_high	yes	yes	yes
3	?	no	no	yes
4	high	yes	yes	yes
5	high	?	yes	yes
6	normal	yes	no	yes
7	normal	no	yes	no
8	*	yes	*	no

in $X \subseteq U$ will be called a *lower approximation* of X. The smallest definable set $\overline{X}$ containing $X \subseteq U$ will be called an *upper approximation* of X. Therefore in the topological space (U, I), the lower approximation of the set is its *interior* and the upper approximation of the set is its *closure*. Thus in any discussion on definability and approximations of the set, we may restrict ourselves to elements of the space (U, I_A). This is an important property since the set I_A is easy to compute.

In the topological space (U, I), the set $X \subseteq U$ will be called *J-definable* if it can be presented as a union of some elements of the family J, where $J \subseteq I$. Obviously, if $J_1 \subseteq J_2$, then every set J_1-definable is also J_2-definable, [16,17].

3 Incomplete Data Sets

In practice, input data for data mining are frequently affected by missing attribute values. In other words, the corresponding function ρ is incompletely specified (partial). A decision table with an incompletely specified function ρ will be called *incomplete*.

For the rest of the paper we will discuss incomplete data sets such that for each case at least one attribute value is specified and all decision values are specified. In this paper we will distinguish two types of missing attribute values.

The first type of missing attribute value will be called *lost*. A missing attribute value is lost when for some case (example, object) the corresponding attribute value was mistakenly erased or forgotten to enter into the data set. The original value existed but for a variety of reasons now it is not accessible.

The second type of missing attribute values, called *"do not care" conditions*, are based on an assumption that missing attribute values were initially, when the data set was created, irrelevant. For example, in a medical setup, patients were subjected to preliminary tests. Patients whose preliminary test results were negative were diagnosed as not affected by a disease. They were perfectly well diagnosed in spite of the fact that not all tests were conducted on them. Thus some test results are missing because these tests were redundant. In different words, a missing attribute value of this type may be potentially replaced by any value typical for that attribute. This type of a missing attribute value will be called a "do not care" condition.

Note that both types of missing attribute values are universal (or standard), since they can be used for any incomplete data set. Obviously, if we are familiar with the reason why some attribute values are missing, we should apply the appropriate interpretation: lost values or "do not care" conditions.

For the rest of the paper we will denote lost values by "?" and "do not care" conditions by "*". An example of incomplete decision table is shown in Table 1.

For incomplete decision tables there are two special cases: in the first case, all missing attribute values are lost, in the second case, all missing attribute values are "do not care" conditions. Incomplete decision tables in which all attribute values are lost, from the viewpoint of rough set theory, were studied for the first time in [8], where two algorithms for rule induction, modified to handle lost attribute values, were presented. This approach was studied later, e.g., in [21] and [22], where the indiscernibility relation was generalized to describe such incomplete decision tables.

On the other hand, incomplete decision tables in which all missing attribute values are "do not care" conditions, from the view point of rough set theory, were studied for the first time in [2], where a method for rule induction was introduced in which each missing attribute value was replaced by all values from the domain of the attribute. Originally such values were replaced by all values from the entire domain of the attribute, later, by attribute values restricted to the same concept to which a case with a missing attribute value belongs. Such incomplete decision tables, with all missing attribute values being "do not care conditions", were extensively studied in [10], [11], including extending the idea of the indiscernibility relation to describe such incomplete decision tables.

Other types of missing attribute values are possible as well, see, e.g., [6]. Moreover, note that some other rough-set approaches to missing attribute values were presented in, e.g., [1,2,15].

4 Blocks of Attribute-Value Pairs

An important tool to analyze decision tables is a block of the attribute-value pair. Let a be an attribute, i.e., $a \in A$ and let v be a specified value of a for some case. A block of an attribute-value pair is defined in the following way:

- If for a specified attribute a there exists a case x such that $\rho(x, a) = ?$, i.e., the corresponding value is lost, then the case x should not be included in any blocks$[(a, v)]$ for all values v of attribute a,
- If for a specified attribute a there exists a case x such that the corresponding value is a "do not care" condition, i.e., $\rho(x, a) = *$, then the case x should be included in blocks $[(a, v)]$ for all specified values v of attribute a.

Alternatively, a block of the pair (a, v) is defined according to the following formula:

$$[(a, v)] = \{x \in U | \rho(x, a) = v \text{ or } \rho(x, a) = *\}.$$

Thus,

[(Temperature, high)] = {1, 4, 5, 8},
[(Temperature, very_high)] = {2, 8},
[(Temperature, normal)] = {6, 7, 8},
[(Headache, yes)] = {2, 4, 6, 8},
[(Headache, no)] = {3, 7},
[(Nausea, no)] = {1, 3, 6, 8},
[(Nausea, yes)] = {2, 4, 5, 7, 8},

For data sets with other types of missing attribute values, the definition of the attribute-value block is modified, see, e.g., [6].

5 Definability

As it was mentioned in Section 1, for complete data sets the family I_A is the set of all elementary sets. Additionally, the cardinality of the set I_A is smaller than or equal to the cardinality of the set U. Thus testing whether a set X is definable is—computationally—a simple task. For incomplete data the situation is different. For a case $x \in U$ the *characteristic set* $K_B(x)$ is defined as the intersection of the sets $K(x, a)$, for all $a \in B$, where the set $K(x, a)$ is defined in the following way:

- If $\rho(x, a)$ is specified, then $K(x, a)$ is the block $[(a, \rho(x, a))]$ of attribute a and its value $\rho(x, a)$,
- If $\rho(x, a) =?$ or $\rho(x, a) = *$ then the set $K(x, a) = U$.

Characteristic set $K_B(x)$ may be interpreted as the set of cases that are indistinguishable from x using all attributes from B and using a given interpretation of missing attribute values. Thus, $K_A(x)$ is the set of all cases that cannot be distinguished from x using all attributes. In [25] $K_A(x)$ was called a successor neighborhood of x, see also [12,13,14,19,24,26,27].

Let $K = \{K_B(x) | x \in U, B \subseteq A\}$ and $K_B = \{K_B(x) \mid x \in U\}$, for $B \subseteq A$. For Table 1 members of the family K_A are:

$K_A(1) = \{1, 4, 5, 8\} \cap U \cap \{1, 3, 6, 8\} = \{1, 8\}$,
$K_A(2) = \{2, 8\} \cap \{2, 4, 6, 8\} \cap \{2, 4, 5, 7, 8\} = \{2, 8\}$,
$K_A(3) = U \cap \{3, 7\} \cap \{1, 3, 6, 8\} = \{3\}$,
$K_A(4) = \{1, 4, 5, 8\} \cap \{2, 4, 6, 8\} \cap \{2, 4, 5, 7, 8\} = \{4, 8\}$,
$K_A(5) = \{1, 4, 5, 8\} \cap U \cap \{2, 4, 5, 7, 8\} = \{4, 5, 8\}$,
$K_A(6) = \{6, 7, 8\} \cap \{2, 4, 6, 8\} \cap \{1, 3, 6, 8\} = \{6, 8\}$,
$K_A(7) = \{6, 7, 8\} \cap \{3, 7\} \cap \{2, 4, 5, 7, 8\} = \{7\}$, and
$K_A(8) = U \cap \{2, 4, 6, 8\} \cap U = \{2, 4, 6, 8\}$.

The characteristic relation $R(B)$ is a relation on U defined for $x, y \in U$ as follows

$$(x, y) \in R(B) \; if \; and \; only \; if \; y \in K_B(x).$$

The characteristic relation $R(B)$ is reflexive but—in general—does not need to be symmetric or transitive. Also, the characteristic relation $R(B)$ is known if we know characteristic sets $K_B(x)$ for all $x \in U$. In our example, $R(A)$ = {(1, 1), (1, 8), (2, 2), (2, 8), (3, 3), (4, 4), (4, 8), (5, 4), (5, 5), (5, 8), (6, 6), (6, 8), (7, 7), (8, 2), (8, 4), (8, 6), (8, 8)}. The most convenient way to define the characteristic relation is through the characteristic sets.

For decision tables, in which all missing attribute values are lost, a special characteristic relation was defined in [21], see also, e.g., [20,22].

For decision tables where all missing attribute values are "do not care" conditions a special characteristic relation was defined in [10], see also, e.g., [11].

For incomplete data sets, a set X will be called *B-globally definable* if it is K_B-definable, i.e., if X is a union of members of the family K_B. A set that is A-globally definable will be called *globally definable.* Obviously, the cardinality of the set K_A is smaller than or equal to the cardinality of U, so checking whether a set is globally definable is computationally easy. However, in general K_A is not a base of the approximation space (U, K). Thus X may be K_B-definable in this space (for some $B \subseteq A$) in spite of the fact that it is not K_A-definable. Moreover, a set may be K-definable and be not K_B-definable for any fixed $B \subseteq A$. The family K may have much greater cardinality than the family K_A. As a consequence, the problem of checking whether a set is definable in the space (U, K) is computationally complex. Similarly, searching for a base of the space (U, K) may be computationally complex.

For incomplete data set it is advantageous to define a local definability. A set T of attribute-value pairs, where all attributes belong to set B and are distinct, will be called a *B-complex*. Any A-complex will be called—for simplicity—a *complex.* Obviously, any set containing a single attribute-value pair is a complex. For the rest of the paper we will discuss only *nontrivial complexes*, i.e., such complexes that the intersection of all attribute-value blocks from a given complex is not the empty set.

Set X *depends* on a complex T if and only if

$$\emptyset \neq [T] = \bigcap \{[t] \mid t \in T\} \subseteq X.$$

For an incomplete decision table and a subset B of the set A of all attributes, a union of intersections of attribute-value pair blocks of attribute-value pairs from some B-complexes, will be called a *B-locally definable* set. A-*locally definable* sets will be called locally definable. If (U, A, d, V, ρ) is an incomplete decision table, then the space (U, L) is an approximation space, where L is a family of all subsets of the set of all possible intersections of attribute-value blocks, members of complexes. Obviously, $L \supseteq K$ for any decision table. Thus, the computational complexity of the problems of looking for a minimal base of the space (U, L) and checking whether a set is locally definable is exponential.

Any set X that is B-globally definable is B-locally definable, the converse is not true. In the example of Table 1, the set {7, 8} is a A-locally definable since it is equal to the intersection of [(Temperature, normal)] and [(Nausea, yes)]. Nevertheless, the set {7, 8} is not A-globally definable.

The importance of the idea of local definability is a consequence of the following fact: A set is locally definable if and only if it can be expressed by decision rule sets. This is why it is so important to distinguish between locally definable sets and those that are not locally definable.

For decision tables in which all missing attribute values are lost, local definability is reduced to global definability. The proof of this fact will be given in Section 7.

Note that definability, introduced in [19], differs from our definitions. For example, the set {1, 2, 4, 6, 8}, globally definable according to our definition, is not definable in [19]. Additionally, sets that are definable in [19], are not even locally definable according to our definition.

6 Local Approximations

Let X be any subset of the set U of all cases. The set X is called a *concept* and is usually defined as the set of all cases defined by a specific value of the decision. In general, X is not a B-definable set, locally or globally.

Let $B \subseteq A$. The *B-local lower* approximation of the concept X, denoted by $L\underline{B}X$, is defined as follows

$$\bigcup\{[T] \mid T \textit{ is a complex of } X, \ [T] \subseteq X\}.$$

The *B-local upper* approximation of the concept X, denoted by $L\overline{B}X$, is defined as the minimal set containing X and defined in the following way

$$\bigcup\{[T] \mid \exists \textit{ a family } \mathcal{T} \textit{ of complexes } T \textit{ of } X \textit{ with } \forall\ T \in \mathcal{T},\ [T] \cap X \neq \emptyset\}.$$

Obviously, the B-local lower approximation of X is unique and it is the largest B-locally definable set contained in X. Any B-local upper approximation of X is B-locally definable, it contains X, and is, by definition, the smallest.

For Table 1

$$L\underline{A}\{1,2,3,4,5,6\} = ([(Headache, no)] \cap [(Nausea, no)]) = \{3\},$$

so one complex, {(Headache, no), (Nausea, no)}, describes $L\underline{A}\{1,2,3,4,5,6\}$,

$$L\underline{A}\{7,8\} = [(Temperature, normal)] \cap [(Nausea, yes)] = \{7,8\},$$

so again, one complex, {(Temperature, normal), (Nausea, yes)}, describes $L\underline{A}$ $\{7,8\}$.

$$L\overline{A}\{1,2,3,4,5,6\} =$$
$$[(Temperature, high)] \cup [(Headache, yes)] \cup [(Nausea, no)] =$$
$$\{1,2,3,4,5,6,8\},$$

therefore, to describe $L\overline{A}\{1,2,3,4,5,6\}$ three complexes are necessary: {(Temperature, high)}, {(Headache, yes)}, {(Nausea, no)}. Finally,

$$L\overline{A}\{7,8\} = L\underline{A}\{7,8\} = \{7,8\}.$$

For the incomplete decision table from Table 1 the local lower approximations for both concepts, {1, 2, 3, 4, 5, 6} and {7, 8}, as well as the upper local approximations for these concepts, are unique. Though the local lower approximations are always unique, the local upper approximations, in general, are not unique. For example, let us consider an incomplete decision table from Table 2.

Table 2. An incomplete decision table

	Attributes			Decision
Case	Age	Complications	Hypertension	Delivery
1	*	alcoholism	mild	pre-term
2	>35	obesity	severe	pre-term
3	>35	obesity	?	pre-term
4	*	none	none	pre-term
5	>35	none	none	full-term
6	<25	none	none	full-term
7	25..35	none	none	full-term

For Table 2

[(Age, <25)] = {1, 4, 6},
[(Age, 25..35)] = {1, 4, 7},
[(Age, >35)] = {1, 2, 3, 4, 5},
[(Complications, alcoholism)] = {1},
[(Complications, obesity)] = {2, 3},
[(Complications, none)] = {4, 5, 6, 7},
[(Hypertension, mild)] = {1}.
[(Hypertension, severe)] = {2}.
[(Hypertension, none)] = {4, 5, 6, 7}.

Moreover, for Table 2

$$L\underline{A}\{1,2,3,4\} = \\ [(Complications, alcoholism)] \cup [(Complications, obesity)] = \\ \{1,2,3\},$$

$$L\underline{A}\{5,6,7\} = \emptyset,$$

However,

$$L\overline{A}\{1, 2, 3, 4\}$$

is not unique, any of the following sets

$$[(Age, > 35)] = \{1, 2, 3, 4, 5\},$$

$$[(Age, < 25)] \cup [(Complications, obesity)] = \{1, 2, 3, 4, 6\},$$

or

$$[(Age, 26..35)] \cup [(Complications, obesity)] = \{1, 2, 3, 4, 7\}.$$

may serve as local upper approximations of {1, 2, 3, 4}.

Lastly,

$$L\overline{A}\{5, 6, 7\} = [(Complications, none)] = \{4, 5, 6, 7\}.$$

Algorithms to compute local lower or upper approximations are NP-hard, since the corresponding problems may be presented in terms of prime implicants, monotone functions, and minimization. A similar result for reducts of complete decision tables is well known [18].

7 Global Approximations

For incomplete decision tables global lower and upper approximations may be defined in a few different ways, see, e.g., [3,4,5]. In this paper we suggest yet another definition of global approximations. Note that our definition of global approximations is based on characteristic sets, as oppose to local approximations, introduced in the previous section, where attribute-value blocks were used.

Again, let $B \subseteq A$. Then *B-global lower* approximation of the concept X, denoted by $G\underline{B}X$, is defined as follows

$$\bigcup\{K_B(x) \mid x \in X, K_B(x) \subseteq X\}.$$

Let us observe that the definition of global lower approximation is identical with the definition of subset (or concept) lower approximation [3,4,5]. The *B-global upper* approximation of the concept X, denoted by $G\overline{B}X$, is defined as the minimal set containing X and defined in the following way

$$\bigcup\{K_B(x) \mid \exists\, Y \subseteq U\ \forall\, x \in Y,\ K_B(x) \cap X \neq \emptyset\}.$$

Similarly as for local approximations, a global lower approximation for any concept X is unique. Additionally, both B-global approximations, lower and upper, are B-globally definable. On the other hand, global upper approximations do not need to be unique. For Table 1,

$$G\underline{A}\{1, 2, 3, 4, 5, 6\} = K_A(3) = \{3\},$$

$$G\underline{A}\{7,8\} = K_A(7) = \{7\},$$

$$G\overline{A}\{1,2,3,4,5,6\} =$$
$$K_A(1) \cup K_A(2) \cup K_A(3) \cup K_A(5) \cup K_A(6) = \{1,2,3,4,5,6,8\}.$$

Furthermore,

$$G\overline{A}\{7,8\}$$

may be computed in four different ways:

(1) as $K_A(1) \cup K_A(7) = \{1,7,8\}$,

(2) as $K_A(2) \cup K_A(7) = \{2,7,8\}$,

(3) as $K_A(4) \cup K_A(7) = \{4,7,8\}$,

(4) or as $K_A(6) \cup K_A(7) = \{6,7,8\}$,

all four sets are global upper approximations of the concept {7, 8}.

In general, local approximations are more precise than global approximations. For any concept X and a subset B of A,

$$L\underline{B}X \supseteq G\underline{B}X$$

and

$$L\overline{B}X \subseteq G\overline{B}X.$$

It is not difficult to find a simple algorithm to compute global lower approximation in polynomial time. Nevertheless, algorithms to compute global upper approximations are NP-hard as well.

On the other hand, determining local and global approximations is quite simple for incomplete data sets with all missing values being *lost*. For decision tables with all missing values being lost, the following results hold:

Lemma 1. Let the only missing attribute values in a decision table be lost. Let $x, y \in U$, let B be a subset of the attribute set A, and let $y \in K_B(x)$. Then $K_B(y) \subseteq K_B(x)$.

Proof. Let $a_1, a_2, ..., a_n$ be all attributes from B such that $\rho(x, a_i) = v_i$ is specified (i.e., $v_i \neq ?$) for all $i = 1, 2, ..., n$. Then $K_B(x)$ is equal to

$$[(a_1, v_1)] \cap [(a_2, v_2)] \cap ... \cap [(a_n, v_n)].$$

If $y \in K_B(x)$ then $y \in [(a_1, v_1)]$, $y \in [(a_2, v_2)]$,... $y \in [(a_n, v_n)]$. Moreover, $\rho(y, a_i)$ are all specified, for $i = 1, 2, ..., n$, since all missing attribute values are lost. Obviously, it is possible that for some $a \in A - B$, $\rho(y, a)$ is specified as well. Thus $K_B(y)$ is a subset of the following set $[(a_1, v_1)] \cap [(a_2, v_2)] \cap ... \cap [(a_n, v_n)]$, or,

$$K_B(y) \subseteq K_B(x).$$

Lemma 2. Let the only missing attribute values in a decision table be lost. Let T be a nontrivial complex. Let B be the set of all attributes involved in T. There exists $x \in U$ such that $[T] = K_B(x)$.

Proof. Let $\{a_1, a_2, ..., a_n\} = B$. Let $T = \{(a_1, v_1), (a_2, v_2), ..., (a_n, v_n)\}$, where $v_1, v_2, ...v_n$ are values of $a_1, a_2, ..., a_n$, respectively. Then for any $x \in [T]$, $\rho(x, a_1) = v_1, \rho(x, a_2) = v_2, ..., \rho(x, a_n) = v_n$, since all missing attribute values are lost. Therefore, $[T] = K_B(x)$.

Lemma 3. Let the only missing attribute values in a decision table be lost. Let B be a subset of the set A of all attributes. For every nontrivial complex T such that all attributes involved in T are in B there exists a subset X of U such that

$$[T] = \bigcup\{K_B(x) \mid x \in X\}.$$

Proof. Let C be the set of all attributes involved in T, i.e., if $T = \{(a_1, v_1), (a_2, v_2), ..., (a_n, v_n)\}$, where $v_1, v_2, ...v_n$ are values of $a_1, a_2, ..., a_n$, respectively, then $C = \{a_1, a_2, ..., a_n\}$. There exists $y \in U$ such that $[T] = K_C(y)$, by Lemma 2. Let $X = [T]$. Thus, $\cup\{K_B(y) \mid y \in X\} \subseteq [T]$ since $C \subseteq B$. On the other hand, $[T] \subseteq \cup\{K_B(y) \mid y \in X\}$ since $y \in K_B(y)$. Thus $\cup\{K_B(y) \mid y \in X\} = [T]$.

Due to our last result, we may observe that for data sets with the only missing attribute values being lost, sets $K_B(x)$, for $x \in U$ and $B \subseteq A$, are finer granules (subsets of U) than any nontrivial B-complexes, since any nontrivial B-complex is a union of some sets $K_B(x)$.

Theorem 1. Let the only missing attribute values in a decision table be lost and let B be a subset of the attribute set A. Then every subset $X \subseteq U$ is B-locally definable if and only if it is B-globally definable.

Proof. Straightforward, due to Lemma 3.

Theorem 2. Let the only missing attribute values in a decision table be lost and let B be a subset of the attribute set A. Then for every concept $X \subseteq U$, its B-local approximations are equal to its B-global approximations. Moreover, computing such approximations is of polynomial computational complexity.

Proof. For lower approximations, the proof is obvious since $L\underline{B}X \supseteq G\underline{B}X$ and Lemma 3. Similarly for upper approximations since $L\overline{B}X \subseteq G\overline{B}X$ and Lemma 3. Thus, we may compute the lower approximation (B-local or B-global) using the following formula

$$G\underline{B}X = \bigcup\{K_B(x) \mid x \in X, K_B(x) \subseteq X\},$$

and, by analogy, the upper approximation (also, B-local or B-global) using the following formula

$$G\overline{B}X = \bigcup\{K_B(x) \mid x \in X\}.$$

The last formula needs some explanation. The B-global upper approximation $G\overline{B}X$ is defined as a minimal set satisfying the following formula: $\bigcup\{K_B(x) \mid \exists Y$

$\subseteq U \; \forall \, x \in Y, \; K_B(x) \cap X \neq \emptyset\}$. Any $y \in U - X$ such that $K_B(y) \cap X \neq \emptyset$ may be ignored as a member of Y since if $x \in K_B(y) \cap X$ then $K_B(x) \subseteq K_B(y)$ by Lemma 1, i.e., any element from $K_B(y) \cap X$ can be covered by some $K_B(x)$, where $x \in X$. Thus we may assume that $Y \subseteq X$.

Moreover, we may also assume that $Y = X$. Indeed, let us suppose that Y should be a proper subset of X and let $x \in X - Y$. Then $x \in K_B(y)$ for some $y \in Y$. However, $K_B(x) \subseteq K_B(y)$, by Lemma 1. Therefore, if we assume that $Y = X$, the set $G\overline{B}X$ will be not affected.

Computing both $G\underline{B}X$ and $G\overline{B}X$ using such formulas requires an algorithm with time computational complexity, in the worst case, of $O(n^2 \cdot m)$, where n is the cardinality of U and m is the cardinality of A.

Corollary. Let the only missing attribute values in a decision table be lost and let B be a subset of the attribute set A. For every concept $X \subseteq U$, all B-local upper approximations of X are unique, all B-global upper approximations of X are unique, and are equal to one another.

8 Conclusions

In this paper we introduced two new kinds of approximations: local and global. These approximations describe optimally approximated sets (lower approximations are largest, upper approximations are smallest and, at the same time, local approximations are locally definable while global approximations are globally definable).

Note that our global approximations may be used to describe behavior of systems defined by relations that are not equivalence relations, as in [12,13,14,19,24,25,26,27].

As a final point, optimality comes with the price: in a general case algorithms to compute both local upper approximations and global upper approximations are NP-hard.

References

1. Greco, S., Matarazzo, B., Slowinski, R.: Dealing with missing data in rough set analysis of multi-attribute and multi-criteria decision problems. In: Decision Making: Recent developments and Worldwide Applications, pp. 295–316. Kluwer Academic Publishers, Dordrecht (2000)
2. Grzymala-Busse, J.W.: On the unknown attribute values in learning from examples. In: Raś, Z.W., Zemankova, M. (eds.) ISMIS 1991. LNCS, vol. 542, pp. 368–377. Springer, Heidelberg (1991)
3. Grzymala-Busse, J.W.: Rough set strategies to data with missing attribute values. In: Workshop Notes, Foundations and New Directions of Data Mining, the 3-rd International Conference on Data Mining, Melbourne, FL, USA, November 19–22, pp. 56–63 (2003)

4. Grzymala-Busse, J.W.: Data with missing attribute values: Generalization of idiscernibility relation and rule induction. In: Peters, J.F., Skowron, A., Grzymała-Busse, J.W., Kostek, B.z., Świniarski, R.W., Szczuka, M.S. (eds.) Transactions on Rough Sets I. LNCS, vol. 3100, pp. 78–95. Springer, Heidelberg (2004)
5. Grzymala-Busse, J.W.: Characteristic relations for incomplete data: A generalization of the indiscernibility relation. In: Tsumoto, S., Słowiński, R., Komorowski, J., Grzymała-Busse, J.W. (eds.) RSCTC 2004. LNCS (LNAI), vol. 3066, pp. 244–253. Springer, Heidelberg (2004)
6. Grzymala-Busse, J.W.: Incomplete data and generalization of indiscernibility relation, definability, and approximations. In: Ślęzak, D., Wang, G., Szczuka, M., Düntsch, I., Yao, Y. (eds.) RSFDGrC 2005. LNCS (LNAI), vol. 3641, pp. 244–253. Springer, Heidelberg (2005)
7. Grzymala-Busse, J.W., Rzasa, W.: Local and global approximations for incomplete data. In: Greco, S., Hata, Y., Hirano, S., Inuiguchi, M., Miyamoto, S., Nguyen, H.S., Słowiński, R. (eds.) RSCTC 2006. LNCS (LNAI), vol. 4259, pp. 244–253. Springer, Heidelberg (2006)
8. Grzymala-Busse, J.W., Wang, A.Y.: Modified algorithms LEM1 and LEM2 for rule induction from data with missing attribute values. In: Proc. of the Fifth International Workshop on Rough Sets and Soft Computing (RSSC 1997) at the Third Joint Conference on Information Sciences (JCIS 1997), Research Triangle Park, NC, March 2–5, 1997, vol. 5, pp. 69–72 (1997)
9. Hong, T.P., Tseng, L.H., Chien, B.C.: Learning cevrage rules from incomplete data based on rough sets. In: Proc. of the IEEE International Conference on Systems, Man and Cybernetics, Hague, the Netherlands, October 10–13, 2004, pp. 3226–3231 (2004)
10. Kryszkiewicz, M.: Rough set approach to incomplete information systems. In: Proceedings of the Second Annual Joint Conference on Information Sciences, Wrightsville Beach, NC, September 28–October 1, 1995, pp. 194–197 (1995)
11. Kryszkiewicz, M.: Rules in incomplete information systems. Information Sciences 113, 271–292 (1999)
12. Lin, T.Y.: Neighborhood systems and approximation in database and knowledge base systems. In: Fourth International Symposium on Methodologies of Intelligent Systems (Poster Sessions), Charlotte, North Carolina, October 12–14, pp. 75–86 (1989)
13. Lin, T.Y.: Chinese Wall security policy—An aggressive model. In: Proceedings of the Fifth Aerospace Computer Security Application Conference, Tucson, Arizona, December 4–8, 1989, vol. 8, pp. 286–293 (1989)
14. Lin, T.Y.: Topological and fuzzy rough sets. In: Slowinski, R. (ed.) Intelligent Decision Support. Handbook of Applications and Advances of the Rough Sets Theory, pp. 287–304. Kluwer Academic Publishers, Dordrecht (1992)
15. Nakata, M., Sakai, H.: Rough sets handling missing values probabilistically interpreted. In: Ślęzak, D., Wang, G., Szczuka, M., Düntsch, I., Yao, Y. (eds.) RSFDGrC 2005. LNCS (LNAI), vol. 3641, pp. 325–334. Springer, Heidelberg (2005)
16. Pawlak, Z.: Rough Sets. International Journal of Computer and Information Sciences 11, 341–356 (1982)
17. Pawlak, Z.: Rough Sets. In: Theoretical Aspects of Reasoning about Data, Kluwer Academic Publishers, Dordrecht (1991)
18. Skowron, A., Rauszer, C.: The discernibility matrices and functions in information systems. In: Slowinski, R. (ed.) Handbook of Applications and Advances of the Rough Sets Theory, pp. 331–362. Kluwer Academic Publishers, Dordrecht (1992)

19. Slowinski, R., Vanderpooten, D.: A generalized definition of rough approximations based on similarity. IEEE Transactions on Knowledge and Data Engineering 12, 331–336 (2000)
20. Stefanowski, J.: Algorithms of Decision Rule Induction in Data Mining. Poznan University of Technology Press, Poznan (2001)
21. Stefanowski, J., Tsoukias, A.: On the extension of rough sets under incomplete information. In: Zhong, N., Skowron, A., Ohsuga, S. (eds.) RSFDGrC 1999. LNCS (LNAI), vol. 1711, pp. 73–81. Springer, Heidelberg (1999)
22. Stefanowski, J., Tsoukias, A.: Incomplete information tables and rough classification. Computational Intelligence 17, 545–566 (2001)
23. Wang, G.: Extension of rough set under incomplete information systems. In: Proc. of the IEEE International Conference on Fuzzy Systems (FUZZ_IEEE 2002), Honolulu, HI, May 12–17, 2002, vol. 2, pp. 1098–1103 (2002)
24. Yao, Y.Y.: Two views of the theory of rough sets in finite universes. International J. of Approximate Reasoning 15, 291–317 (1996)
25. Yao, Y.Y.: Relational interpretations of neighborhood operators and rough set approximation operators. Information Sciences 111, 239–259 (1998)
26. Yao, Y.Y.: On the generalizing rough set theory. In: Wang, G., Liu, Q., Yao, Y., Skowron, A. (eds.) RSFDGrC 2003. LNCS (LNAI), vol. 2639, pp. 44–51. Springer, Heidelberg (2003)
27. Yao, Y.Y., Lin, T.Y.: Generalization of rough sets using modal logics. Intelligent Automation and Soft Computing 2, 103–119 (1996)

A Rough Set Approach to Multiple Criteria ABC Analysis

Ye Chen[1], Kevin W. Li[2], Jason Levy[3], Keith W. Hipel[1], and D. Marc Kilgour[4]

[1] Department of Systems Design Engineering, University of Waterloo, Waterloo, Ontario, N2L 3G1, Canada
{y3chen,kwhipel}@uwaterloo.ca

[2] Odette School of Business, University of Windsor, Windsor, Ontario, N9B 3P4, Canada
kwli@uwindsor.ca

[3] Huxley College of the Environment, Western Washington University, Washington, 98225, US
Jason.Levy@wwu.edu

[4] Department of Mathematics, Wilfrid Laurier University, Waterloo, Ontario, N2L 3C5, Canada
mkilgour@wlu.ca

Abstract. A dominance-based rough set approach (DRSA) to multiple criteria ABC analysis (MCABC) is designed and compared to other approaches using a practical case study. ABC analysis is a well-known inventory planning and control approach, which classifies inventory items, or stock-keeping units (SKUs), based solely on their annual dollar usage. Recently, it has been suggested that MCABC can provide more managerial flexibility by considering additional criteria such as lead time and criticality. This paper proposes an MCABC method that employs DRSA to generate linguistic rules to represent a decision maker's preferences based on the classification of a test data set. These linguistic rules are then applied to classify other SKUs. A case study is used to compare the DRSA with other MCABC approaches to demonstrate the applicability of the proposed method.

Keywords: Inventory management, ABC analysis, multiple criteria decision analysis, rough set theory, dominance-based rough set approach.

1 Introduction

In response to demand for mass customization, firms often increase inventories of components, work-in-progress, and spare parts [30]. The different items in an inventory system, referred to as stock-keeping units (SKUs), typically number in the thousands. Corner convenience stores, for instance, may have several thousand SKUs. In such a large inventory system, specific control schemes for individual SKUs are simply not practical, as they would leave no resources for other management activities [5]. Instead, a general practice in industry is to

J.F. Peters and A. Skowron (Eds.): Transactions on Rough Sets VIII, LNCS 5084, pp. 35–52, 2008.

aggregate SKUs into several groups and apply control policies that are uniform across each group [1].

One commonly used approach to classifying SKUs is ABC analysis. In the traditional ABC analysis, SKUs are ranked in descending order of annual dollar usage, the product of unit price and annual demand. The top few SKUs, with the highest annual dollar usage, are placed in group A, which will receive the most management attention; the SKUs with least annual dollar usage are placed in group C and will receive the least management attention; and the remaining SKUs are placed in group B. Figure 1 captures the essence of this rule.

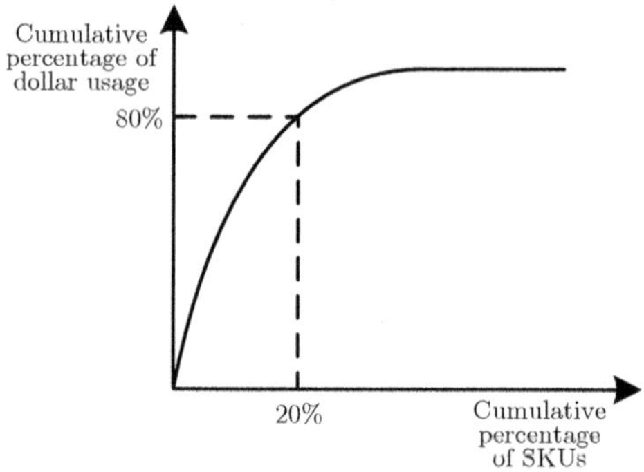

Fig. 1. Example of Dollar Usage Distribution Curve [2]

Traditional ABC analysis can be viewed as an implementation of Pareto's famous observation about the uneven distribution of national wealth [19]: the majority of national wealth is controlled by a few, and the majority of the population controls only a small portion of the wealth. Applications similar to ABC analysis are found in many managerial areas [32]; for instance, in marketing it is often observed that the majority of sales come from a few important customers, while a significant proportion of total sales is due to a large number of very small customers.

Classical ABC analysis has been criticized because of the amount of attention that management pays to an SKU depends on a single criterion, the annual dollar usage of the SKU at the time of classification [9]. However, other attributes of an SKU sometimes play a significant role in prioritization. For instance, suppose that two SKUs are virtually identical except that one is easy to replace while the other is unique and has only one specific supplier. Understandably the SKU with higher substitutability should receive less management attention. Other criteria that could be accounted for include obsolescence, repairability, criticality, and lead time [7], [8].

To carry out multiple criteria classification of SKUs, a variety of approaches has been proposed. One of the first attempts was the Flores and Whybark's bi-criteria matrix method [7]. This approach begins by selecting another critical criterion, in addition to dollar usage, depending on the nature of the industry. Some examples are obsolescence, lead time, substitutability, reparability, criticality and commonality [7]. Next, the model requires that SKUs be divided into three levels of importance, A, B, and C, for each of the two criteria, respectively. The model then reclassifies SKUs into three categories, AA, BB, and CC, representing the three new groups, according to some rules jointly determined by the new criterion and the dollar usage. The structure of the model can be conveniently represented as a joint criteria matrix as shown in Figure 2, adapted from [7]. A general guideline as indicated by the arrows is to regroup AB and BA as AA, AC and CA as BB, and BC and CB as CC.

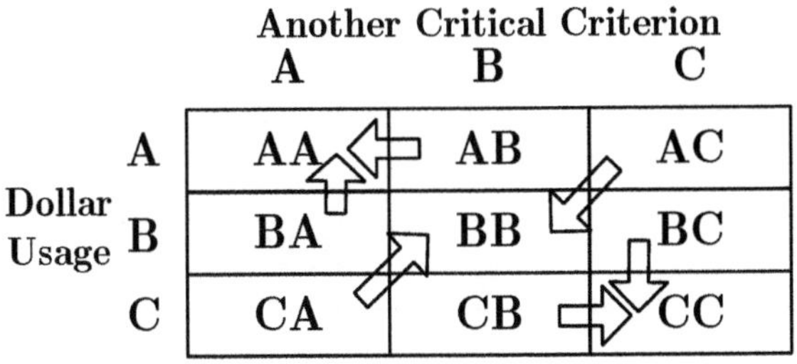

Fig. 2. The Joint Matrix for Two Criteria

Other approaches include the analytic hierarchical process (AHP) [9,21], genetic algorithm [15] and artificial neural networks [20]. Recently, based upon the same case study as described in [9], Ramanathan [27], Ng [17], Zhou and Fan [33], and Chen et al. [2,4] proposed various new approaches to MCABC. For example, Chen et al. [4] proposed a multiple criteria ABC analysis (MCABC) method that employs DRSA to generate linguistic rules for representing a decision maker's preferences based on the classification of a test data set.

In this paper, we refine the previous work in [4] and provide a comprehensive analysis procedure to demonstrate how DRSA can be applied to MCABC. Our results are then compared with other approaches using a practical case study. More specifically, we show how DRSA, a recent advance in rough set theory [11], can be applied to extract information about a decision maker's preferences from the classification of test data and then generate a set of decision rules to classify other SKUs. In addition, the compatibility of DRSA to generate decision rules with other methods are tested and the comparison of classification ability is explored.

The reminder of the paper is organized as follows. Section 2 provides some background pertaining to multiple criteria decision analysis, while Section 3 describes the DRSA in the context of MCABC. An illustrative example is furnished in Section 4, followed by some concluding remarks in Section 5.

2 Multiple Criteria Decision Analysis

Multiple criteria decision analysis (MCDA) is a set of techniques to assist a single decision maker (DM) to *choose*, *rank*, or *sort* a finite set of alternatives according to two or more criteria [28]. The first step of MCDA is to establish a basic structure of the decision problem: define the objectives, arrange them into criteria, identify all possible alternatives, and measure the consequences of each alternative on each criterion. A consequence is a direct measurement of the success of an alternative against a criterion (e.g. cost in dollars). Note that a consequence is usually a physical measurement or estimate; it should not include preferential information.

Figure 3, adapted from [3], shows the basic structure of an MCDA problem. In this figure, $\mathbf{N} = \{N^1, N^2, \cdots, N^i, \cdots, N^n\}$ is a set of alternatives, and $\mathbf{Q} = \{1, 2, \cdots, j, \cdots, q\}$ is a set of criteria. The consequence of alternative N^i over criterion j is denoted $c_j(N^i)$, which can be shortened to c_j^i when there is no possibility of confusion. Note that there are $n > 1$ alternatives and $q > 1$ criteria.

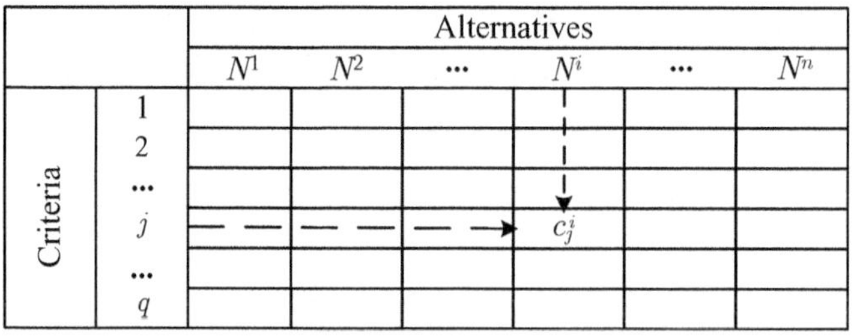

Fig. 3. The Structure of MCDA

Several approaches are available for a DM to structure a decision problem as per Figure 3. Roy [28] suggested that MCDA can be organized into three **problématiques**, or fundamental problems, as follows:

- α, **Choice problématique**. Choose the best alternative from **N**.
- β, **Sorting problématique**. Sort the alternatives of **N** into predefined, relatively homogeneous groups, arranged in preference order.
- γ, **Ranking problématique**. Rank the alternatives of **N** from best to worst.

MCABC is a special kind of sorting problématique: the alternatives are SKUs, and they are to be arranged into three groups, **A**, **B** or **C**. The preference order $\mathbf{A} \succ \mathbf{B} \succ \mathbf{C}$ signifies that an SKU in **A** is to receive more management attention than an SKU in **B**, for instance. It is understood that SKUs in the same group are to receive equal management attention; in this sense, they are indifferent.

The DM's preferences are crucial to the solution of any MCDA problem; moreover, different ways of expressing them may lead to different results. Pareto-Superiority [19] may be used to identify some inferior alternatives, but almost always a more elaborate preference construction is needed to carry out any of the problématiques. Generally speaking, there exist two kinds of preference expressions: *v*alues, which are preferences on consequences, and *w*eights, which are preferences on criteria.

After the structure of an MCDA problem is determined and the DM's preferences are acquired, a model must be constructed to aggregate preferences, thereby permiting the chosen problématique to be investigated. Some methods, such as multiattribute utility theory (MAUT) [18], are direct models in which explicit numerical functions are constructed to evaluate alternatives; others, including Outranking methods [28] and AHP [29], employ pair-wise comparison procedures rather than explicit functions to conduct the evaluation; still others, such as rough set theory [11], tackle the MCDA problem implicitly using linguistic rules.

3 A Rough Set Approach to MCABC

Pawlak [22,23] introduced Rough Sets as a tool to describe dependencies among attributes and to evaluate the significance of individual attributes. Because of its ability to handle the inherent uncertainty or vagueness of data, rough set theory complements probability theory, evidence theory, fuzzy set theory, and other approaches. Recent advances in rough set theory have made it a powerful tool for data mining, pattern recognition, and information representation. For example, Pawlak and Skowron [24] provided a comprehensive literature review of rough set theory including different research directions and various applications. Some theoretical extensions of rough set theory are proposed in [25], and the hybrid of rough set theory and Boolean reasoning with different applications are discussed in [26].

An important principle of rough sets is that all relevant information about alternatives, which may include both condition and decision attributes, can be expressed in a data set [22]. Condition attributes refer to the characteristics of the alternatives; for instance, condition attributes describing a firm can include size, financial characteristics (profitability, solvency, liquidity ratios), market position, and so on. Decision attributes define a partition of the alternatives into groups reflecting the condition attributes in some way. In terms of MCDA, condition and decision attributes are regarded as criteria and decision choices, respectively.

3.1 A Dominance-Based Rough Set Theory for MCABC

As pointed out in [11,14], the original rough set approach cannot efficiently extract knowledge from the analysis of a case set. In MCDA problems, preferences over groups and indiscernibility or similarity must be replaced by the *dominance* relation [14] (also see [10,12] for a detailed discussion of the relationship between the classical rough set approach and DRSA).

To apply rough set theory to MCABC, we treat SKUs as alternatives and relevant data about SKUs as criteria (conditions). We select a non-empty case set $\mathbf{T} \subseteq \mathbf{N}$ and ask the DM to decide how to partition the case set into three non-overlapping classes, $\mathbf{A}'$, $\mathbf{B}'$ and $\mathbf{C}'$, with a preference order $\mathbf{A}' \succ \mathbf{B}' \succ \mathbf{C}'$. (Typically, $\mathbf{T}$ is much smaller than $\mathbf{N}$. For convenience, we assume that $\mathbf{T} = \{N^1, \ldots, N^m\}$.) Then we use rough set theory to extract a set of linguistic rules, $\mathbf{R}$, that capture preferential information in the case set classification, and apply $\mathbf{R}$ to all elements of $\mathbf{N}$ to extend $\mathbf{A}'$ to $\mathbf{A}$, $\mathbf{B}'$ to $\mathbf{B}$, and $\mathbf{C}'$ to $\mathbf{C}$. Thus, $\mathbf{N}$ is

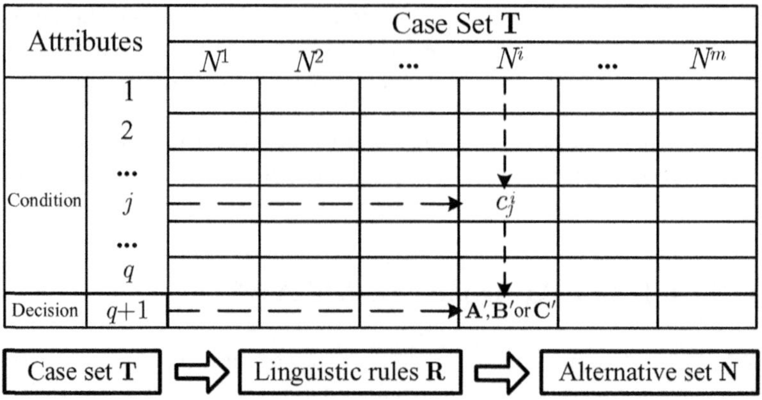

Fig. 4. The Structure of the Case Set

sorted into three classes, **A**, **B**, and **C**, with a preference order $\mathbf{A} \succ \mathbf{B} \succ \mathbf{C}$. This classification procedure is illustrated in Figure 4.

Let S_j be a binary preference relation with respect to criterion $j \in \mathbf{Q}$, such that $N^i S_j N^l$ means that "N^i is at least as good as N^l with respect to criterion j", where $N^i, N^l \in \mathbf{T}$ are alternatives. We assume that S_j is a complete preorder, i.e. a strongly complete and transitive binary relation, and that $\mathbf{S} = (S_1, S_2, \ldots, S_q)$ is a comprehensive preference relation on $\mathbf{N}$, i.e. $N^i \mathbf{S} N^l$ means $N^i S_j N^l$ for every criterion $j \in Q$, for $N^i, N^l \in \mathbf{N}$.

The *upward union* and *downward union* [11,14] with respect to the classes in the test set is defined next. Upward unions are denoted by subscript "$\geq$", and downward unions by subscript "$\leq$".

- $\mathbf{C}'_{\geq} = \mathbf{C}' \cup \mathbf{B}' \cup \mathbf{A}'$; $\mathbf{C}'_{\leq} = \mathbf{C}'$.
- $\mathbf{B}'_{\geq} = \mathbf{B}' \cup \mathbf{A}'$; $\mathbf{B}'_{\leq} = \mathbf{C}' \cup \mathbf{B}'$.
- $\mathbf{A}'_{\geq} = \mathbf{A}'$; $\mathbf{A}'_{\leq} = \mathbf{C}' \cup \mathbf{B}' \cup \mathbf{A}'$.

For example, $\mathbf{C}'_{\geq}$ consists of those test items that at least belong to group $\mathbf{C}'$, and $\mathbf{C}'_{\leq}$ those test items that at most belong to group $\mathbf{C}'$.

N^i dominates N^l with respect to criterion set $\mathbf{P} \subseteq \mathbf{Q}$ and is written as $N^i D_{\mathbf{P}} N^l$, iff $N^i S_j N^l$ for all $j \in \mathbf{P}$. Relative to N^i, the **P**-dominating set is defined by

$$D^{+}_{\mathbf{P}}(N^i) = \{N^l \in \mathbf{T} : N^l D_{\mathbf{P}} N^i\},$$

and the **P**-dominated set by

$$D^{-}_{\mathbf{P}}(N^i) = \{N^l \in \mathbf{T} : N^i D_{\mathbf{P}} N^l\}.$$

With respect to $\mathbf{P} \subseteq \mathbf{Q}$, we say that N^i belongs to $\mathbf{G}'_{\geq}$ unambiguously, where $\mathbf{G}' = \mathbf{A}'$, $\mathbf{B}'$ or $\mathbf{C}'$, iff $N^i \in \mathbf{G}'_{\geq}$ and, for any $N^l \in D^{+}_{\mathbf{P}}(N^i)$, $N^l \in \mathbf{G}'_{\geq}$. More generally, the **P**-lower approximation to $\mathbf{G}'_{\geq}$ is

$$\underline{\mathbf{P}}(\mathbf{G}'_{\geq}) = \{N^i \in \mathbf{T} : D^+_{\mathbf{P}}(N^l) \subseteq \mathbf{G}'_{\geq}\},$$

and the **P**-upper approximation to $\mathbf{G}'_{\geq}$ is

$$\overline{\mathbf{P}}(\mathbf{G}'_{\geq}) = \bigcup_{A^l \in \mathbf{G}'_{\geq}} D^+_{\mathbf{P}}(N^l).$$

Similarly, the **P**-lower approximation to $\mathbf{G}'_{\leq}$ is

$$\underline{\mathbf{P}}(\mathbf{G}'_{\leq}) = \{N^l \in \mathbf{N}' : D^-_{\mathbf{P}}(N^l) \subseteq \mathbf{G}'_{\leq}\},$$

and the **P**-upper approximation to $\mathbf{G}'_{\leq}$ is

$$\overline{\mathbf{P}}(\mathbf{G}'_{\leq}) = \bigcup_{N^l \in \mathbf{G}'_{\leq}} D^-_{\mathbf{P}}(N^l).$$

The **P**-boundaries (**P**-doubtful regions) of $\mathbf{G}'_{\leq}$ and $\mathbf{G}'_{\geq}$ are

$$BN_{\mathbf{P}}(\mathbf{G}'_{\leq}) = \overline{\mathbf{P}}(\mathbf{G}'_{\leq}) - \underline{\mathbf{P}}(\mathbf{G}'_{\leq}),$$

$$BN_{\mathbf{P}}(\mathbf{G}'_{\geq}) = \overline{\mathbf{P}}(\mathbf{G}'_{\geq}) - \underline{\mathbf{P}}(\mathbf{G}'_{\geq}).$$

The quality of the sorting of the case set **T** with respect to $\mathbf{P} \subseteq \mathbf{Q}$ is

$$\gamma_{\mathbf{P}}(\mathbf{G}') = \frac{\left|\mathbf{N} - \left\{\left(\bigcup_{\mathbf{I}'=\mathbf{A}',\mathbf{B}',\mathbf{C}'} BN_{\mathbf{P}}(\mathbf{I}'_{\leq})\right) \cup \left(\bigcup_{\mathbf{I}'=\mathbf{A}',\mathbf{B}',\mathbf{C}'} BN_{\mathbf{P}}(\mathbf{I}'_{\geq})\right)\right\}\right|}{m},$$

where m is the size (cardinality) of the case set **T**. Thus, $\gamma_{\mathbf{P}}(\mathbf{G}')$ represents the proportion of alternatives in the case set **T** that are accurately sorted using only the criteria in **P**.

Each minimal subset $\mathbf{P} \subseteq \mathbf{Q}$ such that $\gamma_{\mathbf{P}}(\mathbf{T}) = \gamma_{\mathbf{Q}}(\mathbf{T})$ is called a *reduct* of **Q**. A case set **T** can have more than one reduct; the intersection of all reducts is called the *core* [11,14].

3.2 Decision Rules for MCABC

The approximations obtained through dominance can be used to construct decision rules capturing preference information contained in the classification of a case set [11]. Assume that all criteria are benefit criteria, i.e. that $c_j(N^i) \geq c_j(N^l)$ implies $N^i S_j N^l$ for all $j \in \mathbf{Q}$ and $N^i, N^l \in \mathbf{N}$. Then three types of decision rules can be generated from a non-empty set of criteria $\mathbf{P} \subseteq \mathbf{Q}$ and are used to sort **N** into **G** and **H**, respectively, where $\mathbf{G} \neq \mathbf{H}$ and $\mathbf{G}, \mathbf{H} \in \{\mathbf{A}, \mathbf{B}, \mathbf{C}\}$, as required.

- $\mathbf{R}_{\geq}$ decision rules, which have the syntax

  $$\text{If } c_j(N^i) \geq r_j \text{ for all } j \in \mathbf{P}, \text{ then } N^i \in \mathbf{G}_{\geq},$$

  where, for each $j \in \mathbf{P}$, $r_j \in \mathbb{R}$ is a consequence threshold for criterion j. Rules of this form are supported only by alternatives from the **P**-lower approximations of class $\mathbf{G}'_{\geq}$.
- $\mathbf{R}_{\leq}$ decision rules, which have the syntax

$$\text{If } c_j(N^i) \leq r_j \text{ for all } j \in \mathbf{P}, \text{ then } N^i \in \mathbf{G}_{\leq},$$

where, for each $j \in \mathbf{P}$, $r_j \in \mathbb{R}$ is a consequence threshold for criterion j. Rules of this form are supported only by alternatives from the **P**-lower approximations of class $\mathbf{G}'_{\leq}$.

- $\mathbf{R}_{\geq\leq}$ decision rules, which have the syntax

$$\text{If } c_j(N^i) \geq r_j \text{ for all } j \in \mathbf{O} \text{ and } c_j(N^i) \leq r_j \text{ for all } j \in \mathbf{P} - \mathbf{O},$$

$$\text{then } N^i \in \mathbf{G} \cup \mathbf{H},$$

where $\mathbf{O} \subseteq \mathbf{P}$ such that both **O** and $\mathbf{P} - \mathbf{O}$ are non-empty, and $r_j \in \mathbb{R}$ is a consequence threshold for criterion j for each $j \in \mathbf{P}$. Rules of this form are supported only by alternatives from the **P**-boundaries of the unions of the classes $\mathbf{G}'_{\geq}$ and $\mathbf{H}'_{\leq}$.

A set of decision rules is *complete* if, when it is applied to alternatives in the case set **T**, all of them can be reclassified to one or more groups and there is no alternative for which rules cannot be applied for classification. Furthermore, alternatives are consistent when they are classified to the original groups; alternatives are inconsistent when they are assigned to a different group or more than one group. A set of decision rules is *minimal* if it is complete and non-redundant, i.e. exclusion of any rule makes the set incomplete [11]. Fortunately, software is available (see below) that produces sets of minimal decision rules.

4 Application

4.1 Background

We now employ a case study on a hospital inventory system, based on data in [9], to demonstrate the proposed procedure. In the reference, 47 disposable SKUs used in a respiratory therapy unit are classified using AHP-based method [29] for MCABC analysis. Table 1 lists data on the SKUs, referred to as S1 through S47. Four criteria as listed in Column 2-5 of Table 1 are considered to be relevant to the MCABC analysis: (1) average unit cost (\$), ranging from \$5.12 to \$210.00; (2) annual dollar usage (\$), ranging from \$25.38 to \$5840.64; (3) criticality, described by numerical values (1, for high or very critical, 0.5, for moderate or important, and 0.01, for low or non-critical); (4) lead time (weeks), the normal time to receive replenishment after an order is placed, ranging from 1 to 7 weeks. The last column of Table 1 shows the AHP-based classification results.

As indicated earlier, in addition to the initial work of Flores et al. [9], several MCABC methods, including Chen et al. [2,4], Ramanathan [27], Ng [17], and Zhou and Fan [33], have been proposed and used the same data set as listed above for demonstration purposes. Here, two types of comparisons of DRSA with other approaches are conducted to show the applicability of DRSA in MCABC:

- *Decision rule generation comparison*: The approaches by Flores et al.[9], Ramanathan [27], Ng [17], and Zhou and Fan [33] rely on direct sorting in which explicit numerical functions are constructed and employed to classify SKUs. To carry out the comparison, DRSA is used to generate decision rules based on the data set including the condition attributes as shown in Table 1 and the decision attributes calculated by other methods. Then, employing the generated decision rules, the reclassification of the training set is done to examine the compatibility of DRSA with other approaches. To a certain extent, this comparison employs a practical example to validate the conclusions drawn by Slowinski et al. [31] and Greco et al. [13] that DRSA is the most general MCDA methodology and other MCDA approaches can be represented in terms of decision rules.
- *Comparison of classification results*: Next, the classification results obtained by using the AHP-based approach in [9] are adopted as benchmark data and are compared with those generated by the herein proposed DRSA approach. To apply the DRSA procedure, a training set, consisting of three **A** items, five **B** items, and seven **C** items, is randomly selected from the 47 SKUs. The training set is then fed into the DRSA-based MCABC procedure to generate decision rules for classifying all 47 SKUs in the inventory system. The sampling is conducted 20 times and the final classification results are then reconciled with those in [9] to examine how well our proposed approach can extract the inherent knowledge imbedded in the training set.

4.2 Decision Rule Generation Comparison

Firstly, based on the classification results obtained by an AHP-based approach [9] as shown in the last column of Table 1, DRSA is utilized to generate decision rules to reflect the DM's subjective judgement, and these rules are employed to reclassify the data set to verify the compatibility of these two approaches. Then, a summary of similar comparisons is provided with other MCABC models. The purpose of these comparisons is to examine whether the reclassification can reproduce the results obtained with other MCABC approaches, thereby confirming the claim in [31] and [13] that DRSA is the most general MCDA methodology and other MCDA approaches can be represented in terms of decision rules. Note that, in general, the case (training, test) set information can be provided by the expert directly and, hence, the generated rules should express the knowledge of the expert used to give his/her classification information.

Analysis Procedures. The software 4eMka2 [16] is employed to conduct the calculations and the analysis procedures are given as follows:

(1) Criteria specification

The detailed criteria specification using 4eMka2 is shown in Figure 5. All of the criteria are interpreted to be benefit criteria: for example, lead time is a gain criterion since the greater the lead time, the higher the level of management attention required. Hence, their preferences are all set as "Gain" as shown in the fifth column of the figure. Note that A1, A2, A3 and A4 represent the criteria of average unit cost, annual dollar usage, criticality and lead time, respectively.

Table 1. Listing of SKUs with multiple criteria, adapted from [9]

SKUs	Criteria				
	Average unit cost ($)	Annual dollar usage ($)	Critical factor	Lead time (week)	Group
S1	49.92	5840.64	1	2	**A**
S2	210.00	5670.00	1	5	**A**
S3	23.76	5037.12	1	4	**A**
S4	27.73	4769.56	0.01	1	**C**
S5	57.98	3478.80	0.5	3	**B**
S6	31.24	2936.67	0.5	3	**C**
S7	28.20	2820.00	0.5	3	**C**
S8	55.00	2640.00	0.01	4	**C**
S9	73.44	2423.52	1	6	**A**
S10	160.50	2407.50	0.5	4	**B**
S11	5.12	1075.20	1	2	**B**
S12	20.87	1043.50	0.5	5	**B**
S13	86.50	1038.00	1	7	**A**
S14	110.40	883.20	0.5	5	**B**
S15	71.20	854.40	1	3	**A**
S16	45.00	810.00	0.5	3	**C**
S17	14.66	703.68	0.5	4	**B**
S18	49.50	594.00	0.5	6	**A**
S19	47.50	570.00	0.5	5	**B**
S20	58.45	467.60	0.5	4	**B**
S21	24.40	463.60	1	4	**A**
S22	65.00	455.00	0.5	4	**B**
S23	86.50	432.50	1	4	**A**
S24	33.20	398.40	1	3	**A**
S25	37.05	370.50	0.01	1	**C**
S26	33.84	338.40	0.01	3	**C**
S27	84.03	336.12	0.01	1	**C**
S28	78.40	313.60	0.01	6	**C**
S29	134.34	268.68	0.01	7	**B**
S30	56.00	224.00	0.01	1	**C**
S31	72.00	216.00	0.5	5	**B**
S32	53.02	212.08	1	2	**B**
S33	49.48	197.92	0.01	5	**C**
S34	7.07	190.89	0.01	7	**C**
S35	60.60	181.80	0.01	3	**C**
S36	40.82	163.28	1	3	**B**
S37	30.00	150.00	0.01	5	**C**
S38	67.40	134.80	0.5	3	**C**
S39	59.60	119.20	0.01	5	**C**
S40	51.68	103.36	0.01	6	**C**
S41	19.80	79.20	0.01	2	**C**
S42	37.70	75.40	0.01	2	**C**
S43	29.89	59.78	0.01	5	**C**
S44	48.30	48.30	0.01	3	**C**
S45	34.40	34.40	0.01	7	**B**
S46	28.80	28.80	0.01	3	**C**
S47	8.46	25.38	0.01	5	**C**

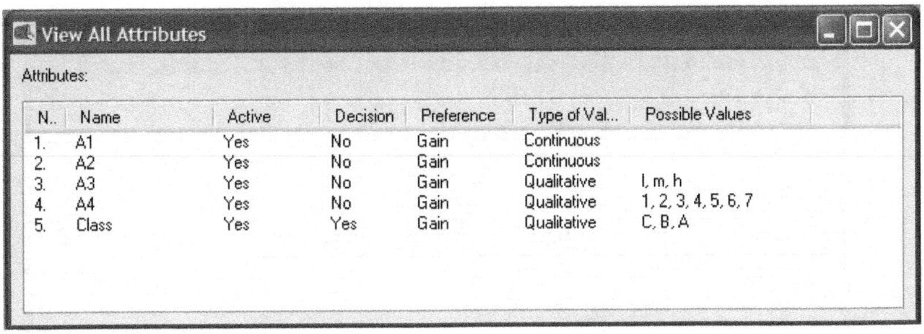

N..	Name	Active	Decision	Preference	Type of Val...	Possible Values
1.	A1	Yes	No	Gain	Continuous	
2.	A2	Yes	No	Gain	Continuous	
3.	A3	Yes	No	Gain	Qualitative	l, m, h
4.	A4	Yes	No	Gain	Qualitative	1, 2, 3, 4, 5, 6, 7
5.	Class	Yes	Yes	Gain	Qualitative	C, B, A

Fig. 5. The Criterion Settings

Average unit cost and annual dollar usage are identified as continuous criteria shown in the sixth column of Figure 5, while criticality and lead time are identified as discrete criteria along with all possible values shown in the last column of the figure. Note that criticality in Flores et al. [9] is represented using numerical values, 1, 0.5 and 0.01, for high critical, moderate and low critical, respectively. Considering the ordinal nature of this criterion, in the DRSA, the linguistic expressions, h, m and l are used instead of 1, 0.5 and 0.01. The same setting is applied to other comparisons. The last row of the figure is the decision attribute, class, which indicates three sorting groups, A, B, and C, for MCABC.

(2) Input data

All data in Table 1 are input into the software for training as shown in Figure 6.

(3) Calculation of unions

All upward unions, downward unions, and boundaries for each class, $\mathbf{A}'$, $\mathbf{B}'$, and $\mathbf{C}'$, are calculated by the software and shown in Figure 7. There are no cases in each group boundary, indicating that the case set has been classified consistently.

(4) Rule generation

As shown in Figure 8, 17 rules are generated based on the algorithm, DomLEM as described in [12], to construct a minimal cover. These rules can be regarded as experts' knowledge in linguistic expressions generated by rough set theory and may help a DM to identify and explain his or her preferences using natural languages. The DM can check and update them as necessary and then apply them to classify any remaining SKUs. For convenience, these 17 rules are reproduced below:

- Rule 1. (A1 $\leq$ 7.07) & (A2 $\leq$ 197.92) $\Rightarrow$ (Class at most **C**);
- Rule 2. (A2 $\leq$ 150) & (A4 $\leq$ 6) $\Rightarrow$ (Class at most **C**);
- Rule 3. (A3$\leq$1) & (A4$\leq$ 6) $\Rightarrow$ (Class at most **C**);
- Rule 4. (A1$\leq$ 31.24) & (A3 $\leq$ m) & (A4$\leq$3) $\Rightarrow$ (Class at most **C**);
- Rule 5. (A2$\leq$ 2936.670000) & (A3 $\leq$ m) & (A4 $\leq$ 3) $\Rightarrow$ (Class at most **C**);
- Rule 4. (A2$\leq$ 2936.670000) & (A3 $\leq$ m) & (A4 $\leq$ 3) $\Rightarrow$ (Class at most **C**);
- Rule 5. (A1$\leq$ 45) & (A2 $\leq$ 810) & (A3 $\leq$ m) & (A4 $\leq$ 3) $\Rightarrow$ (Class at most **C**);
- Rule 6. (A2 $\leq$ 370.5) $\Rightarrow$ (Class at most **B**);

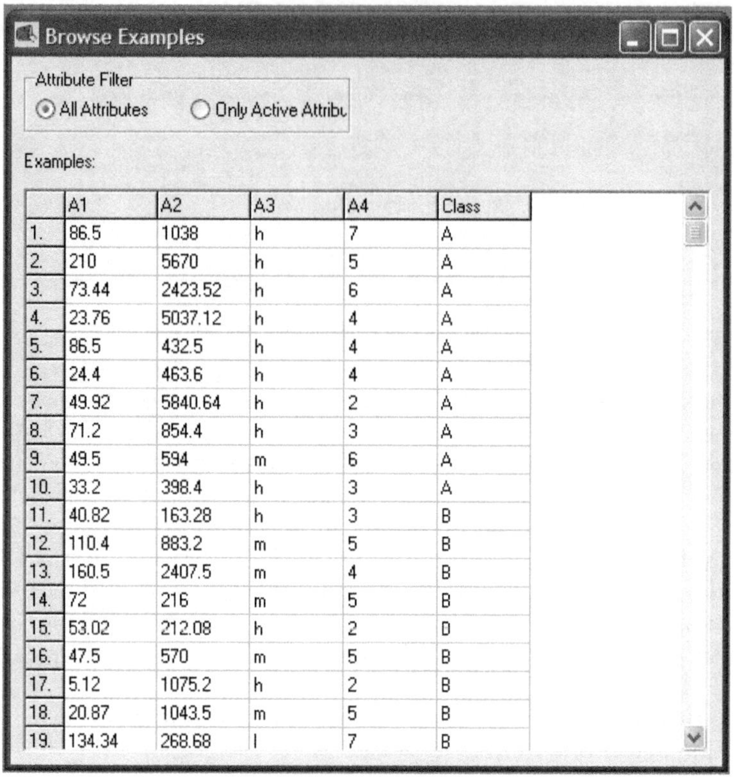

	A1	A2	A3	A4	Class
1.	86.5	1038	h	7	A
2.	210	5670	h	5	A
3.	73.44	2423.52	h	6	A
4.	23.76	5037.12	h	4	A
5.	86.5	432.5	h	4	A
6.	24.4	463.6	h	4	A
7.	49.92	5840.64	h	2	A
8.	71.2	854.4	h	3	A
9.	49.5	594	m	6	A
10.	33.2	398.4	h	3	A
11.	40.82	163.28	h	3	B
12.	110.4	883.2	m	5	B
13.	160.5	2407.5	m	4	B
14.	72	216	m	5	B
15.	53.02	212.08	h	2	D
16.	47.5	570	m	5	B
17.	5.12	1075.2	h	2	B
18.	20.87	1043.5	m	5	B
19.	134.34	268.68	l	7	B

Fig. 6. The Case Set Input

- Rule 7. (A3 $\leq m$) & (A4 ≤ 4) $\Rightarrow$ (Class at most **B**);
- Rule 8. (A1 ≤ 20.87) $\Rightarrow$ (Class at most **B**);
- Rule 9. (A2 ≤ 883.2) & (A3 $\leq m$) & (A4 ≤ 5) $\Rightarrow$ (Class at most **B**);
- Rule 10. (A2 ≥ 5037.12) $\Rightarrow$ (Class at least **A**);
- Rule 11. (A2 ≥ 398.4) & (A3 $\geq h$) & (A4 ≥ 3) $\Rightarrow$ (Class at least **A**);
- Rule 12. (A3 $\geq m$) & (A4 ≥ 6) $\Rightarrow$ (Class at least **A**);
- Rule 13. (A3 $\geq h$) $\Rightarrow$ (Class at least **B**);
- Rule 14. (A2 ≥ 455) & (A3 $\geq m$) & (A4 ≥ 4) $\Rightarrow$ (Class at least **B**);
- Rule 15. (A1 ≥ 34.4) & (A4 ≥ 7) $\Rightarrow$ (Class at least **B**);
- Rule 16. (A1 ≥ 57.98) & (A2 ≥ 3478.8) $\Rightarrow$ (Class at least **B**);
- Rule 17. (A1 ≥ 72) & (A3 $\geq m$) $\Rightarrow$ (Class at least **B**);

(5) Classification precision

All items in the case set are then reclassified using the generated rules. The reclassification results are used to assess classification precision. The generated rules successfully reclassified all items in the case study into corresponding "correct" groups. Therefore, the generated decision rules can accurately capture the DM's preferences, as represented in the classification results by using the AHP-based approach [9].

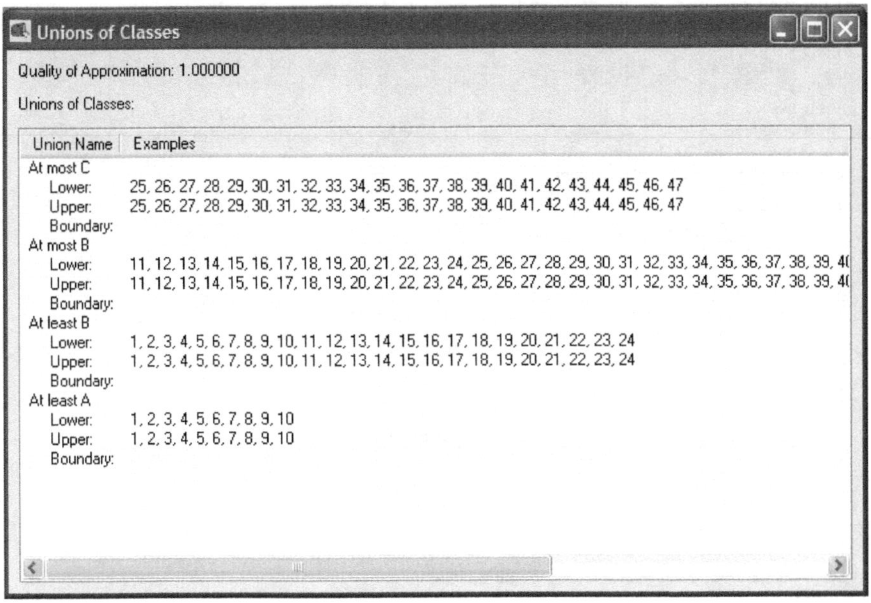

Fig. 7. The Unions in the Case Set

Comparison Summary. Similar procedures are employed to generate decision rules based on the classification information provided by other approaches including Ramanathan's [27], Ng's [17] (Note that the criterion, critical factor, is dropped in Ng's paper and hence, DRSA only analyzes the data set without the condition attribute of critical factor.), and Zhou and Fan's [33]. The detailed analytical steps are skipped here and the compatibility of DRSA with other methods is summarized in Table 2.

Table 2. Summary of Decision Rule Generation Comparisons

Approach Name	Reclassification Results		
	Correct Answers	Incorrect Decisions	Ambiguous Decisions
Flores et al. (AHP)	47	0	0
Ramanathan	17	0	30
Ng	47	0	0
Zhou and Fan	47	0	0

Conclusions : With the approaches of Flores et al. [9], Ng [17], and Zhou and Fan [33], DRSA successfully reclassified all SKUs into the relevant "correct" groups, and there are no incorrect or ambiguous decisions. However, the precision of reclassification using Ramanathan's approach [27] is not so promising, since there are 17 correct, but 30 ambiguous decisions. This large number of ambiguous

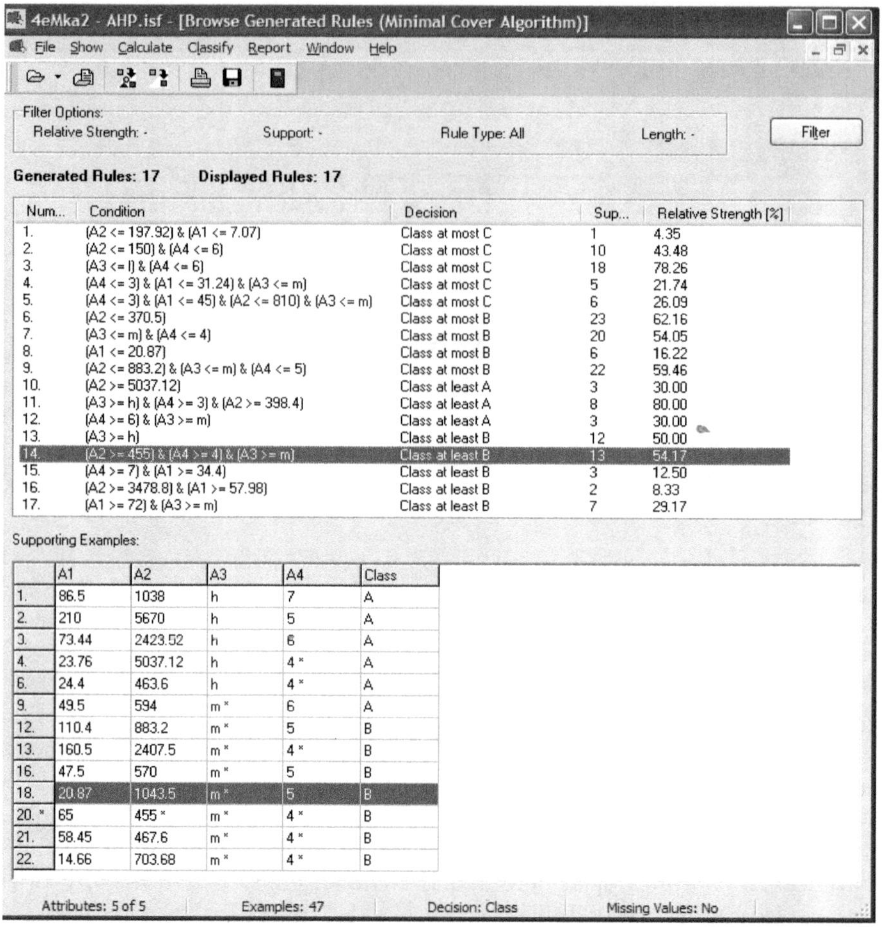

Fig. 8. The Rules Generated by 4eMka2

decisions seemingly resonates with Zhou and Fan's [33] call for an improvement of Ramanathan's model [27] as the classification results may be skewed by extreme values in less important criteria [33]. In short, these comparisons demonstrate that DRSA can successfully generate decision rules that reclassify SKUs into corresponding "correct" groups. This experiment confirms the main conclusion as drawn by Slowinski et al. [31] and Greco et al. [13] that DRSA is the most general MCDA methodology and other MCDA approaches can be represented in terms of decision rules.

4.3 Comparison of Classification Results

Now, a sample of 15 SKUs, consisting of three **A**, five **B**, and seven **C** items, is randomly drawn from the classification results of the AHP-based approach. This sample is input into the DRSA procedure as a training set to generate a list of decision rules. These rules are then applied to all 47 SKUs for classifying them into appropriate groups. This sampling process is repeated 20 times in order to draw statistically significant inferences. The summary of the experiment is given in Table 3. The first column from the left identifies the random training set (the actual lists of T_i, $i = 1, 2, ..., 20$ are not elaborated for the sake of space, but they are available upon request). The second, third, and fourth columns summarize the results from the output of 4eMka2 [16], specifying the number of correct, incorrect, and ambiguous classification results out of the 47 SKUs. The last column indicates the number of "extreme errors", where original **A** items in [9] are classified into **C**, or **C** items in [9] are classified into group **A** as per the generated rules. From a managerial point of view, this kind of "errors" is severe and should always be avoided.

Table 3. Summary of Comparison of Classification Results

Test Set	Reclassification Results			
	Correct Answers	Incorrect Decisions	Ambiguous Decisions	Extreme Errors
T_1	36	9	2	3
T_2	29	13	5	0
T_3	33	3	11	0
T_4	39	6	2	0
T_5	35	6	6	3
T_6	35	10	2	0
T_7	30	15	2	0
T_8	33	10	4	2
T_9	39	7	1	0
T_{10}	41	6	0	0
T_{11}	31	7	9	0
T_{12}	36	5	6	2
T_{13}	35	9	3	0
T_{14}	41	6	0	0
T_{15}	32	8	7	7
T_{16}	37	10	0	0
T_{17}	39	6	2	0
T_{18}	36	4	7	0
T_{19}	28	11	8	0
T_{20}	37	7	3	0

Table 3 demonstrates the applicability of our proposed DRSA approach for MCABC. Firstly, it is rare to have extreme errors of classifying **A** items into group **C**, or **C** items into group **A**: our sample gives a 95% confidence interval of 0.850 ± 0.835. Secondly, most of the 47 SKUs can be categorized into corresponding "correct" groups (the 95% confidence interval is 35.100 ± 1.777). The remaining ambiguity is largely due to the lack of effective mechanisms in DRSA to prioritize different criteria in classifying SKUs while the AHP-based approach allows a DM to determine the weights of criteria. If some conflicting rules generated from the training set are examined and properly removed by the experts who conducted the AHP-based analysis [9], one can expect an increasing number of "correct decisions" and a lower number of "ambiguous decisions".

5 Conclusions

Classical ABC analysis is a straightforward technique to achieve cost-effective inventory management by categorizing SKUs into three groups according to annual dollar usage and then applying similar inventory management procedures throughout each group. However, management can often be made more effective by classifying SKUs under additional criteria, such as lead time and criticality. MCABC furnishes an inventory manager with the flexibility of accounting for more factors when an SKU is categorized.

This paper proposes a dominance-based rough set approach to solve MCABC problems under the umbrella of MCDA theory. Two comparison experiments are conducted based upon a case study. The first experiment, *decision rule generation comparison*, examines whether the DRSA can reproduce the results obtained by other decision models. It is shown that, in most situations, the results are comparable with those obtained using other decision analysis methods such as the AHP-based approach, thereby confirming the applicability of this approach.

In the second experiment, *comparisons of classification results*, the classification result obtained by using the AHP-based approach is adopted as a benchmark and is compared with the one generated by the DRSA. It demonstrates that the decision rules obtained by the DRSA can provide a good approximation of the decision analysis conducted by the AHP method. Future research is needed to compare the classification abilities of this method in various situations with other case-based classification methods, such as methods described by Doumpos and Zopounidis [6].

References

1. Chakravarty, A.K.: Multi-item inventory aggregation into groups. Journal of Operational Research Society 32, 19–26 (1981)
2. Chen, Y., Li, K.W., Kilgour, D.M., Hipel, K.W.: A Case-based distance model for multiple criteria ABC analysis. Computers and Operations Research 35, 776–796 (2008)
3. Chen, Y., Kilgour, D.M., Hipel, K.W.: Multiple criteria classification with an application in water resources planning. Computers and Operations Research 33, 3301–3323 (2006)

4. Chen, Y., Li, K.W., Levy, J., Kilgour, D.M., Hipel, K.W.: Rough-Set multiple-criteria ABC analysis. In: Greco, S., Hata, Y., Hirano, S., Inuiguchi, M., Miyamoto, S., Nguyen, H.S., Słowiński, R. (eds.) RSCTC 2006. LNCS (LNAI), vol. 4259, pp. 328–337. Springer, Heidelberg (2006)
5. Cohen, M.A., Ernst, R.: Multi-item classification and generic inventory stock control policies. Production and Inventory Management Journal 29, 6–8 (1988)
6. Doumpos, M., Zopounidis, C.: Multicriteria decision aid classification methods. Kluwer, Dordrecht (2002)
7. Flores, B.E., Whybark, D.C.: Multiple criteria ABC analysis. International Journal of Operations and Production Management 6, 38–46 (1986)
8. Flores, B.E., Whybark, D.C.: Implementing multiple criteria ABC analysis. Journal of Operations Management 7, 79–84 (1987)
9. Flores, B.E., Olson, D.L., Dorai, V.K.: Management of multicriteria inventory classification. Mathematical and Computer Modeling 16, 71–82 (1992)
10. Greco, S., Matarazzo, B., Slowinski, R.: The use of rough sets and fuzzy sets in MCDM. In: Gal, T., Stewart, T., Hanne, T. (eds.) Advances in Multiple Criteria Decision Making, ch. 14, pp. 14.1–14.59. Kluwer Academic Publishers, Dordrecht (1999)
11. Greco, S., Matarazzo, B., Slowinski, R.: Rough set theory for multicriteria decision analysis. European Journal of Operational Research 129, 1–47 (2001)
12. Greco, S., Matarazzo, B., Slowinski, R.: Rough approximation by dominance relations. International Journal of Intelligent Systems 17(2), 153–171 (2002)
13. Greco, S., Matarazzo, B., Slowinski, R.: Axiomatic characterization of a general utility function and its particular cases in terms of conjoint measurement and roughset decision rules. European Journal of Operational Research 158, 271–292 (2004)
14. Greco, S., Matarazzo, B., Slowinski, R.: Decision rule approach. In: Figueira, J., Greco, S., Erghott, M. (eds.) Multiple Criteria Decision Analysis: State of the Art Surveys, pp. 507–563. Springer, Berlin (2005)
15. Guvenir, H.A., Erel, E.: Multicriteria inventory classification using a genetic algorithm. European Journal of Operational Research 105, 29–37 (1998)
16. Institute of Computing Sceience, Poznan University of Technology, Poland, 4eMka2 software (accessed on March 18, 2007), `http://idss.cs.put.poznan.pl/site/4emka.html`
17. Ng, W.L.: A simple classifier for multiple criteria ABC analysis. European Journal of Operational Research 177, 344–353 (2007)
18. Keeney, R.L., Raiffa, H.: Decision with multiple objectives: preferences and value tradeoffs. Wiley, New York (1976)
19. Pareto, V.: Mannual of Political Economy (English translation). A. A. M. Kelley Publishers, New York (1971)
20. Partovi, F.Y., Anandarajan, M.: Classifying inventory using an artificial neural network approach. Computers and Industrial Engineering 41, 389–404 (2002)
21. Partovi, F.Y., Hopton, W.E.: The analytic hierarchy process as applied to two types of inventory problems. Production and Inventory Management Journal 35, 13–19 (1994)
22. Pawlak, Z.: Rough sets. International Journal of Computer and Information Sciences 11, 341–356 (1982)
23. Pawlak, Z.: Rough sets: theoretical aspects of reasoning about data. Kluwer Academic Publishers, Dordrecht (1991)
24. Pawlak, Z., Skowron, A.: Rudiments of rough sets. Information Sciences 177, 3–27 (2007)

25. Pawlak, Z., Skowron, A.: Rough sets: some extensions. Information Sciences 177, 28–40 (2007)
26. Pawlak, Z., Skowron, A.: Rough sets and boolean reasoning. Information Sciences 177, 41–73 (2007)
27. Ramanathan, R.: ABC inventory classificition with multiple-criteria using weighted linear optimization. Computers and Operations Research 33, 695–700 (2006)
28. Roy, B.: Multicriteria methodology for decision aiding. Kluwer, Dordrecht (1996)
29. Saaty, T.L.: The Analytic Hierarchy Process. McGraw Hill, New York (1980)
30. Silver, E.A., Pyke, D.F., Peterson, R.: Inventory management and production planning and scheduling, 3rd edn. Wiley, New York (1998)
31. Slowinski, R., Greco, S., Matarazzo, B.: Axiomatization of utility, outranking and decision-rule preference models for multiple-criteria classification problems under partial inconsistency with the dominance principle. Control and Cybernetics 31, 1005–1035 (2002)
32. Swamidass, P.M.: ABC analysis or ABC classification. In: Swamidass, P.M. (ed.) Encyclopedia of production and manufacturing management, pp. 1–2. Kluwer Academic Publishers, Boston (2000)
33. Zhou, P., Fan, L.: A note on multi-criteria ABC inventory classification using weighted linear optimization. European Journal of Operational Research 182, 1488–1491 (2007)

Partially Ordered Monads and Rough Sets

Patrik Eklund[1] and M. Ángeles Galán[2,⋆]

[1] Department of Computing Science, Umeå University, Sweden
peklund@cs.umu.se
[2] Department of Applied Mathematics, University of Málaga, Spain
magalan@ctima.uma.es

Abstract. In this paper we will show that partially ordered monads contain appropriate structure for modeling rough sets in a generalized relational setting. Partially ordered monads are further shown to be useful in topological and generalized convergence frameworks. The paper thus demonstrates the use of monad constructions for applications to rough sets and even further towards entirely new types of applications of these generalized rough sets. In doings so, the paper opens up previously unknown research directions for rough sets both towards applications within other theoretical research areas but also for real world applications.

1 Introduction

We aim in this paper to provide a categorical setting for rough sets, connecting them with categorical structures so that rough sets can be described in a more generalized setting providing an innovative approach to categorical rough sets. The paper is an extension of [6] where properties of generalized relations were studied with respect to techniques as provided by operations within partially ordered monads.

A major advantage of category theory is its 'power of abstraction' in the sense that many mathematical structures can be characterized in terms of relatively few categorical ones. This fact enables to pursue a more general study towards generalizations of the structures. Category theory has been successfully applied in different areas as topology, algebra, geometry or functional analysis. In recent years, category theory has also contribute on the development of computer science: the abstraction of this theory has brought the recognition of some of the constructions as categories. This growing interest towards categorical aspects can be found in, for instance, term rewriting systems, game semantics, concurrency. In a gross manner we one can say a *category* is given by a class of *object* and a class of *morphisms* between the objects under certain mathematical conditions. Examples of categories come not only from mathematics (the category of groups and group homomorphisms, the category of topological spaces and continuous functions, etc.) but also from computer science. Deductive systems is a category where the objects are formulas and morphisms are proofs. Partially ordered sets form a category where objects are partially ordered sets and morphisms are monotone mappings. A particular partially ordered set also forms a category where objects are its elements and there is exactly one morphism from an element x to an element y if and only if $x \leq y$.

⋆ Partially supported by Spanish projects P06-FQM-02049 and TIN2006-15455-C03-01.

J.F. Peters and A. Skowron (Eds.): Transactions on Rough Sets VIII, LNCS 5084, pp. 53–74, 2008.

We can go beyond categories and wonder if there is a category of categories. The answer is yes (once properly respecting the underlying selected set theory). In this category of categories the objects are categories and the morphisms are certain structure-preserving maps between categories, called functors. Examples of functors are for instance the list functor, the powerset functor and the term functor. The concept of naturality is important in many of the applications of category theory. Natural transformation are certain structure-preserving maps from one functor to another. It might seem abstract to consider morphisms between morphisms of categories, but natural transformations appear in a natural way very frequently both in mathematics as well as in computer science. Natural transformations are cornerstones in the concept of *m*onads.

Category theory, being an abstract and general theory, enables us to describe and present concepts like *fuzzy sets* or *rough sets* in a unified way. Fuzzy set theory is founded on the idea that many nonmathematical properties cannot be adequately described in terms of crisp sets comprising those elements that fulfil a given property. Often the notion of membership is considered as a gradual property for fuzzy sets. Fuzzy set theory offers a more expressive mathematical language and has many applications in a very wide variety of fields. Originally, fuzzy sets were introduced by Zadeh [38] in 1965 as a need to increase the expressiveness of classical mathematics to deal with information systems that are incomplete, uncertain and imprecise. A fuzzy set is a class with unsharp boundaries. Zadeh introduced the idea of grade of membership of an object in a fuzzy set as a number in the unit interval: Given a set X, a fuzzy set A in X is characterized by a membership function which associates to each element in X a real number in the interval $[0, 1]$, representing the grade of membership of x in A. A fuzzy set, therefore, can be represented as a mapping $A : X \to [0, 1]$, with $A(x) \in [0, 1]$ being the membership degree of x in A. Crisp sets can now be seen as those fuzzy sets A such that $A(x) = 1$ if $x \in A$ or 0 otherwise. This definition correspond exactly of that given by the *crisp powerset monad*.

In 1967, Goguen, [17], extended the idea of fuzzy sets to L-fuzzy sets, considering order structures, L, beyond the unit interval. An L-fuzzy set A is now represented as a mapping $A : X \to L$. Note here that considering the two element lattice, denoted by $L = 2$, we obtain the crisp situation, i.e. we obtain essentially set theory. The case of $L = [0, 1]$, represent the case of fuzzy sets as introduced by Zadeh. Goguen remarked also that the set of all L-fuzzy sets, L^X, can be given whatever operation L has, and these operations in L^X will obey any law valid in L which extends point by point.

The extended notion of L-fuzzy sets can also be represented extending the crisp powerset monad and considering a completely distributive lattice L. Therefore, we can say that powerset monads are a categorical way to represent fuzzy sets. Beyond fuzzy sets, we will show in this paper how considering some particular types of monads, *partially ordered monads*, we can also represent rough sets. In particular, the crisp partially ordered monad is used to represent classical rough sets. The different generalizations of the powerset functor become important in order to extend the classical notions of rough sets.

The paper is organized as follows. In Section 2 we introduce monads and provide the main examples of monads in Section 3. Section 4 extends monads to partially ordered monads, again followed by Section 5 providing examples of partially ordered monads. Section 6 gives an overview of related applications of partially ordered monads to topology and convergence. In Section 7, partially ordered monads are then applied to rough sets, laying out the framework for generalized rough sets and demonstrating the power of partially ordered monad constructions for generalized rough sets and their properties. Finally, Section 8 concludes the paper.

2 Monads and Submonads

As remarked in [2], the naming and identification of monads, in particular as associated with adjoints, can be seen initiated around 1958. Godement was at that time one of the very first authors to use monads, even if then only named *standard constructions*. Huber in 1961 shows that adjoint pairs give rise to monads. Kleisli [26] and also Eilenberg and Moore [4] proved the converse in 1965. The construct of a Kleisli category was thus made explicit in those contributions. Lawvere [29] introduced universal algebra into category theory. This can be seen as the birth of the term monad. These developments then contain all categorical elements for substitution theories. The exploitation of terms and unifications thereof within logic programming is formally described in [33] as early as in 1965. It is therefore somewhat surprising that the categorical connection to unification was not found until twenty years later by Rydeheard and Burstall in [34].

Let $\mathtt{C}$ be a category. A monad (or triple, or algebraic theory) over $\mathtt{C}$ is written as $\mathbf{\Phi} = (\varphi, \eta, \mu)$, where $\varphi : \mathtt{C} \to \mathtt{C}$ is a (covariant) functor, and $\eta : id \to \varphi$ and $\mu : \varphi \circ \varphi \to \varphi$ are natural transformations for which $\mu \circ \varphi\mu = \mu \circ \mu\varphi$ and $\mu \circ \varphi\eta = \mu \circ \eta\varphi = id_\varphi$ hold. A Kleisli category $\mathtt{C}_{\mathbf{\Phi}}$ for a monad $\mathbf{\Phi}$ over a category $\mathtt{C}$ is given with objects in $\mathtt{C}_{\mathbf{\Phi}}$ being the same as in $\mathtt{C}$, and morphisms being defined as $hom_{\mathtt{C}_{\mathbf{\Phi}}}(X,Y) = hom_{\mathtt{C}}(X, \varphi Y)$. Morphisms $f\colon X \rightharpoonup Y$ in $\mathtt{C}_{\mathbf{\Phi}}$ are thus morphisms $f\colon X \to \varphi Y$ in $\mathtt{C}$, with $\eta^\varphi_X\colon X \to \varphi X$ being the identity morphism. Composition of morphisms in $\mathtt{C}_{\mathbf{\Phi}}$ is defined as

$$(X \overset{f}{\rightharpoonup} Y) \diamond (Y \overset{g}{\rightharpoonup} Z) = X \overset{\mu^\varphi_Z \circ \varphi g \circ f}{\longrightarrow} \varphi Z. \tag{1}$$

Composition in the case of the term monad comes down to substitution, and this brings us immediately to substitution theories in general for monads. Monads can be composed and especially the composition of the powerset monad with the term monad provides groundwork for a substitution theory as a basis for many-valued logic [16]. In the following we will provide detail for powerset monads.

The concept of *subfunctors* and *submonads* can be used to provide a technique for constructing new monads from given ones.

Definition 1. *Let φ be a set functor. A set functor φ' is a subfunctor of φ, written $\varphi' \leq \varphi$, if there exists a natural transformation $e\colon \varphi' \to \varphi$, called the inclusion transformation, such that $e_X\colon \varphi' X \to \varphi X$ are inclusion maps, i.e., $\varphi' X \subseteq \varphi X$. The conditions on the subfunctor imply that $\varphi f \mid_{\varphi' X} = \varphi' f$ for all mappings $f : X \to Y$. Further, $\leq$ is a partial ordering.*

Proposition 1 ([8]). *Let $\mathbf{\Phi} = (\varphi, \eta, \mu)$ be a monad over* Set, *and consider a subfunctor φ' of φ, with the corresponding inclusion transformation $e : \varphi' \to \varphi$, together with natural transformations $\eta' : id \to \varphi'$ and $\mu' : \varphi'\varphi' \to \varphi'$ satisfying the conditions*

$$e \circ \eta' = \eta, \tag{2}$$

$$e \circ \mu' = \mu \circ \varphi e \circ e\varphi'. \tag{3}$$

Then $\Phi' = (\varphi', \eta', \mu')$ *is a monad, called the submonad of* Φ*, written* $\Phi' \preceq \Phi$.

3 Examples of Monads

In this section we develop some examples of powerset monads. These examples have an important role in many applications. Powerset monads and their many-valued extensions are in close connection to fuzzification and are good candidates to represent situations with incomplete or imprecise information. With respect to topological application, the fuzzy filter monad is a key construction when studying convergence structures from a more general point of view.

Unless otherwise stated, we assume L to be a completely distributive lattice. For $L = \{0, 1\}$ we write $L = 2$.

3.1 The Powerset Monad

The covariant powerset functor L_{id} is obtained by $L_{id}X = L^X$, i.e. the set of mappings (or L-fuzzy sets) $A : X \to L$, and following [17], for a morphism $f\colon X \to Y$ in $\mathtt{Set}$, the category of sets and functions, by defining

$$L_{id}f(A)(y) = \bigvee_{f(x)=y} A(x).$$

Further, define $\eta_X : X \to L_{id}X$ by

$$\eta_X(x)(x') = \begin{cases} 1 & \text{if } x = x' \\ 0 & \text{otherwise} \end{cases} \tag{4}$$

and $\mu : L_{id} \circ L_{id} \to L_{id}$ by

$$\mu_X(\mathcal{M})(x) = \bigvee_{A \in L_{id}X} A(x) \wedge \mathcal{M}(A).$$

It was shown in [30] that $\mathbf{L}_{id} = (L_{id}, \eta, \mu)$ indeed is a monad. Note that $\mathbf{2}_{id}$ is the usual covariant powerset monad $\mathbf{P} = (P, \eta, \mu)$, where PX is the set of subsets of X, $\eta_X(x) = \{x\}$ and $\mu_X(\mathcal{B}) = \bigcup \mathcal{B}$. Further, note that the transitivity condition, relationally viewed as $f \circ f \subseteq f$, translates to $\bigcup Pf(f(x)) \subseteq f(x)$ for all $x \in X$.

Remark 1. The category of 'sets and relations', i.e. where objects are sets and morphisms $f : X \to Y$ are ordinary relations $f \subseteq X \times Y$ with composition of morphisms being relational composition, is isomorphic to the Kleisli category $\mathtt{Set}_{\mathbf{2}_{id}}$. Indeed, relations $f \subseteq X \times Y$ are morphisms $f : X \rightharpoonup Y$ in $\mathtt{Set}_{\mathbf{2}_{id}}$, i.e. morphisms $f : X \to PY$ in $\mathtt{Set}$, and relational composition corresponds exactly to composition according to (1).

Remark 2. Extending functors to monads is not trivial, and unexpected situations may arise. Let the id^2 functor be extended to a monad with

$$\eta_X(x) = (x, x) \text{ and } \mu_X((x_1, x_2), (x_3, x_4)) = (x_1, x_4).$$

Further, the proper powerset functor P_0, where $P_0X = PX \setminus \{\emptyset\}$, as well as $id^2 \circ P_0$ can, respectively, be extended to monads, even uniquely. However, as shown in [13], $P_0 \circ id^2$ cannot be extended to a monad.

Remark 3. The interaction between monads and algebras is well-known. The tutorial example is the isomorphism between the Kleisli category of the powerset monad and the category of 'sets and relations'. The Eilenberg-Moore category of the powerset monad is isomorphic to the category of complete lattices and join-preserving maps. The Kleisli category of the term monad coincides with its Eilenberg-Moore category and is isomorphic to the category of Ω-algebras. A rather intrepid example, although still folklore, is the isomorphism between the Eilenberg-Moore category of the ultrafilter monad and the category of compact Hausdorff spaces. Here is where "algebra and topology meet".

3.2 Powerset Monads with Fuzzy Level Sets

In [9], a number of set functors extending the powerset functor together with their *extension principles* are introduced. By *extension principles* we mean the two possible generalizations of a mapping $f: X \to Y$ where X, Y are sets, when working in the fuzzy case according to an optimistic or pessimistic interpretation of the fuzziness degree.

1. Maximal extension principle: $Ff_M: FX \to FY$,

$$Ff_M(A)(y) = \begin{cases} \sup\{A(x) \mid f(x) = y \text{ and } A(x) > 0\} & \text{if the set is nonempty} \\ 0 & \text{otherwise} \end{cases}$$

2. Minimal extension principle: $Ff_m: FX \to FY$,

$$Ff_m(A)(y) = \begin{cases} \inf\{A(x) \mid f(x) = y \text{ and } A(x) > 0\} & \text{if the set is nonempty} \\ 0 & \text{otherwise} \end{cases}$$

Both extensions Ff_M and Ff_m coincide with the direct image extension in the case of crisp subsets, that is, given $A \in PX$, then $Pf_M(A) = Pf_m(A) = f(A) \in PY$. These maximal and minimal extension principles can be further generalized to the L-fuzzy powersets, just changing the calculations of suprema and infima by the lattice join and meet operators. We will use the set $I = \{x \in X \mid f(x) = y \text{ and } A(x) > 0\}$:

1. Maximal L-fuzzy extension principle: $Lf_M: LX \to LY$ is

$$Lf_M(A)(y) = \begin{cases} \bigvee_I A(x) & \text{if } I \neq \emptyset \\ 0 & \text{otherwise} \end{cases}$$

2. Minimal L-fuzzy extension principle: $Lf_m: LX \to LY$,

$$Lf_m(A)(y) = \begin{cases} \bigwedge_I A(x) & \text{if } I \neq \emptyset \\ 0 & \text{otherwise} \end{cases}$$

We can now extend the definition of powersets to powersets with fuzzy level sets. Functors for α-upper L-fuzzy sets and α-lower L-fuzzy sets, denoted L_α and L^α, respectively, are given as follows:

$$L_\alpha X = \{A \in L_{id}X \mid A(x) \geq \alpha \text{ or } A(x) = 0, \text{ for all } x \in X\}$$
$$L^\alpha X = \{A \in L_{id}X \mid A(x) \leq \alpha \text{ or } A(x) = 1, \text{ for all } x \in X\}.$$

For mappings $f: X \to Y$, we define $L_\alpha f : L_\alpha X \to L_\alpha Y$ as the restriction of the mapping given by the minimal L-fuzzy extension principle to the L-fuzzy set $L_\alpha X$. Similarly, $L^\alpha f : L^\alpha X \to L^\alpha Y$ is given as the restriction of the mapping given by the maximal L-fuzzy extension principle.

L-fuzzy set categories are defined for each of these extended power set functors and the rationality of the extension principle is proved in the categorical sense, i.e. the associated L-fuzzy set categories are shown to be equivalent to the category of sets and mappings. Each of these new set functors, L_α and L^α can be extended to be monads. Since (L_{id}, η, μ) is a monad, we can easily generalize this result to the case of $(L^\alpha, \eta^\alpha, \mu^\alpha)$. To provide L_α with the monad structure, the unit and multiplications are defined as follows:

$$\eta_{\alpha X}(x)(x') = \begin{cases} 1 & \text{if } x = x' \\ 0 & \text{otherwise} \end{cases}$$

$$\mu_{\alpha X}(\mathcal{A})(x) = \begin{cases} \bigwedge_{A \in I} A(x) \wedge \mathcal{A}(A) & \text{if } I = \{A \in L_\alpha X \mid A(x) \wedge \mathcal{A}(A) > 0\} \neq \emptyset \\ 0 & \text{otherwise} \end{cases}$$

Remark 4. For mappings $f: X \to Y$, we could obtain $L^\alpha f$ as $L_{id} f_{|L^\alpha X}$. Thus, L^α become subfunctor of L_{id} and $\mathbf{L}^\alpha = (L^\alpha, \eta^{L^\alpha}, \mu^{L^\alpha})$ is a submonads of $\mathbf{L_{id}}$.

Remark 5. For $L = 2$, $\mathbf{L}_\alpha = \mathbf{L}^\alpha = \mathbf{2_{id}}$.

3.3 The Covariant Double Contravariant Powerset Monad

The contravariant powerset functor L^{id} is the contravariant hom-functor related to L, i.e. $L^{id} = hom(-, L) : \texttt{Set} \to \texttt{Set}$, which to each set X and mapping $f : X \to Y$ assigns the set L^X of all mappings of X into L, and the mappings $hom(f, L)(g) = g \circ f \ \ (g \in L^Y)$, respectively. Note that 2^{id} is the usual contravariant powerset functor, where $2^{id} X = PX$, and morphisms $X \xrightarrow{f} Y$ in $\texttt{Set}$ are mapped to $2^{id} f$ representing the mapping $M \mapsto f^{-1}[M]$ $(M \in PY)$ from PY to PX.

For double powerset functors it is convenient to write $L_{L_{id}} = L_{id} \circ L_{id}$ and $L^{L^{id}} = L^{id} \circ L^{id}$. Note that $L^{L^{id}}$ is a covariant functor. It may be interesting also to note that the *filter*[1] functor is a subfunctor of $2^{2^{id}}$, but not a subfunctor of $2_{2_{id}}$.

In the case of $L^{L^{id}}$, for $X \xrightarrow{f} Y$ in $\texttt{Set}$ and $\mathcal{M} \in L^{L^X}$, we have $L^{L^{id}} f(\mathcal{M}) = \mathcal{M} \circ L^{id} f$, and hence, $L^{L^{id}} f(\mathcal{M})(g) = \mathcal{M}(g \circ f)$.

Proposition 2 ([13]). *The covariant set functor $LL = L^{id} \circ L^{id}$ can be extended to a monad, considering the following definitions of the natural transformations η^{LL} and μ^{LL}:*

$$\eta_X^{LL}(x)(A) = A(x), \qquad \mu_X^{LL}(\mathcal{U}) = \mathcal{U} \circ \eta_{LX}^{LL}.$$

[1] A *filter* on a set X is a nonempty set $\mathcal{F}$ of subsets of X such that: (i) $\emptyset \notin \mathcal{F}$, (ii) $A, B \in \mathcal{F} \Rightarrow A \cap B \in \mathcal{F}$, (iii) $A \in \mathcal{F}$ $A \subseteq B \Rightarrow B \in \mathcal{F}$.

It is well-known that the proper[2] filter functor F_0 becomes a monad where $\eta^{F_0} : id \to F_0$ is the unique natural transformation and $\mu^{F_0} : F_0 \circ F_0 \to F_0$ is given by

$$\mu_X^{F_0}(\mathcal{U}) = \bigcup_{R \in \mathcal{U}} \bigcap_{\mathcal{M} \in R} \mathcal{M}$$

i.e. the contraction mapping suggested in [28].

Remark 6. In relation with the functor $2^{2^{id}}$, it can easily be seen that $\mu_X^{2^{2^{id}}}(\mathcal{U}) = \mu_X^{F_0}(\mathcal{U})$.

3.4 The Fuzzy Filter Monad

In the case of fuzzy filter monads we assume L to be a complete chain with different least and last element 0 and 1, respectively. A a mapping $\mathcal{M} : L_{id}X \to L$ is called an *L-fuzzy filter* on X [13] if

(F1) $\mathcal{M}(\bar{\alpha}) = \alpha$ for all $\alpha \in L$ and
(F2) $\mathcal{M}(f \wedge g) = \mathcal{M}(f) \wedge \mathcal{M}(g)$ for all $f, g \in L_{id}X$.

where $\bar{\alpha}$ denote the constant mapping from X into L with value α.

Let POSet denote the category of partially ordered sets. Moreover, let $(\mathtt{F}_L, \leq) :$ Set $\to$ POSet be the covariant functor defined as follows: For each set X, $(\mathtt{F}_L, \leq)X$ is the partial ordered set $(\mathtt{F}_L X, \leq)$, where $\mathtt{F}_L X$ consists of all L-fuzzy filters on X and $\leq$ is defined argumentwise by

$$\mathcal{M} \leq \mathcal{N} \iff \mathcal{M}(f) \geq \mathcal{N}(f) \text{ for all } f \in L_{id}X.$$

For each mapping $f : X \to Y$, $(\mathtt{F}_L, \leq)f$ is the mapping $\mathtt{F}_L f : \mathtt{F}_L X \to \mathtt{F}_L Y$ defined by

$$\mathtt{F}_L f(\mathcal{M})(g) = \mathcal{M}(g \circ f).$$

for all L-fuzzy filters $\mathcal{M}$ on X and all $g \in L_{id}X$.

The underlying set functor $\mathtt{F}_L$: Set $\to$ Set of $(\mathtt{F}_L, \leq)$ is called the *L-fuzzy filter functor*. The *L-fuzzy filter monad* $(\mathtt{F}_L, \eta, \mu)$ (see [13]) consists of natural transformations $\eta_X : X \to \mathtt{F}_L X$ and $\mu_X : \mathtt{F}_L \mathtt{F}_L X \to \mathtt{F}_L X$ given by

$$\eta_X(x)(f) = f(x) \tag{5}$$

for all $x \in X$ and $f \in L_{id}X$ and

$$\mu_X(\mathcal{L}) = \mathcal{L} \circ E_X \tag{6}$$

for all $\mathcal{L} \in \mathtt{F}_L \mathtt{F}_L X$, where E_X is the mapping of $L_{id}X$ into $L^{\mathtt{F}_L X}$ which assigns to each $f \in L_{id}X$ the mapping $\mathcal{M} \mapsto \mathcal{M}(f)$ $(\mathcal{M} \in \mathtt{F}_L X)$, written $E_X f$. For each mapping $f \in L_{id}X$,

$$\sup E_X f = \sup f. \tag{7}$$

[2] $F_0 X = FX \setminus \{\emptyset\}$

4 Basic Triples and Partially Ordered Monads

Partially ordered monads are generalizations derived from convergence structures, originally involving filters [28]. In [18], convergence spaces were empowered with functors extended to monads. The development of partially ordered monads is due to [20], with applications e.g. for compactifications in [21]. Topology and convergence were driving forces in the development of partially ordered monads and the demonstration of their power. However, these monads are useful also in other areas. We will show that they contain sufficient structure for modelling rough sets [31] in a generalized setting with set functors. This generalization builds upon a more general powerset functor setting far beyond just strings [25] and relational algebra [37]. Kleene algebras are widely used e.g. in formal languages [35] and analysis of algorithms [1].

Unless otherwise stated, throughout this section we assume that the underlying category is $\mathtt{acSLAT}$, the category of almost complete semilattices, i.e. partially ordered sets $(X, \leq)$ such that the suprema $\sup \mathcal{M}$ of all non-empty subsets $\mathcal{M}$ of X exists. Morphisms $f : (X, \leq) \to (Y, \leq)$ satisfy $f(\sup \mathcal{M}) = \sup f[\mathcal{M}]$ for non-empty subsets $\mathcal{M}$ of X.

A *basic triple* ([20]) is a triple $\Phi = (\varphi, \leq, \eta)$, where $(\varphi, \leq) : \mathtt{Set} \to \mathtt{acSLAT}$, $X \mapsto (\varphi X, \leq)$ is a covariant functor, with $\varphi : \mathtt{Set} \to \mathtt{Set}$ as the underlying set functor, and $\eta : \mathrm{id} \to \varphi$ is a natural transformation. If $(\varphi, \leq, \eta^\varphi)$ and $(\psi, \leq, \eta^\psi)$ are basic triples, then also $(\varphi \circ \psi, \leq, \eta^\varphi \psi \circ \eta^\psi)$ is a basic triple. For each set X and all $x, y \in X$, the infimum $\eta_X(x) \wedge \eta_X(y)$ only exists in case of $x = y$, i.e. η_X injective.

For each set X the partial ordering of $(\varphi X, \leq)$ is considered as a *finer relation*. The elements of φX are called *φ-objects* on X, the minimal elements of $(\varphi X, \leq)$ are called *ultra φ-objects*. For each set X all non-empty suprema of $(\varphi X, \leq)$ exist and for each mapping $f : X \to Y$ the mapping $\varphi f : \varphi X \to \varphi Y$ assigns non-empty suprema to non-empty suprema. The behaviour of infima of subsets of φX may have an essential influence on the topological properties of the general topological structures defined by means of Φ.

A *basic subtriple* of Φ is a basic triple $\Phi' = (\varphi', \leq, \eta')$ such that

1. φ' is a subfunctor of φ and for each set X, $(\varphi' X, \leq)$ is an almost complete subsemilattice of $(\varphi X, \leq)$ (hence the supremum in $(\varphi X, \leq)$ of a non-empty subset of $\varphi' X$ is an element of $\varphi' X$).
2. $\eta'_X(x) = \eta_X(x)$ for each set X and each $x \in X$.

A *partially ordered monad* is a quadruple $\Phi = (\varphi, \leq, \eta, \mu)$, such that

(i) $(\varphi, \leq, \eta)$ is a basic triple.
(ii) $\mu : \varphi\varphi \to \varphi$ is a natural transformation such that (φ, η, μ) is a monad.
(iii) For all mappings $f, g : Y \to \varphi X$, $f \leq g$ implies $\mu_X \circ \varphi f \leq \mu_X \circ \varphi g$, where $\leq$ is defined argumentwise with respect to the partial ordering of φX.
(iv) For each set X, $\mu_X : (\varphi\varphi X), \leq) \to (\varphi X, \leq)$ preserves non-empty suprema.

In the following we introduce the notion of partially ordered submonads. Let $\Phi = (\varphi, \leq, \eta, \mu)$ and $\Phi' = (\varphi', \leq, \eta', \mu')$ be partially ordered monads. Φ' is called a *partially ordered submonad* of Φ, provided that

1. $(\varphi', \leq, \eta')$ is a basic subtriple of $(\varphi, \leq, \eta)$ and
2. $e \circ \mu' = \mu \circ \varphi e \circ e_{\varphi'}$, where $e : \varphi' \to \varphi$ is the natural transformation consisting of all inclusion mappings $e_X : \varphi' X \to \varphi X$.

5 Examples of Partially Ordered Monads

Some of the examples presented in Section 3 are now shown to be extendable to partially ordered monads.

5.1 The Crisp Powerset Monad

The usual covariant powerset monad $\mathbf{P} = (P, \eta, \mu)$, can be extended to a partially ordered monad, $(P, \subseteq, \eta, \mu)$, considering as the partial ordering the inclusion, $\subseteq$.

Clearly by the properties of the monad, $(P, \subseteq, \eta)$ is a basic triple, μ is a natural transformation and $\mu_X : (PPX), \subseteq) \to (PX, \subseteq)$ preserves non-empty suprema.

Given $f, g : Y \to PX$ with $f \subseteq g$ e.g. $f(y) \subseteq g(y)$ for all $y \in Y$ implies $\mu_X \circ Pf \subseteq \mu_X \circ Pg$:

$$(\mu_X \circ Pf)(B) = \bigcup_{y \in B \subseteq Y} f(y) \subseteq \bigcup_{y \in B \subseteq Y} g(y) = (\mu_X \circ Pg)(B)$$

5.2 The Fuzzy Powerset Monad

The powerset monad, (L_{id}, η, μ) can also be extended to a partially ordered monad, considering the partial order defined as $A \leq A'$, with $A, A' \in L_{id}X$ if $A(x) \leq A'(x)$ for all $x \in X$.

Let us see that $\mu_X \circ L_{id} f \leq \mu_X \circ L_{id} g$: provided that $f \leq g$ where $f, g : Y \to L_{id}X$.

$$\begin{aligned}
\mu_X^{L_{id}}(L_{id}f(B))(x) &= \bigvee_{A \in L_{id}X} A(x) \wedge L_{id}f(B)(A) \\
&= \bigvee_{A \in L_{id}X} A(x) \wedge \bigvee_{f(y)=A} B(y) \\
&= \bigvee_{A \in L_{id}X} \bigvee_{f(y)=A} A(x) \wedge B(y) \\
&= \bigvee_{y \in Y} f(y)(x) \wedge B(y) \\
&\leq \bigvee_{y \in Y} g(y)(x) \wedge B(y) \\
&= \mu_X^{L_{id}}(L_{id}g(B))(x).
\end{aligned}$$

5.3 Powerset Monads with Fuzzy Level Sets

The monad $(L^\alpha, \eta^\alpha, \mu^\alpha)$ can also be extended to a partially ordered monad. This result is a generalization of L_{id} being extendable to a partially ordered monad.

To provide L_α with the partially ordered monad structure we need to check that if $f, g : Y \to L_\alpha X$ are such that $f \leq g$ then $\mu_X \circ L_\alpha f \leq \mu_X \circ L_\alpha g$. In the same way as the case of L_{id}, the partial order is defined as $A \leq A'$, with $A, A' \in L_\alpha X$ meaning $A(x) \leq A'(x)$ for all $x \in X$.

$$\begin{aligned}
\mu_X^{L_\alpha}(L_\alpha f(B))(x) &= \bigwedge_{A \in L_\alpha X, A(x)>0, L_\alpha f(B)(A)>0} A(x) \wedge L_\alpha f(B)(A) \\
&= \bigwedge_{A \in L_\alpha X, A(x)>0, L_\alpha f(B)(A)>0} A(x) \wedge \bigwedge_{y \in Y, f(y)=A, B(y)>0} B(y) \\
&= \bigwedge_{A \in L_\alpha X, A(x)>0, f(y)=A, B(y)>0} A(x) \wedge B(y) \\
&= \bigwedge_{B(y)>0} f(y)(x) \wedge B(y) \\
&\leq \bigwedge_{B(y)>0} g(y)(x) \wedge B(y) \\
&= \mu_X^{L_\alpha}(L_\alpha g(B))(x).
\end{aligned}$$

Note that $f \leq g$ implies $f(y)(x) \wedge B(y) \leq g(y)(x) \wedge B(y)$ for all $x \in X$ and therefore $\mu_X^{L_\alpha}(L_\alpha f(B))(x) \leq \mu_X^{L_\alpha}(L_\alpha g(B))(x)$.

5.4 The Covariant Double Contravariant Powerset Monad and the Partially Ordered Fuzzy Filter Monad

Consider $f, g : Y \to LLX$, $f \leq g$. To see if $\mu_X \circ LLf \leq \mu_X \circ LLg$ we have

$$(\eta_{LX}(A) \circ f)(y) = f(y)(A)$$

Therefore, $f \leq g$ implies $(\eta_{LX}(A) \circ f)(y) \leq (\eta_{LX}(A) \circ g)(y)$. By definition of μ_X,

$$\mu_X \circ LLf(\mathcal{M}) = LLf(\mathcal{M}) \circ \eta_{LX}$$

For $A \in LX$

$$\mu_X \circ LLf(\mathcal{M})(A) = \mathcal{M}(\eta_{LX}(A) \circ f)$$

To drop any conclusion of the type $\mathcal{M}(\eta_{LX}(A) \circ f) \leq \mathcal{M}(\eta_{LX}(A) \circ g)$ we need to add conditions on the definition of $\mathcal{M} : LY \to L$.

Clearly to get the partially ordered monad, we need to add conditions on the definition of $\mathcal{M}$. In particular this is the situation for the fuzzy filter monad, where the definition of $\mathcal{M}$ has to fulfil certain conditions, being those conditions the reason why the we can extend the monad to a partially ordered monad.

Indeed, the quadrupel $(\mathcal{F}_L, \leq, \eta, \mu)$, is a partially ordered monad [14].

6 Previous Work on Partially Ordered Monads for Fuzzy Convergence

Compactness for ordinary topologies, viewed as convergence spaces, comes down to ultrafilters always converging. For convergence spaces in general compactification theories are developed based on corresponding Cauchy spaces and their completions. The completion given by [28] is particularly interesting as it turns out that the completion construction makes use of the multiplication in the underlying filter monad. This observation was indicated in [5] and further fully developed in [14]. The Kowalsky completion of an associated Cauchy space, obtained with respect to one suitable choice function, is indeed a compactification ans is frequently called the Richardson compactification since in the filter case it coincides with the usual Richardson compactification [32]. The underlying set functor (and monad) of convergence structures is the filter functor (that extends to a monad). Convergence structures in a more general view, i.e. without particular binding to underlying monads, was introduced as extension structures in [19]. This notion contains important structural properties of Cauchy structures and it is basically for a general completion theory.

6.1 Extension Structures

Let X be a set. By a *φ-extension structure* on X we mean a triple $(S, T, \sim)$ such that

$$S \subseteq \varphi X\,,\ T \subseteq \varphi X \times X$$

holds and $\sim$ is an equivalence relation on S which fulfills the following conditions, where $\mathcal{M} \xrightarrow{T} x$ is written instead of $(\mathcal{M}, x) \in T$:

(E1) $\mathcal{M} \xrightarrow{T} x$ implies $\mathcal{M} \in S$,
(E2) $\mathcal{M} \xrightarrow{T} x\,,\ \mathcal{M} \sim \mathcal{N}$ imply $\mathcal{N} \xrightarrow{T} x$,
(E3) $\mathcal{M} \xrightarrow{T} x\,,\ \mathcal{N} \xrightarrow{T}$ imply $\mathcal{M} \sim \mathcal{N}$.

X equipped with a φ-extension structure on X is called a *φ-extension space*. A *morphism* $f : (X, (S_1, T_1, \sim_1)) \to (Y, (S_2, T_2, \sim_2))$ between φ-extension spaces is a mapping $f : X \to Y$, such that the following conditions are fulfilled:

(M1) $\mathcal{M} \in S_1$ implies $\varphi f(\mathcal{M}) \in S_2$,
(M2) $\mathcal{M} \xrightarrow{T_1} x$ implies $\varphi f(\mathcal{M}) \xrightarrow{T_2} f(x)$,
(M3) $\mathcal{M} \sim_1 \mathcal{N}$ implies $\varphi f(\mathcal{M}) \sim_2 \varphi f(\mathcal{N})$.

Let $\varphi\text{-}\mathtt{EXT}$ be the category of the φ-extension spaces and their related morphisms. A φ-extension space $(X, (S, T, \sim))$ is called *separated* provided that $\mathcal{M} \xrightarrow{T} x$ and $\mathcal{M} \xrightarrow{T} y$ imply $x = y$, which means that T is a partial mapping, that is, a mapping of a subset of φX into X, and $(X, (S, T, \sim))$ is called *complete* provided that for each $\mathcal{M} \in S$ we have $\mathcal{M} \xrightarrow{T} x$ for some $x \in X$. A φ-extension space $(X, (S, T, \sim))$ and also its φ-extension structure are called *η-stable* provided that $\eta_X(x) \xrightarrow{T} x$ holds for all $x \in X$. For each η-stable φ-extension space $(X, (S, T, \sim))$ we have

$$T \;=\; \{\, (\mathcal{M}, x) \mid \mathcal{M} \sim \eta_X(x) \,\}\,.$$

Hence, T is completely fixed by means of S and $\sim$.

6.2 Φ-Cauchy Structures and Completions

Let $\Phi = (\varphi, \leq, \eta)$ be a basic triple. A *Φ-Cauchy structure* on a set X is a subset S of φX with the following properties:

(C1) $\eta_X(x) \in S$ holds for all $x \in X$.
(C2) $\mathcal{M} \in S$ and $\mathcal{N} \leq \mathcal{M}$ imply $\mathcal{N} \in S$.
(C3) For all $\mathcal{M}, \mathcal{N} \in S$, for which the infimum $\mathcal{M} \wedge \mathcal{N}$ exists, we have $\mathcal{M} \vee \mathcal{N} \in S$.

In the filter case, Φ-Cauchy structures are Cauchy structures in the usual sense. A pair consisting of a set X and a Φ-Cauchy structure on X, is called a *Φ-Cauchy space*. If (X, S) is a Φ-Cauchy space, then the elements of S are called the *Φ-Cauchy objects* of this space. A mapping $f : (X, S) \to (Y, S')$ between Φ-Cauchy spaces is called *Φ-Cauchy continuous*, provided that φf assigns Φ-Cauchy objects to Φ-Cauchy objects. Each Φ-Cauchy structure S on a set X can be identified with the η-stable φ-extension structure $(S, T, \sim)$, where the equivalence relation on S is defined by

$$\mathcal{M} \sim \mathcal{N} \iff \mathcal{M} \vee \mathcal{N} \in S \tag{8}$$

and T is given by

$$\mathcal{M} \xrightarrow{T} x \iff \mathcal{M} \vee \eta_X(x) \in S\,. \tag{9}$$

A mapping between Φ-Cauchy spaces is Φ-Cauchy continuous if and only if it is a morphism between the associated η-stable φ-extension spaces.

Proposition 3. *For each Φ-Cauchy structure S on X the set T defined by* (8), *is a* Φ-limit structure, *that is, the following conditions are fulfilled:*

(L1) $\eta_X(x) \xrightarrow{T} x$ *holds for all* $x \in X$.
(L2) $\mathcal{M} \xrightarrow{T} x$ *and* $\mathcal{N} \leq \mathcal{M}$ *imply* $\mathcal{N} \xrightarrow{T} x$.
(L3) *From* $\mathcal{M} \xrightarrow{T} x$ *and* $\mathcal{N} \xrightarrow{T} x$ *it follows* $\mathcal{M} \vee \mathcal{N} \xrightarrow{T} x$.

For each Φ-Cauchy space the set T defined by (9), is called the *Φ-limit structure of S* and (X, T) is called the *related Φ-limit space of (X, S)*.

Proposition 4. *Let $f : (X, S) \to (Y, S')$ be a Φ-Cauchy continuous mapping between Φ-Cauchy spaces. Then the mapping $f : (X, T) \to (Y, T')$ between the related Φ-limit spaces is* continuous, *that is, $\mathcal{M} \xrightarrow{T} x$ implies $\varphi f(\mathcal{M}) \xrightarrow{T'} x$.*

Of course, the notions of *separatedness* and *completeness* introduced for extension spaces, we also use for Φ-Cauchy spaces. Clearly, separatedness is the usual separation axiom of the associated Φ-limit spaces. For each set X a mapping $p : X \to \varphi X$ for which $\eta_X \leq p$ holds, is called a *Φ-pretopology*. Each Φ-pretopology on X can be identified with the Φ-limit structure $T = \{\, (\mathcal{M}, x) \in \varphi X \times X \mid \mathcal{M} \leq p(x)\,\}$.

For the completions of Φ-Cauchy spaces the following conditions are required.

(R) If for a set X and for $\mathcal{M}, \mathcal{N} \in \varphi X$ and $x, y \in X$ the infimum

$$(\mathcal{M} \vee \eta_X(x)) \wedge (\mathcal{N} \vee \eta_X(y))$$

exists, then at least one of the infima $\mathcal{M} \wedge \mathcal{N}$, $\mathcal{M} \wedge \eta_X(y)$ and $\mathcal{N} \wedge \eta_X(x)$ exists or $x = y$.

(V) If for a mapping $f : X \to Y$ and $\mathcal{M} \in \varphi X$ and $y \in Y$ the infimum $\eta_Y(y) \wedge \varphi f(\mathcal{M})$ exists, then there is an element x of X such that $y = f(x)$.

Proposition 5. *[21] Let $\Phi = (\varphi, \leq, \eta)$ be a basic triple. If Φ fulfills the conditions* (R) *and* (V)*, then the category Φ-$\mathtt{CHY}_{sep\,cpl}$ of all separated and complete Φ-Cauchy spaces is an epireflective subcategory of the category Φ-$\mathtt{CHY}_{sep}$ of all separated Φ-Cauchy spaces.*

6.3 Compactness as Completeness and Monadic Compactifications

Let $\Phi = (\varphi, \leq, \eta)$ be a basic triple. A Φ-limit space (X, T) is called *compact* provided that all ultra φ-objects $\mathcal{M}$ on X converge with respect to T. A Φ-limit space (X, T) is called *weakly separated* if for all $x, y \in X$, whenever there is a φ-object on X converging to x and to y with respect to T, then

$$\{\mathcal{M} \mid \mathcal{M} \xrightarrow{T} x\} = \{\mathcal{M} \mid \mathcal{M} \xrightarrow{T} y\}.$$

Proposition 6. *[21] Let (X, T) be a weakly separated Φ-limit space and let*

$$S = \{\mathcal{M} \mid \mathcal{M} \xrightarrow{T} x \text{ for some } x\} \cup \{\mathcal{M} \mid \mathcal{M} \text{ ultra } \varphi\text{-object}\}.$$

Then (X, S) is a Φ-Cauchy space and (X, T) is the associated Φ-limit space of (X, S).

We have that (X, T) is compact if and only if (X, S) is complete.

The Čech-Stone-compactification gives that the category of compact Hausdorff spaces is an epireflective subcategory of the category of all Hausdorff spaces. In [3] it was shown that this does not hold for the category of separated compact filter limit spaces which is not an epireflective subcategory of the category of all separated filter limit spaces. For fuzzy filters the situation can obviously not be improved, but a special type of compactification proposed in [32] is possible by applying the notion of partially ordered monads.

Let $\Phi = (\varphi, \leq, \eta, \mu)$ be a partially ordered monad. We assume that Φ fulfills the following conditions

(U1) For each φ-object $\mathcal{M}$ on a set X there exists an ultra φ-object $\mathcal{U}$ on X such that $\mathcal{U} \leq \mathcal{M}$.
(U2) For each mapping $f : X \to Y$ and each ultra φ-object $\mathcal{U}$ on X, $\varphi f(\mathcal{U})$ is an ultra φ-object on Y.

Because of (U1) and (U2) there is a *subfunctor* φ^u of φ such that for each set X, $\varphi^u X$ is the set of all ultra φ-objects on X. Let $e : \varphi^u \to \varphi$ be the natural transformation such that for each set X, $e_X : \varphi^u X \to \varphi X$ is an inclusion mapping. A further condition is required.

(U3) For each set X and each ultra φ-object $\mathcal{U}$ on $\varphi^u X$, $(\mu_X \circ \varphi e_X)(\mathcal{U})$ is an ultra φ-object on X.

Condition (U3) implies that there is a *natural transformation* $\mu^u : \varphi^u \varphi^u \to \varphi^u$ such that $e \circ \mu^u = \mu \circ \varphi e \circ e_\varphi$.

In the following let a separated Φ-limit space (X,T) be fixed, let S be the Φ-Cauchy structure on X defined as in Proposition 6, let $\sim$ be the related equivalence relation on S defined by (1) and let $X^\wedge = \{ \mathcal{M}^\sim \mid \mathcal{M} \in S \}$ be the set of all equivalence classes with respect to $\sim$. We define a mapping $\kappa : X^\wedge \to \varphi X$ by

$$\kappa(\mathcal{M}^\sim) = \begin{cases} \eta_X(x) & \texttt{if } \mathcal{M} \xrightarrow{T} x \\ \mathcal{M} & \text{if } \mathcal{M} \text{ is a non-converging ultra } \varphi\text{-object.} \end{cases}$$

Proposition 7. *[21] Given that* (U1), (U2) *and* (U3) *hold, each ultra φ-object $\mathcal{K}$ on $X^\wedge$, $(\mu_X \circ \varphi\kappa)(\mathcal{K})$ is an ultra φ-object on X. Further, $(X^\wedge, T^\wedge)$ with*

$$T^\wedge = \{ (\mathcal{K}, \mathcal{M}^\sim) \in \varphi X^\wedge \times X^\wedge \mid (\mu_X \circ \varphi\kappa)(\mathcal{K}) \in \mathcal{M}^\sim \}$$

is a separated and compact Φ-limit space.

$(X^\wedge, T^\wedge)$ is called the *monadic Richardson compactification* of (X,T). Let $\iota : X \to X^\wedge$ be the mapping $x \mapsto \eta_X(x)^\sim$.

Proposition 8. *[21] Assume that* (U1), (U2) *and* (U3) *are fulfilled. Then $\iota : (X,T) \to (X^\wedge, T^\wedge)$ is an epimorphism of the category Φ-$\texttt{LIM}_{sep}$ and for each continuous mapping $f : (X,T) \to \mathcal{Y}$ of (X,T) into a separated, compact and regular Φ-limit space $\mathcal{Y}$, there exists one and only one continuous mapping $f^\wedge : (X^\wedge, T^\wedge) \to \mathcal{Y}$ such that $f = f^\wedge \circ \iota$.*

7 Applications to Rough Sets

The structure offered by partially ordered monads allows us to model rough sets in a generalized settings. Rough sets build upon relations, and categorically, relations are presented here as mappings in a category. Their extension to fuzzy relations are studied in the same context. Upper and lower approximations for rough sets are introduced in terms of the natural transformations of the partially ordered monads and generalized powerset partially ordered monads are used to extend the classical view of rough sets.

Partially ordered submonads are also studied in this section as a way to reduce data in a given information system.

7.1 Relations, Fuzzy Relations and Kleisli Categories

Relations play a fundamental role in many areas. Particular properties of the relations are on demand for classifying data and mathematical foundations are needed to provide bases for developing methods. Uncertain situations or information systems with lack of information is an important application on the study of rough sets.

Rough sets are, traditionally, based on indistinguishable relations and given an information system, one of the first criteria for reducing data is by means of equivalent relations. Some extensions, think over relations where the transitivity property is dropped (*tolerance relations*). There are also other situations where, due to a more imprecise and uncertain nature of the situation considered, some extended relations, fuzzy relations, are to be considered.

A classical (n-ary) relation, R is defined over crisp sets $X_1, X_2, \ldots, X_n$ as a subset of the cartesian product $X_1 \times X_2 \times \ldots \times X_n$. If $X = X_1 = \ldots = X_n$, the relation is said to be defined on X.

Let us consider a binary relation $R \subseteq X \times Y$. We will use the notation xRy to represent that the element $(x, y) \in R$. Considering P, the crisp powerset functor, we can represent the relation as a mapping $\rho : X \to PY$, where

$$\rho(x) = \{y \in Y \text{ such that } xRy\}.$$

As regarded as mappings, considering the composition of two relations, $\rho : X \to PY$ and $\rho' : Y \to PZ$ we clearly see that the conventional composition of mappings can not be done since the domain of ρ' and codomain of ρ are different. To find the appropriate definition of this composition we have to consider the Kleisli composition as defined previously in the paper (1), i.e. we need to use that P is a monad and has a "flattering" operator, μ:

$$(X \overset{\rho}{\rightharpoondown} Y) \diamond (Y \overset{\rho'}{\rightharpoondown} Z) = X \overset{\mu_Z^P \circ P\rho' \circ \rho}{\longrightarrow} PZ.$$

This, indeed, works since the Kleisli category associated to the crisp powerset monad is equivalent to the category `SetRel`, where the objects are sets and the morphisms are relations: $\rho : X \to PY$ corresponds to a relation $R \subseteq X \times Y$ by the observation $(x, y) \in R$ if and only if $y \in \rho(x)$; and Kleili composition, due to its definition in the case of the powerset monad $\mu_Z^P \circ P\rho'(\rho(x)) = \bigcup_{y \in \rho(x)} \rho'(y)$, corresponds to the usual composition of relations $R \subseteq X \times Y$, $R' \subseteq Y \times Z$, $(x, z) \in R' \circ R$ if and only if $\exists y, y \in \rho(x), z \in \rho'(y)$.

Relations can now be extended to fuzzy relations. Let X and Y be nonempty sets. A fuzzy relation R is a fuzzy subset of the cartesian product $X \times Y$. If $X = Y$ we say that R is a binary fuzzy relation on X. $R(x, y)$ is interpreted as the degree of membership of the pair (x, y) in R.

If we consider now the generalized powerset monad, $L_{id}X$ is the set of all L-fuzzy sets. An L-fuzzy set A is nothing but a mapping $A : X \to L$. As a first step, and in the same way as before we can extend the concept of relation to a fuzzy relation, i.e. a mapping $\rho : X \to L_{id}Y$, $\rho(x)$ is nothing but an element in $L_{id}Y$, a mapping $\rho(x) : Y \to L$. An element $y \in Y$ will be assigned a membership degree, $\rho(x)(y)$ representing, as a value in L, the degree on which the elements x and y are fuzzy related. Note that this situation extend the classical relations (crisp powerset situation) in the sense that membership values are 1 if the elements are related and 0 otherwise.

With respect to the Kleisli category associated to the powerset monad $\mathbf{L_{id}}$, the objects are sets and homomorphism are given as mappings $X \to L_{id}Y$ in Set.

Proposition 9. *The Kleisli category associated to* $\mathbf{L_{id}}$ *is equivalent to the category of set and fuzzy relations,* `SetFuzzRel`.

Proof. To see that $Set_{L_{id}}$ is equivalent to $SetFuzzRel$ we need to see that there is one to one correspondence between the membership valued of the fuzzy relations. And this is clear since we define $\rho(x) : Y \to L$ as the mapping assigning to an element y, the degree of membership on the fuzzy relation of the pair (x, y).

Proposition 10. *Kleisli composition associated to* $\mathbf{L_{id}}$ *is given by:*

$$\mu_Z^{L_id}(L_{id}\rho'(\rho(x)))(z) = \bigvee_{y \in Y} \rho'(y)(z) \wedge \rho(x)(y)$$

Proof. By the definition of the multiplication of the monad it follows:

$$\begin{aligned}\mu_Z^{L_id}(L_{id}\rho'(\rho(x)))(z) &= \bigvee_{A \in L_{id}X} A(z) \wedge L_{id}\rho'(\rho(x))(A) \\ &= \bigvee_{A \in L_{id}X} A(z) \wedge \bigvee_{\rho'(y)=(A)} \rho(x)(y) \\ &= \bigvee_{A \in L_{id}X} \bigvee_{\rho'(y)=(A)} A(z) \wedge \rho(x)(y) \\ &= \bigvee_{y \in Y} \rho'(y)(z) \wedge \rho(x)(y).\end{aligned}$$

The previous proposition tells which membership grade we should assign to the composition of two fuzzy relations, i.e. the suprema of the membership grades on the fuzzy relations. This Kleisli composition of fuzzy relations can be connected to situations where we want to combine different information systems and study rough approximations.

7.2 Ordinary Relations and Rough Sets

Let R be a relation on X, i.e. $R \subseteq X \times X$. We represent the relation as a mapping $\rho_X : X \to PX$, where $\rho_X(x) = \{y \in X | xRy\}$. The corresponding inverse relation R^{-1} is represented as $\rho_X^{-1}(x) = \{y \in X | xR^{-1}y\}$. Based on indistinguishable relations, *rough sets* are introduced by defining the upper and lower approximation of sets. These approximations represent uncertain or imprecise knowledge. To be more formal, given a subset A of X, the lower approximation of A correspond to the objects that surely (with respect to an indistinguishable relation) are in A. The lower approximation of A is obtained by

$$A^{\downarrow} = \{x \in X | \rho(x) \subseteq A\}$$

and the upper approximation by

$$A^{\uparrow} = \{x \in X | \rho(x) \cap A \neq \emptyset\}.$$

In what follows we will assume that the underlying almost complete semilattice has finite infima, i.e. is a join complete lattice. Considering P as the functor in its corresponding partially ordered monad we then immediately have

Proposition 11. *[6] The upper and lower approximations of a subset A of X are given by*

$$A^{\uparrow} = \bigvee_{\rho_X(x) \wedge A > 0} \eta_X(x) = \mu_X \circ P\rho_X^{-1}(A)$$

and

$$A^{\downarrow} = \bigvee_{\rho_X(x) \leq A} \eta_X(x),$$

respectively.

The corresponding R-weakened and R-substantiated sets of a subset A of X are given by

$$A^{\Downarrow} = \{x \in X | \rho^{-1}(x) \subseteq A\}$$

and

$$A^{\Uparrow} = \{x \in X | \rho_X^{-1}(x) \cap A \neq \emptyset\}.$$

Proposition 12. *[6] The R-weakened and R-substantiated sets of a subset A of X are given by*

$$A^{\Uparrow} = \mu_X \circ P\rho_X(A)$$

and

$$A^{\Downarrow} = \bigvee_{\rho_X^{-1}(x) \leq A} \eta_X(x),$$

respectively.

Proposition 13. *If $A \subseteq B$ then $A^{\uparrow} \subseteq B^{\uparrow}$, $A^{\downarrow} \subseteq B^{\downarrow}$, $A^{\Uparrow} \subseteq B^{\Uparrow}$, $A^{\Downarrow} \subseteq B^{\Downarrow}$.*

The upper and lower approximations, as well as the R-weakened and R-substantiated sets, can be viewed as $\uparrow_X, \downarrow_X, \Uparrow_X, \Downarrow_X \colon PX \to PX$ with $\uparrow_X (A) = A^{\uparrow}$, $\downarrow_X (A) = A^{\downarrow}$, $\Uparrow_X (A) = A^{\Uparrow}$ and $\Downarrow_X (A) = A^{\Downarrow}$.

Considering the crisp powerset monad we define equivalent relations (reflexive, symmetric and transitive) by

Definition 2. $\rho_X : X \to PX$ *is* reflexive *if* $\eta_X \subseteq \rho_X$, symmetric *if* $\rho = \rho^{-1}$ *and* transitive *if $y \in \rho(x)$ implies $\rho(y) \subseteq \rho(x)$.*

In what follows, equivalence relations are now connected to upper and lower approximations.

Proposition 14. *The following properties hold:*

(i) If ρ_X is reflexive $A^{\downarrow} \subseteq A$ and $A \subseteq A^{\uparrow}$.
(ii) If ρ_X is symmetric $A^{\downarrow\uparrow} \subseteq A$ and $A \subseteq A^{\uparrow\downarrow}$.
(iii) If ρ_X is transitive $A^{\uparrow\uparrow} \subseteq A^{\uparrow}$ and $A^{\downarrow} \subseteq A^{\downarrow\downarrow}$.

Corollary 1. *If ρ_X is an equivalence relation, $A^{\downarrow\uparrow} = A^{\downarrow}$ and $A^{\uparrow\downarrow} = A^{\uparrow}$.*

7.3 Inverse Relations

Inverse relations in the ordinary case means to mirror pairs around the diagonal. The following propositions relate inverses to the multiplication of the corresponding monads.

Proposition 15. *[6] In the case of P,*

$$\bigvee_{\rho_X(x)\wedge A>0} \eta_X(x) = \mu_X \circ P\rho_X^{-1}(A)$$

if and only if

$$\rho_X^{-1}(x) = \bigcup_{\eta_X(x)\leq\rho_X(y)} \eta_X(y).$$

Concerning inverse relations, in the case of $\varphi = L_{id}$ we would accordingly define $\rho_X^{-1}(x)(x') = \rho_X(x')(x)$.

Proposition 16. *[6] In the case of L_{id},*

$$\mu_X \circ L_{id}\rho_X^{-1}(A)(x) = \bigvee_{x'\in X} (\rho_X(x)\wedge A)(x').$$

Consider now the powerset monads with fuzzy level sets, $\mathbf{L}^\alpha$ and $\mathbf{L}_\alpha$. For L^α is similar to L_{id} situation. Let us see how is the situation for L_α

Proposition 17. *In the case of L_α,*

$$\mu_X \circ L_\alpha\rho_X^{-1}(A)(x) = \bigwedge_{x'\in X} (\rho_X(x)\wedge A)(x').$$

Proof.

$$\begin{aligned}
\mu_X^{L_\alpha}(L_\alpha\rho_X^{-1}(A))(x) &= \bigwedge_{B\in L_\alpha X, B(x)>0, L_\alpha\rho_X^{-1}(A)(B)>0} B(x)\wedge L_\alpha\rho_X^{-1}(A)(B)\\
&= \bigwedge_{B\in L_\alpha X, B(x)>0, L_\alpha\rho_X^{-1}(A)(B)>0} B(x)\wedge \bigwedge_{x'\in\rho_X(B), A(x')>0} A(x')\\
&= \bigwedge_{B\in L_\alpha X, B(x)>0, x'\in\rho_X(B), A(x')>0} B(x)\wedge A(x')\\
&= \bigwedge_{x'\in X} \rho_X^{-1}(x')(x)\wedge A(x')\\
&= \bigwedge_{x'\in X} \rho_X(x)(x')\wedge A(x')\\
&= \bigwedge_{x'\in X} (\rho_X(x)\wedge A)(x').
\end{aligned}$$

We have made use of the definition of the inverse relation $\rho_X^{-1}(x')(x) = \rho_X(x)(x')$.

Note that in the case of $L = 2$, for the functor 2_α we obtain the classical definition of the upper approximation of a set A.

Generalizing from the ordinary power set monad to a wide range of partially ordered monads requires attention to relational inverses and complement. The role of the diagonal clearly changes, and the representation of inverses is an open question. Inverses and complements must be based on negation operators as given by implication operators within basic many-valued logic [22].

7.4 Monadic Relations and Rough Monads

Let $\Phi = (\varphi, \leq, \eta, \mu)$ be a partially ordered monad. We say that $\rho_X : X \to \varphi X$ is a *Φ-relation* on X, and by $\rho_X^{-1} : X \to \varphi X$ we denote its *inverse*. The inverse must be specified for the given set functor φ. For any $f\colon X \to \varphi X$, the following condition is required:

$$\varphi f(\bigvee_i a_i) = \bigvee_i \varphi f(a_i)$$

This condition is valid both for P as well as for L_{id}.

Remark 7. Let ρ_X and ρ_Y be relations on X and Y, respectively. Then the mapping $f : X \to Y$ is a congruence, i.e. $x' \in \rho_X(x)$ implies $f(x') \in \rho_Y(f(x))$, if and only if $Pf \circ \rho_X \leq \rho_Y \circ f$. Thus, congruence is related to kind of weak naturality.

Let $\rho_X : X \to \varphi X$ be a Φ-relation and let $a \in \varphi X$. The Φ-ρ-upper and Φ-ρ-lower approximations, and further the Φ-ρ-weakened and Φ-ρ-substantiated sets, now define rough monads using the following monadic instrumentation:

$$\Uparrow_X (a) = \mu_X \circ \varphi\rho_X(a)$$
$$\downarrow_X (a) = \bigvee_{\rho_X(x)\leq a} \eta_X(x)$$
$$\uparrow_X (a) = \mu_X \circ \varphi\rho_X^{-1}(a)$$
$$\Downarrow_X (a) = \bigvee_{\rho_X^{-1}(x)\leq a} \eta_X(x)$$

Proposition 18. *[6] If $a \leq b$, then $\Uparrow_X a \leq \Uparrow_X b$, $\downarrow_X a \leq \downarrow_X b$, $\uparrow_X a \leq \uparrow_X b$, $\Downarrow_X a \leq \Downarrow_X b$.*

Definition 3. *$\rho_X : X \to \varphi X$ is* reflexive *if $\eta_X \leq \rho_X$, and* symmetric *if $\rho = \rho^{-1}$.*

Note that in the case of relations for P and L_{id}, if the relations are reflexive, so are their inverses.

Proposition 19. *[6]*

(i) If ρ is reflexive, $a \leq \Uparrow_X (a)$.
(ii) ρ is reflexive iff $\downarrow_X (a) \leq a$.
(iii) ρ_X^{-1} is reflexive iff $a \leq \uparrow_X (a)$.
(iv) If ρ is symmetric, then $\uparrow_X (\downarrow_X (a)) \leq a$.

In the particular case $a = \eta_X(x)$ we have $a \leq \downarrow_X \circ \uparrow_X (a)$.

The idea of submonad is similar to the idea of subsets.In this sense, the calculations related to submonads is a way to reduce data in a given information system.

Let $\Phi' = (\varphi', \leq, \eta', \mu')$ be a partially ordered submonad of $\Phi = (\varphi, \leq, \eta, \mu)$. Given $a' \in \varphi' X$ we have the following proposition:

Proposition 20. *For* $a' \in \varphi' X$,

$$\uparrow_X (a') = \mu_X \circ \varphi\rho_X^{-1}(a')$$

Proof. Since $a' \in \varphi' X$, $\uparrow_X (a') = \mu'_X \circ \varphi' \rho_X^{'-1}(a')$. For defining submonad we use the mapping $e : \varphi' \to \varphi$ that is the natural transformation consisting of all inclusion mappings $e_X : \varphi' X \to \varphi X$. $\rho_X^{'-1}$ is the restriction of the relation ρ_X^{-1} on φX. Therefore $e_X \circ \rho_X^{'-1} = \rho_X^{-1}$ and $a' = e_X(a')$. The by naturality of e_X and submonad properties we obtain:

$$\begin{aligned}\uparrow_X (a') = \uparrow_X (e_X(a')) &= e_X \circ \mu'_X \circ \varphi' \rho_X^{'-1}(a') \\ &= \mu_X \circ \varphi e_X \circ e_{\varphi' X} \circ \varphi' \rho_X^{'-1}(a') \\ &= \mu_X \circ \varphi e_X \circ \varphi \rho_X^{'-1} \circ e_X(a') \\ &= \mu_X \circ \varphi (e_X \circ \rho_X^{'-1})(e_X(a')) \\ &= \mu_X \circ \varphi \rho_X^{-1}(e_X(a')) \\ &= \mu_X \circ \varphi \rho_X^{-1}(a').\end{aligned}$$

Similarly for the lower approximation we have

Proposition 21. *For* $a' \in \varphi' X$,

$$\downarrow_X (a') = \bigvee_{\rho_X(x) \leq a'} \eta_X(x)$$

These propositions show us that rough approximations are well defined wrt submonads, i.e. their definition in the submonad correspond to the one for the monad.

8 Conclusions

Partially ordered monads enrich monads with structure for generalizing rough sets. Various semantic viewpoints become available for the purpose of modelling uncertainties in many-valued logic. This paper can, on the one hand, provide a solid categorical background and, on the other hand, due to its mathematical core, establish connections with other categorical structures with the objective to enrich the theory. This is for instance the situation for topological and convergence like structures. Their applications to rough sets is future work and need to be based on suitable connections within the categorical framework. Further, algebraic structures of rough sets should be investigated, not only in direction towards topological notions but also involving logical structures. Substructures are always important as they provide additional examples.

Acknowledgements

The authors want to thanks the anonymous referees for their valuable suggestions that significantly improved the paper.

References

1. Aho, A.V., Hopcroft, J.E., Ullman, J.D.: The Design and Analysis of Computer Algorithms. Addison-Wesley, Reading (1975)
2. Barr, M., Wells, C.: Toposes, Triples and Theories. Springer, Heidelberg (1985)
3. Butzmann, H.P., Kneis, G.: On Čech-Stone compactifications of pseudo-topological spaces. Math. Nachr. 128, 259–264 (1986)
4. Eilenberg, S., Moore, J.C.: Adjoint functors and triples. Illinois J. Math. 9, 381–398 (1965)
5. Eklund, P.: A class of completely regular spaces. Internat. J. Math. & Math. Sci. 7, 197–200 (1984)
6. Eklund, P., Galán, M.A.: Monads can be rough. In: Greco, S., Hata, Y., Hirano, S., Inuiguchi, M., Miyamoto, S., Nguyen, H.S., Słowiński, R. (eds.) RSCTC 2006. LNCS (LNAI), vol. 4259, pp. 77–84. Springer, Heidelberg (2006)
7. Eklund, P., Galán, M.A., Ojeda-Aciego, M., Valverde, A.: Set functors and generalised terms. In: Proc. 8th Information Processing and Management of Uncertainty in Knowledge-Based Systems Conference (IPMU), pp. 1595–1599 (2000)
8. Eklund, P., Galán, M.A., Medina, J., Ojeda-Aciego, M., Valverde, A.: Composing submonads. In: Proc. 31st IEEE Int. Symposium on Multiple-Valued Logic (ISMVL 2001), Warsaw, Poland, May 22-24, 2001, pp. 367–372 (2001)
9. Eklund, P., Galán, M.A., Medina, J., Ojeda-Aciego, M., Valverde, A.: Set functors, L-fuzzy set categories and generalized terms. Computers and Mathematics with Applications 43, 693–705 (2002)
10. Eklund, P., Galán, M.A., Medina, J., Ojeda-Aciego, M., Valverde, A.: A categorical approach to unification of generalised terms. Electronic Notes in Theoretical Computer Science, vol. 66(5) (2002), `http://www.elsevier.nl/locate/entcs/volume66.html`
11. Eklund, P., Galán, M.A., Gähler, W., Medina, J., Ojeda-Aciego, M., Valverde, A.: A note on partially ordered generalized terms. In: Proc. of Fourth Conference of the European Society for Fuzzy Logic and Technology and Rencontres Francophones sur la Logique Floue et ses applications (Joint EUSFLAT-LFA 2005), pp. 793–796 (2005)
12. Eklund, P., Gähler, W.: Generalized Cauchy spaces. Math. Nachr. 147, 219–233 (1990)
13. Eklund, P., Gähler, W.: Fuzzy Filter Functors and Convergence. In: Applications of category theory to fuzzy subsets, Theory and Decision Library B, pp. 109–136. Kluwer, Dordrecht (1992)
14. Eklund, P., Gähler, W.: Completions and Compactifications by Means of Monads. In: Fuzzy Logic, State of the Art, pp. 39–56. Kluwer Academic Publishers, Dordrecht (1993)
15. Eklund, P., Gähler, W.: Partially ordered monads and powerset Kleene algebras. In: Proc. 10th Information Processing and Management of Uncertainty in Knowledge Based Systems Conference (IPMU 2004) (2004)
16. Galán, M.A.: Categorical Unification, Dissertation, Umeå University (2004) ISBN 91-7305-642-1
17. Goguen, J.A.: L-fuzzy sets. J. Math. Anal. Appl. 18, 145–174 (1967)
18. Gähler, W.: Monads and convergence. In: Proc. Conference Generalized Functions, Convergences Structures, and Their Applications, Dubrovnik (Yugoslavia) 1987, pp. 29–46. Plenum Press (1988)

19. Gähler, W.: Completion theory. In: Convergence Structures in Topology and Analysis, Mathematisches Forschungsinstitut Oberwolfach, Tagungsbericht 48/1991, 8
20. Gähler, W.: General Topology – The monadic case, examples, applications. Acta Math. Hungar. 88, 279–290 (2000)
21. Gähler, W., Eklund, P.: Extension structures and compactifications. In: Categorical Methods in Algebra and Topology (CatMAT), pp. 181–205 (2000)
22. Hájek, P.: Metamathematics of Fuzzy Logic. Kluwer Academic Publishers, Dordrecht (1998)
23. Järvinen, J.: On the structure of rough approximations. Fundamenta Informaticae 53, 135–153 (2002)
24. Keller, H.H.: Die Limesuniformisierbarkeit der Limesräume. Math.Ann. 176, 334–341 (1968)
25. Kleene, S.C.: Representation of events in nerve nets and finite automata. In: Shannon, C.E., McCarthy, J. (eds.) Automata Studies, pp. 3–41. Princeton University Press, Princeton (1956)
26. Kleisli, H.: Every standard construction is induced by a pair of adjoint functors. Proc. Amer. Math. Soc. 16, 544–546 (1965)
27. Kortelainen, J.: A Topological Approach to Fuzzy Sets, Ph.D. Dissertation, Lappeenranta University of Technology, Acta Universitatis Lappeenrantaensis 90 (1999)
28. Kowalsky, H.-J.: Limesräume und Komplettierung. Math.Nachr. 12, 301–340 (1954)
29. Lawvere, F.W.: Functorial semantics of algebraic theories. Dissertation, Columbia University (1963)
30. Manes, E.G.: Algebraic Theories. Springer, Heidelberg (1976)
31. Pawlak, Z.: Rough sets. Int. J. Computer and Information Sciences 5, 341–356 (1982)
32. Richardson, G.D.: A Stone-Čech compactification for limit spaces. Proc. Amer. Math. Soc 25, 403–404 (1970)
33. Robinson, J.A.: A machine-oriented logic based on the resolution principle. J. ACM 12, 23–41 (1965)
34. Rydeheard, D.E., Burstall, R.M.: A categorical unification algorithm. In: Poigné, A., Pitt, D.H., Rydeheard, D.E., Abramsky, S. (eds.) Category Theory and Computer Programming. LNCS, vol. 240, pp. 493–505. Springer, Heidelberg (1986)
35. Salomaa, A.: Two complete axiom systems for the algebra of regular events. J.ACM 13, 158–169 (1966)
36. Serra, J.: Image Analysis and Mathematical Morphology, vol. 1. Academic Press, London (1982)
37. Tarski, A.: On the calculus of relations. J. Symbolic Logic 6, 65–106 (1941)
38. Zadeh, L.A.: Fuzzy sets. Information and Control 8, 338–353 (1965)

A Rough Sets Approach to the Identification and Analysis of Factors Affecting Biological Control of Leafy Spurge

Mohamed T. Elhadi[1] and Wojciech Ziarko[2]

[1] Department of Computer Science, Sultan Qaboos University,
Alkhod 123, Oman
elhadi@squ.edu.om
[2] Department of Computer Science, University of Regina
Regina, Saskatchewan, Canada S4S 0A2
ziarko@cs.uregina.ca

Abstract. This paper constitutes an account of the authors' experiences and a presentation of results obtained in a real-life application of rough set theory's methods and techniques in the identification and analysis of data dependencies and relationships on an empirical data. The data was collected in the course of an experiment on the biological control of the Leafy Spurge [13] weed in the prairies of Western Canada using an agent beetle known as *Aphthona nigriscutis (A. n.)*. The rough set theory was applied to the data in order to identify and analyze the different factors affecting the success of the biological control of the host weed. This led to the discovery and confirmation of meaningful patterns and the computation of a set of rules for the critical application of agent beetle *A. n* in the control of Leafy Spurge weed.

1 Introduction

Plants that spread accidentally or intentionally become serious and undesirable weeds. Losses from weeds are believed to be equal to the combined losses from insects and diseases, and rank second only to losses from soil erosion [1]. Cultural and chemical practices constitute the main methods of weed control. Such methods are very costly, bring limited relief, increase soil erosion and, worst of all, they contaminate underground water and pollute the environment [1], [5]. For those reasons, such methods are considered undesirable and environmentally unsafe, especially on uncultivated and graze lands. Other alternatives are pursued. One alternative method is a Biological Control (Biocontrol). It is based on the identification and manipulation of weeds' natural enemies. Such enemies can be utilized to influence the abundance and existence of their host plants.

The transfer and manipulation of natural enemies of weeds, primarily plant feeding insects, have resulted in various levels of success of control in a wide range of weeds around the world and all over the prairies of Canada and the USA [2], [3], [4]. There are many examples of successful iocontrol programs. One of the earliest successes was with the cottony cushion scale, a pest that was devastating

J.F. Peters and A. Skowron (Eds.): Transactions on Rough Sets VIII, LNCS 5084, pp. 75–92, 2008.

the California citrus industry in the late 1800s. A predatory insect, the Vidalia beetle, and a parasitoid fly were introduced from Australia. In a short time the cottony cushion scale was completely controlled by these introduced agents. Damage from the alfalfa weevil was substantially reduced by the introduction of natural enemies. Less than two decades after their introduction, the alfalfa acreage treated for alfalfa weevil in the northeastern United States was reduced by 75 percent. A small wasp, *Trichogramma ostriniae*, brought from China to help control the European corn borer, is another example of biocontrol efforts. Many biocontrol programs for insect pests and weeds are under way across the United States and Canada [2].

Despite all the success, biocontrol still faces many challenges and obstacles, with many trials failing [6]. One critical factor in this respect is to do with a scientist's inability to make sense of the relationships and dependencies that exist among agents, weeds, level of success of the projects on one hand and the multifactors. Our research is addressing this issue in the context of rough set theory [8]. Work presented here is based on work conducted by Agriculture Canada dealing with control of Leafy Spurge weed using, among others agents, a beetle known as *Aphthona nigriscutis* (A. n.). Leafy Spurge [13] is an herbaceous perennial of Eastern European origin. It has dominated and excluded most other herbaceous plants on uncultivated land on the North American prairies since its introduction around 1865 with alarming increases in the last few decades [7]. Major projects to study and demonstrate the applicability and effectiveness of the biocontrol on the reduction of the weed have been undertaken by Agriculture Canada. It has been discovered that the growth of the A. n. agent and its effectiveness as biocontrol was determined by the interaction of a variety factors with incomplete understanding of the nature of the relationships between those factors. In our approach, a data analysis methodology, the subject of this paper, based on the application of learning and data discovery techniques derived within the framework of rough set theory, was introduced and used. The methodology was used to analyze the available data, explore hidden relationships, define new patterns, confirm existing dependencies and helpfully provide a way to predict the suitability of sites before the release of the beetle.

In summary, the objectives of this project (and consequently the subject of this paper) were:

- The identification of the major factors contributing to the effectiveness of biological control of Leafy Spurge;
- The analysis of the relationships and dependencies between the factors and obtained control results;
- The creation of general classification rules that are representative of patterns contained in the data. Such rules are to be used for the prediction of the outcome for sites with predetermined factors;
- The investigation of the use of rough set techniques and the verification of their usefulness and effectiveness as an analytical, data discovery and development tool for the biological control of leafy spurge in particular and similar tasks in general.

The rest of the paper contains a general description of the basics of rough set theory, techniques used in our experiments, the description of the experiments performed on collected data, results and analysis followed by the conclusions.

2 The Rough Set Approach

In this section we briefly review the relevant concepts contained in rough set theory as introduced by Pawlak [8][9]. It is worth noting here that many of the techniques and methods of rough set theory presented here and elsewhere are backed up with practical implementations of toolkits and systems [14]. DataQuest [15], one of the early rough set-based systems was used in our work.

2.1 Approximation Space

Given a finite non-empty set U of objects, called a universe. Typically, in applications, the objects are represented by collected observations expressed in terms of data vectors. The limits of discernibility of objects are formally expressed by an equivalence relation over a set of objects. Each object of U is characterized by a description, in the form of a set of attribute-values. In its original form as first introduced by Pawlak [8][9], an equivalence relation on the universe of objects is defined based on their attribute-values. The equivalence relation is called an *indiscernibility relation* $R(C)$, where C is a set of *condition attributes* used to represent objects belonging to the domain of interest U. The attributes are discrete and finite-valued properties of objects. Each attribute a belonging to C is a mapping $a : U \rightarrow V_a$, where V_a is a finite set of values called the *domain* of the attribute a. The indiscernibility relation represents prior classification knowledge about the universe of interest U. It is expressed in terms of identity of values of the condition attributes C on objects.

The pair (U, R) is called an approximation space and the equivalence classes of R are called *elementary sets* or *classes*. It is assumed, as it is the case in applications, that the approximation space consists of a finite number of elementary sets.

2.2 Rough Approximations

If we let R^* be a collection of all elementary sets, then any *definable* subset of the universe U is a set union of some elementary sets. All other subsets are *undefinable* or *rough*. For any definable set X there exits an uncertainty-free criterion for determining the membership status in the set of any object belonging to the universe U. The criterion normally is a logical formula specifying combinations of conditions to be satisfied by all objects belonging to the definable set. The criterion is referred to as a *description* of the set X, denoted as *des(X)*. If the set X is rough, the defining description does not exist and the membership status of some objects with respect to the set X cannot be determined with certainty. The rough set X of interest in our analysis will be referred to as *target set*. Rough sets can be approximately characterized by the following two definable sets (called lower and upper approximations respectively):

The lower approximation, or the positive region, of a rough set $X \subseteq U$ in the approximation space (U, R) is defined as

$$\underline{R}(X) = \bigcup\{E \in R^* : E \subseteq X\}. \quad (1)$$

The lower approximation of X is a union of elementary classes totally included in X, that is, this is the largest definable set contained in the rough set X. Objects belonging to the lower approximation with *certainty* belong to the set X.

The upper approximation of X, denoted as $R(X)$, is a union of these elementary classes which have some overlap with X, i.e.

$$\overline{R}(X) = \bigcup\{E \in R^* : E \cap X \neq \emptyset\}. \quad (2)$$

The upper approximation is the smallest definable set containing the rough set X. Objects belonging to the upper approximation *possibly* belong to the set X.

In addition, based on the upper and lower approximations, the *boundary area* consisting of objects whose membership status with respect to the set X is uncertain can be defined as follows:

$$BNR(X) = \overline{R}(X) - \underline{R}(X). \quad (3)$$

That is, the boundary area is a union of such elementary classes which have only partial overlap with the set X.

The union of all elementary classes which are completely disjoint from the set X, is called the negative region of X, denoted $NEGR(X)$:

$$NEGR(X) = \bigcup\{E \in R^* : E \cap X = \emptyset\} \quad (4)$$

The negative region is a largest definable collection of objects which with certainty do not belong to X.

2.3 Information Tables

The attributes used to form the approximation space are called *condition attributes C*, whereas the prediction target is called *decision attribute D*.

We will assume the target set X is represented by a single binary-valued attribute d called a *decision attribute*. The sets of objects corresponding to values of the decision attribute, that is to X and $\neg X$, are called *decision categories*.

Both condition attributes and the decision attributes define a mapping denoted as

$$\mathbf{C} \cup \{\mathbf{d}\} : U \rightarrow \mathbf{C} \cup \{\mathbf{d}\}(U) \subseteq \otimes_{a \in C \cup \{d\}} V_a, \quad (5)$$

where $\otimes$ denotes Cartesian product operator of all domains of attributes $a \in C \cup \{d\}$. Let $S \subseteq U$ be a finite subset of the universe referred to as *sample*. The mapping $\mathbf{C} \cup \{\mathbf{d}\} : S \rightarrow \mathbf{C} \cup \{\mathbf{d}\}(S)$ can be represented by an *information table* consisting of *tuples*, corresponding to elements of the collection $\mathbf{C} \cup \{\mathbf{d}\}(S)$. An example information table, based on the biocontrol data, is shown in Table 1.

Table 1. Sample partial information table for weed control data

SIZE	DATE	SP	ORG	PH	CLAY	SILT	SAND	TEXT	ECO	SLO	ASPE	REL	SHD	SUR	SHR	SV	PP	BI	EVAL
LRG	EARL	1	MID	MID	LOW	LOW	MID	SCL	SG	YES	STH	SLP	YES	PP	NO	YES	YES	YES	GOOD
LRG	EARL	2	MID	MID	LOW	LOW	MID	SCL	SG	YES	STH	SLP	YES	PP	NO	YES	YES	YES	GOOD
MID	EARL	2	MID	MID	LOW	LOW	HI	SL	SG	YES	STH	SLP	NO	FC	YES	NO	NO	NO	GOOD
MID	EARL	2	MID	MID	LOW	LOW	HI	SND	ASP	YES	STH	CONX	YES	FC	YES	NO	NO	NO	GOOD
MID	EARL	2	MID	MID	LOW	LOW	HI	LS	ASP	NO	NO	FLAT	YES	FC	YES	NO	NO	NO	FAIL
SML	MID	1	HI	HI	MID	MID	LOW	SICL	MO	YES	STH	SLP	YES	FC	YES	NO	NO	NO	FAIL
SML	MID	1	HI	MID	LOW	MID	MID	LOM	SG	YES	STH	SLP	YES	PP	YES	NO	NO	YES	FAIL
MID	MID	1	LOW	HI	LOW	LOW	HI	LS	SG	YES	EW	CONX	NO	FC	YES	NO	NO	YES	FAIL
MID	MID	1	LOW	HI	LOW	LOW	HI	LS	SG	YES	EW	SLP	NO	MB	YES	NO	NO	NO	FAIL
MID	MID	1	MID	HI	LOW	LOW	HI	SND	SG	NO	NO	CONX	NO	FC	YES	NO	NO	YES	FAIL
MID	MID	1	HI	HI	LOW	MID	MID	LOM	FG	NO	NO	FLAT	NO	FC	YES	NO	NO	NO	GOOD
MID	LATE	1	MID	HI	LOW	LOW	HI	SL	ASP	YES	STH	CONX	YES	PP	YES	NO	NO	YES	FAIL
MID	LATE	1	LOW	HI	LOW	LOW	HI	LS	SG	YES	NOR	CONV	NO	MB	YES	NO	NO	NO	GOOD
MID	LATE	1	HI	HI	MID	MID	LOW	SICL	SG	YES	STH	CONV	YES	PP	YES	NO	YES	NO	FAIL

With the above notation, the elementary set corresponding to the tuple t of values of condition attributes C, that is for $t \in \mathbf{C}(U)$, is given by:

$$E_t = \mathbf{C}^{-1}(t) = \{e \in U : \mathbf{C}(e) = t\}. \tag{6}$$

The example partial information table derived from Leafy Spurge biocontrol data is shown in Table 2 (end of section 2.5).

2.4 Dependency Analysis and Data Reduction

One of the major tasks taken by the rough sets methodology is the analysis of data dependencies, identification and elimination of redundancies in information tables. In this project, the rough sets methodology is adapted and used for the determination of the usefulness of a given information table, identification of important attributes, and assessment of their degree of importance, and relevance with respect to a predefined decision action attribute. In the context of the DataQuest system used in our experiments, the overall process involved the following steps:

- **Selection of condition and decision attributes.** A decision must be made on the factors to be used in the determination of the target (decision) attribute value.
- **Selection of the target decision value.** A choice must also be made on the decision value(s) of the decision attribute to evaluate for.
- **Computation of dependencies.** Computation of dependencies involves measuring degree of determinism in representation of decision categories in terms of combinations of values of condition attributes. Technically, the dependency measure is a relative size of all decision categories obtained by dividing the cardinality of lower approximations of the categories by the number of observations [9]. It reflects the degree of functional dependency

between condition and decision attributes . The computed degree of dependency can assume any value between 0 and 1. The system DataQuest can provide for individual or total dependency. Individual dependency refers to the degree to which individual condition attribute can be used to determine the decision attribute. Whereas total dependency refers to the degree to which the whole set of condition attributes can be used to determine the decision attribute.

- **Computation of relative reducts.** This is a process of redundancy identification and elimination [9] involving finding a minimal subset of condition attributes, which have the same degree of dependency with the decision attribute as all condition attributes. By computing the set of possible reducts, we decide on which set of condition attributes to be used for the computation of rules. All reducts preserve total dependency. The main consideration in making the selection is that of the user preferences and domain characteristics. For instance, a user may use cost of data sampling or accuracy of measurements for the selection of attributes to be used.
- **Attribute significance determination.** This is the evaluation of the relative power of individual condition attribute, with respect to total dependency between conditions and decision attributes. This is, to determine which factors are dominant and which are irrelevant, with respect to the decision. The *core* attributes [9], that is the ones that are contained in each reduct, are considered to be the most important ones as far as the dependency is concerned. In addition, the system DataQuest allows for computation of the significance factor for each condition reduct attribute, defined as the relative degree of dependency decrease due to elimination of the attribute from the reduct [15].
- **Computation of minimal rules.** Once a reduct has been selected, the system can generate rules based on suggested sets of attributes. These rules can then be used for consultation and prediction of the suitability of sites before the release of the agent. The description of the rule computation algorithm is presented in the next section.

2.5 Computation of Rules

Rule computation based on the identified dependency is one of the most fundamental aspects of the rough set theory. In the context of rough set theory, rules are expressions representing relationships between subsets of the universe U. For any definable set Y and the target set X of the universe U, the *rule* is a formal statement $des(Y) \rightarrow X$, where $des(Y)$ is a defining description of the set Y. The set Y is referred to as *rule support set.* The rule is interpreted as a statement which says that if an object matches the description of the set Y then it belongs to the target set X.

In our experiments, rules were computed by the system DataQuest according to the algorithm based on the notions of reducts of categories introduced in [9]. The algorithm produces a set of rules forming locally minimal cover of the lower approximation, or of the boundary area, of the target set X. The lower approximation-based rules are *deterministic*, or *certain*, with a unique outcome, whereas the boundary area-based rules are *non-deterministic*, or *uncertain* with several possible outcomes, in general.

Let $F = \{X_1, X_2, ...X_n\}$ be a family of subsets of U and let Y be a subset of U such that $\bigcap F \subseteq Y$. The family of sets $H \subseteq F$ is a relative *join-reduct* of categories of F with respect to the set Y if:

- $\bigcap H \subseteq Y$;
- the family of sets H is non-redundant with respect to preservation of the relation $\bigcap H \subseteq Y$, that is for every subset $X \in H$, the relation $\bigcap(H - X) \subseteq Y$ does not hold.

The join-reduct of categories H is a *locally minimal* sub-family of the family of sets F preserving the relation $\bigcap H \subseteq Y$ since no component set of H can be removed from it without affecting the relation.

The other kind of reduct of categories used in the computation of rules is the *union-reduct*. The union-reduct is a locally minimal subset of the family F preserving the union $\bigcup F$ of its members. In detail, the family of sets $H \subseteq F$ is a union reduct of F if:

- $\bigcup H = \bigcup F$;
- the family of sets H is non-redundant with respect to preservation of the relation $\bigcup H = \bigcup F$, that is for every subset $X \in H$, $\bigcup(H - \{X\}) \neq \bigcup F$.

The rules are computed from the information table obtained based on pre-declared condition and decision attributes. They can be computed for each decision category. The main stages of the rule computation algorithm of the DataQuest system are sketched as follows.

Algorithm (Computation of Locally Minimal Rules).

1. Identify descriptions of all elementary sets forming lower approximation of the selected decision category or boundary area.
2. Based on the identified descriptions, express each elementary set as an intersection of sets corresponding to symbols appearing in the description of each elementary set.
3. For each elementary set of the lower approximation (or boundary area), compute relative join-reduct of categories.
4. Form the set union of the join-reducts obtained in step (3);
5. To obtain locally minimal cover of the lower approximation (or the boundary area) of the target set, compute union-reduct of join-reducts obtained in step (4).
6. The join-reducts obtained in step (5) are definable sets whose descriptions are conjunctions of conditions defining the intersecting sets appearing in each join-reduct. The descriptions form preconditions of the rules for the lower approximation of the target set or the boundary area.

Table 2. Rules in tabular form computed from the weed control data

ID	CNO	DNO	SZE	SPN	ORG	SLT	TXT	ECO	ASP	REL	SHD	PP	BI	AF	EVL
1	2	46	–	2	–	–	SND	–	–	FLT	NO	–	–	–	GD
2	8	46	–	2	LOW	LOW	–	–	–	–	NO	–	–	–	GD
3	3	46	LRG	–	–	MID	–	–	–	–	YES	–	–	–	GD
4	4	46	–	1	–	–	–	MG	STH	–	–	–	–	–	GD
5	3	46	–	–	HI	–	SND	–	–	–	–	–	–	YES	GD
6	1	46	–	1	HI	LOW	–	–	–	–	–	–	YES	–	GD
7	2	46	–	–	LOW	–	SND	MG	–	–	–	–	–	NO	GD
8	7	46	–	–	–	–	–	–	–	SLP	–	–	–	YES	GD
9	3	46	–	–	–	–	–	–	–	FLT	YES	–	YES	–	GD
10	4	46	–	–	–	–	SND	–	STH	COX	–	–	NO	–	GD
11	2	46	–	–	–	–	SND	–	EW	SLP	–	–	–	–	GD
12	3	46	–	–	LOW	–	–	ASP	–	FLT	NO	–	–	–	GD
13	5	46	–	–	–	LOW	–	–	–	SLP	–	–	YES	–	GD
14	5	46	LRG	–	LOW	–	OTH	–	–	FLT	NO	–	–	–	GD
15	3	46	–	–	–	–	OTH	–	STH	SLP	NO	–	–	–	GD
16	1	46	–	–	–	–	–	MG	EW	–	–	–	–	–	GD
17	7	46	–	–	–	MID	SND	–	–	–	–	–	NO	NO	GD
18	1	46	–	–	–	–	–	–	–	CAV	NO	–	–	–	GD
19	1	34	–	2	–	–	–	MG	–	–	–	YES	–	–	FL
20	3	34	–	2	–	–	OTH	–	–	–	–	NO	–	–	FL
21	3	34	–	–	–	–	- -	ASP	NTH	–	–	–	–	–	FL
22	1	34	–	–	–	–	–	ASP	STH	–	–	YES	–	–	FL
23	5	34	–	–	–	MID	OTH	–	–	–	–	–	–	–	FL
24	1	34	–	–	–	–	–	MG	NON	–	–	–	–	YES	FL
25	2	34	–	1	–	LOW	SND	–	–	SLP	–	NO	NO	–	FL
26	1	34	–	–	–	LOW	–	ASP	EW	–	–	NO	–	–	FL
27	2	34	–	–	–	–	–	MG	NTH	–	–	–	–	NO	FL
28	2	34	–	1	–	LOW	SND	SG	–	–	–	NO	–	–	FL
29	8	34	–	–	–	LOW	–	ASP	NON	–	–	NO	NO	–	FL
30	4	34	–	1	–	–	–	SG	–	COX	–	–	–	–	FL
31	2	34	–	–	–	–	–	SG	EW	–	–	–	–	–	FL
32	3	34	–	–	–	LOW	–	–	–	COX	–	–	YES	–	FL
33	3	34	SMLL	–	–	–	–	–	–	–	–	–	–	–	FL

Quite independent, sometimes overlapping, but never redundant production rules are computed by the algorithm. They can be based on the whole set of data or on manually, or automatically, selected attribute reducts [9]. The rules computed are displayed in a table format, with each row corresponding to a single rule. Values that do not matter to the decision are left blank and are dealt with as do not care values. Computation of rules based on less than full functional dependency means that the system will lack sufficient information to be able to discriminate between every decision category and some boundary area rules will be produced.

A number of measures are provided by the system to help establish confidence in the computed rules include:

- **The number of rows in the information table that support each given rule (CNo).** That is, this is the number of observations that match rule condition part. This value corresponds to the rule *support* or *generality* parameter reflecting the percentage of cases in the universe matching rule condition part. The percentage of cases matching rule condition part and also matching the rule's decision part, the rule *accuracy* measure is always at maximum value of 1 for all lower approximation-based (deterministic) rules. (for more details on rule evaluation measures see, for example, [17],[16]).

- **The number of cases with a particular decision value (DNo).** That is, this is the number of observations in the information table with the same value of the decision attribute as that of the rule. This value, jointly with CNo, corresponds to the rule *coverage*, or *resolution* [17] parameter reflecting the percentage of the target set covered by objects matching rule condition part, which is given by the ratio of $\frac{CNo}{Dno}$ for deterministic rules.

The DataQuest system provides for adaptive classification environment and consultation interface. A set of condition attributes can be entered into the system to be used to predict the outcome. Incorrect decisions, different multiple decisions, or no decisions at all, are not the results of the systems failure but a lack of sufficient cases. To have the systems experience grow, every failure case can be added to the respective table with corrected and verified decision value. The task of building classification system using DataQuest reduces to an accumulation of sufficient number of representative cases.

3 Experiments

In this section, we provide a comprehensive overview of the application of the rough set methodology to the analysis of data reflecting the effectiveness of biological control of growth of weeds.

3.1 Experimental Procedure

The investigation method is based on the computer analysis of empirical data acquired from a number of test sites. Test sites are locations where the A. n. beetles are released and monitored. The data has been collected in the course of research carried out at the Regina Research Station of Agriculture Canada.

In the context of this research, the observable and measurable data and the effects on the present status of the site are expressed as attributes and attribute values for each release site. A number of experiments were done each based on a set of data with particular properties and measurements of different aspects of release sites. These runs were used at first for tuning the knowledge base so as to arrive at a final representation. The final data set was then used in identifying the most important attributes, rule generation and prediction.

3.2 Data Collection and Usage

Data from release cases were collected, tabulated and refined to arrive at a workable set of data that is suitable for computers and representable as an information system. Two important aspects were kept in mind:

- inclusion of as many factors as possible and
- inclusion of as many complete cases as possible.

Cases with incomplete data factors were excluded. A final data table representing non-uniform cases from releases made in Canadian provinces of Saskatchewan,

Table 3. Factors used in the experiments

Name	Description	Name	Description
Size	Number of beetles released in the site	Aspect	Direction of slopes
Date	Part of the summer when site created	Relief	Site's relief:concave, convex
Span	Time from site creation till evaluation	Shade	Presence of shade
Organic C	Level of Organic Carbon in soil	Cover	Presence of bare ground
pH	Ph level of the soil.	Shrubs	Presence of shrubs
Clay	Percentage level of Clay in the soil	S. c.	Presence of Stipa comata
Silt	Percentage level of Silt in the soil	S. v.	Presence of Stipa viridulav
Sand	Percentage level of Sand in the soil	P. p.	Presence of Poa Pratensis
Texture	Soil texture in terms of Sandy, loamy	B. i.	Presence of Bromus inermis
Eco Region	Ecological region type where the site is	A. f.	Presence of Artemisia frigida
Slope	Indication of the existence of slopes	E. A.	Presence of Equisetum arvensis
Evaluation	Effectiveness of the control in the site	–	–

Manitoba and Alberta was compiled. The data included an incomplete table of over 150 cases and 23 attributes. After the exclusion of agents other than the A. n., data was reduced to 128 cases of which 80 were non-redundant, complete and were used for training and testing the system. Table 3 shows the list of selected factors used in the experiment.

The following two sections contain the results of the experiments performed on the final and complete set of data [7], [11], along with their analysis and testing results.

4 Results and Analysis

Four major categories of factors, Release, Physical, Ecological and Vegetation were first analyzed separately. Then the attributes were combined and analyzed producing results as shown in Table 4 and described in the next few sections.

4.1 Data Modelling and Analysis

Two important considerations used in the evaluation of release sites were supplied by the experts. They included the amount of weed depression introduced by the control agent and density of the beetles' presence. The former is expressed as diameter of the control area whereas the latter is expressed by the number of beetles in five sweeps. The two factors were combined according to the following formula supplied by the field experts:

$$q = (diameter/2)^2(number - of - beetles) \tag{7}$$

The resulting q values were then mapped into one of GOOD or FAIL decision categories using predefined ranges as supplied by the domain expert [11]. The survival of the colony and the reduction of the target weed are thought to be a function of many factors related to Release, Physical, Ecological or Vegetation considerations. Previous studies [12] have revealed that:

Table 4. Summary of significance of factors

No	Sets	Attribute Name	Sub-sets Sig.	Comb.Set Core
1	Release Factors	Size	75	Yes
2	–	Date	25	–
3	–	Span	25	Yes
4	Physical Factors	pH	63.0	–
5	–	Texture	59.3	–
6	–	Silt	48.1	–
7	–	Organic C	44.4	–
8	–	Clay	25.9	–
9	–	Sand	0.0	Yes
10	Ecological Factors	Region	44.7	–
11	–	Cover	44.7	Yes
12	–	Shade	42.1	Yes
13	–	Relief	31.6	–
14	–	Aspect	28.9	–
15	–	Slope	0.0	–
16	Vegetation Factors	Shrubs	33.3	–
17	–	Bis	33.3	–
18	–	S.c	20.8	–
19	–	P.p	20.8	–
20	–	S.v	16.7	–
21	–	A.f	12.5	–
22	–	E.a	0.0	–
23	Effect and Evaluation	Evaluation	–	–

- The controlling agent thrives well in OPEN, DRY, and COARSE soils.
- Presence of Stipa comata (S.c), southerly slopes and bare grounds are indications of favorable conditions that should evaluate to GOOD.
- On the other hand, presence of shade, northern slopes and swales are indication of more moisture and characteristic of poor sites and should evaluate to POOR.

The initial set of data was first used and it yield 93% total dependency indicating strong patterns in the data. Some inconsistencies resulting from attribute Span produced less than 100% dependency on the table led to the elimination of the earlier releases of the inconsistent cases. It is observed that the finer the qualitative values, the stronger the dependency. Based on this idea, the database in the initial runs contained all possible values. This resulted in a 100% dependency. Some of the categories proved to be unnecessary and thus were eliminated or merged with each other while maintaining maximal dependency of (100%) [11]. This resulted in coarser values for the attributes Span, Size, Date, Organic-carbon, pH, Clay, and Sand. The values for each of those attributes were re-modelled and made coarser by merging of values. The resulting set was then used by the system with results as discussed next.

4.2 Release Factors

This subset includes Size, Date, and Span factors. Running the system on the release data revealed the following :

- A total dependency of 5.0% was obtained meaning that approximately 5% of the cases could be unambiguously classified to their correct decision categories. Obviously a very weak dependency suggesting that release data on its own may not be very detrimental to the characterizations of relationship between evaluation of success and release factors.
- Rules generated based on the release data concluded that all cases with releases of less than 50 beetles did poorly, in line with expert's expectations [7][11][12].
- The significance factor of each parameter was established and is shown in Table 4 with Size as the most important of the three (75%), followed by Date and Span (25% each). The core of this group of data is made up of Size, Date, and Span.

4.3 Physical Factors

This subset contained the attributes of Organic carbon, pH, Clay, Sand, Slit and Texture. A number of interesting results obtained as outlined next:

- A much greater dependency (33%) was obtained. That is, one third of the cases were unambiguously classified into their correct decision category. This suggested that physical data is an important component inline with expert's findings [7][11] [12].
- Except for Sand, all other attributes were considered as important. This was expected and understandable in the case of Texture, Slit and Clay since they indicated the moisture holding capacity of the soil. the PH and Organic carbon's importance were not previously known to the expert, but were satisfactorily explained since soil with high organic carbon develops in poorly drained sites. As far as pH is concerned, two aspects might explain its importance. Soils with low pH values are not good for the beetles. High pH values are an indication of alkaline soils. In turn, alkalis are indicative of dry sandy soils and are better for the beetles. This is true in the prairies where factors such as low rain fall, warm weather and long periods of sun exposure contribute to high evaporation and efficient drainage on sandy soils.
- Rules generated by the system were consistent with the above observations in that high organic carbon content was a characteristic of poor sites; 27% of the relevant cases did poorly even though they were on sandy soil with low silt content. And 47% of the sites did poorly because they were on non-sandy soil but contained high organic carbon and low pH values. On the other hand, 67% of the relevant cases that did well were releases on sandy soils with high pH values.
- The attribute Sand was given 0% significance value while Texture was rated at 59.3%. This is a very interesting result, since texture includes the sand content. Thus, the attribute Sand was not necessary and was eliminated as redundant. The core of this set of data included the attributes pH, Texture, Silt, Organic Carbon, and Clay.

4.4 Ecological Factors

The six attributes considered were Ecological region, Relief, Aspect, Slope, Surface cover and Shade. Results were as follows:

- A greater total dependency of 47.5% was obtained for this set of ecological factors. This is quite useful, since factors of topological nature require less effort in collecting and sampling when compared to physical data for example.
- All of the attributes, except for Slope, were judged important with very close values of importance suggesting that they are all equally important.
- Rules generated have confirmed the above observations quite well. For example, 26% of the relevant cases that did poorly were released on Aspen shaded-areas with no slope or bare ground (highly moist and considered bad for the beetles). The 17% of sites that did poorly were released in the Aspen region with northerly aspects. On the other hand, 21% that did well were on short grass areas with bare ground, and 17% were on southerly aspects even though they were on Aspen regions and had no bare ground. The set of attributes of Ecological region, Surface cover, Shade, Relief and Aspect were considered the most important.The interaction of such factors can result in differing degrees of moisture which can accordingly affect the beetles [7].

4.5 Vegetation Factors

This sub-set of data accounts for the occurrence of certain plant species in a site including Shrubs, Bromus inermis (Bis), Stipa comata (S.c) , Artemisia fridida (A.f), Poa pratensis (P.p) and Equisetum arvense (E.a). The occurrence of each plant is an indication of certain conditions, but its absence may not indicate the opposite conditions. Results were as follows:

- A total dependency of 30.0% was obtained for vegetation factors, higher than that of release factors but lower than physical and ecological factors.
- Interestingly enough Shrubs had the highest importance (33.3%) suggesting that groups of plants of common characteristics may be far better in characterizing the relationships than single plants. S.c had the same significance of 33.3% in accordance with expert's observations that S.c seems to indicate favorable sites. Only E. a. was left out of the most important attributes.
- Rules generated confirmed the above observations as well as previous findings [7][11]. For instance, 21% of the relevant cases that did well had Stipa comat (S.c) and 26% had no Shrubs. The core of this set is made up of the attributes Shrubs, S.c, P.p, Bis, S.v and A.f.

4.6 Combined Factors

The final run of the system was on the combined set of 23 factors. The following are some of the results obtained:

1. A total dependency of 100% was obtained. This means that all of the cases in the knowledge base were correctly and unambiguously classified. A 100% total dependency represents a very strong relationship between the condition factors and decision attributes. However, one has to be cautions with the interpretation of an extreme result like that. The high value of the dependency my be just the reflection of the relatively high number of attributes used and of the small number of observations which results in a classification containing only the singleton elementary classes leading to full (100%) dependency.
2. The list of most important factors contributing to the success or failure of the site contained Size, Span, Aspect, Texture and Relief. This was a very interesting result, since each attribute of the core set is a good indicator of moisture level, openness and soil texture. This in turn makes the list of factors quite important to the characterization of good and poor sites. This has a direct effect on the cost and efficiency of running the experiments as all core factors except for Texture are observable and easy to measure factors. Effect of each CORE attribute can be explained as follows:

   - **Size:** Small numbers of beetles will probably end up in failure whereas large ones will be successful even in less favorable sites.
   - **Span:** It seems that evaluations after one year may not indicate whether the beetle will succeed. Cases for sites of longer periods are needed to tell if the Span affects the sites.
   - **Texture:** It was found that soil types represented as sandy, which include sand, sandy loam, sandy clay loam and loamy soil types, are characteristics of good sites. Analysis of these types using a table provided by the Research Station showed that sandy soils contain about 50 % or more sand and about 50% or less silt and clay. The reverse is true for poor sites. Soils rich in silt or clay and low in sand are characteristics of unfavorable sites. This is in line with findings by [7] in that the two types are used in characterizing how coarse and dry soil is. Moist and fine soils (Silty or Clay) are poor for the beetle whereas dry and coarse (Sandy to Loam) are good for the beetle.
   - **Ecological region:** Good sites tend to be in the short grass prairies regions and are rare in the Aspen, Montane or Fescue grass regions. The opposite is true for the poor sites. However, there are exceptions. Since, the ecological region is a broad categorization of huge areas; moist sites may exist in dry regions and vise versa.
   - **Aspect:** Southerly to easterly aspects are characteristics of good sites whereas northerly aspects are characteristics of poor sites. This is consistent with findings and is expected since southerly aspects are subject to more sun exposure than northerly aspects. Thus, they are dryer and warmer [7][12]. For example, 7% of the cases did well even though they were on a non-sandy soil due to southerly aspect. Also 9% did well even though they were on unfavorable ecological region due to southerly aspect.

- **Relief:** Though flat sites can be successful, sloped and convex sites generally have a better chance of success. They tend to have better drainage, more bare grounds and thus are generally dryer. In contrast, concave sites and northerly sloped which are moister. This is evident from results obtained. For instance, 15%, that did well had slopes, whereas 25% which did poorly were flat.

3. As far as reduction is concerned, the system was able to identify 24 different combinations of attributes that can be used to discriminate good sites from poor ones. Each subset generated by the system corresponds to a combination of factors that include, along with the core, one or more of the factors from the original set. Each reduct represents a reduction from 46% to 55%. Any subset can be used to generate rules used to predict the success or failure of sites. These many different reducts represent some of the flexibility provided for the user in selecting a cost-effective, accurate and easy to sample set of factors, while maintaining the same ability of characterizing good sites from poor ones.
4. A set of rules was generated based on combined set of data. Examples of such rules are shown here, for the complete set of rules generated see [11].
5. In an attempt to verify the usefulness of the dependency expressed by the rules, 80 cases were tested out. Leave-one-out method of testing was used. That is, the test data was basically the same as the training data except, that the rules used to test one case were based on the knowledge base not containing the test case. A 100% total dependency was obtained on all 80 runs that were made. Testing results are shown in Table 6. It contains the test results for each decision and for all decision categories.
   The test results do confirm the usefulness of the rules that can be generated from the developed knowledge base. It is important however to note the following:
   - The degree of correctness (incorrectness) of the prediction made by the system varies from one decision category to another. Best results of 73% were obtained for good sites. This may be attributed to the fact that the knowledge base contained more good cases than poor ones. It is also possible that sites may fail for any considerations other than being in unfavorable places and conditions. This is certainly true for cases reported to be grazed by cattle, where no beetles were found and thus the site was judged poor rather than good. This meant that sites judged as poor by the experts were predicted good by the system. This is like saying that the site has favorable conditions but may have failed due to other reasons.
   - The system was not able to suggest any answers for 6% of the test cases. This is like saying that the knowledge base was not representative enough and that more cases and/or factors are needed to reach a maximum effectiveness.
   - All of the cases that were not predictable fall in the good decision category and none in the fail category. This confirms that more good cases need to be included in the knowledge base so as to make it representative enough.

Table 5. Sample rules for each decision category

Combined
Site is **GOOD** IF
Size = Large and Silt = Medium and Texture = Sandy and Shrubs = Yes
Size=Large and Organic Carbon = Low and Texture = Non-Sandy and Relief = Flat and Shrubs = No
Texture = Sandy and Aspect = STHerly and Relief = Convex and B.I = No
Site is **FAIL** IF
Span = two years and Ecological Region = Aspen and Aspect = North P.p =No
Sitl Content = Low and Ecological Region = Aspen and Aspect = North and P.p.= No and B. i.= No
Vegetation Factors only
Site is **GOOD** IF
Shrubs =Y es and P. P. = Yes and A. f. = Yes
Site is **FAIL** IF
Shrubs = No and S. c. =Y es and S. v. = No and A. f. = No
Ecological Factors only
Site is **GOOD** IF
Ecological Region = Aspen and Aspect = STHly = Relief = Convex and Surface = Fully covered
Site is **FAIL** IF
Ecological Region = Mixed Grass and Shade = No and Surface = Bare
Physical Factors only
Site is **GOOD** IF
Ph = High and Silt = Medium and Texture = Sandy
Site is **FAIL** IF
Organic Carbon = High and Ph = Low and Clay = Low and Texture = non-sandy
Release factors only
Site is **GOOD** IF
Size = Large and Date = Early
Site is **FAIL** IF
Size = Small

Table 6. Results of tests

Decision	—	Total	Correct	Incorrect	Unable-To
ALL	Raw	80	51	24	5
	Percent	100%	64	30%	6%
	Raw	75	51	24	-
	Percent	94%	68%	32%	-
GOOD	Raw	46	30	11	5
	Percent	58%	65%	24%	11
	Raw	41	30	11	-
	Percent	51%	73%	27%	-
POOR	Raw	34	18	16	0
	Percent	42%	53%	47%	0
	Raw	34	8	16	-
	Percent	42%	68%	32%	-

In summary, the system did well in correctly predicting the success of 73% of the good cases and 68% of all cases.

5 Conclusion

The basic elements of the original rough set model of rough sets as introduced by Pawlak were presented and applied to the biological control of Leafy Spurge by establishing its utility in analysis, discovery and rule generation. Using rough set's tools and techniques, data factors were analyzed, some dependencies were identified, and rules describing these dependencies were computed and tested.

By applying the theory in the biological control of weeds, the effects of different parameters on the success or failure of this application were analyzed. Major data groups relating to different aspects of the prairies environment were collected. The surveyed factors included four important aspects of release physical, topological characteristics and vegetation.

Results obtained confirm strongly the usefulness and need for rough set in domains where no formulas exist and little data is available.

Not only have the obtained results confirmed the findings of respective researchers but also suggested the importance of factors some of which researchers were not aware. Different combinations of factors, including the most important ones, were presented as alternatives that can be used to evaluate the appropriateness of sites before new releases are made.

Test runs on the generated rules assert the validity of such rules, as a representation of strong dependencies existent in the knowledge base. Analysis of the work done indicate that the theory of rough sets can be fruitfully applied in the field of biological control of weeds and similar fields.

Acknowledgments. Sincere thanks and appreciation are due to Dr. Peter Harries of Agriculture Canada for his help and expertise support. Also thanks to the people in the Agriculture Station for their help and support. The research reported in the article was partially supported by a research grant awarded to the second author by the Natural Sciences and Engineering Research Council of Canada.

References

1. Huffaker, C.B., Messenger, P.S. (eds.): Theory and practice of biological control, pp. 481–496. Academic Press, London (1976)
2. Julien, M.H., Griffiths, M.W. (eds.): Biological Control of Weeds: A World Catalogue of Agents and Their Target Weeds, 4th edn. CABI Publishing and the Australian Centre for International Agricultural Research, Antony Rowe, Chippenham, United Kingdom (1999)
3. Julien, M.H.: Biological control of weeds worldwide: Trends, rates of success and the future. Biocontrol News Inform. 10, 299–306 (1989)
4. Harris, P.: Biological control of weeds. In: Franz, J.M. (ed.) Biological Plant and Health Protection. Fortschr. Zool., vol. 32, pp. 123–138 (1986)
5. Harris, P.: Environmental impact of introduced biological control agents. In: Mackauer, M., Ehler, L., Roland, J. (eds.) Critical Issues in Biological Control Intercept, Andover, UK, pp. 289–299 (1990)
6. Harris, P.: Classical biocontrol of weeds: its definitions, selection of effective agents, and administrative-political problems. The Canadian Entomologist 123, 827–849 (1991)
7. Biological control of spurge: Progress report 1990-95. Regina Agriculture Station, Regina, Canada (1990)

8. Pawlak, Z.: Rough sets. International Journal of Computer and Information Sciences 11(5), 341–356 (1982)
9. Pawlak, Z.: Rough Sets: Theoretical Aspects of Reasoning About Data. Kluwer Academic Publishers, Boston (1991)
10. Pawlak, Z.: Information systems-theoretical foundations. Information Systems 6, 205–218 (1981)
11. Elhadi, P.: An experimental analysis of rough sets theory as applied to biological control of weeds. M. Sc. Thesis, Department of Computer Science, University of Regina (1991)
12. Harris, P.: Classical biological control with insects and pathogens. Paper outlining results. Regina Agriculture Station, Regina (1989)
13. Gassman, A.: Aphthona nigriscutis Foudras (Coleoptera: Chrysomelidae): a candidate for the biological control of cypress spurge and leafy spurge in North America. Final screening report. International Institute of Biological Control, Delmont, Switzerland (1985)
14. Polkowski, L., Skowron, A. (eds.): A Rough Sets Knowlege Discovery 2: Applications, Case studies and Software Systems. Physica-Verlag, Heidelberg (1998)
15. Ziarko, W.: Systems: DataQuest, DataLogic and KDD-R. In: Proc. of the Fourth International Workshop on Rough Sets, Fuzzy Sets, and Machine Discovery (RSFD 1996), Tokyo, Japan, pp. 441–442 (1996)
16. Ziarko, W.: Probabilistic rough sets. In: Ślęzak, D., Yao, J., Peters, J.F., Ziarko, W., Hu, X. (eds.) RSFDGrC 2005. LNCS (LNAI), vol. 3642, pp. 283–293. Springer, Heidelberg (2005)
17. Wong, S.K.M., Ziarko, W.: On learning and evaluation of decision rules in the context of rough sets. In: Proceedings of the International Symposium on Methodologies for Intelligent Systems, Knoxville, Tennessee, pp. 224–308 (1986)

Interpretation of Extended Pawlak Flow Graphs Using Granular Computing

Huawen Liu[1,2], Jigui Sun[1,2], and Huijie Zhang[1,3]

[1] College of Computer Science, Jilin University, Changchun 130012, China
[2] Key Laboratory of Symbolic Computation and Knowledge Engineering of Ministry of Education, Changchun 130012, China
Huaw.Liu@gmail.com, JgSun@jlu.edu.cn
[3] College of Computer, Northeast Normal University, Changchun 130021, China
Zhanghj167@nenu.edu.cn

Abstract. In this paper, we mainly discuss the relationship between the extended Pawlak flow graph (EFG) with granular computing (GrC), and develop a both simple and concrete model for EFG using GrC. The distinct advantage is that we can resort to merits of GrC to benefit us in analyzing and processing data using flow graph, for its structure is inherently consistent with GrC, which provides us with both structured thinking at the philosophical level and structured problem solving at the practical level. In pursuit of our purpose, at first, EFG will be mainly discussed in three aspects under GrC, namely, granulation of EFG, some relationships and operations of granules. Under the framework of GrC model, inference and reformation in EFG can be easily implemented in virtue of decomposition and composition of granules, respectively. Based on this scheme, two efficient reduction algorithms about EFG are also proposed.

Keywords: Pawlak Flow graph, Granular computing, Rough set, Information System, Reduct.

1 Introduction

Since Zdzisław Pawlak proposed a flow graph (FG) in his pioneer paper [18], which is a new graphical model for representing knowledge and reasoning data, series of relative papers about FG, such as [19,20,21,22,23,24,25,26,27], have been continuously put forward to place emphasis upon its importance in data analysis. In these literatures, Pawlak investigated that FG is tightly related with several theories (e.g., rough sets, decision systems, Bayes' theorem, data mining and decision tree). In addition, these works also pave the way for its application in many fields [26]. For example, Palwak firstly discussed the relations between probability theory and FG in [20]. Then FG was linked up with decision systems in sources [21,22,24], and tied up with rough sets in his recent paper [25].

Unlike optimal flow, FG mainly concerns information flow distribution in the view of quantity. In a FG, each node denotes one element set and each branch

J.F. Peters and A. Skowron (Eds.): Transactions on Rough Sets VIII, LNCS 5084, pp. 93–115, 2008.

describes flow distribution between nodes. In addition, each branch is considered as a rule. Thus, every path from the root to a leaf denotes a decision rule, where leaf represents decision class or label. As a result, a FG is a set of decision rules. In other words, FG is a kind of graph representation of decision algorithm in some ways [18]. Under this context, each branch or path associates with three coefficients, i.e., the strength, certainty and coverage factors. Moreover, these information flow distributions in FG are governed by Bayes' formula and abide by flow conservation equations [25].

As a mathematical model of finding and mining knowledge, FG has some advantages, such as intuitional representation, straightforward computation, explicit relations and parallel processing. Just owing to these, a considerable attention from many researchers has been paid on FG since it has been proposed. For example, Butz et al. argued that rough set FGs can be transformed into Bayesian networks in polynomial times [3]. Subsequently, they figured out that the FG inference algorithm in the past has exponential complexity and then presented a polynomial time complexity algorithm for inference in FG [4,5]. While Kostek and Czyzewski successfully applied the inference technology about FG in musical metadata retrieval, in order to improve retrieval efficiency [9,10].

Even so, some undesirable effects also reside in FG. For example, it cannot exactly or precisely depict the relationships among nodes in network, because FG is only based on information flow distribution and represents the relationships among nodes in quantity of flow. To tackle with this problem, Sun et al. introduced an extension of flow graph (in short, EFG) in [33] on the ground of the fact that qualitative analysis is no less important than quantification measures in data mining. This extension, however, not only has the capability of FG in the quantification aspect, but also can be interpreted by information systems or granular computing (GrC) from qualitative view.

Since information granulation has been addressed by L.A. Zadeh in his paper [40], many fruitful researches have been made (see [11], [16], [39] and [41]). To label studies on information granulation and computations, T.Y. Lin introduced a term, called GrC, which is more about a philosophical way of thinking and a practical methodology of problem solving deeply rooted in human mind [38]. Its emerging is primarily motivated by the practical needs for simplification, clarity, low cost, approximation, and tolerance of uncertainty. By effectively using levels of granularity, GrC provides a systematic, natural way to analyze, understand, represent, and solve real world problems [38]. The predominance of GrC lies in that complicated problem can be divided into several simple and easily managed sub-problems, so as to cut the solution cost down and assist users to get far away from unnecessary and trivial matters.

Ascribing to its well-structural network, EFG is intrinsically consistent with GrC [34,14]. In this paper, the relationships between EFG and GrC will be revealed from three aspects: granulation of the universe, relationships of granules and computing with granules. These elements, however, are three fundamental issues of granular computing [39]. Without loss of generality, we will first discuss granulation of EFG in details, and then involve some relationships among granules. The transition among granules describe the capability of model in solving problem among different levels of granularity. Further, an approximation reduction algorithm about EFG based on the model of GrC will be represented after decomposition and composition of granules are introduced. Furthermore, some issues of inference and reformation in EFG will be concerned later, and corresponding algorithms will also be given.

The structure of the rest is organized as follows. Section 2 briefly reviews related work about the state of the art of FG. Section 3 presents some notations of flow graph and its extension. In Section 4, the model of EFG under the interpretation of GrC is highlighted. Section 5 provides the inference and reformation procedures in EFG according to decomposition and composition on granules, respectively. Moreover, the corresponding algorithms are also given in this section. Section 6 introduces two kinds of reduction algorithms about EFG under the framework of GrC and some simulated experiments for the proposed reduction algorithms have been performed in Section 7. Finally, several concluding remarks are summarized in Section 8.

2 Related Work

Since flow graph has been introduced, many promising works have been down on it in the past years. In this section, we will briefly review the state of the art of flow graphs.

After addressed the concept of flow graph in [18], Pawlak put his great endeavor to its theoretic foundation and application. As an illustration, he pointed out that FG is closely related with rough sets and decision algorithms in [25], and then discussed in details its application in data mining [26], such as production quality and voting analysis. More works about FG proposed by Pawlak can be found in the related literatures listed in the reference.

Butz et al. in [3] figured out that FG is a special case of Bayesian network. Under this context, they demonstrated that the time complexity of the traditional inference method in FG is exponential. In addition, they investigated an efficient inference algorithm in [4] to reduce its computational cost. This method eliminates variables (i.e., layers) one by one, not all variables at a time. As a result, the reasoning process will be ended in polynomial time. To further improve its efficiency, a variable elimination order, which is measured by the number of input and output for each node of variable, was introduced in [5].

To remedy the deficiency of FG in qualitative aspect, Sun et al. [33] proposed the concept of EFG. Subsequently, they integrated EFG with granular computing together, for they shares many common characteristics in structure [34,14]. Just owing to the advantages of GrC, the inference and reformation processes in EFG can be complemented freely and easily under the partition model of GrC. Meanwhile, a reduction of EFG can be achieved in a hierarchy manner.

In a FG, each node denotes a pair of attribute and its value. This implies that FG can not be utilized to represent fuzzy information systems effectively. To alleviate this problem, Mieszkowicz-Rolka and Rolka [15] generalized FG into a fuzzy one, called fuzzy flow graph, to describe decision tables with fuzzy attributes. Under this representation, node in fuzzy FG refers to a linguistic value of attribute. In addition, a T-norm operator was chosen to calculate the information flow between nodes. Correspondingly, the certainty and strength of path were also extended.

As a graphical framework, FG is more intuitional and comprehensible in comparison with rule representation form. However, rule is widely used in data mining ascribing to its easy interpretation. Thus, several scholars attempt to build the relationship between rule and FG. For example, Pattaraintakorn et al. [17] considered graphical representation of rules for ordinal prediction, and adopted the measurements in FG to scale prediction rules. Moreover, Chitchareon and Pattaraintakorn [7] recently revealed the relationship between association rules and flow graphs, that is, a path in FG is an association rule if its confidence and certainty are all determinative. Besides, Chan and Tsumoto [6] elaborated the relationship between decision rule and FG under the context of multiset decision tables. In multiset decision table, each row denotes one path in FG and the whole table is a minimal representation of FG. Thus, the learning procedure of decision rules from inconsistent data becomes easy, and the obtained rule set is a minimal one.

For the application of FG, Kostek and Czyzewski [9,10] successfully applied FG to analyze the dependent relation between meta-data in musical databases. Its purpose is to improve the effectiveness and efficiency of access to such information and release the limitation of online resources. In temporal data mining, succeeding data is usually determined by the anterior models. To predict future behavior of modeled systems, Suraj and Pancerz [32] exploited FG to obtain this kind of dependent relation and represented it as prediction rule, so as to describe the changes of components in the consecutive time windows of a temporal information system.

3 Flow Graphs and Its Extension

In this section, some concepts of flow graph and its extension will be recalled briefly. More relevant notations can be consulted [23] and [33].

3.1 Flow Graph

A *flow graph* (FG) is a *directed, acyclic, finite* graph $G = (N, B, \varphi)$, where N is a set of *nodes*, $B \subseteq N \times N$ is a set of *directed branches*, $\varphi : B \to R^+$ is a *flow function* and R^+ is non-negative real [23].

In a FG, if $(n, n') \in B$ then n is called an *input* of n' and n' is an *output* of n reversely. $I(n)$ and $O(n)$ are the sets of all inputs and outputs of n, respectively. That is to say, $I(n) = \{n' \in N | (n', n) \in B\}$ and $O(n) = \{n' \in N | (n, n') \in B\}$. $\varphi(n, n')$ denotes the *throughflow* from node n to n'. For each node n, its *inflow* and *outflow* are $\varphi_+(n) = \sum_{n' \in I(n)} \varphi(n', n)$ and $\varphi_-(n) = \sum_{n' \in O(n)} \varphi(n, n')$, respectively.

A *normalized* FG is a *directed, acyclic, finite* graph $G = (N, B, \sigma)$, where N and B are denoted as above. $\sigma : B \to (0, 1]$ is *normalized flow* of $(n, n') \in B$ such that $\sigma(n, n') = \varphi(n, n')/\varphi(G)$. Usually, $\sigma(n, n')$ refers to the *strength* of the branch (n, n'). Correspondingly, the *inflow* and *outflow* of node n are $\sigma_+(n) = \sum_{n' \in I(n)} \sigma(n', n)$ and $\sigma_-(n) = \sum_{n' \in O(n)} \sigma(n, n')$, respectively. Obviously, the normalized throughflow of G is $\sigma(G) = 1$. The *certainty* and *coverage* of (n, n') are defined as $cer(n, n') = \sigma(n, n')/\sigma(n)$ and $cov(n, n') = \sigma(n, n')/\sigma(n')$, where $\sigma(n) \neq 0$ and $\sigma(n') \neq 0$.

Cohering with decision table closely, FGs can be directly constructed from decision tables. Furthermore, FG is a graphical representation of decision algorithms by organizing the decision rules obtained from decision table as a directed acyclic path [18]. For instance, each node $n \in N$ in FG depicts a pair of attribute-value $n = (a, v)$ and branch $(n, n') \in B$ shows the correlation between n and n'. Moreover, $\varphi(n)$ represents the meaning of its corresponding attribute-value pair $n = (a, v)$, and $\sigma(n, n')$ denotes the probability of co-occurrence about n and n'. In other words, a FG is a set of decision rules, where each decision rule is a path from root to leaf with strength, certainty and coverage. Moreover, these coefficients are accord with Bayes' theorem and flow conservation equations [25].

3.2 An Extension of Flow Graph

As mentioned above, we know that FG is a quantification graph, that is, it represents simply the relations among nodes by virtue of flow distribution of information. Despite that some valuable results can be achieved using quantitative factors, however, it is not sufficient to depict concretely and exactly the relationships among nodes. In addition, qualitative factors, as well as quantitative ones, play very important roles in data mining, for they can bring more reasonable outcomes to data analysis. Therefore, an extension of FG has been proposed in [33] on the ground of the information or objects flowing in the network.

An *extension* of flow graph (EFG) is a *directed, acyclic, finite* graph $G = (E, N, B, \varphi, \alpha, \beta)$, where E and N are the sets of *objects* and *nodes* respectively. $B \subseteq N \times N$ is the set of *directed branches*, $\varphi : B \to 2^E$ is the set of *objects* which flow through branches and $\alpha, \beta : B \to [0, 1]$ are thresholds of *certainty* and *decision*, respectively.

In the same way as the tight relation between FGs and decision tables, an EFG $G = (E, N, B, \varphi, \alpha, \beta)$ can also be illustrated as a decision table $DT = (U, C, D)$. To be more specific, each node $n \in N$ in G is an attribute-value pair (a, v) in DT, and its throughflows refers to the meaning of the pair, namely, $\varphi(n) = \{o \in U | a(o) = v\}$, where $a \in C \cup D$ and $v \in V_a$. Additionally, branch $(n, n') \in B$ means that its corresponding attribute-value pairs associate with each other, i.e., $\varphi(n, n') = \{o \in U | a(o){=}v \wedge a'(o){=}v'\}$, where U is the universe in DT, (a, v) and (a', v') are pairs with respect to n and n' respectively.

For the sake of simplicity, hereafter we assume that nodes with same attribute form a layer and an EFG are arranged in several layers, denoted by a set L. In the same layer, there does not exist any branch among nodes and one object only belongs to one node. The throughflow of (n, n') is not empty if $(n, n') \in B$.

In analogy to FG, node n is an *input* (*father*) of n', if $(n, n') \in B$, and n' is an *output* (*child*) of n reversely. Likewise, $I(n)$ and $O(n)$ are the sets of *fathers* and *children* of node n respectively. In EFG, node n is called the *root* if $I(n) = \emptyset$ holds, and n is a *leaf* if $O(n) = \emptyset$. In addition, n is an *internal* node if n is neither the *root* nor a *leaf*. The *inflow* and *outflow* of node n are respectively defined as

$$\varphi_+(n) = \bigcup\nolimits_{n' \in I(n)} \varphi(n', n) \quad and \quad \varphi_-(n) = \bigcup\nolimits_{n' \in O(n)} \varphi(n, n'). \tag{1}$$

Moreover, $\varphi(n) = \varphi_+(n) = \varphi_-(n)$ if $n \in N$ is an *internal* node. For branch $(n, n') \in B$, its input and output are nodes n and n' respectively. Thus, an object flowing through (n, n') also pass through both n and n', and vice versa. This means that if $\varphi(n)$ and $\varphi(n')$ are given, we can obtain the throughflow of $(n, n') \in B$ by

$$\varphi(n, n') = \varphi_-(n) \cap \varphi_+(n') = \varphi(n) \cap \varphi(n'), \tag{2}$$

in virtue of Eq. (1). The *input* and *output* of the EFG G are denoted as $I(G) = \{n \in N | I(n) = \emptyset\}$ and $O(G) = \{n \in N | O(n) = \emptyset\}$, respectively.

For any layer $l \in L$ in EFG, its total throughflow is $\varphi(l) = \bigcup_{n \in l} \varphi(n) = E$. This implies that $\varphi(l)$ is a partition on E, because $\varphi(n) \cap \varphi(n') = \emptyset$ for any $n, n' \in l$.

The *certainty* and *coverage* factors of each branch $(n, n') \in B$ in an EFG are

$$cer(n, n') = |\varphi(n, n')| / |\varphi(n)| \quad and \quad cov(n, n') = |\varphi(n, n')| / |\varphi(n')|, \tag{3}$$

respectively, where $|X|$ is the cardinality of X, $\varphi(n) \neq \emptyset$ and $\varphi(n') \neq \emptyset$. Meanwhile, $|\varphi(n, n')|$ is also called the *strength* of (n, n').

A *directed path* from n to n', denoted by $[n...n']$, is a sequence of nodes $n_1, ..., n_m$, where $(n_i, n_{i+1}) \in B$ for $1 \leq i \leq m-1$, $n_1 = n$, $n_m = n'$ and $\bigcap_{i=1}^{m-1} \varphi(n_i, n_{i+1}) \neq \emptyset$. In addition, the *support*, *certainty* and *coverage* of the path $[n_1...n_m]$ are

$$\begin{aligned} \varphi(n_1...n_m) &= \bigcap\nolimits_{i=1}^{m-1} \varphi(n_i, n_{i+1}), \\ cer(n_1...n_m) &= |\varphi(n_1...n_m)| / |\varphi(n_1...n_{m-1})|, \\ cov(n_1...n_m) &= |\varphi(n_1...n_m)| / |\varphi(n_m)| \end{aligned} \tag{4}$$

respectively, where $\varphi(n_1...n_{m-1}) \neq \emptyset$ and $\varphi(n_m) \neq \emptyset$. Readers interested in more detailed description should refer to the paper written by the authors.

For an EFG $G = (E, N, B, \varphi, \alpha, \beta)$, if we only cast our lights on quantity of objects flowing through branches in G, rather than concrete objects, the EFG can be transformed into a FG $G' = (N, B, \varphi')$. That is, $\varphi'(n, n') = |\varphi(n, n')| / |E|$, $\alpha = 0$ and $\beta = 0$. Additionally, an approximate FG [23] can be achieved from the EFG by adjusting the value of α or β. This implies a fact that EFG has more powerful capabilities than FG.

One probably observe that EFG is similar to information map proposed by Skowron and Synak [30]. Information map is defined by means of transition relations on set of states and every state consists of information label and information. These states is similar to nodes in EFG. However, EFG is distinguished

Table 1. Discretized Iris database

id	Petalwidth	Petallength	Sepalwidth	Sepallength	Class
o_0	≤1.65	≤2.45	(2.25, 3.05]	≤6.05	Iris-Setosa
o_1	≤1.65	≤2.45	>3.05	≤6.05	Iris-Setosa
o_2	≤1.65	(2.45, 4.95]	≤2.25	≤6.05	Iris-Versicolour
o_3	≤1.65	(2.45, 4.95]	≤2.25	>6.05	Iris-Versicolour
o_4	≤1.65	(2.45, 4.95]	(2.25, 3.05]	≤6.05	Iris-Versicolour
o_5	≤1.65	(2.45, 4.95]	(2.25, 3.05]	>6.05	Iris-Versicolour
o_6	≤1.65	(2.45, 4.95]	>3.05	≤6.05	Iris-Versicolour
o_7	≤1.65	(2.45, 4.95]	>3.05	>6.05	Iris-Versicolour
o_8	(1.65, 1.75]	(2.45, 4.95]	(2.25, 3.05]	≤6.05	Iris-Virginica
o_9	>1.75	>4.95	(2.25, 3.05]	>6.05	Iris-Virginica
o_{10}	>1.75	>4.95	>3.05	>6.05	Iris-Virginica

from information map at two aspects, i.e., binary relation and information function. The binary relation is symmetrical in EFG, which is a hierarchy structure, and can be constructed if only both nodes in adjoining layers have shared objects, whereas the relation among states is transition one and has partial orders in many cases [29]. Moreover, the functions in information map among branches shared the same father probably include common information. However, as we stated above, EFG is a partition model and there has nothing to the throughflow of branches shared the same input.

In the interest of having a better comprehension to EFG, a well-known database, *Iris Plants* (Iris), in machine learning community will be served as a running example to illuminate our ideas.

Example 1. The Iris database, which is available from the UCI machine learning repository [2], has four numerical predictive attributes (i.e., *petalwidth, petallength, sepalwidth, sepallength*) and the class. After numerical attributes have been discretized and then the same rows have been removed, Iris is shown in Table 1.

According to the relation between decision table and EFG, an EFG $G = (E, N, B, \varphi, \alpha, \beta)$, which is depicted in Fig. 1, can be built from the above decision table. It is organized as six layers in the light of five attributes: *Petalwidth, Petallength, Sepalwidth, Sepallength* and *Class*, where $E = \{o_1, ..., o_{10}\}$ denotes eleven objects observed, $N = \{n_{01}, n_{11}, n_{12}, n_{13}, n_{21}, n_{22}, n_{23}, n_{31}, n_{32}, n_{33}, n_{41}, n_{42}\} \cup \{n_{51}, n_{52}, n_{53}\}$ represents corresponding features, $\alpha = 0, \beta = 0, CL = \{l_0, l_1, l_2, l_3, l_4\}, O(G) = l_5$ and B, φ as shown in Fig. 1.

In this EFG, the root of G is n_{01} and leaves are n_{51}, n_{52} and n_{53}. The input and output of node n_{22} are $I(n_{22}) = \{n_{11}, n_{12}\}$ and $O(n_{22}) = \{n_{31}, n_{32}, n_{33}\}$, respectively. The throughflow of branch $(n_{11}, n_{21}) \in B$ is $\varphi(n_{11}, n_{21}) = \{o_0, o_1\}$. For node n_{22}, its inflow is $\varphi_+(n_{22}) = \varphi(n_{11}, n_{22}) \cup \varphi(n_{12}, n_{22}) = \{o_2, ..., o_8\}$ and outflow is $\varphi_-(n_{22}) = \varphi(n_{22}, n_{31}) \cup \varphi(n_{22}, n_{32}) \cup \varphi(n_{22}, n_{33}) = \{o_2, ..., o_8\}$, i.e., $\varphi(n_{22}) = \varphi_+(n_{22}) = \varphi_-(n_{22})$.

The sequence $n_{01}, n_{11}, n_{22}, n_{31}$ is a path and its degrees of certainty and coverage are $cer(n_{01}, n_{11}, n_{22}, n_{31}) = 2/7$ and $cov(n_{01}, n_{11}, n_{22}, n_{31}) = 1$, respectively.

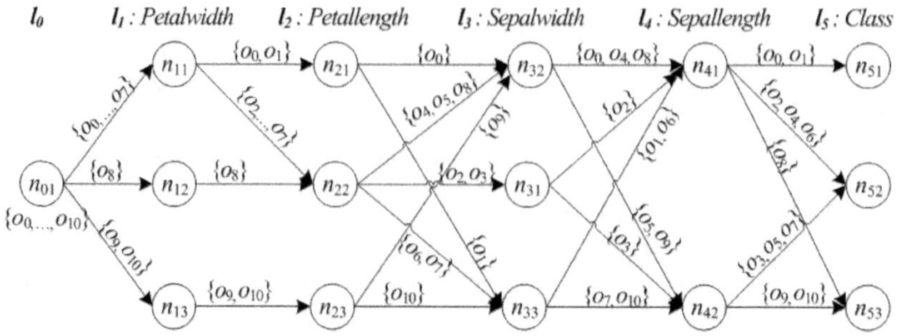

Fig. 1. An EFG G generated from the discretized Iris database, where n_{11}, n_{12}, n_{13}:$\leq$ 1.65, (1.65, 1.75], >1.75; n_{21}, n_{22}, n_{23}:$\leq$2.45, (2.45, 4.95], >4.95; n_{31}, n_{32}, n_{33}:$\leq$2.25, (2.25, 3.05], >3.05; n_{41}, n_{42}:$\leq$6.05, >6.05; n_{51}, n_{52}, n_{53}: *Iris-Setosa, Iris-Versicolour, Iris-Virginica*

In addition, sequence $n_{01}, n_{11}, n_{22}, n_{31}, n_{41}, n_{52}$ denotes a decision rule $n_{01} \wedge n_{11} \wedge n_{22} \wedge n_{31} \wedge n_{41} \rightarrow n_{52}$ in its corresponding decision table. □

4 Relationship Between EFG and GrC

As a graphical tool of data analysis in data mining, FG has been interpreted by decision algorithms, probability and rough sets [25]. However, we will investigate the relationship between EFG and GrC in this section, for EFG shares many common features with GrC in structural facet.

4.1 Granulation

With respect to a layer $l \in L$ in an EFG G, two objects $x, y \in E$ probably flow through the same node $n \in l$, i.e., $x, y \in \varphi(n)$. At this point, x is indistinguishable from y with respect to n. This means that x, y can be grouped into a *granule*.

Definition 1. *Let $G = (E, N, B, \varphi, \alpha, \beta)$ be an EFG, if $x, y \in E$ flow through $n \in N$, i.e., $x, y \in \varphi(n)$, then we will say x, y belong to the same granule $g(n)$, denoted as a pair $(n, m(n))$, where n is its descriptor and $m(n)$ is its meaning of the granule respectively, and $x, y \in m(n)$.*

This indicates a fact that the meaning of the granule $g(n)$ represents those objects which flow through the node n. Hence, the equation

$$m(n) = \{x \in E | x \in \varphi(n)\} = \varphi(n), \tag{5}$$

holds for $\forall n \in N$ in EFG.

Definition 2. *Let $G = (E, N, B, \varphi, \alpha, \beta)$ be an EFG, a granule $g(n)$ will be called an element granule if n is a single node in G, i.e., $n \in N$.*

On the basis of assumption, each object $x \in E$ only flows through one node in each layer l in EFG G. As a result, the family of granules $F(\{l\}) = \{g(n) | m(n) \neq$

$\emptyset, \forall n \in l\}$ forms a partition over E, denoted by E/l, and its corresponding equivalence relation R_l on E is

$$xR_ly \Leftrightarrow x \in m(n) \wedge y \in m(n) \wedge n \in l. \quad (6)$$

Since the equivalence class of x with reference to R_l is $[x]_{R_l} = \{y \in E|xR_ly\}$, this equivalence class is also a granule according to Def. 1 and represented as $g(n) = (n, [x]_{R_l})$, where n is one of nodes in layer l and $x \in \varphi(n)$.

Definition 3. *Let $g(n), g(n')$ be two granules, the combined granule $g(n \wedge n')$ of the granules is $g(n \wedge n') = g(n \wedge n', m(n \wedge n'))$, where $\wedge$ is combined operation.*

Proposition 1. *Let $g(n), g(n')$ be two granules and $g(n \wedge n')$ be their combined granule, then $m(n \wedge n') = m(n) \cap m(n')$ holds.*

Proof. By assuming that $x, y \in E$ are indiscernible with respect to the combined granule $g(n \wedge n')$, i.e., $x, y \in m(n \wedge n')$. This implies that x flow through both nodes n and n'. That is to say, $x \in \varphi(n)$ and $x \in \varphi(n')$. Meanwhile, y also belongs to the meanings of n and n'. Thus $x, y \in \varphi(n) \cap \varphi(n')$. In terms of Eq.(5), we have $x, y \in m(n) \cap m(n')$.

Likely, if $x, y \in m(n) \cap m(n')$, both x and y flow through nodes n and n' at the same time. As a result, $x, y \in \varphi(n \cap n')$, and $x, y \in m(n \cap n')$ on the ground of Def. 1. □

Def. 3 and Proposition 1 depict the manner of composition among granules. In EFG, each node is an element granule. Thus, the combined granule of two element granules (e.g. $g(n)$ and $g(n')$) is the granule $g(n \wedge n')$ corresponding to the branch (n, n') and its meaning $m(n \wedge n')$ is the throughflow $\varphi(n, n')$. In other words, every node or branch in EFG is a granule. Furthermore, each path is also considered as a granule by combining the nodes and branches and its throughflow is the meaning of the granule. Assume that $[n...n', n'']$ is a path, then $\varphi(n...n', n'') = \varphi(n...n') \cap \varphi(n'')$ holds. This means that the granule $g(n \wedge ... \wedge n' \wedge n'')$ consists of the granules $g(n \wedge ... \wedge n')$ and $g(n'')$ and its meaning is $m(n \wedge ... \wedge n' \wedge n'') = \varphi(n...n', n'')$. Moreover, in terms of Proposition 1 and Eq. (5), we have the following proposition.

Proposition 2. *In GrC model of EFG, if $g(n \wedge ... \wedge n' \wedge n'')$ is the granule of path $[n...n', n'']$, the meaning of this granule is*

$$m(n \wedge ... \wedge n' \wedge n'') = \varphi(n...n') \cap \varphi(n''). \quad (7)$$

Generally, for a subset of layers L' in EFG, the equivalence relation is $R_{L'} = \bigcap_{l \in L'} R_l$ such that $xR_{L'}y \Leftrightarrow \wedge_{l \in L'} x \in m(n) \wedge y \in m(n) \wedge n \in l$, and each granule is $g(\wedge_{n \in l, l \in L'} n)$, whose meaning is $m(\wedge_{n \in l, l \in L'} n) = \bigcap_{n \in l, l \in L'} m(n)$.

Moreover, if $l, l' \in L$ are different layers in the same EFG, the family of granules $F(\{l, l'\}) = \{g(n \wedge n')|m(n \wedge n') \neq \emptyset, \forall n \in l \wedge \forall n' \in l'\}$ also forms a partition $E/\{l, l'\}$ over E, for $m(n \wedge n') = \varphi(n) \cap \varphi(n')$ holds for any $n \in l, n' \in l'$.

Definition 4. *Let $g(n), g(n')$ be two granules, $g(n)$ is finer than $g(n')$, denoted as $g(n) \subseteq g(n')$, if $m(n) \subseteq m(n')$. A family of granules $F(L)$ is finer that another $F(L')$, denoted as $F(L) \subseteq F(L')$, if there exists a granule $g(n')$ in $F(L')$ for any granule $g(n) \in F(L)$, such that $g(n) \subseteq g(n')$.*

Obviously, if $L' \subseteq L$, $F(L) \subseteq F(L')$. That is, the more layers, the finer partition over E and granules generated by the layers in the light of Def. 3. What's more, all paths with the same length, which start from the root in EFG, compose a partition over E.

Example 2. **(cont.)** In the EFG G illustrated as Fig. 1, $g(n_{ij})$ is an element granule and its corresponding meaning is $\varphi(n_{ij})$, where $n_{ij} \in N$. Branch $(n_{11}, n_{22}) \in B$ forms a combined granule $g(n_{11} \wedge n_{22}) = (n_{11} \wedge n_{22}, \varphi(n_{11}) \cap \varphi(n_{22}))$. Furthermore, l_1:*Petalwidth*, l_2:*Petallength*, l_3:*Sepalwidth*, l_4:*Sepallength* and l_5:*Class* generate respectively different partitions over E and their corresponding granules are shown in the following:

$l_1 : g(n_{11}){=}(n_{11}, \{o_0, ..., o_7\}),\quad g(n_{12}){=}(n_{12}, \{o_8\}),\quad g(n_{13}){=}(n_{13}, \{o_9, o_{10}\});$
$l_2 : g(n_{21}){=}(n_{21}, \{o_0, o_1\}),\quad g(n_{22}){=}(n_{22}, \{o_2, ..., o_8\}),\quad g(n_{23}){=}(n_{23}, \{o_9, o_{10}\});$
$l_3 : g(n_{31}){=}(n_{31}, \{o_2, o_3\}),\quad g(n_{32}){=}(n_{32}, \{o_0, o_4, o_5, o_8, o_9\}),$
$\quad g(n_{33}){=}(n_{33}, \{o_1, o_6, o_7, o_{10}\});$
$l_4 : g(n_{41}){=}(n_{41}, \{o_0, o_1, o_2, o_4, o_6, o_8\}),\quad g(n_{42}){=}(n_{42}, \{o_3, o_5, o_7, o_9, o_{10}\});$
$l_5 : g(n_{51}){=}(n_{51}, \{o_0, o_1\}),\quad g(n_{52}){=}(n_{52}, \{o_2, ..., o_7\}),$
$\quad g(n_{53}){=}(n_{53}, \{o_8, o_9, o_{10}\});$

The meanings of granules with respect to $L' = \{l_0, l_1\}$ and $L'' = \{l_0, l_1, l_2\}$ are
$L' : m(n_{01} \wedge n_{11}) = \{o_0, ..., o_7\},\quad m(n_{01} \wedge n_{12}) = \{o_8\},$
$\quad m(n_{01} \wedge n_{13}) = \{o_9, o_{10}\};$
$L'' : m(n_{01} \wedge n_{11} \wedge n_{21}) = \{o_0, o_1\},\quad m(n_{01} \wedge n_{11} \wedge n_{22}) = \{o_2, ..., o_7\},$
$\quad m(n_{01} \wedge n_{12} \wedge n_{22}){=}\{o_8\},\quad m(n_{01} \wedge n_{13} \wedge n_{23}){=}\{o_9, o_{10}\}.$
Obviously, the granules in $F(L'')$ are finer than those in $F(L')$. □

4.2 Decomposition and Composition of Granules

Decomposition and composition are two necessary operations in problem-solving for granules, because these operations can traverse views among different levels of granularity at any moment. However, there is no exception to the GrC model of EFG.

Granule decomposition deals with the transition from a coarse granule to finer ones in order to provide more details for data analysis, whereas composition operation tackles with the shift from several fine granules to a coarser one to make distinct granules no longer differentiable by discarding some trivial matters [38]. Decomposition operation also means that the problem will been divided into a series of more manageable and smaller subtasks to reduce an overall computing cost, while combination operation integrates several sub-problems into one, in order to provide with a better insight into its essence rather than get buried in unnecessary details.

According to the analysis in subsection 4.1, we observe that the granule model of EFG is a partition one [37] and the conversion process from one granularity to another can be easily achieved under the framework of quotient space theory [42]. More details or extra information are needed, when a granule is decomposing into several ones, which are usually finer than their father. However, they carry more information. In EFG, information is often encoded as layers and each layer divides E into a partition on the basis of Def. 2. Hence, granule decomposition is in fact the process that breaks granules in one partition down to finer ones in another partition. Further, the finer granules in EFG are, the more layers are required.

Definition 5. *In GrC model of EFG, granules decomposition function is a mapping* $Dec : \mathcal{F}\times\mathcal{F} \to \mathcal{F}$ *such that* $Dec(\mathcal{F}', \mathcal{F}'') = \{g(n\wedge n')|m(n\wedge n') \neq \emptyset \wedge g(n) \in \mathcal{F}' \wedge g(n') \in \mathcal{F}''\}$, *where* $\mathcal{F}$, $\mathcal{F}'$ *and* $\mathcal{F}''$ *are the families of granules which compose different partitions over* E, *that is,* $\bigcup_{g(n_i)\in\mathcal{F}} m(n_i) = E$ *and* $m(n_i) \cap m(n_i')=\emptyset$ *for* $\forall g(n_i), g(n_i') \in \mathcal{F}$.

Proposition 3. *In Grc model of EFG,* $Dec(\mathcal{F}, \mathcal{F}')$ *is a partition over* E, *if* $\mathcal{F}$ *and* $\mathcal{F}'$ *are two different families of granules.*

Proof. This proposition can be proved in a straightforward way according to Def. 1, 3 and Proposition 1. □

Example 3. **(cont.)** Assume that two partitions $F(\{l_0, l_1\})$ and $F(\{l_2\})$ in *Example* 2 are given. If the family of granules $F(\{l_0, l_1\})$ is too coarse, however, a finer one can be obtained by decomposition operation $Dec(F(\{l_0, l_1\}), F(\{l_2\}))$, and it is represented as $F(L'')$ in *Example* 2. □

In contrast with decomposition, granule composition is the procedure that extracts the common information from granules regardless of distinct ones for the purpose of generalization from specificity. As described above, information is represented as layer in EFG. Thus, the common information owned by several granules refers to the common layers which they all share. For example, granule $g(n_{11})$ in *Example* 2 is the composition of granules $g(n_{11} \wedge n_{21})$ and $g(n_{11} \wedge n_{22})$, for they share the common information (i.e., *Petalwidth*), where $n_{11} \in l_1$ is a node in the *Petalwidth* layer.

Definition 6. *In GrC model of EFG, granules composition function is a mapping* $Com : \mathcal{F} \to \mathcal{F}\times\mathcal{F}$ *such that* $Com(\mathcal{F}) = \{(g(n), g(n'))|m(n) \neq \emptyset \wedge g(n\wedge n') \in \mathcal{F}\}$, *where* $\mathcal{F}$ *is the family of granules which constitutes a partition over* E, *and* $g(n)$ *and* $g(n')$ *are the common and unnecessary knowledge, respectively.*

Proposition 4. *In Grc model of EFG, if* $\mathcal{F}$ *is a partition consisting of granules, all granules* $g(n)$ *also constitute another partition over* E, *where* $(g(n), g(n')) \in Com(\mathcal{F})$.

Proof. Let $\mathcal{F}'$ be the family of granules $g(n)$. This means that $\mathcal{F}'$ is one common knowledge of $\mathcal{F}$ by $Com(\mathcal{F})$. From Def. 6, we know any granule $g(n \wedge n')$ in $\mathcal{F}$ is finer its corresponding one $g(n)$ in $\mathcal{F}'$, namely, $\mathcal{F} \subseteq \mathcal{F}'$. However, $\mathcal{F}$ is a partition over E, thus $\bigcup_{g(n)\in\mathcal{F}'} m(n) = E$. Furthermore, for any $x \in E$, there is only a granule $g(n \wedge n') \in \mathcal{F}$ such that $x \in m(n \wedge n')$. As a result, only the corresponding granule $g(n) \in \mathcal{F}'$ of $g(n \wedge n')$ satisfies $x \in g(n)$, for $g(n \wedge n')$ is finer than $g(n)$. In conclusion, this proposition holds. □

In GrC model of EFG, the size of a partition $F(L \cup L')$ over E denotes how much the knowledge we have. So the function *Com* gets the shared knowledge $F(L)$ from all granules in $F(L \cup L')$ and changes the former knowledge into a coarser one $F(L')$ at the same time.

Example 4. **(cont.)** Let the partition $F(\{l_0, l_1, l_2\})$ in *Example* 2 be given. If we only need the knowledge about *Petalwidth* (namely, l_1), rather than *Petallength* l_2. That is to say, the information about l_2 is useless and can be ignored. Thus, the common knowledge $F(\{l_0, l_1\})$ can be achieved by combining all granules in $F(\{l_0, l_1, l_2\})$. □

Since *Dec* and *Com* work under the partitions of model, they are the special cases of binary neighborhood relations [11]. Based on Def. 4, the following property is immediately accessible.

Property 1. Let G be a GrC model of EFG, if $\mathcal{F}'$ and $\mathcal{F}''$ are partitions over E in G, then
1). $Dec(\mathcal{F}', \mathcal{F}'') \subseteq \mathcal{F}'$, $Dec(\mathcal{F}', \mathcal{F}'') \subseteq \mathcal{F}''$;
2). $\mathcal{F}' \subseteq \mathcal{F}''$, where $\mathcal{F}''$ is the common knowledge generated by $Com(\mathcal{F}')$.

5 Inference and Reformation

As mentioned above, each directed path in EFG represents a granule and the longer the path, the finer the granule. Thus a kind of nested granulations hierarchy is constituted by all granules corresponding to paths starting from the root. For convenience, in the rest of this paper, granules denote those paths stemming from the root and *leaves* are arranged in *decision layer DL* (i.e., $O(G)$) and others in *condition layers CL*.

Since an EFG is the graph model of decision rules in some way, where each path from the root to a leaf is a decision rule, inference of EFG in fact is a procedure of granule decomposition. Based on Def. 3, 4 and 5, we can directly obtain the following proposition.

Proposition 5. *In GrC model of EFG,* $F(L \cup \{l\})$=$Dec(F(L), F(\{l\}))$, *where* $L \subseteq CL$, $l \in CL$ *and* $F(L), F(\{l\})$ *are families of granules generated by* L *and* $\{l\}$, *respectively.*

In the granules hierarchy, the root lies in the highest layer and the leaves is in the lowest one. The granules in the same layer form a partition over E and the granules in the i-th layer are finer than those in the j-th layer if $i > j$. Moreover, one granule in one high layer can be split into several disjoint finer granules in its next layer by granule decomposition. In terms of this principle, the data inference in EFG can be easily implemented by employing the granule decomposition operation. Meanwhile, more details (or information) about the granule can be obtained. As a result, granules decomposition (or composition) can be accessed in a top-down (or bottom-up) manner.

The main idea of the inference of EFG is that E has been separated by the root firstly, and then the partition $F(L)$ is divided by layer $\{l\} \in CL$, step by step, into a new one $Dem(F(L), F(\{l\}))$. If a granule $g(n) \in Dem(F(L), F(\{l\}))$ is finer than a decision granule $g(n') \subseteq F(DL)$, then $g(n)$ will be removed from $Dem(F(L), F(\{l\})))$, otherwise it would be processed further. The algorithm will be terminated when all granules are classified or the layers in CL are all used out. More details about inference in EFG are given in Alg. 1.

Algorithm 1. Inference algorithm in EFG

Input : An EFG $G = (E, N, B, \varphi, \alpha, \beta)$.
Output: A new EFG $G' = (E, N', B', \varphi, \alpha, \beta)$.
$F(L) = \{E\}; F(DL) = \{g(n)|n \in N \text{ is a leaf}\}; B' = \emptyset; N' = \{n|n \in N \text{ is a leaf}\}$;
while $F(L) \neq \emptyset$ *and* $CL \neq \emptyset$ **do**
  $F(L) = Dem(F(L), F(\{l\})); CL = CL - \{l\}$; //Select l from CL ;
  for $\forall g(n) \in F(L)$ **do**
    If $\exists g(n') \in F(DL)$ *and* $g(n) \subseteq g(n')$ **then**
      $F(L) = F(L) - \{g(n)\}; N' = N' \cup \{n\}; B' = B' \cup \{(n, n')\}$;
  end
end
if $F(L) \neq \emptyset$ *and* $CL = \emptyset$ **then**
  //In this case, there exists inconsistent path in EFG ;
  for $\forall g(n) \in F(L)$ **do**
    If $\exists g(n') \in F(DL)$ *and* $g(n) \cap g(n') \neq \emptyset$ **then** $B' = B' \cup \{(n, n')\}$;
    $N' = N' \cup \{n\}; F(L) = F(L) - \{g(n)\}$;
  end
end

On the contrary, the common knowledge $F(\{l\})$ can also be drawn from the specified knowledge $Dem(F(L), F(\{l\}))$ by granule composition. Similarly, the following fact holds in the light of Def. 3, 4 and 6.

Proposition 6. *In GrC model of EFG,* $Com(F(L \cup \{l\}))=(F(L), F(\{l\}))$, *where* $L \subseteq CL, l \in CL, F(\{l\})$ *is the family of coarse granules (i.e., common knowledge) generated by* $\{l\}$ *and* $F(L)$ *is the knowledge without* $\{l\}$.

This proposition illustrates that a hierarchy graph can be constructed by continuously composing granules, where each layer denotes a kind of common knowledge and different knowledge lies in different layers. At granules composition stage, coarser granules can be created and put into a higher abstract level regardless of some inessential information from several ones in the same layer. However, this is also the thought of reformation algorithm in EFG (Alg. 2). The reformation begins from the finest granules $F(L \cup \{l\})$, and then extracts continuously their common knowledge $F(\{l\})$ which forms a new layer l in EFG. The reformation algorithm will end if there is no different knowledge in $F(L)$, i.e., $|L| = 1$, and its pseudo-code is shown in Alg. 2.

Algorithm 2. Reformation algorithm in EFG

Input : An EFG $G = (E, N, B, \varphi, \alpha, \beta)$.
Output: A reformed EFG $G' = (E, N', B', \varphi, \alpha, \beta)$.
$B' = B; N' = N; F(CL) = \{g(n)|n \in CL\}$;
while $|CL| \neq 1$ **do**
    $l = first(CL); CL = CL - \{l\}$; // i.e., $CL = \{l\} \cup L$;
    for $\forall g(n \wedge n') \in F(CL)$ **do**
        if $n \in l$ **then**
            $F(CL) = F(CL) - \{g(n \wedge n')\}; F(CL) = F(CL) \cup \{g(n)\} \cup \{g(n')\}$;
            $N' = N' \cup \{n\}; B' = B' \cup \{(n, n')\}$;
            for $\forall(n'', n \wedge n') \in B'$ **do** $B' = B' \cup \{(n'', n)\}$;
        end
    end
end

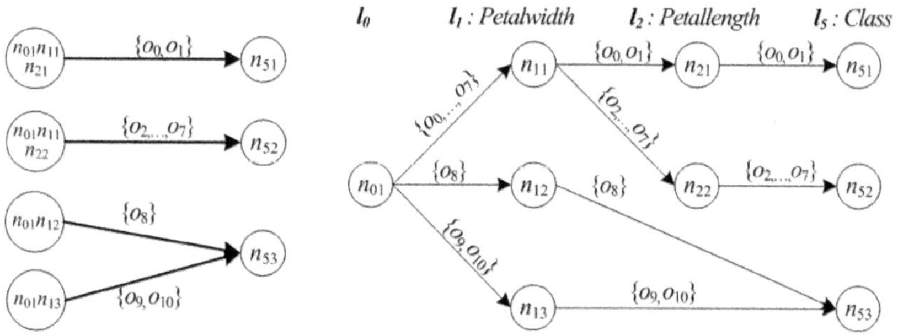

Fig. 2. An inferred EFG G' obtained from the original EFG G in *Example* 1 (left) and a reconstructed EFG G''(right) yielded by Alg. 2 on G'

Since the number of granules in each layer is at most $|E|$, the cost of granule decomposition is $|E|^2$, and Alg. 1 takes less $|CL|$ times iteration before it stops. Thus, the time complexity of Alg. 1 is $O(|E|^2|CL|)$. Likewise, the complexity of Alg. 2 is $O(|E|^2|CL|)$.

Example 5. **(cont.)** Let G be the EFG in Fig. 1. After the inference algorithm (Alg. 1) is performed over G, an inferential EFG G' is available and shown as the left of Fig. 2. Furthermore, a reconstructed EFG G'' (in the right of Fig. 2) is achieved, if the Alg. 2 is carried out on the inferential EFG G'. □

6 Reduction of EFG

An simplification of EFG can bring some advantages, such as low costs and rapid reasoning, in data analysis and reasoning. In this section, two kinds of EFG reduction algorithms based on GrC will be discussed. Above all, some definitions about reduction of EFG and path are given as follows.

Definition 7. *Let G be an EFG and $[n_1, ..., n_k]$ be a path in G, we will say n_i is dispensable in the path with respect to $O(G)$ if $\varphi(n_1, ..., n_k) = \varphi(n_1, ..., n_{i-1}, n_{i+1}, ..., n_k)$, where $1 \leq i \leq k$, otherwise n_i is indispensable. If there is no dispensable nodes in the path $[n_1, ..., n_k]$, then $[n_1, ..., n_k]$ is called a minimal reduction one.*

As we known, an EFG is a set of paths from the root to leaves. If each path in an EFG is indispensable, however, the EFG is a simplified one.

Definition 8. *Let G be an EFG, we will say G is a minimal reduction EFG if all paths in G are minimal reduction paths.*

The relationship between flow graphs and decision algorithms is first presented by Pawlak in [21], where every branch $(n, n') \in B$ is interpreted as a rule $n \rightarrow n'$, and each path $[n_1,...,n_k]$ is a sequence of rules $n_1 \rightarrow n_2,..., n_{k-1} \rightarrow n_k$, in short $n_1, ..., n_{k-1} \rightarrow n_k$. As mentioned above, a rule refers to a path in EFG. Hence, in this paper measures of rules are slightly different from those in [21].

Under the framework of GrC, each path in EFG from the root to other node is denoted as a granule, and any element contained in the meaning of the granule flows through the path. An object x satisfies granule $g(n)$, if x flows through its corresponding path n, namely, $x \in \varphi(n)$. On the ground of this interpretation, the meaning of granule $g(n_1 \wedge ... \wedge n_k)$ is the throughflow of the path $[n_1, ..., n_k]$ determining a rule $n_1, ..., n_{k-1} \rightarrow n_k$. Hence, the *support* of rule $n_1, ..., n_{k-1} \rightarrow n_k$ is equal to the throughflow $\varphi(n_1...n_k)$ of the path $[n_1, ..., n_k]$(i.e., the meaning of granule $g(n_1 \wedge ... \wedge n_k)$). Further, the *certainty* and *coverage* of $n_1, ..., n_{k-1} \rightarrow n_k$ are those certainty and coverage of the path $[n_1, ..., n_k]$, i.e.

$$cer(n_1...n_{k-1} \rightarrow n_k) = \frac{|m(n_1 \wedge ... \wedge n_{k-1}) \cap m(n_k)|}{|m(n_1 \wedge ... \wedge n_{k-1})|} = cer(n_1...n_k), \quad (8)$$

$$cov(n_1...n_{k-1} \rightarrow n_k) = \frac{|m(n_1 \wedge ... \wedge n_{k-1}) \cap m(n_k)|}{|m(n_k)|} = cov(n_1...n_k), \quad (9)$$

respectively, where $m(n_1 \wedge ... \wedge n_{k-1}) \neq \emptyset$ and $m(n_k) \neq \emptyset$. However, these quantities consistent with those in [39]. If n_k belongs to the decision layer DL, rule $n_1...n_{k-1} \rightarrow n_k$ is a decision rule.

6.1 Reformation Method

As stated in Def. 7, whether node n_i in path $[n_1, ..., n_k]$ is dispensable or not is determined by its contributions to decision-making. If its occurrence in the path does not affect the original results, it means that the node n_i is useless and can be safely removed from the path. In the view of GrC, n_i is a redundant one, if the meaning of the path $[n_1, ..., n_{i-1}]$ will not be changed after n_i has been inserted into the path. However, this gives us a good indication to eliminate superfluous nodes by adopting a top-down manner. In addition, the inference algorithm in Section 5 can be utilized to achieve our purpose after some minor modifications have been carried, for the inference procedure has the capability of producing all decision rules.

Algorithm 3. Revised Decomposition Operation

Input : Two families of granules (i.e., partitions) $F(L)$ and $F(\{l\})$ over E.
Output: The new family of granules $F(L)$.
$F(L') = F(L)$; $F(L) = Dem(F(L'), F(\{l\}))$;
for $\forall g(n) \in F(L)$ **do**
  if $\exists g(n') \in F(L')$ *and* $m(n) = m(n')$ **then**
    $F(L) = F(L) - \{g(n)\}$; $F(L) = F(L) \cup \{g(n')\}$;
  end
end

It is observed that the '$F(L) = Dem(F(L), F(\{l\}))$' statement in Alg. 1 is used to generate finer granules by decomposition operation. That is to say, some redundant information(i.e., nodes) will not be considered in this statement, if more strict constraints are imposed on it. The details of revised process is illustrated in Alg. 3.

The '$F(L) = F(L) - \{g(n)\}$' statement in Alg. 3 is applied to discard the new granule $g(n)$, while the aim of '$F(L) = F(L) \cup \{g(n')\}$' is to add the original granule $g(n')$ of $g(n)$ into $F(L)$. This indicates that the corresponding path n' of $g(n')$ will not benefit from the knowledge $F(\{l\})$ because of $m(n) = m(n')$. After the Alg. 1 with the revised decomposition operation is conducted on an EFG, a more simply inferential EFG can be obtained. Moreover, a reduction of the original EFG is available when the Alg. 2 is employed later. Alg. 4 is the reduction algorithm about EFG.

Algorithm 4. Reduction algorithm of EFG

Input : An EFG $G = (E, N, B, \varphi, \alpha, \beta)$.
Output: A reduction of EFG $G = (E, N, B, \varphi, \alpha, \beta)$.
1) Performing the revised algorithm Alg. 1 with Alg. 3 on G ;
2) Running Alg. 2 on G generated in step 1);

6.2 Inference Method

As previous illustration, Alg. 4 consists of two steps. The first one is the inference procedure from the top to bottom, while the other is the reformation about the EFG reversely. Besides this compound method, this subsection introduces another reduction algorithm which only adopts a manner of top-down.

A rule $n \to n'$ in EFG is *certainty* if $cer(n \to n')$=1, namely, $\varphi(n) \subseteq \varphi(n')$, and $m(n) \subseteq m(n')$ in its GrC model. However, it is too rigid to the real-life world filled with noise data. To cope with this problem, an variable degree of certainty rule is introduced in a naive way.

Definition 9. *[43] A rule $n \to n'$ is approximation certainty if $cer(n \to n') \geq \beta$, where 0< $\beta \leq$1.*

In GrC model of EFG, if granule $g(n_1 \wedge ... \wedge n_k)$ corresponding to $[n_1, ..., n_k]$ is finer than decision granule $g(n_d) \in F(DL)$, all sub-granules $g(n_1 \wedge ... \wedge n_k \wedge n_{k+1})$

Algorithm 5. Approximation reduction algorithm of EFG

Input : An EFG $G = (E, N, B, \varphi, \alpha, \beta)$.
Output: An approximation reduction of EFG $G' = (E, N', B', \varphi, \alpha, \beta)$.

```
F(L) = {E}; F(DL) = {g(n)|n ∈ N is a leaf};
B' = ∅; N' = {n|n ∈ N is a leaf or the root};
for ∀g(n1 ∧ ... ∧ ni) ∈ F(L) do
    IsExistDispensableNode = False ;
    if ∃g(n) ∈ F(DL) and |m(n1∧...∧ni) ∩ m(n)|/|m(n1∧...∧ni)| ≥ β then
        // g(n1 ∧ ... ∧ ni) → g(n) is an approximation rule ;
        B' = B' ∪ {(ni, n)} ;
    else
        for ∀(ni, nj) ∈ B and m(n1∧...∧ni) ∩ m(nj) ≠ ∅ do
            if m(n1∧...∧ni) ∩ m(nj) = m(n1∧...∧ni) then
                IsExistDispensable = True ; // nj is dispensable ;
                for ∀(nj, nt) ∈ B do B = B ∪ (ni, nt) ;
            else
                B' = B' ∪ {(ni, nj)}; N' = N' ∪ {nj} ;
                if g(nj) ∉ F(DL) then
                    m(n1∧...∧ni ∧ nj) = m(n1∧...∧ni) ∩ m(nj) ;
                    F(L) = F(L) ∪ g(n1 ∧ ... ∧ ni ∧ nj);
                end
            end
        end
    end
    if IsExistDispensableNode = False then F(L) = F(L) − {g(n1 ∧ ... ∧ ni)};
end
```

are also finer than $g(n_d)$. At this case, other knowledge $F(l)$ is useless for $g(n_1 \wedge ... \wedge n_k)$ with respect to $F(DL)$, where $g(n_{k+1}) \in F(l)$. Based on this fact, an approximation reduction algorithm about EFG is described as Alg. 5. Alg. 5 is slightly distinguish from the inference algorithm (Alg. 1). The main distinctions lie in the manner of processing dispensable nodes and inserting branches and nodes into new EFG G'. In Alg. 5, redundant nodes will not be accessed by updating their related branches (the third *For* statement), whereas Alg. 1 will not tackle useless nodes unless their corresponding rules have been generated. Moreover, branches and nodes are inserted into new EFG G', once rules have formed. However, all branches and nodes will be preserved in G' except those related with superfluous nodes in the approximation reduction algorithm. The time complexity of Alg. 5 is $O(|E|^2|CL|)$, where $|E|$ is the number of objects occurring in G and $|CL|$ is the maximal number of conditional layers. Under the context of GrC, however, the reduction algorithms (Alg. 4 and Alg. 5) are all lower than those in [33] where the time complexity of nodes reduction algorithm of EFG is $O(|E|^{2|CL|})$.

One may consider that different layer orders will lead to different EFG and for a given EFG G, there are probably many reductions. In order to achieve better results, different methods should be adopted to suit for specific problems at hand. Therefore, heuristic strategies, such as the priority of layers with most

Table 2. The steps of the Alg. 5

Loop	*State*
0	$F(L)=\{g(n_{01})\}$; $N'=\{n_{01},n_{51},n_{52},n_{53}\}$; $B'=\{\}$;
1	$F(L)=\{g(n_{01}\wedge n_{13}),g(n_{01}\wedge n_{12}),g(n_{01}\wedge n_{11})\}$; $N'=\{n_{01},n_{51},n_{52},n_{53},n_{13},n_{12},n_{11}\}$; $B'=\{(n_{01},n_{13}),(n_{01},n_{12}),(n_{01},n_{11})\}$;
2	$F(L)=\{g(n_{01}\wedge n_{12}),g(n_{01}\wedge n_{11})\}$; $N'=\{n_{01},n_{51},n_{52},n_{53},n_{13},n_{12},n_{11}\}$; $B'=\{(n_{01},n_{13}),(n_{01},n_{12}),(n_{01},n_{11}),(n_{13},n_{53})\}$;
3	$F(L)=\{g(n_{01}\wedge n_{11})\}$; $N'=\{n_{01},n_{51},n_{52},n_{53},n_{13},n_{12},n_{11}\}$; $B'=\{(n_{01},n_{13}),(n_{01},n_{12}),(n_{01},n_{11}),(n_{13},n_{53}),(n_{12},n_{53})\}$;
4	$F(L)=\{g(n_{01}\wedge n_{11}\wedge n_{22}),g(n_{01}\wedge n_{11}\wedge n_{21})\}$; $N'=\{n_{01},n_{51},n_{52},n_{53},n_{13},n_{12},n_{11},n_{22},n_{21}\}$; $B'=\{(n_{01},n_{13}),(n_{01},n_{12}),(n_{01},n_{11}),(n_{13},n_{53}),(n_{12},n_{53}),(n_{11},n_{22}),(n_{11},n_{21})\}$;
5	$F(L)=\{g(n_{01}\wedge n_{11}\wedge n_{21})\}$; $N'=\{n_{01},n_{51},n_{52},n_{53},n_{13},n_{12},n_{11},n_{22},n_{21}\}$; $B'=\{(n_{01},n_{13}),(n_{01},n_{12}),(n_{01},n_{11}),(n_{13},n_{53}),(n_{12},n_{53}),(n_{11},n_{22}),(n_{11},n_{21}),$ $(n_{22},n_{52})\}$;
6	$F(L)=\{\}$; $N'=\{n_{01},n_{51},n_{52},n_{53},n_{13},n_{12},n_{11},n_{22},n_{21}\}$; $B'=\{(n_{01},n_{13}),(n_{01},n_{12}),(n_{01},n_{11}),(n_{13},n_{53}),(n_{12},n_{53}),(n_{11},n_{22}),(n_{11},n_{21}),$ $(n_{22},n_{52}),(n_{21},n_{51})\}$;

nodes [12], need to be taken into account to obtain a minimal reduction of EFG. In fact, the order of importance of attributes is also a crucial problem to classification in data mining. For the sake of simplicity, we only employed a greedy method in Alg. 4 and Alg. 5.

Example 6. **(cont.)** Let G be the EFG depicted in Fig. 1 and $\beta = 1$, the steps of the approximation reduction algorithm are shown in Table 2 and the result (namely, the approximation reduction EFG G'' of G) is illustrated as Fig. 2, after Alg. 5 has been performed. For the simplified EFG G'', only four decision rules are available if required. While eleven decision rules will be generated by the LEM2 from the Iris database. □

7 Simulation Experiments

In order to validate the reduction algorithms presented in the above section is feasible in practice, eleven benchmark data sets, which are all available from the UCI Machine Learning Repository [2], were used in our experiments. Most of the datasets are frequently used in literatures and the machine learning community. The appropriate references of the origin can be obtained from the UCI website. Table 3 shows some general information about these datasets. The number of objects (i.e., records) in each dataset is shown as the second column and the third is the number of attributes (namely, features) with respect to the datasets.

All experiments were conducted on a Pentium IV, with a CPU clock rate of 2.8 GHz, 512 MB main memory. The proposed reduction algorithm[1] has been

[1] The code is available at http://ccst.jlu.edu.cn/ai/hwliu/publications/EFG07.zip

Table 3. Data sets used in experiments

Data sets	Records	Attributes	Data sets	Records	Attributes
Autos	205	26	Heart-disease	270	14
Breast-cancer	286	10	Mushroom	8124	23
Credit-a	690	16	Sonar	208	61
Diabetes	768	9	Vehicle	846	19
Flags	194	30	Zoo	101	18
Glass	214	10			

implemented in Borland Delphi 7. In experiments, the last attribute in each dataset was taken as decision class. Since EFG can only process discrete and complete data, some pre-processing steps must be performed on datasets. Currently, various tools are available and capable of dealing with this problem. However, we choose RSES software (version 2.2), which has been developed by Skowron et al. [28] and is now widely applied in the rough set community and data mining domain. The main reason is that RSES not only has implemented many preprocessing and classification algorithms, but also is competent for exploring the intermediate results to end-users, so as to help users to analyze the experimental data. Our experiments briefly consist of the following steps:

- **Step one: Handling with missing values**. Up to now, several typical approaches have been applied in practice to cope with the missing value, such as ignoring or deleting objects with missing values, or treating it with a special value. Among these methods, an interposition approach, which replaces missing values with some values in the same attribute, perhaps is the most often used. In our experiments, we filled the missing values with most common values or with mean ones in RSES.
- **Step two: Discretization**. Discretization refers to transform attributes with continuous values into categorical ones, in order to reduce time complexity in rule induction and to prevent over-fitting in classification. Numerous

Table 4. The experimental results of the reduction algorithms

Data sets	Original EFG		Reduced EFG		Elapsed time (s)
	Nodes	Layers	Nodes	Layers	
Autos	96	25	47	14	0.064
Breast-cancer	44	9	32	8	0.032
Credit-a	76	15	44	11	0.126
Diabetes	34	8	28	7	0.148
Flags	326	29	216	5	0.342
Glass	33	9	25	6	0.031
Heart-disease	28	13	20	8	0.031
Mushroom	119	22	48	9	1.25
Sonar	73	60	21	10	0.031
Vehicle	49	18	39	13	0.281
Zoo	144	17	86	6	0.031

Notes: the number of layers only refers to the condition layers.

discretization methods have been proposed till now [13]. However, a simple discretization in RSES, known as "a global discretization" [1], was taken into consideration in our experiments.
- **Step three: Reduct and inference in EFG**. After being performed preprocessing, data can be used to construct EFG. We have been implemented reduct and inference algorithms in Delphi 7.0. Since only dataset with TAB format has been recognized in RSES, our program also permit datasets with this format.

Intuitionally, the reduction result of the proposed reduction algorithm is determined by the order of condition layers in EFGs. That is to say, different arranged orders of condition layers in an EFG would yield different results. To obtain fair results, we carried out experiments repeatedly 10 times, and the orders of condition layers was arranged randomly at each time. Ultimately, the mean values were obtained and presented in Table 4.

From the experimental results, we know that lots of condition layers and nodes were reduced effectively in many cases. For example, in the Sonar database, only ten condition layers and twenty-one nodes were preserved in the reduction EFG, while its original one had 60 layers and 73 nodes. This means that the reduction algorithm works well. However, its limitation is also evident. In our experiments, the numbers of layers and nodes in the reduction EFGs varied from each other, under the different arranged orders of the condition layers. For instance, the least number of nodes was twenty-four with seven layers, while the least number of layers was two and its corresponding nodes were 109 in the case of the Zoo database.

8 Conclusion

Owing to its well-structured representation, EFG is closely relative with GrC. In this paper, we firstly discussed the relation between EFG and GrC from three aspects in GrC, and then gave a GrC model of EFG. Under this interpretation, each node, branch or path in EFG represents a granule, and the meaning of the granule is the throughflow of its corresponding path. Moreover, the granules, whose corresponding path starts from the root of EFG, form a hierarchical structure. In this hierarchy, the granules with respect to the same length paths are arranged in the same layer and getting finer along with the depth of the hierarchy.

Benefiting from structured ideas in GrC, the inference and reformation procedures in EFG can be easily acquired through granule composition and decomposition operation. Since the GrC model of EFG is a partition one, however, the transform of granularity between different layers can be freely achieved without loss of knowledge. Further, their corresponding algorithms have also been implemented in polynomial time. Moreover, two kinds of efficient reduction algorithms about EFG have been proposed.

Since EFGs require a great deal of storage, however, this is unpractical in processing mass data. Moreover, the arranged order of condition layers will perform great effect on the reduction algorithms. Thus, our future works will be carried on how to reduce EFG's storage and arrange the condition layers in EFG heuristically.

Acknowledgment

The authors would like to thank anonymous reviewers, Prof. J.F. Peters and A. Skowron for their valuable comments and constructive suggestions which have helped to improve the clarity and the completeness of the paper. This work is supported by the National NSF of China(60773097) and Doctor Point Founds of Educational Department(20050183065) and Science Foundation for Young Teachers of Northeast Normal University(20081003).

References

1. Bazan, J., Nguyen, H.S., Nguyen, S.H., Synak, P., Wróblewski, J.: Rough set algorithms in classification problem. In: Polkowski, L., Tsumoto, S., Lin, T. (eds.) Rough Set Methods and Applications, pp. 49–88. Physica, Heidelberg (2000)
2. Blake, C.L., Merz, C.J.: UCI Repository of Machine Learning Databases, University of California (1998), `http://www.ics.uci.edu/~mlearn/MLRepository.html`
3. Butz, C.J., Yan, W., Yang, B.: The Computational Complexity of Inference using Rough Set Flow Graphs. In: [31], pp. 335–344 (2005)
4. Butz, C.J., Yan, W., Yang, B.: An Efficient Algorithm for Inference in Rough Set Flow Graphs. Transaction on Rough Sets V, 102–122 (2006)
5. Butz, C.J., Yan, W.: Current Trends in Rough Set Flow Graphs. In: Hassanien, A.E., Suraj, Z., Ślęzak, D., Lingras, P. (eds.) Rough Computing: Theories, Technologies and Applications, Idea Group Inc. (2007)
6. Chan, C.-C., Tsumoto, S.: On Learning Decision Rules From Flow Graphs. In: Proceedings of North American Fuzzy Information Processing Society (NAFIPS 2007), pp. 655–658 (2007)
7. Chitchareon, D., Pattaraintakorn, P.: Knowledge Discovery by Rough Sets mathematical Flow Graphs and Its Extension. In: Proceedings of the IASTED International Conference on Artificial Intelligence and Applications (AIA 2008), Innsbruck, Austria, pp. 340–345 (2008)
8. Greco, S., Hata, Y., Hirano, S., Inuiguchi, M., Miyamoto, S., Nguyen, H.S., Słowiński, R. (eds.): RSCTC 2006. LNCS (LNAI), vol. 4259. Springer, Heidelberg (2006)
9. Czyzewski, A., Szczerba, M., Kostek, B.: Musical Metadata Retrieval with Flow Graphs. In: [35], pp. 691–698 (2004)
10. Kostek, B., Czyzewski, A.: Processing of Musical Metadata Employing Pawlak's Flow Graphs. In: Peters, J.F., Skowron, A. (eds.) Transactions on Rough Sets I, pp. 279–298. Springer, Heidelberg (2004)
11. Lin, T.Y.: Granular Computing on Binary Relations I: Data Mining and Neighborhood Systems. In: Skoworn, A., Polkowski, L. (eds.) Rough Sets In Knowledge Discovery, pp. 107–121. Springer, Heidelberg (1998)
12. Lin, T.Y., Yin, P.: Heuristically Fast Finding of the Shortest Reducts. In: [35], pp. 465–470 (2004)

13. Liu, H., Hussain, F., Tan, C.L., Dash, M.: Discretization: An Enabling Technique. Data Mining and Knowledge Discovery 6, 393–423 (2002)
14. Liu, H., Sun, J., Qi, C., Bai, X.: Inference and Reformation in Flow Graph using Granule Computing. In: Kryszkiewicz, M., Peters, J.F., Rybinski, H., Skowron, A. (eds.) RSEISP 2007. LNCS (LNAI), vol. 4585, pp. 261–270. Springer, Heidelberg (2007)
15. Mieszkowicz-Rolka, A., Rolka, L.: Flow Graphs and Decision Tables with Fuzzy Attributes. In: Rutkowski, L., Tadeusiewicz, R., Zadeh, L.A., Żurada, J.M. (eds.) ICAISC 2006. LNCS (LNAI), vol. 4029, pp. 268–277. Springer, Heidelberg (2006)
16. Nguyen, H.S., Skowron, A., Stepaniuk, J.: Granular Computing, A Rough Set approach. Computational Intelligence 17(3), 514–544 (2001)
17. Pattaraintakorn, P., Cercone, N., Naruedomkul, K.: Rule learning: Ordinal prediction based on rough sets and soft-computing. Applied Mathematics Letters 19(12), 1300–1307 (2006)
18. Pawlak, Z.: Decision algorithms, Bayes' Theorem and Flow Graphs. In: Proceeding of the 6th Conference on Neural Networks and Soft Computing (2002)
19. Pawlak, Z.: The Rough Set View on Bayes' Theorem. In: Pal, N.R., Sugeno, M. (eds.) AFSS 2002. LNCS (LNAI), vol. 2275, pp. 106–116. Springer, Heidelberg (2002)
20. Pawlak, Z.: Probability, truth and flow graphs. In: Proceedings of the Workshop on Rough Sets in Knowledge Discovery and Soft Computing at ETAPS, pp. 1–9 (2003)
21. Pawlak, Z.: Flow graphs and decision algorithms. In: [36], pp. 1–11 (2003)
22. Pawlak, Z.: Decision Networks. In: [35], pp. 1–7 (2004)
23. Pawlak, Z.: Some Issues on Rough Sets. In: Peters, J.F., Skowron, A., Grzymała-Busse, J.W., Kostek, B.z., Świniarski, R.W., Szczuka, M. (eds.) Transactions on Rough Sets I. LNCS, vol. 3100, pp. 1–58. Springer, Heidelberg (2004)
24. Pawlak, Z.: Decisions rules and flow networks. European Journal of Operational Research 154, 184–190 (2004)
25. Pawlak, Z.: Rough Sets and Flow Graphs. In: [31], pp. 1–11 (2005)
26. Pawlak, Z.: Flow Graphs and Data Mining. In: Peters, J.F., Skowron, A. (eds.) Transactions on Rough Sets III. LNCS, vol. 3400, pp. 1–58. Springer, Heidelberg (2005)
27. Pawlak, Z.: Decision Trees and Flow Graphs. In: [8], pp. 1–11 (2006)
28. Skowron, A., Bazan, J.G., Latkowski, R.: RSES 2.2 User's Guide. Institute of Mathematics, Warsaw University (2002), `http://logic.mimuw.edu.pl/~rses/`
29. Skowron, A., Synak, P.: Patterns in Information Maps. In: Alpigini, J.J., Peters, J.F., Skowron, A., Zhong, N. (eds.) RSCTC 2002. LNCS (LNAI), vol. 2475, pp. 453–460. Springer, Heidelberg (2002)
30. Skowron, A., Synak, P.: Reasoning Based on Information Changes in Information Maps. In: [36], pp. 229–236 (2003)
31. Ślęzak, D., et al. (eds.): Rough Sets, Fuzzy Sets, Data Mining and Granular Computing. Springer, Berlin (2005)
32. Suraj, Z., Pancerz, K.: Flow Graphs as a Tool for Mining Prediction Rules of Changes of Components in Temporal Information Systems. In: Yao, J., Lingras, P., Wu, W.-Z., Szczuka, M., Cercone, N.J., Ślęzak, D. (eds.) RSKT 2007. LNCS (LNAI), vol. 4481, pp. 468–475. Springer, Heidelberg (2007)
33. Sun, J., Liu, H., Zhang, H.: An Extension of Pawlak's Flow Graphs. In: Proceedings of the 1st International Conference on Rough Sets and Knowledge Technology, Chongqing, pp. 191–199 (2006)

34. Sun, J., Liu, H., Qi, C., Zhang, H.: An Interpretation of Flow Graphs by Granular Computing. In: [8], pp. 448–457 (2006)
35. Tsumoto, S., Słowiński, R., Komorowski, J. (eds.): RSCTC 2004. LNCS (LNAI), vol. 3066. Springer, Heidelberg (2004)
36. Wang, G.Y., et al. (eds.): RSFDGrC 2003. LNCS (LNAI), vol. 2639. Springer, Heidelberg (2003)
37. Yao, Y.Y.: A partition model of granular computing. In: Peters, J.F., Skowron, A. (eds.) Transactions on Rough Sets I. LNCS, vol. 3100, pp. 232–253. Springer, Heidelberg (2004)
38. Yao, Y.Y.: Perspectives of Granular Computing. In: Proceedings of 2005 IEEE International Conference on Granular Computing, vol. 1, pp. 85–90 (2005)
39. Yao, Y.Y., Zhong, N.: Granular computing using information tables. In: Lin, T.Y., Yao, Y.Y., Zadeh, L.A. (eds.) Data Mining, Rough Sets and Granular Computing, pp. 102–124. Physica, Heidelberg (2002)
40. Zadeh, L.: Towards a theory of fuzzy in formation granulation and its centrality in human reasoning and fuzzy logic. Fuzzy Sets and Systems 90, 111–121 (1997)
41. Zadeh, L.: Some reflections on soft computing, granular computing and their roles in the conception, design and utilization of information/intelligent systems. Soft Computing 2, 23–25 (1998)
42. Zhang, L., Zhang, B.: The quotient space theory of problem solving, In: [36], pp. 11–15 (2003)
43. Zhang, C., Zhang, S. (eds.): Association Rule Mining: models and algorithms. LNCS (LNAI), vol. 2307. Springer, Heidelberg (2002)

Generalized Indiscernibility Relations: Applications for Missing Values and Analysis of Structural Objects

Wojciech Jaworski

Faculty of Mathematics, Computer Science and Mechanics
Warsaw University, Banacha 2, 02-07 Warsaw, Poland
wjaworski@mimuw.edu.pl

Abstract. In this paper, we discuss an approach to structural objects based on a generalisation of indiscernibility relation used in rough set theory. The existing results in rough set theory are based on the assumption that objects are perceived by attribute value vectors.

We propose the new point of view on rough set theory. We replace information systems with the knowledge representation models that incorporate information relative to the structure of objects. We redefine the indiscernibility relation as a relation on objects characterised by some axioms. Such a definition can be naturally applied for information systems with missing values and multivalued attributes. We extend the approach on structural objects. We introduce the meaning representation language for expressing properties of structural objects and we show how to select relevant formulae from this language for sequential data.

Keywords: Indiscernibility relation, rough sets, knowledge representation, sequential data, missing values, multivalued attributes, structural objects, model of reality, data analysis, information system.

1 Introduction

Rough set theory [28,29] is based on the analysis of information systems consisting of sets of objects characterised by attribute value vectors. This data representation found successful application in many fields [30,31,32]. However, such information systems have some drawbacks when we deal with multivalued attributes or attributes with missing values. Moreover, the representation by means of vector of feature values should be treated as a consequence of a more primitive representation of structural objects. This representation is next used for computing the values of relevant attributes.

For the above reasons, we postulate a new approach to knowledge representation. We model the reality and our domain knowledge using an approach based on logic, in particular on model theory [22]. We describe our knowledge by a set of axioms written in a formal language. Such an approach solves semantic ambiguities which are present, for example, in information systems with missing values: the missing value of an attribute may be interpreted as originally

J.F. Peters and A. Skowron (Eds.): Transactions on Rough Sets VIII, LNCS 5084, pp. 116–145, 2008.

specified for a given object, yet unknown under the present knowledge or the attribute may be not applicable for a certain case or the attribute may have any value from its domain for this object. In order to solve these ambiguities, an additional domain knowledge, not represented in a given information system, is needed.

We define rough set concepts such as indiscernibility, definability and set approximations in terms of axioms. We prove that, in thecase of complete information systems, the proposed approach is equivalent to the approach used so far in rough set theory [28,29]. Then we compare our idea with extensions of rough set theory for information systems with missing values and multivalued attributes presented in [8,9,14,15,16,17,18,19,20,4,27].

Finally, we extend the rough set approach on structural objects. We introduce the meaning representation language for expressing properties of structural objects represented by data. Our language provides description of features of data irreducible relative to the attribute value vectors. The idea of knowledge representation by means of formal language was thoroughly studied in Artificial Intelligence [36].

Structural objects are often represented by means of the sequential data generated as the result of purposeful actions, for example, textual documents, voice signals or recorded parameters of cars on road. Unfortunately, the sequential representation of structural objects is computationally opaque — the structure is hidden in the sequence and the fair deal of knowledge is needed to extract it [3,34].

We show how to extract object descriptions from the sequential data using a relevant rule system (grammar of regular language) equipped with semantic attachments. Our approach is motivated by Structural Pattern Recognition [5,24,26], Natural Language Processing [12] and Information Extraction [21] as well as by Knuth's Attributed Grammars [13]. We already studied this feature extraction task in [11].

Structural objects are composed out of subobjects connected by relational constraints. These relations together with properties of objects create an ontology [2,7]. The ontology defining the structure of objects, is acquired from an expert. It helps to select the object properties relevant for the further applications. The ontology is representing hints showing how to translate data into the meaning representation language formula and how to identify objects.

The novelty of our approach is that it makes possible the generalisation of the definition of indiscernibility, making it independent from the existence of missing values, multivalued attributes and other characteristics of data. This definition is driven by semantics and it depends on the meaning of features. This allows us to merge indiscernibility with domain knowledge, introduce structure on the attribute values, express the fact that one value may be a specific case of another and deal with the problem that one object may have many different descriptions.

The indiscernibility classes are defined on the basis of the structure of objects. They provide us patterns that approximate high level concepts relevant for classification and knowledge discovery. These patterns are expressed by logical formulae.

Missing values and multivalued attributes are examples of ambiguity in data. These are simple cases illustrating ambiguity arising during the sequential data processing. Due to the ambiguity we obtain various contradicting interpretations of data. Each of them represents a possible model of reality. We may not determine which of these interpretations is correct without acquiring further information such as a domain knowledge or more data.

This paper is structured as follows. In Section 2, we define our knowledge representation model. In Section 3, we present the application of our knowledge representation model to the information systems. In Section 4, we show how to represent structural objects as the meaning representation language formulae. In Section 5, we describe the transformation of sequential data into the meaning representation language formulae. In Section 6, we define the set approximations. In Section 7, we conclude our paper presenting applications and extensions of our ideas.

2 Knowledge Representation

The goal of data analysis is to discover information about reality from a given data. However, the data itself can also be treated as a kind of information. In this section we present the knowledge representation model and we consider the problem of a relationship between this representation and the reality.

The latter question is known in the philosophy of language as the reference problem. Some ideas presented below corresponds to picture theory of language developed by Wittgenstein in [37].

Overview of various theories of reference in the computer science context is presented in [10] and an example of the basic data model for rough set theory is reported in [6].

2.1 Reality Perceived by Sensors

We do not possess the direct insight into the nature of reality. We perceive it by means of sensors. Sensors generate structural data on the basis of reality, by extracting objects and relations among them. Sensors recognise properties of objects as well as relations between them. As examples of sensors, one can consider a thermometer, a camera as well as an expert. Generally, every analysis of the reality that results in structural data may be understood as obtained by means of sensor measurements.

We represent sensor measurements by means of relational structures developed using model theory [23]. In model theory structures are composed out of a set of individuals, and some relations among them. The set of all individuals is called as the universe of the structure. We also refer to the individuals as objects or entities.

Individuals and relations from a relational structure are represented by symbols. The set of such symbols is called a signature. The interpretation is a function that maps symbols from a given signature to individuals and relations. The interpretation assigns objects or relations to names. It assures that the object or relation represent the aspect of sensor activity stated by its name. For example, the symbol `red` is interpreted as an object that represents the red colour. The signature is a lexicon for every syntactic construct (i.e. language formula, information system, image) by means of which we describe sensor measurements.

Let the structure $\mathcal{P}$ be a model of the reality observed by our sensors [1]. We assume that $\mathcal{P}$ is a relational structure of signature $\Sigma_{\mathcal{P}}$ and with the universe P. $\Sigma_{\mathcal{P}}$ consists of symbols of constants $u_1, u_2, \ldots$ and relational symbols $a_1, a_2, \ldots$. In symbols, $\mathcal{P}$ can be described by:

$$\mathcal{P} = \langle P, u_1^{\mathcal{P}}, u_2^{\mathcal{P}}, \ldots, a_1^{\mathcal{P}}, a_2^{\mathcal{P}}, \ldots \rangle,$$

where P includes objects as well as attribute values for objects $u_1^{\mathcal{P}}, u_2^{\mathcal{P}}, \ldots$, and $a_1^{\mathcal{P}}, a_2^{\mathcal{P}}, \ldots$ are interpretations of the symbols from $\Sigma_{\mathcal{P}}$. Constants are mapped to individuals and relational symbols to relations.

Let us consider the following example: Our goal is to describe persons. Our sensors recognise person properties such as `name`, `age`, `colour of hair`, `number of hairs`, `known languages`. A person may be a `parent` for another person and the `colour of hair` of some person may be `similar` to the `colour of hair` of another person.

In our example of a world, we are given 10 persons: $p_1, p_2, \ldots, p_{10}$. We know the `names`: `Alice, Bob, Charlie, David`. The `age` may be `young` or `old`. The possible `colours of hair` are `blond`, `brown`, `black` or non specified directly h_1, h_2, h_3. The `number of hairs` varies from `none`, `little`, `many`. The `known languages` are `Pascal`, `Ocaml` and `Cobol`.

We denote our example world as $\mathcal{P}_1$. The signature $\Sigma_{\mathcal{P}_1}$ consists of

- constants $p_1, p_2, \ldots, p_{10}$, `Alice`, `Bob`, `Charlie`, `David`, `young`, `old`, `blond`, `brown`, `black`, h_1, h_2, h_3, `none`, `little`, `many`, `Pascal`, `Ocaml`, `Cobol`,
- unary relations `is a person`, `is a hair colour` that define categories of objects
- binary relations `name`, `age`, `colour of hair`, `number of hairs`, `known languages`, `parent` that define person attributes .
- binary relation `similar` that define the similarity between `colours of hair`.

The structure $\mathcal{P}_1$ is defined as follows:

$$\mathcal{P}_1 = \langle P_1, p_1^{\mathcal{P}_1}, p_2^{\mathcal{P}_1}, \ldots, p_{10}^{\mathcal{P}_1}, \texttt{Alice}^{\mathcal{P}_1}, \texttt{Bob}^{\mathcal{P}_1}, \texttt{Charlie}^{\mathcal{P}_1}, \texttt{David}^{\mathcal{P}_1},$$

$$\texttt{young}^{\mathcal{P}_1}, \texttt{old}^{\mathcal{P}_1}, \texttt{blond}^{\mathcal{P}_1}, \texttt{brown}^{\mathcal{P}_1}, \texttt{black}^{\mathcal{P}_1}, h_1^{\mathcal{P}_1}, h_2^{\mathcal{P}_1}, h_3^{\mathcal{P}_1}$$

$$\texttt{none}^{\mathcal{P}_1}, \texttt{little}^{\mathcal{P}_1}, \texttt{many}^{\mathcal{P}_1}, \texttt{Pascal}^{\mathcal{P}_1}, \texttt{Ocaml}^{\mathcal{P}_1}, \texttt{Cobol}^{\mathcal{P}_1},$$

$$\texttt{is a person}^{\mathcal{P}_1}, \texttt{is a hair colour}^{\mathcal{P}_1},$$

$$\texttt{name}^{\mathcal{P}_1}, \texttt{age}^{\mathcal{P}_1}, \texttt{colour of hair}^{\mathcal{P}_1}, \texttt{number of hairs}^{\mathcal{P}_1},$$

$$\texttt{known languages}^{\mathcal{P}_1}, \texttt{parent}^{\mathcal{P}_1}, \texttt{similar}^{\mathcal{P}_1} \rangle.$$

[1] We mean the *model* in a sense of mathematical modelling, and the *structure* in a sense of model theory.

There is no clear distinction between individuals and values of individual attributes. Therefore, we consider interpretation of every constant as an object. For example, property values like names of colours are objects. In our example, the hair colours are values of the attribute `colour of hair` and individuals on whose the relation `similar` is defined.

Objects may be abstract entities. The constant `Alice` refers to the abstract entity which is a *name*, while each p_i points to a certain person.

Let us consider the digital camera as an another example. We may define photos, data produced by this sensor, as a structure whose universe is composed out of pixels, colours and objects that represent the picture itself. We introduce binary relations `horizontal neighbours` and `vertical neighbours` that defines the topology of pixels and the 3-argument relation `colour of pixel` whose arguments are a picture, pixel and colour. The latter relation provide us an access to the picture contents.

Assume that we would like to recognise people on photos. We add new sensor which provides data structuralised by concepts such as `man`, `face`, `eye`, `hand`. The problem of people recognition is equivalent to the problem of describing one sensor concept in terms of some other concepts.

We consider objects in the universe as atomic entities. We describe their structure by means of relations yet we do not asume that objects are explicitely represented by vectors of feature values. There are two reasons for such a decision. First, in order to define structure of a given object one must possess its unambiguous decomposition into essential components. And the problem is whether such essential components exist. The second reason is that that such a structure is accessible only on the metalanguage level and therefore is useless. Instead of this, we represent the internal structure of objects by means of relations between them, for examle the *meronymy relation*.

Objects perceived by sensors need not necessarily be the real things. Objects may be epiphenomena, created by sensors (for example by human perception), nevertheless we need them in order to operate on reality. While modelling the reality we do not recognise existing objects. We define them.

2.2 Semantics of Knowledge

Knowledge, or information, provides us an insight into $\mathcal{P}$, the model of the reality observed by sensors. Information about $\mathcal{P}$ is represented by means of symbols connected by syntactic rules. Languages, information systems, images, tables, time series etc. are examples of such representations.

Besides the symbolic representation we do not possess any insight into $\mathcal{P}$ and this representation describes the results of sensor measurements in an incomplete and ambiguous way. It includes information only for a part of interesting us sensors and objects. For a given information there exists lots of different models consistent with this information.

We introduce semantics for all symbolic representations. This semantics allows us to define the symbolic representations in the formal way and translate information between such representations.

Definition 1. *We describe semantics of symbolic representation by the class of structures, which we name as the class* possible worlds *and denote as* $\mathbb{P}$. *Every* $\mathcal{Q} \in \mathbb{P}$ *is a model of possible reality, i.e., the world that does not contradict our knowledge. Each* $\mathcal{Q} \in \mathbb{P}$ *posses signature* $\Sigma_{\mathcal{Q}}$ *and interpretation* $I_{\mathcal{Q}}$.

The class of structures $\mathbb{P}$ defines knowledge independently from the syntactic medium. Every $\mathcal{Q} \in \mathbb{P}$ describes sensor measurements in a precise way ($\mathcal{Q}$ describes also the possible reality with the precision relative to the sensor precision). Imprecise description provided by the symbolic representation is interpreted as a collection of precise descriptions. Possible worlds consists of all the possible extensions of the set of given sensor measurements without any estimation or inductive reasoning over unknown measurements. Even as the multicipity of possible worlds refer to the incompleteness of knowledge, the $\mathcal{P} \in \mathbb{P}$ statement defines its correctness: the knowledge is correct if the real world belongs to the class of possible ones.

For each $\mathcal{Q} \in \mathbb{P}$ its signature $\Sigma_{\mathcal{Q}}$ is a set of atomic symbols available for symbolic representation. Signature differs across the possible worlds since some symbols may be unknown to us or we may assume existence of something that in fact does not really exist.

The interpretation is a link between sensor measurements and symbols. A part of symbols refers to the sensor construction. Interpretations of these symbols should be correlated among the possible worlds. In the digital camera example constants that represent pixels and colours as well as relations `horizontal neighbours` and `vertical neighbours` should be identical in all possible worlds.

In order to formalise the above we introduce *primitives.*

Definition 2. Primitives *are symbols such that*

- *For each primitive symbol* σ
$$\sigma \in \Sigma_{\mathcal{P}}.$$
- *For each constant primitive symbol* σ *and for each* $\mathcal{Q} \in \mathbb{P}$ *if* $\sigma \in \Sigma_{\mathcal{Q}}$ *then*
$$I_{\mathcal{P}}(\sigma) = I_{\mathcal{Q}}(\sigma).$$
- *For each n-ary relational primitive symbol* σ *and for each* $\mathcal{Q} \in \mathbb{P}$ *if* $\sigma \in \Sigma_{\mathcal{Q}}$ *then*
$$\forall_{u_1,\ldots,u_n \in P \cap Q}\, I_{\mathcal{P}}(\sigma)(u_1,\ldots,u_n) = I_{\mathcal{Q}}(\sigma)(u_1,\ldots,u_n).$$

In other words, each primitive has the same interpretation for all real objects in every possible world. The primitives for unreal objects may be defined in an arbitrary way. We require from the interpretations of all possible worlds to preserve the constraints defined by primitives.

Primitives connect names with specific objects. They allow us state that objects have diverse properties, belong to distinct sorts. We provide the *meaningful names* for primitives in order to represent their metalanguage definitions.

In our example the constants `Alice`, `Bob`, `Charlie`, `David`, `young`, `old`, `blond`, `brown`, `black`, `none`, `little`, `many`, `Pascal`, `Ocaml`, `Cobol` and the categories of objects `is a person`, `is a hair colour` are primitives

2.3 Language

Signature symbols denote basic concepts given by sensors. Complex concepts are defined out of basic ones by means of language. The language is a set of syntactic rules for connecting signature symbols. Each syntactic rule generate a language formula. For each rule there is provided a method of calculation the interpretation of the formula generated by the rule .

Definition 3. *Let $\mathbb{A}$ be a set of language formulae without free variables. We say that $\mathbb{A}$ is* valid *in the structure $\mathcal{Q}$ iff for each formula $a \in \mathbb{A}$, truth is an interpretation of a in the structure $\mathcal{Q}$. We denote the above as*

$$\mathcal{Q} \models \mathbb{A}.$$

We use language for the knowledge representation. In terms of language we define sensor properties.

We create constraints for symbols whose interpretation depend on the nature of the sensor. The requirement of category for the relation argument is defined by the formula that states that if a relation is true for a certain objects as its arguments, then these objects must belong to a certain category. The category itself is determined by a primitive relation. For example

$$\forall_{x,y}\ \texttt{similar}(x,y) \Longrightarrow \texttt{is a hair colour}(x) \wedge \texttt{is a hair colour}(y)$$

and

$$\forall_{x,y}\ \texttt{name}(x,y) \Longrightarrow \texttt{is a person}(x) \wedge$$
$$\wedge\ (y = \texttt{Alice} \vee y = \texttt{Bob} \vee y = \texttt{Charlie} \vee y = \texttt{David}).$$

In the same way we may state that for each object the attribute has exactly one value. For example

$$\forall_x \exists!_y\ \texttt{name}(x,y).$$

We provide the meaningful names for symbols whose interpretation is restricted by the properties of sensors

Such an approach assures us that the given symbol represents corresponding relations in all possible worlds. For example, the relation denoted by symbol `colour of hair` may vary in different possible worlds, yet it is desired for it to point in each world to the colour of hair.

The second use of language is to define values of sensor measurements.

We define the class $\mathbb{P}(\mathbb{A})$ of possible realities by a set of language formulae $\mathbb{A}$, which we call axioms:

$$\mathbb{P}(\mathbb{A}) = \{\mathcal{Q} : \mathcal{Q} \models \mathbb{A}\}.$$

Definition 4. *We say that the set of language formula φ is a* semantic consequence *of axioms $\mathbb{A}$ iff for each structure $\mathcal{Q}$, such that $\mathcal{Q} \models \mathbb{A}$,*

$$\mathcal{Q} \models \varphi.$$

We denote the above as

$$\mathbb{A} \models \varphi.$$

Table 1. Notation

Symbol	Interpretation
$\mathcal{P}$	a structure that is a real world
$\Sigma_{\mathcal{P}}$	the signature of $\mathcal{P}$
P	the universe of $\mathcal{P}$
$I_{\mathcal{P}}$	the interpretation of $\mathcal{P}$
$\mathcal{Q}$	an arbitrary structure
$\Sigma_{\mathcal{Q}}$	the signature of $\mathcal{Q}$
Q	the universe of $\mathcal{Q}$
$I_{\mathcal{Q}}$	the interpretation of $\mathcal{Q}$
$\mathbb{P}$	the class of possible worlds
$\mathcal{Q} \in \mathbb{P}$	a possible world
$\mathbb{A}$	the set of axioms
$\models$	the relation of semantic consequence
$\mathcal{Q} \models \mathbb{A}$	the axioms $\mathbb{A}$ are satisfied in the structure $\mathcal{Q}$
$\mathbb{A} \models \varphi$	the formula φ is the semantic consequence of axioms $\mathbb{A}$
U	the universe of information system
A	the set of attributes of information system
$m(u, a)$	the set of values of the attribute a for the object u in the information system.
$\underline{\mathbb{A}}X$	lower approximation of X
$\overline{\mathbb{A}}X$	upper approximation of X

The knowledge is provided in three ways: Primitives defined in metalanguage assure the connection between symbols and elements of sensor measurements. Axioms expressed as language formulae describe the properties of sensors. Language is used also to formulate axioms that define values of sensor measurements.

The Table 1 contains the summary of notation introduced in this and the following sections.

3 Information Systems

In this section, we consider data sets presented in a form of *information system* [27]. We propose axiomatic representation of information systems, define indiscernibility and set approximations for complete information systems, then we extend the definitions on the case of missing values and multivariate attributes. We compare our approach with the literature.

3.1 Complete Data

An example of information system is presented in Table 2, we denote this system by I_1. The system I_1 contains information about structure $\mathcal{P}_1$ from Section 2, yet not the whole information. Rows of the information system represent known objects, elements of $\mathcal{P}$ universe, while columns are labelled by known attributes. Attributes are relations in $\mathcal{P}$. The set of labels of objects described in the information system will be denoted by U. In Table 2, $U = \{p_1, p_2, p_3, p_4, p_5\}$. The set of all attributes included in the information system will be denoted by

Table 2. A complete information system

	name	age
p_1	Alice	young
p_2	Alice	old
p_3	Bob	young
p_4	Bob	old
p_5	Bob	young

A. In Table 2, $A = \{\texttt{name}, \texttt{age}\}$. We assume that both U and A are finite. Each attribute a defines a relation between the set of objects and the set of attribute values V_a. In Table 2, $V_{\texttt{name}} = \{\texttt{Alice}, \texttt{Bob}\}$ and $V_{\texttt{age}} = \{\texttt{young}, \texttt{old}\}$. Any information system defines attribute values for given objects. Let

$$m(u, a)$$

denote the set of values of the attribute a for the object u in the information system. Usually each attribute has exactly one value for each object, i.e $m(u, a)$ contains one element for every u and a. In such a case information system is called *complete*.

We consider an information system as a sensor. Constants that represent attribute values are primitives. We introduce the primitive relation `is an object` which recognise objects measured by sensor described in the information system. For each $u \in U$ we the statement `is an object`(u) is true, yet the relation `is an object` is broader: it contains all objects that the sensor perceived, perceive and will perceive. The information system provides also the structural information about the domains of the attributes. We encode this information in the following way: for each attribute a we state

$$\mathcal{P} \models \forall_{x,y}\ a(x, y) \Longrightarrow \texttt{is an object}(x) \wedge y \in V_a.$$

The complete information system also states that every attribute has exactly one value for each object: for each attribute a we state

$$\mathcal{P} \models \forall_x \left(\texttt{is an object}(x) \Longrightarrow \exists!_y\ a(x, y)\right).$$

We encode the information system as a set of axioms $\mathbb{A}$ in the following way: For each $u \in U$, for each $a \in A$ we state

$$\mathcal{P} \models a(u, v),$$

where $v \in m(u, a)$ in the information system.

The above transformation treats both an object etiquette and an attribute value as constants. The attributes are considered as binary relations.

For Table 2, our knowledge about $\mathcal{P}_1$ provided by an information system I_1 is restricted to the following axiom:

$$\mathcal{P}_1 \models \texttt{name}(p_1, \texttt{Alice}) \wedge \texttt{age}(p_1, \texttt{young}) \wedge \texttt{name}(p_2, \texttt{Alice}) \wedge \texttt{age}(p_2, \texttt{old}) \wedge$$
$$\wedge \texttt{name}(p_3, \texttt{Bob}) \wedge \texttt{age}(p_3, \texttt{young}) \wedge \texttt{name}(p_4, \texttt{Bob}) \wedge \texttt{age}(p_4, \texttt{old}) \wedge$$
$$\wedge \texttt{name}(p_5, \texttt{Bob}) \wedge \texttt{age}(p_5, \texttt{young}).$$

Axioms derived for Table 2 allow us to define many different structures. The set of possible worlds $\mathbb{P}(\mathbb{A})$ consists of all the possible extensions of information system presented in Table 2. $\mathbb{P}(\mathbb{A})$ will include

$$\mathcal{Q}_1 = \langle Q_1, p_1^{\mathcal{Q}_1}, p_2^{\mathcal{Q}_1}, \ldots, p_5^{\mathcal{Q}_1}, \texttt{Alice}^{\mathcal{Q}_1}, \texttt{Bob}^{\mathcal{Q}_1}, \texttt{young}^{\mathcal{Q}_1}, \texttt{old}^{\mathcal{Q}_1}, \texttt{name}^{\mathcal{Q}_1}, \texttt{age}^{\mathcal{Q}_1} \rangle$$

as well as

$$\mathcal{Q}_2 = \langle Q_2, p_1^{\mathcal{Q}_2}, p_2^{\mathcal{Q}_2}, \ldots, p_{10}^{\mathcal{Q}_2}, \texttt{Alice}^{\mathcal{Q}_2}, \texttt{Bob}^{\mathcal{Q}_2}, \texttt{young}^{\mathcal{Q}_2}, \texttt{old}^{\mathcal{Q}_2}, \texttt{name}^{\mathcal{Q}_2}, \texttt{age}^{\mathcal{Q}_2} \rangle$$

or

$$\mathcal{Q}_3 = \langle Q_3, p_1^{\mathcal{Q}_3}, p_2^{\mathcal{Q}_3}, \ldots, p_5^{\mathcal{Q}_3},$$
$$\texttt{Alice}^{\mathcal{Q}_3}, \texttt{Bob}^{\mathcal{Q}_3}, \texttt{young}^{\mathcal{Q}_3}, \texttt{old}^{\mathcal{Q}_3}, \texttt{name}^{\mathcal{Q}_3}, \texttt{age}^{\mathcal{Q}_3}, \texttt{parent}^{\mathcal{Q}_3} \rangle.$$

Axioms do not provide any information about objects $p_6, \ldots, p_{10}$. They may have arbitrary properties. Relations name, age and parent are specified in any way that satisfy $\mathbb{A}$. They do not need to be consistent with their definition in $\mathcal{P}$.

Rough set theory [28,29] is based on the idea of an indiscernibility relation. Let B be a nonempty subset of the set A of all attributes. The indiscernibility relation $IND(B)$ is a relation on objects in a complete information system defined for $x, y \in U$ as follows

$$(x, y) \in IND(B) \text{ iff } \forall a \in B \big(m(x, a) = m(x, a)\big)$$

or equivalently

$$(x, y) \in IND(B) \text{ iff } \forall a \in B\ \forall v \in V_a\ \big(a(x, v) \Longleftrightarrow a(y, v)\big).$$

For example, for Table 2, p_3 and p_5 are indiscernible with respect to the attributes name and age.

The indiscernibility is an equivalence relation. We will denote its equivalence class generated by object u as

$$[u]_{IND(B)}.$$

Definition 5. *By a* query *over the set of attributes A we denote any formula*

$$\bigwedge_{i=1}^{n} a_i(x, v_i),$$

where $a_i \in A$, $v_i \in V_{a_i}$ and $n \leq |A|$. x is a free variable, which is valuated to an object.

Consider the query:

$$\varphi(x) = \texttt{name}(x, \texttt{Bob}) \wedge \texttt{age}(x, \texttt{young}).$$

This formula is satisfied either if p_3 is the value for x or its value is p_5. p_3 and p_5 cannot be distinguished by formula $\varphi(x)$.

We postulate the following definition of indiscernibility:

Definition 6. *Let $\mathbb{A}$ be a set of axioms. Let $\varphi(x)$ be a query with free variable x. Let u_1 and u_2 be constants. We say that u_1 and u_2 are* indiscernible by the query $\varphi(x)$ *if*

$$(\mathbb{A} \models \varphi(u_1)) \Longleftrightarrow (\mathbb{A} \models \varphi(u_2)).$$

Theorem 1. *Let $I = (U, A)$ be a complete information system with the set of objects' labels U and the set of attributes A. Let B be a subset of A. Objects $u_1 \in U$ and $u_2 \in U$ are indiscernible with respect to attribute set B iff they are indiscernible with respect to every query over the set of attributes B.*

Proof. Let $\mathbb{A}$ be axioms derived from I.

If $(u_1, u_2) \in IND(B)$, for every $a \in B$ there exists v_a such that

$$m(u_1, a) = m(u_2, a) = \{v_a\}.$$

Then the set of formulae $\{a(u_1, v_a) : a \in B\}$ is a subset of $\mathbb{A}$ and the set of formulae $\{a(u_2, v_a) : a \in B\}$ is a subset of $\mathbb{A}$. So for all $a \in B$ and $v \in V_a$ we have

$$\mathbb{A} \models a(u_1, v) \Leftrightarrow \mathbb{A} \models a(u_2, v).$$

Thus for every query $\varphi(x)$ we obtain $\mathbb{A} \models \varphi(u_1) \Leftrightarrow \mathbb{A} \models \varphi(u_2)$.

If $(u_1, u_2) \notin IND(B)$ we have $a \in B$ and v_1, v_2 such that $v_1 \neq v_2$, $m(u_1, a) = \{v_1\}$ and $m(u_2, a) = \{v_2\}$. Thus

$$\mathbb{A} \models a(u_1, v_1) \wedge \mathbb{A} \not\models a(u_2, v_1)$$

and the query $\varphi(x) = a(x, v_1)$ distinguishes u_1 and u_2.

Assume that we have two sensors. Measurements of both of them are represented by information systems and these information systems shares their sets of objects, i.e. the relation `is an object` is identical for both sensors. We wish to describe the measurements of one sensor by means of measurements of the second one. Alternatively we say the that we are describing the value of attribute in an information system (which we call the decision attribute) by the values of the rest of attributes. We may reduce this problem to the problem of description of the set objects in the information system for which the decision attribute has a certain value. Such a set is either *definable* or *indefinable* by other attributes.

Definition 7. *Let X be a subset of U. We say that X is* definable *by $\mathbb{A}$ iff there exist queries $\varphi_1(x), \ldots, \varphi_n(x)$ such that*

$$\forall u \in U \; \big(u \in X \Longleftrightarrow \mathbb{A} \models \varphi_1(u) \vee \cdots \vee \varphi_n(u)\big).$$

Each definable set is a sum of objects that satisfy at least one of a given queries.

Proposition 1. *Let $I = (U, A)$ be a complete information system with set of objects' labels U and set of attributes A. Let $\mathbb{A}$ be axioms derived from I. X is* definable *by $\mathbb{A}$ iff*

$$X = [u_1]_{IND(A)} \cup \cdots \cup [u_n]_{IND(A)}$$

for some $u_1, \ldots, u_n \in U$.

Proof. X is definable by $\mathbb{A}$ if and only if

$$X = \bigcup_{i=1}^{n} \{u \in U : \mathbb{A} \models \varphi_i(u)\}.$$

Theorem 1 states that $u_1, u_2 \in \{u \in U : \mathbb{A} \models \varphi_i(u)\}$ iff $(u_1, u_2) \in IND(A)$.

An indefinable set $X \subset U$ may be approximated by two definable sets. The first one is called the *lower approximation* of X, denoted by $\underline{\mathbb{A}}X$, and is defined by

$$\bigcup \{Y \mid Y \subset X \wedge Y \text{ is definable by } \mathbb{A}\}.$$

The second set is called the *upper approximation* of X, denoted by $\overline{\mathbb{A}}X$, and is defined by

$$\bigcap \{Y \mid X \subset Y \wedge Y \text{ is definable by } \mathbb{A}\}.$$

$\overline{\mathbb{A}}X \subset U$ because every definable set is a subset of U.

Proposition 2. *The lower and the upper approximations of any set $X \subset U$ are definable.*

Proof. For a given information system, there is a finite number of definable sets. Thus

$$\bigcup \{Y \mid Y \subset X \wedge Y \text{ is definable}\} = Y_1 \cup \cdots \cup Y_n,$$

where Y_i is defined by a formula $\varphi_i(x)$. Hence, $Y_1 \cup \cdots \cup Y_n$ is defined by $\varphi_1(x) \vee \cdots \vee \varphi_n(x)$. Similarly

$$\bigcap \{Y \mid X \subset Y \wedge Y \text{ is definable}\} = Y_1 \cap \cdots \cap Y_n,$$

where every Y_i is defined by a formula $\varphi_i(x)$. Hence, $Y_1 \cap \cdots \cap Y_n$ is defined by $\varphi_1(x) \wedge \cdots \wedge \varphi_n(x)$. The last formula may be transformed into a form of an alternative of queries.

Theorem 2. *Let $I = (U, A)$ be a complete information system with set of objects' labels U and set of attributes A, such that U and A are finite. Let $\mathbb{A}$ be the set of axioms derived from I. Then*

$$\underline{A}X = \underline{\mathbb{A}}X \text{ and } \overline{A}X = \overline{\mathbb{A}}X.$$

Proof. According to Prop. 1

$$\underline{\mathbb{A}}X = \bigcup \{[u_1]_{IND(A)} \cup \cdots \cup [u_n]_{IND(A)} \subset X\} =$$

$$= \bigcup \{[u]_{IND(A)} \mid [u]_{IND(A)} \subset X\} = \underline{A}X,$$

$$\overline{\mathbb{A}}X = \bigcap \{X \subset [u_1]_{IND(A)} \cup \cdots \cup [u_n]_{IND(A)}\}.$$

$IND(A)$ is an equivalence relation, so

$$\overline{\mathbb{A}}X = \bigcup \{[u]_{IND(A)} \mid [u]_{IND(A)} \cap X \neq \emptyset\} = \overline{\mathbb{A}}X.$$

Let the set X be the subset of universe of information system for which the decision attribute d takes the value v. Let $\underline{\varphi}(x)$ and $\overline{\varphi}(x)$ be the formulae that define the lower and upper approximation of X. $\underline{\varphi}(x)$ is equivalent to $\overline{\varphi}(x)$ when X is definable. Symbolically:

$$\{x \mid x \in U \wedge \underline{\varphi}(x)\} \subseteq \{x \mid x \in U \wedge d(x,v)\} \subseteq \{x \mid x \in U \wedge \overline{\varphi}(x)\}.$$

The claim that

$$\{x \mid \texttt{is an object}(x) \wedge \underline{\varphi}(x)\} \subseteq \{x \mid \texttt{is an object}(x) \wedge d(x,v)\} \subseteq$$
$$\subseteq \{x \mid \texttt{is an object}(x) \wedge \overline{\varphi}(x)\}.$$

we denote as *inductive reasoning*.

Inductive reasoning bases on the assumption that the definition generated for the specific data is still valid in the general case. $\underline{\varphi}(x)$ and $\overline{\varphi}(x)$ constitutes a classifier that assigns values of decision attribute to new samples.

Since a sample of objects included in the information system is not representative enough to define the bounds correctly, statistical methods are used in order to obtain bounds that are correct with the high probability. Such as method are for example: the limit of the number of queries in the bound definition, the minimal support for each query in the bound and so on.

3.2 Incomplete Data

Real-life data are frequently incomplete, i.e. values for some attributes are missing. We will assume three different interpretations of missing values:

- missing attribute values that are *lost*, i.e they are specified, yet their value are unknown
- attributes *not applicable* for a certain case, e.g. the colour of hair for a completely bald person
- *do not care values*: the attribute may have any value from its domain.

We will extend the definition of $m(u,a)$. $m(u,a) =?$ will mean that the value of attribute a for object u is lost, $m(u,a) = \star$ that it is 'do not care' and $m(u,a) = -$ that it is not applicable.

The problem of missing values was thoroughly studied (see e.g. [8,9,14,15]). The presented ideas were based on various modifications of indiscernibility relation so it could handle missing values and remain definable in terms of attributes.

The definitions of indiscernibility, definability, lower and upper approximation we stated in the above section may do not need to be modified for information systems with missing values. They are equivalent to the definitions proposed in the cited papers.

We express the various types of missing value semantics using axioms:

- for each $u \in U$, for each $a \in A$ we state

$$\mathcal{P} \models a(u,v),$$

where $v \in m(u,a)$ in the information system.

Table 3. An information system with missing values

	number of hairs	colour of hair
p_1	none	-
p_2	little	brown
p_3	?	blond
p_4	$\star$	brown

- 'lost' values we define as follows: for each $u \in U$, for each $a \in A$ we state

$$\mathcal{P} \models a(u, v_1) \vee \cdots \vee a(u, v_n),$$

  where $v_1, \ldots, v_n$ are all possible values of attribute a.
- for each $u \in U$, for each $a \in A$ whose value is not applicable we state

$$\mathcal{P} \models \forall_x \neg a(u, x),$$

- for each $u \in U$, for each $a \in A$, for each v from the domain of a we state

$$\mathcal{P} \models a(u, v),$$

  when the value of a is 'do not care' for object u.

We may describe contents of Table 3 using the following formula [2]:

$$\mathcal{P}_1 \models \texttt{number of hairs}(p_1, \texttt{none}) \ \wedge \forall_x \neg \texttt{colour of hair}(p_1, x) \wedge$$

$$\wedge \ \texttt{number of hairs}(p_2, \texttt{little}) \wedge \texttt{colour of hair}(p_2, \texttt{brown}) \wedge$$

$$\wedge \ \big(\texttt{number of hairs}(p_3, \texttt{none}) \vee \texttt{number of hairs}(p_3, \texttt{little}) \vee$$

$$\vee \ \texttt{number of hairs}(p_3, \texttt{many})\big) \wedge \texttt{colour of hair}(p_3, \texttt{blond})$$

$$\wedge \ \texttt{number of hairs}(p_4, \texttt{none}) \wedge \texttt{number of hairs}(p_4, \texttt{little}) \wedge$$

$$\wedge \ \texttt{number of hairs}(p_4, \texttt{many}) \wedge \texttt{colour of hair}(p_4, \texttt{brown}).$$

Since indiscernibility with respect to the set of attributes does not work for incomplete information systems authors extended it or replaced it by another concepts.

The extension proposed in [14] for the information systems with 'do not care' missing values is the relation

$$(x, y) \in SIM(B) \text{ iff } \forall a \in B\big(m(x, a) = \star \vee m(y, a) = \star \vee m(x, a) = m(y, a)\big).$$

[2] Since it is impossible to have none, little and many hairs at the same time, the formula $\texttt{number of hairs}(p_4, \texttt{none}) \wedge \texttt{number of hairs}(p_4, \texttt{little}) \wedge \texttt{number of hairs}(p_4, \texttt{many})$ is contradictory. Yet, for the purpose of example, we do not take this fact into account.

Theorem 3. *Let $I = (U, A)$ be an information system with 'do not care' missing vales. Let U be the set of objects and A be the set of attributes. Let B be a subset of A. If objects $u_1 \in U$ and $u_2 \in U$ are indiscernible with respect to every query over the set of attributes B then*

$$(u_1, u_2) \in SIM(B).$$

The reverse implication is not valid for information systems with nontrivial missing values.

Proof. Let $\mathbb{A}$ be axioms derived from I. If $(u_1, u_2) \notin SIM(B)$ we have $a \in B$ and $v_1, v_2 \in V_a$ such that $v_1 \neq v_2$, $m(u_1, a) = \{v_1\}$ and $m(u_2, a) = \{v_2\}$. Thus

$$\mathbb{A} \models a(u_1, v_1) \text{ and } \mathbb{A} \not\models a(u_2, v_1)$$

and the query $\varphi(x) = a(x, v)$ distinguishes u_1 and u_2.

For the case of reverse implication let us consider Table 3. We have

$$(p_2, p_4) \in SIM(\{\texttt{number of hairs}\}),$$

yet the query

$$\texttt{number of hairs}(x, \texttt{none})$$

distinguish them.

In [8] an another approach for 'do not care' and 'lost' missing values is presented. The indiscernibility with respect to the set of attributes is replaced by the concept of characteristic set:

Definition 8. *For an object $u \in U$ the* characteristic set $K_A(u)$ *is defined as*

$$K_A(u) = \bigcap_{a \in A} K(u, a),$$

where $K(u, a)$ is defined in the following way

- *if $m(u, a) = \{v\}$ then*

$$K(u, a) = \{u' \in U \mid m(u', a) = \{v\} \vee m(u', a) = \star\}.$$

- *if $m(u, a) = ?$ or $m(u, a) = \star$ then $K(u, a) = U$.*

Lemma 1. *Let $I = (U, A)$ be an information system with 'lost' and 'do not care' missing vales. Let U be the set of objects and A be the set of attributes. Let $\mathbb{A}$ be axioms derived from I. For every $a \in A$ and for each $u \in U$ such that $m(u, a) = \{v\}$*

$$x \in K(u, a) \Longleftrightarrow \mathbb{A} \models a(x, v).$$

Proof. Let x be an element of $K(u, a)$. Then $m(u, a) = \{v\}$ or $m(u, a) = \star$. In both cases $a(x, v)$ is satisfied by $\mathbb{A}$.

If $\mathbb{A} \models a(x, v)$, then either the value of a on x was specified as v either it was 'do not care' missing value. In both cases $x \in K(u, a)$.

Theorem 4. *Let $I = (U, A)$ be an information system with 'lost' and 'do not care' missing vales. Let U be the set of objects and A be the set of attributes. Let $\mathbb{A}$ be axioms derived from I. The set $X \subset U$ is definable iff X is the union of characteristic sets.*

Proof. Let $u_1, \ldots, u_n$ be such that

$$X = \bigcup_{i=0}^{n} K_A(u_i) = \bigcup_{i=0}^{n} \bigcap_{a \in A} K(u_i, a) = \bigcup_{i=0}^{n} \bigcap_{a \in A_i} K(u_i, a),$$

where A_i is the set of all attributes specified for u_i. Let $m(u_i, a) = \{v_{a,i}\}$. According to Lemma 1

$$K(u_i, a) = \{x \in U \mid \mathbb{A} \models a(x, v_{a,i})\}.$$

Thus $x \in X$ iff

$$\mathbb{A} \models \bigvee_{i=0}^{n} \bigwedge_{a \in A_i} a(x, v_{a,i}).$$

Theorem 5. *Let $I = (U, A)$ be an information system with 'lost' and 'do not care' missing vales. Let U be the set of objects and A be the set of attributes. Let $\mathbb{A}$ be axioms derived from I. For each $X \subset U$*

$$\underline{\mathbb{A}}X = \bigcup \{K_A(x) \mid K_A(x) \subset X\},$$

$$\overline{\mathbb{A}}X = \bigcup \{K_A(x) \mid x \in U, K_A(x) \cap X \neq \emptyset\}.$$

Lower and upper approximations are equivalent to subset *lower and upper approximations (defined in [8]).*

Proof. $\underline{\mathbb{A}}X$ is definable, so according to Thm. 4

$$\underline{\mathbb{A}}X = \bigcup_{i=1}^{n} K_A(u_i)$$

for some $u_1, \ldots, u_n \in U$. Since $\underline{\mathbb{A}}X \subset X$, we obtain $K_A(u_i) \subset X$. If $K_A(x) \subset X$ then $K_A(x) \subset \bigcup_{i=1}^{n} K_A(u_i)$, because $K_A(x)$ is definable and $\underline{\mathbb{A}}X$ is the largest definable subset of X.

$\overline{\mathbb{A}}X$ is definable, so according to Thm. 4

$$\overline{\mathbb{A}}X = \bigcup_{i=1}^{n} K_A(u_i)$$

for some $u_1, \ldots, u_n \in U$. Since $\overline{\mathbb{A}}X$ is the smallest definable set such that $X \subset \overline{\mathbb{A}}X$, we obtain $K_A(u_i) \cap X \neq \emptyset$.

3.3 Multivalued Attributes

Multiple valued attributed (introduced in [27] and studied in [20]) may reflect our incomplete knowledge about their values, what makes them similar to 'lost' missing values. The may also represent attributes that have a few values simultaneously, in which case the are like 'do not care' missing values.

- 'lost' multiple values we define as follows: for each $u \in U$, for each $a \in A$ we state
$$\mathcal{P} \models a(u, v_1) \vee \cdots \vee a(u, v_n),$$
where $v_1, \ldots, v_n$ are all possible values of attribute a for object u mentioned in information system.
- for each $u \in U$, for each $a \in A$, for each value v of attribute a for object u in information system
$$\mathcal{P} \models a(u, v),$$
when the value of a is 'do not care' multiple value for object u.

Table 4. A multiple valued information system

	name	known languages
p_5	Bob	Pascal, Ocaml, Cobol
p_6	David, Alice	Ocaml

For example objects in Table 4 will be described by the following formula:

$$\mathcal{P}_1 \models \texttt{name}(p_5, \texttt{Bob}) \wedge \texttt{known languages}(p_5, \texttt{Pascal}) \wedge$$

$$\wedge\ \texttt{known languages}(p_5, \texttt{Ocaml}) \wedge \texttt{known languages}(p_5, \texttt{Cobol}) \wedge$$

$$\wedge\ (\texttt{name}(p_6, \texttt{David}) \vee \texttt{name}(p_6, \texttt{Alice})) \wedge \texttt{known languages}(p_6, \texttt{Ocaml}).$$

Multivalued attributes is a simple extension of the 'missing values' case and the whole theory derived for the information systems with missing attributes is applicable here.

4 Structural Objects

Information systems are devoted to representation of simple objects described by a vector of attributes. What makes compound objects different from the simple ones is the internal structure. Structural objects are composed of subobjects connected by relations.

We shown in Section 3 that the idea of representing knowledge in terms of axioms provides us a flexible and extendable framework for coherent theory of data analysis. Now, we formally define the language for knowledge representation, which allow us to describe properties of structured objects. We call it a *meaning representation language*.

Table 5. A set of data sequences

s_1	`Alice has brown hair.`
s_2	`Charlie and David know Ocaml and Cobol.`
s_3	`Bob's hair colour is black,` `similar to David's hair colour.`
s_4	`Parents of Alice and Bob are old.`
s_5	`Alice, Bob, Charlie and David are old.`

The meaning representation language represents concepts included in data and dependencies between them. It is an extension of the formulae used to describe axioms derived from information systems in Section 3.

Syntax of the language is defined as follows: We have the set of constants and the set of predicate names.

Constants play the role of labels for entities described in data. Constant names may be meaningful (see Section 2) or may not carry any information about pointed entity. The anonymity of constants reflects the fact that we do not posses direct access to the entities. We know only the relations between entities and these relation does not define entities in an exact way. It reflects the incompleteness of our knowledge.

Predicates posses lists of one or more arguments. Number of arguments for a given predicate is not fixed. Predicates represent relations on finite sequences of entities. The predicates have meaningful names.

Atomic formula is a predicate. Formula is composed of one or more atomic formulae connected by means of conjunction or alternative. We do not use quantifiers, functions or negation.

The semantics of the meaning representation language is based on the concepts presented in Section 2. The structure $\mathcal{P}$ plays the role of the reality model. Information about $\mathcal{P}$ is represented by axioms $\mathbb{A}$. The axioms are written using the meaning representation language. $\mathbb{P}(\mathbb{A})$ is the set of possible world defined by $\mathbb{A}$. We assume that data are consistent; in other words:

$$\mathcal{P} \models \mathbb{A}.$$

For example, Table 5 provides us knowledge representation of s_1 by the following axioms:

$$\mathcal{P}_1 \models \mathtt{name}(u_1, \mathtt{Alice}) \wedge \mathtt{colour\ of\ hair}(u_1, \mathtt{brown}).$$

Note that object identifiers are not sequence identifiers s_i. One sequence may describe many objects. The same object may be mentioned in several sequences, yet we must use the domain knowledge in order to assure that different constants denote the same object. Let us consider now s_2:

$$\mathcal{P}_1 \models \mathtt{name}(u_2, \mathtt{Charlie}) \wedge \mathtt{name}(u_3, \mathtt{David}) \wedge \mathit{and}(u_4, u_2, u_3) \wedge$$

$$\wedge \mathtt{known\ languages}(u_4, \mathtt{Ocaml}) \wedge \mathtt{known\ languages}(u_4, \mathtt{Cobol}).$$

The conjunction and in the sequence s_2 has two meanings. The second time it is used as logical '$\wedge$', while in the first case of use it forks the sequence. We represent this operation using symbol *and* defined as

$$and(a, a_1, \dots, a_n) \wedge \varphi(a, a_1, \dots, a_n) \iff$$

$$\iff \varphi(a_1, a_1, \dots, a_n) \wedge \dots \wedge \varphi(a_n, a_1, \dots, a_n).$$

In sequence s_3, the `colour of hair` is an object and the property of person in the same time:

$$\mathcal{P}_1 \models \texttt{name}(u_5, \texttt{Bob}) \wedge \texttt{colour of hair}(u_5, \texttt{black}) \wedge$$

$$\wedge \texttt{name}(u_6, \texttt{David}) \wedge \texttt{colour of hair}(u_6, u_7) \wedge \texttt{similar}(\texttt{black}, u_7).$$

The sequence s_4 is ambiguous: in first interpretation `Parents of Bob are old` and in the second `Bob is old`. We use '$\vee$' in order to represent both possibilities.

$$\mathcal{P}_1 \models \texttt{name}(u_8, \texttt{Alice}) \wedge \texttt{parent}(u_9, u_8) \wedge \texttt{parent}(u_{10}, u_8) \wedge \texttt{name}(u_{11}, \texttt{Bob}) \wedge$$

$$\wedge \big(and(u_{12}, u_9, u_{10}, u_{11}) \vee \big(\texttt{parent}(u_9, u_{11}) \wedge \texttt{parent}(u_{10}, u_{11}) \wedge$$

$$\wedge and(u_{12}, u_9, u_{10})\big)\big) \wedge \texttt{age}(u_{12}, \texttt{old}).$$

In sequence s_5 we have the list of objects that could be arbitrary long:

$$\mathcal{P}_1 \models \texttt{name}(u_{13}, \texttt{Alice}) \wedge \texttt{name}(u_{14}, \texttt{Bob}) \wedge \texttt{name}(u_{15}, \texttt{Charlie}) \wedge$$

$$\wedge \texttt{name}(u_{16}, \texttt{David}) \wedge and(u_{17}, u_{13}, u_{14}, u_{15}, u_{16}) \wedge \texttt{age}(u_{17}, \texttt{old}).$$

5 Sequential Data Processing

Now, we show how to obtain structural object description written in form of axioms. We assume that the source information is given as the sequence of symbols, for example textual data, recorded sound, sequence of some measurements etc. We will carefully study the process of translation of sequential data into our meaning representation language in the following sections. This process is similar to the segmentation of images [35]. As we will see it is tightly connected with the syntax of the meaning representation language.

Sequential data are a description of some world $\mathcal{P}$. This description is not a precise definition, rather the theory of the set of possible worlds $\mathbb{P}(\mathbb{A})$. Our goal is to transform this description into the set of axioms $\mathbb{A}$ that would define the same theory of $\mathbb{P}(\mathbb{A})$. Thanks to this processing the data obtain the description that has a formal semantics and identifies objects and their properties.

The sequential data processing is an example of complex translation from one sensor into another one. We show in Section 6 that this process may be considered as rough set approximation.

The classical approach to sequential data analysis consists in splitting the data into subsequences of constant length denoted as windows. Then each window is treated as a vector of attribute values. The advantage of this approach is a simple translation into an information system. The disadvantage is that it does not reflect the semantics of the data. If the windows are small, they cut object descriptions. If they are large, they do not distinguish objects. When the sequence length used for describing objects varies, the proper window size does not exist. In addition windows does not allow to express the properties of structured objects described by the sequential data.

In our approach, we divide sequences into the windows that vary in size and merge windows into larger structures. We transform data sequences into axioms using the methodology of the attributed grammars [13]. The basic idea is to perform the syntactic decomposition of the sequence using generative grammar and add the semantic value for each grammar symbol. In our case, these semantic values are formulae of meaning representation language. The semantic values are calculated by means of semantic attachments assigned to grammar rules. We extract concepts explicitly stated in sequence, not the ones that can be deduced from it.

5.1 Syntactic Rules

First, we define grammar which we will use for describing syntactic structure of data sequences.

We decided that our grammar would recognise regular languages. Yet we may replace our grammar with Context-Free Grammar without deep modification in the system.

We represent syntactic rules using a modification of context-free grammars by adding some special rule, called, a *term accumulation rule*. Formally let

$$G = (\Sigma, N, X_I, R, +, \prec)$$

be such that

- Σ is a finite set of terminal symbols,
- N is a finite set of non-terminal symbols.
- $X_I \in N$ is the start-symbol of grammar.
- R is a finite set of production rules. Each production has the form $A \to \alpha$ or $A \to \beta+$, where A is a non-terminal and α is a sequence of terminals and non-terminals and $\beta \in \Sigma \cup N$. $A \to \beta+$ is a shortcut for set of rules: $A \to \beta, A \to \beta\beta, A \to \beta\beta\beta, \ldots$.
- $\prec$ is binary relation of $\Sigma \cup N$ such that $A \prec B$ if and only if there is a rule $A \to \alpha$ in R such that B belongs to α or there is a rule $A \to B+$.
- $\prec$ is a irreflexive and transitive partial order.

We will denote every subsequence parsed to a grammar symbol as a phrase.

Proposition 3. *Language L can be recognised by grammar defined above if and only if L is regular language.*

For example, for sequences from Table 5 the following grammar may be generated:

```
[name] ::= Alice | Bob | Charlie | David
[age] ::= young | old
[colour of hair] ::= brown | black
[number of hairs] ::= none | little | many
[known language] ::= Pascal | Ocaml | Cobol
[known languages] ::= [known language] |
     [known language] and [known language]
[name,] ::= [name] ,
[name list] ::= [name,] +
[names] ::= [name] | [name] and [name] | [name list] and [name]
[parents] ::= [names] | parents of [names]
[person] ::= [parents] | [parents] and [parents]
X_I ::= [person] are [age] |
     [person] has [colour of hair] hair |
     [person] know [known languages]
```

Names of the symbols in the grammar reflect the concept names. The grammar is ambiguous. The sequence s_4 may be parsed in two different ways which reflect two possible interpretations of sequence.

5.2 Data Sequence Representation

We are looking for all the possible derivation trees for a given data sequence and grammar.

We need representation that can describe ambiguous, partially parsed data. We represent it as directed acyclic graph whose edges are labelled by grammar symbols. We call it *the graph of syntactic decomposition.*

We represent data sequence as a graph that is a list. Formally, let $\{\sigma_i\}_1^n$, $\sigma_i \in \Sigma$ be the sequence. We create graph with vertexes $V = \{v_0, \ldots, v_n\}$ and set of edges $E = \{v_0 \xrightarrow{\sigma_1} v_1, \ldots, v_{n-1} \xrightarrow{\sigma_n} v_n\}$.

While applying the rule we find path in the graph with edge labels that match to the rule. Then we add to graph a new edge from beginning to end of the path labelled with rule production.

In order to apply the rule $A \to \alpha_1, \ldots, \alpha_k$ we find all paths

$$v_{a_0} \xrightarrow{\alpha_1} v_{a_1} \xrightarrow{\alpha_2} v_{a_2} \ldots v_{a_{k-1}} \xrightarrow{\alpha_k} v_{a_k}$$

and we add for each of them the edge

$$v_{a_0} \xrightarrow{A} v_{a_k}$$

to the graph.

While applying the $A \to \beta+$ rule, we find all paths

$$v_{a_0} \xrightarrow{\beta} v_{a_1} \xrightarrow{\beta} v_{a_2} \ldots v_{a_{k-1}} \xrightarrow{\beta} v_{a_k}$$

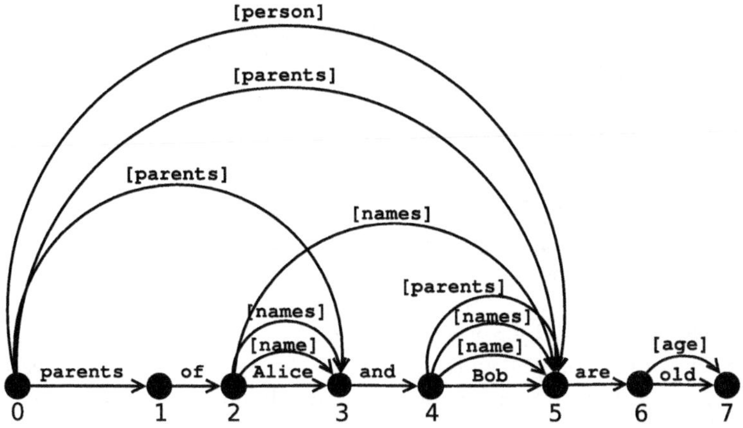

Fig. 1. Part of the graph of syntactic decomposition for sequence s_4

and we add for each of them the edge

$$v_{a_0} \xrightarrow{A} v_{a_k}$$

to the graph.

We will denote the edge labelled α such that $v_i \xrightarrow{\alpha} v_j$ by $\alpha_{i,j}$.

As a result of parsing process we obtain the edge from the beginning to the end of graph labelled by the start symbol of grammar.

5.3 Parser Algorithm

Having defined the data representation, we describe the parser algorithm.

We divide the set of symbols into layers: Let $N_0 = \Sigma$ and let

$$N_{n+1} = \{A : \exists A \to \alpha_1 \ldots \alpha_k \forall i(\alpha_i \in N_n) \cup \exists A \to \beta+(\beta \in N_n)\}.$$

Now we divide the rules set R into layers. Let $R_{-1} = \emptyset$ and

$$R_n = \{A \to \alpha_1 \ldots \alpha_k : \forall i \alpha_i \in N_n\} \cup \{A \to \beta+ : \beta \in N_n\} \setminus R_{n-1}.$$

Since we do not allow recurrent symbol to occur there is finite number of layers.

For example, for grammar created for sequences from Table 5 we obtain:

$$\begin{aligned}
N_0 &= \{\texttt{Alice}, \texttt{Bob}, \texttt{Charlie}, \texttt{David}, \texttt{young}, \texttt{old}, \texttt{brown}, \texttt{black}, \\
&\quad \texttt{none}, \texttt{little}, \texttt{many}, \texttt{Pascal}, \texttt{Ocaml}, \texttt{Cobol}, \\
&\quad \texttt{and}, \texttt{,}, \texttt{parents}, \texttt{of}, \texttt{are}, \texttt{has}, \texttt{know}, \texttt{hair}\} \\
N_1 &= \{\texttt{[name]}, \texttt{[age]}, \texttt{[colour of hair]}, \\
&\quad \texttt{[number of hairs]}, \texttt{[known language]}\} \\
N_2 &= \{\texttt{[known languages]}, \texttt{[name,]}\} \\
N_3 &= \{\texttt{[name list]}\} \\
N_4 &= \{\texttt{[names]}\} \\
N_5 &= \{\texttt{[parents]}\} \\
N_6 &= \{\texttt{[person]}\} \\
N_7 &= \{X_I\}
\end{aligned}$$

and

$$R_0 = \begin{array}{l} \texttt{[name]} ::= \texttt{Alice} \mid \texttt{Bob} \mid \texttt{Charlie} \mid \texttt{David} \\ \texttt{[age]} ::= \texttt{young} \mid \texttt{old} \\ \texttt{[colour of hair]} ::= \texttt{brown} \mid \texttt{black} \\ \texttt{[number of hairs]} ::= \texttt{none} \mid \texttt{little} \mid \texttt{many} \\ \texttt{[known language]} ::= \texttt{Pascal} \mid \texttt{Ocaml} \mid \texttt{Cobol} \end{array}$$

$$R_1 = \begin{array}{l} \texttt{[known languages]} ::= \texttt{[known language]} \mid \\ \qquad \texttt{[known language] and [known language]} \\ \texttt{[name,]} ::= \texttt{[name] ,} \end{array}$$

etc.

Rules belonging to each layer are independent. Hence we may go through the sequence once for each layer and apply all matching rules simultaneously.

For each path in graph, the algorithm finds all rules that match to the path and add their production to graph. We begin with graph (V, E_0), where $E_0 = E$. We obtain graph (V, E_{n+1}) by applying to (V, E_n) rules from R_n. For each text's subsequence we find all its possible syntactic consequences.

In order to do it efficiently we create prefix tree out of every layer: For each rule $A \to \alpha_1 \dots \alpha_k$ in R_n we create path in the tree from the root labelled by symbols α_1 till α_k and we label the leaf tree node by A. For each node we merge paths that have identical labels.

Using this data structure we can apply all $A \to \alpha$ rules in layer in $\mathcal{O}(|E_n| l \log | \Sigma \cup E| + |E_n||R_n^+|)$ time, where

$$l = \max_{R_n} \{k : A \to \alpha_1 \dots \alpha_k \in R_n\}.$$

Since l, $\log |\Sigma \cup E|$ and number of layers is relatively small $|E_n|$ is crucial for parser performance.

For the different kinds of grammar parser may be replaced with other known in literature parsers [12].

5.4 Semantic Values of Grammar Symbols

In case of ambiguous grammar, the number of possible syntax derivation trees may be exponential to the sequence length. The concept of the graph of syntactic decomposition is their compact representation. The number of possible semantic values of the sequence is equal to the number of syntax derivation trees. That is why we cannot represent them directly. Instead, we distribute the semantic values across the graph of syntactic decomposition.

The meaning representation language formulae must have syntax coherent with the graph of syntactic decomposition. This requirement creates the dependence between the syntax of the meaning representation language and the process of translating data into axioms.

The formulae are spread across the graph in a way presented below.

Consider the edge $\alpha_{i,j}$ of the graph. This edge was created as the result of parsing a phrase. The phrase described an entity. We represent this entity by means of constant $a_{\alpha,i,j}$. We describe its properties derived from the phrase by the formula of meaning representation language. We name this formula *semantic value of grammar symbol* and denote it as $[\![\alpha]\!]_{i,j}$. We assign the formula $[\![\alpha]\!]_{i,j}$ to the edge $\alpha_{i,j}$ on the implementation level.

The formula $[\![\alpha]\!]_{i,j}$ has the following structure:

$$[\![\alpha]\!]_{i,j} := \bigvee_{k=1}^{n} p_k(a_{\alpha,i,j}, a_{\alpha_1^k,i_1^k,j_1^k}, \dots, a_{\alpha_{m_k}^k,i_{m_k}^k,j_{m_k}^k}) \wedge \bigwedge_{l=1}^{m_k} [\![\alpha_l^k]\!]_{i_l^k,j_l^k}.$$

Each $[\![\alpha_l^k]\!]_{i_l^k,j_l^k}$ is assigned to the edge ${\alpha_l^k}_{i_l^k,j_l^k}$, so only the set of atomic formulae

$$\{p_1(a_{\alpha,i,j}, a_{\alpha_1^1,i_1^1,j_1^1}, \dots, a_{\alpha_{m_k}^1,i_{m_k}^1,j_{m_k}^1}), \dots$$
$$\dots, p_n(a_{\alpha,i,j}, a_{\alpha_1^n,i_1^n,j_1^n}, \dots, a_{\alpha_{m_k}^n,i_{m_k}^n,j_{m_k}^n})\}$$

must by associated with graph edge on the implementation level.

Semantics for a terminal symbol $\alpha_{i,j}$ is an one-argument predicate whose name is α and whose argument is the variable $a_{\alpha,i,j}$.

For example, for graph of syntactic decomposition presented on Fig. 1 we will obtain the following semantic values:

$$\begin{aligned}
[\![\texttt{[name]}]\!]_{2,3} &= \texttt{[name]}(a_{\texttt{[name]},2,3}, \texttt{Alice})\\
[\![\texttt{[names]}]\!]_{2,5} &= \mathit{and}(a_{\texttt{[names]},2,5}, a_{\texttt{[name]},2,3}, a_{\texttt{[name]},4,5}) \wedge [\![\texttt{[name]}]\!]_{2,3} \wedge\\
&\quad \wedge [\![\texttt{[name]}]\!]_{4,5}\\
[\![\texttt{[parents]}]\!]_{0,3} &= \texttt{[parents]}(a_{\texttt{[parents]},0,3}, a_{\texttt{[names]},2,3}) \wedge [\![\texttt{[names]}]\!]_{2,3}\\
[\![\texttt{[person]}]\!]_{0,5} &= \big(\mathit{equal}(a_{\texttt{[person]},0,5}, a_{\texttt{[parents]},0,5}) \wedge [\![\texttt{[parents]}]\!]_{0,5}\big) \vee\\
&\quad \vee \big(\mathit{and}(a_{\texttt{[person]},0,5}, a_{\texttt{[parents]},0,3}, a_{\texttt{[parents]},4,5}) \wedge\\
&\quad \wedge [\![\texttt{[parents]}]\!]_{0,3} \wedge [\![\texttt{[parents]}]\!]_{4,5}\big).
\end{aligned}$$

5.5 Semantic Attachments

Semantic values of grammar symbols are constructed using semantic attachments of grammar rules. Semantic attachment are functions that compose semantics of greater objects out of semantics of smaller ones.

Let $A \to \alpha_1 \dots \alpha_k$ be a syntactic rule and $f_{A \to \alpha_1 \dots \alpha_k}$ be a semantic attachment assigned to it. Assume that the rule was matched to the path $\alpha_{1,i_0,i_1}, \dots, \alpha_{k,i_{k-1},i_k}$. As the rule was applied the symbol A_{i_0,i_k} was created.

The semantic value for A_{i_0,i_k} is constructed as follows: first we calculate the value of

$$f_{A \to \alpha_1 \dots \alpha_k}([\![\alpha_1]\!]_{i_0,i_1}, \dots, [\![\alpha_k]\!]_{i_{k-1},i_k}).$$

We demand from the values of the semantic attachments to be predicate. Let

$$p(a_{A,i_0,i_k}, a_{\beta_1,j_1,k_1}, \dots, a_{\beta_n,j_n,k_n}) := f_{A \to \alpha_1 \dots \alpha_k}([\![\alpha_1]\!]_{i_0,i_1}, \dots, [\![\alpha_k]\!]_{i_{k-1},i_k}),$$

where every β_{i,j_i,k_i} belongs to $\{\alpha_{1,i_0,i_1}, \dots, \alpha_{k,i_{k-1},i_k}\}$. Now we define semantics of A_{i_0,i_k} as

$$[\![A]\!]_{i_0,i_k} := p(a_{A,i_0,i_k}, a_{\beta_1,j_1,k_1}, \dots, a_{\beta_n,j_n,k_n}) \wedge \bigwedge_{1 \le i \le n} [\![\beta_i]\!]_{j_i,k_i}) \vee [\![A]\!]_{i_0,i_k}.$$

$[\![A]\!]_{i_0,i_k}$ on the right side of assignment is the semantic value of the edge A_{i_0,i_k} before the rule application. The semantic value for nonexistent edge is falsity.

The first argument of the predicate p is the constant a_{A,i_0,i_k} whose value is the entity described by phrase parsed to A_{i_0,i_k}.

The semantic attachment manipulates on formulae considering them as terms. It extracts parts of formulae and constructs the predicate p using them. In most cases it uses the constant pointing to the entity described by the predicate. The semantic attachment often concatenates the names of predicates that are the semantic values of terminal symbols

For example, the grammar created for sequences from Table 5 may have the following semantic attachments:

$\texttt{[name]} ::= \texttt{Alice}$
$\quad\texttt{name}(u_{\texttt{[name]}}, \texttt{Alice})$
$\texttt{[age]} ::= \texttt{old}$
$\quad\texttt{age}(u_{\texttt{[age]}}, \texttt{old})$
$\texttt{[known language]} ::= \texttt{Pascal}$
$\quad\texttt{[known language]}(u_{\texttt{[name]}}, \texttt{Pascal})$
$\texttt{[known languages]} ::= \texttt{[known language]}$
$\quad equal(u_{\texttt{[known languages]}}, u_{\texttt{[known language]}})$
$\texttt{[known languages]} ::= \texttt{[known language]}_1 \texttt{ and } \texttt{[known language]}_2$
$\quad equal(u_{\texttt{[known languages]}}, u_{\texttt{[known language]}_1}, u_{\texttt{[known language]}_2})$
$\texttt{[names]} ::= \texttt{[name]}$
$\quad equal(u_{\texttt{[names]}}, u_{\texttt{[name]}})$
$\texttt{[names]} ::= \texttt{[name]}_1 \texttt{ and } \texttt{[name]}_2$
$\quad and(u_{\texttt{[names]}}, u_{\texttt{[name]}_1}, u_{\texttt{[name]}_2})$
$\texttt{[parents]} ::= \texttt{[names]}$
$\quad equal(u_{\texttt{[parents]}}, u_{\texttt{[names]}})$
$\texttt{[parents]} ::= \texttt{parents of [names]}$
$\quad\texttt{parents}(u_{\texttt{[parents]}}, u_{\texttt{[names]}})$
$\texttt{[person]} ::= \texttt{[parents]}$
$\quad equal(u_{\texttt{[person]}}, u_{\texttt{[names]}})$
$\texttt{[person]} ::= \texttt{[parents]}_1 \texttt{ and } \texttt{[parents]}_2$
$\quad and(u_{\texttt{[person]}}, u_{\texttt{[parents]}_1}, u_{\texttt{[parents]}_2})$
$X_I ::= \texttt{[person] are [age]}$
$\quad equal(u_{X_I}, u_{\texttt{[person]}}, u_{\texttt{[age]}})$
$X_I ::= \texttt{[person] know [known languages]}$
$\quad equal(u_{X_I}, u_{\texttt{[person]}}, u_{\texttt{[known languages]}})$.

The space complexity determines the representation of semantics. Each rule application add one predicate, so space complexity of semantics is proportional to the number of applied rules.

Thank to such representation of formulae we omit combinatorial explosion during the analysis of data with high ambiguity.

Various predicates generated for a given subsequence are possible descriptions of an entity. That is why we point that entity by the same constant in each predicate.

We connect different possible subsequence interpretations by means of alternative. Only one of them may be correct because only one of contradicting formulae may be consistent, ensuring that only one clause of alternative will be true. In case when text interpretations do not contradict, it could happen that a few clauses will be true (consistent with the facts described in the document) at the same time, despite the fact that only one of them could be meant do be written.

For for each $A \to \beta+$ rule we assign semantic action $f_{A\to\beta+}$ such that

$$[\![A]\!] = f_{A\to\beta+}([\![\beta_1]\!], [\![\beta_2]\!], \ldots).$$

We represent semantic value of the symbol generated by accumulation rule as a graph, whose vertexes are constants that are arguments of predicate. Each path from beginning to ending vertex in such a graph represents a list of predicate arguments. The predicates are connected by alternative. We denote such a graph as *graph of logic structure* of accumulation symbol.

Size of semantics for accumulation rule is smaller than $\frac{n(n+1)}{2}$, where n is number of vertexes in the graph of syntax decomposition.

We may add a few different semantic attachments to a syntactic rule. Obtaining rules that are grammatically identical but differ on semantic level.

In the end of parsing process we obtain an edge labelled by start symbol of grammar. Its semantic value is a formula that contains every possible translation for the entire text into meaning representation language.

6 Set Approximations

Now, when we studied the process of generating axioms for a given data, we define rough sets for objects described by axioms written as the meaning representation language formulae.

The most important difference with the case of information systems is the extension of the definition of query that makes it suitable for data represented in a form of meaning representation language formulae.

Definition 9. *By a* query *we denote any formula* $\varphi(x)$ *of the form*

$$\exists_{x_1,\ldots,x_n} \bigwedge_{i=1}^{k} p_i(a_1^i, \ldots, a_{k_i}^i),$$

where each a_j^i *either is a variable belonging to* $\{x, x_1, \ldots, x_n\}$ *or a constant.* x *is a free variable and* p_i *are predicates.*

For example, the query

$$\varphi_1(x) = \exists_{x_1} \texttt{colour of hair}(x, x_1) \wedge \texttt{similar}(x_1, \texttt{brown})$$

refers to the people whose hair has colour similar to brown. In Table 6, the formula $\varphi_1(x)$ is satisfied either if p_1 is the value of x or its value is p_2. p_1 and p_2 cannot be distinguished by formula $\varphi_1(x)$.

Table 6. A set of axioms

$$\mathcal{P}_1 \models \texttt{colour of hair}(p_1, h_1) \wedge \texttt{similar}(h_1, \texttt{brown})$$
$$\mathcal{P}_1 \models \texttt{colour of hair}(p_2, h_2) \wedge \texttt{similar}(h_2, \texttt{brown})$$

The indiscernibility is defined in the same way as for information systems (compare with Definition 6 in Section 3):

Definition 10. *Let $\mathbb{A}$ be a set of axioms. Let $\varphi(x)$ be a query with free variable x. Let u_1 and u_2 be constants. We say that u_1 and u_2 are* indiscernible by the query $\varphi(x)$ *if*

$$(\mathbb{A} \models \varphi(u_1)) \iff (\mathbb{A} \models \varphi(u_2)).$$

In case of information systems, the set of objects' labels U were given. When data are represented as a set of axioms, we define the set of objects' labels U as the set of all constant symbols included in axioms.

Definition 11. *Let X be a subset of U. We say that X is* definable *by $\mathbb{A}$ iff there exist queries $\varphi_1(x), \ldots, \varphi_n(x)$ such that*

$$\forall u \in U \; \big(u \in X \iff \mathbb{A} \models \varphi_1(u) \vee \cdots \vee \varphi_n(u)\big).$$

Nondefinable X may be approximated by two definable sets. The first one is called *lower approximation* of X, denoted by $\underline{\mathbb{A}}X$ and defined as

$$\bigcup\{Y \mid Y \subset X \wedge Y \text{ is definable by } \mathbb{A}\}.$$

The second set is called *upper approximation* of X, denoted by $\overline{\mathbb{A}}X$ and defined as

$$\bigcap\{Y \mid X \subset Y \wedge Y \text{ is definable by } \mathbb{A}\}.$$

$\overline{\mathbb{A}}X \subset U$ because every definable $Y \subset U$.

Lower and upper approximations are definable, so

$$\underline{\mathbb{A}}X = \{u \in U \mid \underline{\varphi}(u)\},$$

$$\overline{\mathbb{A}}X = \{u \in U \mid \overline{\varphi}(u)\}.$$

We consider rough set approximation as a pair of formula

$$\forall_{u \in U} \; \big(\underline{\varphi}(u) \implies u \in X\big),$$

$$\forall_{u \in U} \; \big(\neg\overline{\varphi}(u) \implies \neg u \in X\big),$$

providing that $\underline{\varphi}$ and $\overline{\varphi}$ are the strongest formulae for which the above implications holds.

The sequential data processing methodology presented in the Section above may be interpreted as rough set approximation. Syntactic rules are queries and semantic attachments define the measurements of the approximated sensor. As the effect of sequential data processing we obtain the upper approximation of structural sensor by means of means of sequential sensor measurements.

7 Conclusions

When we process data into meaning representation language formulae we obtain knowledge base, which we may use for various data mining applications.

We can look for information using concepts from the documents. We describe properties of desired objects by means of queries and then we find the set of objects that satisfy the query.

The meaning representation language formulae provide us features for clustering and classification of the structural objects. If we define a decision attribute, we may construct rule based classifiers that use queries as selectors in decision rules. We plan to adapt the classical rule generation algorithms, so they could analyse the information contained in the properties of structurally described objects.

One of the classification tasks is an object identification. The goal of object identification is to determine the object's ontological category. We use properties of objects extracted from the sequential data for that purpose.

Object descriptions extracted from the sequential data are often incomplete. Part of attributes is missing. When we identify the object's ontological category, we add the values of missing attributes to the set of axioms as 'lost' missing values.

We plan also to study the process of high level concept extraction: the algorithms for defining the concepts that are not included in data but only approximated.

We will search for automatic methods for generating ontologies and determining the structure of objects in a way that would provide features useful for further application.

The ontology is most important during the process of feature selection. In our case the development of grammar and semantics for processing the sequential data into the meaning representation language formulae. This suggest that both problems are tightly connected and should be studied together.

On the other hand we plan to derive methods of extracting features from visual data and to extend our system on numerical data sequences.

Acknowledgments

I would like to thank Professor Andrzej Skowron for the inspiration and comprehensive support during the writing of this paper.

The research has been supported by the grant N-N206-400234 from Ministry of Science and Higher Education of the Republic of Poland.

References

1. Ballesteros, S., Shepp, B.E.: Object Perception: Structure and Process. Lawrence Erlbaum Associates, Hillsdale (1989)
2. Bazan, J., Nguyen, S.H., Nguyen, H.S., Skowron, A.: Rough set methods in approximation of hierarchical concepts. In: Tsumoto, S., Słowiński, R., Komorowski, J., Grzymała-Busse, J.W. (eds.) RSCTC 2004. LNCS (LNAI), vol. 3066, pp. 346–355. Springer, Heidelberg (2004)

3. Bazan, J., Skowron, A., Peters, J.F., Synak, P.: Spatio-temporal approximate reasoning over complex objects. Fundamenta Informaticae 67(1-3), 249–269 (2005)
4. Demri, S., Orłowska, E.: Incomplete Information: Structure, Inference, Complexity. Springer, Heidelberg (2002)
5. Duda, O., Hart, P.E., Stark, D.G.: Pattern Classification. A Wiley-Interscience Publication John Wiley & Sons, Chichester (2001)
6. Düntsch, I., Gediga, G.: Rough set data analysis: A road to non-invasive knowledge discovery. Methodos, Bangor (2000)
7. Gruber, T.R.: A translation approach to portable ontologies. Knowledge Acquisition 5(2), 199–220 (1993)
8. Grzymała-Busse, J.W., Grzymała-Busse, W.J.: An Experimental Comparison of Three Rough Set Approaches to Missing Attribute Values. In: Peters, J.F., Skowron, A., Düntsch, I., Grzymała-Busse, J.W., Orłowska, E., Polkowski, L. (eds.) Transactions on Rough Sets VI. LNCS, vol. 4374, pp. 31–50. Springer, Heidelberg (2007)
9. Grzymała-Busse, J.W.: A Rough Set Approach to Data with Missing Attribute Values. In: Wang, G.-Y., Peters, J.F., Skowron, A., Yao, Y. (eds.) RSKT 2006. LNCS (LNAI), vol. 4062, pp. 58–67. Springer, Heidelberg (2006)
10. Hanseth, O., Monteiro, E.: Modelling and the representation of reality: some implications of philosophy on practical systems development. Scandinavian Journal of Information Systems 6(1), 25–46 (1994)
11. Jaworski, W.: Learning Compound Decision Functions for Sequential Data in Dialog with Experts. In: Greco, S., Hata, Y., Hirano, S., Inuiguchi, M., Miyamoto, S., Nguyen, H.S., Słowiński, R. (eds.) RSCTC 2006. LNCS (LNAI), vol. 4259, Springer, Heidelberg (2006)
12. Jurafsky, D., Martin, J.H.: Speech and Language Processing: An Introduction to Natural Language Processing, Computational Linguistics, and Speech Recognition. Prentice-Hall, Englewood Cliffs (2000)
13. Knuth, D.E.: The Genesis of Attribute Grammars. In: WAGA 1990, pp. 1–12 (1990)
14. Kryszkiewicz, M.: Rough Set Approach to Incomplete Information Systems. Inf. Sci. 112(1-4), 39–49 (1998)
15. Kryszkiewicz, M.: Properties of incomplete information systems in the framework of rough sets. In: Polkowski, L., Skowron, A. (eds.) Studies in Fuzziness and Soft Computing. Rough Sets in Knowledge Discovery 1, vol. 18, pp. 422–450. Physica-Verlag, Heidelberg (1998)
16. Latkowski, R.: On decomposition for incomplete data. Fundamenta Informaticae 54(1), 1–16 (2003)
17. Latkowski, R.: On indiscernibility relations for missing attribute values. In: Proceedings of the Workshop on Concurrency, Specification and Programming (CSP 2004), Caputh, Germany, September 24-26, 2004, Informatik-Bericht 170, strony pp. 330–335. Humboldt Universität, Berlin (2004)
18. Latkowski, R., Mikołajczyk, M.: Data decomposition and decision rule joining for classification of data with missing values. In: Peters, J.F., Skowron, A., Grzymała-Busse, J.W., Kostek, B.z., Świniarski, R.W., Szczuka, M.S. (eds.) Transactions on Rough Sets I. LNCS, vol. 3100, pp. 299–320. Springer, Heidelberg (2004)
19. Latkowski, R.: Flexible indiscernibility relations for missing attribute values. Fundamenta Informaticae 67(1-3), 131–147 (2005)
20. Lipski Jr., W.: On Databases with Incomplete Information. J. ACM 28(1), 41–70 (1981)

21. Moens, M.-F.: Information Extraction: Algorithms and Prospects in a Retrieval Context. The Information Retrieval Series, vol. 21. Springer, Heidelberg (2006)
22. Marker, D.: Model Theory: An Introduction. Springer, Heidelberg (2002)
23. Mendelson, E.: Introduction to Mathematical Logic, 4th edn. International Thomson Publishing (1997)
24. Mohr, R., Pavlidis, T., Sanfeliu, A. (eds.): Structural Pattern Analysis. World Scientific, Teaneck (1990)
25. Nguyen, S.H., Bazan, J., Skowron, A., Nguyen, H.S.: Layered learning for concept synthesis. In: Peters, J.F., Skowron, A., Grzymała-Busse, J.W., Kostek, B.z., Świniarski, R.W., Szczuka, M.S. (eds.) Transactions on Rough Sets I. LNCS, vol. 3100, pp. 187–208. Springer, Heidelberg (2004)
26. Pavlidis, T.: Structural Pattern Recognition. Springer, Heidelberg (1977)
27. Pawlak, Z.: Information systems — theoretical foundations. Inf. Syst. 6(3), 205–218 (1981)
28. Pawlak, Z.: Rough Sets. International Journal of Computer and Information Sciences 11, 341–356 (1982)
29. Pawlak, Z.: Rough Sets. Theoretical Aspects of Reasoning about Data. Kluwer Academic Publishers, Dordrecht (1991)
30. Pawlak, Z., Skowron, A.: Rudiments of rough sets. Information Sciences 177(1), 3–27 (2007)
31. Pawlak, Z., Skowron, A.: Rough sets: Some extensions. Information Sciences 177(1), 28–40 (2007)
32. Pawlak, Z., Skowron, A.: Rough sets and Boolean reasoning. Information Sciences 177(1), 41–73 (2007)
33. Peters, J.F., Skowron, A. (eds.): Transactions on Rough Sets III. LNCS, vol. 3400. Springer, Heidelberg (2005)
34. Roddick, J.F., Hornsby, K., Spiliopoulou, M.: YABTSSTDMR - Yet Another Bibliography of Temporal, Spatial and Spatio-Temporal Mining Research. In: Uthurusamy, R., Unnikrishnan, K.P. (eds.) SIGKDD Temporal Data Mining Workshop, San Francisco, CA, pp. 167–175. ACM Press, Springer (2001)
35. Russ, J.C.: The Image Processing Handbook, Fourth Edition, 4th edn. CRC Press, Inc., Boca Raton (2002)
36. Russell, S.J., Norvig, P.: Artificial Intelligence. A Modern Approach. Prentice-Hall, Englewood Cliffs (1995)
37. Wittgenstein, L.: Tractatus Logico-Philosophicus. Routledge, London (1974)

A Categorical Approach to Mereology and Its Application to Modelling Software Components*

Dai Tri Man Lê** and Ryszard Janicki

McMaster University, Department of Computing and Software
Hamilton, ON, L8S 4K1 Canada
{ledt,janicki}@mcmaster.ca

Abstract. *Mereocat* (mereological category), a theory of part-whole relations based on category theory, is proposed and its basic properties are discussed. The concepts of mereological sum is redefined in categorical framework. The theory is then applied to model some properties of component software architecture.

Keywords: mereology, category theory, software component, "part of" relation.

1 Introduction

A correct construction of complex entities from the more primitive ones is one of the basic problems in many parts of science and virtually all engineering. This problem is especially important for the young field of software engineering, since software systems have grown much larger and more complex each year. The need of precise rules for both composition and decomposition in software constructions had been recognised more than thirty years ago (see [25]), but widely accepted formal techniques have not yet been found.

Attempts to formalise the concept of *"part of"* and *"fusion"* (composition) of parts go back to S. Leśniewski (1916-37, [20, 36]), and H. Leonard, N. Goodman (1940-50, [9, 19]). Leśniewski invented Mereology as an alternative to what is now called "standard set theory" (i.e. based on Zermelo-Fraenkel axioms[1]). Leonard and Goodman formulated Mereology within set theory, which makes the theory more accessible to applications. Both models of Mereology have been substantially extended [4,9,6,32,33,36], and recently many new concepts have been added. For example, *Mereotopology* [35] tackles the problem of axiomatising various topological properties; a new generalisation of a mereological sum, called *mereological supremum* has been introduced in [15], and the problem of *equivalent parts* is analysed in [13]. Most of known applications came from philosophy, cognitive science and pure mathematics [4,6,23,33,36].

* Partly supported by NSERC of Canada Grant.
** Partly supported by Ontario Graduate Scholarship.

[1] The fundamental difference is that in Leśniewski's systems, the operator called "*is*" and usually denoted by "ε", corresponds to both set theory membership "$\in$" and subset "$\subseteq$" operators, i.e. we would write $a\,\varepsilon A$ for $a \in A$ and $B\,\varepsilon A$ for $B \subseteq A$ [20, 32, 36].

J.F. Peters and A. Skowron (Eds.): Transactions on Rough Sets VIII, LNCS 5084, pp. 146–174, 2008.

However since mid nineties, mereological ideas started to be applied in other parts of science (and engineering) as for instance: industrial engineering [31], approximate reasoning [28], knowledge engineering [24], and recently in computer science and software engineering [12, 13, 14, 15].

In both Leśniewski's and Leonard-Goodman's mereologies, the "parthood" is considered to be a *global* relation, i.e. the relationship "a is part of b" does not have any external name. However, "a is part of b" and "a is part of c" are often very different relationships. In particular, when the same parts might be used to construct different objects, having unique names would often help both the understanding of relationships and part manipulation. A natural solution is to name the relationship "a is a part of b" as for instance f_{ab}, or, to say that the relationship "a is a part of b" is a *morphism* $f_{ab} : a \to b$. This leads us to category theory [22].

In this paper, which is a substantial extension of some ideas from [18], we propose a categorical mereology, called *mereocat*. The work was inspired by *Basic Mereology* of [13, 15] and Goguen's *General System Theory* [7, 8].

We will next use mereocat to model some hierarchical structures of *Component Based Software*. Component Based Software development [1] focuses on building a large software system by composing pre-existing *parts*, i.e. software components. The paradigm arises as an answer for the disadvantage of object-oriented paradigm that does not have an architectural method to separate the *computational* and *compositional* aspects. However it appears that both the formal and intuitive sense of parthood in Component Based Software is fuzzier and more confusing than the one in classical engineering. Although "part-whole" relationship is considered to be one of the most important UML modelling concepts within object-oriented modelling through the notion of *aggregation* and *composition* (see [24, 29]) and very recently within component based modelling through the UML notion of *"structure classifiers"* (see [11]), according to the authors' knowledge, no formal and complete parthood model does exist. We hope to contribute a little to a solution of this problem.

In the next two chapters we recall the basic concepts and results of Standard Mereology and Category Theory. In Chapter 4 we will define our categorical mereology, *mereocat*. Mereocat sums will be discussed in Chapter 5 and mereocat product in a very short Chapter 6. Elements of *CommUnity*, a component design framework [5, 21] will be discussed in Chapter 7. We will use CommUnity as a design tool in Chapter 8, where we show how mereocat can help in modelling parthood properties of Component Based Software. Chapter 9 contains our final comments.

2 Standard Mereology

Before introducing our categorical mereology, to make our ideas and concepts better understood we start with basic notions of "standard mereology" [4, 33]. This is a mereology based on the Calculus of Individuals by Leonard and Goodman [9, 19]. From a mathematical point of view it is a part of the theory of partially ordered sets. We will not discuss here the mathematics of Leśniewski's systems[2] as they cannot easily be formulated within standard set theory [32, 33, 36].

To make the papers self-sufficient, we start with a survey on the principal properties of *partial orders*.

Let X is a set. A relation $\preceq \subseteq X \times X$ is called a *partial order* iff it is *reflexive* ($x \preceq x$), *anti-symmetric* ($x \preceq y \wedge y \preceq x \Rightarrow x = y$), and *transitive* ($x \preceq y \wedge y \preceq z \Rightarrow x \preceq z$). If $\preceq$ is a partial order then the pair $(X, \preceq)$ is called a *partially ordered set* or *poset*. A relation $\prec$ defined as $x \prec y \iff x \preceq y \wedge x \neq y$ is called a *strict partial order*.

Let $(X, \preceq)$ be a poset and let $A \subseteq X$. An element $a \in X$ is called an *upper bound* (a *lower bound*) of A iff $\forall x \in A.\ x \preceq a$ ($\forall x \in A.\ a \preceq x$). The sets of all *upper bounds* and *lower bounds* of A are denoted by $ub(A)$ and $lb(A)$ respectively.

The element $\top \in X$ satisfying $\forall x \in X.\ x \preceq \top$ is called the *top* of X, and the element $\bot$ satisfying $\forall x \in X.\ \bot \preceq x$ is called the *bottom* of X.

An element $a \in A$ is a *minimal* (*maximal*) element of A iff $\forall x \in A.\ \neg(x \prec a)$ ($\forall x \in A.\ \neg(a \prec x)$). The set of all *minimal* (*maximal*) elements of A will be denoted by $min(A)$ ($max(A)$).

An element $a \in X$ is called the *least upper bound* (*supremum*) of A, denoted $sup(A)$, iff $a \in ub(A)$ and $\forall x \in ub(A).\ a \preceq x$, and it is called the *greatest lower bound* (*infimum*) of A, denoted $inf(A)$, iff $a \in lb(A)$ and $\forall x \in lb(A).\ x \preceq a$.

The minimal elements of the set $X \setminus \{\bot\}$ are called *atoms* of the poset $(X, \preceq)$, and $\mathcal{A}toms$ denotes the set of all atoms of X.

The relation $\hat{\prec}$ defined as $x \hat{\prec} y \iff x \prec y \wedge \neg(\exists z.\ x \prec z \prec y)$ is called the *cover relation* for $\preceq$.

Now we will begin with mereological axioms, but to do so we need some definitions. Let $(X, \preceq)$ be a poset (with or without $\bot$). The relation $\preceq$ is now interpreted as *"part of"*; a is a *part of* b iff $a \preceq b$, and a is a *proper part of* b iff $a \prec b$. Notice that "a is a *part of* b" is equivalent to saying that "b is a *whole of* a". The element $\bot$ is interpreted as an empty part. The relation $\circ$, $\dagger$ and $\diamond$ on $X \setminus \{\bot\}$ defined as

$$x \circ y \iff \exists z \in X \setminus \{\bot\}.\ z \preceq x \wedge z \preceq y \qquad \text{(overlap)}$$
$$c \dagger y \iff \neg(x \circ y) \qquad \text{(disjoint)}$$
$$x \diamond y \iff \exists z \in X \setminus \{\bot\}.\ x \preceq z \wedge y \preceq z \qquad \text{(underlap)}$$

are called *overlapping, disjointness* and *underlapping* respectively. Two element x and y *overlap* iff they have a common non-empty part, they are disjoint iff they do not have a common non-empty part, and they underlap if they are both parts of another element (see [4, 33] for more properties).

There are many mereological axioms, we will only discuss those we consider the most important for our purposes [4, 19, 13, 33]. The first three axioms below deal with the meaning of "part of" relation, the remaining five describe global properties of mereologies.

---

[2] Conceptually the ideas of Leśniewski and Leonard-Goodman appear to be similar, but mathematical results about this relationship are hard to find and the major one [10] has not been widely accepted among Leśniewski's disciples (see [36]). The paper [10] is right in the case of Leśniewski's elementary mereology, however Leśniewski's non-elementary mereology is a much richer system, so [10] is often considered heresy for some Leśniewski's followers [36].

$$\neg(y \preceq x) \Rightarrow \exists z. (z \preceq y \wedge x \dagger z) \quad \text{(SSP)}$$
$$x \prec y \Rightarrow (\exists z \in X. z \prec y \wedge x \dagger z) \vee x = \bot \quad \text{(WSP)}$$
$$x \prec y \Rightarrow (\exists z \in X. z \prec y \wedge \neg(z \preceq x)) \vee x = \bot \quad \text{(SCP)}$$
$$\forall x \in X \setminus \{\bot\}. \exists y \in \mathcal{A}toms. y \preceq x \quad \text{(ATM)}$$
$$\bot \in X \quad \text{(BOT)}$$
$$\top \in X \quad \text{(TOP)}$$
$$x \preceq y \Longleftrightarrow x(\widehat{\prec})^{*}y \quad \text{(CCL)}$$
$$\forall x \in X. \exists y \in max(X). \ x \preceq y \quad \text{(WUB)}$$

The axiom SSP is called *Strong Supplementation Principle.* It implies quite regular properties, among others *it guarantees that different objects have different sets of proper parts*. Mereologies satisfying SSP are called *Extensional Mereologies*. Detailed discussion of SSP and its consequences can be found in [4, 33]. We think SSP is too restricted for our purposes. In mathematics and computer science different objects are very often built from the same proper parts. For instance, $A \times B \neq B \times A$ (unless $A = B$), and both $A \times B$ and $B \times A$ are intuitively built from the same parts.

The axiom WSP, called *Weak Supplementation Principle*, is much weaker and very popular. Among others, it guarantees that if an element has a proper non-empty part, it has more than one. Mereologies satisfying WSP are often called *Minimal Mereologies* [4, 33]. For example, a totally ordered set is not a minimal mereology. It is often believed that any reasonable mereology must conform to this axiom [33], however we *strongly disagree* with this opinion. The hidden assumption behind WSP is that some parts, including the most elementary ones, are unique. Consider the set $S = \{(a,1),(a,2),(b,1)\}$. Intuitively $\{a\}$, $\{b\}$, $\{1\}$, $\{2\}$, $\{(a,1)\}$, $\{(a,2)\}$, $\{(b,1)\}$, and $\{(a,1),(a,2)\}$, $\{(a,2),(b,1)\}$, $\{(a,1),(b,1)\}$ are proper parts of it. WSP does not hold as $\{(a,2),(b,1)\}$ and $\{(a,1)\}$, both proper parts of S, overlap on $\{a\}$ and $\{1\}$. A possible solution is to "tag" all elementary components, i.e. to consider the set $S_{WSP} = \{(a_1,1_a),(a_2,2_a),(b_1,1_b)\}$ instead of S, but this usually leads to unnecessary complexity of a model. We will come back to this issue in our Example 1.

The axiom SCP, is called *Strong Company Principle* [4]. It is weaker than WSP and, intuitively, it says that a proper part x has to be supplemented by some other proper part z which is not "included" in the part x. For example, a totally ordered set does not satisfy SCP. However even SCP may not hold if the repetition of components is allowed. Consider a mereology of multisets, the only part of $\{a,a,a,a\}$ is $\{a\}$ (or a). Again, "tagging" elementary elements seems to solve the problem but may result in unnecessarily complex model. Nevertheless, we believe this axiom seems to be helpful in many "software oriented" mereologies.

Some modern mereologies, for instance *Rough Mereology* of Polkowski and Skowron [28] do not assume even SCP. The exact meaning of a "part" is left for particular applications.

The axiom ATM (*Atomistic Mereology*) says that all objects (except the empty part) are built from elementary elements called *atoms*. The axiom BOT simply says that the empty part does exist. Most of classical mereological theories assume the empty part does not exist. The argument is that the empty part (empty element) is not needed except for completeness properties [4,33]. We believe, that empty part, as empty set or empty string, is a very useful concept that eventually will make our theory simpler. The top $\top$ is called *Universe* in mereology, and it plays either the role of "universe of concourse" or it simply represents the most complex object [33]. The axiom CCL (*Cover Closed*) seems to be implicitly assumed in many mereological theories, however it not usually openly discussed [4,33]. It was formally introduced in [13] and it states that the part-of relation is the reflexive and transitive closure of the cover relation for $\preceq$. Every finite poset satisfies CCL. The final axiom WUB (*Weakly Upper Bounded*) means that the set $max(X)$ is a roof that cover the whole set. This axiom is crucial when the concept of equivalent parts is introduced [13].

Posets satisfying ATM and SSP are often called *Hyperextensional Mereologies* [4,9, 33]. The name follows from the fact that in such cases the objects are identical if they are built from the identical sets of atoms. Hyperextensional mereologies usually lead to very elegant theories, but they appear to be too restricted for our purposes.

A mereology that satisfies BOT, WSP, CCL, WUB and ATM is called *Basic Mereology* in [13], however, detailed analysis of examples from [13] and [15] indicates that the author of [13] actually meant SCP *not* WSP.

We believe a search for a "universal" Mereology does not lead very far, so for any particular application a specific Mereology should be constructed . In mereocat analysed in the rest of this paper, we will only assume BOT and SCP, with a possibility of adding additional axioms when needed.

Example 1 ([13, 15]). For every set A, let $\widehat{A} = \{\{a\} \mid a \in A\}$ be the set of all singletons generated by A, i.e. if $A = \{a,b\}$, then $\widehat{A} = \{\{a\},\{b\}\}$.

Let $D_1 = \{a,b\}$, $D_2 = \{1,2\}$ be sets and let $X = 2^{D_1} \cup 2^{D_2} \cup 2^{D_1 \times D_2}$.

Define the relation $\preceq$ in $X \times X$ as follows:

$$A \preceq B \iff A \subseteq B \vee A \subseteq \pi_i(B),\ i = 1,2$$

where $\pi_i(B)$ is the projection of B on i-th coordinate, i.e. $\pi_1(B) = \{x_1 \mid (x_1,x_2) \in B\}$, $\pi_2(B) = \{x_2 \mid (x_1,x_2) \in B\}$.

One can show by inspection that the pair $(X,\preceq)$ satisfies BOT, SCP, CCL, WUB and ATM with $\mathcal{A}toms = \widehat{D_1} \cup \widehat{D_2}$, $\bot = \emptyset$ and $\top = \{(a,1),(a,2),(b,2),(b,2)\}$. This example is a special case of a more general model that is discussed in detail in [12, 13, 15]. A Hasse diagram of the relation $\preceq$ is presented in Figure 1. The axiom WSP is not satisfied, as for instance, $\{(a,2),(b,1)\}$ and $\{(a,1)\}$, are both proper parts of $\{(a,1),(a,2),(b,1)\}$, and they overlap on $\{a\}$ and $\{1\}$. □

The operations $\oplus$ and $\odot$ defined by

$$z = x \oplus y \iff (\forall w \in X.\, w \circ z \Leftrightarrow w \circ x \vee w \circ y), \quad \text{(sum)}$$
$$z = x \odot y \iff (\forall w \in X.\, w \preceq z \Leftrightarrow w \preceq x \wedge w \preceq y) \wedge z \neq \bot \quad \text{(product)}$$

are called the *mereological sum* and *mereological product* respectively [9, 4, 33]. It is implicitly assumed for both definitions that in order to exist, z must be unique. Both concepts can easily be extended from two elements to any set in a standard way [4, 33]. The sum of elements of the set A (if exists) will be denoted by $\bigoplus A$, and the product of elements of the set A (if exists) will be denoted by $\bigodot A$. There

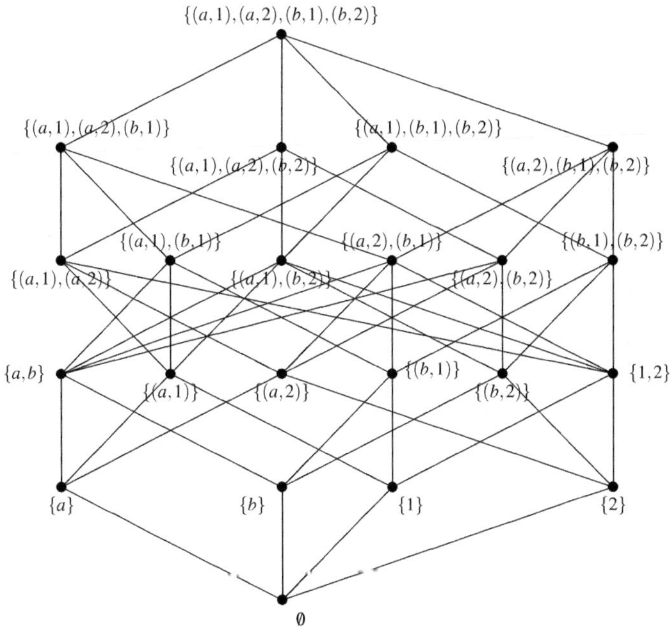

Fig. 1. A Hasse diagram of the relation $\preceq$ from Example 1

is an obvious relationship between mereological sum and least upper bound and between mereological product and greatest lower bound, however those concepts are not identical. Let $X = \{\emptyset, \{a\}, \{b\}, \{c\}, \{a,b,c\}\}$. The tuple $(X, \subseteq)$ is clearly a Minimal Mereology, $sup(\{\{a\},\{b\}\}) = \{a,b,c\}$, $\{a\} \oplus \{b\}$ does not exist. On the other hand $sup(\{\{a\},\{b\},\{c\}\}) = \bigoplus\{\{a\},\{b\},\{c\}\} = \{a,b,c\}$. The idea is that if $\preceq$ represents "part-of" relation and the element $\{\{a\},\{b\},\{c\}\}$ is built from all three parts $\{a\}$, $\{b\}$ and $\{c\}$. Consider the relation $\preceq$ from Figure 1. The least upper bound of $\{(b,2)\}$ and $\{1,2\}$ does not exist while $\{(b,2)\} \oplus \{1,2\} = \{(b,1),(b,2)\}$. However for $\{a,b\}$ and $\{1,2\}$, neither $sup(\{a,b\},\{1,2\})$ nor $\{a,b\} \oplus \{1,2\}$ does exist.

Many mereologies assume that $x \odot y$ implies the existence of $x \oplus y$ [33], which results in very elegant models similar to semi-lattices or, when additional assumptions are made, to quasi boolean algebras [33]. However, for our purposes such assumption is too strong, most of the models we are interested in do not have this property (including systems from [12, 13, 15]). If different objects are allowed to have identical proper parts, then the sum $x \oplus y$ often does not exist. The concept of *mereological supremum* proposed in [15] is a partial (but only partial) solution to this problem. The mereological supremum of $\{a,b\}$ and $\{1,2\}$ is $\{(a,1),(a,2),(b,1),(b,2)\}$, the top of this mereology and the most complex object that can be constructed from parts $\{a,b\}$ and $\{1,2\}$. Unfortunately the definition of mereological supremum is rather complex and has not been tested for many examples.

For more details on Standard and Basic Mereologies the reader is referred to [4, 13, 15, 33].

As we have mentioned in the Introduction, the mereologies discussed in this chapter consider the "parthood" as a *global relation*, and particular relationship "*a* is part of *b*" does not have any specific external name. By providing such a name[3] and interpreting it as a morphism from a to b we move to the domain of category theory.

3 Elements of Category Theory

In order to make this paper self-sufficient, and to set up a uniform notation, we will now introduce the basic concepts of category theory [5, 22].

A *category* $\mathbf{C}$ consists of

1. a class $ob(\mathbf{C})$, which denotes the class of $\mathbf{C}$-*objects* of $\mathbf{A}$,
2. a class $mor(\mathbf{C})$, which denotes the class of $\mathbf{C}$-*morphisms* (or $\mathbf{C}$-*arrows*).
3. for each morphism $f \in mor(\mathbf{C})$ has a unique source object A and target object B where $A, B \in ob(\mathbf{C})$. We write $A \xrightarrow{f} B$ or $f : A \to B$ to denote "f is a morphism from A to B",
4. for each pair (A, B) of $\mathbf{C}$-objects, a class $\mathbf{C}(A, B)$, which denotes the *hom-class* of all $\mathbf{C}$-morphisms from A to B,
5. for each $\mathbf{C}$-object A, a morphism $A \xrightarrow{id_A} A$, called the $\mathbf{C}$-*identity* on A,
6. a *composition law* associating with two $\mathbf{C}$-morphisms $A \xrightarrow{f} B$ and $B \xrightarrow{g} C$ a $\mathbf{C}$-morphism $A \xrightarrow{g \circ f} C$, called the *composition* of f and g,

such that the following axioms hold:

- composition is associative; i.e., for morphisms $A \xrightarrow{f} B$, $B \xrightarrow{g} C$, and $C \xrightarrow{h} D$, the equation $h \circ (g \circ f) = (h \circ g) \circ f$ holds,
- $\mathbf{C}$-identities act as identities with respect to composition; i.e., for $\mathbf{C}$-morphisms $A \xrightarrow{f} B$, we have $id_B \circ f = f$ and $f \circ id_A = f$.

The following types of morphisms will often be used throughout this paper. Let $\mathbf{C}$ be a category and $A, B \in ob(\mathbf{C})$.

A morphism $A \xrightarrow{f} B$ of $\mathbf{C}$ is:

- an *isomorphism* iff there is a morphism $B \xrightarrow{g} A$ of $\mathbf{C}$ such that $g \circ f = id_A$ and $f \circ g = id_B$. Under these conditions, A and B are said to be isomorphic, which is denoted by $A \cong B$.
- a *monomorphism* if $f \circ g = f \circ h$ implies $g = h$ for all $\mathbf{C}$-morphisms $g, h : C \to A$. In this case, we say f is *monomorphic* (or *monic*).

Finally the *functor* is defined as follows. For the categories $\mathbf{C}$ and $\mathbf{D}$, a *functor* F from $\mathbf{C}$ to $\mathbf{D}$ is a mapping that:

- associates to each object $X \in ob(\mathbf{C})$ an object $F(X) \in ob(\mathbf{D})$,

[3] Naming relationship looks trivial from a theoretical point of view, but it is extremely important, often underestimated part of any formal software specification procedure [26].

- associates to each morphism $f : X \to Y \in mor(\mathbf{C})$ a morphism $F(f) : F(X) \to F(Y) \in mor(\mathbf{D})$

and the following two properties hold:

- $F(id_X) = id_{F(X)}$ for every object $X \in ob(\mathbf{C})$
- $F(g \circ f) = F(g) \circ F(f)$ for all morphisms $f : X \to Y$ and $g : Y \to Z$ of $\mathbf{C}$.

We denote the indentity functor from a category $\mathbf{C}$ to itself by $I_{\mathbf{C}}$, (or simply, by $\mathbf{C}$).

4 Introduction to Mereocat

Mereocat is a theory whose structures possess mereological properties and are defined in terms of category theory. Goguen's work on *inclusive categories* [7] provided a major motivation. The concept of *inclusive morphism* from [7] appears to be close to the concept of "part of" as understood in software engineering (see [12]). In mathematics, the categorical approach to set theory can be found in Lawvere and Rosebrugh [17], where the whole set theory is presented as the algebra of mappings. In [17] the phrase "part of" which denotes monomorphism represents both "subset of" and "member of", but no reference to existing papers on Mereology is given. Nevertheless, these ideas are very close to fundamental Leśniewski's concepts [20, 32, 36], even though their formulations use categorical notions.

Before giving the definition of a *mereocat*, we recall the notion of a *comma category*[4] [22].

Suppose that $\mathbf{A}$, $\mathbf{B}$, and $\mathbf{C}$ are categories, and two functors $\mathbf{A} \xrightarrow{S} \mathbf{C} \xleftarrow{T} \mathbf{B}$, we can form the comma category $(S \downarrow T)$ as follows:

- The objects are triples (D, E, f), with $D \in ob(\mathbf{A})$, $E \in ob(\mathbf{B})$, and $f : T(D) \to S(E)$ a morphism in $\mathbf{C}$.
- The morphisms from (D, E, f) to (D', E', f') are pairs (g, h) where $g : D \to D'$ and $h : E \to E'$ are morphisms in $\mathbf{A}$ and $\mathbf{B}$ respectively making the following diagram commutes

$$\begin{array}{ccc} S(D) & \xrightarrow{S(g)} & S(D') \\ \Big\downarrow f & & \Big\downarrow f' \\ T(E) & \xrightarrow{T(h)} & T(E') \end{array}$$

Morphisms are composed by taking $(g, h) \circ (g', h')$ to be $(g \circ g', h \circ h')$.

A special case of comma category is of an *arrow category* [2,22]. From any category $\mathbf{C}$, the arrow category $\mathcal{A}r(\mathbf{C})$ of arrows of $\mathbf{C}$ is the category $(I_C \downarrow I_C)$. In detail, $\mathcal{A}r(\mathbf{C})$ has objects as $\mathbf{C}$-morphisms and a morphism between from an object $A \xrightarrow{f} A'$ to an

[4] The name comes from the notation originally used in [17], which involved the comma punctuation mark.

object $B \xrightarrow{g} B'$ is a pair of part-morphisms $A \xrightarrow{h} B$ and $A' \xrightarrow{k} B'$ making following diagram commute

$$\begin{array}{ccc} A & \overset{h}{\dashrightarrow} & B \\ f\downarrow & & \downarrow g \\ A' & \overset{k}{\dashrightarrow} & B' \end{array}$$

This morphism will be denoted as $(k,l) : f \to g$.

We can now provide the main definition of this paper, our *categorical mereology*.

Definition 1. *A* ***mereocat MC*** *is a category consisting of*

- *its object in* $ob(\mathbf{MC})$*, which are called* ***mereo-objects*** *(or* ***objects****),*
- *its morphisms which are called* ***part-morphisms*** *(or* ***parts****),*
- *an initial object denoted by* $\perp$ *(every morphism* $\_ : \perp \to A$ *is called* ***empty-part****),*

such that the following conditions hold

1. *if* $A \xrightarrow{f} A$ *is a* $\mathbf{MC}$*-morphism then* $f = id_A$*,*
2. *for every pair of part-morphisms* (f,g)*, the hom-class* $\mathcal{A}r(MC)(f,g)$ *has at most one element, and*
3. *for every part-morphism* $A \xrightarrow{f} B$ *of* $\mathbf{MC}$ *such that* $A \not\cong B$ *and* $A \not\cong \perp$*, there exists a part-morphism* $C \xrightarrow{g} B$ *such that* $C \not\cong \perp$ *and* $C \not\cong B$ *and* $\mathcal{A}r(\mathbf{MC})(g,f) = \emptyset$. □

The first very important thing is that we do *not* call mereo-objects "parts" as in the traditional mereologies. An AA battery is a part of a camera, only when it is placed into it, not in general. The same battery might be a part of a clock, remote control, etc., or it might be a top of an abstract sculpture. In this model, the morphisms $(battery \xrightarrow{in} camera)$, $(battery \xrightarrow{in'} clock)$, $(battery \xrightarrow{in''} remote\text{-}control)$, $(battery \xrightarrow{in'''} abstract\text{-}sculpture)$, etc., are *parts*. When saying some object is a part of a whole, we assign implicitly a *role* the object plays with respect to the whole, and the *nature* of the whole as well.

However, every mereo-object A itself can be considered as an *external* part (a whole) because every object A is equipped with an identity map i_A. Since i_A and A are categorically the same, in this paper, we will abuse the notation by also calling an object A "part", or "part of B", if the morphism $A \xrightarrow{f} B$ is obvious from the context.

The conditions (1), (2) and (3) in principle state that this category can be interpreted as a mereology.

The condition (1) makes sure we have a unique way to identify a whole as an *external* part using its identity morphism.

The condition (2) makes sure that **MC** and $\mathcal{A}r(\mathbf{MC})$ behave like a *posets*. Note that the concept of poset in this categorical definition is not "strict" since antisymmetry is not assumed to resolve into *extensional equality* [17].

The condition (3) is the categorical formulation of SCP axiom. Intuitively, it says that a "proper" part f has to be supplemented by some other "proper" part g which is not "included" in the part f.

Proposition 1. *Let **MC** be a mereocat, then*

1. *All part-morphisms of **MC** are monomorphic.*
2. *All morphisms of $\mathcal{A}r(\mathbf{MC})$ are monomorphic.*

Proof. Follows from condition (2) of Definition 1. □

For example, a category of finite sets whose morphisms are inclusive mappings, i.e. function of the form $i(x) = x$, is a mereocat. And so is the mereology described in Figure 1, with morphisms $A \xrightarrow{f_{AB}} B$ iff $A \prec B$, and $A \xrightarrow{id_A} A$ for all A.

Some examples of structures which are not mereocats are:

- Finite sets and functions (because functions are not necessarily monomorphic).
- The following simple category:

$$\bot \xrightarrow{f} B \xrightarrow{g} C$$

since, for instance, $B \neq C$ but we cannot find another morphism with the target C to satisfy the condition (3) of Definition 1.

Since in this model, parts are morphisms, we need to define what is "part of a *part*". The *part-of* relationship between parts is defined as follows:

Definition 2. *Let $f : A \to B$ and $g : A' \to B'$ be two parts of a mereocat **MC**. We say f is a part of g iff there exists a morphism $(k,l) : f \to g$ in the category $\mathcal{A}r(\mathbf{MC})$.*

$$\begin{array}{ccc} A & \overset{k}{\dashrightarrow} & A' \\ {\scriptstyle f}\downarrow & & \downarrow{\scriptstyle g} \\ B & \underset{l}{\dashrightarrow} & B' \end{array}$$

*Under this condition, we say f is **part-of** g.* □

Note that we allow $f = id_A : A \to A$ and $g = id_{A'} : A' \to A'$, which further justify considering occasionally mereo-object as parts, if this does not lead to any ambiguity.

By using morphisms we may provide very exact and unambiguous specification of various aspects of parthood, which we believe cannot easily be achieved using the traditional relational models alone. Every mereocat **MC** automatically generates its companion category $\mathcal{A}r(\mathbf{MC})$, which actually is *the* part-of category, and can be used to model more elaborate parthood properties.

Definition 3. *For a mereocat **MC**, the category of $\mathcal{A}r(\mathbf{MC})$ is called the **part-of category** of **MC**.* □

The Definitions 2 and 3, and their properties are illustrated in the below example.

Example 2. Let **BBlock** be a mereocat where objects are house building blocks and a part-morphism $f : A \to C$ means A is used to build C. Suppose the following square commutes:

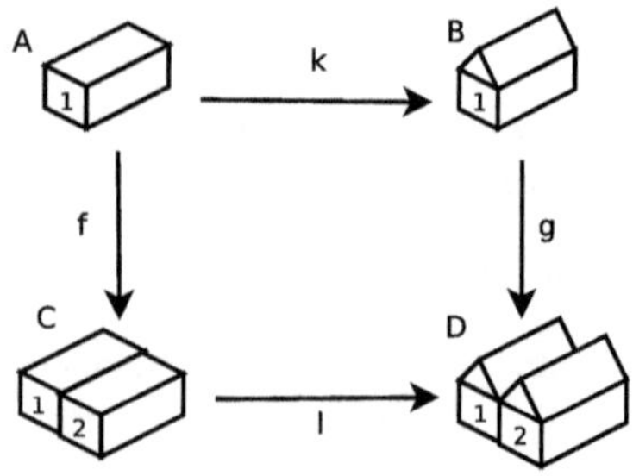

This implies that $(k,l) : f \to g$ is a morphism of $\mathcal{A}r(\mathbf{BBlock})$. Hence, the part-morphism f, which specifies how the block (1) of A become block (1) of C, is part-of the part-morphism g, which specifies how the house of B becomes one of the houses of D as follows:

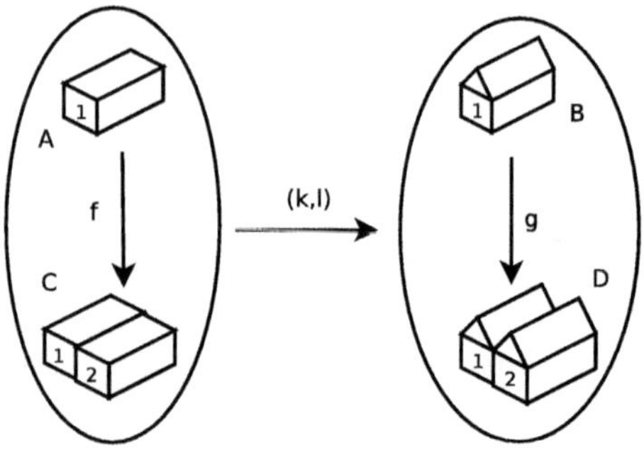

Furthermore, the commutativity of the square also implies the part-of relationship between k and l through the existence of the morphism (f,g) as shown in the next diagram:

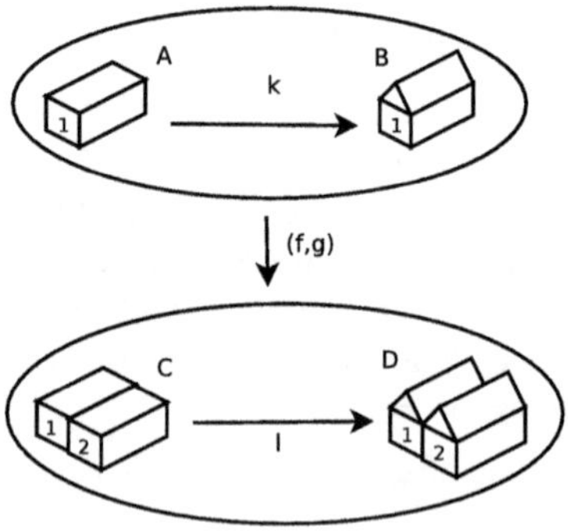

□

The next, rather abstract, example illustrates both Definition 1 and Definition 2.

Example 3. Let us consider a mereocat **M** specified using the following diagram[5] where the square $BCFE$ and the triangle ABC commute,

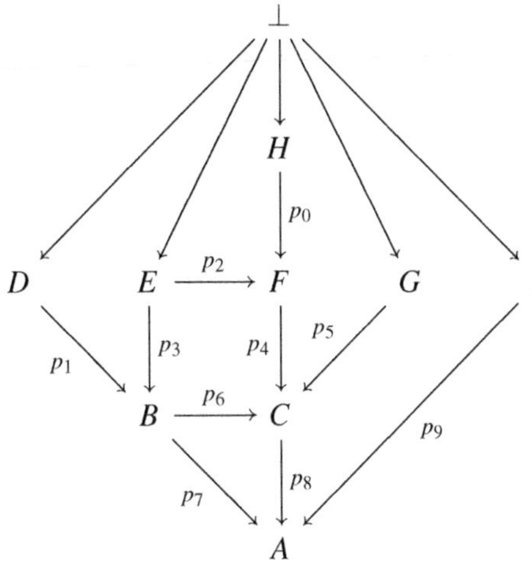

We have:

- for any part morphism $p : X \to Y$ in the diagram, $(p, p) : id_X \to id_Y$ is a morphism of $\mathcal{A}r(\mathbf{M})$. Hence, if there an arrow between any two objects of **M**, we say id_X is part-of id_Y (or we can abuse the notation by saying X is part-of Y). For instance, B is part of A, C is part of A, D is part-of A and $\perp$ is part-of everything in the diagram.
- p_7 is part-of p_8 because $(p_6, id_A) : p_7 \to p_8$ is a morphism of $\mathcal{A}r(\mathbf{M})$ (the triangle ABC commutes)
- p_3 is part-of p_4 because $(p_2, p_6) : p_3 \to p_4$ is a morphism of $\mathcal{A}r(\mathbf{M})$ (the square $BCFE$ commutes), but also p_2 is part-of p_6 because $(p_3, p_4) : p_2 \to p_6$ is a morphism of $\mathcal{A}r(\mathbf{M})$. □

The next example deals with a more realistic situation.

Example 4. Assume that we have a *filled car* with two passengers in it. We can model two possible parts of it as the collection of two people and the car with empty seats.

In Figure 2, **People** models the collections of two people; **Empty Car** and **Filled Car** model the empty and filled car respectively. The arrows from **People** to **Filled Car** describe the morphism **People**$\xrightarrow{f}$**Filled Car**, which associates the people to the appropriate car seats. The arrows from **Empty Car** to **Filled Car** describe the morphism **Empty Car**$\xrightarrow{g}$**Filled Car**, which *associates* the seats of the empty car to the seats of filled car. Notice that the arrows are needed since in category theory *the mappings are used to characterise the objects*. This is different from theory, e.g. set theory, where the members of a collection define the collection itself. As a result, category theory is often considered a "component-free" or "blackbox" approach. □

[5] In category theory, to simplify the diagram, identity arrows and arrows that can be inferred from composition of arrows are not drawn.

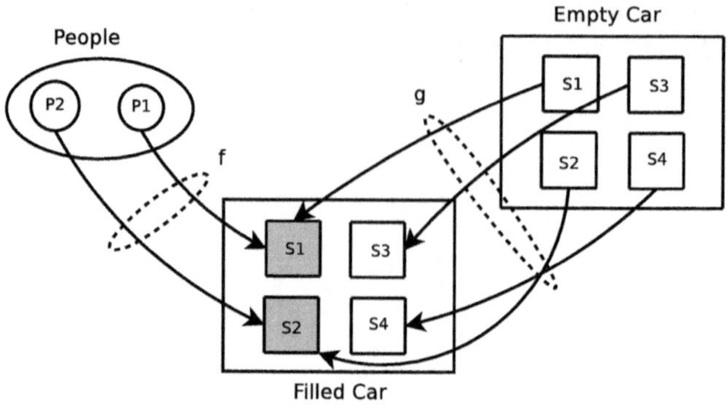

Fig. 2. The car model with part mophisms f and g from Example 4

We will now show that a mereocat can be interpreted as a mereology which is defined in Section 2. Since we consider mereocat to be a generic rather than specific concept, the mereological assumptions are very weak. We assume only equivalents of BOT and SCP, the other axioms may be added when needed. The following results show formal relationship among a mereocat, its arrow category companion, and a standard mereology.

Proposition 2. *Let* $\boldsymbol{MC}$ *be a mereocat satisfying* $A \cong B \Rightarrow A = B$ *for all* $A, B \in ob(\boldsymbol{MC})$, *and let* $\preceq$ *be a relation on* $ob(\boldsymbol{MC})$ *defined as:*

$$A \preceq B \Longleftrightarrow \exists f \in hom(\boldsymbol{MC}).\ f : A \to B$$

Then $(ob(\boldsymbol{MC}), \preceq)$ *is a mereology that satisfies BOT and SCP.*

Proof. It is a straightforward consequence of the conditions (1), (2) and (3) of Definition 1. □

Proposition 3. *Let* $\boldsymbol{MC}$ *be a mereocat such that*

1. *for any two objects* $A, B \in ob(\boldsymbol{MC})$, *if* $A \cong B$ *then* $A = B$, *and*
2. *for every part-morphism* $A \xrightarrow{f} B$ *of* $\boldsymbol{MC}$ *with* $A \neq B$ *and* $A \neq \bot$, *there exists a part-morphism* $C \xrightarrow{g} B$ *with* $C \neq B$ *and* $C \neq \bot$ *such that no* $\boldsymbol{MC}$*-object* $D \neq \bot$ *and two part-morphisms* $D \xrightarrow{k} A$ *and* $D \xrightarrow{l} C$ *making the following diagram commute*

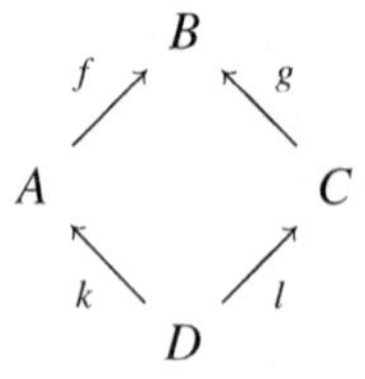

Let $\preceq$ *be a relation on* $ob(\boldsymbol{MC})$ *defined as:*

$$A \preceq B \Longleftrightarrow \exists f \in hom(\boldsymbol{MC}).\ f : A \to B$$

Then $(ob(\boldsymbol{MC}), \preceq)$ *is a minimal mereology with BOT.*

Proof. Form Proposition 2 and the condition (2) above, which is a categorical equivalence of WSP. □

Proposition 4. *Let* $\boldsymbol{BM} = (X, \preceq)$ *be a mereology satisfying BOT and SCP. Define* $\mathbf{C}$ *as a category such that* $ob(\mathbf{C}) = X$ *and there is a morphism* $f_{AB} : A \to B$ *iff* $A \preceq B$*. The category* $\mathbf{C}$ *is a mereocat.*

Proof. Proof follows from the fact **BM** is a poset so there is at most one morphism from one object to another in $ob(\mathbf{C})$, while BOT and SCP imply the remaining mereocat properties. □

Proposition 5. *Let* $\boldsymbol{MC}$ *be a mereocat such that for any two part-morphisms* f *and* g*,* $f \cong g$ *implies* $f = g$*. Then the arrow category* $\mathcal{A}r(\boldsymbol{MC})$ *is a mereology with BOT and SCP.*

Proof. – Reflexivity: any part $f : A \to B$ is a part of itself, since the following diagram commutes

$$\begin{array}{ccc} A & \xrightarrow{i_A} & A \\ {\scriptstyle f}\downarrow & & \downarrow{\scriptstyle f} \\ B & \xrightarrow{i_B} & B \end{array}$$

- Antisymmetry: if $f : A \to B$ is part-of $g : A' \to B'$ and g is also part-of f, then hom-classes $\mathcal{A}r(MC)(f,g)$ and f $\mathcal{A}r(MC)(f,g)$ and $\mathcal{A}r(MC)(g,f)$ are not empty then there is a unique morphism from f to g and a unique morphism from g to f. In other words, $f \cong g$. Hence, we conclude $f = g$.

- Transitivity: suppose $f : A \to B$ is part-of $g : A' \to B'$ and g is a part-of $h : A'' \to B''$. Since $\mathcal{A}r(MC)$ is a category, we have following commutative diagram

$$\begin{array}{ccccc} A & \xrightarrow{k} & A' & \xrightarrow{k'} & A'' \\ {\scriptstyle f}\downarrow & & \downarrow{\scriptstyle g} & & \downarrow{\scriptstyle h} \\ B & \xrightarrow{l} & B' & \xrightarrow{l'} & B'' \end{array}$$

and hence f is part-of h. Since the BOT and SCP are also satisfied, we are done. □

Sometimes we want to look at the categorical structure consisting of only the parts of some specific whole. This can be done using the concept of *slice category*. Given a category $\mathbf{C}$ and a fixed object $A \in ob(\mathbf{C})$, the *slice category* (see [2]) over A, denoted by $(\mathbf{C} \downarrow A)$, is a category whose objects are morphisms with the same target A and a morphism from $f : B \to A$ to $g : C \to A$ in $(\mathbf{C} \downarrow A)$ is a commutative triangle

$$\begin{array}{ccccc} & & B & & \\ & {\scriptstyle f}\nearrow & & \nwarrow{\scriptstyle g} & \\ A & & \xrightarrow{h} & & C \end{array}$$

Definition 4. *Let* **MC** *be a mereocat and A be an object of* **MC***, the part-of category* ***restricted to*** *A is the slice category* $(\mathbf{MC} \downarrow A)$. □

In Example 3, the part-of category restricted to C is

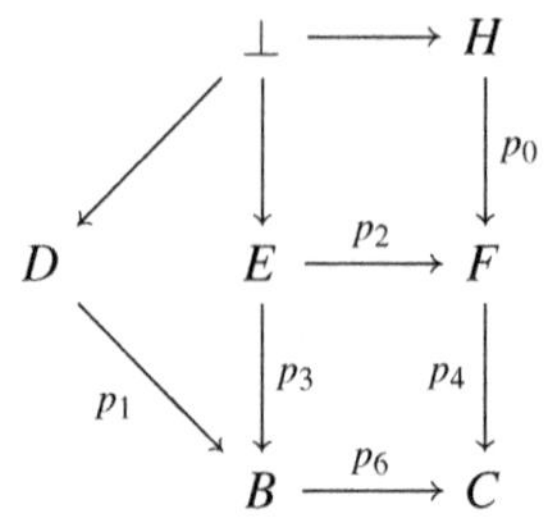

The concept of *atomic part* can be defined categorically as follows:

Definition 5. *Let* **MC** *be a mereocat, an object A of* **MC** *is called an* ***atom*** *in* **MC** *iff the slice category* $(\mathbf{MC} \downarrow A)$ *consists only two objects* id_A *and* $\bot \to A$*. In this case, the part* id_A *(or A) is called an* ***atomic part***. □

In Example 3, D, E, H, G and I are atomic parts. Intuitively, a part is atomic if they do not have any sub-parts except the empty part.

5 Mereocat Sums

The traditional concepts of fusion in mereology through the use of mereological sum as in [36, 33] are always problematic. The problems are carefully discussed by Meirav [23]. According to Meirav, the root of some most noteworthy difficulties has been the traditional tendency to focus on the horizontal dimension of wholes (i.e. the part-whole relations) and neglect the vertical dimension (i.e. relations among the whole itself and its parts). In fact, in software engineering, the very popular object modelling method UML [30] mainly provides only mechanisms to model the vertical relationships with both their static and dynamic behaviours, which are extremely important as well.

However, the solutions proposed in [23], mainly based on the linguistic differences on the meaning of "part of", does not seem to suit computer science and software engineering needs.

Another solution, the notion of *mereological supremum* of [15], is formally rather complex, and occasionally not very intuitive and does not solve many problems discussed in [23].

In our model, fusion will be defined based on the concepts of part-morphism, mereological sum and categorical universal constructions. Instead of using only one mereological sum definition for every kinds of composition as in Standard Mereology, we use separate concepts of sums which depends on how the sub-parts are related. Four different types of sum and their generalisation will be provided and discussed.

5.1 Independent Sum

In many cases, a whole is constructed by putting two or more parts together and all of these parts contribute to the behaviours of the whole *independently*. This form of composition is defined as follows:

Definition 6. *In a mereocat* **MC**, *a whole C is an* **independent sum** *of two parts A and B iff*

- *there exists two part morphisms* $f_A : A \to C$ *and* $f_B : B \to C$, *and*
- *for any part morphism* $f_E : E \to C$ *and* $E \ncong \bot$, *at least one of the following conditions satisfies:*
  1. *there exist an object* $F \ncong \bot$ *and two parts* $h_1 : F \to E$ *and* $h_2 : F \to A$ *making the following diagram commute:*

$$\begin{array}{ccc} F & \xrightarrow{h_2} & A \\ {\scriptstyle h_1}\downarrow & & \downarrow{\scriptstyle f_A} \\ E & \xrightarrow{f_E} & C \end{array}$$

  2. *there exist an object* $G \ncong \bot$ *and two parts* $i_1 : G \to E$ *and* $i_2 : G \to B$ *making the following diagram commute:*

$$\begin{array}{ccc} G & \xrightarrow{i_2} & B \\ {\scriptstyle i_1}\downarrow & & \downarrow{\scriptstyle f_B} \\ E & \xrightarrow{f_E} & C \end{array}$$

□

The first part of definition says C is a whole such that f_A and f_B are part-of it and the second part of the definition says that any part overlaps the whole iff it overlaps at least one of the part of the whole. The second part of this definition is based on the overlapping requirement of the mereological sum of Standard Mereology but for the case when A and B are disjoint.

It is worth to notice that the whole C is only defined to be *an* independent sum, because there might be several way to build a whole satisfying this definition of the sum as demonstrated the following example.

Example 5. Recall the mereocat of **BBlock** from Example 2. In Figure 3, the house D is a sum of B and E, but D' is also a sum of B and E.

□

Hence, to make sure that the sum is unique up to an isomorphism in a sense if two objects C and C' are both sums of set of objects, then $C \cong C'$, we need to impose the *universal property* [22]. This restricted form of independent sum can be defined in mereocat as follows:

Definition 7. *In a mereocat* **MC**, *a whole C is a* **strict** *independent sum of two parts A and B iff there exist part morphisms* $f_A : A \to C$ *and* $f_B : B \to C$ *such that*

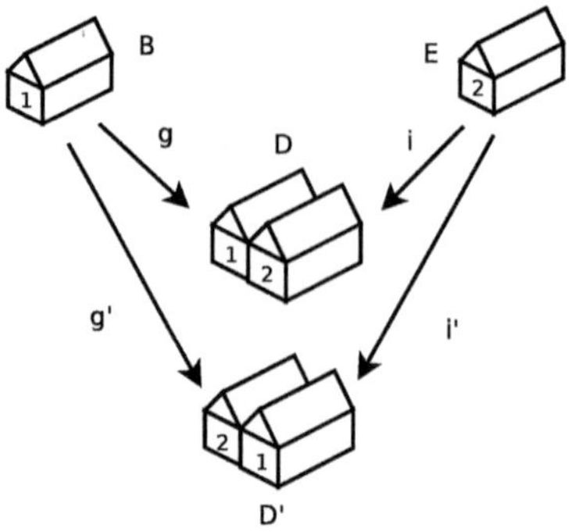

Fig. 3. Both D and D' are indepedent sums of B and E

- *C an independent sum of A and B, and*
- *for any object C' and a pair of morphisms $g_A : A \to C'$ and $g_B : B \to C'$ of* ***MC*** *there is a unique morphism $k : C \to C'$ in* ***MC*** *making the following diagram commute*

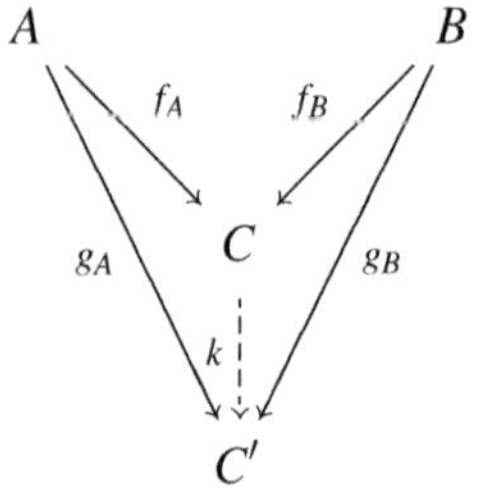

□

The above definition says C is a whole such that f_A and f_B are part-of it, and for any other whole C' containing A and B, C is a part of that whole through the existence of k. Intuitively, this means C is the "minimal" whole containing both A and B. This requirement of the existence of a unique k imposes the universal property needed. This definition of strict independent sum can be seen as a special form of *coproduct* in category theory [22].

For instance, in the Example 5, if D and D' are both strict independent sums of B and E then we will require that the existence of part mophisms j and j' as in Figure 4. Intuitively, this existence of j and j' says that D and D' are "the same", but the fact whether they should be considered the same or not depends mainly on a designer's specification purpose which affects how morphisms are defined. Notice that if we only require, for instance, D to be the strict independent sum then we only need a unique arrow $j : D \to D'$.

5.2 Interactive Sum

In many cases, it might not be possible to build any generic structures by only using the independent sum operation on parts. Consider the following example:

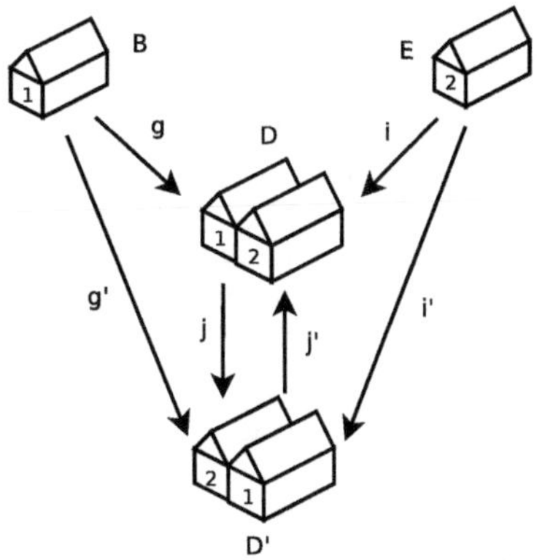

Fig. 4. There exist arrows $j : D \to D'$ and $j' : D' \to D$

Example 6. In the mereocat **BBlock**, we want to be able to specify explicitly that the house D is a sum of B and C and the block A is the common part of them as shown in Figure 5. To clarify things, we use the dotted arrows to specify each morphisms f, g, h and i by associating each block from each source to the correspondent block in the target house.

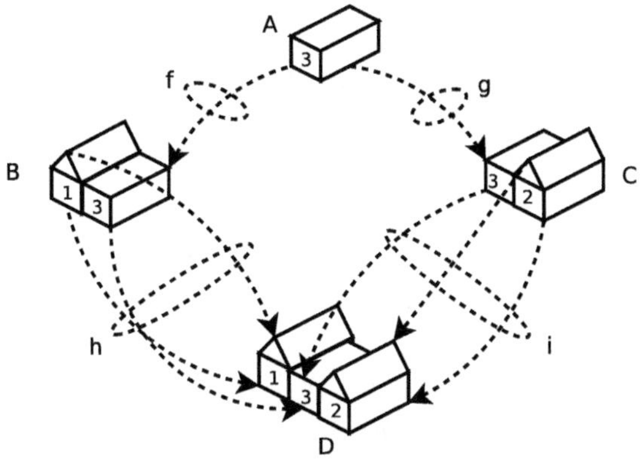

Fig. 5. The block A is the common part of two houses B and C

However, having the common part A, D is not an independent sum of B and C.

□

Intuitively, the block A acts as some form of "connector" to connect B and C together to create D. Hence, we need another kind of sum which can be defined as follows:

Definition 8. *In a mereocat* ***MC****, a whole S is an* ***interactive sum*** *of two parts $f : A \to B$ and $g : A \to C$ in* ***MC*** *iff there exist two part-morphisms $i_B : B \to S$ and $i_C : C \to S$ in* ***MC*** *such that:*

– $i_B \circ f = i_C \circ g$, *and*

– *for any part morphism $i_E : E \to S$ and $E \neq \bot$, at least one of the following conditions satisfies:*
  1. *there exist an object $F \ncong \bot$ and two parts $h_1 : F \to E$ and $h_2 : F \to B$ making the following diagram commute:*

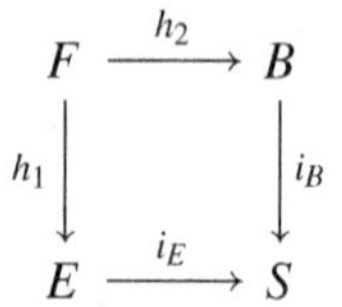

  2. *there exist an object $G \ncong \bot$ and two parts $l_1 : G \to E$ and $l_2 : G \to C$ making the following diagram commute:*

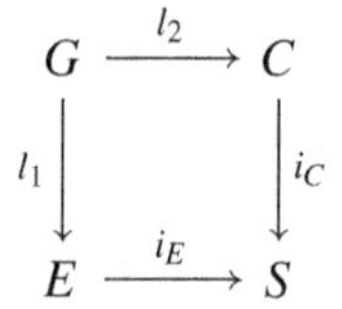

□

The first part of the definition says S is a whole such that i_B and i_C are part-of it and A is the common part of S, i_B and i_C. The second part imposes the mereological sum requirement from Standard Mereology (i.e any part overlaps the whole if it overlaps at least one of the parts of the whole).

In Example 6, we can conclude that the whole D is an interactive sum of two parts $f : A \to B$ and $g : A \to C$.

The strict version of interactive sum which requires to satisfy the universal property is defined as follows:

Definition 9. *In a mereocat **MC**, a whole S is the **strict** interactive sum of two parts $f : A \to B$ and $g : A \to C$ in **MC** iff there exist two part-morphisms $i_B : B \to S$ and $i_C : C \to S$ in **MC** such that:*

– *S is an interactive sum of $f : A \to B$ and $g : A \to C$, and*
– *for any object S' and two part-morphisms $i'_B : B \to S'$ and $i'_C : C \to S'$ in **MC** such that $i'_B \circ f = i'_C \circ g$, there is a unique morphism $k : S \to S'$ in **MC** making the following diagram commute*

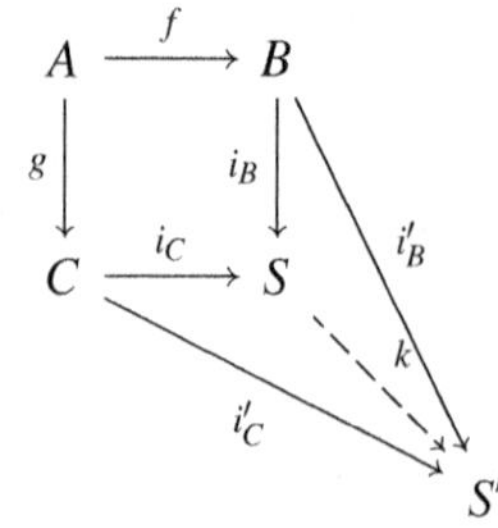

□

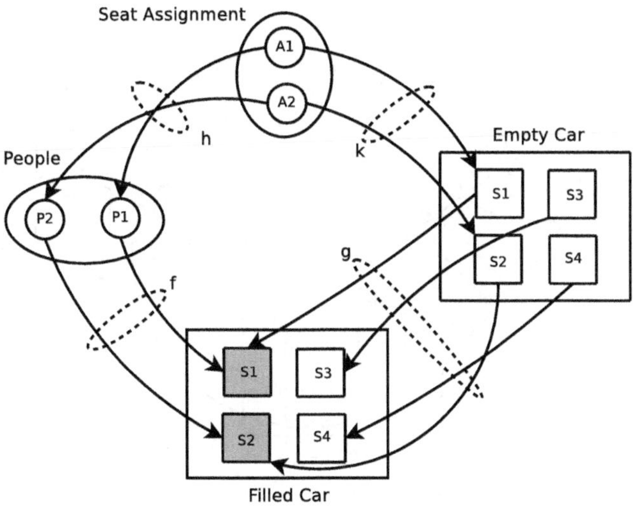

Fig. 6. Car made of two parts h and k from Example 7

Intuitively, this definition guarantees that S is the "minimal" whole containing both f and g. When the strict interactive sum exists, it is unique up to an isomorphism. This definition of strict interactive sum can be seen as a special form of *pushout* in category theory [22].

Example 7. In Example 4, we can also build **Filled Car** from the the object **Seat Assignment** with two part-morphisms **Seat Assignment**$\xrightarrow{h}$**People** and **Seat Assignment**$\xrightarrow{k}$**Empty Car** as in Figure 6. Two morphisms h and k associate explicitly each person with his/her corresponding seat. Intuitively, we can think of each seat assignment as some form of *common attribute* between each person and an empty seat. Hence, **Filled Car** is an interactive sum of two parts h and k.

We can change the choices of seats by using a new morphism **Seat Assignment'** $\xrightarrow{k'}$**Empty Car** in the place of k' as shown in Figure 7.

It is worth to notice in the sum **Filled Car'** how the morphism **People**$\xrightarrow{f'}$**Filled Car'** is different from f to reflect the change. □

Hence, our generic definition of mereological sums provides a simple but very explicit way to specify how the sum is built. The information of how the each sum is composed is not just contained in each object itself, but also the morphisms with respect to it. For example, how **Filled Car'** is different from **Filled Car** is reflected in the differences between $\{f, g\}$ and $\{f', g'\}$.

The use of morphisms to specify the sum has a strong advantage here. Together with the version of non-strict independent/interactive sum, we allow different non-isomorphic wholes to be built from a same set of parts. Hence, the morphisms equipped to each resulting whole will help distinguishing one whole from another.

Notice that although part-morphisms in the previous mereocat example are mostly functions, part-morphisms need not be functions. It is also clear that (strict) independent

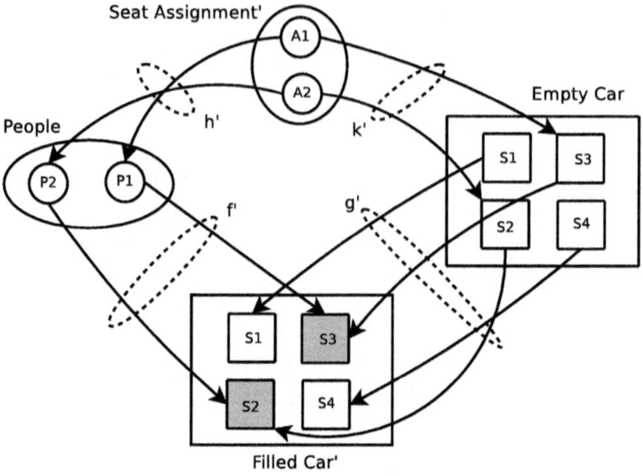

Fig. 7. Car made of two parts h' and k' from Example 7

sum is a special case of (strict) interactive sum when the common part in the (strict) independent sum is the $\perp$ part.

5.3 Generalised Sum

We first recall the concept of a *diagram* in category theory. A diagram in a category **C** is a graph homomorphism $\delta : \mathbf{I} \to |\mathbf{C}|$ for some (directed) graph **I** where $|\mathbf{C}|$ denotes the *underlying graph* of **C**. For simplicity, we only write $\delta : I \to \mathbf{C}$ to denote the diagram δ. Such diagram is called a *diagram of type* **I** and **I** is called the *index graph* of δ. δ is called a *finite diagram* if the index graph has only finite nodes and arrows.

Let $\delta : \mathbf{I} \to \mathbf{C}$ be a diagram in a category **C**. A *cocone* [22] with base δ is an object Z of **C** together with a family $\{p_A : \delta(A) \to Z\}_{A \in ob(\mathbf{I})}$ of morphisms of **C**, usually denoted by $p : \delta \to Z$. The object Z is said to be the vertex of the cocone, and for each $A \in ob(\mathbf{I})$, the morphism p_A is said to be the edge of the cocone at point A. The cocone p is said to be commutative iff for every arrow $s : A \to B$ of graph **I**, the following diagram commutes:

$$\begin{array}{ccc} & Z & \\ {\scriptstyle p_A}\nearrow & & \nwarrow{\scriptstyle p_B} \\ \delta(A) & \xrightarrow{\delta(s)} & \delta(B) \end{array}$$

In a mereocat, a commutative cocone can be interpreted as the whole which contains all the parts in a diagram. Clearly, not every commutative cocone is a sum of a diagram in a mereocat, since a commutative cocone might contain more parts than the parts of the diagram itself. Hence, we can define the notion of *generalised sum* which generalises both the independent and interactive sum concepts as following:

Definition 10. *In a mereocat* ***MC****, a* ***generalised sum*** *of a diagram* $\delta : \boldsymbol{I} \to \boldsymbol{MC}$ *is an object* Z *together with a family* $\{p_A : \delta(A) \to Z\}_{A \in ob(\boldsymbol{I})}$ *of part-morphisms such that*

– $\{p_A : \delta(A) \to Z\}_{A \in ob(\boldsymbol{I})}$ *is a commutative cocone of* δ*, and*

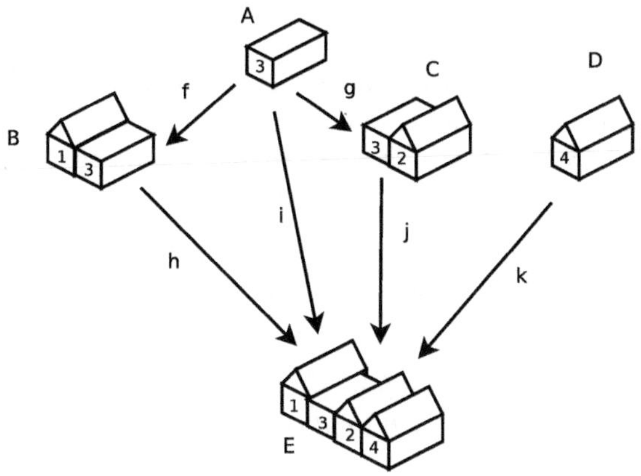

Fig. 8. The commutative cocone $\{h,i,j,k\}$ with vertex E from Example 8

– *for any part-morphism $i_E : E \to S$ and $E \not\cong \bot$, there exists some $A \in ob(\boldsymbol{I})$ and two part morphisms $f : F \to \delta(A)$ and $g : F \to E$ and $F \not\cong \bot$ making the following diagram commute:*

$$\begin{array}{ccc} F & \xrightarrow{f} & \delta(A) \\ {\scriptstyle g}\downarrow & & \downarrow{\scriptstyle p_A} \\ E & \xrightarrow{i_E} & S \end{array}$$

□

Just like the notions of *strict* independent and interactive sums, we also want to guarantee the minimality and the uniqueness up to an isomorphism of a generalised sum. This is where the notion of *colimit* in category theory [22] is needed.

Let $\delta : \mathbf{I} \to \mathbf{C}$ be a diagram in category $\mathbf{C}$. A colimit of δ is a commutative cocone $p : \delta \to Z$ such that for every other commutative cocone $p' : \delta' \to Z'$, there is a unique morphism $f : Z \to Z'$ such that $f \circ p = p'$.

Definition 11. *In a mereocat **MC**, the* **strict** *generalised sum of a diagram $\delta : \boldsymbol{I} \to \boldsymbol{MC}$ is an object Z together with a family $\{p_A : \delta(A) \to Z\}_{A \in ob(\boldsymbol{I})}$ of part-morphisms such that*

- *Z is a generalised sum of δ, and*
- *$\{p_A : \delta(A) \to Z\}_{A \in ob(\boldsymbol{I})}$ is the colimit of δ.* □

Example 8. Recall the mereocat **BBlock**. In Figure 8, the house E is a generalised sum of the diagram created by A, B, C, D and the morphisms between them. The commutative cocone in this case is $\{h,i,j,k\}$.

□

The generalised sum might not always exist for any finite arbitrary diagram of a mereocat. When the strict generalised sum always exists for any arbitrary diagram of a mereocat then following the category theory terminology, we will call that mereocat a **cocomplete** mereocat. When the strict generalised sum only exists for finite diagram, we will call the mereocat **finitely cocomplete**.

6 Mereocat Product

Although not as carefully discussed as the notions of sum, we also include a brief discussion on the **product** construction in mereocat. The product notion which seems most interesting to us is the *dual* notion of the strict interactive sum by reversing the direction of the arrows. The definition can be given in detail as follows:

Definition 12. *In a mereocat* **MC**, *a whole P is the* **product** *of two parts* $f_A : A \rightarrow C$ *and* $f_B : B \rightarrow C$ *iff there exist two part-morphisms* $p_A : P \rightarrow A$ *and* $p_B : P \rightarrow B$ *such that*

- $f \circ p_A = g \circ p_B$ *and*
- *for any other pair of part-morphisms* $p'_A : P' \rightarrow A$ *and* $p'_B : P' \rightarrow B$, *there exists a unique part-morphism k making the following diagram commute*

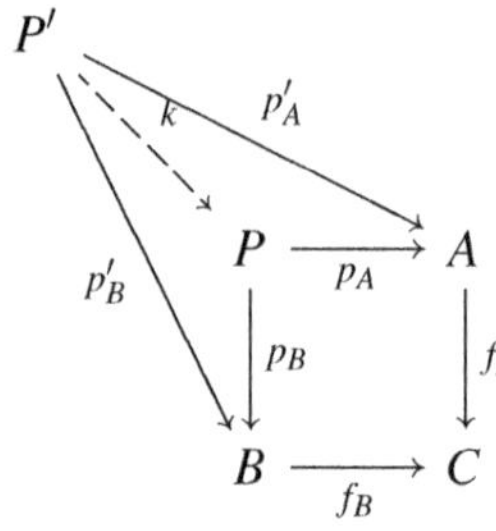

□

The universal property guarantees that if the product exists then it is unique up to some isomorphism. Intuitively, it also means that *P* is the *maximal* common part of *A* and *B*. Clearly, this concept of product is similar to the merelogical one in Standard Mereology. The product concept in mereocat corresponds exactly to the *pullback* notion in category theory [22].

For instance, in Example 6, the block *A* is the product of *B* and *C*.

7 Categorical Connector Framework

We assume the reader has rudimentary knowledge of basic notions of Component Based Software paradigm [1].

Mereocat is a relatively general theory and in order to be used to model component-based softwares, it needs to be equipped with some formal architectural framework. For this role we adopt CommUnity, the architectural design framework invented by Fiadeiro et al. in [21,5]. CommUnity is flexible and quite general, and most importantly, it does not restrict us to any specific *architecture description language*. It also allows easy modelling of different aspects of parallel design. Last but not least, the categorical formalisation of CommUnity can be easily merged with the concepts of mereocat. To keep the paper as self-sufficient as possible we will give a brief overview of CommUnity's three architectural elements: components, configurations and connectors.

Components, which can be thought in a sense of Component Based Software, are the model entities that perform computation and are able to synchronise with their environments and exchange information through channels. Hence, components are given in terms of their channels and actions in a form of "designs". For example, component design ***print*** below consists of input channel *i*, output channel *po* and private channel *rd*. Actions of ***print*** are given in CommUnity as a special form of "guarded commands", except satisfying the guards only means the actions can be executed but does not force the action to be executed right away. In ***print***, if *rd=false* then action *print* is allowed to be executed and change *rd* to *true*. Action *prod* of ***print*** does the "opposite" of action *print* and also assign input *i* to output *po*. The ***convert*** component does the task of of a conversion module which convert a MSWord document to a PS document.

design print
in *i:ps*
out *po:ps*
prv *rd:bool*
do *print[rd]:* $\neg$*rd* $\rightarrow$ *po:=i* $\|$ *rd:= true*
[] *prod[rd]: rd* $\rightarrow$ *rd:= false*

design convert
out *o:ps*
prv *w:MSWord*
do *to_ps[o]: true, false* $\rightarrow$ *o:=ps(w)*

Configurations are diagrams in a category of designs where objects are designs and morphisms are *superposition*, also called design morphisms. A configuration describes the coordination and interaction of component designs when they are instantiated. A design morphism $\sigma : P_1 \rightarrow P_2$ identifies that P_2 can be obtained from P_1 by "augmenting" additional behaviours to P_1 while still preserving properties of P_1 in P_2.

From a *meaningful* configuration (e.g. an output channel is not connected to other output channels) [21], a new design can be constructed using colimit construction. For example, we want to build a new useful design from the previous designs ***print*** and ***convert***, using the configuration in the diagram below where *cable*, ***convert***, ***print*** are objects and each arrow represents a morphism between them.

Notice that explicit names are not given to the action and channel of *cable* used for interconnection, but • symbols are used instead [5,21]. The reason is the interconnection does not rely on the global naming but precisely on associations (name binding), for example, we need to explicitly specify that *o*, *to_ps* are bound to *i* and *prod* respectively. In other words, the action and channel names of *cable* do not play any important role with respect to the resulting design, for instance, ***user*** in this case.

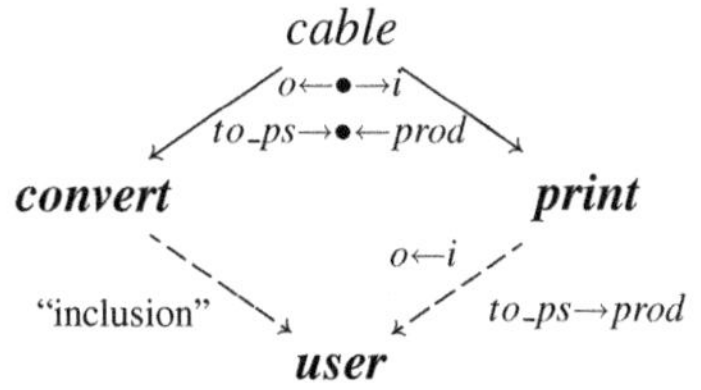

design user
out *o,po:ps*
prv *rd:bool, w: MSWord*
do *print[rd]:* $\neg$*rd* $\rightarrow$ *po:=o* $\|$*rd:= true*
[] *to_ps[o,rd]: rd* $\rightarrow$ *o:=ps(w)*$\|$ *rd:= false*

Using colimit construction, the new object ***user*** and two arrows from ***convert*** and ***print*** to ***user*** are introduced into the diagram. Here the colimit, as the "interactive sum" in our mereocat terminology, returns the minimal single design representing the whole configuration.

The design objects and design morphisms constitute category ***C-DSGN***.

Connectors are model entities independent from components whose purpose is to coordinate interactions between components as in the spirit of [1]. Connectors are given in CommUnity in terms of a "glue" design and collection of "role" designs. Since the formal concept of connectors is quite lengthy, readers are referred to [21].

8 Software Components and Mereocat

In real world, a Component Based Software development for complex systems is more than just composing a system from pre-existing components together using connectors. When a system becomes larger and larger, it helps to understand the architectural structure of the system better by analysing different architectural views, which are the different abstractions of the same software system. The first kind of views is by partitioning a system vertically into subsystems, which *aggregate* modules implementing related functionalities. The second kind of views is by looking at the horizontal sections that may have different scope within the system. Layers may belongs to a single subsystem, a part of subsystem or across different subsystems [16].

The categorical framework discussed previously are designed to support composition of subsystems from component designs. The framework is very successful in separating and clarifying the roles of coordination vs. computation in component architecture development. However, it does not seem obviously how the framework supports the layer and subsystem views of software architecture. Their approach mathematically treats all the designs as categorical objects and strongly emphasise the properties preserved by morphisms, but also "flattens" down the whole architectural structure. Our goal is to complement their framework by bringing back the depth to the architectural structure using part-of relation from mereocat. Fortunately, the task is trivial since CommUnity has already defined precisely the design and design morphism concepts. We only need a proper interpretation of parts and the part-of relationship.

Proposition 6. *Assume that a design S, which can be software system or subsystem, is constructed using colimit construction from a configuration diagram*[6] $\delta : \boldsymbol{I} \rightarrow$ ***C-DSGN*** *upon instantiation. Let* $\perp$ *denote the empty design which has a unique empty design morphism to every design and let* $\{p_A : \delta(A) \rightarrow S\}_{A \in ob(\boldsymbol{I})}$ *denote the colimit which constitutes S, then the category* $\boldsymbol{C}$ *such that*

- $ob(\boldsymbol{C}) = \{\perp\} \cup \{\delta(A)\}_{A \in ob(\boldsymbol{I})} \cup \{S\}$, *and*
- $hom(\boldsymbol{C}) = \{\perp \rightarrow \delta(A)\}_{A \in ob(\boldsymbol{I}) \cup \{S\}} \cup \{\delta(A)\}_{A \in hom(\boldsymbol{I})} \cup \{p_A : \delta(A) \rightarrow S\}_{A \in ob(\boldsymbol{I})}$ $\cup \{id_A\}_{A \in ob(\boldsymbol{C})}$

is a mereocat.

Proof. Follows from the properties of the colimit construction and the fact that in the base diagram δ of instantiation configuration there is at most one morphism from one design to another. □

This proposition presents the proper definition of the parthood relation in Component Based Software in one layer of the compositional architecture where the designs in

[6] See the definition of *diagram* at the beginning of subsection Generalised Sum.

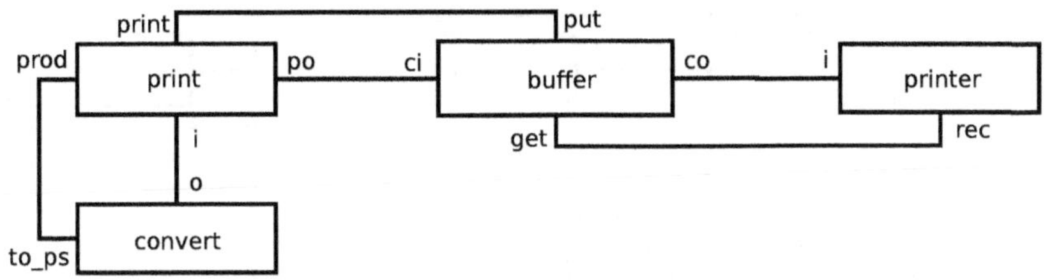

Fig. 9. A non-hierachical way of specifying the ***User-Printer*** system

$\{\delta(A)\}_{A\in ob(\mathbf{I})}$ contribute directly to the collective behaviour of the system (or subsystem) S. In this view, the design S and the cocone $\{p_A : \delta(A) \to S\}_{A\in ob(\mathbf{I})}$ is exactly the generalised sum of δ in the mereocat $\mathbf{C}$.

For example, in the case of ***user*** system, the commutative diagram clearly implies that ***convert*** is part-of ***user***, ***print*** is part-of ***user***, *cable* is part-of ***convert***, *cable* is part-of ***print*** and *cable* is part-of ***user***. Moreover, due to how we define the part-of relation in mereocat, we can also conclude that *cable* $\to$ ***convert*** is part-of ***print*** $\to$ ***user*** and *cable* $\to$ ***print*** is part-of ***convert*** $\to$ ***user***.

We can use the tools of mereocat, to analyse only one subsystem, or we can also apply the construction of Proposition 6 again carefully to build a new mereocat from the mereocats of subsystem. This in turn provides us an effective way of modularisation as shown in the next example.

Example 9. Using the previous components ***convert*** and ***print***, we can design a ***User-Printer*** application where, a user application send a PS document to a "printing server" component ***printer*** to print the document. All the communication is done through a bounded buffer ***buffer***, which prevents ***user*** from sending a new document when there is no space and prevents ***printer*** from reading a new message when no new message has been sent. The designs of ***buffer*** and ***printer*** are given as follows:

design buffer
in *ci:ps*
out *co:ps*
prv *rd: bool; q:queue*
do *put:* $\neg$*full(q)* $\to$ *q:= enqueue (i,q)*
[]*prv* *next:* $\neg$*empty(q)* $\wedge \neg$ *rd*
$\to$ *o:=head(q)* $\|$ *q:= tail(q)* $\|$ *rd:=true*
[] *get: rd* $\to$ *rd:= false*

design printer
in *i:ps*
prv *busy:bool*
do *rec:* $\neg$*busy* $\to$ *busy:=true*
[] *end_print:busy* $\to$ *busy:=false*

We can specify the configuration of the situation in two different ways.

The first method is to specify the architectural configuration as in Figure 9.

Notice that instead of drawing the categorical diagram as in previous section, we use the "syntactic sugar" of name-binding to associate the correspondent methods and channels of designs. According to the name-binding method of [21], we bind ***print*** component with the input of ***buffer***.

The second method, by using mereocat, is to specify the subsystem ***user*** and its sub-parts as a way to modularise the configuration as in Figure 10. We connect the resulting *subsystem* ***user*** to ***buffer*** where ***user*** is the strict interactive sum of two parts

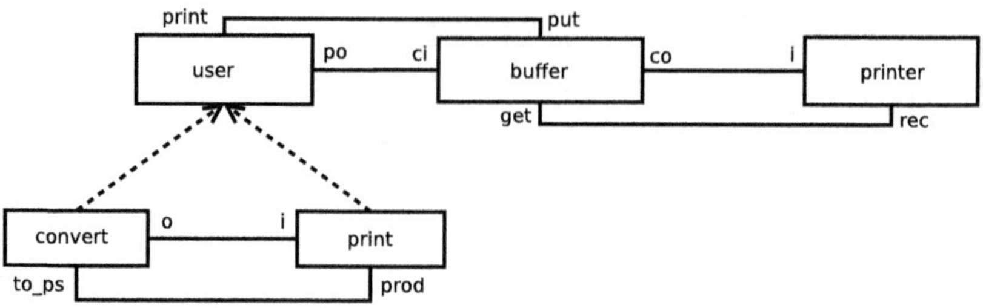

Fig. 10. Specifying the ***User-Printer*** system using mereocat

cable $\rightarrow$ ***convert*** and *cable* $\rightarrow$ ***print***. The dotted arrows denote the morphisms ***convert*** $\rightarrow$ ***user*** and ***print*** $\rightarrow$ ***user*** of the interactive sum. As a result, we have a more hierarchical view of the whole ***User-Printer*** system. □

Obviously, we can recursively apply this method to different parts of the system when the system grows larger and larger. Notice that ***buffer*** (the "glue" design) and its associations (name-bindings) constitute a *connector* according to the connector definition in [21].

9 Conclusion

The main contribution of this paper is an introduction of a categorical mereology - *mereocat*. The major philosophical difference between mereocat and traditional mereologies is the treatment of "parts". In traditional mereologies parthood is defined globally as a relation (partial order) among all the elements. In mereocat parts are defined locally, as morphisms $f : a \rightarrow b$ interpreted "a is a part of b and the *name* of this relation is f". Defining parts as morphisms gives us more flexibility and among others allows to define more realistic and useful sum and product operations. As a result, we have found a categorical setting that seems more intuitive and applicable for specification purposes than traditional mereologies. We then applied mereocat to model some parthood properties of Component Based Software architecture.

The paper provides a first complete version of our categorical mereology, but it also leaves many questions unanswered and problems untouched. We will mention only a few of them. Can we define a "universal sum" such that the notion of mereological supremum of [13] is a special case? How a categorical setting can help in defining "constructors" and "destructors" proposed in [12] and discussed in [15]? How the fact that parts are morphisms affect the entire concept of "constructors" and "destructors"? Can we easily incorporate equivalence relations discussed in [13] into mereocat? We have not exploited categorical funtor-based constructions in this paper.

We are currently working on a mereological refinement notion for mereocat. We believe mereocat with refinement would be more useful modelling tool than a flattened architecture of [5, 21]. For example, we can substitute not just one of components ***convert*** or ***print*** but the whole subsystem ***user*** in the evolution of the system ***User-Printer***.

On the surface there seems to be a link between the model presented in this paper and the category of *Chu spaces*, where objects are *classifications* and morphisms are *infomorphisms*, and infomorphisms are used to capture some kind of part-whole relationship [3, 34]. However, a more detailed study indicates that the Chu spaces model does not relate to the concept of mereology as understood in this paper. Moreover, we do not describe things using states [3] and we do not have the concept of 'time' built in. However, we plan to study the possible relationships more thoroughly in the future. There also seems to be some sort of relationship between our *interactive sum* and the *sum with constraints* of [34], however a detailed discussion requires introducing many detailed concepts of *Rough Sets* [27, 28, 34], and is beyond the scope of this paper.

Acknowledgements: The first author thanks Tom Maibaum and Gregory H. Moore for many stimulating discussions which motivated the categorical approach of this paper. Both authors gratefully acknowledge the anonymous referees, whose comments significantly contributed to the final form of this paper.

References

1. Allen, R., Garlan, D.: A Formal Basis for Architectural Connection. ACM Transactions on Software Engineering and Methodology 6(3), 213–249 (1997)
2. Barr, M., Wells, C.: Toposes, Triples and Theories. Springer, Heidelberg (1985)
3. Barwise, J., Seligman, J.: Information Flow: The Logic of Distributed Systems. Cambridge University Press, Cambridge (1997)
4. Casati, R., Varzi, A.: Parts and places: the structures of spatial representation. MIT Press, Cambridge (1999)
5. Fiadeiro, J.L.: Categories for Software Engineering. Springer, Heidelberg (2004)
6. Forrest, P.: Nonclassical mereology and its application to sets. Notre Dame Journal of Formal Logic 43(2), 79–94 (2002)
7. Goguen, J.A.: Categorical foundations for general systems theory. In: Pichler, F., Trappl, R. (eds.) Advances in Cybernetics and Systems Research, pp. 121–130, Transcripta Books (1973)
8. Goguen, J.A., Rosu, G.: Institution morphisms. Formal Aspects Computing 13(3-5), 274–307 (2002)
9. Goodman, N.: The structure of appearance. Dordrecht (1977)
10. Grzegorczyk, A.: The system of Leśniewski in relation to contemporary logical research. Studia Logica 3, 77–95 (1955)
11. Ivers, J., Clements, P., Garlan, D., Nord, R., Schmerl, B., Oviedo, J.: Documenting component and connector views with UML 2.0. Technical Report CMU-CS-2004-TR-008, School of Computer Science, Carnegie Mellon University (April 2004)
12. Janicki, R.: On a mereological system for relational software specifications. In: Diks, K., Rytter, W. (eds.) MFCS 2002. LNCS, vol. 2420, pp. 375–386. Springer, Heidelberg (2002)
13. Janicki, R.: Basic mereology with equivalence relations. In: Jedrzejowicz, J., Szepietowski, A. (eds.) MFCS 2005. LNCS, vol. 3618, pp. 507–519. Springer, Heidelberg (2005)
14. Janicki, R., Khedri, R.: On a formal semantics of tabular expressions. Science of Computer Programming 39(2–3), 189–213 (2001)
15. Janicki, R., Lê, D.T.M.: Towards a pragmatic mereology. Fundamenta Informaticae 75(1-4), 295–314 (2007)
16. Jazayeri, M., Ran, A., van der Linden, F.: Software architecture for product families: principles and practice. Addison-Wesley, Reading (2000)
17. Lawvere, F.W., Rosebrugh, R.: Sets for Mathematics. Cambridge University Press, Cambridge (2003)

18. Lê, D.T.M., Janicki, R.: On a parthood specification method for component software. In: Greco, S., Hata, Y., Hirano, S., Inuiguchi, M., Miyamoto, S., Nguyen, H.S., Słowiński, R. (eds.) RSCTC 2006. LNCS (LNAI), vol. 4259, pp. 537–546. Springer, Heidelberg (2006)
19. Leonard, H.S., Goodman, N.: The calculus of individuals and its uses. Journal of Symbolic Logic 5(2), 45–55 (1940)
20. Leśniewski, S.: Grundzüge eines neuen systems der grundlagen der mathematik. Fundamenta Mathematicae 24, 1–81 (1929)
21. Lopes, A., Wermelinger, M., Fiadeiro, J.L.: High-order architectural connectors. ACM Transactions on Software Engineering and Methodology 12(1), 64–104 (2003)
22. MacLane, S.: Categories for the Working Mathematician. Springer (1998)
23. Meirav, A.: Wholes, Sums and Unities (Philosophical Studies). Springer (2003)
24. Motschnig-Pitrik, R., Kaasbøll, J.J.: Part-whole relationship categories and their application in object-oriented analysis. IEEE Transactions on Knowledge and Data Engineering 11(5), 779–797 (1999)
25. Parnas, D.L.: On the criteria to be used in decomposing systems into modules. Communications of the ACM 15(12), 1053–1058 (1972)
26. Parnas, D.L.: On a buzzword: Hierarchical structure. In: Proc. of IFIP 1974 Congress, pp. 336–339. North-Holland, Amsterdam (1974)
27. Pawlak, Z.: Rough Sets. Kluwer Academic Publishers, Dordrecht (1991)
28. Polkowski, L., Skowron, A.: Rough mereology: A new paradigm for approximate reasoning. Journal of Approximate Reasoning 15, 316–333 (1997)
29. Rumbaugh, J., Eddy, F.: Object-Oriented Modeling and Design. Prentice-Hall, Englewood Cliffs (1991)
30. Rumbaugh, J., Jacobson, I., Booch, G.: The Unified Modeling Language Reference Manual. Addison-Wesley, Reading (2004)
31. Salustri, F.A., Lockledge, J.C.: Towards a formal theory of products including mereology. In: Proc. of 12th Int'l Conf. on Engineering Design, pp. 1125–1130 (1999)
32. Simons, P.: On understanding Leśniewski. History and Philosophy of Logic 3(2), 165–191 (1982)
33. Simons, P.: Parts: A Study in Ontology. Oxford University Press, Oxford (1987)
34. Skowron, A., Stepaniuk, J., Peters, J.F.: Rough Sets and Infomorphisms: Towards Approximation of Relations in Distributed Environments. Fundamenta Informaticae 54(1-3), 263–277 (2003)
35. Smith, B.: Mereotopology: a theory of parts and boundaries. Data and knowledge engineering 20(3), 287–303 (1996)
36. Srzednicki, J.T.J., Rickey, V.F. (eds.): Leśniewski's Systems. Kluwer, Dordrecht (1984)

Esoteric Rough Set Theory: Algebraic Semantics of a Generalized VPRS and VPFRS

A. Mani

Calcutta Mathematical Society
9B, Jatin Bagchi Road
Kolkata(Calcutta)-700029, India
a_mani_sc_gs@yahoo.co.in

Abstract. In different theories involving indiscernibility, it is assumed that at some level the objects involved are actually assignable distinct names. This can prove difficult in different application contexts if the main semantic level is distinct from the semantic-naming level. Set-theoretically too this aspect is of much significance. In the present research paper we develop a framework for a generalized form of rough set theory involving partial equivalences on different types of approximation spaces. The theory is also used to develop an algebraic semantics for variable precision rough set and variable precision fuzzy rough set theory. A quasi-inductive concept of relativised rough approximation is also introduced in the last section. Its relation to esoteric rough sets is considered.

Keywords: Esoteric Rough Set Theory, Rough Sets, Variable Precision Rough Set Theory, Algebraic Semantics of Generalized Rough Set Theory, Irreflexive Rough Set Theory, VPRFS, GVPRFS.

1 Introduction

In both rough set theory and generalizations thereof involving relations explicitly, it is assumed that the relation in the underlying approximation or information space is at least reflexive. This amounts to a degree of distinguishability at a higher meta-level. Naturally we assume familiarity with rough sets [1], generalizations thereof [2,3,4,5,6], VPRS [7], VPRFS [8] and algebraic semantics thereof (where possible) [2,9,10].

Suppose we have a finite set of objects, but have forgotten (as a functor), all the maps that can prove the finiteness and suppose we know that for each function symbol there exists an object and a partial interpretation of the function on the object which gives the last value. The last condition is optional. We can as well allow indeterminate finiteness or even transfiniteness. In this situation suppose we have no mechanism of distinguishing between any two of the objects, because the interpretation of names or the counting functions is a problem. If we allow real time modifications of the number of objects, we have other things to consider. Models of such can be constructed by weakening extensionality or

J.F. Peters and A. Skowron (Eds.): Transactions on Rough Sets VIII, LNCS 5084, pp. 175–223, 2008.

by modifying the underlying logic with a suitable modified concept of equality of sets. Intuitionist versions of such indeterminacy inclusive set theory may be seen as mixing intuitionist things with indeterminacy of the kind. In the present paper we will concern ourselves with rough set theory.

Rough set theory in this situation requires deep modifications. Modifications which effectively break much of it's usual possible logics and semantics. Nor does any of the current generalized versions of rough set theory allow a reasonable treatment of the situation. All this in spite of the naturalness of partial equivalences in the situation. Though the concepts of approximability by definite objects and the concept of being between the lower and upper approximation of an object are possible at least partially, embedding such in some kind of 'completed model' is yet to be investigated on the errors involved.

There are statistical properties of the information associated. These are known to be really bad. But much improvement is possible by improving on the partitionability of the set...which amounts to requiring additional information...obtainable by special procedures. We deal with all this in particular in the *esoteric rough set theory* developed in the present research.

The following notions are stated to make the notation and terminology precise. An *approximation space* $X = \langle \underline{X}, R \rangle$ is a pair with $\underline{X}$ being a set in ZFC and R an equivalence relation. For any $A \in \wp(X)$, the *lower approximation* A^l of A is defined via,

$$A^l = \bigcup \{Y; Y \in X \mid R, Y \subset A\}$$

($X \mid R$ being the set of classes of R), while its *upper approximation* A^u is defined via,

$$A^u = \bigcup \{Y; Y \in X \mid R, Y \cap A \neq \phi\}.$$

A^l (resp. A^u) can be seen to be the collection of objects that *R-definitely* (resp. *R-possibly*) belong to A or as the collection of objects of X whose types are fully included in (resp. intersect), the set of types of objects of A. The triple (X, R, A) is called a *rough set*. A is *roughly included* in B, $A, B \subset X$, $A \prec B$ if and only if $A^l \subset B^l$ and $A^u \subset B^u$. A and B are *roughly equal* iff $A \prec B$ and $B \prec A$ if and only if $A^l = B^l$ and $A^u = B^u$.

In contrast to the above, an *Information System* is a tuple of the form $\langle \underline{U}, A, \nu_a : a \in A \rangle$, with $\underline{U}$ being the universe, A is a set of attributes and ν_a is a valuation map : $U \mapsto U$ which is defined for each $a \in A$.

Algebras used in the paper will be written in the form

$$S = \langle \underline{S}, f_1, f_2, \ldots f_n, (v_1, v_2, \ldots v_n) \rangle$$

with $\underline{S}$ being a set, $f_1, f_2, \ldots f_n$ being operation symbols and $v_1, v_2, \ldots v_n$ being their corresponding arities. The operation symbols together with their arities constitute the signature of the algebra. The interpretation of the signature on the set $\underline{S}$ will be kept implicit, so that operations will be written like operation symbols. Distinguished elements will be treated as 0-place operations.

A *pre-rough algebra* is an algebra of the form

$$S = \langle \underline{S}, \sqcap, \sqcup, \Rightarrow, L, \neg, 0, 1, (2, 2, 2, 1, 1, 0, 0) \rangle$$

which satisfies

(i) $\langle \underline{S}, \sqcap, \sqcup, \neg \rangle$ is a De Morgan lattice.

(ii) $\neg\neg a = a$; $L(a) \sqcap a = L(a)$

(iii) $LL(a) = L(a)$; $L(1) = 1$; $L(a \sqcap b) = L(a) \sqcap L(b)$

(iv) $\neg L \neg L(a) = L(a)$; $\neg L(a) \sqcup L(a) = 1$

(v) $L(a \sqcup b) = L(a) \sqcup L(b)$

(vi) $(L(a) \sqcap L(b) = L(a), \neg L(\neg(a \sqcap b)) = \neg L(\neg a) \longrightarrow a \sqcap b = a)$

(vii) $a \Rightarrow b = (\neg L(a) \sqcup L(b)) \sqcap (L(\neg a) \sqcup \neg L(\neg b))$.

A distributive lattice is said to be *completely distributive* if and only if the infinite supremum operation distributes over meet and dually the infinite meet operation distributes over join. Naturally the lattice must be complete. A completely distributive pre-rough algebra is called a *rough algebra*. In all these algebras it is possible to define an operation M by setting $M(x) = \neg L \neg(x)$ for each element x. M corresponds to the upper approximation operator. The operation $\Rightarrow$ is weaker than classical implication and corresponds to rough inclusion by way of $a \Rightarrow b = 1$ iff $a \leq b$ in the associated lattice order. Bi-implication naturally corresponds to rough equality.

1.1 Generalized Covers Approach

In [3], a theory of generalized rough sets based on covers of subsets of a given set $\underline{S}$ is considered. Let S be a set and $\mathcal{K} = \{K_i\}_1^n$ be a collection of subsets of it. If $X \subseteq S$, then consider the sets (with $K_0 = \emptyset$, $K_{n+1} = S$)

(i) $X^{l1} = \bigcup\{K_i : K_i \subseteq X, \ i \in \{0, 1, ..., n\}\}$

(ii) $X^{l2} = \bigcup\{\cap(S \setminus K_i) : \cap_I(S \setminus K_i) \subseteq X, \ I \subseteq \{1, ..., n+1\}\}$

(iii) $X^{u1} = \bigcap\{\cup_{i \in I} K_i : X, \subseteq \cup K_i, \ I \subseteq \{1, ..., n+1\}\}$

(iv) $X^{u2} = \bigcap\{S \setminus K_i : X, \subseteq S \setminus K_i, \ i \in \{0, ..., n\}\}$

The pair (X^{l1}, X^{u1}) is called a *AU-rough set* by union, while (X^{l2}, X^{u2}) a *AI-rough set* by intersection (in the notation of [3] these are $(\mathcal{F}_*^{\cup}(X), \mathcal{F}_{\cup}^*(X))$ and $(\mathcal{F}_*^{\cap}(X), \mathcal{F}_{\cap}^*(X))$ respectively).

Theorem 1. *The following are true :*

(i) $X^{l1} \subseteq X \subseteq X^{u1}$

(ii) $X^{l2} \subseteq X \subseteq X^{u2}$

(iii) $\emptyset^{l1} = \emptyset^{l2} = \emptyset$

(iv) $(\cup\mathcal{K} = S \longrightarrow S^{u1} = S^{u2} = S)$

(v) $(\cup\mathcal{K} = S \longrightarrow \emptyset^{u2} = \emptyset, S^{l1} = S)$

(vi) $(\cap\mathcal{K} = \emptyset \longrightarrow \emptyset^{u1} = \emptyset, S^{l2} = S)$

(vii) $(X \cap Y)^{l1} \subseteq X^{l1} \cap Y^{l1}, (X \cap Y)^{l2} = X^{l2} \cap Y^{l2}$

(viii) $(X \cup Y)^{u1} = X^{u1} \cup Y^{u1}, X^{u2} \cup Y^{u2} \subseteq (X \cup Y)^{u2}$

(ix) $(X \subseteq Y \longrightarrow X^{l1} \subseteq Y^{l1}, X^{l2} \subseteq Y^{l2})$

(x) *If $\mathcal{K}$ is pairwise disjoint then* $(X \cap Y)^{l1} = X^{l1} \cap Y^{l1}, (X \cup Y)^{u2} = X^{u2} \cup Y^{u2}$

(xi) $(X \subseteq Y \longrightarrow X^{u1} \subseteq Y^{u1}, X^{u2} \subseteq Y^{u2})$

(xii) $X^{l1} \cup Y^{l1} \subseteq (X \cup Y)^{l1}$

(xiii) $X^{l2} \cup Y^{l2} \subseteq (X \cup Y)^{l2}$

(xiv) $(X \cap Y)^{u1} \subseteq X^{u1} \cap Y^{u1}$

(xv) $(X \cap Y)^{u2} \subseteq X^{u2} \cap Y^{u2}$

(xvi) $(S \setminus X)^{l1} = S \setminus X^{u2}$

(xvii) $(S \setminus X)^{l2} = S \setminus X^{u1}$

(xviii) $(S \setminus X)^{u1} = S \setminus X^{l2}$

(xix) $(S \setminus X)^{u2} = S \setminus X^{l1}$

(xx) $(X^{l1})^{l1} = X^{l1}, (X^{l2})^{l2} = X^{l2}$

(xxi) $(X^{u1})^{u1} = X^{u1}, (X^{u2})^{u2} = X^{u2}$

(xxii) $(X^{l1})^{u1} = X^{l1}, (X^{u2})^{l2} = X^{u2}$

(xxiii) $X^{l2} \subseteq (X^{l2})^{u2}, (X^{u1})^{l1} \subseteq X^{u1}$

(xxiv) $(\mathcal{K}_j^{\cap}(X))^{u2} = \mathcal{K}_j^{\cap}(X)$, $j = 1, 2, ..., t_1$

(xxv) $(\mathcal{K}_j^{\cup}(X))^{l1} = \mathcal{K}_j^{\cup}(X)$, $j = 1, 2, ..., t_2$

In this, $(\mathcal{K}_j^{\cup}(X))$ is the minimal union for j being in the indicated range and $(\mathcal{K}_j^{\cap}(X))$ is the maximal intersection.

Proof. The properties are essentially set-theoretical as shown in the following proofs:

(i) $X^{l1} \subseteq X \subseteq X^{u1}$. Suppose X is non-empty. If $x \in X^{l1}$, then x is in some of those K_i) for which $K_i) \subseteq X$ holds. So $x \in X$ as well. If $x \in X$, x is in all those unions of the form $\cup_{i \in I} K_i$ that include X for some $I \subseteq \{0, 1, \ldots, n\}$. So $x \in X^{u1}$.

(ii) If X is nonempty and $x \in X^{l2}$, then x is in some of the intersections of complements of K_is that are contained in X. So $X^{l2} \subseteq X$. If $x \in X$, then if x is in a K_i that is contained in X, then there must exist a K_i for some $i \in \{0, 1 \ldots n\}$ such that its complement contains X. So also if x is in a K_i that is not contained in X. When x is in no K_i, then x is in every complement of K_is. So we have $X \subseteq X^{u2}$.

(iii) This follows from the definition.

(iv) If $\cup\mathcal{K} = S$ then $S^{u1} = \bigcap\{\cup K_i : S = \cup K_i\} = S$. $S^{u2} = \bigcap\{S \setminus K_i : S = S \setminus K_i, i \in \{0, 1, \ldots n\}\} = S$ in the situation too.

(v) If $\cup\mathcal{K} = S$, then $\emptyset^{u2} = \bigcap\{S \setminus K_i : i \in \{0, \ldots, n\} = \emptyset$, while $S^{l1} = \bigcup\{K_i : K_i \subseteq S\} = \cup\mathcal{K} = S$.

(vi) If $\cap\mathcal{K} = \emptyset$, then $\emptyset^{u1} = \bigcap\{K_i : i \in \{1, 2, \ldots, n\}\} = \emptyset$ and $S^{l2} = \bigcup\{\cap_I(S \setminus K_i) : I \subseteq \{1, 2, \ldots, n+1\}\} = S$.

(vii) If $x \in (X \cap Y)^{l1}$ then x is in some of the $K_i \subseteq (X \cap Y)$, but these will be subsets of both X and Y respectively. So x will be in $X^{l1} \cap Y^{l1}$. For the second part it is clear that $X^{l2} \cap Y^{l2} \subseteq (X \cap Y)^{l2}$. Suppose that the reverse inclusion is false. Then there must exist $x \in (X \cap Y)^{l2}$ such that $x \notin X^{l2}$ and $x \notin Y^{l2}$. The latter means that there exist no intersection of complements of K_is that are included in X and Y, but are included in $X \cap Y$. This obvious contradiction proves that $(X \cap Y)^{l2} = X^{l2} \cap Y^{l2}$.

(viii) From the definitions it is clear that $X^{u1} \cup Y^{u1}$ will be a subset of $(X \cup Y)^{u1}$ as the latter is the intersections of the unions of K_is that contain $(X \cup Y)$. If $x \in (X \cup Y)^{u1}$ then x is in all of the unions of K_is that contain $(X \cup Y)$. Now each of these unions will contain X and Y respectively. If x is in X or Y, then there is nothing to prove. So suppose that x is in neither. Now if x is neither in X^{u1} and Y^2, then we will be able to form a collection of K_i that contains $X \cup Y$, contradicting our original assumption that $x \in (X \cup Y)^{u1}$.

(ix) Let $X \subset Y$ and let $K_i \subseteq X$, then K_i is also included in Y. The union of all such K_i's is the corresponding lower approximation. So $X^{l1} \subseteq Y^{l1}$. Again if $X \subset Y$ and if $\cap_{i \in I} S \setminus K_i \subseteq X$, then it is a subset of Y as well, in fact some subsets of I may also have the property for Y and not for X. And as the unions of these intersections is the second lower approximation, so $X^{l2} \subseteq Y^{l2}$.

(x) We already have $(X \cap Y)^{l1} \subseteq X^{l1} \cap Y^{l1}$. Let $x \in X^{l1} \cap Y^{l1}$. As the constituent K_i's in X^{l1} and Y^{l1} are respectively disjoint, suppose $x \in K_j$ for a fixed j. If x is not in $X \cap Y$, then we have an obvious contradiction to the existence of such a K_j. Therefore we have $X^{l1} \cap Y^{l1} \subseteq (X \cap Y)^{l1}$

(xi) Let $X \subset Y$ and let $Y \subseteq \cup_{i \in I} K_i$ for some $I \subseteq \{1, 2, \dots n+1\}$ then $X \subseteq \cup_{i \in I} K_i$. X^{u1} and Y^{u1} are formed by the intersection of such unions of K_i's. So $X^{u1} \subseteq Y^{u1}$.
Again if $X \subset Y$, and $Y \subseteq S \setminus K_i$ for some i, then X will also be a subset of the same. But there may exist some K_j for which, X is a subset of it's complement and Y is not so. So $X^{u2} \subseteq Y^{u2}$.

(xii) If $x \in X^{l1} \cup Y^{l1}$, then x is in at least one of the K_i's contained in X^{l1} or Y^{l1}. But that K_i must be contained in $X \cup Y$ and therefore in $(X \cup Y)^{l1}$ as well. So $X^{l1} \cup Y^{l1} \subseteq (X \cup Y)^{l1}$.

(xiii) If $x \in X^{l2} \cup Y^{l2}$, then x is in at least one of the intersections of the form $\cap_{i \in I}(S \setminus K_i)$ that is included in X or Y. Therefore x is in $X \cup Y$. But $\cap_{i \in I}(S \setminus K_i) \subseteq X \cup Y$. So $X^{l2} \cup Y^{l2} \subseteq (X \cup Y)^{l2}$.

(xiv) The intersection of the unions of K_is that contain X and Y respectively will intersect in a set containing $(X \cap Y)^{u1}$ (as a larger number of unions of K_is will contain $(X \cap Y)^{u1}$ and as their intersection is precisely $(X \cap Y)^{u1}$).

(xv) If $x \in (X \cap Y)^{u2}$, then x is present in all those $S \setminus K_i$ that contain $X \cap Y$. It is not present in any of those K_is that are included in $S \setminus (X \cap Y) = (S \setminus X) \cup (S \setminus Y)$. Suppose x is not in $(X^{u2} \cap Y^{u2}$, then in each of the three cases we have a contradiction to our original assumption. So $(X \cap Y)^{u2} \subseteq X^{u2} \cap Y^{u2}$.

(xvi) $(S \setminus X)^{l1} = \bigcup\{K_i : K_i \subseteq S \setminus X\}$. Now this is the same as $\bigcup\{K_i : X \subseteq S \setminus K_i\} = S \setminus X^{u2}$

(xvii) $(S \setminus X)^{l2} = S \setminus X^{u1}$ is proved in the same way as the above.

(xviii) $(S \setminus X)^{u1} = S \setminus X^{l2}$ is proved in the same way as the above.

(xix) $(S \setminus X)^{u2} = S \setminus X^{l1}$ is proved in the same way as the above.

The other statements are easy to prove through standard set-theoretic arguments. □

1.2 Concrete Katrinak Algebras

This is a streamlined modification of the approach due to Duntsch [11]. A *double Stone algebra*
$\langle \underline{L}, +, \cdot, *, ^+, 0, 1\rangle$ is an algebra of type $(2, 2, 1, 1, 0, 0)$ such that

(i) $\langle \underline{L}, +, \cdot, 0, 1\rangle$ is a bounded distributive lattice.

(ii) x^* is the pseudo-complement of x, i.e. $y \leq x^* \Leftrightarrow y \cdot x = 0$,

(iii) x^+ is the dual pseudo-complement of x, i.e. $x^+ \leq y \Leftrightarrow y + x = 1$,

(iv) $x^* + x^{**} = 1$, $x^+ \cdot x^{++} = 0$.

It is possible to replace the second and the third condition by the equations,

- $x \cdot (x \cdot y)^{*} = x \cdot y^{*}, \quad x + (x + y)^{+} = x + y^{+}$
- $x \cdot 0^{*} = x, \quad x + 1^{+} = x$
- $0^{**} = 0, \quad 1^{(++)} = 1$

.

A double Stone algebra is *regular* if and only if $x \cdot x^{+} \leq y + y^{*}$ if and only if

$$(x^{+} = y^{+}, x^{*} = y^{*} \longrightarrow x = y).$$

Let B be a Boolean algebra and F a filter on it, then let

$$[B, F] = \{(a, b) : a, b \in B, a \leq b, (a \vee b^{c}) \in F\}.$$

On this,the operations $+, \cdot, {}^{*}, {}^{+}$ are definable via

- $(a, b) + (c, e) = (a \vee c, b \vee e)$
- $(a, b) \cdot (c, e) = (a \wedge c, b \wedge e)$
- $(a, b)^{*} = (b^{c}, b^{c})$
- $(a, b)^{+} = (a^{c}, a^{c})$

An algebra $\mathsf{K} = \langle [B, F], +, \cdot, {}^{*}, {}^{+} \rangle$ of this form is called a *Katrinak algebra* in [11].In such an algebra we can identify B with $\{(a, a) : a \in B\}$ and F with $\{(a, 1) : (a, 1) \in \mathsf{K}\}$.

If instead we start from an arbitrary set S and construct the collection of all fields of subsets $\mathcal{F}(S)$ of it, then the Katrinak algebras formed from the elements of $\mathcal{F}(S)$ will be called *concrete Katrinak algebras.* The following theorem is an adaptation of theorem that had been originally proved in [12].

Theorem 2. *Each concrete Katrinak algebra is a regular double Stone algebra and conversely every regular double stone algebra is isomorphic to a concrete Katrinak algebra.*

Proof. Let K be a concrete Katrinak algebra constructed over a set S, so that its universal set is $[B, F]$, $B = \langle B, +, \cdot, -, 0, 1 \rangle$ being the Boolean algebra and F being a filter on B such that

$$[B, F] = \{(a, b) \in B \times B : a \leq b \,\text{and}\, -b + a \in F\}.$$

Then, $L = \langle [B, F], +, \cdot, {}^{*}, {}^{+} \rangle$ is a regular double Stone algebra if we set

$$(a, b)^{*} = (-b, -b), \; (a, b)^{+} = (-a, -a)$$

Let $\langle \underline{L}, +, \cdot, *, ^+, 0, 1\rangle$ be a regular double Stone algebra. Define the center $B(L)$ and the dense set of L via, $B(L) = \{x^* : x \in \underline{L}\}$ and $\Delta(L) = \{x : x^* = 0\}$ respectively. Then $B(L)$ is a Boolean algebra isomorphic to B in which the operations * and $^+$ coincide with the complementation operation. Let $H = (\Delta(L))^{++}$. If we define a map $\tau : L \mapsto (B(L), H)$, under $\tau(x) = (x^{++}, x^{++})$ then this is an isomorphism. □

In the associated algebraization, Duntsch uses a language $\mathfrak{L}$ of rough set logic consisting of a nonempty set of propositional variables P, two binary connectives $\vee$, $\wedge$, two unary connectives *, $^+$ (representing negations) and a constant $\mathbf{T}$ for truth. Formulas are constructible in the usual way, so that the set $\mathcal{F}(\mathfrak{L})$ of formulas is a free algebra of type $(2, 2, 1, 1, 0)$ generated over P. A model of $\mathfrak{L}$ then is a pair of the form (W, v), where W is a set and $v : P \mapsto \wp(W) \times \wp(W)$ is a valuation, such that if $v(p) = (A, B)$ then $A \subseteq B$.

Given a model $\mathfrak{M} = (W, v)$, it's meaning function σ is defined as an extension of the valuation function $\sigma : \mathcal{F}(\mathfrak{L}) \longmapsto \wp W \times \wp W$ such that,

- $\sigma(\mathbf{T}) = (W, W)$
- $\forall p \in P \; \sigma(p) = v(p)$
- If $\sigma(\varphi) = (A, B)$ and $\sigma(\psi) = (C, E)$, then
  - $\sigma(\varphi \wedge \psi) = (A \cap C, B \cap E)$
  - $\sigma(\varphi \vee \psi) = (A \cup C, B \cup E)$
  - $\sigma(\varphi^*) = (-B, -B)$
  - $\sigma(\varphi^+) = (-A, -A)$, $-A$ being the complement of A in $\wp(W)$.

Now on $Ran(\sigma) = \{\sigma(\varphi) : \varphi \in \mathcal{F}(\mathfrak{L})\}$, if we define the operations $+, \cdot, ^*, ^+$ via,

- $\sigma(\varphi) \cdot \sigma(\psi) = \sigma(\varphi \wedge \psi)$
- $\sigma(\varphi) + \sigma(\psi) = \sigma(\varphi \vee \psi)$
- $(\sigma(\varphi))^* = \sigma(\varphi^*)$
- $(\sigma(\varphi))^+ = \sigma(\varphi^+)$.

With these operations $Ran(\sigma)$ is a Katrinak algebra and σ is a morphism. The variety generated by it coincides with the variety of regular double stone algebras. Moreover the associated logic has a finitely complete strongly sound inferential base and fails the Beth definability property.

2 Esoteric Rough Set Theory

We intend the term *esoteric rough set theory* to qualify all those generalized rough set theories that involve a lack of consistent naming scheme in the domain of approximation. In rough set theory along with its classicalist meta-level interpretation, we have at least two levels of discernibility. At the meta-level we have enough discernibility to distinguish between all the elements of the set. In esoteric rough set theory too the intended interpretation will be by the transfer of all discernibility to the main relation of the approximation space. The eventual logic corresponding to the theory may be the most faithful representation of the intended interpretation.

Definition 1. *By a partial equivalence R on a set S we mean a binary relation satisfying all of the following :*

(i) $R \cap \Delta_S \neq \Delta_S$

(ii) $R^c = R$

(iii) $((x, y), (y, z) \in R, x \neq y \neq z \longrightarrow (x, z) \in R)$

The third condition will be referred to as weak transitivity, while the condition

$$((x, y), (y, z) \in R \longrightarrow (x, z) \in R)$$

will be called strong transitivity. *For each $x \in S$, $[x] = \{y : (x, y) \in R\}$ will be the* pseudoclass *formed by x. A pseudoclass $[x]$ will be said to be an* essential class *if and only if $\forall y \in [x]\, [x] = [y]$.*

Proposition 1. *The set of pseudo-classes $\mathcal{E}_R$ induced by a partial equivalence R on a set S satisfy all of the folllowing :*

(i) $\bigcup \mathcal{E}_R \subseteq S$

(ii) $x \neq y, [x] \cap [y] \neq \emptyset \longrightarrow \{x, y\} \in [x] \cap [y]$

(iii) $[x] \cap [y] \neq \emptyset \longrightarrow [x] = [y]$

Proof. (i) If $[x] = \{y : (x, y) \in R\}$, then $\mathcal{E}_R = \{[x] : x \in S\}$. As it is possible that $x \notin [x]$, and that $(\forall y)\, (x, y) \notin R$ so in general we have $\bigcup \mathcal{E}_R \subseteq S$.

(ii) If the intersection of $[x]$ and $[y]$ is nonempty, then there must exist at least one $z \in S$ such that $(x, z), (y, z) \in R$. It follows that in such a case $\{x, y\} \in [x] \cap [y]$.

(iii) If $z \in [x] \cap [y]$, then $(\forall a \in [x])(\forall b \in [y])(a \neq b \longrightarrow (a, z), (b, z) \in R)$. So it follows that $(\forall a \in [x])(\forall b \in [y])(a \neq b \longrightarrow (a, b) \in R)$. This ensures $[x] = [y]$.

□

It is easy to prove the following proposition by any form of transfinite induction, but in our approach this extension is not particularly useful at the object level as we do not accept extensionality there!

Proposition 2. *Every partial equivalence ς on a set S is extensible to a unique smallest equivalence relation $\varsigma^{\oslash}$ (in ZFC).*

Proof. Let $pEQ(S)$ be the set of all partial equivalence relations on the set S. The inclusion order $\subseteq$ on the set is a partial order on it. By this order we will be able to find a set of equivalences that contain the partial equivalence ς. As the set of equivalences on a set forms a complete lattice (in the induced order), the intersection of the set of equivalences that contain ς will be an equivalence. This must necessarily be the least equivalence containing ς. Note that we can always obtain $\varsigma^{\oslash}$ as the reflexive, transitive and symmetric completion of ς.

Definition 2. *A* partial approximation space *S will be a pair of the form $\langle \underline{S}, R\rangle$, where R is a partial equivalence relation on the set $\underline{S}$. If A is a subset of $\underline{S}$, then the lower approximation A^l of A can be defined via*

$$A^l = \bigcup\{[x] : [x] \subseteq A\}$$

. By $[x]$ of course we mean the set $\{y : (x, y) \in R\}$. Similarly we can define the upper approximation via

$$A^u = \{[x] ; [x] \cap A \neq \emptyset\}$$

Proposition 3. *If there exists an element $x \in \underline{S}$ such that $\neg((x, x) \in R)$, then it is possible that $S^l \subset S$, but if x is not R-related to all other elements in S, then it is necessary that $S^l \subset S$ and $S^u \subset S$.*

Proof. Suppose it is the case that (x, x) is not in R, but $(x, y) \in R$ for some $y \neq x$, then $x \in [y]$ will hold. In this case if for example all other elements are R-related to themselves, then $S^l = S$ will hold.

If on the other hand if we have a single x, that is not R-related to any other element, then x will not be in any of the pseudo-classes, so that $S^l \subset S$ will necessarily hold. □

Proposition 4. *If A is a subset of a partial approximation space S, then it is possible that $A \nsubseteq A^u$. If however each element of A is R-related to some other element in S, then $A \subseteq A^u$ will hold.*

Proof. Suppose A contains an element x that is not R-related to any other element in S, then by definition $x \notin A^u$. Hence the result. □

Definition 3. *In the light of the above pathology, we will say that a subset A of a partial approximation space is a* well-related *subset if and only if A does not contain any element that is not related to every other element of S. A partial approximation space that satisfies the condition $S^l = S$ will be said to be a*well partial approximation space *(ASW). Otherwise we will term it an* ill-posed partial approximation space *(ASI)*

For the greater part of this research paper we will avoid this crucial distinction.

Proposition 5. *A partial approximation space S will be a well partial approximation space if and only if each subset of S is well-related.*

Proof. There is nothing much to prove here. If S is an ASW, then each element of S must be related to some other element of it. As partial equivalences are symmetric, every subset of S will have the property. The converse is obvious. □

Definition 4. *If $S = \langle \underline{S}, R \rangle$ is a partial approximation space, let*

(i) $\underline{S}^* = \underline{S} \cup \{\bot, T\}$

(ii) Extend R to R^* by the condition $R^* = R \cup \{(T, x) : x \neq \bot\}$

$S^* = \langle \underline{S}^*, R^* \rangle$ *will be said to be the* twisted approximation space.

It should be noted that it is often possible to define many partial and total algebraic operations on partial approximation spaces with the intended meaning being of partial or total causal dependence. This is easier to effect in twisted approximation spaces. In general we may end up with multiple terms with no sensible interpretation. In a twisted approximation space, we will have the option of assigning these to the elements $\bot$, T. Another reason for defining the twisted approximation space is in the following :

Proposition 6. *The set $\mathcal{E}_R^*$ of pseudo-classes of R is a lattice with respect to the set-theoretic inclusion.*

Proof. On $\mathcal{E}_R^*$ we can define the following operations naturally :

(i) $[x] \vee [y] = [y]$ if and only if $[x] \subseteq [y]$

(ii) $[x] \wedge [y] = [x]$ if and only if $[x] \subseteq [y]$

(iii) $[x] \wedge [y] = \emptyset = [\bot]$ if and only if $[x]$ is not comparable with y.

(iv) $[x] \vee [y] = [T]$ if and only if $[x]$ is not comparable with $[y]$ and there is no $c \in S$ such that $[x], [y] \subset [c]$.

(v) $[x] \vee [y] = [c]$ for some $c \neq T$ if and only if $[x] \cap [y] \neq \emptyset$ and $[x] \cup [y] = [c]$

(vi) $[x] \wedge [T] = [T]$

(vii) $[x] \vee [\bot]\ [x]$

Given this it is easy to check that it is a lattice. Note that it is possible that $[x] \subset [y]$.

Proposition 7. *On $\mathcal{E}_R^*$ it is provable that* *(i)* $(a \vee b = [T] \longrightarrow a \wedge b = [\perp])$

(ii) $a \vee (b \wedge c) = (a \vee b) \wedge (a \vee c)$

Proof. (i) If two pseudoclasses are not comparable, only then will their join be $[T]$ and then it must be the case that the two pseudoclasses have empty or nonempty intersection. The nonempty intersection cannot be a pseudo-class unless it is $[\perp]$. So the result holds.

(ii) This part is direct.

□

Definition 5. *The* degree $\eta(x)$ *of a pseudo-class* $[x]$ *is the cardinality of the set*

$$\{y : y \in [x], (y, y) \notin R, x \neq y\}.$$

The cardinality of the set $\{y : y \in [x], (y, y) \notin R\}$ *will be called the* full degree $\gamma([x])$ *of the pseudo-class. The structure* $\Sigma(S) = \langle \mathcal{E}_R^*, \vee, \wedge, \perp, T, \eta, \gamma \rangle$ *will be said to be the* discernibility spectra *of the partial approximation space.*

Definition 6. *Given two partial approximation spaces* $S = \langle \underline{S}, R \rangle$ *and* $K = \langle \underline{K}, P \rangle$*, a mapping* $\xi : \Sigma(S) \mapsto \Sigma(K)$ *(from a discernibility spectra into another) will be said to be a injective morphism if and only if it satisfies :*

- ξ *induces a injective lattice morphism :* $\langle \mathcal{E}_R^*, \vee, \wedge, \perp, T \rangle \mapsto \langle \mathcal{E}_P^*, \vee, \wedge, \perp, T \rangle$ *(forgetfully)*
- $\eta(x) \leq \eta(\xi(x))$
- $\gamma(x) \leq \gamma(\xi(x))$

ξ *will be called an isomorphism if it induces a lattice isomorphism and corresponding* η *and* γ *values are respectively equal.*

Definition 7. *Let* $S = \langle \underline{S}, R \rangle$ *and* $T = \langle \underline{Y}, K \rangle$ *be two partial approximation spaces such that there is an injective relation morphism* $\zeta : \underline{S} \mapsto \underline{Y}$ *and* $(x, y) \in R$ *implies* $(\zeta(x), \zeta(y)) \in K$*, then we will say that* Y *is an* injective extension *of* S*.*

In the following two theorems we can assume that the partial approximation spaces involved are finite. The concepts of degree and full degree of pseudo classes can be used to arrive at notions of equivalent partial approximation spaces. However, we can prove

Theorem 3. *If a partial approximation space* Y *is an injective extension of a well partial approximation space* S *with the injective relation morphism* $\zeta : \underline{S} \mapsto \underline{Y}$ *being such that for any* $x \in S$ $(\zeta(x)\zeta(x)) \in P$ *implies* $(x, x) \in R$*, then we have an embedding of* $\Sigma(S)$ *into* $\Sigma(T)$*.*

Proof. Let $\zeta : \underline{S} \mapsto \underline{Y}$ be the injective relation morphism associated. For any $x \in S$, we have $\zeta([x]) \subseteq [\zeta(x)]$ in general. So let ζ^* be a map from $\Sigma(S)$ into $\Sigma(Y)$ such that

$$(\forall x \in S)\, \zeta^*[x] = [\zeta(x)] .$$

Clearly ζ^* is well defined provided there are no isolated singletons (which is ensured by requiring that S is a well partial approximation space).

In $\langle \mathcal{E}_R^*, \vee, \wedge, \perp, T \rangle$, equal pseudoclasses are of course identified and the meet of two incomparable elements is the bottom $\perp$ and their join is the top T. If $\zeta^*[x] = \zeta^*[y]$ then $[x] = [y]$ is essential precisely because of the condition for any $x \in S$ $(\zeta(x)\,\zeta(x)) \in P$ implies $(x, x) \in R$. ζ^* serves as the required embedding as the lattice embedding part is a standard result. The rest is easy to verify.

Theorem 4. *If two partial approximation spaces have isomorphic discernibility spectras, then they themselves are isomorphic to each other (relation morphically).*

Proof. Let $X = \langle \underline{X}, R\rangle$ and $Y = \langle \underline{Y}, P\rangle$ be two partial approximation spaces with isomorphic discernibility spectras $\Sigma(X)$ and $\Sigma(Y)$ respectively. If the spectras are isomorphic then there exists a map $\varphi : Sigma(X) \longmapsto \Sigma(Y)$ such that the conditions mentioned in the definition hold. The equality of corresponding η and γ values along with the finiteness assumptions ensures the result.

Proposition 8. *The set $\mathcal{E}_R$ of pseudo-classes of R is endowable with a partial lattice structure using the set-theoretic inclusion order. It satisfies all of*

(i) $[x] \vee ([y] \vee [z]) \stackrel{s}{=} ([x] \vee [y]) \vee [z]$ *(and dually)*
(ii) If $[x] \subseteq [y]$ *then* $[x] \vee [y] = [y]$

(iii) If $[x] \vee [y] = [z]$ *and* $[x] \nsubseteq [y]$ *and conversely, then* $[x] \wedge [y]$ *is not defined.*
(iv) $\mathcal{E}_R$ *is a tree (a poset in which each principal order ideal is a well-ordered set.*

$\stackrel{s}{=}$ *is the strong weak equality (if either side is defined then the other is and the two are equal).*

Proof. The first thing that is to be noted is that if $[x] \subset [y]$, then $[y] = [x] \cup \{x\}$. If $[x]$ and $[y]$ have nonempty intersection and either is not included in the other, then it is necessary that $(x, x), (y, y) \notin R$. In this case their intersection is not a pseudoclass and the largest pseudoclass contained in will not be defined unless their intersection is a singleton. This proves the third part that can happen in the only one situation. The strong weak associativity in the first part is induced by the set theoretic property. □

Proposition 9. *If $S^* = \langle \underline{S}^*, R^*\rangle$ is the twisted approximation space corresponding to S and $\mathcal{E}_R^*$ the set of its pseudoclasses, then $\mathcal{E}_R$ (as a partial lattice) is embeddable in the lattice $\mathcal{E}_R^*$.*

Proof. The proof is easy as it based on the natural identification of corresponding pseudo classes. Note that we can construct $\mathcal{E}_R^*$ from $\mathcal{E}_R$ as a special two point completion (for the notion see [13] for example). □

Proposition 10. *Given a partial approximation space S there exists a unique minimal approximation space $S^{\oslash}$ that contains it in the following sense; the relation of S is contained in the relation of $S^{\oslash}$ and the underlying set of S is also a subset of that of $S^{\oslash}$.*

Proof. $S = \langle \underline{S}, R\rangle$ is a partial approximation space with R being a partial equivalence. So the closure $R^{\oslash}$ of R under reflexivity and transitivity will be well defined. $S^{\oslash}$ can be taken to be the set along with the closure of the relation. The class of all approximation spaces containing S is partially orderable by the lexicographic order induced by set-inclusion. It is easy to show by any form of general transfinite induction that the above defined $S^{\oslash}$ is minimal and unique as well. □

For proceeding with the algebraization procedure, it is very important to have answers for the following question(s) in particular (in the power set of the partial approximation space):

- What is the fine structure of lower and upper approximation of sets and how do they interact with derived set-theoretic operations?
- Given a set in the form A^y, does there exist a set B such that $B^f = A^y$ and in what form can B be in (y, f being variables over the set $\{l, u, lu, ul, ll, uu\}$ or subsets thereof)?
- What is the relationship between the different types of approximations ?

The interaction with derived set-theoretic operations is considered in the section on generalized esoteric covers.

We shall show that if approximations of a set or sets satisfies a property that is not satisfied in an approximation space, then it must be due to the existence of certain kinds of subsets. This allows us a particular approach to the main questions. But we will consider other approaches too.

3 Exceptional Sets

Definition 8. *A set K of a partial approximation pace S will be said to R-*isolated *if and only if $((x, y) \in R, x \in K \longrightarrow y \in K)$.*

The relationship of the collection $\mathcal{F}(A)$ of all isolated sets which are not disjoint from a given set A can be expected to actually define the approximations of the set A. We shall show that it suffices to restrict ourselves to one, two and three element isolated sets alone. These are classified into different types on the basis of the relationship of the elements to A, R and S. This is done in coherence with the different types of possible approximations.

Classification of Three Element Isolated Sets not Disjoint from a Given Set A

Definition 9. *A three element isolated set* $\{x, y, z\}$ *not contained in a set* A *will be said to be* $\frac{3}{1}$-exceptional for A *if and only if all of the following are true.*

(i) $x \in A$, $y, z \in S \setminus A$

(ii) $(x, x) \notin R$, $(y, y) \in R$, $(z, z) \notin R$

(iii) $(x, y), (x, z) \in R$

Definition 10. *A three element isolated set* $\{x, y, z\}$ *not contained in a set* A *will be said to be* $\frac{3}{2}$-exceptional for A *if and only if all of the following are true.*

(i) $x, y \in A$, $z \in S \setminus A$

(ii) $(x, x) \notin R$, $(y, y) \notin R$, $(z, z) \notin R$

(iii) $(x, y), (x, z) \in R$

Definition 11. *A three element isolated set* $\{x, y, z\}$ *not contained in a set* A *will be said to be* $\frac{3}{3}$-exceptional for A *if and only if all of the following are true.*

(i) $x, y \in A$, $z \in S \setminus A$

(ii) $(x, x) \notin R$, $(y, y) \in R$, $(z, z) \notin R$

(iii) $(x, y), (x, z) \in R$

Definition 12. *A three element isolated set* $\{x, y, z\}$ *not contained in a set* A *will be said to be* $\frac{3}{4}$-exceptional for A *if and only if all of the following are true.*

(i) $x, y \in A$, $z \in S \setminus A$

(ii) $(x, x) \notin R$, $(y, y) \notin R$, $(z, z) \in R$

(iii) $(x, y), (x, z) \in R$

Definition 13. *A three element isolated set* $\{x, y, z\}$ *not contained in a set* A *will be said to be* $\frac{3}{5}$-exceptional for A *if and only if all of the following are true.*

(i) $x \in A$, $y, z \in S \setminus A$

(ii) $(x, x) \in R$, $(y, y) \notin R$, $(z, z) \notin R$

(iii) $(x, y), (x, z) \in R$

Definition 14. *A three element isolated set* $\{x, y, z\}$ *not contained in a set* A *will be said to be* $\frac{3}{6}$-exceptional for A *if and only if all of the following are true.*

(i) $x \in A$, $y, z \in S \setminus A$

(ii) $(x, x) \notin R$, $(y, y) \in R$, $(z, z) \in R$

(iii) $(x, y), (x, z) \in R$

Classification of Two Element Isolated Sets (non-disjoint from a given subset A)

Definition 15. *A two element isolated set* $\{x, y\}$ *not contained in a set A will be said to be* $\frac{2}{1}$-exceptional for A *if and only if all of the following are true.*

(i) $x \in A, \;\; y \in S \setminus A$

(ii) $(x, x) \notin R, \;\; (y, y) \notin R$

(iii) $(x, y) \in R$

Definition 16. *A two element isolated set* $\{x, y\}$ *not contained in a set A will be said to be* $\frac{2}{2}$-exceptional for A *if and only if all of the following are true.*

(i) $x \in A, \;\; y \in S \setminus A$

(ii) $(x, x) \notin R, \;\; (y, y) \in R$

(iii) $(x, y) \in R$

Definition 17. *A two element isolated set* $\{x, y\}$ *not contained in a set A will be said to be* $\frac{2}{3}$-exceptional for A *if and only if all of the following are true.*

(i) $x \in A, \;\; y \in S \setminus A$

(ii) $(x, x) \in R, \;\; (y, y) \notin R$

(iii) $(x, y) \in R$

Classification of Isolated Singletons in a Given Set

There are two possible types of singleton isolated sets. In the first type the element must be related to itself. In the other type the element must not be related to itself. Naturally the existence of the second type of sets will make the partial approximation space an ASI.

Definition 18. *A set K will be said to be exceptional for a set A if and only if all of the following hold:*

(i) K *is isolated.*
(ii) $(\exists x \in K)\, (x, x) \notin R$

(iii) $K \not\subseteq A,\ \ K \cap A \neq \emptyset$

Proposition 11. *If K is a subset of a partial approximation space S with the properties that*

(i) $(\forall x \in K)(\forall y)((x, y) \in R \longrightarrow (y, y) \in R)$

(ii) $(\forall x \in K)(x, x) \in R,$

then

$$K^{ll} = K^{l} = K^{lu} \subseteq K \subseteq K^{u} = K^{ul} = K^{uu}.$$

Proof. Let S^* be the reflexive completion of the partial approximation space. It is easy to check that the lower approximation of K in both the spaces are identical. The case is similar for upper approximations and all possible successive approximations too. So the result follows. □

From the above we see that any deviation from the usual approximation properties must be due to exceptional sets. In the next few theorems we

- Determine the general relation between approximations in the presence of different kinds of exceptional sets.
- Prove that the above defined exceptional sets are essentially independent in their determination of the relation between approximations and
- Show that for determining the most general possible set-theoretic relations between the different approximations, it suffices to restrict ourselves to at most three element exceptional sets.

On an exceptional set K for A, for any $x, y \in K$, let $x \sim_A y$ if and only if one of the following is true :

(i) $x, y \in A$ and $(x, x), (y, y) \in R$

(ii) $x, y \in A$ and $(x, x), (y, y) \notin R$

(iii) $x, y \in S \setminus A$ and $(x, x), (y, y) \in R$

(iv) $x, y \in S \setminus A$ and $(x, x), (y, y) \notin R$.

Then $\sim_A$ is an equivalence on K and we have exactly four classes (some of which may be empty). Note that elements in the first of these classes remain in A and all approximations thereof and have no new effect essentially. So we have the following theorem :

Theorem 5. *If K is an exceptional set for A, then the relative set-theoretic relationship between the different approximations (with respect to the inclusion order) will remain invariant when K is replaced with a suitable finite subset F in S. In more clear terms we mean if A, A^l, A^{lu}, A^u, A^{uu}, A^{ll} are related within themselves by the relations $\{\subset, =\}$ in the way Φ, then B, B^l, B^{lu}, B^u, B^{uu}, B^{ll} will be related in the way Φ, where $B = (A \setminus K) \cup (F \cap A)$. All interpretations are assumed to be in S.*

A subset A of a partial approximation space S will be fully featured if it includes all types of elements relevant for the approximation process. This is possible because the existence of distinct elements of distinct types are essentially independent as far as the implications on possible properties are concerned. We have arrived at the following definition by a careful analysis of all possible types of existence of exceptional sets for a given set.

Definition 19. *A set of the form*

$$K = K_0 \cup \{x_1, y_1, y_2, z_1, z_2, a_1, b_1, c_1, f_1, f_2, g_1, h_1\}$$

satisfying all of

(i) K_0 *is a nonempty subset of* S *with at least two elements that are not subsets of any other exceptional subsets of* K.
(ii) Elements of K_0 *are distinct from the others listed.*

(iii) $(\forall x \in K_0)(x, x) \in R$

(iv) $(\forall x \in K_0)(\forall y)((x, y) \in R \longrightarrow (y, y) \in R)$.

(v) $\exists x_2, x_3 \in S \setminus K$

(vi) $\{x_1, x_2, x_3\}$ *is a* $\frac{3}{1}$*-exceptional set for* K.

(vii) $\exists y_3 \in S \setminus K$

(viii) $\{y_1, y_2, y_3\}$ *is a* $\frac{3}{2}$*-exceptional set for* K.

(ix) $\exists z_3 \in S \setminus K$

(x) $\{z_1, z_2, z_3\}$ *is a* $\frac{3}{3}$*-exceptional set for* K.

(xi) $\exists f_3 \in S \setminus K$

(xii) $\{f_1, f_2, f_3\}$ *is a* $\frac{3}{4}$*-exceptional set for* K.

(xiii) $\exists g_2, g_3 \in S \setminus K$

(xiv) $\{g_1, g_2, g_3\}$ *is a* $\frac{3}{5}$*-exceptional set for* K.

(xv) $\exists h_2, h_3 \in S \setminus K$

(xvi) $\{h_1, h_2, h_3\}$ *is a* $\frac{3}{6}$*-exceptional set for* K.

(xvii) $\exists a_2 \in S \setminus K$

(xviii) $\{a_1, a_2\}$ *is a* $\frac{2}{1}$*-exceptional set for* K.

(xix) $\exists b_2 \in S \setminus K$

(xx) $\{b_1, b_2\}$ *is a* $\frac{2}{2}$*-exceptional set for* K.

(xxi) $\exists c_2 \in S \setminus K$

(xxii) $\{c_1, c_2\}$ *is a* $\frac{2}{3}$*-exceptional set for* K.

will be said to be fully featured *in* S.

Proposition 12. *If* F, K *are two fully featured sets in a partial approximation space, then* $F \mid\sim$ *is bijective to* $K \mid\sim$.

Proof. They will have the same number of classes with respect to the equivalence $\sim$. □

Proposition 13. *Any subset* K *of a partial approximation space* S *is representable as a disjoint union of* R*-isolated sets.*

Proof. Suppose $\mathcal{K}$ is the set of pseudoclasses of K. We can form non-disjoint subcollections $\{P_\lambda\}_{\lambda \in \Lambda}$ of $\mathcal{K}$. For each λ in the indexing set Λ, let $A_\lambda = \bigcup P_\lambda$. Each A_λ is R-isolated and

$$S = \bigcup A_\lambda .$$

□

Proposition 14. *Let* A *be a subset of a partial approximation space* S *with the only exceptional set for it being the single* $\frac{2}{1}$*-exceptional set* $\{x, y\}$, *then the following hold in general*

(i) $A^{ll} = A^{lu} = A^l \subseteq A^u = A^{ul} = A^{uu}$

(ii) A^l, A^{ul}, A^u *are not unions of pseudoclasses of elements in them.*

(iii) None of the approximations include the exceptional set.

Proof. $\{x, y\}$ is the only $\frac{2}{1}$-exceptional set for A, so $x \in A$ and $y \in S \setminus A$ and $[y] = \{x\} \in A^l$. $A^l \setminus \{x\}$ consists of essential-classes alone. So the first part of the result follows. The second and third part are obvious as $y \notin A^u$. □

Proposition 15. *Let* A *be a subset of a partial approximation space* S *with the only exceptional set for it being the single* $\frac{2}{2}$*-exceptional set* $\{x, y\}$, *then the following hold in general*

(i) $A^{ll} = A^l = A^{lu} \subset A^u = A^{ul} = A^{uu}$

(ii) A^l *and* A^u *are unions of pseudoclasses of elements in them.*

(iii) A^u *contains the exceptional set* $\{x, y\}$ *for* A.

Proof. (i) Clearly $[x] = \{y\}$ and $[y] = \{x, y\}$. So both of these pseudoclasses are not present in A^l. The rest of the pseudoclasses in A^l are essential-classes. But the two pseudoclasses $[x]$, $[y]$ are subsets of A^u. So we have $A^l \subset A^u$ in general.
(ii) This follows from the proof of the first part.
(iii) This follows from the proof of the first part.

□

Proposition 16. *Let* A *be a subset of a partial approximation space* S *with the only exceptional set for it being the single* $\frac{2}{3}$*-exceptional set* $\{x, y\}$*, then the following hold in general*

(i) $A^{ll} = A^l \subset A^{lu} \subseteq A^u = A^{ul} = A^{uu}$

(ii) A^{ul} *and* A^u *are unions of pseudoclasses of elements in them.*

(iii) A^u *contains the exceptional set* $\{x, y\}$ *for* A.

Proof. (i) Clearly $[x] = \{x, y\}$ and $[y] = \{x\}$. So $[y]$ is included in A, but $[x]$ is not included in it. Apart from $[y]$ all other pseudoclasses included in A^l are essential-classes. So $A^{ll} = A^l \subset A^{lu}$ (as $[x] \subset A^{lu}$).
(ii) Note that $[x] \subset A^u$, so the result follows.
(iii) Follows from the first part.

□

Proposition 17. *Let* A *be a subset of a partial approximation space* S *with the only exceptional set for it being the single* $\frac{3}{1}$*-exceptional set* $\{x, y, z\}$*, then the following hold in general*

(i) $A^l \subset A$, $A^{lu} \subset A$ *and* $A^{lu} \subset A^{ul} = A^u = A^{uu}$

(ii) A^u *is a union of pseudoclasses of elements (but not in* A*).* A^l *is a union of essential-classes.*

(iii) A^u *contains the exceptional set* $\{x, y, z\}$ *for* A.

Proof. (i) Clearly $[x] = \{y, z\}$, $[y] = \{x, y, z\}$, $[z] = \{x, y\}$. So none of them are included in A and only $[y]$, $[z]$ intersect it to form the nonempty the set $\{x\}$. So it is essential that $A^l \subset A$. As A^l is made up of essential-classes alone, so $A^{lu} = A^l$. The rest follows.
(ii) We need to include $[y]$ or $[z]$ for getting A^u.
(iii) Follows from the proof of the first part.

□

Proposition 18. *Let* A *be a subset of a partial approximation space* S *with the only exceptional set for it being the single* $\frac{3}{2}$*-exceptional set* $\{x, y, z\}$*, then the following hold in general* *(i)* $A^{ll} = A^l \subseteq A$ *and* $A^l \subset A^{lu} \subseteq A^u = A^{ul} = A^{uu}$

(ii) A^l *is not a union of pseudoclasses of elements in it, while* A^u *is a union of pseudoclasses of elements in* A.

(iii) $\{x, y, z\}$ *is still exceptional for* A^l *and is contained in* A^u.

Proof. (i) In this case $[x] = \{y, z\}$, $[y] = \{x, z\}$ $[z] = \{x, y\}$. Clearly $[z]$ is included in A and all three pseudoclasses have nonempty intersection with A. A^l is a not a union of essential-classes but still we have $A^{ll} = A^l$. $A^{lu} = A^l \cup \{z\}$ and in general this will be a subset of A^u or maybe equal to it. The equality $A^u = A^{ul} = A^{uu}$ is obvious as there exist no elements in $S \setminus A^u$, that form pseudoclasses intersecting A^u.
(ii) A^l includes $[z]$ and this pseudoclass cannot be replaced by any other.
(iii) Follows from the above two parts.

□

Proposition 19. *Let* A *be a subset of a partial approximation space* S *with the only exceptional set for it being the single* $\frac{3}{3}$*-exceptional set* $\{x, y, z\}$*, then the following hold in general*

(i) $A^{ll} = A^l \subset A^{lu} \subseteq A^{ul} = A^u = A^{uu}$

(ii) A^l *is not a union of pseudoclasses of elements in it, while* A^u *is a union of pseudoclasses in* A *itself.*

(iii) $\{x, y, z\}$ *is contained in* A^u.

Proof. (i) In this case $[x] = \{y, z\}$, $[y] = \{x, y, z\}$ $[z] = \{x, y\}$. $x, y \in A$ and $z \in S \setminus A$. So $x, y \in A^l$. Apart from these A^l is a union of essential classes. But A^{lu} will be $A^l \cup \{z\}$ precisely, while A^u will include A^{lu} and some other essential classes in general. So the result follows.
(ii) Clearly $[x]$ is not included in A^l, but both A^u and A^{lu} are unions of pseudoclasses of some elements in it.
(iii) This has been shown in the proof of the first part.

□

Proposition 20. *Let* A *be a subset of a partial approximation space* S *with the only exceptional set for it being the single* $\frac{3}{4}$*-exceptional set* $\{x, y, z\}$*, then the following hold in general*

(i) $A^l = A^{lu} \subset A^{ul} = A^u = A^{uu}$

(ii) A^l, A^u *are a union of pseudoclasses of some elements in it.*

(iii) $\{x, y, z\}$ *is contained in* A^u*, but not in* A^{lu}.

Proof. (i) In this case $[x] = \{y, z\}$, $[y] = \{x, z\}$ $[z] = \{x, y, z\}$. So A^l will be the union of essential classes alone and A^{lu} will be also be the union of just these. A^u will contain $[z]$, these essential classes and some other essential classes possibly. So the result holds.
(ii) None of $[x]$, $[z]$ and $[y]$ are included in A^l in the first place and A^l is a union of essential classes. The argument for A^u is in the proof of the first part.
(iii) Has been proved in the first part.

□

Proposition 21. *Let A be a subset of a partial approximation space S with the only exceptional set for it being the single $\frac{3}{5}$-exceptional set $\{x, y, z\}$, then the following hold in general*

(i) $A^l = A^{lu} \subset A^{ul} = A^u = A^{uu}$

(ii) Both A^l and A^u are unions of pseudoclasses of some elements in it.

(iii) $\{x, y, z\}$ is contained in A^u, but not in A^{lu}.

Proof. (i) In this case $[x] = \{x, y, z\}$, $[y] = \{x, z\}$ $[z] = \{x, y\}$. As only $x \in A$, so none of these pseudoclasses can be contained in A^l and it is a union of essential classes contained in A. A^{lu} will therefore be equal to A^l. A^u will contain $[x]$, these essential classes and some other essential classes possibly. So the result follows.
(ii) Follows from the first part.
(iii) Follows from the first part. This is what ensures the strict inclusion in the first part.

□

Proposition 22. *Let A be a subset of a partial approximation space S with the only exceptional set for it being the single $\frac{3}{6}$-exceptional set $\{x, y, z\}$, then the following hold in general*

(i) $A^l = A^{lu}$ *and* $A^{ul} = A^u = A^{uu}$

(ii) A^u is a union of pseudoclasses of elements in itself but not in A.

(iii) $\{x, y, z\}$ is contained in A^u, but not in A^{lu}.

Proof. (i) In this case $[x] = \{y, z\}$, $[y] = \{x, y, z\}$ $[z] = \{x, y, z\}$. So both A^l and A^{lu} will contain none of the elements $\{x, y, z\}$, but A^u and A^{uu} will contain all of them. A^l will consequently be a union of some essential classes in A, but A^u will be a union of pseudoclasses in A^u.
(ii) Clearly the pseudoclasses of elements of A do not contain x.
(iii) Follows from the first part.

□

Remark 1. In the above, equality between sets is to be understood in the sense of conjunction of $\subseteq$ and $\supseteq$. But we can use a different contextual logic of combining interpretations in the following:

- If $A \subset B$ in case X alone applies and $A = B$ in case Y alone applies, then $A \subset B$ holds when both case X and Y apply.
- If $A \subset B$ in case X alone applies and $B \subset A$ in case Y alone applies, then $A \parallel B$ when both case X and Y apply.
- If $A \subseteq B$ in case X alone applies and $A = B$ in case Y alone applies, then $A \subset B$ holds when both case X and Y apply.
- If $A \parallel B$ in case X alone applies and $A \subseteq B$ in case Y alone applies, then $A \parallel B$ holds when both case X and Y apply.
- If $A \parallel B$ in case X alone applies and $A \parallel B$ in case Y alone applies, then $A \parallel B$ holds when both case X and Y apply.

X, Y will be one of the exclusive exceptional sets existing for A or B or some set whose approximation is A or B.

Theorem 6. *The following hold in any partial approximation space :*

(i) In each of the nine propositions, we can use an arbitrary number of exceptional sets of the same corresponding type with no change in the content of the propositions.
(ii) Given an arbitrary set A with any number of exceptional sets of any of the above nine types, the relationship between the approximations A^l, A^{ll}, A^u, A^{uu}, A^{lu}, A^{ul} *is generable by the contextual logic of the individual relationships.*

Relationship Diagrams : The first diagram below summarizes the relation between the different approximations in ASW. The abbreviations correspond to applications of the lower and upper approximation and the identity operator.

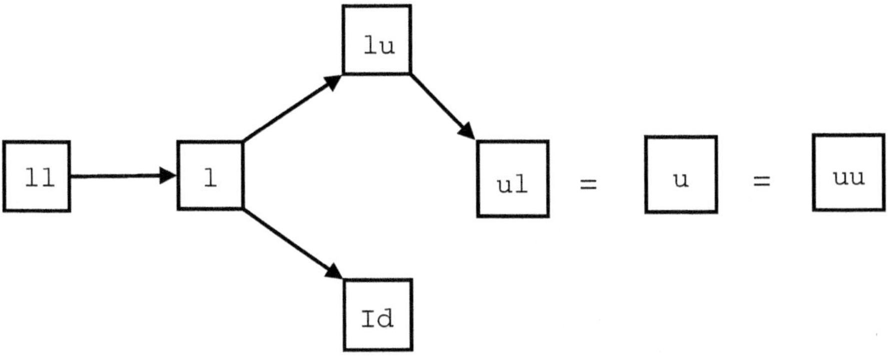

The following summarizes the relation between the different approximations in ASI.

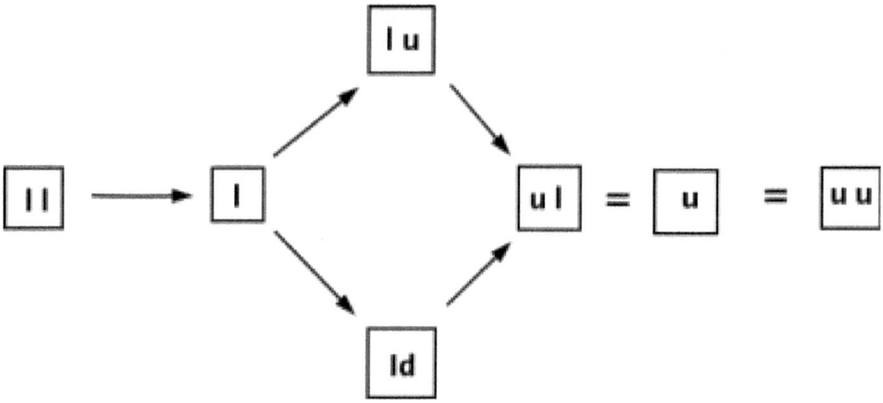

Definition 20. *In a partial approximation space, tuples of the form* $\langle A^l, A^{lu} A^u \rangle$ *will be called* esoteric rough tuples.

Definition 21. *Two subsets* A, B *of* S, *will be said to be* roughly pseudo equal, $A \doteq B$ *if and only if*

(i) $A^l = B^l,\ A^u = B^u$

(ii) $A^{lu} = B^{lu}$

Proposition 23. *The rough pseudo-equality relation* $\doteq$ *defined is an equivalence relation on* $\wp(S)$. *The class generated by a subset* A *will be denoted by* $[A]_r$.

Proposition 24. *If a subset* B *of* S *is of the form* A^l, *then it is possible to find a set* C, *such that* $C^u = B$.

Proof. $B = A^l = \cup\{[x] : [x] \subseteq A\, x \in S\}$. We cannot in general take C to be the set A^l itself. Let the union of the essential classes contained in A (and therefore A^l) be P. If $[z]$ is a pseudoclass contained in $A^l \setminus P$ with $z \in S \setminus A$, then we will exclude all elements of the pseudoclass from C and include z alone. We will need to repeat the procedure for similar pseudoclasses. We can safely include other types of pseudoclasses in A^l and P in our required C. That is it. □

Proposition 25. *Let* A, B *be two subsets of a partial approximation space* S, *then there exists a subset* C *such that* $A^{lu} \cup B^{lu} = C^{lu}$.

Proof. If we prove this for fully featured sets, then we are done. But we will need to wade through all the possible intersections. The result is obvious when there are no exceptional sets for A and B. As in this case the intersection will include some or no pseudo-classes (and with no exceptional sets being present in them).

But it suffices to note that the operation of lu on a fully featured set of the form (where the symbols mean as in the definition of a fully featured set)

$$K_0 \cup \{x_1, y_1, y_2, z_1, z_2, a_1, b_1, c_1, f_1, f_2, g_1, h_1\}$$

results in one of the form

$$(K_0)^l \cup \{y_1, y_2, y_3, z_1, z_2, z_3, a_1\}$$

In the most general case A and B can have

(i) a common set which does not contain exceptional subsets for either of A or B.
(ii) uncommon parts under the above condition
(iii) a common exceptional subset.
(iv) disjoint exceptional subsets.
(v) and exceptional parts of one being contained in the other as isolated sets.

And in all cases the construction of C is possible. □

4 Equalities in Esoteric Rough Set Theory

In this brief section we shall clarify on the different possible weak equivalences definable on a set and on a related algebra. These are directly related to the very definition of possible partial approximation spaces and their semantics. The main theorem in this section is due to [14,15].

Definition 22. *A weak partial congruence ϑ on an algebra of the form $L = \langle \underline{L}, f_1, \ldots, f_n \rangle$ is a weak equivalence relation which is compatible with the operations of L in the sense*

$$\bigwedge (x_i, y_i) \in \vartheta \longrightarrow (f(x_1, \ldots, x_n), f(y_1, \ldots, y_n)) \in \vartheta$$

for every operation of the corresponding arity. The set of all weak partial congruences on L will be denoted by $C_w(L)$.

Definition 23. *The algebra*

$$K_w(L) = \langle C_w(L), \vee, \wedge, \circ, ^{-1}, , \triangle, \sigma, L^2 \rangle$$

is called the weak partial congruence algebra *where the lattice operations, relation composition (whenever it is defined) $\circ$, inverse and the diagonal relation (corresponding to the least subalgebra on the lattice of subalgebras) are defined in the usual way.*

Theorem 7. *A weak partial congruence algebra satisfies all of the following :*

(i) $\triangle \circ (\triangle \vee \rho) = \triangle \vee \rho$

(ii) $\triangle \circ (\triangle \wedge \rho) = \triangle \wedge \rho$

(iii) $\triangle \wedge (\rho \vee \theta) = (\triangle \wedge) \vee (\triangle \wedge \theta)$

(iv) $\rho \circ \theta \in C_w(L)$ *if and only if* $\rho \circ \theta = \theta \circ \rho$

(v) If $\rho \circ \theta \in C_w(L)$ *and* $\rho \vee \theta = \rho \circ \theta$ *if and only if* $\triangle \wedge \rho = \triangle \wedge \theta$

Given an abstract weak congruence algebra, it need not happen that there exists an algebra L, such that its weak congruence algebra is isomorphic to it. In spite of this, the algebra is useful in the selection of improved forms of partial weak congruences or equivalences given particular ones.

4.1 Connections with Equalities in the Rough Context and Dynamic Extensions

Given a partial approximation space $S = \langle \underline{S}, R \rangle$, we can always construct an approximation space $S^m = \langle \underline{S}, R^m \rangle$, with R^m being the reflexive completion of R. It turns out that the possible concepts of rough equality on the respective spaces do not correspond in any natural way. This consequently bears on the semantics too. But again note that we can construct the greatest approximation space S^g contained in the given partial approximation space (This is the approximation space on the subset $\underline{T} = \{x : x \in \underline{S}, (x, x) \in R\}$ with the equivalence $R^g = R \cap T^2$). We will refer to the former approximation space as the *supra-approximation space* and the latter as the *infra-approximation space* corresponding to the partial approximation space S. By $=_m$ and $=_g$ we will mean the rough equalities induced by S^m and S^g respectively.

A sequence of approximation spaces ordered by a concept of subrelational structure is the primary structure of interest in dynamic rough contexts and temporal rough contexts. If $A = \langle \underline{A}, \rho \rangle$ and $B = \langle \underline{B}, \sigma \rangle$ are two approximation spaces, then let $A \lhd B$ if and only if $\underline{A} \subseteq \underline{B}$ and $\rho = \sigma \cap \underline{A}^2$. A single partial approximation space can serve as a representation of such a sequence. This order can be extended to partial approximation spaces as well. Unless we impose additional conditions on the combining process, the sequence in general cannot determine the partial approximation space uniquely. But this part is less important.

Theorem 8. *Every partial approximation space S uniquely determines an interval of approximation spaces under the $\lhd$-order. This interval is given by*

$$[S^g, S^m] = \{X : S^g, \lhd X \lhd S^m\}$$

Proof. By definition S^g and S^m are uniquely determined by S. These in turn uniquely determine the interval of approximation spaces. □

The converse question is of natural interest. But it is actually dependent on the procedure adopted. If we are given an approximation space and some sub structures of it then we can always define an associated partial approximation space. But that partial approximation will correspond to the interval of approximation spaces generated by the original set of approximation spaces. The following two theorems roughly specify the converse context. A more thorough study of the connections with dynamic versions of rough set theory and temporal extensions thereof will be considered in a separate paper by the present author.

Theorem 9. *Any pair of approximation spaces A, B satisfying $A \lhd B$ determines a partial approximation space S uniquely. In this situation A, B are respectively isomorphic (relation theoretically) to S^g and S^m respectively.*

Proof. Given $A = \langle \underline{A}, R_A \rangle$ and $B = \langle \underline{B}, R_B \rangle$ satisfying the subobject relationship, let $\underline{S} = \underline{B}$ and $R = R_B \setminus \Delta_{B \setminus A}$, where $\Delta_{B \setminus A} = \{(x, x) : x \in B \setminus A\}$, is the required partial approximation space.

The converse construction will obviously lead to A being relation morphically isomorphic to S^g and B to S^m respectively.

We consider the relationship by way of correspondences between the different types of equalities next.

Proposition 26. *If X, Y are approximation spaces such that $X \lhd Y$, then $A \approx B$ in Y implies $A \approx B$ in X.*

Proof. The proof is well known. In words the result says that if the means of distinguishing between entities is reduced then that which could not be distinguished before will remain indistinguishable. □

Proposition 27. *If A, B are two subsets of $\underline{S}$, such that $A \triangleq B$ in the partial approximation space S, then it is not necessary that they are roughly equal in the space S^m.*

Proof. Suppose in addition that $(\exists a \in (S \setminus A) \cap B)\, [a] \subset A \cap B$ in the partial approximation space S, then it will be the case that $A^l = B^l$ in S and $A^l \subset B^l$ in S^m. (In this situation the required models that satisfy $A^u = B^u$ and $A^{lu} = B^{lu}$ in S can be constructed, by assuming that no further exceptional sets exist). This will ensure $\neg(A =_m B)$. □

Theorem 10. *If A, B are two subsets of $\underline{S}$, such that $A \triangleq B$ in the partial approximation space S, then it is necessary that their intersection with the universe of S^g be roughly equal in the space S^g.*

Proof. If the underlying universe of S^g is $\underline{T} = \{x : x \in \underline{S}, (x,x) \in R\}$ with the equivalence relation being $R^g = R \cap T^2$, then let $A_T = A \cap \underline{T}$ and $B_T = B, \cap \underline{T}$. If $[x]$ is an essential class contained in A, then $[x] \cap \underline{T}$ is a class contained in A_T and similarly for B. Suppose $[z]$ is a maximal pseudoclass (in the usual inclusion order) contained in A with $z \notin A$, then the class generated by any member of $[z]$ (if any) in S^g will be contained in A. It is necessary that this $z \notin B$ and $[z] \subset B$ in S and the class generated by any member of $[z]$ in S^g will be in $[B]$ (else we will have a contradiction). This will ensure that $A_T^l = B_T^l$. Continuing the argument we have the result. □

The above result can be proved using fully featured sets and restrictions thereof. Avoiding it does make it nonconstructive.

Theorem 11. *If A, B are two subsets of $\underline{S}$, such that $A =_m B$ in the approximation space S^m, then it is not necessary that they are roughly pseudo equal in the space S.*

Proof. We can obtain a model affirming the situation by essentially expanding the lower approximation of one of the sets A in S with an extra pseudoclass $[z]$ (say) such that $z \notin A \cap B$ and $[z] \subset A$, while $[z] \not\subseteq B$. This will ensure that the lower approximations differ in the partial approximation space. In fact it is easy to show that if we are given a partial approximation space S and S^m, then we can find sets A, B with the required properties. □

5 Generalized Esoteric Covers

The method of generalized covers mentioned in the introduction can be adapted to the esoteric rough context in different ways. We consider the direct adaptation, a generalization thereof and formulate its main properties.

If we directly start from a partial approximation space S and take the collection of sets $\mathcal{K}$ to be the same as the collection of pseudo-classes, then the result proved in the section on generalized covers remains valid (provided the number of pseudoclasses is finite). In general the collection of all pseudoclasses $\mathcal{K}$ if finite in number need not be such that their union is S and neither need they be pairwise disjoint. If we take only the maximal pseudoclasses alone then also they need not be pairwise disjoint. For well-related partial approximation spaces however, it is the case that $\cup\mathcal{K} = S$.

Definition 24. *A partial approximation space will be said to* finitary *if and only if it has a finite number of pseudoclasses.*

As before if $X \subseteq S$, then consider the sets (with $K_0 = \emptyset$, $K_{n+1} = S$)

(i) $X^{l1} = \bigcup\{K_i : K_i \subseteq X, i \in \{0, 1, ..., n\}\}$

(ii) $X^{l2} = \bigcup\{\cap(S \setminus K_i) : \cap_I(S \setminus K_i) \subseteq X, I \subseteq \{1, ..., n+1\}\}$

(iii) $X^{u1} = \bigcap\{\cup_I K_i : X, \subseteq \cup_{i \in I} K_i I \subseteq \{1, ..., n+1\}\}$

(iv) $X^{u2} = \bigcap\{S \setminus K_i : X, \subseteq S \setminus K_i i \in \{0, ..., n\}\}$

The pair (X^{l1}, X^{u1}) will be called a *PAU-rough set* by union, while (X^{l2}, X^{u2}) will be called a *PAI-rough set* by intersection We will deal with the connections with the lower and upper approximations introduced in the esoteric approach.

Theorem 12. *In a finitary partial approximation space* $S = \rangle \underline{S}, R\langle$, *for any* $X \subseteq S$,

- $X^l = X^{l1}$
- $X^{u1} \subseteq X^u$, $X^{u1} \subseteq X^{u2}$
- $X^{l1} \subseteq X^{l2}$

Proof. – By our assumptions the definitions of X^l and X^{l1} are the same.

- The inequality of X^u and X^{u1} in general can be shown with a two element counterexample. Suppose $\{x, y\}$ is a set such that (x, x), (x, y), $(y, x) \in R$ alone then $\{x\}^u = \{x, y\}$, but $\{x\}^{u1} = \{x\}$. The $\subseteq$ part is set-theoretic. If $x \in X^{u1}$ then it is in all the unions of pseudoclasses containing X. So it will remain in the complements (of pseudoclasses) which, contain X . Possible equality can be contradicted by considering sets containing isolated singletons. The element that is not related to anything will be present in X^{u2}, but not in X^{u1}.
- The existence of an irreflexive element $x \in S \setminus X$ whose pseudoclass is contained in X, will cause $X^{l1} \subset X^{l2}$. Other types of exceptional sets do not disturb the equality of the two.

□

Theorem 13. *The following hold in any finitary partial approximation space S under the assumption that $\mathcal{K}$ is the set of pseudoclasses of S. X, Y being arbitrary subsets of S.*

(i) $X^{l1} \subseteq X \subset X^{u1} \subseteq X^{u2}$

(ii) $X^{l1} \subseteq X^{l2} \subseteq X \subseteq X^{u2}$

(iii) $\emptyset^{l1} = \emptyset^{l2} = \emptyset$

(iv) $(\cup\mathcal{K} = S \longrightarrow S^{u1} = S^{u2} = S)$

(v) $(\cup\mathcal{K} = S \longrightarrow \emptyset^{u2} = \emptyset,\ S^{l1} = S)$

(vi) $\emptyset^{u1} = \emptyset,\ S^{l2} = S$

(vii) $(X \cap Y)^{l1} = X^{l1} \cap Y^{l1},\ (X \cap Y)^{l2} = X^{l2} \cap Y^{l2}$

(viii) $(X \cup Y)^{u1} = X^{u1} \cup Y^{u1} \subseteq X^{u2} \cup Y^{u2} = (X \cup Y)^{u2}$

(ix) $(X \subseteq Y \longrightarrow X^{l1} \subseteq Y^{l1},\ X^{l2} \subseteq Y^{l2})$

(x) *If $\mathcal{K}$ is pairwise disjoint then* $(X \cap Y)^{l1} = X^{l1} \cap Y^{l1},\ (X \cup Y)^{u2} = X^{u2} \cup Y^{u2}$

(xi) $(X \subseteq Y \longrightarrow X^{u1} \subseteq Y^{u1},\ X^{u2} \subseteq Y^{u2})$

(xii) $X^{l1} \cup Y^{l1} \subseteq (X \cup Y)^{l1}$

(xiii) $X^{l2} \cup Y^{l2} \subseteq (X \cup Y)^{l2}$

(xiv) $(X \cap Y)^{u1} \subseteq X^{u1} \cap Y^{u1}$

(xv) $(X \cap Y)^{u2} \subseteq X^{u2} \cap Y^{u2}$

(xvi) $(S \setminus X)^{l1} = S \setminus X^{u2}$

(xvii) $(S \setminus X)^{l2} = S \setminus X^{u1}$

(xviii) $(S \setminus X)^{u1} = S \setminus X^{l2}$

(xix) $(S \setminus X)^{u2} = S \setminus X^{l1}$

(xx) $(X^{l1})^{l1} = X^{l1}$, $(X^{l2})^{l2} = X^{l2}$

(xxi) $(X^{u1})^{u1} = X^{u1}$, $(X^{u2})^{u2} = X^{u2}$

(xxii) $(X^{l1})^{u1} = X^{l1}$, $(X^{u2})^{l2} = X^{u2}$

(xxiii) $X^{l2} \subseteq (X^{l2})^{u2}$, $(X^{u1})^{l1} \subseteq X^{u1}$

(xxiv) $(\mathcal{K}_j^{\cap}(X))^{u2} = \mathcal{K}_j^{\cap}(X)$, $j = 1, 2, ..., t_1$

(xxv) $(\mathcal{K}_j^{\cup}(X))^{l1} = \mathcal{K}_j^{\cup}(X)$, $j - 1, 2, ..., t_2$

Proof. The theorem follows immediately from the proof of the above theorem and the theorem in the generalized cover section. □

It must be noted that the generalized covers approach crucially depends on the choice of the collection $\mathcal{K}$. Already when we take it as the collection of all pseudoclasses we have seen that the approximations are quite distinct from the approximations that we introduced except for $l1$ and the lower approximation.

Below we introduce a new way of using multiple collections to arrive at a rationalized approximation operations on partial approximation spaces. The associated heuristics may be seen as being essentially rough-set theoretical.

If L is a subset of pseudoclasses with the property that it is maximal with respect to being pairwise nondisjoint, then let $\cap L = L_0$ and $\cup L = L^+$. Since the collection $\mathcal{K}$ of pseudoclasses can be represented as a union of such maximal pseudo-classes, so it makes sense to consider new collections $\mathcal{L}$ of sets of the form L_0 and collections $\mathcal{M}$ of sets of the form L^+. The collections $\mathcal{L}$ and $\mathcal{M}$ are respectively pairwise disjoint. In fact, for each element $x \in S$, we can associate the empty set or an element of the collection $\mathcal{L}$ in a unique way. We can do the same with $\mathcal{M}$.

Definition 25. *The eight approximation operators generated respectively by the collections $\mathcal{L}$ and $\mathcal{M}$ will be termed* rationalized approximation operators.

In general these do not coincide with the approximation operators in the spaces S^g and S^m, but simulate an approximation-space like semantics over the partial approximation space. In conjunction with the operators generated from the collection of pseudoclasses we have an enhanced set of operators. It is here that some of the compositions of operators behave like *approximation operators over approximation spaces.* We will deal with the details of this in a separate paper [16].

6 Three Algebraic Semantics

A semantics of a generalized form of rough set theoretical reasoning need not necessarily be based on *the behaviour of the indistinguishable in the face of approximation operations on them.* Even classical rough set theory can be differently modelled (see [17] for example). We define more than three different algebraic semantics for the set theoretical semantics of esoteric rough set theory.

Definition 26. *A subset A of a partial approximation space S will be said to be an* almost definite set *if and only if it is the case that*

$$A^l = A^{lu} = A^u$$

Definition 27. *A subset A of a partial approximation space S will be said to be a* definite set *if and only if it is the case that*

$$A^l = A^{lu} = A^u = A.$$

Let $\mathbf{E}(S)$ and $\mathbf{F}(S)$ be the set of all almost definite and definite subsets of S. In general,

$$\mathbf{E}(S) \subseteq \mathbf{F}(S).$$

Theorem 14. *The set* $\mathbf{F}(S)$ *is endowable with a boolean algebra structure.*

Proof. For any $\alpha, \beta \in \mathbf{F}(S)$ if we let

$$\alpha \wedge \beta = \bigcup\{[x] : [x] \subseteq \alpha, [x] \subseteq \beta\}$$

$$\alpha \vee \beta = \bigcup\{[x] : [x] \subseteq \alpha \cup \beta\}$$

$$\alpha^c = \bigcup\{[x] : [x] \subseteq S \setminus \alpha\}$$

then the structure $\mathcal{F}(S) = \left\langle \underline{\mathbf{F}(S)}, \vee, \wedge, {}^c, 0, 1 \right\rangle$ is a boolean algebra, where $0 = \emptyset$ and $1 = 0^c$. The last operation is not a partial operation as the elements of $\mathbf{F}(S)$ are definite sets. Note that 1 need not coincide with S itself. The distributive lattice structure under the defined operations follows from purely set-theoretic considerations. □

Theorem 15. *The set* $\mathbf{E}(S)$ *is not necessarily a lattice under operations defined in the same way as in the above proof.*

Proof. It is possible that elements of $\mathbf{E}(S)$ contain isolated elements. For such subsets $\wedge$ and $\vee$ as defined above will not be lattice operations (idempotency will also fail). Counterexamples are easy to construct. □

Let $\underline{K}(S)$ be the set of all subsets of the form A^l, A^{lu} and A^u, A being a subset of $\underline{S}$ (S being the partial approximation space). Now let

$$\Pi(S) = \{(x, y) : x, y \in \underline{K}(S), x \subseteq y\}.$$

On $\Pi(S)$ we can define the following operations and relations:

- $(x, y) \leq (a, b)$ if and only if $x \subseteq a$ and $y \subseteq b$.
- $(a, b) \sqcap (x, y) \triangleq (a \cap x, b \cap y)$
- $(a, b) \sqcup (x, y) \triangleq (a \cup x, b \cup y)$
- $\sim (a, b) \triangleq (b^c, a^c)$ if defined.
- $L(a, b) \triangleq (a, a)$
- $0 \triangleq (\emptyset, \emptyset)$
- $T \triangleq (S^u, S^u)$
- $\neg(a, b) \triangleq (S^u \setminus b, S^u \setminus a)$
- $\Box(a, b) \triangleq (b, b)$

Definition 28. *The partial algebra*

$$\Pi(S) = \left\langle \underline{\Pi(S)}, \sqcap, \sqcup, \sim, \neg, L, \Box, 0, T, (2, 2, 1, 1, 1, 1, 0, 0) \right\rangle$$

will be said to be the esoteric definite algebra *corresponding to the partial approximation space S.*

A binary operation f is said to be *weakly commutative* if and only if, if both the terms $f^{\underline{S}}(x, y)$ and $f^{\underline{S}}(y, x)$ are defined, then the two are equal in value. $f^{\underline{S}}$ is the interpretation of the symbol f over S. This can also be written as $f^{\underline{S}}(x, y) \stackrel{w}{=} f^{\underline{S}}(y, x)$. $\rightsquigarrow$ is the directional equality; if the LHS is defined, then the RHS is and is equal to it. For more details refer to [13,18].

Theorem 16. *An esoteric definite algebra satisfies all of the following :*

(i) $\sqcup, \sqcap$ *are idempotent, commutative and associative operations.*

(ii) $(\sim x = a \longrightarrow \sim\sim x = x$

(iii) $(a \sqcup b) \sqcap a = a$

(iv) $(a \sqcap b) \sqcup a = a$

(v) $\neg\neg(x) = \neg(x)$

(vi) $\sim (a \sqcap b) \stackrel{w}{=} \sim (a) \sqcup \sim (b)$

(vii) $(a \sqcap b) \sqcup c = (a \sqcup c) \sqcap (b \sqcup c)$

(viii) $LL(x) = L(x)$

(ix) $\Box\Box(x) = \Box(x)$

(x) $\Box L(x) = L(x)$

(xi) $L(\Box(x)) = \Box(x)$

Proof. (i) Idempotency, commutativity and associativity are induced by the properties of the corresponding set theoretical operations on the components.

(ii) $\sim x = a$ in the premise essentially means that $\sim x$ is defined. The conclusion is ensured by the property of set-theoretic complements. It is easy to show that $\sim\sim x$ need not be equal to x always by considering isolated elements and their set complements.

(iii) Set intersection and union operation are being performed on each of the components in the $\sqcap$ and $\sqcup$ operations respectively. So the statement is essentially about absorption.

(iv) Same as the above.

(v) $\neg x$ is defined for any x in the first place as $S^u \setminus A^l$ for any set A will be representable as a set of the form X^l or X^u or X^{lu} and similarly for the upper and mixed approximations of A. The rest is obvious.

(vi) Suppose both sides of the weak inequality are defined, then the complements of a and b are of the required form and so also is the RHS. The equality is purely set theoretic property though.

(vii) This follows from the distribution of $\cup$ over $\cap$ and its dual.

(viii) If $x = (a, b)$, then $LL(x) = LL(a, b) = L(a, a) = (a, a)$, while $L(a, b) = (a, a)$.

(ix) If $x = (a, b)$, then $\Box\Box(x) = \Box\Box(a, b) = \Box(b, b) = (b, b)$, while $\Box(a, b) = (b, b)$.

(x) If $x = (a, b)$, then $\Box L(x) = \Box L(a, b) = \Box(a, a) = (a, a)$, while $L(a, b) = (a, a)$.

(xi) If $x = (a, b)$, then $L\Box(x) = L\Box(a, b) = L(b, b) = (b, b)$, while $\Box(a, b) = (b, b)$.

All this is in sharp contrast to the structure of definite sets in rough set theory. Note that the main differences are in the behaviour of negation and the top. The other point is that if we ignore sets of the form A^{lu} in the above considerations, then the structure improves. We can expect this to be a partial semantics of esoteric rough set theory. In case of rough sets, the corresponding structure is a topological quasi-boolean algebra (see [9]).

Theorem 17. *If S is an approximation space, then $\Pi(S)$ without the operations $\Box, \neg$ is isomorphic to the topological quasi boolean algebra generated by its definite elements.*

Proof. Under the conditions, the set $K(S)$ consists of only definite elements of the form A^u and A^l alone. So the result follows by the theorem proved in [9] [19].

Different concepts of rough equalities are definable for collections of subsets of a partial approximation space. These include the rough pseudo-equality defined in the earlier section and the ones definable on the basis of different types of esoteric covers of sets. We get different partial algebraic semantics of the version of esoteric rough set theory corresponding to these.

Let $\wp(\underline{S})$ be the powerset of the underlying set of a partial approximation space S. Then $\hat{=}$ is an equivalence relation on it. We can define different interesting operations on the quotient (see [16]).

Neo BZ-Lattices

Given a partial approximation space $\langle \underline{S}, R\rangle$, form its power set $\wp(S)$. Let $\neg((x,y) \in R)$ if and only if $(x, y) \in F$, then if R is a partial equivalence, then F is a partially reflexive and symmetric relation. For any set $H \subseteq S$, let

- $H^\circ = \{x;\, (\forall y \in H)\,(x, y) \in F\}$
- $L(H) = H^{c\circ\circ c}$
- $U(H) = H^{\circ\circ}$

Definition 29. *The algebra*

$$\mathcal{P} = \left\langle \underline{\wp(S)}, \vee, \wedge, {}^\circ, {}^c, l, u, L, U\, \emptyset, S, (2, 2, 1, 1, 1, 1, 1, 1 0, 0)\right\rangle$$

will be called a Neo BZ-Lattice. *The operations l, u are the lower and upper approximations due to R respectively.*

Theorem 18. $\mathcal{P} = \left\langle \underline{\wp(S)}, \vee, \wedge, {}^\circ, {}^c, l, u, L, U\, \emptyset, S\right\rangle$ *is of type* $(2, 2, 1, 1, 1, 1, 1, 1 0, 0)$ *and is boolean algebra with extra operations which satisfies all of the following :*

(i) $\left\langle \underline{\wp(S)}, \vee, \wedge, {}^c, \emptyset, S, (2, 2, 1, 0, 0)\right\rangle$ *is a boolean algebra.*

(ii) $H^\circ \subseteq H^{\circ\circ\circ}$

(iii) $(H \subseteq K \longrightarrow (K^\circ \setminus K^*) \subseteq .H^\circ)$, *where* $K^* = K \wedge K^\circ$.

(iv) $H^\circ \wedge K^\circ \subseteq (H \vee K)^\circ$

(v) $(H \wedge K)^\circ \subseteq H^\circ \vee K^\circ$

(vi) $H^\circ \setminus H^* \subseteq H^c$

Proof. (i) $\vee$ and $\wedge$ are the same as the usual set-theoretic operations of union and intersection.

(ii) If $(a, a) \in F$, $a \in H$ and $\{a\} \cup K$ is an isolated set for some $K \subseteq H^c$, then a is in H° and elements R-related to a cannot be in H°. But a subset of K will in general be included in H°. a will then be in $H^{\circ\circ}$. But if no part of K is included in $H^{\circ\circ}$, then $a \notin H^{\circ\circ}$. It is easy to construct examples. $H^\circ \subseteq H^{\circ\circ\circ}$ can be verified by considering fully featured sets.

Proposition 28. *In general for a subset H, $H^{\circ\circ}$ is not comparable with H.*

Proof. Let $x_0 \in H$ be such that

- $(\forall x \in H) \neg (x_0, x) \in R$
- $(x_0, x_0) \notin R$

Then it is that $x_0 \in H^\circ$, but it is possible that the element may or may not be in $H^{\circ\circ}$ (depending on how the elements related to x_0 are). The required counterexample is easy.

We can extend the theory to a more abstract level by essentially using a BZ-lattice abstraction as in [2]. But we have reservations on the generality and the bias inherent in the approach especially in case of similarity spaces [16].

An Abstract Algebraic Approach

The basic idea in this approach is to capture a set theory in the light of the truths offered by esoteric rough set theory. It differs severely from the Katrinak algebras approach in rough set theory, but is motivated by similar considerations. The intended valuation is into sets of tuples of the form (A^l, A^{lu}, A^u). But we need to refine this to restrict our considerations to unions of pseudo-classes alone. For this purpose we use the additional operation τ, (this essentially removes the pseudoclasses that are not contained within the set on which it is operating).

In this algebraization, our language $\mathfrak{L}$ of general rough set logic will consist of a nonempty set of propositional variables P, two binary connectives $\vee$, $\wedge$, three unary connectives $*$, $+$, $\flat$ and three constants $\mathbf{T}$, $\mathbf{T}_0$, $\mathbf{F}$ for truth. Formulas are constructible in the usual way, so that the set $\mathcal{F}(\mathfrak{L})$ of formulas is a free algebra of type $(2, 2, 1, 1, 1, 0, 0)$ generated over P. A model of $\mathfrak{L}$ then is a pair of the form (W, v), where W is a set and $v : P \mapsto \wp(W) \times \wp(W) \times \wp(W)$ is a valuation, such that if $v(p) = (A, B, C)$ then $A \subseteq B \subseteq C$. Further we will assume that an operation $\tau : \wp(W) \mapsto \wp(W)$ satisfying all of the following is given:

Inclusion $\tau(A) \subseteq A$

Idempotence $\tau(\tau(A)) = \tau(A)$

Monotonicity $(A \subseteq B \longrightarrow \tau(A) \subseteq \tau(B))$

Empty-set $\tau(\emptyset) = \emptyset$

$\perp$ $A^{\perp} = \tau(A^c)$, c being the complementation operation.

Given a model $\mathfrak{M} = (W, \upsilon)$, it's meaning function σ will be an extension of the valuation function $\sigma : \mathcal{F}(\mathfrak{L}) \longmapsto \wp(W) \times \wp(W) \times \wp(W)$ such that,

1. $\sigma(\mathbf{T}) = (\tau(W), \tau(W), \tau(W)) = 1$
2. $\sigma(\mathbf{T}_0) = (W, W, W) = 2$
3. $\sigma(\mathbf{F}) = (\emptyset), \emptyset, \emptyset) = 0$
4. $\forall p \in P\ \sigma(p) = \upsilon(p)$
5. If $\sigma(\varphi) = (A, B, C)$ and $\sigma(\psi) = (E, F, G)$, then
   - $\sigma(\varphi \wedge \psi) = (A \cap E, B \cap F, C \cap G)$
   - $\sigma(\varphi \vee \psi) = (A \cup E, B \cup F, C \cup G)$
   - $\sigma(\varphi^*) = (C^{\perp}, C^{\perp}, C^{\perp})$
   - $\sigma(\varphi^+) = (A^{\perp}, A^{\perp}, A^{\perp})$
   - $\sigma(\varphi^{\flat}) = (B^{\perp}, B^{\perp}, B^{\perp})$

Now on $Ran(\sigma) = \{\sigma(\varphi) : \varphi \in \mathcal{F}(\mathfrak{L})\}$, if we define the operations $\oplus, \cdot, {}^*, {}^+, {}^{\flat}$ via,

1. $\sigma(\varphi) \cdot \sigma(\psi) = \sigma(\varphi \wedge \psi)$
2. $\sigma(\varphi) \oplus \sigma(\psi) = \sigma(\varphi \vee \psi)$
3. $(\sigma(\varphi))^* = \sigma(\varphi^*)$
4. $(\sigma(\varphi))^+ = \sigma(\varphi^+)$
5. $(\sigma(\varphi)^{\flat} = \sigma(\varphi^{\flat}))$.

Theorem 19. *$Ran(\sigma)$ with the defined operations is an algebra satisfying all of the following :*

(i) $\langle Ran(\sigma), +, \cdot, 0, 1\rangle$ *is a bounded distributive lattice.*

(ii) $(a \leq b^* \longrightarrow (a \cdot b) = 0)$

(iii) $a^{***} = a^*$

(iv) $a^{+++} = a^+$

(v) $a^{\flat\flat} \cdot a = a$

(vi) $x^* \oplus x^{**} = 1$

(vii) $x^+ \cdot x^{++} = 0$

(viii) $(x^+ = y^+, x^* = y^*, x^\flat = y^\flat \longrightarrow x \oplus y^* = 1)$

(ix) $x^+ \cdot x^* = x^*$

(x) $2^* = 0$

(xi) $1^{**} = 1$

Proof. The following proof can be understood in a purely abstract way with no explicit reference to sets. We have retained the connection with sets for a more visual presentation of the proof.

(i) $\oplus$ and $\cdot$ are clearly distributive lattice operations on $Ran(\sigma)$. It is bounded by 0 and 2 and it is essential that 2 covers 1 (lattice-theoretically).
(ii) If $a \leq b^*$, then a is contained in the image of the complement (by τ) of the upper approximation of b, so that the *meet* $\cdot$ of a and b is the image 0 of the triple of empty sets.
(iii) a^{***} is obtained by three applications of an upper approximation followed by a complementation and then by the τ operation in order on the components. But τ essentially forms the union of the largest collection of pseudoclasses that are contained within the complement (component-wise). We are using most of the results regarding upper approximations that have been developed in the section on exceptional sets.
(iv) $a^{+++} = a^+$ can be proved in the same way as the above.
(v) $a^\flat$ will consist of pseudoclasses included in complement of the lu applications on the components of a. An application lu on the resulting components will have no effect. The complements of these will contain a component-wise. An application of the τ operation on the components will still contain a. So $a^{\flat\flat} \cdot a = a$.
(vi) x^* is a subset of the complement of the upper approximation of x. x^{**} is essentially the largest union of pseudoclasses contained in the complement of x^*. Now their disjunction ($\oplus$) will not contain the singleton isolated sets alone. This is precisely 1.

(vii) $x^+ \cdot x^{++} = 0$ follows by an argument similar to the one above.
(viii) If $(x^+ = y^+, x^* = y^*, x^\flat = y^\flat$, then obviously it does not mean that $x = y$. But given an arbitrary nonempty pseudoclass, it must be the case that it is either in x or y^*. So we have $x \oplus y^* = 1$.
(ix) x^* is included in x^+

(x) The complement of 2 is 0 and τ of that is still τ.
(xi) τ of the 2 is 1.

□

Definition 30. *$Ran(\sigma)$ along with the defined operations will be called an* esoteric 2SA algebra.

Definition 31. *An* abstract esoteric 2SA algebra *will be an algebra of the form*

$$A = \langle \underline{A}, \vee, \wedge, \flat, *, +, 0, 1, (2, 2, 1, 1, 1, 0, 0) \rangle$$

that satisfies all of the following:

(i) $Ran(\sigma), +, \cdot, 0, 1$ *is a bounded distributive lattice.*

(ii) $a \leq b^* \longrightarrow (a \cdot b) = \theta$

(iii) $a^{***} = a^*$

(iv) $a^{+++} = a^+$

(v) $a^{\flat\flat} \cdot a = a$

(vi) $x^* \oplus x^{**} = 1$

(vii) $x^+ \cdot x^{++} = 0$

(viii) $(x^+ = y^+, x^* = y^*, x^\flat = y^\flat \longrightarrow x \oplus y^* = 1)$

(ix) $x^+ \cdot x^* = x^*$

(x) $2^* = 0$

(xi) $1^{**} = 1$

Definition 32. *If $S = \langle \underline{S}, R \rangle$ and $F = \langle \underline{F}, H \rangle$ are two partial approximation spaces such that the associated lattices of pseudo-classes $\mathcal{E}^*(R)$ and $\mathcal{E}^*(H)$ are respectively isomorphic, then we will say that the partial approximation spaces are* p-equivalent.

These algebras are investigated in greater detail by the present author in [16]. In particular we prove an abstract representation theorem and that, if the esoteric 2SA algebras associated with S and F are isomorphic, then the two partial approximation spaces must be p-equivalent.

7 Examples

We consider some examples for the general esoteric rough set theory context in this section.

Example 1: Let $S = \{1, 2, 3, 4, 5\}$ be a set and

$$R = \{(1, 2), (2, 1), (3, 3), (4, 4), (5, 5), (4, 5), (5, 4)\}$$

a partial equivalence on it. Consider the subsets $A = \{2, 3\}$ and $F = \{1, 4\}$. We have,

- $F^l = \{1\}$, $F^{ll} = \{1\}$
- $F^{lu} = \{1\}$
- $F^u = \{1, 4, 5\}$, $F^{uu} = \{1, 4, 5\}$
- $A^l = \{2, 3\}$, $A^{ll} = \{2, 3\}$
- $A^u = \{2, 3\}$, $A^{uu} = \{2, 3\}$
- $A^{lu} = \{2, 3\}$

This shows that it is possible that sets like A can actually exist in particular. Note that the classes of elements of A are not contained in A.
Now consider S^m, here we have

- $A^l = \{3\}$ and $F^l = \emptyset$
- $A^u = \{1, 2, 3\}$ and $F^u = \{1, 2, 4, 5\}$

Example 2: Fishes in a Pond Suppose we have a pond of fishes of different type. Suppose also that our problem is to determine some biological information matrix relating to each of the fishes on a dynamic rule-evolution scheme. If the experimenter observes streams of fishes and stops after a certain *correct* time. If it is possible that the same fish is considered more than once and the attributes measured are subject to much imprecision, then we have a context for the present theory.

Example 3: Any VPRS or VPFRS context can be interpreted as a context involving partially reflexive relations. This is developed further in the next section.

Example 4: Contexts involving tolerance relations (similarity spaces) do not require the esoteric approach, but if we have problems of defining a minimal identification map for the set of elements under consideration, then the approach is certainly warranted. Suppose we have a context involving searching for a particular data set from a huge collection of data sets with limited computational resources (as of a needle in a haystack problem). Examples include detecting cancerous tissues in sequences of mammograms. Here the methods of esoteric rough set theory can be effectively used.

Example 5: In clustering of very large sets of documents available from multiple sources too, the theory can be effective. Often the same document may be found at multiple points. Rough set theory has been considered for document clustering in [20] for example.

Example 6: Let $S = \{1, 2, 3, 4, 5\}$ be a set and

$$R = \{(1, 2), (2, 1), (1, 1), (3, 3), (4, 4), (5, 5), (4, 5), (5, 4)\}$$

a partial equivalence on it. Consider the subsets $A = \{2, 3\}$ and $F = \{1, 4\}$. We have,

- $F^l = \{1\}$, $F^{ll} = \{1\}$
- $F^{lu} = \{1\}$
- $F^u = \{1, 4, 5\}$, $F^{uu} = \{1, 4, 5\}$
- $A^l = \{3\}$, $A^{ll} = \{3\}$
- $A^u = \{1, 2, 3\}$, $A^{uu} = \{1, 2, 3\}$
- $A^{lu} = \{3\}$

This shows that it is possible that sets like A can actually exist in particular. Note that the classes of elements of A are not contained in A.

Now consider S^m, here we have

- $A^l = \{3\}$ and $F^l = \emptyset$
- $A^u = \{1, 2, 3\}$ and $F^u = \{1, 2, 4, 5\}$

8 Esoteric Rough Set Theory, VPRS and VPRFS

What is actually done in VPRS and VPRFS is a process of trimming which results in simpler or more complex degree based elimination of elements from approximations (see [21,22,7,8]). In case of VPRFS of course the same trimming process is applied to (interpreted over) fuzzy rough sets. The main issue is of the inclusion of elements in the approximations with very low degree of membership in the different approximations. The problems of noise in the data come next. Our proposed solution for these consists in using preference orders on subsets in conjunction with esoteric rough set theory. In our approach we start from an approximation space together with a set of preferred subsets of it and derive a partial approximation space with an esoteric semantics for it. This way the problems relating to the application contexts of variable precision rough and rough-fuzzy set theory can be approached differently through esoteric rough set theory (ESRT).

In what follows we will develop the main transformation process for the VPRS context. It should be noted that it is possible to arrive at a suitable esoteric semantics without sufficient information about the membership functions too.

Let $S = \langle \underline{S}, R \rangle$ be an approximation space with R being an equivalence relation on $\underline{S}$. Let A, B be crisp subsets of S, then $A \overset{\beta}{\subseteq} B$ if and only if $e(A, B) = \beta$, where

$$e(A, B) = 1 - \frac{o(A \cap B)}{o(A)}.$$

In more general forms of VPRS two numbers a, b are selected under $0 \leq a \leq b \leq 1$ and the b-*lower* and a-*upper* approximations are defined via

$$A^{l_b} = \{x \,;\, [x] \overset{1-b}{\subseteq} A\}$$

and

$$A^{u_a} = \{x \,;\, e([x]_R, A) \leq 1 - a\}.$$

Given this we will require a set ζ of subsets of S to be specified as a set of preferred sets. This procedure is not part of the different VPRS variants. Given the set of preferred sets, we can construct a new partially reflexive, weakly transitive and symmetric relation through certain rules. We will refer to these rules as the *Modified VPR Trimming Rules* :

- $((\forall A \in \zeta)\, [x]_R \overset{1-b}{\nsubseteq} A) \Leftrightarrow (x, x) \notin R^+$
- $((\forall A \in \zeta)\, [x]_R \overset{1-b}{\nsubseteq} A), (x, y) \in R \Rightarrow (x, y), (y, x) \notin R^+$
- R^+ is the largest subset of R also satisfying symmetry and weak transitivity

Definition 33. *Given an approximation space and a set of preferred sets ζ, the relation R^+ will be called* the relation determined by ζ

Definition 34. *In the above context if ζ_1 and ζ_2 are sets of preferred sets satisfying $A \in \zeta_1$ then there is a $B \in \zeta_2$ such that $B \subseteq A$, then ζ_2 will be said to be* pre-finer *than ζ_1. Note that we assume nothing about ζ being a partition.*

Proposition 29. *In the above context $S_+ = \langle \underline{S}, R^+ \rangle$ is a partial approximation space.*

Proof. Clearly R^+ is a partially reflexive relation by construction. Weak transitivity is ensured by the last step. So S_+ is a partial approximation space. □

What is the nature of a preferred set in the context(s) ?

A preferred set is a set which, we expect to be more severely influenced by classes with low degrees of membership in it. It is possible to involve expert heuristics in the situation. In case of the VPRS and VPRFS approaches such a provision is not admissible. The existence of very nice measures is presumed in the VPRS and to a lesser extent in the VPRFS case.

Instead of using the above procedure of modifying VPRS and VPRFS, we can instead modify a given approximation space in the following way. This will be called the *Generatively Trimmed VPRS* procedure.

GTVPRS We start with an approximation space $S = \langle \underline{S}, R \rangle$ and a collection of subsets $\mathcal{F}$ of $\underline{S}$.

Definition 35. *For a subset X of $\underline{S}$ it's* cautious upper approximation *by $\mathcal{F}$ will be the set*

$$X^{uf} = \bigcup\{[x] : \exists K \in \mathcal{F},\, [x] \subseteq K,\, [x] \cap X \neq \emptyset,\, x \in \underline{S}\}.$$

The collection of all cautious upper approximations of subsets of S will be denoted by $UC(S)$. This cautious upper approximation may not include the original set X, but will hopefully intersect it !

Definition 36. *For a subset X of $\underline{S}$ it's* cocautious upper approximation *by $\mathcal{F}$ will be the set*

$$X^{ue} = \bigcup\{[x] : \forall K \in \mathcal{F},\, [x] \nsubseteq K,\, [x] \cap X \neq \emptyset\}.$$

The collection of all cocautious upper approximations of subsets of S will be denoted by $UCO(S)$.

Definition 37. *For a subset X of $\underline{S}$ it's* cautious lower approximation *by $\mathcal{F}$ will be the set*

$$X^{lf} = \bigcup\{[x] : \exists K \in \mathcal{F},\, [x] \subseteq K,\, [x] \subseteq X\ x \in \underline{S}\}.$$

The collection of all cautious lower approximations of subsets of S will be denoted by $LC(S)$.

Definition 38. *For a subset X of $\underline{S}$ it's* cocautious lower approximation *by $\mathcal{F}$ will be the set*

$$X^{le} = \bigcup\{[x] : \forall K \in \mathcal{F},\, [x] \nsubseteq K,\, [x] \subseteq X\}.$$

The collection of all cocautious lower approximations of subsets of S will be denoted by $LCO(S)$.

The following definition is concrete in it's dependence on the whole process.

Definition 39. *By a* generatively cautiously trimmed variable precision rough space *$GCTVPRS$, we will mean the tuple*

$$S^{+} = \langle \underline{S},\, LC(S),\, UC(S),\, \mathcal{F}\rangle\,.$$

Definition 40. *By a* generative cocautiously trimmed variable precision rough space *$GCCTVPRS$, we will mean the tuple*

$$S^{co} = \langle \underline{S},\, LCO(S),\, UCO(S),\, \mathcal{F}\rangle\,.$$

Now given the above definition sequence we can formulate the trimming portion of the construction as follows :

(i) If $S = \langle \underline{S},\, R\rangle$ is an approximation space and $\mathcal{F}$ of $\underline{S}$, a collection of subsets characterized by a concept of being *R-well-defined* or *ill-defined* or *preferred.*

(ii) Depending on the type that we select in the previous step we will define our concept of *generatively trimmed rough set*. If it is R-well-defined or preferred, then any tuple of the form $\langle A, A^{lf}, A^{uf} \rangle$ with A being any subset of $\underline{S}$ will be such a set.

(iii) A *full generatively trimmed rough set* will be a tuple of the form

$$\langle A, A^{lf}, A^{uf}, A^{le}, A^{ue} \rangle$$

with A being any subset of $\underline{S}$.

Given the above we can proceed by a reconstruction procedure which uses the esoteric rough set theory mechanism or attempt a direct semantic approach. The co-cautious approach apparently has the edge in application contexts, while from the algebraization viewpoint, they are similar.

Esoteric Reconstruction Procedure

In this the basic aim is to obtain esoteric rough set semantics from the trimmed structures. Clearly this is always possible and we have some choice here.

Definition 41. *By the* LC-Respace *we will mean the structure*

$$K(S^{+l}) = \langle \underline{S}, P \rangle$$

with P being a new relation defined by the following

(i) From $LC(S)$ form the collection of all minimal intersections $LC_m(S)$ and maximal unions $LC_M(S)$ and similarly from $UC(S)$, form $UC_m(S)$ and $UC_M(S)$ respectively.

(ii) $x \in \bigcup LC_m(S)$ if and only if $(x, x) \in P$

(iii) If $x, y, z \in \bigcup LC_m(S)$ and $(\exists B \in LC_m(S))\, x, y, z \in B$ then $(x, y), (y, z), (y, x) \in P$

(iv) If $x, y \in \bigcup LC_m(S)$, $(\exists B \in LC(S))\, x, y, z \in B$ and $z \notin \bigcup LC_m(S)$ then $(x, z), (z, x) \notin P$ and $(x, y) \in P$.

Proposition 30. *Any LC-Respace is a partial approximation space.*

Proof. $LC_m(S)$ is the collection of minimal intersections generated by $LC(S)$. The second condition ensures the partial reflexivity of P. Symmetry is also immediate from the form of the definition. For weak transitivity, note that the elements of $LC(S)$ are all unions of classes, while $LC_m(S)$ is also similar. □

Definition 42. *By the* UC-Respace *we will mean the structure*

$$K(S^{+u}) = \langle \underline{S}, Q \rangle$$

with Q being a new relation defined by the following :

(i) $x \in \bigcup UC_m(S)$ *if and only if* (x, x) *in* Q

(ii)

$$\text{If } x,\, y,\, z \in \bigcup UC_m(S) \text{ and } (\exists B \in UC_m(S))x,\, y,\, z \in B$$

$$\text{then}(x,\, y)\,(y,\, z),\, (y,\, x) \in Q$$

(iii)

$$\text{If } x,\, y \in \bigcup UC_m(S),\ (\exists B \in UC(S))\, x,\, y,\, z \in B \text{ and } z \notin \bigcup UC_m(S)$$

$$\text{then } (x,\, z),\, (z,\, x) \notin Q \text{ and } (x, y) \in Q.$$

Proposition 31. *Any UC-Respace is a partial approximation space.*

Proof. $UC_m(S)$ is the collection of minimal intersections generated by $UC(S)$. The second condition ensures the partial reflexivity of Q. Symmetry is also immediate from the form of the definition. For weak transitivity, note that the elements of $UC(S)$ are all unions of classes, while $UC_m(S)$ is also similar. □

Definition 43. *By the* ULC-Respace *we will mean the structure*

$$K(S^{+ul}) = \langle \underline{S},\, P \rangle$$

with P *being a new relation defined by the following :*

(i) $x \in \bigcup LC_m(S)$ *if and only if* (x, x) *in* P
(ii) If $x,\, y,\, z \in \bigcup UC_m(S)$ *and* $(\exists B \in UC_m(S))x,\, y,\, z \in B$ *then* $(x,\, y)\,(y,\, z),\, (y,\, x) \in P$
(iii) If $x,\, y \in \bigcup LC_m(S)$, $(\exists B \in LC(S))\, x,\, y,\, z \in B$ *and* $z \notin \bigcup LC_m(S)$ *then* $(x,\, z),\, (z,\, x) \notin P$ *and* $(x, y) \in P$.

Proposition 32. *Any ULC-Respace is a partial approximation space.*

Proof. The second condition ensures the partial reflexivity of P. Symmetry is also immediate from the form of the definition. For weak transitivity, note that the elements of $UC(S)$ and $LC(S)$ are all unions of classes, while $UC_m(S)$ and $LC_m(S)$ are similar. □

Remark 2. These will not be extensible to the original approximation space in general. A fuller characterization of this will be considered separately. The notions of cocautiously trimmed spaces yield quite different partial approximation spaces.

A third procedure for dealing with VPRS and VPRFS is by obtaining partially ordered partial approximation spaces directly. The associated computational aspects maybe difficult.

9 Relativised Approximations

We introduce different new concepts of relativised rough approximations with some quasi-inductive import in the following. Important questions on the connections with these relativised approximations arise naturally in the above context. Computationally the relativised approach may prove difficult if used directly.

Definition 44. *Let A, B be two subsets of a partial approximation space S then by the B-*lower approximation *of A we will mean the set*

$$A^l_B = \bigcup\{[x] : [x] \subseteq A, x \in B\}$$

Definition 45. *For a subset A of a partial approximation space S by a* lower plus approximation *(LPA) of A, we will mean a set A^{l+} for which*

$$A^{l+} = \bigcup\{[x] : [x] \subseteq A, x \in A^{l+}\}$$

holds.

Definition 46. *For a subset A of a partial approximation space S by the* lower minus approximation *(LM0) of A, we will mean the set*

$$A^{l-} = \bigcup\{[x] : [x] \subseteq A^{l-}, x \in A\}.$$

Proposition 33. *(i) If S is an approximation space then for a subset A, $A^l_A = A^l$.*

(ii) If A is a subset of a partial approximation space S, then generally $A^l_A \subseteq A^l$.

(iii) If in a partial approximation space S, $A^l_A = A^l$, then there exist no exceptional sets for A of type $\frac{2}{1}$, $\frac{2}{3}$, $\frac{3}{2}$ and $\frac{3}{3}$ alone.

Proof. (i) Obviously the two are identical as classes of elements contained in A cannot be generated by elements outside A.

(ii) For example, if $\{x, y\}$ is a $\frac{2}{1}$-exceptional set for A, then A^l_A will not contain x, while A^l will contain the element x. So in general $A^l_A \subseteq A^l$.

(iii) This is the full proof of how $A^l_A \subset A^l$ can happen. It can be deduced from the definition. Other types of exceptional sets will not affect the relation between A^l and A^l_A.

□

Proposition 34. *The following are all true for a subset A of a partial approximation space S:*

(i) $A^{l+} \subseteq A^l_A$.

(ii) $A^{l-} \subseteq A^l_A$

(iii) $A^{l-} \subseteq A^l$

Proof. The construction of A^{l+} is well-defined and contradictions must be actively used in determining it. Suppose we have a $x \in S \setminus A$ such that $[x] \subset A$, then $[x] \nsubseteq A^{l+}$. If $[y] \subseteq A^{l+}$, then $[y] \subseteq A$, and $y \in A$, so $[y] \subseteq A^{l}_{A}$. This proves the first assertion.

The proofs of the other parts follow from the definitions and the required counterexamples (for possible strict inclusion) can be generated directly using the different types of exceptional sets. For example note that if $\{j_1, j_2, j_3\}$ is an isolated set such that $j_1 \in A$ and $j_2, j_3 \in S \setminus A$ and $((j_3, j_3) \in R$, $((j_i, j_k) \in R$ only when $i \neq k$. Then $j_1, j_2 \notin A^{l-}$, but $j_1, j_2 \in A^{l}$.

Proposition 35. *The following hold for a subset A of an approximation space S:*

(i) $A^{l-} = A^{u}$.

(ii) $A^{l+} = A^{l}$

Proof. (i) The class $[x]$ generated by an element $x \in A$ will be a subset of A^{l-} and A^{u} and both of them consist of precisely the union of such classes.
(ii) A^{l+} is also a form of B-lower approximation with $B = S$ in approximation spaces.

□

Proposition 36. *Given a subset A of a partial approximation space, there exists a least subset* A^{*}, *such that* $A^{u}_{A^{*}} = A^{u}$ *and a least subset* A^{+} *such that* $A^{l}_{A^{+}} = A^{l}$.

Proof. By Zorn's Lemma or any other form of transfinite induction.

In subjective terms reducts are minimal sets of attributes that preserve the quality of classification. These are defined in the context of decision tables which, in turn have a dual relationship with approximation spaces. The main problem is in getting at good scalable algorithms for the computation of the different types of reducts (or supersets that are close to them). In actual applications the appropriate concept of reduct is to be decided from the context especially when the size of the decision table necessitates the use of approximate reducts (see [23,24]).

Different modified concepts of decision tables may be defined as a proper generalization of the corresponding concept for approximation spaces. Using an additional distinguished attribute for example, we can redefine the indiscernibility relation generated. This way we can get the required partial reflexivity. Another possibility is to proceed through the decision tables corresponding to the supra and infra approximation spaces associated. A useful version is

Definition 47. *By a* p-decision table *we mean a tuple of the form*

$$T = \langle \mathcal{O}, AT \cup \{\delta, \tau\} \rangle$$

with $\mathcal{O}$ *being a set of objects,* AT *a set of attributes,* δ *a decision* ($\delta \in AT$) *and* $\tau \in AT$ *being a distinguished element (possibly an extra dummy element).*

Moreover each object a must map $\mathcal{O}$ to the value set V_a, τ being the sole partial map. Atomic descriptors will be attribute-value pairs. The partially reflexive indiscernibility relation $\mathcal{I}$ generated by a subset X of attributes can then be defined via

$$\mathcal{I}(X) = \{(x, y) \in \mathcal{O}^2 \,;\, \forall a \in X(x \neq y \rightarrow a(x) = a(y), \text{ or } \tau(x) \text{ and } x = y\}$$

Algorithms using relative approximations have also been considered by the present author. These will appear separately.

Problems over partial approximation space also lead to relative approximations. This is the case when we need to consider the partial approximations given an approximation perspective determined by equivalences or tolerances. This aspect is considered along with an application of rough sets to Bayesian belief networks in [5]

10 Conclusion

In this research we have developed a theory of esoteric rough sets for contexts involving problems with self identification of elements and the contexts of VPRS and VPFRS. It is also shown that sequences of rough sets that occur in dynamic contexts can be approached through the methods developed. We have also proposed three different algebraic semantics for the theory. The proposed algebraic semantics will be of natural interest in the development of associated logics.

A partial approximation space in a real context actually ends up capturing situations of the form: *we know that two things are indistinguishable, but at the same time one of them is not indistinguishable from others in the same way as the other.* It can be argued that this amounts to admitting a dialethic or paraconsistent scheme of interpretation at the meta-level. From a classical perspective of discernibility this can still be resolved by claiming that the concept of discernibility afforded by partial approximation spaces is itself distinct from the one provided by approximation spaces. So the logics that can be associated can be quite different in semantic content.

The importance of dealing with sequences of rough sets in a unified way under different conditions is implicit in the proposed model for dealing with layered learning in [25]. The methods developed in the section on dynamic extensions need to be explored in this light too. The connections with similarity spaces is another aspect to be investigated in more detail. Some problems on reducts and variations thereof are also being investigated through relative approximations and esoteric methods by the present author.

Acknowledgements

I would like to thank the anonymous referee for useful suggestions that lead to considerable improvement of the research paper and its readability.

References

1. Komorowski, J., Pawlak, Z., Polkowski, L., Skowron, A.: Rough sets – a tutorial. In: Pal, S.K. (ed.) Rough Fuzzy Hybridization, pp. 3–98. Springer, Heidelberg (1999)
2. Cattaneo, G., Ciucci, D.: Algebras for rough sets and fuzzy logics. In: Peters, J.F., Skowron, A., Grzymała-Busse, J.W., Kostek, B.z., Świniarski, R.W., Szczuka, M. (eds.) Transactions on Rough Sets 2. LNCS, vol. 3100, pp. 208–252. Springer, Heidelberg (2004)
3. Inuiguchi, M.: Generalisation of rough sets and rule extraction. In: Peters, J.F., Skowron, A., Grzymała-Busse, J.W., Kostek, B.z., Świniarski, R.W., Szczuka, M. (eds.) Transactions on Rough Sets I. LNCS, vol. 3100, pp. 96–116. Springer, Heidelberg (2004)
4. Mani, A.: Rough equalities from posets and rough difference orders. Fundamenta Informaticae 53, 321–333 (2002)
5. Mani, A.: Relativised generalized rough sets and bayesian belief networks (to be Submitted, 2007)
6. Lin, T.Y., Yao, Y.Y., Wong, S.K.: A review of rough set models. In: Lin, Y.Y., et al. (eds.) Analysis of Information databases,Rough Sets and Data Mining, Kluwer Academic Publishers, Dordrecht (1997)
7. Ziarko, W.: Variable precision rough set model. J. of Computer and System Sciences 46, 39–59 (1993)
8. Rolka, A.M., Rolka, L.: Variable precision fuzzy rough sets. In: Peters, J.F., Skowron, A., Grzymała-Busse, J.W., Kostek, B.z., Świniarski, R.W., Szczuka, M. (eds.) Transactions on Rough Sets I. LNCS, vol. 3100, pp. 144–160. Springer, Heidelberg (2004)
9. Banerjee, M., Chakraborty, M.K.: Algebras from rough sets – an overview. In: Pal, S.K., et al. (eds.) Rough-Neural Computing, pp. 157–184. Springer, Heidelberg (2004)
10. Yao, G.T.: Constructions and algebraic methods of the theory of rough sets. Information Science 109, 21–47 (1998)
11. Düntsch, I.: Rough sets and algebras of relations. In: Orlowska, E. (ed.) Incomplete Information and Rough Set Analysis, pp. 109–119. Physica, Heidelberg (1998)
12. Katrinak, T.: Construction of regular double p-algebras. Bull. Roy. Soc. Sci. Liege 43, 294–301 (1974)
13. Burmeister, P.: A Model-Theoretic Oriented approach to Partial Algebras. Akademie-Verlag (1986) (2002)
14. Vojvodic, G.: A note on weak partial congruence algebras. Rev. of Res. Fac. Sci. Univ of Novisad 22, 89–94 (1992)
15. Seselja, B., Vojvodic, G.: On the lattice of weak congruence relations. Algebra Universalis 25, 121–130 (1988)
16. Mani, A.: Esoteric rough algebras, p. 24 (submitted, 2006)
17. Mani, A.: Super rough semantics. Fundamenta Informaticae 65, 249–261 (2005)
18. Ljapin, E.S.: Partial algebras and their applications. Academic, Kluwer (1996)
19. Banerjee, M., Chakraborty, M.K.: Rough sets through algebraic logic. Fundamenta Informaticae 28, 211–221 (1996)
20. Bleyberg, M., Elumalai, A.: Using rough sets to construct sense type decision trees for text categorisation. Transactions of IEEE, 19–24 (2001)
21. Katzberg, T., Ziarko, W.: Variable precision extension of rough sets. Fundamenta Informaticae 27, 1–12 (1996)

22. Pawlak, Z., Skowron, A.: Rough membership functions. In: Yager, R., et al. (eds.) Advances in Dempster-Schafer Theory of Evidence, Wiley, Chichester (1994)
23. Kryszkiewicz, M., Cichon, K.: Towards scalable algorithms for discovering rough set reducts. In: Peters, J.F., Skowron, A. (eds.) Transactions of Rough Sets-1, vol. LNCS-3100, pp. 120–143. Springer, Heidelberg (2004)
24. Bazan, J., Nguyen, S., Nguyen, H., Synak, P., Wroblewski, J.: Rough set algorithms in classification problem. In: Polkowski, L., et al. (eds.) Rough Set Methods and Applications, pp. 49–88. Physica Verlag, Heidelberg (2000)
25. Bazan, J., Nguyen, S.H., Skowron, A., Szczuka, M.: A view on rough set concept approximations. In: Wang, G., Liu, Q., Yao, Y., Skowron, A. (eds.) RSFDGrC 2003. LNCS (LNAI), vol. 2639, pp. 181–188. Springer, Heidelberg (2003)

Domain Knowledge Assimilation by Learning Complex Concepts

Tuan Trung Nguyen

Polish-Japanese Institute of Information Technology
ul. Koszykowa 86, 02-008 Warsaw, Poland
nttrung@pjwstk.edu.pl

Abstract. Domain, or background, knowledge has proven to be a key component in the development of high-performance classification systems, especially when the objects of interest exhibit complex internal structures, as in the case of images, time series data or action plans. This knowledge usually comes in extrinsic forms such as human expert advices, often contains complex concepts expressed in quasi-natural descriptive languages and need to be assimilated by the classification system. This paper presents a framework for the assimilation of such knowledge, equivalent to matching different ontologies of complex concepts, using rough mereology theory and rough set methods. We show how this framework allows a learning system to acquire complex, highly structured concepts from an external expert in an intuitive and fully interactive manner. We also argue the needs to focus on expert's knowledge elicited from outlier or novel samples, which we deem have a crucial impact on the classification process. Experiment results show that the proposed methods work well on a large collection of handwritten digits, though they are by no means limited to this particular type of data.

Keywords: Rough mereology, concept approximation, ontology matching, handwritten digit recognition, outlier samples.

1 Introduction

A machine learning problem can be viewed as a search within a space of hypotheses H for a hypothesis h that best fits a set of training samples T. Amongst the most popular approaches to such problems are e.g. statistical learning, decision trees, neural networks or genetic algorithms, commonly referred to as *inductive* learning methods, i.e. methods that generalize from observed training examples by finding features that empirically distinguish positive from negative training examples. Though these methods allow for highly effective learning systems, there often exist proven bounds on the performance of the classifiers they can construct, especially when the samples involved exhibit complex internal structures, such as optical characters, facial images or time series data. It is believed that *analytical* learning methods based on structural analysis of training examples are more suitable in dealing with such samples. In practice, best performances are obtained using a combination of the two learning methods [1].

J.F. Peters and A. Skowron (Eds.): Transactions on Rough Sets VIII, LNCS 5084, pp. 224–236, 2008.

An analytical learning algorithm, in addition to the training set T and a hypothesis space H assume a *domain theory* D which carries prior knowledge about the samples being learned. The search is now for a hypothesis h that best fits T and at the same time conforms to D. In other words, a background or domain knowledge is available to the learning system and may help facilitate the search for the target hypothesis. One of the widely used approach to analytical learning is the *Explanation Based Learning* (EBL) method, which uses specific training examples to analyze, or explain, which features are relevant or irrelevant to the target classification function. The explanations therefore can serve as *search control knowledge* by establishing initial search points or by subsequent altering search directions. Domain (or background) knowledge can serve as additional search control tools. Usually fast and efficient greedy searches have limits in the patterns they can discover, while complex and more elaborated, more exhaustive strategies typically display high computational costs. The trade-off between the two groups might be greatly refined with domain knowledge in order to steer the search process to more promising areas more quickly or to fine tune the construction of components patterns that would be difficult to find greedily.

In this paper, we investigate an architecture in which the explanation comes from an external, possibly human, expert. Moreover, the explanations will not come as *a priori*, but will be provided by the expert in a two way dialog along with the evolution of the learning system. It is worthy to note that while humans sometimes may not be able to explicitly explain how they perform certain tasks, they often find it easy to correct things that "went wrong" on specific examples. Incorporating this knowledge into the learning process is an effective way to improve its overall performance. Learning from external domain knowledge sources constitutes an integral part of the intensively pursued research over Knowledge-rich Data Mining, as stipulated in, e.g. [2]

One of the first and major challenges of this approach is that the knowledge employed by the external expert is often expressed in a descriptive language, called a *foreign language* L_f, which may contain natural language constructs, like, for example *"Ed has a square face"* or *"The Sun is in eclipse"*. This language is usually alien to the learning system, which has its own knowledge encoded using a different *domestic language* L_d describing, for example, physical sensor measurements. This is because the expert and the system have different knowledge *ontologies*, meaning they rely on different *concepts* and *relations* [3]. An *ontology matching*, i.e. a mapping between concepts and, in a further step, relations used by the expert and the learning system is needed.

The expert knowledge ontology, similarly to the samples to which it applies, will be highly structured. More specifically, it has the form of a *lattice*, or acyclic tangled trees of concepts, representing different aspects of the expert's perception about training samples. One can view these concepts as abstract information granules which, together with binding relations amongst them, form the expert's reasoning about the samples. These concepts and, in a further steps, their binding relations have to be translated, or in other words, *approximated* by the learning system by means of its domestic expressions. Examples:

- *SquareFace*(Ed)≡ (Ed.getFace().Width - Ed.getFace().Height $\leq$ 2.0 cm)
- *IsEclipse*(p)≡ (s=p.GetSun())$\wedge$(m=p.GetMoon())$\wedge$(s$\cap$m.Area$\geq$ s.Area$\cdot$0.9)

A key issue is that although the concepts and relations get approximated, their hierarchical structure remains intact in translation. This aims to allow parent concepts be approximated using the approximations of children concepts, essentially building an *approximate reasoning scheme.* We will show how this multi layered approximation can be performed using rough inclusion measures, rough set decision rules and how to ensure the quality of approximation using tools based on rough mereology theory.

The principal requirements for a satisfactory translation/approximation at a given level in the hierarchy are the following:

- A flexible matching of a variations of similar domestic patterns to a foreign concept, i.e. the translation result should not be a single patterns, but rather a collection or cluster of patterns.
- It should find approximations for the foreign concepts and relations, while preserving their hierarchical structure. In other words, inherent structure of the provided knowledge should be passed intact.
- Robustness, which means it is proof to noisy input data and incidental underperformance of approximation on lower levels
- Stability, which guarantees that any input pattern matching concepts on a lower level to a satisfactory degree will result in a satisfactory target pattern on the next level.

The expert's advices are based, in a natural way, on his perception on training samples. Human perception and behavior are subject of extensive research of Cognitive Science [4]. We will discuss resemblances and common points of interest between complex concepts' approximation and popular cognitive architectures.

Another important issue is the focus we place on the analysis of atypical, or outlier samples. Recent developments in pattern recognition clearly indicate they are crucial to search refining. They allow to better understand the interclass dependencies of the sample collection and help to steer the search process through vital points in the search spaces. Together with the explanation based learning approach these outliers, borderline samples often prove to be key in forming effective domain reasoning schemes.

2 Knowledge Elicitation from External Expert

We assume an architecture that allows a learning recognition system to consult a human expert for advices on how to analyze a particular sample or a set of samples. Typically this is done in an iterative process, with the system subsequently incorporating knowledge elicited on samples that could not be properly classified in previous attempts.

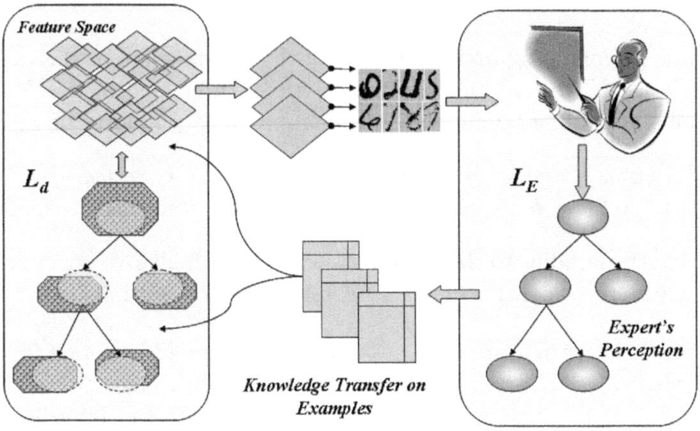

Fig. 1. System's Overview

2.1 Ontology Matching

The knowledge on training samples that comes from an expert obviously reflects his perception about the samples. The language used to describe this knowledge is a component of the expert's ontology which is an integral part of his perception. In a broad view, an ontology consists of a vocabulary, a set of concepts organized in some kind of structures, and a set of binding relations amongst those concepts [3]. We assume that the expert's ontology when reasoning about complex structured samples will have the form of a multi-layered hierarchy, or a lattice, of concepts. A concept on a higher level will be synthesized from its children concepts and their binding relations. The reasoning thus proceeds from the most primitive notions at the lowest levels and work bottom-up towards more complex concepts at higher levels.

As the human perception is inherently tolerant to variation and deviation, concepts and relations in his ontology are approximate by design. To use the terms of granular computing, they are information granules that encapsulate the autonomous yet interdependent aspects of human perception.

The knowledge elicitation process assumes that samples for which the learning system deems it needs additional explanations are submitted to the expert, which returns not only their correct class identity, but also an explanation on *why*, and perhaps more importantly, *how* he arrived at his decision. This explanation is passed in the form of a rule:

$$[CLASS(u) = k] \equiv \Im(EFeature_1(u), ..., EFeature_n(u))$$

where $EFeature_i$ represents the expert's perception of some characteristics of the sample u, while synthesis operator $\Im$ represents his perception of some relations between these characteristics. In a broader view, $\Im$ constitutes of a *relational structure* that encompasses the hierarchy of experts' concepts expressed by $EFeature_i$.

The ontology matching aims to translate the components of the expert's ontology, such as $EFeature_i$ and binding relations embedded in the $\Im$ structure, expressed in the foreign language L_f, which may have the form of, e.g.

"A six is a digit that has a closed belly below a slanted neck."
$[CLASS(u) =$'6'$] \equiv a, b$ are parts of u; "Below"(b,a); "SStroke"(a);"CBelly"(b)

into the patterns familiar to the learning system, which involve, e.g. pixels counting or calculations of density or mass center of pixel collections.

Single concept approximation. A foreign concept C is approximated by a domestic pattern (or a set of patterns) p in term of a rough inclusion measure $Match(p, C) \in [0, 1]$. The measure tells how well the patterns describe the subset of samples which fit the concept. Such measures take root in the theory of rough mereology [5], and are designed to deal with the notion of inclusion to a degree. An example of concept inclusion measures would be:

$$Match(p, C) = \frac{|\{u \in T : Found(p, u) \wedge Fit(C, u)\}|}{|\{u \in T : Fit(C, u)\}|}$$

where T is a common set of samples used by both the system and the expert to communicate with each other on the nature of expert's concepts, $Found(p, u)$ means a pattern p is present in u and $Fit(C, u)$ means u is regarded by the expert as fit to his concept C.

Our principal goal is, for each expert's explanation, find sets of patterns Pat, Pat_1,...,Pat_n and a relation $\Im_d$ so as to satisfy the following *quality requirements*:

$$\textbf{if } (\forall i : Match(Pat_i, EFeature_i) \geq p_i) \wedge (Pat = \Im_d(Pat_1, ..., Pat_n))$$
$$\textbf{then } Quality(Pat) \geq \alpha$$

where $p, p_i : i \in \{1, .., n\}$ and α are certain cutoff thresholds, while the *Quality* measure, intended to verify if the target pattern Pat fits into the expert's concept of sample class k, can be any, or combination, of popular quality criteria such as *support*, *coverage*, or *confidence* [6].

–

$$Support_{CLASS=k}(Pat) = |\{u \in U : Found(Pat, u) \wedge CLASS(u) = k\}|$$

–

$$Confidence_{CLASS=k}(Pat) = \frac{Support(Pat)}{|\{u \in U : Found(Pat, u)\}|}$$

–

$$Coverage_{CLASS=k}(Pat) = \frac{Support(Pat)}{|\{u \in U : CLASS(u) = k\}|}$$

where U is the training set.

In other words, we seek to translate the expert's knowledge into the domestic language so that to generalize the expert's reasoning to the largest possible number of training samples. More refined versions of the inclusion measures would involve the granulation of the $Found$ and Fit test functions

$$Match(p, C) = \frac{|\{u \in T : Found(p, u) \geq c_1 \wedge Fit(C, u) \geq c_2\}|}{|\{u \in T : Fit(C, u) \geq c_2\}|}$$

where c_1 and c_2 constitute adjustable threshold values.

We also may consider additional weight coefficients attached to the said functions in order to a more flexible construction of $Match(p, C)$. Adjustment of these coefficients based on feedback from actual training data may help optimize the approximation quality.

The use of rough inclusion measures allows for a very flexible approximation of foreign concept. For instance, a stroke at 85 degree to the horizontal in an image can still be regarded as a vertical stroke, though obviously not a 'pure' one. Instead of just answering in a '*Yes*/*No*' fashion, the expert may express his degrees of belief using such terms as '*Strong*', '*Fair*', or '*Weak*'.

Domestic patterns satisfying the defined quality requirement can be quickly found, taking into account that sample tables submitted to experts are usually not very large. The most effective strategies seem to be genetic algorithms equipped with some greedy heuristics. For example, [7] reported using this kind of tools and methods for a similar problem.

Relations between features. Relations between expert's features may include concepts such as '*Above*', '*Below*' or simply '*Near*'. They express not only expert's perceptions about particular concepts, but also the interdependencies among them. Similarly to the stand-alone features, these relations can also be described by the expert with a degree of tolerance.

The approximation of these relations has been formalized within the framework of perception structures developed by Skowron [8]. A *perception structure* S, in a simpler form, is defined as:

$$S = (U, M, F, \models, p)$$

where U is a set of samples, F is a family of formulas expressed in domestic language that describe certain features of the samples and M is a family of relational structures in which these formulas can be evaluated, while $p\colon U \rightarrow M \times F$ is a *perception function* such that $\forall u {\in} U : p_1(u) {\models} p_2(u)$ (p_1 and p_2 are

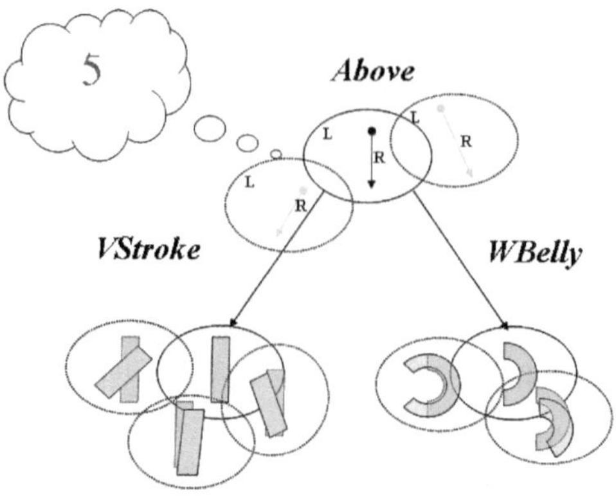

Fig. 2. Tolerant matching by expert

the first and second component projections of p) which means that $p_2(u)$ is satisfied (is true) in the relational structure $p_1(u)$. This may express that some relations among features within samples are observed.

Perception structures, following natural constructs in the expert's foreign language, can involve tolerant matching. Two relational structures might be considered approximately *the same* if they allow for similar formulas to yield similar results in majority of cases when these formulas are applicable.

Layered approximate reasoning paradigm. Let's observe that the approximation quality requirement previously introduced yields a powerful feature of the multi-layered approximation scheme. First, as the target pattern Pat retains its quality regardless of deviations of input patterns, the approximation is *robust* with regards to noisy input data or imperfect performances on lower levels. This also means high reusability of the same framework on changing or evolving data. Second, we have the *global stability*, which guarantees that if only some input patterns Pat'_i are equally "close" or "similar" to $EFeature_i$, then the target pattern $Pat' = \Im_d(Pat'_1, ..., Pat'_n)$ will meet the same quality requirements as Pat to a satisfactory degree.

These issues are illustrated in Fig. 3. Any variances (presented as polygonal shapes) of the concepts approximated by Pat_1 or Pat_2 (inner circles), as long as they are sufficiently close to these standard concepts, i.e. they do not surpass the outer, dot-lined circles, can be used to approximate the upper level concept within a guaranteed degree of precision. The same applies to variances of the bounding relation R.

This leads to an approximation of $EFeature_i$ which is *independent* from particular patterns Pat_i. The hierarchy scheme itself therefore becomes a high level search knowledge control mechanism that allow for the classifier system, when conditions are met, to bypass intermediate levels of reasoning without sacrificing too much on approximation quality.

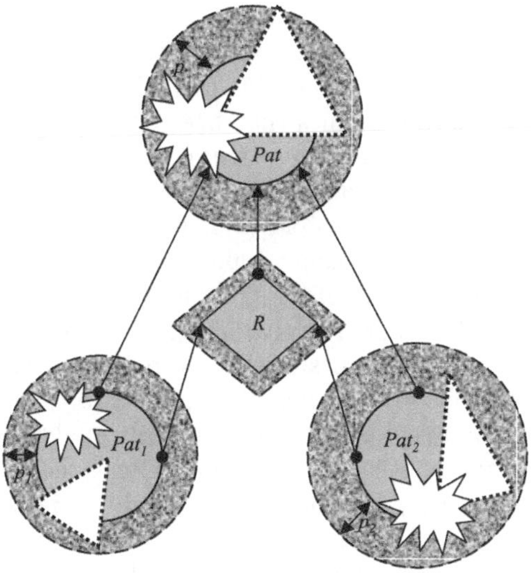

Fig. 3. Quality constraints

The preservation of approximation quality allows retaining of subsequent advances in search processes. This is especially important in applications pertaining to complex objects, where search processes are generally computationally expensive. It is argued that analyzing and mining complex structured objects heavily rely on the synthesis of more abstract objects' features from more basic ones by way of domain knowledge incorporation. External knowledge transfer is deemed crucial for meaningful data mining tasks [9].

It is noteworthy to observe that our approach, based on approximate reasoning scheme and granular computing, though developed independently, have much in common with theories and methods of Cognitive Science. For example, one of the most fundamental assumption of Unified Theory of Cognition [4] stipulates that human perception are inherently hierarchical and theories on such perception should be deliberately approximate. Most, if not all, cognitive architectures such as SOAR, ACT-R, Prodigy or recently developed ICARUS [10] are based on *knowledge and data chunking*, which follows the hierarchical structure of human perception. Chunking resembles in many ways the layered reasoning paradigm. Many other common issues such as search control, target function learning or external background knowledge assimilation can also be observed.

On the other hand, cognitive architectures seem not to incorporate the approximation of internal predicates or goal seeking strategies to a large extent, while the approximation of concepts and their binding relations is at the core of our approach.

2.2 Analysis of Outlier Cases

Our architecture typically asks for the expert's additional knowledge on samples which escaped previous classification attempts. This eventually results in asking the expert's help on "hard" samples that had defied much of our classification efforts, mainly because they differ to a significant extent from other samples of their class, or belong to a boundary region between several classes.

Outliers are kind of atypical samples that either are markedly different from the rest of their group in terms of some similarity measures, or behave very differently from the norm [11]. These samples previously tended to be treated as bias or noisy input data and were frequently discarded or suppressed in the learning process. However, there is an increasing effort to develop better methods for their analysis, based on the observation that they often carry useful diagnosis on the characteristics of the sample domain and, if properly analyzed, may provide valuable guidance in discovering the causalities underlying the behavior of a learning system. As such, they may prove to be valuable as additional search control knowledge. Most popular measures to detect outliers can be found in [12].

While outlier detection does not pose significant computation problems, their effective use in eliciting additional domain knowledge is believed difficult without support of a human expert.

Our approach to outlier detection and analysis will assume a somewhat different perspective. It focuses on two main issues:

1. Elicitation of intensional knowledge from outliers by approximating the perception of external human experts.
2. Evaluation of suspicious samples by verification the performance of classifiers constructed using knowledge elicited from these samples.

Having established a mechanism for eliciting expert's knowledge as described above, we can develop outlier detection tests that might be completely independent from the existing similarity measures within the learning system as follows. For a given training samples u^*, we ask the expert for his explanation on u^* and received a foreign knowledge structure $\Im(u^*)$. Next, we approximate $\Im(u^*)$ under restrictive matching degrees to ensure only the immediate neighborhood of u^* is investigated. Let's say the result of such an approximation is a pattern (or set of pattern) p_u^*. It is now sufficient to check $Coverage(p_u^*)$. If this coverage is high, it signifies that u^* may bear significant information that is also found in many other samples. The sample u^* therefore cannot be regarded as an outlier despite the fact that there may not be many other samples in its vicinity in terms of existing domestic distance measures of the learning system. This test shows that outlier analysis and expert's elicited knowledge are complementary to each other.

In our architecture, outliers may be detected as samples that defied previous classification efforts, or samples that pass the above described outlier test, but may also be selected by the expert himself. In this way, we can benefit from the best of both sources of knowledge.

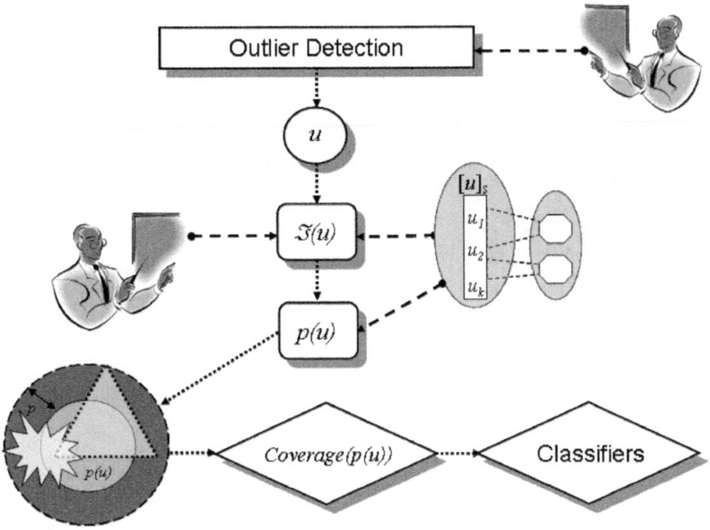

Fig. 4. Outlier analysis scheme

3 Implementation

The proposed framework and methods have been verified with a OCR system working on the NIST SD 19 handwritten digits. The domestic representational language of digit images involves various simple pixel evaluation functions and the Loci coding scheme, which reflects the local and global topological morphology or strokes in an image.

The expert's advices employ concepts such as *'Circle'*, *'Slanted Strokes'* or *'West Open Belly'*. The expert will explain what he means when he says, e.g. *'Circle'*, by providing a decision table (U, d) with reference samples, where d is the expert decision to which degree he considers that *'Circle'* appears in samples $u \in U$. The samples in U may be provided by the expert, or may be picked up by him among samples explicitly submitted by the system, e.g. those that had been misclassified in previous attempts.

Table 1. Perceived features

	Circle
u_1	*Strong*
u_2	*Weak*
...	...
u_n	*Fair*

Table 2. Translated features

	#NESW	Circle
u_1	252	*Strong*
u_2	4	*Weak*
...	...	...
u_n	90	*Fair*

We then attempt to find domestic feature(s) that approximates these degrees of belief using, among other means, genetic algorithms. In this particular

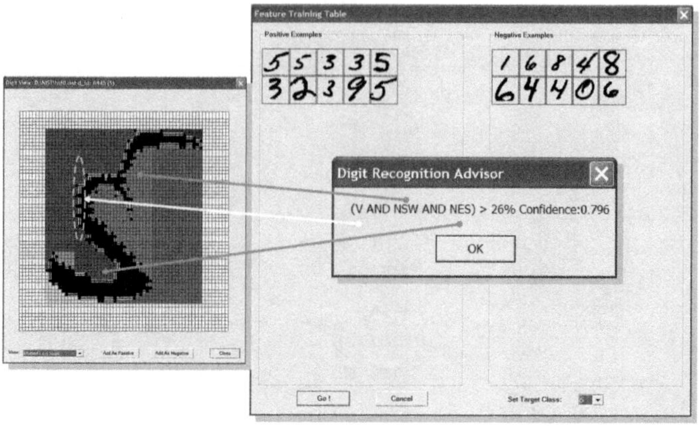

Fig. 5. Expert Concepts Approximated by Patterns

example, such feature may be the number of pixels that have black neighbors in all four directions (See Tab. 2).

Having approximated the expert's features $EFeature_i$, we can try to translate his relation $\Im$ into our $\Im_d$ by asking the expert to go through U and provide us with the additional attributes of how strongly he considers the presence of $EFeature_i$ and to what degree he believes the relation $\Im$ holds (See Tab. 3).

Table 3. Perceived relations

	VStroke	*WBelly*	*Above*
u_1	*Strong*	*Strong*	*Strong*
u_2	*Fair*	*Weak*	*Weak*
...	...	...	...
u_n	*Fair*	*Fair*	*Weak*

Table 4. Translated relations

	#V_S	#NES	$S_y < B_y$	*Above*
u_1	0.8	0.9	(*Strong*,1.0)	(*Strong*, 0.9)
u_2	0.9	1.0	(*Weak*, 0.1)	(*Weak*, 0.1)
...	...	...	...	...
u_n	0.9	0.6	(*Fair*, 0.3)	(*Weak*, 0.2)

We then replace the attributes corresponding to $EFeature_i$ with the rough inclusion measures of the domestic feature sets that approximate those concepts (computed in the previous step). In the next stage, we try to add other features, possibly induced from original domestic primitives, in order to approximate the decision d. Such a feature may be expressed by $S_y < B_y$, which tells whether the median center of the stroke is placed closer to the upper edge of the image than the median center of the belly. (See Tab. 4) Again, this task should be resolved by means of adaptive or evolutionary search strategies without too much computing burden, although it is more time-expensive compared with single concept approximation.

The expert's perception *"A '6' is something that has a 'vertical stroke' 'above' a 'belly open to the west'"* is eventually approximated by a classifier in the form of a rule:

Table 5. Comparison of performances

	No domain knowledge	With domain knowledge	Gain
Total learning time	205s	168s	22%
Negative classifier learning time	3.7s	2.2s	40%
Positive classifier learning time	28.2s	19.4s	31%
Skeleton graph size	3-5 nodes	2-5 nodes	

if $S(\#BL\_SL > 23)$ **AND** $B(\#NESW > 12\%)$ **AND** $S_y < B_y$ **then** CL='6',

where S and B are designations of pixel collections, #BL_SL and #NESW are numbers of pixels with particular Loci codes, and $S_y < B_y$ reasons about centers of gravity of the two collections.

In this way, the knowledge transfer from the human expert to the recognition engine is conducted within a fully interactive process, which essentially performs a refined, supervised, multilayered learning scheme, taking full advantage of the introduced rough set and rough mereology methodologies.

We compared the performances gained by a standard learning approach with and without the aid of the domain knowledge. The additional knowledge, passed by a human expert on popular classes as well as some atypical samples allowed to reduce the time needed by the learning phase from 205 minutes to 168 minutes, which means an improvement of about 22 percent without loss in classification quality. In case of screening classifiers, i.e. those that decide a sample *does not* belong to given classes, the improvement is around 40 percent. The representational samples found are also slightly simpler than those computed without using the background knowledge.

4 Conclusion

A formal framework based on multi-layered approximate reasoning schemes for the domain knowledge assimilation problem is proposed. We demonstrated that rough mereology theory and granular computing can be successfully used to transfer domain knowledge expressed in quasi-natural languages into domestic languages of computer learning system. A universal, robust and stable scheme for human-computer ontology matching in a clear, friendly interactive manner is also presented. We also argue that outlier analysis is key to successful domain knowledge elicitation whence elicited domain knowledge can help detect new outlier. Comparison of selected common aspects with cognitive theories and architectures has been outlined. Proposed methods have been verified by an OCR system working on a large handwritten digit dataset.

Acknowledgment. The author would like to express gratitude to Professor Andrzej Skowron for his insightful comments and his invaluable support during the work on this paper. This work has been supported by the grant N N516 368334 from Ministry of Science and Higher Education of the Republic of Poland and by the grant Innovative Economy Operational Programme 2007-2013 (Priority

Axis 1. Research and development of new technologies) managed by Ministry of Regional Development of the Republic of Poland.

References

1. Mitchell, T.M.: Machine Learning. McGraw-Hill, New York (1997)
2. Domingos, P.: Toward knowledge-rich data mining. Data Minning and Knowledge Discovery 15, 21–28 (2007)
3. Fensel, D.: Ontologies: A Silver Bullet for Knowledge Management and Electronic Commerce. Springer, New York (2003)
4. Newell, A.: Unified theories of cognition. Harvard University Press, Cambridge (1994)
5. Polkowski, L., Skowron, A.: Rough mereology: A new paradigm for approximate reasoning. Journal of Approximate Reasoning 15, 333–365 (1996)
6. Polkowski, L., Skowron, A.: Constructing rough mereological granules of classifying rules and classifying algorithms. In: B.B.M., et al. (eds.) Technologies for Constructing Intelligent Systems I, pp. 57–70. Physica-Verlag, Heidelberg (2002)
7. Oliveira, L., Sabourin, R., Bortolozzi, F., Suen, C.: Feature selection using multiobjective genetic algorithms for handwritten digit recognition. In: International Conference on Pattern Recognition (ICPR 2002), vol. I, pp. 568–571 (2002)
8. Skowron, A.: Rough sets in perception-based computing. In: Pal, S.K., Bandyopadhyay, S., Biswas, S. (eds.) PReMI 2005. LNCS, vol. 3776. Springer, Heidelberg (2005)
9. Kriegel, H.-P., Borgwardt, K.M., Kröger, P., Pryakhin, A., Schubert, M., Zimek, A.: Future trends in data mining. Data Min. Knowl. Discov. 15, 87–97 (2007)
10. Langley, P., Laird, J.E.: Cognitive architectures: Research issues and challenges (Technical Report) (2002)
11. Aggarwal, C.C., Yu, P.S.: Outlier detection for high dimensional data. In: Proceedings of the 2001 ACM SIGMOD International Conference on Management of Data, pp. 37–46. ACM Press, New York (2001)
12. Hodge, V., Austin, J.: A survey of outlier detection methodologies. Artif. Intell. Rev. 22, 85–126 (2004)

Information Quanta and Approximation Operators: Once More Around the Track

Marcin Wolski

Department of Logic and Methodology of Science,
Maria Curie-Skłodowska University, Poland
marcin.wolski@umcs.lublin.pl

Abstract. The present paper investigates topological variations of information structures (together with the associated operators) from the theory of Information Quanta, namely Information Quantum Relational Systems (IQRSs) and Property Systems, with special emphasis put on approximation spaces from Rough Set Theory (RST) and contexts from Formal Concept Analysis (FCA), respectively. The main novelty of this study comes from the influence of the theory of Heyting-Brouwer Algebras (HBAs) – some of that influence is reflected in the choice of Alexandroff topological spaces as the main source of both IQRSs and Property Systems. It allows us to establish some new results concerning RST and FCA, and to extend RST by a new type of rough sets, i.e. Heyting-Brouwer rough sets, defined with respect to bitopological spaces. Furthermore, we provide two interpretations of operators from RST and FCA, the first one in terms of HBAs and the second one (by the extended Gödel translation from intuitionistic logic to S4) in terms of temporal logic S4.t.

1 Introduction

The framework of Information Quanta (IQ), explicitly introduced and explored by P. Paliagni and M. Chakraborty [7,8], distinguishes two levels of how information gathered, for example, in the process of observation, comes into account. On the first level, whose information structures are called Property Systems, information is given "locally": objects are regarded as separate elements which possess or not some known properties. On the second level, whose information structures are called Information Quantum Relational Systems (IQRSs), information is encoded "globally": an object a is not viewed separately but is regarded as a "cluster" consisting of other objects which are sufficiently similar to a with respect to a relation R which encodes information from a Property System. In the present paper we apply the framework of IQ to the well-known theories of data mining and knowledge acquisition, namely Rough Set Theory (RST) [9] and Formal Concept Analysis (FCA) [16]. Basically, RST is concerned with IQRSs while FCA deals with Property Systems.

Firstly, we shall discuss how operators from RST and FCA behave when applied to the first and second level information structures. The special case of interest to us is when both Property Systems and IQRSs are induced by Alexandroff spaces. This idea comes from the theory of Heyting-Brouwer Algebras (HBAs) [4,11,12,13,19], which play an important role in our investigation. C. Rauszer introduced two standard methods

J.F. Peters and A. Skowron (Eds.): Transactions on Rough Sets VIII, LNCS 5084, pp. 237–250, 2008.

of obtaining HBAs [11,12,13]. The first method is based on preordered sets; the second method is based on the concept of a bitopological space. A bitopological space is a (nonempty) set X equipped with two topologies τ_1 and τ_2 satisfying some additional requirements. These two topologies are usually represented by their interior Int_{τ_1} and closure Cl_{τ_2} operators, respectively. As a consequence, a bitopological space is typically introduced as a triple (X, Int, Cl).

Each topological approximation space (X, τ) may be represented as a bitopological space (X, Int, Cl), where both operators Int and Cl are induced by the same topology τ. A rough set is then a pair $(Int(A), Cl(A))$, for some $A \subseteq X$. By a simple generalisation of this observation we shall introduce the new concept of a Heyting-Brouwer rough set called a HB-rough set. Given an Alexandroff topological space we introduce its corresponding bitopological space (X, Int, Cl) (where Int and Cl may give rise to two different topologies) and define a HB-rough set as a pair $(Int(A), Cl(A))$, for some $A \subseteq X$. The interior operator Int is defined by means of RST while the closure operator Cl is defined in terms of FCA. For topological approximation spaces the concepts of a rough set and a HB-rough set coincide; thus the notion of a HB-rough set really generalises the concept of a rough set in the sense of Pawlak's definition [9]. At the end of the paper we provide an interpretation of operators employed by the concept of a HB-rough set in terms of temporal logic S4.t. Firstly, we consider topological models and related IQRSs and then present the corresponding Property Systems.

2 Information Structures and Approximation Operators

Following [7,8], we shall assume that the pieces of information collected in the process of observation are basically given by: (a) a set G of objects, (b) a set M of observable properties, and (c) a fulfilment relation $\models$ between G and M, where $g \models m$ reads as "the object g has a property m." It is also assumed, that there are no "dummy" objects and properties, i.e., objects without any property and properties enjoyed by no object.

Definition 1 (Property System). *A triple $(G, M, \models)$, where G and M are finite sets, $\models \subseteq G \times M$ is a relation such that for all $g \in G$ there exists $m \in M$ such that $g \models m$, and for all $m \in M$ there exists $g \in G$ such that $g \models m$, is called a* context *or* Property System.

On this view, the collected pieces of information are given "locally", that is, we have a set of individuals which can be examined if they enjoy (or not) some known properties. However, we generally do not have an access to individual objects because the set of properties may not be sufficiently rich to distinguish any two given elements. In this case information must be encoded "globally" as a relation which "glues" objects into information granulae.

Definition 2 (IQRS). *Let S be a Property System and for all $g \in G, Q_g = \{g' : \forall m \in M$ if $g \models m$ then $g' \models m\}$, then we say that R is induced by S iff*

$$(g, g') \in R \text{ iff } g' \in Q_g.$$

The system (G, R) is called an Information Quantum Relational System *(IQRS) and R is called an* information quantum relation.

Basically, FCA deals with Property Systems and $\models$ whereas RST deals with IQRSs and R. Both theories has its own operators induced by binary relations, which can be described as Galois connections [3].

Definition 3 (Galois Connection). *Let* $(\mathcal{U}, \leq)$ *and* $(\mathcal{V}, \preceq)$ *be partially ordered sets (posets). If* $\pi_* : \mathcal{U} \to \mathcal{V}$ *and* $\pi^* : \mathcal{V} \to \mathcal{U}$ *are functions such that for all* $a \in \mathcal{U}$ *and* $b \in \mathcal{V}$, $a \leq \pi^* b$ *iff* $\pi_* a \preceq b$, *then the quadruple* $\pi = \langle(\mathcal{U}, \leq), \pi_*, \pi^*, (\mathcal{V}, \preceq)\rangle$ *is called a* Galois connection, *where* π_* *and* π^* *are called the coadjoint and adjoint part of* π, *respectively.*

Now we discuss Galois connections induced by binary relations, namely *polarities* and *axialities* (also called residuated mappings or Galois adjunctions) [3].

Proposition 1. *Any relation* $R \subseteq \mathcal{U} \times \mathcal{V}$ *induces a Galois connection called* polarity ${R_+}^+ = \langle(\boldsymbol{P}\mathcal{U}, \subseteq), R_+, R^+, (\boldsymbol{P}\mathcal{V}, \subseteq)\rangle$, *where* $\boldsymbol{P}\mathcal{U}$ *is the power set of* $\mathcal{U}$, R_+ *and* R^+ *are defined as follows: for any* $A \subseteq \mathcal{U}$ *and* $B \subseteq \mathcal{V}$,

$$R_+(A) = \{b \in \mathcal{V} : (\forall a \in A)\langle a, b\rangle \in R\}$$

$$R^+(B) = \{a \in \mathcal{U} : (\forall b \in B)\langle a, b\rangle \in R\}$$

Below, we shall present FCA in a very concise way to give the reader at least a "taste" of this theory. For a detailed exposition of FCA see [16].

A triple $\langle \mathcal{U}, \mathcal{V}, R\rangle$, where $R \subseteq \mathcal{U} \times \mathcal{V}$, is called a *context*. Each context is associated with two operators R_+ and R^+ called *derivation operators*. These operators allows one to build concepts, i.e. meaningful entities which constitute our knowledge about the context.

Definition 4 (Concept). *A concept of a given context* $\langle \mathcal{U}, \mathcal{V}, R\rangle$ *is a pair* (A, B), *where* $A \subseteq \mathcal{U}$ *and* $B \subseteq \mathcal{V}$ *such that* $A = R^+(B)$ *and* $B = R_+(A)$.

Basically, FCA is concerned with hierarchies of concepts induced by formal contexts. However, in the present paper we shall pay attention only to the derivation operators and their compositions.

Proposition 2. *Any relation* $R \subseteq \mathcal{U} \times \mathcal{V}$ *induces a Galois connection (adjunction) called* axiality ${R_\exists}^\forall = \langle(\boldsymbol{P}\mathcal{U}, \subseteq), R_\exists, R^\forall, (\boldsymbol{P}\mathcal{V}, \subseteq)\rangle$, *where* $R_\exists$ *and* $R^\forall$ *are defined as follows: for any* $A \subseteq \mathcal{U}$ *and* $B \subseteq \mathcal{V}$,

$$R_\exists(A) = \{b \in \mathcal{V} : (\exists a \in \mathcal{U})\langle a, b\rangle \in R \;\&\; a \in A\}$$

$$R^\forall(B) = \{a \in \mathcal{U} : (\forall b \in \mathcal{V})\langle a, b\rangle \in R \Rightarrow b \in B\}$$

The theoretical dual of ${R_\exists}^\forall$, defined as ${R^\exists}_\forall = \langle R^\exists, R_\forall\rangle = {(R^{-1})_\exists}^\forall$, is also an axiality but from $(\mathbf{P}\mathcal{V}, \subseteq)$ to $(\mathbf{P}\mathcal{U}, \subseteq)$. R^{-1} means the converse relation of R, that is, $bR^{-1}a$ iff aRb. Now, we recall basic concepts of RST.

Definition 5 (Approximation Operators). *Let* $\mathcal{U}$ *be a set,* E *an equivalence relation on* $\mathcal{U}$, *and* $[a]_E$ *– the equivalence class containing* $a \in \mathcal{U}$. *With each* $A \subseteq \mathcal{U}$, *we can associate its E-lower and E-upper* approximations, $\underline{A}$ *and* $\overline{A}$, *respectively, defined as follows:*

$$\underline{A} = \{a \in \mathcal{U} : [a]_E \subseteq A\},$$
$$\overline{A} = \{a \in \mathcal{U} : [a]_E \cap A \neq \emptyset\}.$$

A pair $(\mathcal{U}, E)$ is called *approximation space*. A *rough set* is a pair $(\underline{A}, \overline{A})$, for some $A \subseteq \mathcal{U}$. A subset $A \subseteq \mathcal{U}$ is called *definable* if $A = \bigcup B$ for some $B \subseteq \mathcal{U}/E$, where $\mathcal{U}/E$ is the family of equivalence classes of E. For any definable set A it holds that $\underline{A} = \overline{A}$. The chief idea of RST is to approximate an undefinable set A by means of two definable sets $\underline{A} \subseteq A \subseteq \overline{A}$. The lower approximation $\underline{A}$ consists of points which necessarily belong to A while the upper approximation $\overline{A}$ consists of points which possibly belong to A.

Each approximation space $(\mathcal{U}, E)$ may be converted into topological space $(\mathcal{U}, \tau_E)$; in literature topological spaces induced by equivalence relations are called *approximation topological spaces* [10].

Definition 6 (Approximation Topologial Space). *A topological space (X, τ) with an interior operation Int induced by an equivalence relation E, that is*

$$Int(A) = \bigcup\{[x] \in \boldsymbol{B} : [x] \subseteq A\},$$

*where **B**, the family of all equivalence classes of E, is a subbasis of τ, and $[x]$ denotes the equivalence class of $x \in X$, is said to be an* approximation topological space.

On this view, the lower approximation $\underline{A}$ is the interior of $A \subseteq \mathcal{U}$ and the upper approximation $\overline{A}$ is the closure of A. A set $A \subseteq \mathcal{U}$ is definable only if $A \in \tau_E$. It is worth emphasising that every topological approximation space satisfies the following *clopen sets property*: every closed set is open and every open set is closed [10]. In topology such sets are called *clopen*. Hence both approximations are clopen and, as a result, definable.

Now let us return to the basic structures introduced in this section, that is, Galois connections, Property Systems, and IQRSs. Any (finite) topological space $\mathcal{S} = (X, \tau)$ may be viewed as a Property System $(X, \tau, \models)$, where $x \models A$ iff $x \in A$, and, in consequence, it gives rise to an IQRS $\mathcal{S} = (X, R)$. Let us recall that for any topological space (X, τ) we can convert the relation of set inclusion on τ into a preorder, called the *specialisation preorder*, on elements of X:

$$x \preceq y \text{ iff } Cl(\{x\}) \subseteq Cl(\{y\}). \tag{1}$$

Furthermore, for arbitrary preordered set $(X, \preceq)$ there is always a topology τ whose specialisation preorder is $\preceq$, and there will in general be many of them. Among these topologies the special role is played by specialisation topologies.

Definition 7 (Specialisation Topology). *Let $(X, \preceq)$ be a preordered set. A* specialisation topology *on X is a topology τ with a specialisation order $\preceq$ such that every automorphism of $(X, \preceq)$ is homeomorphism.*

A topology τ_E induced by an equivalence relation E is an example of a specialisation topology. However, the canonical example of this kind of topology is given by an *Alexandroff* topology.

Definition 8. *A topological space (X, τ) is called an* Alexandroff space *if its topology τ is closed under arbitrary intersections and arbitrary unions.*

In such case, each $x \in X$ has the smallest neighbourhood defined as follows:

$$\nabla(x) = \bigcap\{A \in \tau : x \in A\}. \tag{2}$$

The Alexandroff topology is actually the largest specialisation topology induced by $\preceq$. In this topology the following sets

$$\nabla'(x) = \{y \in X : x \preceq y\} \text{ for all } x \in X. \tag{3}$$

form a sub-basis. Moreover one can prove that $\nabla(x) = \nabla'(x)$, for any x.

Proposition 3. *There is a one-to-one correspondence between Alexandroff topologies on a set X and preorders on X.*

Given an Alexandroff topological space (X, τ) we may produce a preorder (namely its specialisation preorder) by (1). Given a preordered set $(X, \preceq)$ we may obtain its Alexandroff topology by (3). Actually, Alexandroff spaces and preordered sets regarded as categories are dually isomorphic and we may identify them. Let us stress that any finite space is an Alexandroff space.

Proposition 4. *Let $\mathcal{S} = (X, \tau)$ be a (finite) topological space, $\preceq$ its specialisation preorder, and $\mathcal{S} = (X, R)$ its IQRS; then $R = \preceq$.*

Proof. By definition $Q_x = \{y : \forall(A \in \tau) \text{ if } x \in A \text{ then } y \in A\} = \{y : \nabla(y) \subseteq \nabla(x)\} = \{y : y \in \nabla(x)\}$. Now, by definition $\langle x, y\rangle \in R$ iff $y \in \nabla(x)$ iff $x \preceq y$.

In other words for any finite topological space its specialisation preorder coincides with the quantum information relation R induced by its Property System. However, the proof actually does not involve the finiteness of the underlying space – it employs the fact that the space is Alexandroff. Therefore we shall modify a little the concepts of Property Systems and IQRSs: any Alexandroff topological space will be regarded as a valid Property System. That is, for Alexandroff spaces we shall drop out the finiteness condition. As a consequence, by Proposition 4 we have that:

Proposition 5. *Let $(\mathcal{U}, E)$ may be an approximation space and $(\mathcal{U}, \tau_E)$ its approximation topological space; then $(\mathcal{U}, E)$ is an IQRS induced by the Property System $(\mathcal{U}, \tau_E, \models)$.*

Now, we recall and give a new proof of some well-known results concerning Galois connections and approximation operators from RST.

Lemma 1 (P. Pagliani, M. Chakraborty). *Let O_i, O_j be a Galois connection induced by an IQRS (G, R), then $O_i O_j = O_j$.*

Proposition 6 (I. Düntsch, G. Gediga). *Let $(\mathcal{U}, \tau_E)$ be a topological approximation space and $A \subseteq \mathcal{U}$ then:*

$$(i)\ E^{\exists}E_{\forall}(A) = E_{\exists}E^{\forall}(A) = \underline{A}$$

$$(ii)\ E^{\forall}E_{\exists}(A) = E_{\forall}E^{\exists}(A) = \overline{A}$$

Proof. We prove only (i); the proof of (ii) is analogous.

$$\underline{A} = \{a \in \mathcal{U} : [a]_E \subseteq A\} =$$
$$\{a \in \mathcal{U} : \forall(b \in \mathcal{U})\langle a, b\rangle \in E \Rightarrow b \in A\} =$$
$$E^{\forall}(A) = E_{\exists}E^{\forall}(A) \text{ (by Lemma 1)}.$$

Since $E^{-1} = E$, we have $E_{\exists}E^{\forall}(A) = E^{\exists}E_{\forall}(A)$.

Proposition 6 – dressed differently – has been firstly proved (by means of other methods) in [1,2]. The direct relationship between Galois connections and RST has been observed in [7,8,17].

The operator ∇ expressed by means of Equation (3) may be easily defined for sets as well:

$$\nabla(A) = \bigcup_{y \in A} \nabla(y) \tag{4}$$

Proposition 7. *Let (X, τ) be an Alexandroff topological space, $(X, \tau, \models)$ its Property System, (X, R) its IQRS, Int and Cl the interior and closure operators induced by τ, respectively. Then*

$$(i)\ \models^{\forall}\models_{\exists} (A) = Cl(A),$$
$$(ii)\ \models^{\exists}\models_{\forall} (A) = Int(A),$$
$$(iii)\ \models^{+}\models_{+} (A) = \nabla(A),$$
$$(iv)\ R^{\forall}R_{\exists}(A) = \nabla(A),$$
$$(v)\ R_{\forall}R^{\exists}(A) = Cl(A),$$
$$(vi)\ R_{\exists}R^{\forall}(A) = Int(A),$$
$$(vii)\ R^{\exists}R_{\forall}(A) = \bigcup_{x \in X} (Cl(\{x\}) \subseteq A).$$

for all $A \subseteq \mathcal{U}$.

Proof. (i) Let:

$$\mathcal{B} = \models_{\exists} (A) = \{B \in \tau : \exists(x \in A) x \models B\}$$

Then:

$$\models^{\forall} (\mathcal{B}) = \{x \in X : \forall(B \in \tau) \text{ if } x \models B \text{ then } B \in \mathcal{B}\}$$

Furthermore, $x \in Cl(A)$ iff for every $B \in \tau$ such that $x \in B$ we have $B \cap A \neq \emptyset$, i.e. $B \in \models_{\exists} (A)$.
(ii) Now, let:

$$\mathcal{B} = \models_{\forall} (A) = \{B \in \tau : \forall(x \in A) \text{ if } B \models^{-1} x \text{ then } x \in A\}.$$

Then:

$$\models^{\exists} (\mathcal{B}) = \{x \in X : \exists(B \in \mathcal{B}) B \models^{-1} x\} =$$
$$\{x \in X : \exists(B \subseteq A) B \models^{-1} x\} =$$
$$\{x \in X : \exists(B \subseteq A) B \supseteq \{x\}\}$$

It means that $x \in \models^{\exists}\models_{\forall} (A)$ iff there exists $B \in \tau$ such that $x \in B$ and $B \subseteq A$ iff $x \in Int(A)$.
(iii) Let:

$$\mathcal{B} = \models_{+} (A) = \{B \in \tau : \forall(x \in A)\ x \models B\} =$$
$$\{B \in \tau : A \subseteq B\}$$

Then:

$$\models^{+} (\mathcal{B}) = \{x \in X : \forall (B \in \mathcal{B}) \ x \models B\}.$$

Thus

$$\models^{+}\models_{+} (A) = \bigcap \{B \in \tau : A \subseteq B\} = \nabla(A)$$

(iv) By Lemma 1 we have to prove that $R^{\exists}(A) = \nabla(A)$. For any Alexandroff space X, we have:

$$\nabla(A) = \bigcup_{y \in A} \nabla(y) =$$

$$\{x \in X : \exists (y \in A) \ x \in \nabla(y)\} =$$

$$\{x \in X : \exists (y \in A) \ yRx\} = R_{\exists}(A)$$

(v) As above, for any Alexandroff space X we have:

$$Cl(A) = \bigcup_{y \in A} Cl(y) =$$

$$\{x \in X : \exists (y \in A) \ x \in Cl(y)\} =$$

$$\{x \in X : \exists (y \in A) \ xRy\} = R^{\exists}(A)$$

(vi)

$$R^{\forall}(A) = \{x \in X : \forall z \text{ if } xRz \text{ then } z \in A\} =$$

$$\{x \in X : \forall z \text{ if } \nabla z \subseteq \nabla x \text{ then } z \in A\} =$$

$$\{x \in X : \forall z \text{ if } z \in \nabla x \text{ then } z \in A\} =$$

$$\{x \in X : \nabla x \subseteq A\} = Int(A)$$

(vii)

$$R_{\forall}(A) = \{x \in X : \forall z \text{ if } zRx \text{ then } z \in A\} =$$

$$\{x \in X : \forall z \text{ if } Cl(\{z\}) \subseteq Cl(\{x\}) \text{ then } z \in A\} =$$

$$\{x \in X : \forall z \text{ if } z \in Cl(\{x\}) \text{ then } z \in A\} =$$

$$\{x \in X : Cl(\{x\}) \subseteq A\} = \bigcup_{x \in X} (Cl(\{x\}) \subseteq A).$$

The above proposition explains how operators from RST and FCA behave when applied to information structures of the first and second level. Generally speaking, their compositions can be described in terms of the operators of interior, closure, and minimal open neighborhood. Please note, that operators from RST and from FCA may be applied to Property Systems (that is Alexandroff topological spaces); however, only operators from RST (i.e. based on axialities) can be applied to IQRSs. Observe that when R is a specialisation preorder then $R_{+}(A)$ gives all points y which belongs to the minimal open neighborhood $\nabla(x)$ of every element x of A. But, in most cases there is no such point. On the other hand, when R is an equivalence relation then $R^{+}R_{+} = R_{\forall}R^{\exists}$. Thus, polarities either brings the empty set $\emptyset$ or may be recovered from axialities. In this sense the derivation operators of FCA, R^{+} and R_{+}, are not regarded as quantum operators [8].

3 Approximation Operators, Heyting-Brouwer Algebras and Rough Sets

In the present section we study the operators from RST and FCA in terms of Heyting-Brouwer Algebras (HBAs), introduced and developed by C. Rauszer [11,12,13]. It was already proved by P. Pagliani that rough sets can be made into a HBA [6]. In contrast, we shall here restrict our attention to the lower and upper approximation operators, leaving aside the concept of a rough set.

We begin by recalling some basic concepts of the theory of HBAs – a detailed exposition of the properties of HBAs may be found in [4,11,12,13,19]. Although C. Rauszer introduced them under a different name, namely semi-Boolean algebras, they are today usually called HBAs for two reasons: (i) a semi-Boolean algebra is a Heyting algebra which is also Brouwerian, (ii) semi-Boolean algebras provide a semantics for Heyting-Brouwer logic – the term introduced by C. Rauszer in [12]. Summing up, the name of HBA seems to be more natural than the original one.

Definition 9 (Heyting Algebra). *An algebra* $(X, \leq, \vee, \wedge, \overset{\wedge}{\rightarrow}, \top, \bot)$ *is called a* Heyting algebra *iff* $(X, \leq, \vee, \wedge, \overset{\wedge}{\rightarrow}, \top, \bot)$ *is a bounded distributive lattice and* $\overset{\wedge}{\rightarrow}$ *is a relative pseudo-complement of* a *with respect to* c*, i.e.,*

$$a \wedge b \leq c \text{ if and only if } b \leq a \overset{\wedge}{\rightarrow} c$$

for all $a, b, c \in X$

Definition 10 (Heyting-Brouwer Algebra). *A* Heyting-Brouwer algebra *(HBA)* $(X, \leq, \vee, \wedge, \overset{\wedge}{\rightarrow}, \overset{\vee}{\rightarrow}, \top, \bot)$ *is a Heyting algebra* $(X, \leq, \vee, \wedge, \overset{\wedge}{\rightarrow}, \top, \bot)$ *equipped with the operation of* coimplication $\overset{\vee}{\rightarrow}$*, that is:*

$$a \vee b \geq c \text{ if and only if } b \geq a \overset{\vee}{\rightarrow} c$$

for all $a, b, c \in X$

It is worth stressing that a HBA can be alternatively defined by means of the operation of *pseudo-difference* $\div$ (originally used by Rauszer):

$$a \vee b \geq c \text{ if and only if } b \geq c \div a$$

for all $a, b, c \in X$. However, later – following [19] – we shall introduce Heyting-Brouwer (sentential) logic (HBL) whose axiomatisation is based on coimplication and therefore a HBA is defined by means of $\overset{\vee}{\rightarrow}$ instead of $\div$.

The standard method of producing HBAs is based on preorders (or Alexandroff topologies):

Proposition 8 (C. Rauszer). *Let* $S = (X, \leq)$ *be a preordered set,* (X, τ_S) *the induced Alexandroff topological space, and*

$$A \overset{\wedge}{\rightarrow} B = \{a \in X : (\forall b \geq a)(b \in A \Rightarrow b \in B)\}$$

$$A \overset{\vee}{\rightarrow} B = \{a \in X : (\exists b \leq a)(b \notin A \;\&\; b \in B)\}$$

then the algebra $(\tau_S, \subseteq, \cap, \cup, \overset{\wedge}{\rightarrow}, \overset{\vee}{\rightarrow}, \emptyset, G)$ *is a HBA.*

Let us now recall that in RST both lower and upper approximations are open sets (i.e. definable sets). However, for arbitrary Alexandroff topological space the closure of a set A is not necessarily open, thus the upper approximation $\overline{A} = Cl(A)$ may be an indefinable set. What we actually need is to make $Cl(A)$ an Int-open so to generalise the notion of a topological approximation space. The solution to this problem is brought by the concept of a bitopological space.

Definition 11 (Bitopological Space). *A* bitopological space $(X, Int, \mathtt{Cl})$ *is a non-empty set equipped with an interior operation* Int *and a closure operation* $\mathtt{Cl}$ *satisfying:*

$$Int(A) = \mathtt{Cl}(Int(A)) \text{ and } \mathtt{Cl}(A) = Int(\mathtt{Cl}(A)),$$

for all $A \subseteq X$.

Please note that any toplogical approximation space (X, τ) gives rise to a bitopological space $(\mathcal{U}, Int, Cl)$, where both operators, Int and Cl, are induced by τ. But in case of an Alexandroff topological space $\mathcal{S} = (\mathcal{U}, \tau)$, the operator Int is "only one half" of a bitopological space $(X, Int, \mathtt{Cl})$, since $\mathtt{Cl}$ – in general – may be different from Cl. It is clear that $\mathtt{Cl}$ returns τ-open sets, i.e. definable sets, as it is required by RST. The same argument applies to the closure operator Cl and $(X, \mathtt{Int}, Cl)$.

Proposition 9. *Let* (X, τ) *be an Alexandroff topological space,* $(X, \models)$ *its Property System and* (X, R) *its IQRS; then* $(X, R_{\exists}R^{\forall}, \models^{+}\models_{+})$ *and* $(X, R_{\exists}R^{\forall}, R^{\forall}R_{\exists})$ *are bitopological spaces.*

Proof. By Proposition 7 both operators, $R_{\exists}R^{\forall}$ and $R^{\forall}R_{\exists}$, return open sets of (X, τ). Furthermore, $R_{\exists}R^{\forall}$ is actually the interior operator induced by τ. It is easy to see that $R^{\forall}R_{\exists}$ is a closure operator which, as said above, also returns elements of τ. Additionally, by the same proposition $\models^{+}\models_{+}$ is equal to $R^{\forall}R_{\exists}$.

Additionally, $R_{\exists}R^{\forall}(R_{\forall}R^{\exists}(A)) = R_{\forall}R^{\exists}(A)$ by Lemma 1. Indeed, $R_{\exists}R^{\forall} = R^{\forall}$ and $R_{\forall}R^{\exists} = R^{\exists}$.

Proposition 10. *Let* (X, τ) *be an Alexandroff topological space,* R *its specialisation preorder and* $(X, \tau, \models)$ *its Property System; then the algebra* $(\tau, \subseteq, \cup, \cap, \overset{\wedge}{\to}, \overset{\vee}{\to}, \mathcal{U}, \emptyset)$*:*

$$A \overset{\wedge}{\to} B = R_{\exists}R^{\forall}(-A \cup B),$$

$$A \overset{\vee}{\to} B = \models^{+}\models_{+} (-A \cap B) = R^{\forall}R_{\exists}(-A \cap B),$$

for all $A, B \in \tau$*, where* $-$ *is the set complement, is is a HBA.*

Proof. By Lemma 1 $R_{\exists}R^{\forall}(-A \cup B) = R^{\forall}(-A \cup B) = \{x \in X : \forall(y \in X)\ xRy \Rightarrow y \in -A \cup B\}$. By classicall propositional logic we get that $R^{\forall}(-A \cup B) = \{x \in X : \forall(y \in X)\ xRy\ \&\ y \in A \Rightarrow y \in B\}$. Analogously, $R^{\forall}R_{\exists}(-A \cap B) = R_{\exists}(-A \cap B) = \{x \in X : \exists(y \in -A \cap B)\ yRx\} = \{x \in X : \exists(y \in X)\ yRx\ \&\ y \notin A\ \&\ y \in B\}$.

The same result may be obtained for the upper approximation operator $R_{\forall}R^{\exists}$. Let us recall that the compositions $S^{\forall}S_{\exists}$ and $S_{\exists}S^{\forall}$, where $S \subseteq X \times Y$ for X, Y being non-empty sets, give the closure and interior operators, respectively. On the other hand both compostions $S^{+}S_{+}$ and $S_{+}S^{+}$ give two closure operators. Therefore, for every $A \subseteq \mathcal{U}$ we have to define *dual* $\models^{+}\models_{+} (A) = -(\models^{+}\models_{+} (-A))$.

Corollary 1. *Let $(\mathcal{U}, \tau)$ be an Alexandroff topological space, R its specialisation order and $(X, \tau, \models)$ its Property System, then $(\mathcal{U}, R^{\exists}R_{\forall}, R_{\forall}R^{\exists})$ and $(\mathcal{U}, dual \models^{+}\models_{+}, R_{\forall}R^{\exists})$ are bitopological spaces.*

Corollary 2. *Let $X, \tau)$ be an Alexandroff topological space, R its specialisation preorder and $(X, \tau, \models)$ its Property System; then the algebra $(-\tau, \subseteq, \cup, \cap, \overset{\wedge}{\rightarrow}, \overset{\vee}{\rightarrow}, \mathcal{U}, \emptyset)$ is a HBA:*

$$A \overset{\wedge}{\rightarrow} B = dual \models^{+}\models_{+} (-A \cup B) = R^{\exists}R_{\forall}(-A \cup B),$$

$$A \overset{\vee}{\rightarrow} B = R_{\forall}R^{\exists}(-A \cap B),$$

for all $A, B \in -\tau$, where $-\tau = \{A \subseteq X : -A \in \tau\}$.

The above corollary follow from Proposition 9 and Proposition 10 by the duality of operators and topologies.

Now, let (X, τ_E) be an approximation topological space and let (X, Int, Cl) be the induced bitopological space. Since a rough set from a topological approximation space may be represented as $(Int(A), Cl(A))$, for some $A \subseteq X$, Proposition 9 suggests a straightforward generalisation of rough sets by means of the induced bitopological space: a HB-*rough set* of an Alexandroff topological space (X, τ) is as a pair of sets $(R_{\exists}R^{\forall}(A), R^{\forall}R_{\exists}(A))$, for some $A \subseteq X$. Please note that both HB-approximations of A, namely $R_{\exists}R^{\forall}(A)$ and $R^{\forall}R_{\exists}(A)$, are definable in (X, τ).

Corollary 3. *Let (X, τ_E) be a topological approximation space induced by an equivalence relation E. Then, the set of rough sets induced by (X, τ_E) is equal to the set of HB-rough sets induced by this space.*

This corollary shows that HB-rough sets are in fact generalised rough sets in the sense of RST. It is worth spelling out again the topological interpretation of these operators. Let (X, τ) be an Alexandroff topological space and R its specialisation order. As we have already said, the lower approximation $\underline{A}$ of a set $A \subseteq X$ is its interior, that is, the largest open set included in A. On the other hand, $dual R^{+}R_{+}(A)$ is the largest closed set included in $A \subseteq X$. In case of a topological approximation spaces both concepts coincide due to the clopen set property. Similar argument applies to the upper approximation operator and $R^{+}R_{+}$. The upper approximation $\overline{A}$ of $A \subseteq X$ is its closure, that is the smallest closed set which includes A, whereas $R^{+}R_{+}(A)$ is the smallest open set which includes A. Since for any topological approximations space open and closed sets coincide, $R^{\forall}R_{\exists}(A)$ is equivalent to $R^{+}R_{+}(A)$, for all $A \subseteq X$.

Now, we consider logics which are related to the operators employed by HB-rough sets or, better still, to their approximation operators. Firstly we introduce Heyting-Brouwer logic.

The Heyting-Brouwer sentential calculus (HBL) is formulated in the propositional language $\mathcal{L}_{HB}$ with connectives $\wedge, \vee, \overset{\wedge}{\rightarrow}, \overset{\vee}{\rightarrow}, \top, \bot$. Its axiomatisation was given by C. Rauszer in [12]. However, we shall follow the presentation given in [19].

Let us abbreviate $\overset{\wedge}{\neg} p = p \overset{\wedge}{\rightarrow} \bot$, $\overset{\vee}{\neg} p = p \overset{\vee}{\rightarrow} \top$ and define:

$$H = \{p \overset{\wedge}{\rightarrow} (q \wedge (q \overset{\vee}{\rightarrow} p)), (q \overset{\vee}{\rightarrow} p) \overset{\wedge}{\rightarrow} \overset{\vee}{\neg} (p \overset{\wedge}{\rightarrow} q),$$

$$(r \xrightarrow{\vee} (q \xrightarrow{\vee} p)) \xrightarrow{\wedge} ((p \vee q) \xrightarrow{\vee} p),$$
$$\overset{\wedge}{\neg} (q \xrightarrow{\vee} p) \xrightarrow{\wedge} (p \xrightarrow{\wedge} q), \overset{\wedge}{\neg} (p \xrightarrow{\vee} p)\}.$$

The HBL is the smallest logic containing $INT \cup H$, where INT is any axiomatisation of intuitionistic logic, and closed under uniform substitutions, *modus ponens* and

$$\frac{p}{\overset{\wedge}{\neg}\overset{\vee}{\neg} p}$$

Proposition 11 (C. Rauszer). *A formula of $\mathcal{L}_{HB}$ is provable in HBL iff it is valid in all HBAs.*

In view of Proposition 10 approximation operators from both RST and FCA are implicit in the semantics of HBL. However, by means of the extended Gödel translation from intuitionistic logic to S4 [14,5] we can elicit these operators from the model of HBL to the language of tense logic S4.t. Let us recall that temporal logics have two diamonds and two boxes – a pair of operators $\Box_F$, $\Diamond_F$ for "always in the Future" and "some time in the Future", respectively, and a pair of operators $\Box_P$, $\Diamond_P$ for "always in the Past" and "some time in the Past", respectively.

The extended Gödel translation t is defined as follows [14,5]:

$$\alpha^t = \Box_F \alpha$$

$$\top^t = \top$$

$$\bot^t = \bot$$

$$(\alpha \wedge \beta)^t = \alpha^t \wedge \beta^t$$

$$(\alpha \vee \beta)^t = \alpha^t \vee \beta^t$$

$$(\alpha \xrightarrow{\wedge} \beta)^t = \Box_F(\alpha^t \rightarrow \beta^t)$$

$$(\alpha \xrightarrow{\vee} \beta)^t = \Diamond_P(\neg\alpha^t \wedge \beta^t)$$

Proposition 12 (P. Łukowski). *For all $\alpha \in \mathcal{L}_{HB}$*

$$HBL \models \alpha \text{ iff } S4.t \models \alpha$$

Let us recall that for an Alexandroff topological space (X, τ) and its specialisation preorder R the triple $(X, R_{\exists}R^{\forall}, R^{\forall}R_{\exists})$ is a bitopological space. This space gives rise to a HBA which forms a model for HBL. Now, the extended Gödel translation actually provides a temporal version of the well known strong completeness theorem of modal logic S4 with respect to the class of all topological spaces.

Definition 12 (IQR Model). *Let (X, τ) be an Alexandroff topological space, R its specialisation preorder and V the valuation function from propositional letters to subsets of $\mathcal{U}$. The function V is extended to Boolean connectives in the standard way, for modal operators the extension is as follows:*

$$V(\Box_F\alpha) = R_\exists R^\forall(V(\alpha)),$$

$$V(\Diamond_F\alpha) = R_\forall R^\exists(V(\alpha)),$$

$$V(\Box_P\alpha) = R^\exists R_\forall(V(\alpha)),$$

$$V(\Diamond_P\alpha) = R^\forall R_\exists(V(\alpha)).$$

The definition of truth is as usual: $a \vDash \alpha$ *iff* $a \in V(\alpha)$. *The triple* (X, τ, V) *will be called an Information Quantum Relational Model (IQRM).*

Proposition 13. *A formula is provable in S4.t iff it is valid in all IQRMs.*

Please observe that an IQRM is in fact an IQRS. Proposition 7 allows us to deal with topological spaces in a more direct way. As we have already said each Alexandroff topological space is in fact a Property System $(X, \tau, \models)$:

Definition 13 (Property Model). *Let* $(X, \tau, \models)$ *be a Property System induced by an Alexandroff topological space and let* V *be defined as follows:*

$$V(\Box_F\alpha) = \models_\exists\models^\forall (V(\alpha)),$$

$$V(\Diamond_F\alpha) = \models_\forall\models^\exists (V(\alpha)),$$

$$V(\Box_P\alpha) = dual \models^+\models_+ (V(\alpha)),$$

$$V(\Diamond_P\alpha) = \models^+\models_+ (V(\alpha)).$$

The tuple $(X, \tau, \models, V)$ *will be called a property model.*

Proposition 14. *A formula is provable in S4.t iff it is valid in all property models.*

Thus S4.t may be regarded as a logic expressing mutual relationships among operators from RST and FCA [18].

4 Concluding Remarks

In the present paper we have investigated Rough Set Theory (RST) and Formal Concept Analysis (FCA) – both expressed in the conceptual framework of the theory of Information Quanta – against the background of the theory of Heyting-Brouwer Algebras (HBAs). In result, we have established some new relationships between approximation operators from RST and derivation operators from FCA, and provided two interpretations of theses operators: the first one in terms of HBAs and the second one – by the extended Gödel translation from intuitionistic logic to S4 – in terms of temporal system S4.t. Also, the paper has introduced the concept of Heyting-Brouwer rough set defined with respect to bitiopological spaces which generalises the usual notion of a rough set.

Acknowledgment

Many thanks are due to anonymous referees for the careful reading of the manuscript and valuable comments.

References

1. Düntsch, I., Gediga, G.: Modal-style operators in qualitative data analysis. In: Proc. of the 2002 IEEE Conference on Data Mining (ICDM 2002), pp. 155–162 (2002)
2. Düntsch, I., Gediga, G.: Approximation operators in qualitative data analysis. In: de Swart, H., Orłowska, E., Schmidt, G., Roubens, M. (eds.) Theory and Applications of Relational Structures as Knowledge Instruments. LNCS, vol. 2929, pp. 214–230. Springer, Heidelberg (2003)
3. Erne, M., Klossowski, E., Melton, A., Strecker, G.E.: A primer on Galois connections. In: Proceedings of the 1991 Summer Conference on General Topology and Applications in Honor of Mary Elen Rudin and Her Work. Annals of the New York Academy of Sciences, vol. 704, pp. 103–125 (1993)
4. Goré, R.: Dual Intuitionistic Logic Revisited. In: Dyckhoff, R. (ed.) TABLEAUX 2000. LNCS, vol. 1847, pp. 252–267. Springer, Heidelberg (2000)
5. Łukowski, P.: Modal interpretation of Heyting-Brouwer logic. Bulletin of the Section of Logic 2, 80–83 (1996)
6. Pagliani, P.: Intrinsic co-Heyting boundaries and information incompleteness in Rough Set Analysis. In: Polkowski, L., Skowron, A. (eds.) RSCTC 1998. LNCS (LNAI), vol. 1424, pp. 123–130. Springer, Heidelberg (1998)
7. Pagliani, P., Chakraborty, M.: Information quanta and approximation spaces. I: Non-classical approximation operators. In: Hu, X., Liu, Q., Skowron, A., Lin, T.S., Yager, R.R., Zhang, E.B. (eds.) Proceedings of the IEEE International Conference on Granular Computing, vol. 2, pp. 605–610. IEEE, Los Alamitos (2005)
8. Pagliani, P., Chakraborty, M.: Information quanta and approximation spaces. II: Generalised approximation space. In: Hu, X., Liu, Q., Skowron, A., Lin, T.S., Yager, R.R., Zhang, E.B. (eds.) Proceedings of the IEEE International Conference on Granular Computing, vol. 2, pp. 611–616. IEEE, Los Alamitos (2005)
9. Pawlak, Z.: Rough sets. Int. J. Computer and Information Sci. 11, 341–356 (1982)
10. Rasiowa, H.: Algebraic models of logics. University of Warsaw (2001)
11. Rauszer, C.: Representation theorem for semi-Boolean algebras I. Bulletin de l'Académie Polonaise des Sciences XIX(10), 881–887 (1971)
12. Rauszer, C.: Semi-boolean algebras and their applications to intuitionistic logic with dual operations. Fundamenta Mathematicae 83, 81–87 (1974)
13. Rauszer, C.: An algebraic and Kripke-style approach to a certain extensions of intuintionistic logic. Dissertationes Mathematicae CLXVII (1980)
14. Restall, G.: Extending intuitionistic logic with subtraction (1997), `www.phil.mq.edu.au/staff/grestall/files/extendingj.ps`
15. Skowron, A., Stepaniuk, J.: Generalised approximation spaces. In: Proceedings of 3rd International Workshop on Rough Sets and Soft Computing, San Jose, pp. 156–163 (1994)
16. Wille, R.: Reconstructing lattice theory: An approach based on hierarchies of concepts. In: Rival, I. (ed.) Ordered Sets, NATO Advanced Studies Institute, Dordrecht, Reidel, pp. 445–470 (1982)

17. Wolski, M.: Galois connections and data analysis. Fundamenta Informaticae 60, 401–415 (2004)
18. Wolski, M.: Formal Concept Analysis and Rough Set Theory from the perspective of finite topological approximations. In: Peters, J.F., Skowron, A. (eds.) Transactions on Rough Sets III. LNCS, vol. 3400, pp. 230–243. Springer, Heidelberg (2005)
19. Wolter, F.: On logics with coimplication. Journal of Philosophical Logic 27, 353–387 (1998)

On Partial Covers, Reducts and Decision Rules

Mikhail Ju. Moshkov[1], Marcin Piliszczuk[2], and Beata Zielosko[3]

[1] Institute of Computer Science, University of Silesia
39, Będzińska St., Sosnowiec, 41-200, Poland
moshkov@us.edu.pl
[2] ING Bank Śląski S.A., 34, Sokolska St., Katowice, 40-086, Poland
marcin.piliszczuk@ingbank.pl
[3] Institute of Computer Science, University of Silesia
39, Będzińska St., Sosnowiec, 41-200, Poland
zielosko@us.edu.pl

Abstract. In the paper, the accuracy of greedy algorithms for construction of partial covers, reducts and decision rules is considered. Bounds on the minimal complexity of partial covers, reducts and decision rules based on an information about greedy algorithm work are studied. The results of experiments with greedy algorithms are described.

Keywords: Rough sets, partial covers, partial reducts, partial decision rules, greedy algorithms.

1 Introduction

This paper is devoted to the consideration of partial decision reducts and partial decision rules on the basis of partial cover study. Partial reducts and partial decision rules are generalizations of exact reducts and exact decision rules which belong to the main notions of rough set theory [12,17].

Rough set theory often deals with decision tables containing noisy data. In this case exact reducts and exact decision rules can be "over-learned", i.e., depend essentially on the noise. In rough set theory reducts and decision rules are considered often as way for knowledge representation [16]. It is clear that, instead of an exact reduct with many attributes, it is more appropriate to work with a partial reduct containing a small number of attributes which separate almost all pairs of rows with different decisions. The same situation is with decision rules. In [13] Zdzisław Pawlak wrote that "the idea of an approximate reduct can be useful in cases when a smaller number of condition attributes is preferred over accuracy of classification".

Last years in rough set theory partial reducts and partial decision rules are studied intensively [9,10,20,21,22,23,25]. There are a number of approaches to the definition of approximate reducts [22]. In [10,21,22,23] it was proved that for each of the considered approaches the problem of partial reduct minimization (construction of a partial reduct with minimal cardinality) is NP-hard. The approach considered in [10] is the closest to the approach studied in this paper

J.F. Peters and A. Skowron (Eds.): Transactions on Rough Sets VIII, LNCS 5084, pp. 251–288, 2008.

(see also [21,23]). Approximate reducts are also investigated in the extensions of rough set model such as VPRS (variable precision rough sets) [27] and α-RST (alpha rough set theory) [14].

In the paper, we consider theoretical and experimental results on partial decision reducts and partial decision rules. These investigations are based on the study of partial covers. The results for covers and partial covers (including known results listed in Sect. 2.2) will be useful for wider spectrum of problems considered in rough set theory, for example, for the investigation of (i) reducts and rules for information systems, (ii) reducts and rules for decision tables with missing values, (iii) subsystems of a given decision rule system which "cover" the same set of rows, etc.

Based on the technique created by Ślęzak in [21,23], we generalize well known results of Feige [2], and Raz and Safra [15] on the precision of approximate polynomial algorithms for exact cover minimization (construction of an exact cover with minimal cardinality) to the case of partial covers. From obtained results and results of Slavík [18,19] on the precision of greedy algorithm for partial cover construction it follows that, under some natural assumptions on the class NP, the greedy algorithm for partial cover construction is close to the best polynomial approximate algorithms for partial cover minimization.

An information about the greedy algorithm work can be used for obtaining of lower and upper bounds on the minimal cardinality of partial covers. We fix some kind of information, and find the best lower and upper bounds depending on this information.

We obtain a new bound on the precision of greedy algorithm for partial cover construction which does not depend on the cardinality of covered set. This bound generalizes the bound obtained by Cheriyan and Ravi [1] and improves the bound obtained by Moshkov [8]. Based on the results of Slavík [18,19] on the precision of greedy algorithm for partial cover construction, we prove that obtained bound is, in some sense, unimprovable.

We prove that for the most part of set cover problems there exist exact (and, consequently, partial) covers with small cardinality. Experimental results show that, for the most part of randomly generated set cover problems, during each step the greedy algorithm chooses a subset which covers at least one half of uncovered elements. We prove that under some assumption, for the most part of set cover problems, during each step the greedy algorithm chooses a subset which covers at least one half of uncovered elements.

The most part of results obtained for partial covers is generalized to the case of partial reducts and partial decision rules. In particular, we show that

- Under some natural assumptions on the class NP, greedy algorithms are close to the best polynomial approximate algorithms for the minimization of the cardinality of partial reducts and the length of partial decision rules.
- Based on an information received during the greedy algorithm work, it is possible to obtain lower and upper bounds on the minimal cardinality of partial reducts and the minimal length of partial decision rules.

– For the most part of randomly generated binary decision tables, greedy algorithms construct simple partial reducts and partial decision rules with relatively high accuracy.

Obtained results will further to wider use of partial reducts and partial decision rules in rough set theory and applications.

Some similar problems were studied in [9] for the case, when each attribute has its own weight. Usually, results obtained in [9] are weaker and proofs from [9] are more complicated than the corresponding results and proofs from this paper. We must also note that even if all weights are equal to 1, then the results of the work of greedy algorithms considered in this paper can be different from the results of the work of greedy algorithms considered in [9].

The paper consists of five sections. In Sect. 2 partial covers are studied. In Sect. 3 partial tests (partial superreducts) and partial reducts are investigated. In Sect. 4 partial decision rules are considered. Sect. 5 contains short conclusions.

2 Partial Covers

2.1 Main Notions

Let $A = \{a_1, \ldots, a_n\}$ be a nonempty finite set and $S = \{B_i\}_{i \in \{1,\ldots,m\}} = \{B_1, \ldots, B_m\}$ be a family of subsets of A such that $B_1 \cup \ldots \cup B_m = A$. We assume that S can contain equal subsets of A. The pair (A, S) is called a *set cover problem*.

Let I be a subset of $\{1, \ldots, m\}$. The family $P = \{B_i\}_{i \in I}$ is called a *subfamily* of S. The number $|I|$ is called the *cardinality* of P and is denoted by $|P|$. Let $P = \{B_i\}_{i \in I}$ and $Q = \{B_i\}_{i \in J}$ be subfamilies of S. The notation $P \subseteq Q$ means that $I \subseteq J$. Let $P \cup Q = \{B_i\}_{i \in I \cup J}$, $P \cap Q = \{B_i\}_{i \in I \cap J}$, and $P \setminus Q = \{B_i\}_{i \in I \setminus J}$.

A subfamily $Q = \{B_{i_1}, \ldots, B_{i_t}\}$ of the family S is called a *partial cover for* (A, S). Let $\alpha \in \mathbb{R}$ and $0 \leq \alpha < 1$. The subfamily Q is called an *α-cover for* (A, S) if $|B_{i_1} \cup \ldots \cup B_{i_t}| \geq (1 - \alpha)|A|$. For example, 0.01-cover means that we must cover at least 99% of elements from A. Note that a 0-cover is an exact cover. By $C_{\min}(\alpha) = C_{\min}(\alpha, A, S)$ we denote the minimal cardinality of α-cover for (A, S). The notation $C_{\min}(\alpha)$ will be used in cases, when A and S are known.

Let us consider a greedy algorithm with threshold α (see Algorithm 1) which constructs an α-cover for (A, S). By $C_{\text{greedy}}(\alpha) = C_{\text{greedy}}(\alpha, A, S)$ we denote the cardinality of constructed α-cover for (A, S).

2.2 Known Results

First, we consider some known results for exact covers, when $\alpha = 0$.

Theorem 1. (Nigmatullin [11]) $C_{\text{greedy}}(0) \leq C_{\min}(0)(1 + \ln |A| - \ln C_{\min}(0))$.

Theorem 2. (Johnson [3], Lovász [6])

$$C_{\text{greedy}}(0) \leq C_{\min}(0)(1 + \ln(\max_{B_i \in S} |B_i|)) \leq C_{\min}(0)(1 + \ln |A|) .$$

Algorithm 1: Greedy algorithm for partial cover construction

Input : Set cover problem (A, S) with $S = \{B_1, \ldots, B_m\}$, and real number α, $0 \leq \alpha < 1$.
Output: α-cover for (A, S).

```
Q ⟵ ∅;
while Q is not an α-cover for (A, S) do
    select B_i ∈ S with minimal index i such that B_i covers the maximal number
    of elements from A uncovered by subsets from Q;
    Q ⟵ Q ∪ {B_i};
end
return Q;
```

More exact bounds (depending only on $|A|$) were obtained by Slavík [18,19].

Theorem 3. (Slavík [18,19]) *If* $|A| \geq 2$, *then* $C_{\rm greedy}(0) < C_{\rm min}(0)(\ln |A| - \ln \ln |A| + 0.78)$.

Theorem 4. (Slavík [18,19]) *For any natural* $m \geq 2$ *there exists a set cover problem* (A, S) *such that* $|A| = m$ *and* $C_{\rm greedy}(0) > C_{\rm min}(0)(\ln |A| - \ln \ln |A| - 0.31)$.

There are some results on exact and approximate polynomial algorithms for cover minimization.

Theorem 5. (Karp [4]) *The problem of construction of* 0*-cover with minimal cardinality is* NP*-hard.*

Theorem 6. (Feige [2]) *If* $NP \not\subseteq DTIME(n^{O(\log \log n)})$, *then for any* ε, $0 < \varepsilon < 1$, *there is no polynomial algorithm that for a given set cover problem* (A, S) *constructs a* 0*-cover for* (A, S) *which cardinality is at most* $(1-\varepsilon)C_{\rm min}(0) \ln |A|$.

Theorem 7. (Raz and Safra [15]) *If* $P \neq NP$, *then there exists* $\gamma > 0$ *such that there is no polynomial algorithm that for a given set cover problem* (A, S) *constructs a* 0*-cover for* (A, S) *which cardinality is at most* $\gamma C_{\rm min}(0) \ln |A|$.

Note that some results on the minimal exact covers for almost all set cover problems from some classes were obtained by Vercellis [24]. Kuzjurin in [5] investigated the behavior of greedy algorithm during the construction of exact covers for almost all problems from some classes of set cover problems such that each element from A belongs to the same number of subsets from S.

We will now consider some known results for partial covers, where $\alpha \geq 0$.

Theorem 8. (Slavík [18,19]) *Let* $0 \leq \alpha < 1$ *and* $\lceil (1-\alpha)|A| \rceil \geq 2$. *Then* $C_{\rm greedy}(\alpha) < C_{\rm min}(\alpha)(\ln \lceil (1-\alpha)|A| \rceil - \ln \ln \lceil (1-\alpha)|A| \rceil + 0.78)$.

Theorem 9. (Slavík [18,19]) *Let* $0 \leq \alpha < 1$. *Then for any natural* $t \geq 2$ *there exists a set cover problem* (A, S) *such that* $\lceil (1-\alpha)|A| \rceil = t$ *and* $C_{\rm greedy}(\alpha) > C_{\rm min}(\alpha)(\ln \lceil (1-\alpha)|A| \rceil - \ln \ln \lceil (1-\alpha)|A| \rceil - 0.31)$.

Theorem 10. (Slavík [19]) *Let* $0 \le \alpha < 1$. *Then* $C_{\text{greedy}}(\alpha) \le C_{\min}(\alpha)(1 + \ln(\max_{B_i \in S} |B_i|))$.

There are some bounds on $C_{\text{greedy}}(\alpha)$ which does not depend on $|A|$. Note that in the next two theorems we consider the case, where $\alpha > 0$.

Theorem 11. (Cheriyan and Ravi [1]) *Let* $0 < \alpha < 1$. *Then* $C_{\text{greedy}}(\alpha) \le C_{\min}(0) \ln \frac{1}{\alpha} + 1$.

This bound was rediscovered by Moshkov in [7] and generalized in [8].

Theorem 12. (Moshkov [8]) *Let* $0 < \beta \le \alpha < 1$. *Then* $C_{\text{greedy}}(\alpha) \le C_{\min}(\alpha - \beta) \ln \frac{1}{\beta} + 1$.

There is a result on exact polynomial algorithms for partial cover minimization.

Theorem 13. (Ślęzak [21,23]) *Let* $0 \le \alpha < 1$. *Then the problem of construction of* α*-cover with minimal cardinality is* NP*-hard.*

2.3 On Polynomial Approximate Algorithms

In this subsection, using technique created by Ślęzak in [21,23], we generalize the results of Feige, Raz and Safra (Theorems 6 and 7) to the case of partial covers.

When we say about a polynomial algorithm for set cover problems (A, S), it means that the time complexity of the considered algorithm is bounded from above by a polynomial depending on $|A|$ and $|S|$.

When we say about an algorithm, that for a given set cover problem (A, S) constructs an α-cover which cardinality is at most $f(A, S)C_{\min}(\alpha, A, S)$, we assume that in the case $f(A, S) < 1$ the considered algorithm constructs an α-cover for (A, S) which cardinality is equal to $C_{\min}(\alpha, A, S)$.

We consider an arbitrary set cover problem (A, S) with $S = \{B_1, \ldots, B_m\}$. Let $\alpha \in \mathbb{R}$ and $0 < \alpha < 1$. We correspond to (A, S) and α a set cover problem (A_α, S_α). Let $n(\alpha) = \left\lfloor \frac{|A|\alpha}{1-\alpha} \right\rfloor$ and $b_1, \ldots, b_{n(\alpha)}$ be elements which do not belong to the set A. Then $A_\alpha = A \cup \{b_1, \ldots, b_{n(\alpha)}\}$ and $S_\alpha = \{B_1, \ldots, B_m, B_{m+1}, \ldots, B_{m+n(\alpha)}\}$, where $B_{m+1} = \{b_1\}, \ldots, B_{m+n(\alpha)} = \{b_{n(\alpha)}\}$.

It is clear that there exists a polynomial algorithm which for a given set cover problem (A, S) and number α constructs the set cover problem (A_α, S_α).

Lemma 1. *Let* $Q \subseteq S$ *be a* 0*-cover for* (A, S) *and* α *be a real number such that* $0 < \alpha < 1$. *Then* Q *is an* α*-cover for* (A_α, S_α).

Proof. It is clear that $|A_\alpha| = |A| + n(\alpha)$. One can show that

$$|A| - 1 < (1 - \alpha)|A_\alpha| \le |A| \ . \tag{1}$$

It is clear that subsets from Q cover exactly $|A|$ elements from A_α. From (1) we conclude that Q is an α-cover for (A_α, S_α). □

Lemma 2. *Let $Q_\alpha \subseteq S_\alpha$ be an α-cover for (A_α, S_α). Then there exists $Q \subseteq S$ which is a 0-cover for (A, S) and for which $|Q| \leq |Q_\alpha|$. There exists a polynomial algorithm which for a given Q_α constructs corresponding Q.*

Proof. Let $Q_\alpha = Q^0 \cup Q^1$, where $Q^0 \subseteq S$ and $Q^1 \subseteq S_\alpha \setminus S$. If Q^0 covers all elements of the set A, then in the capacity of Q we can choose the set Q^0. Let Q^0 cover not all elements from A, A' be the set of uncovered elements from A, and $|A'| = m$. Taking into account that Q_α covers at least $(1-\alpha)|A_\alpha|$ elements from A_α and using (1) we conclude that Q_α covers greater than $|A|-1$ elements. Thus, Q_α covers at least $|A|$ elements. It is clear that each subset from $S_\alpha \setminus S$ covers exactly one element. Therefore, $|Q^1| \geq m$. One can show that there exists a polynomial algorithm which finds $t \leq m$ subsets $B_{i_1}, \ldots, B_{i_t}$ from S covering all elements from A'. Set $Q = Q^0 \cup \{B_{i_1}, \ldots, B_{i_t}\}$. It is clear that Q is a 0-cover for (A, S), and $|Q| \leq |Q_\alpha|$. □

Corollary 1. *Let $\alpha \in \mathbb{R}$ and $0 < \alpha < 1$. Then $C_{\min}(0, A, S) = C_{\min}(\alpha, A_\alpha, S_\alpha)$.*

Proof. From Lemma 1 it follows that $C_{\min}(\alpha, A_\alpha, S_\alpha) \leq C_{\min}(0, A, S)$. From Lemma 2 it follows that $C_{\min}(0, A, S) \leq C_{\min}(\alpha, A_\alpha, S_\alpha)$. □

Lemma 3. *Let α, b and δ be real numbers such that $0 < \alpha < 1$, $b > 0$ and $\delta > 0$, and let there exist a polynomial algorithm $\mathcal{A}$ that, for a given set cover problem (A, S), constructs an α-cover which cardinality is at most $b \ln |A| C_{\min}(\alpha, A, S)$. Then there exists a polynomial algorithm $\mathcal{B}$ that, for a given set cover problem (A, S), constructs a 0-cover which cardinality is at most $(b+\delta) \ln |A| C_{\min}(0, A, S)$.*

Proof. Let us describe the work of the algorithm $\mathcal{B}$. Let $\beta = 1 + \frac{\alpha}{1-\alpha}$ and $a = \max\left\{\frac{1}{b}, \frac{b \ln \beta}{\delta}\right\}$. If $\ln |A| \leq a$, then, in polynomial time, we consider all subfamilies of S, which cardinality is at most $|A|$, and find among them a 0-cover for (A, S) with minimal cardinality. It is clear that the cardinality of this 0-cover is equal to $C_{\min}(0, A, S)$.

Let $\ln |A| > a$. Then $b \ln |A| > 1$, $(b+\delta) \ln |A| > 1$ and

$$\delta \ln |A| > b \ln \beta \ . \tag{2}$$

In polynomial time, we construct the problem (A_α, S_α), and apply to this problem the polynomial algorithm $\mathcal{A}$. As a result, we obtain an α-cover Q_α for (A_α, S_α) such that $|Q_\alpha| \leq b \ln |A_\alpha| C_{\min}(\alpha, A_\alpha, S_\alpha)$.

It is clear that $|A_\alpha| \leq |A|\beta$. Using Corollary 1 we obtain $C_{\min}(\alpha, A_\alpha, S_\alpha) = C_{\min}(0, A, S)$. Therefore, $|Q_\alpha| \leq b(\ln |A| + \ln \beta) C_{\min}(0, A, S)$.

From (2) we obtain $b(\ln |A| + \ln \beta) = (b+\delta) \ln |A| - \delta \ln |A| + b \ln \beta \leq (b+\delta) \ln |A|$. Therefore, $|Q_\alpha| \leq (b+\delta) \ln |A| C_{\min}(0, A, S)$. From Lemma 2 we conclude that, in polynomial time, we can construct a 0-cover Q for (A, S) such that $|Q| \leq (b+\delta) \ln |A| C_{\min}(0, A, S)$. □

We now generalize Theorem 6 to the case of partial covers.

Theorem 14. *Let $\alpha \in \mathbb{R}$ and $0 \leq \alpha < 1$. If $NP \not\subseteq DTIME(n^{O(\log\log n)})$, then for any ε, $0 < \varepsilon < 1$, there is no polynomial algorithm that for a given set cover problem (A, S) constructs an α-cover for (A, S) which cardinality is at most $(1 - \varepsilon)C_{\min}(\alpha, A, S) \ln |A|$.*

Proof. If $\alpha = 0$, then the statement of the theorem coincides with Theorem 6. Let $\alpha > 0$. Let us assume that the considered statement does not hold: let $NP \not\subseteq DTIME(n^{O(\log\log n)})$ and for some ε, $0 < \varepsilon < 1$, there exist a polynomial algorithm $\mathcal{A}$ that, for a given set cover problem (A, S), constructs an α-cover for (A, S) which cardinality is at most $(1 - \varepsilon)C_{\min}(\alpha, A, S) \ln |A|$.

Applying Lemma 3 with parameters $b = (1 - \varepsilon)$ and $\delta = \frac{\varepsilon}{2}$ we conclude that, under the assumption $NP \not\subseteq DTIME(n^{O(\log\log n)})$, there exists a polynomial algorithm $\mathcal{B}$ that, for a given set cover problem (A, S), constructs a 0-cover for (A, S) which cardinality is at most $(1 - \frac{\varepsilon}{2})C_{\min}(0, A, S) \ln |A|$. Last statement contradicts Theorem 6. □

From Theorem 10 it follows that $C_{\text{greedy}}(\alpha) \leq C_{\min}(\alpha)(1 + \ln |A|)$. From this inequality and from Theorem 14 it follows that, under the assumption $NP \not\subseteq DTIME(n^{O(\log\log n)})$, the greedy algorithm is close to the best polynomial approximate algorithms for partial cover minimization.

We now generalize Theorem 7 to the case of partial covers.

Theorem 15. *Let $\alpha \in \mathbb{R}$ and $0 \leq \alpha < 1$. If $P \neq NP$, then there exists $\rho > 0$ such that there is no polynomial algorithm that for a given set cover problem (A, S) constructs an α-cover for (A, S) which cardinality is at most $\rho C_{\min}(\alpha, A, S) \ln |A|$.*

Proof. If $\alpha = 0$, then the statement of the theorem coincides with Theorem 7. Let $\alpha > 0$. We will now show that in the capacity of ρ we can take the number $\frac{\gamma}{2}$, where γ is the constant from Theorem 7. Let us assume the contrary: let $P \neq NP$, and a polynomial algorithm $\mathcal{A}$ exist that, for a given set cover problem (A, S), constructs an α-cover for (A, S) which cardinality is at most $\frac{\gamma}{2}C_{\min}(\alpha, A, S) \ln |A|$.

Applying Lemma 3 with parameters $b = \frac{\gamma}{2}$ and $\delta = \frac{\gamma}{2}$ we conclude that, under the assumption $P \neq NP$, there exists a polynomial algorithm $\mathcal{B}$ that, for a given set cover problem (A, S), constructs a 0-cover for (A, S) which cardinality is at most $\gamma C_{\min}(0, A, S) \ln |A|$. Last statement contradicts Theorem 7. □

2.4 Bounds on $C_{\min}(\alpha)$ Based on Information About Greedy Algorithm Work

Using information on the greedy algorithm work we can obtain bounds on $C_{\min}(\alpha)$. We consider the two simple examples. It is clear that $C_{\min}(\alpha) \leq C_{\text{greedy}}(\alpha)$. From Theorem 10 it follows that $C_{\text{greedy}}(\alpha) \leq C_{\min}(\alpha)(1 + \ln |A|)$. Therefore, $C_{\min}(\alpha) \geq \frac{C_{\text{greedy}}(\alpha)}{1+\ln |A|}$. Another lower bounds on $C_{\min}(\alpha)$ can be obtained based on Theorems 8 and 12.

In this subsection, we fix some information on the greedy algorithm work, and find the best upper and lower bounds on $C_{\min}(\alpha)$ depending on this information.

Information on Greedy Algorithm Work. Assume that (A, S) is a set cover problem and α is a real number such that $0 \le \alpha < 1$. Let us apply the greedy algorithm with threshold α to the problem (A, S). Assume that during the construction of α-cover the greedy algorithm chooses consequently subsets $B_{j_1}, \ldots, B_{j_t}$. Set $B_{j_0} = \emptyset$ and for $i = 1, \ldots, t$ set $\delta_i = |B_{j_i} \setminus (B_{j_0} \cup \ldots \cup B_{j_{i-1}})|$.

Write $\Delta(\alpha, A, S) = (\delta_1, \ldots, \delta_t)$. As information on the greedy algorithm work we will use the tuple $\Delta(\alpha, A, S)$ and numbers $|A|$ and α. Note that $\delta_1 = \max\{|B_i| : B_i \in S\}$ and $t = C_{\text{greedy}}(\alpha, A, S)$. Let us denote by P_{SC} the set of set cover problems and $D_{SC} = \{(\alpha, |A|, \Delta(\alpha, A, S)) : \alpha \in \mathbb{R}, 0 \le \alpha < 1, (A, S) \in P_{SC}\}$.

Lemma 4. *A tuple $(\alpha, n, (\delta_1, \ldots, \delta_t))$ belongs to the set D_{SC} if and only if α is a real number such that $0 \le \alpha < 1$, and $n, \delta_1, \ldots, \delta_t$ are natural numbers such that $\delta_1 \ge \ldots \ge \delta_t$, $\sum_{i=1}^{t-1} \delta_i < (1 - \alpha)n$ and $(1 - \alpha)n \le \sum_{i=1}^{t} \delta_i \le n$.*

Proof. Let $(\alpha, n, (\delta_1, \ldots, \delta_t)) \in D_{SC}$ and $(\alpha, n, (\delta_1, \ldots, \delta_t)) = (\alpha, |A|, \Delta(\alpha, A, S))$. It is clear that α is a real number, $0 \le \alpha < 1$, and $n, \delta_1, \ldots, \delta_t$ are natural numbers. From the definition of greedy algorithm it follows that $\delta_1 \ge \ldots \ge \delta_t$. Taking into account that α is the threshold for the greedy algorithm we obtain $\sum_{i=1}^{t-1} \delta_i < (1 - \alpha)n$ and $(1 - \alpha)n \le \sum_{i=1}^{t} \delta_i \le n$.

Let $(\alpha, n, (\delta_1, \ldots, \delta_t))$ be a tuple for which α is a real number such that $0 \le \alpha < 1$, and $n, \delta_1, \ldots, \delta_t$ are natural numbers such that $\delta_1 \ge \ldots \ge \delta_t$, $\sum_{i=1}^{t-1} \delta_i < (1-\alpha)n$ and $(1-\alpha)n \le \sum_{i=1}^{t} \delta_i \le n$. We define a set cover problem (A, S) in the following way: $A = \{a_1, \ldots, a_n\}$ and $S = \{\{a_1, \ldots, a_{\delta_1}\}, \ldots, \{a_{\delta_1 + \ldots + \delta_{t-1} + 1}, \ldots, a_{\delta_1 + \ldots + \delta_t}\}, \{a_{\delta_1 + \ldots + \delta_t + 1}\}, \ldots, \{a_n\}\}$ (for simplicity, we omit here notation $B_1 = \{a_1, \ldots, a_{\delta_1}\}, \ldots$). It is not difficult to show that $\Delta(\alpha, A, S) = (\delta_1, \ldots, \delta_t)$. Thus, $(\alpha, n, (\delta_1, \ldots, \delta_t)) \in D_{SC}$. □

Best Upper Bound for $C_{\min}(\alpha)$. We define a function $\mathcal{U}_{SC} : D_{SC} \to \mathbb{N}$. Let $(\alpha, n, (\delta_1, \ldots, \delta_t)) \in D_{SC}$. Then $\mathcal{U}_{SC}(\alpha, n, (\delta_1, \ldots, \delta_t)) = \max\{C_{\min}(\alpha, A, S) : (A, S) \in P_{SC}, |A| = n, \Delta(\alpha, A, S) = (\delta_1, \ldots, \delta_t)\}$. It is clear that $C_{\min}(\alpha, A, S) \le \mathcal{U}_{SC}(\alpha, |A|, \Delta(\alpha, A, S))$ is the best upper bound for $C_{\min}(\alpha)$ depending on α, $|A|$ and $\Delta(\alpha, A, S)$.

Theorem 16. *Let $(\alpha, n, (\delta_1, \ldots, \delta_t)) \in D_{SC}$. Then $\mathcal{U}_{SC}(\alpha, n, (\delta_1, \ldots, \delta_t)) = t$.*

Proof. Let us consider an arbitrary set cover problem (A, S) such that $|A| = n$ and $\Delta(\alpha, A, S) = (\delta_1, \ldots, \delta_t)$. It is clear that $C_{\min}(\alpha, A, S) \le C_{\text{greedy}}(\alpha, A, S)$. Since $C_{\text{greedy}}(\alpha, A, S) = t$, we have $\mathcal{U}(\alpha, n, (\delta_1, \ldots, \delta_t)) \le t$.

We consider the following set cover problem (A, S): $A = \{a_1, \ldots, a_n\}$ and $S = \{\{a_1, \ldots, a_{\delta_1}\}, \ldots, \{a_{\delta_1 + \ldots + \delta_{t-1} + 1}, \ldots, a_{\delta_1 + \ldots + \delta_t}\}, \{a_{\delta_1 + \ldots + \delta_t + 1}\}, \ldots, \{a_n\}\}$ (we omit here notation $B_1 = \{a_1, \ldots, a_{\delta_1}\}, \ldots$). It is clear that $|A| = n$. Lemma 4 now shows that $\Delta(\alpha, A, S) = (\delta_1, \ldots, \delta_t)$. Taking into account that all subsets from S are pairwise disjoint it is not difficult to prove that $C_{\min}(\alpha, A, S) = C_{\text{greedy}}(\alpha, A, S) = t$. Therefore, $\mathcal{U}_{SC}(\alpha, n, (\delta_1, \ldots, \delta_t)) \ge t$. □

Thus, $C_{\min}(\alpha, A, S) \le C_{\text{greedy}}(\alpha, A, S)$ is the best upper bound for $C_{\min}(\alpha)$ depending on α, $|A|$ and $\Delta(\alpha, A, S)$.

Best Lower Bound for $C_{\min}(\alpha)$. We define a function $\mathcal{L}_{SC} : D_{SC} \to \mathbb{N}$. Let $(\alpha, n, (\delta_1, \ldots, \delta_t)) \in D_{SC}$. Then $\mathcal{L}_{SC}(\alpha, n, (\delta_1, \ldots, \delta_t)) = \min\{C_{\min}(\alpha, A, S) : (A, S) \in P_{SC}, |A| = n, \Delta(\alpha, A, S) = (\delta_1, \ldots, \delta_t)\}$. It is clear that $C_{\min}(\alpha, A, S) \geq \mathcal{L}_{SC}(\alpha, |A|, \Delta(\alpha, A, S))$ is the best lower bound for $C_{\min}(\alpha)$ depending on α, $|A|$ and $\Delta(\alpha, A, S)$. For $(\alpha, n, (\delta_1, \ldots, \delta_t)) \in D_{SC}$ and $\delta_0 = 0$ set

$$l(\alpha, n, (\delta_1, \ldots, \delta_t)) = \max\left\{\left\lceil \frac{\lceil (1-\alpha)n \rceil - (\delta_0 + \ldots + \delta_i)}{\delta_{i+1}} \right\rceil : i = 0, \ldots, t-1\right\} .$$

Theorem 17. *Let $(\alpha, n, (\delta_1, \ldots, \delta_t)) \in D_{SC}$. Then $\mathcal{L}_{SC}(\alpha, n, (\delta_1, \ldots, \delta_t)) = l(\alpha, n, (\delta_1, \ldots, \delta_t))$.*

Proof. Let us consider an arbitrary set cover problem (A, S) such that $|A| = n$ and $\Delta(\alpha, A, S) = (\delta_1, \ldots, \delta_t)$. Set $p = C_{\min}(\alpha, A, S)$. It is clear that there exist p subsets from S which cover a subset V of the set A such that $|V| \geq \lceil (1-\alpha)n \rceil$.

Let $i \in \{0, \ldots, t-1\}$. After i steps of the greedy algorithm work, at least $\lceil (1-\alpha)n \rceil - (\delta_0 + \ldots + \delta_i)$ elements from the set V are uncovered. Therefore, in the family S there is a subset which can cover at least $\frac{\lceil (1-\alpha)n \rceil - (\delta_0 + \ldots + \delta_i)}{p}$ of uncovered elements. Thus, $\delta_{i+1} \geq \frac{\lceil (1-\alpha)n \rceil - (\delta_0 + \ldots + \delta_i)}{p}$ and $p \geq \frac{\lceil (1-\alpha)n \rceil - (\delta_0 + \ldots + \delta_i)}{\delta_{i+1}}$. Since p is a natural number, we have $p \geq \left\lceil \frac{\lceil (1-\alpha)n \rceil - (\delta_0 + \ldots + \delta_i)}{\delta_{i+1}} \right\rceil$. Taking into account that i is an arbitrary number from $\{0, \ldots, t-1\}$ we obtain $C_{\min}(\alpha, A, S) \geq l(\alpha, n, (\delta_1, \ldots, \delta_t))$. Thus, $\mathcal{L}_{SC}(\alpha, n, (\delta_1, \ldots, \delta_t)) \geq l(\alpha, n, (\delta_1, \ldots, \delta_t))$.

Let us show that $\mathcal{L}(\alpha, n, (\delta_1, \ldots, \delta_t)) \leq l(\alpha, n, (\delta_1, \ldots, \delta_t))$.

Write $d = l(\alpha, n, (\delta_1, \ldots, \delta_t))$, $r = \lceil (1-\alpha)n \rceil$ and $q = n - (\delta_1 + \ldots + \delta_t)$. Let us consider the following set cover problem (A, S): $A = \{a_1, \ldots, a_n\}$ and $S = \{B_1, \ldots, B_t, B_{t+1}, \ldots, B_{t+q}, B_{t+q+1}, \ldots, B_{t+q+d}\}$, where $B_1 = \{a_1, \ldots, a_{\delta_1}\}$, ..., $B_t = \{a_{\delta_1+\ldots+\delta_{t-1}+1}, \ldots, a_{\delta_1+\ldots+\delta_t}\}$, $B_{t+1} = \{a_{\delta_1+\ldots+\delta_t+1}\}$, ..., $B_{t+q} = \{a_n\}$. Let $D = \{a_1, \ldots, a_r\}$. For $j = 1, \ldots, d$, the set B_{t+q+j} includes all elements from the set D of the kind $a_{r-id-j+1}$, $i = 0, 1, 2, \ldots$, and only such elements.

It is clear that subsets $B_{t+q+1}, \ldots, B_{t+q+d}$ form an α-cover for (A, S). Therefore, $C_{\min}(\alpha, A, S) \leq d$.

We prove by induction on $j = 1, \ldots, t$ that, during the step number j, the greedy algorithm chooses the subset B_j from S. Note that from Lemma 4 it follows that $\delta_1 \geq \ldots \geq \delta_t$.

Let us consider the first step of greedy algorithm. It is clear that the cardinality of B_1 is equal to δ_1, and δ_1 is greater than of equal to the cardinality of each of sets $B_2, \ldots, B_{t+q}$. Let us show that δ_1 is greater than of equal to the cardinality of each of sets $B_{t+q+1}, \ldots, B_{t+q+d}$. We have $\left\lceil \frac{r}{\delta_1} \right\rceil \leq d$. Therefore, $\frac{r}{\delta_1} \leq d$ and $\frac{r}{d} \leq \delta_1$. Let $r = sd + a$, where s is a nonnegative integer and $a \in \{0, 1, \ldots, d-1\}$. Then the cardinality of each of the sets $B_{t+q+1}, \ldots, B_{t+q+d}$ is equal to s if $a = 0$, and is at most $s+1$ if $a > 0$. From the inequality $\frac{r}{d} \leq \delta_1$ it follows that $\delta_1 \geq s$ if $a = 0$, and $\delta_1 \geq s+1$ if $a > 0$. So at the first step the greedy algorithm chooses the set B_1.

Let us assume that during j steps, $1 \leq j \leq t-1$, the greedy algorithm chooses the sets $B_1, \ldots, B_j$. Let us consider the step number $j+1$. It is clear that B_{j+1}

covers δ_{j+1} uncovered elements. One can show that each set from $B_{j+2}, \ldots, B_{t+q}$ covers at most δ_{j+1} uncovered elements. Set $u = r-(\delta_1+\ldots+\delta_j)$. Let $u = sd+a$, where s is a nonnegative integer and $a \in \{0, 1, \ldots, d-1\}$. One can show that each set from $B_{t+q+1}, \ldots, B_{t+q+d}$ covers at most s uncovered elements if $a = 0$, and at most $s+1$ uncovered elements if $a > 0$. It is clear that $\left\lceil \frac{u}{\delta_{j+1}} \right\rceil \le d$. Therefore, $\frac{u}{\delta_{j+1}} \le d$ and $\frac{u}{d} \le \delta_{j+1}$. Hence, $\delta_{j+1} \ge s$ if $a = 0$, and $\delta_{j+1} \ge s+1$ if $a > 0$. So at the step number $j+1$ the greedy algorithm chooses the set B_{j+1}.

Since the greedy algorithm chooses subsets $B_1, \ldots, B_t$, we have $\Delta(\alpha, A, S) = (\delta_1, \ldots, \delta_t)$. Therefore, $C_{\min}(\alpha) \ge d$. As it was proved earlier, $C_{\min}(\alpha) \le d$. Hence, $C_{\min}(\alpha) = d$ and $\mathcal{L}_{SC}(\alpha, n, (\delta_1, \ldots, \delta_t)) \le l(\alpha, n, (\delta_1, \ldots, \delta_t))$. Therefore, $\mathcal{L}_{SC}(\alpha, n, (\delta_1, \ldots, \delta_m)) = l(\alpha, n, (\delta_1, \ldots, \delta_t))$. □

So $C_{\min}(\alpha, A, S) \ge l(\alpha, |A|, \Delta(\alpha, A, S))$ is the best lower bound for $C_{\min}(\alpha)$ depending on α, $|A|$ and $\Delta(\alpha, A, S)$.

Properties of Best Lower Bound for $C_{\min}(\alpha)$. Assume that (A, S) is a set cover problem and α is a real number such that $0 \le \alpha < 1$. Let

$$l_{SC}(\alpha) = l_{SC}(\alpha, A, S) = l(\alpha, |A|, \Delta(\alpha, A, S)) \ .$$

Lemma 5. *Let $\alpha_1, \alpha_2 \in \mathbb{R}$ and $0 \le \alpha_1 < \alpha_2 < 1$. Then $l_{SC}(\alpha_1) \ge l_{SC}(\alpha_2)$.*

Proof. Let $\Delta(\alpha_1, A, S) = (\delta_1, \ldots, \delta_{t_1})$ and $\Delta(\alpha_2, A, S) = (\delta_1, \ldots, \delta_{t_2})$. We have $t_1 \ge t_2$. Let $\delta_0 = 0$, $j \in \{0, \ldots, t_2-1\}$ and $\left\lceil \frac{\lceil |A|(1-\alpha_2) \rceil - (\delta_0+\ldots+\delta_j)}{\delta_{j+1}} \right\rceil = l_{SC}(\alpha_2)$. It is clear that $l_{SC}(\alpha_1) \ge \left\lceil \frac{\lceil |A|(1-\alpha_1) \rceil - (\delta_0+\ldots+\delta_j)}{\delta_{j+1}} \right\rceil \ge l_{SC}(\alpha_2)$. □

Corollary 2. $l_{SC}(0) = \max\{l_{SC}(\alpha) : 0 \le \alpha < 1\}$.

The value $l_{SC}(\alpha)$ can be used for obtaining of upper bounds on the cardinality of partial covers constructed by the greedy algorithm.

Theorem 18. *Let α and β be real numbers such that $0 < \beta \le \alpha < 1$. Then $C_{\mathrm{greedy}}(\alpha) < l_{SC}(\alpha-\beta) \ln\left(\frac{1-\alpha+\beta}{\beta}\right) + 1$.*

Proof. Let $\Delta(\alpha-\beta, A, S) = (\delta_1, \ldots, \delta_t)$, $\delta_0 = 0$, $M = (1-\alpha+\beta)|A|$ and $l = l_{SC}(\alpha-\beta)$. We have $l \ge 1$ and $l \ge \max\left\{\frac{M-(\delta_0+\ldots+\delta_i)}{\delta_{i+1}} : i = 0, \ldots, t-1\right\}$. Therefore, for $i = 0, \ldots, t-1$, $\frac{M-(\delta_0+\ldots+\delta_i)}{\delta_{i+1}} \le l$ and

$$\frac{M-(\delta_0+\ldots+\delta_i)}{l} \le \delta_{i+1} \ . \tag{3}$$

Let us assume that $l = 1$. Then $\delta_1 \ge M$ and $C_{\mathrm{greedy}}(\alpha) = 1$. It is clear that $l_{SC}(\alpha-\beta) \ln\left(\frac{1-\alpha+\beta}{\beta}\right) > 0$. Therefore, if $l = 1$, then the statement of the theorem holds. Let $l \ge 2$. Let us show that for $j = 1, \ldots, t$,

$$M-(\delta_0+\ldots+\delta_j) \le M\left(1-\frac{1}{l}\right)^j \ . \tag{4}$$

For $i = 0$, from (3) it follows that $\delta_1 \geq \frac{M}{l}$. Therefore, (4) holds for $j = 1$. Let us assume that (4) holds for some j, $1 \leq j \leq t-1$. Let us show that

$$M - (\delta_0 + \ldots + \delta_{j+1}) \leq M \left(1 - \frac{1}{l}\right)^{j+1} . \quad (5)$$

Write $Q = M - (\delta_0 + \ldots + \delta_j)$. For $i = j$, from (3) it follows that $\delta_{j+1} \geq \frac{Q}{l}$. Using this inequality and (4) we obtain $M - (\delta_0 + \ldots + \delta_{j+1}) \leq Q - \frac{Q}{l} \leq Q\left(1 - \frac{1}{l}\right) \leq M\left(1 - \frac{1}{l}\right)^{j+1}$. Therefore, (5) holds. Thus, (4) holds.

Let $C_{\mathrm{greedy}}(\alpha) = p$. It is clear that $C_{\mathrm{greedy}}(\alpha) \leq C_{\mathrm{greedy}}(\alpha - \beta) = t$. Therefore, $p \leq t$. It is clear that $\delta_1 + \ldots + \delta_{p-1} < |A|(1-\alpha)$. Using (4) we obtain $M - M\left(1 - \frac{1}{l}\right)^{p-1} \leq \delta_1 + \ldots + \delta_{p-1}$. Therefore, $|A|(1-\alpha+\beta) - |A|(1-\alpha+\beta)\left(1 - \frac{1}{l}\right)^{p-1} < |A|(1-\alpha)$. Hence, $|A|\beta < |A|(1-\alpha+\beta)\left(1-\frac{1}{l}\right)^{p-1} = |A|(1-\alpha+\beta)\left(\frac{l-1}{l}\right)^{p-1}$ and $\left(\frac{l}{l-1}\right)^{p-1} < \frac{1-\alpha+\beta}{\beta}$. If we take the natural logarithm of both sides of this inequality, we obtain $(p-1)\ln\left(1 + \frac{1}{l-1}\right) < \ln\left(\frac{1-\alpha+\beta}{\beta}\right)$. Taking into account that $l-1$ is a natural number, and using the inequality $\ln\left(1+\frac{1}{r}\right) > \frac{1}{r+1}$, which holds for any natural r, we obtain $\ln\left(1 + \frac{1}{l-1}\right) > \frac{1}{l}$. Therefore, $C_{\mathrm{greedy}}(\alpha) = p < l \ln\left(\frac{1-\alpha+\beta}{\beta}\right) + 1 = l_{SC}(\alpha-\beta)\ln\left(\frac{1-\alpha+\beta}{\beta}\right) + 1$. □

Corollary 3. *Let $\alpha \in \mathbb{R}$ and $0 < \alpha < 1$. Then $C_{\mathrm{greedy}}(\alpha) < l_{SC}(0)\ln\left(\frac{1}{\alpha}\right) + 1$.*

If $l_{SC}(0)$ is a small number, then we have a good upper bound on $C_{\mathrm{greedy}}(\alpha)$. If $l_{SC}(0)$ is a big number, then we have a big lower bound on $C_{\min}(0)$ and on $C_{\min}(\alpha)$ for some α.

2.5 Upper Bound on $C_{\mathrm{greedy}}(\alpha)$

In this subsection, we obtain one more upper bound on $C_{\mathrm{greedy}}(\alpha)$ which does not depend on $|A|$, and show that, in some sense, this bound is unimprovable.

Theorem 19. *Let α and β be real numbers such that $0 < \beta \leq \alpha < 1$. Then $C_{\mathrm{greedy}}(\alpha) < C_{\min}(\alpha-\beta)\ln\left(\frac{1-\alpha+\beta}{\beta}\right) + 1$.*

Proof. From Theorem 18 it follows that $C_{\mathrm{greedy}}(\alpha) < l_{SC}(\alpha-\beta)\ln\left(\frac{1-\alpha+\beta}{\beta}\right) + 1$, and from Theorem 17 it follows that $l_{SC}(\alpha-\beta) \leq C_{\min}(\alpha-\beta)$. □

Let us show that obtained bound is, in some sense, unimprovable.

Lemma 6. *Let α be a real number, $0 \leq \alpha < 1$, $j \in \{0, \ldots, |A|-1\}$ and $\frac{j}{|A|} \leq \alpha < \frac{j+1}{|A|}$. Then $C_{\min}(\alpha) = C_{\min}(\frac{j}{|A|})$ and $C_{\mathrm{greedy}}(\alpha) = C_{\mathrm{greedy}}(\frac{j}{|A|})$.*

Proof. Taking into account that $\frac{j}{|A|} \leq \alpha$ we conclude that $C_{\min}(\alpha) \leq C_{\min}(\frac{j}{|A|})$ and $C_{\mathrm{greedy}}(\alpha) \leq C_{\mathrm{greedy}}(\frac{j}{|A|})$.

Let $Q = \{B_{i_1}, \ldots, B_{i_t}\}$ be an arbitrary α-cover for (A, S). Let $M = |B_{i_1} \cup \ldots \cup B_{i_t}|$. It is clear that $M \geq |A|(1-\alpha)$. Therefore, $1 - \frac{M}{|A|} \leq \alpha$. Taking into account that $\alpha < \frac{j+1}{|A|}$ we obtain $|A| - M < j+1$. Hence, $|A| - M \leq j$ and $|A| - j \leq M$. Therefore, $M \geq |A|(1 - \frac{j}{|A|})$, and Q is also an $\frac{j}{|A|}$-cover. Thus, each α-cover is an $\frac{j}{|A|}$-cover. Using this fact it is not difficult to show that $C_{\min}(\alpha) \geq C_{\min}(\frac{j}{|A|})$ and $C_{\text{greedy}}(\alpha) \geq C_{\text{greedy}}(\frac{j}{|A|})$. □

Theorem 20. *There is no real $\delta < 1$ such that for any set cover problem (A, S) and for any real α and β, $0 < \beta \leq \alpha < 1$, the following inequality holds:*

$$C_{\text{greedy}}(\alpha) \leq \delta \left(C_{\min}(\alpha - \beta) \ln \left(\frac{1 - \alpha + \beta}{\beta} \right) + 1 \right) . \tag{6}$$

Proof. Assume the contrary: let such δ exist. We now consider an arbitrary α, $0 < \alpha < 1$, and an arbitrary set cover problem (A, S). Let $j \in \{0, \ldots, |A| - 1\}$ and $\frac{j}{|A|} \leq \alpha < \frac{j+1}{|A|}$. Using (6) we obtain

$$\begin{aligned} C_{\text{greedy}} \left(\frac{j}{|A|} + \frac{1}{2|A|} \right) &\leq \delta \left(C_{\min} \left(\frac{j}{|A|} \right) \ln \left(\frac{1 - \frac{j}{|A|} - \frac{1}{2|A|} + \frac{1}{2|A|}}{\frac{1}{2|A|}} \right) + 1 \right) \\ &= \delta \left(C_{\min} \left(\frac{j}{|A|} \right) \ln \left(|A| - j \right) + C_{\min} \left(\frac{j}{|A|} \right) \ln 2 + 1 \right) . \end{aligned}$$

Lemma 6 now shows $C_{\text{greedy}} \left(\frac{j}{|A|} + \frac{j}{2|A|} \right) = C_{\text{greedy}} \left(\frac{j}{|A|} \right) = C_{\text{greedy}}(\alpha)$ and $C_{\min}(\frac{j}{|A|}) = C_{\min}(\alpha)$. Let us evaluate the number $|A| - j$. We have $j \leq \alpha|A| < j+1$. Therefore, $|A| - j - 1 < |A| - \alpha|A| \leq |A| - j$ and $|A| - j = \lceil (1-\alpha)|A| \rceil$. Finally, we have

$$C_{\text{greedy}}(\alpha) \leq \delta \left(C_{\min}(\alpha) \ln \left(\lceil (1-\alpha)|A| \rceil \right) + C_{\min}(\alpha) \ln 2 + 1 \right) . \tag{7}$$

Using Theorem 9 we conclude that for any natural $t \geq 2$ there exists a set cover problem (A_t, S_t) such that $\lceil (1-\alpha)|A_t| \rceil = t$ and $C_{\text{greedy}}(\alpha, A_t, S_t) > C_{\min}(\alpha, A_t, S_t)(\ln t - \ln \ln t - 0.31)$. Let $C_t = C_{\min}(\alpha, A_t, S_t)$. Using (7) we obtain for any $t \geq 2$, $C_t(\ln t - \ln \ln t - 0.31) < \delta(C_t \ln t + C_t \ln 2 + 1)$. If we divide both sides of this inequality by $C_t \ln t$, we obtain $1 - \frac{\ln \ln t}{\ln t} - \frac{0.31}{\ln t} < \delta + \frac{\delta \ln 2}{\ln t} + \frac{\delta}{C_t \ln t}$.

It is clear that $C_t \geq 1$. Therefore, with growth of t the left-hand side of this inequality tends to 1, and the right-hand side of this inequality tends to δ, which is impossible. □

2.6 On Covers for the Most Part of Set Cover Problems

Assume that (A, S) is a set cover problem, the elements of A are enumerated by numbers $1, \ldots, n$, and sets from S are enumerated by numbers $1, \ldots, m$. It is possible that sets from S with different numbers are equal. There is a one-to-one correspondence between such set cover problems and tables with n rows and m columns filled by numbers from $\{0, 1\}$ and having no rows filled by 0 only. Let

$A = \{a_1, \ldots, a_n\}$ and $S = \{B_1, \ldots, B_m\}$. Then the problem (A, S) corresponds to the table which, for $i = 1, \ldots, n$ and $j = 1, \ldots, m$, has 1 at the intersection of i-th row and j-th column if and only if $a_i \in B_j$.

A table filled by numbers from $\{0, 1\}$ will be called *SC-table* if this table has no rows filled by 0 only.

Lemma 7. *The number of SC-tables with n rows and m columns is at least $2^{mn} - 2^{mn-m+\log_2 n}$.*

Proof. Let $i \in \{1, \ldots, n\}$. The number of tables, in which the i-th row is filled by 0 only, is equal to 2^{mn-m}. Therefore, the number of tables, which are not SC-tables, is at most $n2^{mn-m} = 2^{mn-m+\log_2 n}$. Thus, the number of SC-tables is at least $2^{mn} - 2^{mn-m+\log_2 n}$. □

On Exact Covers for the Most Part of Set Cover Problems. First, we study exact covers for the most part of set cover problems such that $m \geq \lceil \log_2 n \rceil + t$ and t is large enough.

Theorem 21. *Let us consider set cover problems (A, S) such that $A = \{a_1, \ldots, a_n\}$, $S = \{B_1, \ldots, B_m\}$ and $m \geq \lceil \log_2 n \rceil + t$, where t is a natural number. Let $i_1, \ldots, i_{\lceil \log_2 n \rceil + t}$ be pairwise different numbers from $\{1, \ldots, m\}$. Then the fraction of set cover problems (A, S), for which $\{B_{i_1}, \ldots, B_{i_{\lceil \log_2 n \rceil + t}}\}$ is an exact cover for (A, S), is at least $1 - \frac{1}{2^t - 1}$.*

Proof. Let $k = \lceil \log_2 n \rceil + t$. The analyzed fraction is equal to the fraction of SC-tables with n rows and m columns which have no rows with only 0 at the intersection with columns $i_1, \ldots, i_k$. Such SC-tables will be called correct.

Let $j \in \{1, \ldots, t\}$. The number of tables with n rows and m columns filled by 0 and 1, in which the j-th row has only 0 at the intersection with columns $i_1, \ldots, i_k$, is equal to 2^{mn-k}. Therefore, the number of SC-tables which are not correct is at most $n2^{mn-k} = 2^{mn-k+\log_2 n}$. Using Lemma 7 we conclude that the fraction of correct SC-tables is at least $1 - \frac{2^{mn-k+\log_2 n}}{2^{mn} - 2^{mn-m+\log_2 n}} = 1 - \frac{1}{2^{k-\log_2 n} - 2^{k-m}} \geq 1 - \frac{1}{2^t - 1}$. □

For example, if $t = 7$, then for at least 99% of set cover problems (A, S) the subsets $B_{i_1}, \ldots, B_{i_{\lceil \log_2 n \rceil + t}}$ form an exact cover for (A, S).

So if $m \geq \lceil \log_2 n \rceil + t$ and t is large enough, then for the most part of set cover problems there exist exact (and, consequently, partial) covers with small cardinality.

On Partial Covers Constructed by Greedy Algorithm for the Most Part of Set Cover Problems. We will now study the behavior of greedy algorithm for the most part of set cover problems such that $m \geq n + t$ and t is large enough.

Let us consider set cover problems (A, S) such that $A = \{a_1, \ldots, a_n\}$ and $S = \{B_1, \ldots, B_m\}$. A problem (A, S) will be called *saturated* if for any nonempty subset A' of A there exists a subset B_i from S which covers at least one half of elements from A'. For a saturated set cover problem, the greedy algorithm at each step chooses a subset which covers at least one half of uncovered elements.

Let us evaluate the number of saturated set cover problems. First, we prove an auxiliary statement.

Lemma 8. *Let k be a natural number and $\sigma \in \{0,1\}$. Then the number of k-tuples from $\{0,1\}^k$, in which the number of σ is less than $\frac{k}{2}$, is at most 2^{k-1}.*

Proof. Let k be even. Then the number of k-tuples from $\{0,1\}^k$, in which the number of σ is less than $\frac{k}{2}$, is equal to $C_k^0 + \ldots + C_k^{\frac{k}{2}-1}$ which is less than 2^{k-1}. Let k be odd. Then the number of k-tuples from $\{0,1\}^k$, in which the number of σ is less than $\frac{k}{2}$, is equal to $C_k^0 + \ldots + C_k^{\lfloor \frac{k}{2} \rfloor}$ which is equal to 2^{k-1}. □

A table with n rows and m columns filled by numbers from $\{0,1\}$ will be called *saturated* if for any $k \in \{1,\ldots,n\}$ for any k rows there exists a column which has at least $\frac{k}{2}$ numbers 1 at the intersection with considered rows. Otherwise, the table will be called *unsaturated.*

Theorem 22. *Let us consider set cover problems (A,S) such that $A = \{a_1,\ldots, a_n\}$, $S = \{B_1,\ldots,B_m\}$ and $m > n$. Then the fraction of saturated set cover problems (A,S) is at least $1 - \frac{1}{2^{m-n}-1}$.*

Proof. It is clear that analyzed fraction is equal to the fraction of saturated SC-tables.

Let us consider tables with n rows and m columns filled by numbers from $\{0,1\}$. Let $k \in \{1,\ldots,n\}$ and $i_1,\ldots,i_k$ be pairwise different numbers from $\{1,\ldots,n\}$. We now evaluate the number of tables in which at the intersection of each column with rows $i_1,\ldots,i_k$ the number of 1 is less than $\frac{k}{2}$. Such tables will be called unsaturated in rows $i_1,\ldots,i_k$.

From Lemma 8 it follows that the number of k-tuples from $\{0,1\}^k$, in which the number of 1 is less than $\frac{k}{2}$, is at most 2^{k-1}. Therefore, the number of tables, which are unsaturated in rows $i_1,\ldots,i_k$, is at most 2^{mn-m}.

There are 2^n different subsets of rows. Therefore, the number of unsaturated tables is at most 2^{mn+n-m}. Using Lemma 7 we conclude that the fraction of saturated SC-tables is at least $1 - \frac{2^{mn+n-m}}{2^{mn}-2^{mn-m+\log_2 n}} = 1 - \frac{1}{2^{m-n}-2^{\log_2 n-n}} \geq 1 - \frac{1}{2^{m-n}-1}$. □

For example, if $m = n+7$, then at least 99% of set cover problems are saturated. Let us analyze the work of greedy algorithm on an arbitrary saturated set cover problem (A,S). For $i = 1,2,\ldots$, after the step number i at most $\frac{|A|}{2^i}$ elements from A are uncovered. We now evaluate the number $C_{\text{greedy}}(\alpha)$, where $0 < \alpha < 1$. It is clear that $C_{\text{greedy}}(\alpha) \leq i$, where i is a number such that $\frac{1}{2^i} \leq \alpha$. One can show that $\frac{1}{2^{\lceil \log_2 \frac{1}{\alpha} \rceil}} \leq \alpha$. Therefore, $C_{\text{greedy}}(\alpha) \leq \lceil \log_2 \frac{1}{\alpha} \rceil$. Some examples can be found in Table 1.

Let us evaluate the number $C_{\text{greedy}}(0)$. It is clear that all elements from A will be covered after a step number i if $\frac{|A|}{2^i} < 1$, i.e., if $i > \log_2 |A|$. If $\log_2 |A|$

Table 1. Values of $\lceil \log_2 \frac{1}{\alpha} \rceil$ for some α

α	0.5	0.3	0.1	0.01	0.001
Percentage of covered elements	50	70	90	99	99.9
$\lceil \log_2 \frac{1}{\alpha} \rceil$	1	2	4	7	10

is an integer, we can set $i = \log_2 |A| + 1$. Otherwise, we can set $i = \lceil \log_2 |A| \rceil$. Therefore, $C_{\text{greedy}}(0) \leq \log_2 |A| + 1$.

We now evaluate the number $l_{SC}(0)$. Let $\Delta(0, A, S) = (\delta_1, \dots, \delta_m)$. Set $\delta_0 = 0$. Then $l_{SC}(0) = \max\left\{ \left\lceil \frac{|A|-(\delta_0+\ldots+\delta_i)}{\delta_{i+1}} \right\rceil : i = 0, \dots, m-1 \right\}$. Since (A, S) is a saturated problem, we have $\delta_{i+1} \geq \frac{|A|-(\delta_0+\ldots+\delta_i)}{2}$ and $2 \geq \frac{|A|-(\delta_0+\ldots+\delta_i)}{\delta_{i+1}}$ for $i = 0, \dots, m-1$. Therefore, $l_{SC}(0) \leq 2$. Using Corollary 2 we obtain $l_{SC}(\alpha) \leq 2$ for any α, $0 \leq \alpha < 1$.

Results of Experiments. We made some experiments with set cover problems (A, S) such that $|A| \in \{10, 50, 100, 1000, 3000, 5000\}$ and $|S| = 10$. For each value of $|A|$ we generated randomly 10 problems (A, S) such that for each element a_i from A and for each subset S_j from S the probability of a_i to be in S_j is equal to $\frac{1}{2}$. The results of experiments are represented in Tables 2 and 3.

In Table 2 the average percentage of elements covered at the i-th step of greedy algorithm is presented, $i = 1, \dots, 10$. For example, 52.5 means that, on the average, 52.5% of elements remaining uncovered before i-th step are covered at i-th step.

Table 2. Average percentage of elements covered at i-th step of greedy algorithm

	Number i of step									
$\lvert A \rvert$	1	2	3	4	5	6	7	8	9	10
10	71.0	87.5	100.0							
50	62.4	67.5	80.1	100.0						
100	58.9	60.6	62.9	67.8	82.7	95.0	100.0			
1000	52.8	52.4	52.4	53.4	54.7	57.3	64.7	76.2	85.0	100.0
3000	51.2	51.5	52.5	52.6	53.6	54.2	56.9	61.2	72.3	100.0
5000	51.1	51.3	51.5	52.4	52.5	54.3	56.7	63.1	82.0	100.0

In Table 3 for each $\alpha \in \{0.1, 0.01, 0.001, 0.0\}$ the minimal, average and maximal cardinalities of α-covers constructed by the greedy algorithm are presented.

The obtained results show that for the most part of the considered set cover problems (not only for the case, when $|S| > |A|$) during each step the greedy algorithm chooses a subset which covers at least one half of uncovered elements.

It must be also noted that with increase of step number the percentage of elements, covered at this step, grows for the most part of the considered set cover problems.

Table 3. Cardinalities of α-covers for set cover problems (A, S) with $|S| = 10$

$\|A\|$	α											
	0.1			0.01			0.001			0.0		
	min	avg	max	min	avg	max	min	avg	max	min	avg	max
10	2	2.0	2	2	2.4	3	2	2.4	3	2	2.4	3
50	2	2.6	3	4	4.0	4	4	4.0	4	4	4.0	4
100	3	3.0	3	5	5.5	7	5	5.5	7	5	5.5	7
1000	3	3.9	4	6	6.6	7	8	8.9	10	8	8.9	10
3000	4	4.0	4	6	6.9	7	8	9.0	10	9	9.9	10
5000	4	4.0	4	7	7.0	7	9	9.0	9	9	9.9	10

3 Partial Tests and Reducts

3.1 Main Notions

Let T be a table with n rows labeled with nonnegative integers (decisions) and m columns labeled with attributes (names of attributes) $f_1, \ldots, f_m$. This table is filled by nonnegative integers (values of attributes). The table T is called a *decision table.*

By $P(T)$ we denote the set of unordered pairs of different (in at least one column) rows of T with different decisions. We say that an attribute f_i *separates* a pair of rows $(r_1, r_2) \in P(T)$ if rows r_1 and r_2 have different values at the intersection with column f_i.

Let $0 \leq \alpha < 1$. A set of attributes Q is called an *α-test for T* if attributes from Q separate at least $(1 - \alpha)|P(T)|$ pairs from the set $P(T)$. An α-test is called an *α-reduct* if each proper subset of the considered α-test is not an α-test. If $P(T) = \emptyset$, then each subset of $\{f_1, \ldots, f_m\}$ is an α-test, and empty set is an α-reduct only.

For example, 0.01-test means that we must separate at least 99% of pairs from $P(T)$. Note that a 0-reduct is an exact reduct. It must be also noted that each α-test contains at least one α-reduct as subset.

By $R_{\min}(\alpha) = R_{\min}(\alpha, T)$ we denote the minimal cardinality of an α-reduct for T. It is clear that $R_{\min}(\alpha, T)$ coincides with minimal cardinality of an α-test for T.

We will now describe a greedy algorithm with threshold α which constructs an α-test for T.

Let us denote by $R_{\text{greedy}}(\alpha) = R_{\text{greedy}}(\alpha, T)$ the cardinality of constructed α-test for T.

3.2 Relationships between Partial Covers and Partial Tests

Let T be a decision table with m columns labeled with attributes $f_1, \ldots, f_m$, and with a nonempty set $P(T)$. We correspond a set cover problem $(A(T), S(T))$ to the considered decision table T in the following way: $A(T) = P(T)$ and $S(T) = \{B_1, \ldots, B_m\}$, where $B_1 = P(T, f_1), \ldots, B_m = P(T, f_m)$, and for $i = 1, \ldots, m$

Algorithm 2: Greedy algorithm for partial test construction

Input : Decision table T with conditional attributes $f_1, \ldots, f_m$, and real number α, $0 \leq \alpha < 1$.
Output: α-test for T.

$Q \longleftarrow \emptyset$;
while *Q is not an α-test for T* **do**
  select $f_i \in \{f_1, \ldots, f_m\}$ with minimal index i such that f_i separates the maximal number of pairs from $P(T)$ unseparated by attributes from Q;
  $Q \longleftarrow Q \cup \{f_i\}$;
end
return Q;

the set $P(T, f_i)$ coincides with the set of pairs from $P(T)$ separated by the attribute f_i.

Let during the construction of an α-test for T the greedy algorithm choose consequently attributes $f_{j_1}, \ldots, f_{j_t}$. Set $P(T, f_{j_0}) = \emptyset$ and for $i = 1, \ldots, t$ set $\delta_i = |P(T, f_{j_i}) \setminus (P(T, f_{j_0}) \cup \ldots \cup P(T, f_{j_{i-1}}))|$. Let $\Delta(\alpha, T) = (\delta_1, \ldots, \delta_t)$. It is not difficult to prove the following statement.

Proposition 1. *Let α be a real number such that $0 \leq \alpha < 1$. Then $|P(T)| = |A(T)|$, $\Delta(\alpha, T) = \Delta(\alpha, A(T), S(T))$, $R_{\min}(\alpha, T) = C_{\min}(\alpha, A(T), S(T))$, and $R_{\text{greedy}}(\alpha, T) = C_{\text{greedy}}(\alpha, A(T), S(T))$.*

Let (A, S) be a set cover problem, $A = \{a_1, \ldots, a_n\}$ and $S = \{B_1, \ldots, B_m\}$. We correspond a decision table $T(A, S)$ to the set cover problem (A, S) in the following way. The table $T(A, S)$ contains m columns labeled with attributes $f_1, \ldots, f_m$ and $n + 1$ rows filled by numbers from $\{0, 1\}$. For $i = 1, \ldots, n$ and $j = 1, \ldots, m$, the number 1 stays at the intersection of i-th row and j-th column if and only if $a_i \in B_j$. The $(n + 1)$-th row is filled by 0. The first n rows are labeled with the decision 0. The last row is labeled with the decision 1.

For $i = \{1, \ldots, n+1\}$, we denote by r_i the i-th row. It is not difficult to see that $P(T(A, S)) = \{(r_1, r_{n+1}), \ldots, (r_n, r_{n+1})\}$. Let $i \in \{1, \ldots, n\}$ and $j \in \{1, \ldots, m\}$. One can show that the attribute f_j separates the pair (r_i, r_{n+1}) if and only if $a_i \in B_j$. It is not difficult to prove the following statements.

Proposition 2. *Let α be a real number such that $0 \leq \alpha < 1$, and $\{i_1, \ldots, i_t\} \subseteq \{1, \ldots, m\}$. Then $\{f_{i_1}, \ldots, f_{i_t}\}$ is an α-test for $T(A, S)$ if and only if $\{B_{i_1}, \ldots, B_{i_t}\}$ is an α-cover for (A, S).*

Proposition 3. *Let α be a real number and $0 \leq \alpha < 1$. Then $|P(T(A, S))| = |A|$, $R_{\min}(\alpha, T(A, S)) = C_{\min}(\alpha, A, S)$, $R_{\text{greedy}}(\alpha, T(A, S)) = C_{\text{greedy}}(\alpha, A, S)$ and $\Delta(\alpha, T(A, S)) = \Delta(\alpha, A, S)$.*

Proposition 4. *There exists a polynomial algorithm which, for a given set cover problem (A, S), constructs the decision table $T(A, S)$.*

3.3 On Precision of Greedy Algorithm

The following three statements are simple corollaries of the results of Slavík (see Theorems 8–10).

Let T be a decision table with m columns labeled with attributes $f_1, \dots, f_m$.

Theorem 23. *Let* $0 \le \alpha < 1$ *and* $\lceil (1-\alpha)|P(T)| \rceil \ge 2$. *Then* $R_{\text{greedy}}(\alpha) < R_{\min}(\alpha)(\ln \lceil (1-\alpha)|P(T)| \rceil - \ln \ln \lceil (1-\alpha)|P(T)| \rceil + 0.78)$.

Proof. Let $(A, S) = (A(T), S(T))$. From Proposition 1 it follows that $|A| = |P(T)|$. Therefore, $\lceil (1-\alpha)|A| \rceil \ge 2$. Theorem 8 now shows that

$$C_{\text{greedy}}(\alpha, A, S) < C_{\min}(\alpha, A, S)(\ln \lceil (1-\alpha)|A| \rceil - \ln \ln \lceil (1-\alpha)|A| \rceil + 0.78) \ .$$

By Proposition 1, we obtain $R_{\text{greedy}}(\alpha) = C_{\text{greedy}}(\alpha, A, S)$ and $R_{\min}(\alpha) = C_{\min}(\alpha, A, S)$. Taking into account that $|A| = |P(T)|$ we conclude that the statement of the theorem holds. □

Theorem 24. *Let* $0 \le \alpha < 1$. *Then for any natural* $t \ge 2$ *there exists a decision table* T *such that* $\lceil (1-\alpha)|P(T)| \rceil = t$ *and*

$$R_{\text{greedy}}(\alpha) > R_{\min}(\alpha)(\ln \lceil (1-\alpha)|P(T)| \rceil - \ln \ln \lceil (1-\alpha)|P(T)| \rceil - 0.31) \ .$$

Proof. From Theorem 9 it follows that for any natural $t \ge 2$ there exists a set cover problem (A, S) such that $\lceil (1-\alpha)|A| \rceil = t$ and $C_{\text{greedy}}(\alpha, A, S) > C_{\min}(\alpha, A, S)(\ln \lceil (1-\alpha)|A| \rceil - \ln \ln \lceil (1-\alpha)|A| \rceil - 0.31)$.

Let us consider the decision table $T = T(A, S)$. From Proposition 3 it follows that $|P(T)| = |A|$, $C_{\text{greedy}}(\alpha, A, S) = R_{\text{greedy}}(\alpha, T)$ and $C_{\min}(\alpha, A, S) = R_{\min}(\alpha, T)$. Therefore, the statement of the theorem holds. □

Theorem 25. *Let* $0 \le \alpha < 1$ *and* $P(T) \ne \emptyset$. *Then* $R_{\text{greedy}}(\alpha) \le R_{\min}(\alpha)(1 + \ln(\max_{j \in \{1,\dots,m\}} |P(T, f_j)|))$.

Proof. For the set cover problem $(A, S) = (A(T), S(T))$, from Theorem 10 it follows that $C_{\text{greedy}}(\alpha, A, S) \le C_{\min}(\alpha, A, S)(1 + \ln(\max_{j \in \{1,\dots,m\}} |P(T, f_j)|))$.

Proposition 1 shows that $C_{\text{greedy}}(\alpha, A, S) = R_{\text{greedy}}(\alpha)$ and $C_{\min}(\alpha, A, S) = R_{\min}(\alpha)$. Therefore, the statement of the theorem holds. □

3.4 On Polynomial Approximate Algorithms

Theorem 26. (Nguyen and Ślęzak [10], Ślęzak [23]) *Let* $0 \le \alpha < 1$. *Then the problem of construction of* α*-reduct with minimal cardinality is* NP*-hard.*

Let us generalize Theorem 14 to the case of partial tests.

Theorem 27. *Let* $\alpha \in \mathbb{R}$ *and* $0 \le \alpha < 1$. *If* $NP \not\subseteq DTIME(n^{O(\log \log n)})$, *then for any* ε, $0 < \varepsilon < 1$, *there is no polynomial algorithm that, for a given decision table* T *with* $P(T) \ne \emptyset$, *constructs an* α*-test for* T *which cardinality is at most* $(1-\varepsilon)R_{\min}(\alpha, T) \ln |P(T)|$.

Proof. Assume the contrary: let $NP \not\subseteq DTIME(n^{O(\log\log n)})$ and for some ε, $0 < \varepsilon < 1$, a polynomial algorithm $\mathcal{A}$ exist that, for a given decision table T with $P(T) \neq \emptyset$, constructs an α-test for T which cardinality is at most $(1 - \varepsilon)R_{\min}(\alpha, T)\ln|P(T)|$.

Let (A, S) be an arbitrary set cover problem, $A = \{a_1, \ldots, a_n\}$ and $S = \{B_1, \ldots, B_m\}$. From Proposition 4 it follows that there exists a polynomial algorithm which, for a given set cover problem (A, S), constructs the decision table $T(A, S)$. Let us apply this algorithm, and construct the table $T = T(A, S)$. Let us apply to the table T the algorithm $\mathcal{A}$. As a result, we obtain an α-test $\{f_{i_1}, \ldots, f_{i_t}\}$ for T such that $t \leq (1 - \varepsilon)R_{\min}(\alpha, T)\ln|P(T)|$. From Proposition 2 it follows that $\{B_{i_1}, \ldots, B_{i_t}\}$ is an α-cover for (A, S). From Proposition 3 it follows that $|A| = |P(T)|$ and $R_{\min}(\alpha, T) = C_{\min}(\alpha, A, S)$. Therefore, $t \leq (1 - \varepsilon)C_{\min}(\alpha, A, S)\ln|A|$.

Thus, under the assumption $NP \not\subseteq DTIME(n^{O(\log\log n)})$, there exists a polynomial algorithm that, for a given set cover problem (A, S), constructs an α-cover for (A, S) which cardinality is at most $(1 - \varepsilon)C_{\min}(\alpha, A, S)\ln|A|$, but this fact contradicts Theorem 14. □

From Theorem 25 it follows that $R_{\text{greedy}}(\alpha) \leq R_{\min}(\alpha)(1 + \ln|P(T)|)$. From this inequality and from Theorem 27 it follows that, under the assumption $NP \not\subseteq DTIME(n^{O(\log\log n)})$, the greedy algorithm is close to the best polynomial approximate algorithms for partial test minimization.

Let us generalize Theorem 15 to the case of partial covers.

Theorem 28. *Let α be a real number such that $0 \leq \alpha < 1$. If $P \neq NP$, then there exists $\rho > 0$ such that there is no polynomial algorithm that, for a given decision table T with $P(T) \neq \emptyset$, constructs an α-test for T which cardinality is at most $\rho R_{\min}(\alpha, T)\ln|P(T)|$.*

Proof. Let us show that in the capacity of such ρ we can choose ρ from Theorem 15. Assume that the considered statement does not hold: let $P \neq NP$ and a polynomial algorithm $\mathcal{A}$ exist that, for a given decision table T with $P(T) \neq \emptyset$, constructs an α-test for T which cardinality is at most $\rho R_{\min}(\alpha, T)\ln|P(T)|$.

Let (A, S) be an arbitrary set cover problem, $A = \{a_1, \ldots, a_n\}$ and $S = \{B_1, \ldots, B_m\}$. From Proposition 4 it follows that there exists a polynomial algorithm which for a given set cover problem (A, S) constructs the decision table $T(A, S)$. Let us apply this algorithm and construct the table $T = T(A, S)$. Let us apply the algorithm $\mathcal{A}$ to the table T. As a result, we obtain an α-test $\{f_{i_1}, \ldots, f_{i_t}\}$ for T such that $t \leq \rho R_{\min}(\alpha, T)\ln|P(T)|$. From Proposition 2 it follows that $\{B_{i_1}, \ldots, B_{i_t}\}$ is an α-cover for (A, S). From Proposition 3 it follows that $|A| = |P(T)|$ and $R_{\min}(\alpha, T) = C_{\min}(\alpha, A, S)$. Therefore, $t \leq \rho C_{\min}(\alpha, A, S)\ln|A|$.

Thus, under the assumption $P \neq NP$, there exists a polynomial algorithm that for a given set cover problem (A, S) constructs an α-cover for (A, S) which cardinality is at most $\rho C_{\min}(\alpha, A, S)\ln|A|$, but this fact contradicts Theorem 15. □

3.5 Bounds on $R_{\min}(\alpha)$ Based on Information About Greedy Algorithm Work

In this subsection, we fix some information on the greedy algorithm work and find the best upper and lower bounds on $R_{\min}(\alpha)$ depending on this information.

Information on Greedy Algorithm Work. Assume that T is a decision table with m columns labeled with attributes $f_1, \ldots, f_m$, $P(T) \neq \emptyset$, and α is a real number such that $0 \leq \alpha < 1$. Let us apply the greedy algorithm with threshold α to the table T. Let during the construction of α-test the greedy algorithm choose consequently attributes $f_{j_1}, \ldots, f_{j_t}$. Set $P(T, f_{j_0}) = \emptyset$ and for $i = 1, \ldots, t$ set $\delta_i = |P(T, f_{j_i}) \setminus (P(T, f_{j_0}) \cup \ldots \cup P(T, f_{j_{i-1}}))|$. Let $\Delta(\alpha, T) = (\delta_1, \ldots, \delta_t)$. As information on the greedy algorithm work we will use the tuple $\Delta(\alpha, T)$ and numbers $|P(T)|$ and α. Note that $\delta_1 = \max\{|P(T, f_i)| : i = 1, \ldots, m\}$ and $t = R_{\text{greedy}}(\alpha, T)$.

Let us denote by P_{DT} the set of decision tables T with $P(T) \neq \emptyset$. Set $D_{DT} = \{(\alpha, |P(T)|, \Delta(\alpha, T)) : \alpha \in \mathbb{R}, 0 \leq \alpha < 1, T \in P_{DT}\}$.

Lemma 9. $D_{DT} = D_{SC}$.

Proof. Let $\alpha \in \mathbb{R}$, $0 \leq \alpha < 1$ and $T \in P_{DT}$. Then from Proposition 1 it follows that $(\alpha, |P(T)|, \Delta(\alpha, T)) = (\alpha, |A(T)|, \Delta(\alpha, A(T), S(T)))$. Therefore, $D_{DT} \subseteq D_{SC}$. Let $\alpha \in \mathbb{R}$, $0 \leq \alpha < 1$ and $(A, S) \in P_{SC}$. Then from Proposition 3 it follows that $(\alpha, |A|, \Delta(\alpha, A, S)) = (\alpha, |P(T(A, S))|, \Delta(\alpha, T(A, S)))$. Therefore, $D_{SC} \subseteq D_{DT}$. □

Note that the set D_{SC} was described in Lemma 4.

Best Upper Bound for $R_{\min}(\alpha)$. We define a function $\mathcal{U}_{DT} : D_{DT} \to \mathbb{N}$. Let $(\alpha, n, (\delta_1, \ldots, \delta_t)) \in D_{DT}$. Then $\mathcal{U}_{DT}(\alpha, n, (\delta_1, \ldots, \delta_t)) = \max\{R_{\min}(\alpha, T) : T \in P_{DT}, |P(T)| = n, \Delta(\alpha, T) = (\delta_1, \ldots, \delta_t)\}$. It is clear that $R_{\min}(\alpha, T) \leq \mathcal{U}_{DT}(\alpha, |P(T)|, \Delta(\alpha, T))$ is the best upper bound for $R_{\min}(\alpha)$ depending on α, $|P(T)|$ and $\Delta(\alpha, T)$.

Theorem 29. *Let* $(\alpha, n, (\delta_1, \ldots, \delta_t)) \in D_{DT}$. *Then* $\mathcal{U}_{DT}(\alpha, n, (\delta_1, \ldots, \delta_t)) = t$.

Proof. Let T be an arbitrary decision table such that $|P(T)| = n$ and $\Delta(\alpha, T) = (\delta_1, \ldots, \delta_t)$. It is clear that $R_{\min}(\alpha, T) \leq R_{\text{greedy}}(\alpha, T) = t$. Therefore, $t \geq \mathcal{U}_{DT}(\alpha, n, (\delta_1, \ldots, \delta_t))$.

Let us show that $\mathcal{U}_{DT}(\alpha, n, (\delta_1, \ldots, \delta_t)) \geq t$. From Lemma 9 it follows that $(\alpha, n, (\delta_1, \ldots, \delta_t)) \in D_{SC}$. From here and from Theorem 16 it follows that there exists a set cover problem (A, S) such that $|A| = n$, $\Delta(\alpha, A, S) = (\delta_1, \ldots, \delta_t)$ and $C_{\min}(\alpha, A, S) = t$. Let us consider the decision table $T = T(A, S)$. From Proposition 3 it follows that $|P(T)| = n$, $\Delta(\alpha, T) = (\delta_1, \ldots, \delta_t)$ and $R_{\min}(\alpha, T) = t$. Therefore, $\mathcal{U}_{DT}(\alpha, n, (\delta_1, \ldots, \delta_t)) \geq t$. □

Thus, $R_{\min}(\alpha, T) \leq R_{\text{greedy}}(\alpha, T)$ is the best upper bound for $R_{\min}(\alpha)$ depending on α, $|P(T)|$ and $\Delta(\alpha, T)$.

Best Lower Bound for $R_{\min}(\alpha)$. We define a function $\mathcal{L}_{DT} : D_{DT} \to \mathbb{N}$. Let $(\alpha, n, (\delta_1, \ldots, \delta_t)) \in D_{DT}$. Then $\mathcal{L}_{DT}(\alpha, n, (\delta_1, \ldots, \delta_t)) = \min\{R_{\min}(\alpha, T) : T \in P_{DT}, |P(T)| = n, \Delta(\alpha, T) = (\delta_1, \ldots, \delta_t)\}$. It is clear that $R_{\min}(\alpha, T) \geq \mathcal{L}_{DT}(\alpha, |P(T)|, \Delta(\alpha, T))$ is the best lower bound for $R_{\min}(\alpha)$ depending on α, $|P(T)|$ and $\Delta(\alpha, T)$.

Let $(\alpha, n, (\delta_1, \ldots, \delta_t)) \in D_{DT}$. We now remind the definition of the parameter $l(\alpha, n, (\delta_1, \ldots, \delta_t))$. Set $\delta_0 = 0$. Then

$$l(\alpha, n, (\delta_1, \ldots, \delta_t)) = \max\left\{\left\lceil \frac{\lceil (1-\alpha)n \rceil - (\delta_0 + \ldots + \delta_i)}{\delta_{i+1}} \right\rceil : i = 0, \ldots, t-1\right\} .$$

Theorem 30. *Let $(\alpha, n, (\delta_1, \ldots, \delta_t)) \in D_{DT}$. Then $\mathcal{L}_{DT}(\alpha, n, (\delta_1, \ldots, \delta_t)) = l(\alpha, n, (\delta_1, \ldots, \delta_t))$.*

Proof. Let T be an arbitrary decision table such that $|P(T)| = n$ and $\Delta(\alpha, T) = (\delta_1, \ldots, \delta_t)$. We now consider the set cover problem $(A, S) = (A(T), S(T))$. From Proposition 1 it follows that $|A| = n$ and $\Delta(\alpha, A, S) = (\delta_1, \ldots, \delta_t)$. Applying Theorem 17 we obtain $C_{\min}(\alpha, A, S) \geq l(\alpha, n, (\delta_1, \ldots, \delta_t))$. From Proposition 1 it follows that $C_{\min}(\alpha, A, S) = R_{\min}(\alpha, T)$. Therefore, $R_{\min}(\alpha, T) \geq l(\alpha, n, (\delta_1, \ldots, \delta_t))$ and $\mathcal{L}_{DT}(\alpha, n, (\delta_1, \ldots, \delta_t)) \geq l(\alpha, n, (\delta_1, \ldots, \delta_t))$.

Let us show that $\mathcal{L}_{DT}(\alpha, n, (\delta_1, \ldots, \delta_t)) \leq l(\alpha, n, (\delta_1, \ldots, \delta_t))$. From Lemma 9 it follows that $(\alpha, n, (\delta_1, \ldots, \delta_t)) \in D_{SC}$. From here and from Theorem 17 it follows that there exists a set cover problem (A, S) such that $|A| = n$, $\Delta(\alpha, A, S) = (\delta_1, \ldots, \delta_t)$ and $C_{\min}(\alpha, A, S) = l(\alpha, n, (\delta_1, \ldots, \delta_t))$. Let us consider the decision table $T = T(A, S)$. From Proposition 3 it follows that $|P(T)| = n$, $\Delta(\alpha, T) = (\delta_1, \ldots, \delta_t)$ and $R_{\min}(\alpha, T) = l(\alpha, n, (\delta_1, \ldots, \delta_t))$. Thus, $\mathcal{L}_{DT}(\alpha, n, (\delta_1, \ldots, \delta_t)) \leq l(\alpha, n, (\delta_1, \ldots, \delta_t))$. □

So $R_{\min}(\alpha, T) \geq l(\alpha, |P(T)|, \Delta(\alpha, T))$ is the best lower bound for $R_{\min}(\alpha)$ depending on α, $|P(T)|$ and $\Delta(\alpha, T)$.

Properties of Best Lower Bound for $R_{\min}(\alpha)$. Assume that T is a decision table from P_{DT} and α is a real number such that $0 \leq \alpha < 1$. Let $l_{DT}(\alpha) = l_{DT}(\alpha, T) = l(\alpha, |P(T)|, \Delta(\alpha, T))$.

Lemma 10. *Let α_1 and α_2 be real numbers such that $0 \leq \alpha_1 < \alpha_2 < 1$. Then $l_{DT}(\alpha_1) \geq l_{DT}(\alpha_2)$.*

Proof. Let $\Delta(\alpha_1, T) = (\delta_1, \ldots, \delta_{t_1})$ and $\Delta(\alpha_2, T) = (\delta_1, \ldots, \delta_{t_2})$. We have $t_1 \geq t_2$. Let $\delta_0 = 0$, $j \in \{0, \ldots, t_2 - 1\}$ and $\left\lceil \frac{\lceil |P(T)|(1-\alpha_2) \rceil - (\delta_0 + \ldots + \delta_j)}{\delta_{j+1}} \right\rceil = l_{DT}(\alpha_2)$. It is clear that $l_{DT}(\alpha_1) \geq \left\lceil \frac{\lceil |P(T)|(1-\alpha_1) \rceil - (\delta_0 + \ldots + \delta_j)}{\delta_{j+1}} \right\rceil \geq l_{DT}(\alpha_2)$. □

Corollary 4. $l_{DT}(0) = \max\{l_{DT}(\alpha) : 0 \leq \alpha < 1\}$.

The value $l_{DT}(\alpha)$ can be used for obtaining of upper bounds on the cardinality of partial tests constructed by the greedy algorithm.

Theorem 31. *Let α and β be real numbers such that $0 < \beta \leq \alpha < 1$. Then $R_{\text{greedy}}(\alpha) < l_{DT}(\alpha - \beta) \ln \left(\frac{1-\alpha+\beta}{\beta} \right) + 1$.*

Proof. Let $(A, S) = (A(T), S(T))$. Theorem 18 now shows $C_{\text{greedy}}(\alpha, A, S) < l_{SC}(\alpha - \beta, A, S) \ln \left(\frac{1-\alpha+\beta}{\beta} \right) + 1$. By Proposition 1, $l_{DT}(\alpha - \beta) = l_{DT}(\alpha - \beta, T) = l_{SC}(\alpha - \beta, A, S)$ and $R_{\text{greedy}}(\alpha) = R_{\text{greedy}}(\alpha, T) = C_{\text{greedy}}(\alpha, A, S)$. Therefore, the statement of the theorem holds. □

Corollary 5. *Let α be a real number and $0 < \alpha < 1$. Then $R_{\text{greedy}}(\alpha) < l_{DT}(0) \ln \left(\frac{1}{\alpha} \right) + 1$.*

If $l_{DT}(0)$ is a small number, then we have a good upper bound on $R_{\text{greedy}}(\alpha)$. If $l_{DT}(0)$ is a big number, then we have a big lower bound on $R_{\min}(0)$ and on $R_{\min}(\alpha)$ for some α.

3.6 Upper Bound on $R_{\text{greedy}}(\alpha)$

Let T be a decision table from P_{DT}. In this subsection, we obtain an upper bound on $R_{\text{greedy}}(\alpha) = R_{\text{greedy}}(\alpha, T)$ which does not depend on $|P(T)|$, and show that, in some sense, this bound is unimprovable.

Theorem 32. *Let α and β be real numbers such that $0 < \beta \leq \alpha < 1$. Then $R_{\text{greedy}}(\alpha) < R_{\min}(\alpha - \beta) \ln \left(\frac{1-\alpha+\beta}{\beta} \right) + 1$.*

Proof. From Theorem 31 it follows that $R_{\text{greedy}}(\alpha) < l_{DT}(\alpha - \beta) \ln \left(\frac{1-\alpha+\beta}{\beta} \right) + 1$, and from Theorem 30 it follows that $l_{DT}(\alpha - \beta) \leq R_{\min}(\alpha - \beta)$. □

Let us show that obtained bound is, in some sense, unimprovable.

Theorem 33. *There is no real $\delta < 1$ such that for any decision table $T \in P_{DT}$ and for any real α and β, $0 < \beta \leq \alpha < 0$, the following inequality holds: $R_{\text{greedy}}(\alpha) \leq \delta \left(R_{\min}(\alpha - \beta) \ln \left(\frac{1-\alpha+\beta}{\beta} \right) + 1 \right)$.*

Proof. Assume the contrary: let such δ exist. Let us consider an arbitrary set cover problem (A, S) and arbitrary $\alpha, \beta \in \mathbb{R}$ such that $0 < \beta \leq \alpha < 0$. Set $T = T(A, S)$. Then $R_{\text{greedy}}(\alpha, T) \leq \delta \left(R_{\min}(\alpha - \beta, T) \ln \left(\frac{1-\alpha+\beta}{\beta} \right) + 1 \right)$.

From Proposition 3 it follows that $R_{\text{greedy}}(\alpha, T) = C_{\text{greedy}}(\alpha, A, S)$ and $R_{\min}(\alpha - \beta, T) = C_{\min}(\alpha - \beta, A, S)$. Therefore, there exists real $\delta < 1$ such that for any set cover problem (A, S) and for any real α and β, $0 < \beta \leq \alpha < 1$, the inequality $C_{\text{greedy}}(\alpha, A, S) \leq \delta \left(C_{\min}(\alpha - \beta, A, S) \ln \left(\frac{1-\alpha+\beta}{\beta} \right) + 1 \right)$ holds, which contradicts Theorem 20. □

3.7 On Tests for the Most Part of Binary Decision Tables

On Tests for the Most Part of Binary Information Systems. A *binary information system* is a table with n rows (corresponding to objects) and m columns labeled with attributes $f_1, \ldots, f_m$. This table is filled by numbers from $\{0, 1\}$ (values of attributes). There are exactly 2^{mn} different binary information systems with n rows and m columns.

A set $Q \subseteq \{f_1, \ldots, f_m\}$ is called a *test* for the considered information system if attributes from Q separate any two rows with different numbers i and j, where $i, j \in \{1, \ldots, n\}$. It is clear that if we add a decision attribute to the considered information system, then each test for this information system is an exact test for obtained binary decision table.

A decision attribute is a tuple $(d_1, \ldots, d_n)$ of decisions corresponding to the rows of information system. Let us fix a set D of decision attributes. Then the number of different decision tables with decision attribute from D, which can be obtained from a given information system, is equal to $|D|$, and the number of binary decision tables with n rows, m columns and decision attribute from D is equal to $2^{mn}|D|$.

If we prove the existence of good tests for the most part of binary information systems with n rows and m columns, then it means the existence of good tests for the most part of binary decision tables with n rows, m columns and decision attributes from D.

The following theorem is very close to a similar result obtained in test theory (see [26] for details).

Theorem 34. *Let us consider binary information systems with n rows and m columns labeled with attributes $f_1, \ldots, f_m$. Let $m \geq \lceil 2\log_2 n \rceil + t$, where t is a natural number, and $i_1, \ldots, i_{\lceil 2\log_2 n \rceil + t}$ be different numbers from $\{1, \ldots, m\}$. Then the fraction of information systems, for which $\{f_{i_1}, \ldots, f_{i_{\lceil 2\log_2 n \rceil + t}}\}$ is a test, is at least $1 - \frac{1}{2^{t+1}}$.*

Proof. Let $k = \lceil 2\log_2 n \rceil + t$, $j, l \in \{1, \ldots, n\}$ and $j \neq l$. The number of information systems, for which j-th and l-th rows are equal at the intersection with columns $f_{i_1}, \ldots, f_{i_k}$, is equal to 2^{mn-k}. The number of pairs $j, l \in \{1, \ldots, n\}$ such that $j \neq l$ is at most $\frac{n^2}{2}$. Therefore, the number of information systems, for which $\{f_{i_1}, \ldots, f_{i_k}\}$ is not a test, is at most $\frac{n^2}{2} 2^{mn-k} = 2^{mn-k+2\log_2 n - 1} \leq 2^{mn-t-1}$. Thus, the fraction of information systems, for which $\{f_{i_1}, \ldots, f_{i_k}\}$ is a test, is at least $\frac{2^{mn} - 2^{mn-t-1}}{2^{mn}} = 1 - \frac{1}{2^{t+1}}$. □

For example, if $t = 6$, then for at least 99% of the considered information systems the attributes $f_{i_1}, \ldots, f_{i_{\lceil 2\log_2 n \rceil + t}}$ form a test.

So if $m \geq \lceil 2\log_2 n \rceil + t$ and t is large enough, then for the most part of binary information systems there exist tests with small cardinality. Therefore, for the most part of binary decision tables (with decision attributes from a fixed set D) there exist exact (and, consequently, partial) tests with small cardinality.

On Partial Tests Constructed by Greedy Algorithm for the Most Part of Binary Decision Tables with Binary Decision Attributes. We now study (under some assumption on relationships between m and n) the behavior of greedy algorithm for the most part of binary decision tables with binary decision attributes with values from $\{0,1\}$. To this end we investigate binary information systems of a special kind.

Let t be a natural number. We will study so-called t-separable binary information systems with n rows and m columns labeled with attributes $f_1,\ldots,f_m$.

Let $W=\{(c,\delta): c\in\{1,\ldots,2^t\}, \delta\in\{0,1\}\}$. An arbitrary tuple $((c_1,\delta_1),\ldots,(c_n,\delta_n))\in W^n$ is interpreted as follows: the set of rows of the considered information system is divided into 2^t classes (it is possible that some classes are empty), and each row is labeled with a decision from $\{0,1\}$. For $j=1,\ldots,n$ the number c_j is the number of class to which j-th row belongs, and δ_j is the decision attached to j-th row. It is clear that $|W^n|=2^{n(t+1)}$.

Fix a tuple $\bar\delta\in W^n$, $\bar\delta=((c_1,\delta_1),\ldots,(c_n,\delta_n))$. Let $|\{c_1,\ldots,c_n\}|=k$. It means that the tuple $\bar\delta$ determines a partition of set of rows into k nonempty classes. It is clear that $1\le k\le 2^t$. For $i=1,\ldots,k$ we denote by A_i^0 the set of rows from the i-th class with decision 0, and by A_i^1 – the set of rows from the i-th class with decision 1. Let us define the notion of *$\bar\delta$-attribute*. This is an attribute (column) which for any $i\in\{1,\ldots,k\}$ at the intersection with rows from A_i^0 has at least $\frac{|A_i^0|}{2}$ numbers 0, and at the intersection with rows from A_i^1 has at least $\frac{|A_i^1|}{2}$ numbers 1. A binary information system with m columns and n rows will be called *t-separable* if for any $\bar\delta\in W^n$ this system has a $\bar\delta$-attribute.

Let the considered information system be t-separable, $t\ge 1$. We now show that in this system all rows are pairwise different. Let $i,j\in\{1,\ldots,n\}$. Let us consider such $\bar\delta$ that i-th and j-th rows are in the first class and have different decisions, and all other rows are in the second class. It is clear that $\bar\delta$-attribute must separate i-th and j-th rows.

We now study the work of greedy algorithm for a decision table T obtained from a t-separable binary information system I by adding a binary decision attribute. Let I have n rows and m columns labeled with attributes $f_1,\ldots,f_m$.

Let us consider the tuple $\bar\delta=((1,\delta_1),\ldots,(1,\delta_n))$, where δ_j is the decision attached to j-th row for $j=1,\ldots,n$. Let A_1^0 be the set of rows with decision 0, and A_1^1 be the set of rows with decision 1. Since I is a t-separable information system, there is a $\bar\delta$-attribute f_{p_1} which at the intersection with rows from A_1^0 has at least $\frac{|A_1^0|}{2}$ numbers 0, and at the intersection with rows from A_1^1 has at least $\frac{|A_1^1|}{2}$ numbers 1. Let a_0 be the number of rows from A_1^0 for which the value of f_{p_1} is equal to 0, a_1 – the number of rows from A_1^0 for which the value of f_{p_1} is equal to 1, b_0 – the number of rows from A_1^1 for which the value of f_{p_1} is equal to 0, and b_1 – the number of rows from A_1^1 for which the value of f_{p_1} is equal to 1.

Let us denote by $P(T)$ the set of unordered pairs of different rows from T with different decisions. We know that all rows are pairwise different. Therefore, $|P(T)|=(a_0+a_1)(b_0+b_1)$. The attribute f_{p_1} separates $a_0b_1+a_1b_0$ pairs from $P(T)$. Let us show that $a_0b_1+a_1b_0\ge\frac{|P(T)|}{2}$. To this end we must prove that

$2a_0b_1 + 2a_1b_0 \geq (a_0 + a_1)(b_0 + b_1) = a_0b_0 + a_0b_1 + a_1b_0 + a_1b_1$. This inequality is equivalent to the inequality $a_0b_1 + a_1b_0 \geq a_0b_0 + a_1b_1$. The last inequality is equivalent to the inequality $a_0(b_1 - b_0) \geq a_1(b_1 - b_0)$. It is clear that $a_0 \geq a_1$ and $b_1 \geq b_0$. Therefore, the considered inequality holds.

Thus, during the first step the greedy algorithm chooses an attribute f_{l_1} which separates at least one half of pairs from $P(T)$.

Let the greedy algorithm make $k \leq t$ steps and choose attributes $f_{l_1}, \ldots, f_{l_k}$. These attributes divide the set of rows into $q \leq 2^k$ nonempty classes. In each class attributes $f_{l_1}, \ldots, f_{l_k}$ are constant. Let us consider the tuple $\bar{\delta} = ((c_1, \delta_1), \ldots, (c_n, \delta_n))$, where c_j is the number of class to which j-th row belongs, and δ_j is the decision attached to j-th row, $j = 1, \ldots, n$. It is clear that $\bar{\delta} \in W^n$. For $i = 1, \ldots, q$, we denote by A_i^0 the set of rows from the i-th class with decision 0, and by A_i^1 – the set of rows from the i-th class with decision 1. The number of unseparated pairs from $P(T)$ is equal to $\sum_{i=1}^q |A_i^0||A_i^1|$. If $\sum_{i=1}^q |A_i^0||A_i^1| = 0$, then $\{f_{l_1}, \ldots, f_{l_k}\}$ is an exact test for T. Let $\sum_{i=1}^q |A_i^0||A_i^1| > 0$. Since I is a t-separable, there exists a $\bar{\delta}$-attribute $f_{p_{k+1}}$. As it was made earlier for one class, we can show that $f_{p_{k+1}}$ separates at least one half of unseparated pairs from $P(T)$. Therefore, during the step number $k+1$ the greedy algorithm chooses an attribute $f_{l_{k+1}}$ which separates at least one half of unseparated pairs from $P(T)$.

Thus, for any decision table T, obtained from a t-separable binary information system I by adding a binary decision attribute, during the first $t+1$ steps the greedy algorithm at each step chooses an attribute that separates at least one half of unseparated pairs from $P(T)$ (if unseparated pairs exist).

Let us evaluate the fraction of t-separable binary information systems, under some assumptions on joint behavior of n and m.

Theorem 35. *Let t and k be natural numbers. Let us consider binary information systems with n rows and $m \geq 2^{2^{t+1}}((t+1)n+k)$ columns. Then the fraction of t-separable binary information systems is at least $1 - \frac{1}{2^k}$.*

Proof. Let us fix a tuple $\bar{\delta} = ((c_1, \delta_1), \ldots, (c_n, \delta_n)) \in W^n$. This tuple determines an equivalence relation on the set of rows of information system: two rows with numbers i and j are equivalent if $c_i = c_j$ and $\delta_i = \delta_j$. This equivalence relation divides the set of rows into $p \leq 2^{t+1}$ classes of equivalence. We now consider an arbitrary equivalence class. Let this class contain s rows. To be $\bar{\delta}$-attribute, an attribute for some $\sigma \in \{0, 1\}$ must have at least $\frac{s}{2}$ numbers σ at the intersection with considered s rows. From Lemma 8 it follows that there are at least 2^{s-1} tuples from $\{0,1\}^s$ which have at least $\frac{s}{2}$ numbers σ. Using this fact it is not difficult to prove that there are at least $2^{n-p} \geq 2^{n-2^{t+1}}$ tuples from $\{0,1\}^n$ which are $\bar{\delta}$-attributes. Therefore, the number of tuples which are not $\bar{\delta}$-attributes is at most $2^n - 2^{n-2^{t+1}}$. Hence, the number of information systems without $\bar{\delta}$-attributes is at most $\left(2^n - 2^{n-2^{t+1}}\right)^m$. The number of information systems without at least one $\bar{\delta}$-attribute, $\bar{\delta} \in W^n$, is at most $2^{n(t+1)}\left(2^n - 2^{n-2^{t+1}}\right)^m$. Therefore, the number of t-separable information systems is at least $2^{mn} - 2^{n(t+1)}\left(2^n - 2^{n-2^{t+1}}\right)^m$, and the fraction

Table 4. Average percentage of pairs of rows separated at i-th step of greedy algorithm ($m = 40$ and $c = 10$)

Number of rows n	Number i of step: 1	2	3	4	5	6	7	8	9	10
10	59.3	67	85.7	100						
50	52	53.2	55	58.1	61.9	69.2	89.7	100		
100	50.9	51.5	52.2	54	55.8	59.7	65	73.5	91	100
1000	50.1	50.1	50.2	50.3	50.5	50.8	51.1	51.5	52.2	53.2
3000	50	50	50.1	50.1	50.2	50.2	50.3	50.5	50.8	51
5000	50	50	50	50.1	50.1	50.1	50.2	50.3	50.4	50.6

of t-separable information systems is at least $1 - \frac{2^{n(t+1)}\left(2^n - 2^{n-2^{t+1}}\right)^m}{2^{mn}} = 1 - 2^{n(t+1)}\left(1 - \frac{1}{2^{2^{t+1}}}\right)^m$. Let $r = 2^{2^{t+1}}$. Using well known inequality$\left(\frac{r-1}{r}\right)^r \leq \frac{1}{e}$ we obtain $2^{n(t+1)}\left(1 - \frac{1}{2^{2^{t+1}}}\right)^m = 2^{n(t+1)}\left(\frac{r-1}{r}\right)^{r\frac{m}{r}} \leq 2^{n(t+1)-\frac{m}{r}}$. Therefore, the fraction of t-separable information systems is at least $1 - 2^{n(t+1)-\frac{m}{2^{2^{t+1}}}}$. If $m \geq 2^{2^{t+1}}((t+1)n + k)$, then this fraction is at least $1 - 2^{-k}$. □

For example, if $m = 256(3n+7)$, then at least 99% of binary information systems are 2-separable.

Let us consider the work of greedy algorithm on an arbitrary decision table T obtained from a t-separable binary information system by adding a binary decision attribute. For $i = 1, 2, \ldots, t+1$, after step number i at most $\frac{|P(T)|}{2^i}$ pairs from $P(T)$ are unseparated. Using this fact it is not difficult to prove that $R_{\text{greedy}}(\alpha) \leq \lceil \log_2 \frac{1}{\alpha} \rceil$ and $l_{DT}(\alpha) \leq 2$ for any α such that $\frac{1}{2^{t+1}} \leq \alpha < 1$.

Results of Experiments. We made some experiments with binary decision tables T containing $n \in \{10, 50, 100, 1000, 3000, 5000\}$ rows, $m \in \{10, 40, 100\}$ conditional attributes and one decision attribute with values from the set $\{1, \ldots, c\}$, $c \in \{2, 10, 100\}$. For each triple of values (n, m, c) we generated randomly 10

Table 5. Average cardinality of α-tests for decision tables with 10 conditional attributes

	Number of different decisions c: 2				10				100			
Number of rows n	α: 0.1	0.01	0.001	0.0	0.1	0.01	0.001	0.0	0.1	0.01	0.001	0.0
10	2.1	2.8	2.8	2.8	3.1	4.5	4.5	4.5	3.1	4.5	4.5	4.5
50	3.7	6	7.8	7.8	4	6.1	8.6	9.1	4	6.2	7.9	8.6
100	3.9	6.4	8.7	9.4	4	6.9	9	10	4	6.9	9	9.9
1000	4	7	9.1	10	4	7	9	10	4	7	9	10
3000	4	7	9	10	4	7	9	10	4	7	9	10
5000	4	7	9	10	4	7	9	10	4	7	9	10

decision tables such that each element of table is equal to b, $b \in \{0, 1\}$, with probability $\frac{1}{2}$, and each decision is equal to d, $d \in \{1, \ldots, c\}$, with probability $\frac{1}{c}$. The results of experiments are represented in Tables 4–7.

In Table 4 the average percentage of pairs of rows from $P(T)$ separated at the i-th step of greedy algorithm, $i = 1, \ldots, 10$, is presented for the case, when $m = 40$ and $c = 10$. For example, 50.2 means that, on the average, 50.2% of pairs remaining unseparated before i-th step are separated at i-th step.

Table 6. Average cardinality of α-tests for decision tables with 40 conditional attributes

	Number of different decisions c											
	2				10				100			
Number	α											
of rows n	0.1	0.01	0.001	0.0	0.1	0.01	0.001	0.0	0.1	0.01	0.001	0.0
10	2.1	2.3	2.3	2.3	3	3.8	3.8	3.8	3	4	4	4
50	3.1	5.1	6.2	6.2	3.7	6	7	7.6	4	6	7.4	7.8
100	3.7	6	8	8.5	4	6	8.3	9.4	4	6.5	8.8	9.6
1000	4	7	10	15.2	4	7	10	15.9	4	7	10	16.6
3000	4	7	10	18.4	4	7	10	19.1	4	7	10	19.3
5000	4	7	10	19.9	4	7	10	20.7	4	7	10	20.9

In Table 5 for each $\alpha \in \{0.1, 0.01, 0.001, 0.0\}$ the average cardinality of α-tests constructed by the greedy algorithm is presented for decision tables with 10 conditional attributes.

In Table 6 for each $\alpha \in \{0.1, 0.01, 0.001, 0.0\}$ the average cardinality of α-tests constructed by the greedy algorithm is presented for decision tables with 40 conditional attributes.

In Table 7 for each $\alpha \in \{0.1, 0.01, 0.001, 0.0\}$ the average cardinality of α-tests constructed by the greedy algorithm is presented for decision tables with 100 conditional attributes.

Table 7. Average cardinality of α-tests for decision tables with 100 conditional attributes

	Number of different decisions c											
	2				10				100			
Number	α											
of rows n	0.1	0.01	0.001	0.0	0.1	0.01	0.001	0.0	0.1	0.01	0.001	0.0
10	1.9	1.9	1.9	1.9	2.4	3.3	3.3	3.3	3	4	4	4
50	3	5	6	6	3.5	6	7	7	4	6	7	7.2
100	3.1	6	7.2	8	4	6	8	9.1	4	6.1	8.1	9
1000	4	7	10	14.6	4	7	10	15.2	4	7	10	15.4
3000	4	7	10	17.9	4	7	10	18.6	4	7	10	18.8
5000	4	7	10	19.1	4	7	10	20	4	7	10	20

The obtained results show that for the most part of the considered decision tables (not only for the case, when $m \gg n$, and only binary decision attributes are used) during each step the greedy algorithm chooses an attribute which separates at least one half of unseparated pairs.

It must be also noted that with increase of step number the percentage of pairs of rows, separated at this step, grows for the most part of the considered decision tables.

4 Partial Decision Rules

4.1 Main Notions

Assume that T is a decision table with n rows labeled with nonnegative integers (decisions) and m columns labeled with attributes (names of attributes) $f_1, \ldots, f_m$. This table is filled by nonnegative integers (values of attributes).

Let $r = (b_1, \ldots, b_m)$ be a row of T labeled with a decision d. By $U(T, r)$ we denote the set of rows from T which are different (in at least one column) from r and are labeled with decisions different from d. We will say that an attribute f_i *separates* a row $r' \in U(T, r)$ from the row r if the rows r and r' have different numbers at the intersection with column f_i. The pair (T, r) will be called a *decision rule problem.*

Let $0 \leq \alpha < 1$. A decision rule

$$(f_{i_1} = b_{i_1}) \wedge \ldots \wedge (f_{i_t} = b_{i_t}) \to d \tag{8}$$

is called an *α-decision rule* for (T, r) if attributes $f_{i_1}, \ldots, f_{i_t}$ separate from r at least $(1 - \alpha)|U(T, r)|$ rows from $U(T, r)$. The number t is called the *length* of the considered decision rule. If $U(T, r) = \emptyset$, then for any $f_{i_1}, \ldots, f_{i_t} \in \{f_1, \ldots, f_m\}$ the rule (8) is an α-decision rule for (T, r). The rule (8) with empty left-hand side (when $t = 0$) is also an α-decision rule for (T, r).

For example, 0.01-decision rule means that we must separate from r at least 99% of rows from $U(T, r)$. Note that a 0-decision rule is an exact decision rule. By $L_{\min}(\alpha) = L_{\min}(\alpha, T, r)$ we denote the minimal length of α-decision rule for (T, r).

We will now describe a greedy algorithm with threshold α which constructs an α-decision rule for (T, r).

Let us denote by $L_{\text{greedy}}(\alpha) = L_{\text{greedy}}(\alpha, T, r)$ the length of constructed α-decision rule for (T, r).

4.2 Relationships between Partial Covers and Partial Decision Rules

Let T be a decision table with m columns labeled with attributes $f_1, \ldots, f_m$, r be a row from T, and $U(T, r)$ be a nonempty set.

We correspond a set cover problem $(A(T, r), S(T, r))$ to the considered decision rule problem (T, r) in the following way: $A(T, r) = U(T, r)$ and $S(T, r) =$

Algorithm 3: Greedy algorithm for partial decision rule construction

Input : Decision table T with conditional attributes $f_1, \ldots, f_m$, row $r = (b_1, \ldots, b_m)$ of T labeled with the decision d, and real number α, $0 \leq \alpha < 1$.
Output: α-decision rule for (T, r).

$Q \longleftarrow \emptyset$;
while *attributes from* Q *separate from* r *less than* $(1 - \alpha)|U(T, r)|$ *rows from* $U(T, r)$ **do**
  select $f_i \in \{f_1, \ldots, f_m\}$ with minimal index i such that f_i separates from r the maximal number of rows from $U(T, r)$ unseparated by attributes from Q;
  $Q \longleftarrow Q \cup \{f_i\}$;
end
return $\bigwedge_{f_i \in Q}(f_i = b_i) \rightarrow d$;

$\{B_1, \ldots, B_m\}$ where $B_1 = U(T, r, f_1), \ldots, B_m = U(T, r, f_m)$ and for $i = 1, \ldots, m$ the set $U(T, r, f_i)$ coincides with the set of rows from $U(T, r)$ separated by the attribute f_i from the row r.

Let during the construction of an α-decision rule for (T, r) the greedy algorithm choose consequently attributes $f_{j_1}, \ldots, f_{j_t}$. Set $U(T, r, f_{j_0}) = \emptyset$ and for $i = 1, \ldots, t$ set $\delta_i = |U(T, r, f_{j_i}) \setminus (U(T, r, f_{j_0}) \cup \ldots \cup U(T, r, f_{j_{i-1}}))|$. Let $\Delta(\alpha, T, r) = (\delta_1, \ldots, \delta_t)$. It is not difficult to prove the following statement.

Proposition 5. *Let* α *be a real number such that* $0 \leq \alpha < 1$. *Then* $|U(T, r)| = |A(T, r)|$, $\Delta(\alpha, T, r) = \Delta(\alpha, A(T, r), S(T, r))$, $L_{\min}(\alpha, T, r) = C_{\min}(\alpha, A(T, r), S(T, r))$, *and* $L_{\text{greedy}}(\alpha, T, r) = C_{\text{greedy}}(\alpha, A(T, r), S(T, r))$.

Let (A, S) be a set cover problem, $A = \{a_1, \ldots, a_n\}$ and $S = \{B_1, \ldots, B_m\}$. We correspond a decision rule problem $(T(A, S), r(A, S))$ to the set cover problem (A, S) in the following way. The table $T(A, S)$ contains m columns labeled with attributes $f_1, \ldots, f_m$ and $n+1$ rows filled by numbers from $\{0, 1\}$. For $i = 1, \ldots, n$ and $j = 1, \ldots, m$, the number 1 stays at the intersection of i-th row and j-th column if and only if $a_i \in B_j$. The $(n + 1)$-th row is filled by 0. The first n rows are labeled with the decision 0. The last row is labeled with the decision 1. Let us denote by $r(A, S)$ the last row of table $T(A, S)$. For $i \in \{1, \ldots, n + 1\}$ we denote by r_i the i-th row. It is not difficult to see that $U(T(A, S), r(A, S)) = \{r_1, \ldots, r_n\}$. Let $i \in \{1, \ldots, n\}$ and $j \in \{1, \ldots, m\}$. One can show that the attribute f_j separates the row $r_{n+1} = r(A, S)$ from the row r_i if and only if $a_i \in B_j$. It is not difficult to prove the following statements.

Proposition 6. *Let* $\alpha \in \mathbb{R}$, $0 \leq \alpha < 1$, *and* $\{i_1, \ldots, i_t\} \subseteq \{1, \ldots, m\}$. *Then* $(f_{i_1} = 0) \wedge \ldots \wedge (f_{i_t} = 0) \rightarrow 1$ *is an* α*-decision rule for* $(T(A, S), r(A, S))$ *if and only if* $\{B_{i_1}, \ldots, B_{i_t}\}$ *is an* α*-cover for* (A, S).

Proposition 7. *Let* α *be a real number such that* $0 \leq \alpha < 1$. *Then* $|U(T(A, S), r(A, S))| = |A|$, $L_{\min}(\alpha, T(A, S), r(A, S)) = C_{\min}(\alpha, A, S)$, $L_{\text{greedy}}(\alpha, T(A, S), r(A, S)) = C_{\text{greedy}}(\alpha, A, S)$ *and* $\Delta(\alpha, T(A, S), r(A, S)) = \Delta(\alpha, A, S)$.

Proposition 8. *There exists a polynomial algorithm which for a given set cover problem (A, S) constructs the decision rule problem $(T(A, S), r(A, S))$.*

4.3 On Precision of Greedy Algorithm

The following three statements are simple corollaries of results of Slavík (see Theorems 8–10). Let T be a decision table with m columns labeled with attributes $f_1, \ldots, f_m$, and r be a row of T.

Theorem 36. *Let $0 \leq \alpha < 1$ and $\lceil (1-\alpha)|U(T,r)| \rceil \geq 2$. Then $L_{\text{greedy}}(\alpha) < L_{\min}(\alpha)(\ln \lceil (1-\alpha)|U(T,r)| \rceil - \ln\ln \lceil (1-\alpha)|U(T,r)| \rceil + 0.78)$.*

The proof of this theorem is similar to the proof of Theorem 23.

Theorem 37. *Let $0 \leq \alpha < 1$. Then for any natural $t \geq 2$ there exists a decision rule problem (T, r) such that $\lceil (1-\alpha)|U(T,r)| \rceil = t$ and $L_{\text{greedy}}(\alpha) > L_{\min}(\alpha)(\ln \lceil (1-\alpha)|U(T,r)| \rceil - \ln\ln \lceil (1-\alpha)|U(T,r)| \rceil - 0.31)$.*

The proof of this theorem is similar to the proof of Theorem 24.

Theorem 38. *Let $0 \leq \alpha < 1$ and $U(T,r) \neq \emptyset$. Then $L_{\text{greedy}}(\alpha) \leq L_{\min}(\alpha)(1 + \ln(\max_{j \in \{1,\ldots,m\}} |U(T,r,f_j)|))$.*

The proof of this theorem is similar to the proof of Theorem 25.

4.4 On Polynomial Approximate Algorithms

Theorem 39. *Let $0 \leq \alpha < 1$. Then the problem of construction of α-decision rule with minimal length is NP-hard.*

Proof. From Theorem 13 it follows that the problem of construction of α-cover with minimal cardinality is NP-hard. Using Propositions 6 and 8 we conclude that there exists a polynomial-time reduction of the problem of construction of α-cover with minimal cardinality to the problem of construction of α-decision rule with minimal length. □

Let us generalize Theorem 14 to the case of partial decision rules.

Theorem 40. *Let $\alpha \in \mathbb{R}$ and $0 \leq \alpha < 1$. If $NP \not\subseteq DTIME(n^{O(\log\log n)})$, then for any ε, $0 < \varepsilon < 1$, there is no polynomial algorithm that for a given decision rule problem (T, r) with $U(T,r) \neq \emptyset$ constructs an α-decision rule for (T, r) which length is at most $(1-\varepsilon)L_{\min}(\alpha, T, r) \ln |U(T,r)|$.*

The proof of this theorem is similar to the proof of Theorem 27.

From Theorem 38 it follows that $L_{\text{greedy}}(\alpha) \leq L_{\min}(\alpha)(1 + \ln |U(T,r)|)$. From this inequality and from Theorem 40 it follows that, under the assumption $NP \not\subseteq DTIME(n^{O(\log\log n)})$, the greedy algorithm is close to the best polynomial approximate algorithms for partial decision rule minimization.

Let us generalize Theorem 15 to the case of partial decision rules.

Theorem 41. *Let α be a real number such that $0 \leq \alpha < 1$. If $P \neq NP$, then there exists $\rho > 0$ such that there is no polynomial algorithm that for a given decision rule problem (T, r) with $U(T, r) \neq \emptyset$ constructs an α-decision rule for (T, r) which length is at most $\rho L_{\min}(\alpha, T, r) \ln |U(T, r)|$.*

The proof of this theorem is similar to the proof of Theorem 28.

4.5 Bounds on $L_{\min}(\alpha)$ Based on Information About Greedy Algorithm Work

In this subsection, we fix some information on the greedy algorithm work and find the best upper and lower bounds on $L_{\min}(\alpha)$ depending on this information.

Information on Greedy Algorithm Work. Assume that (T, r) is a decision rule problem, where T is a decision table with m columns labeled with attributes $f_1, \ldots, f_m$, $U(T, r) \neq \emptyset$, and α is a real number such that $0 \leq \alpha < 1$. Let us apply the greedy algorithm with threshold α to the problem (T, r). Let during the construction of α-decision rule the greedy algorithm choose consequently attributes $f_{j_1}, \ldots, f_{j_t}$. Set $U(T, r, f_{j_0}) = \emptyset$ and for $i = 1, \ldots, t$ set $\delta_i = |U(T, r, f_{j_i}) \setminus (U(T, r, f_{j_0}) \cup \ldots \cup P(U(T, r, f_{j_{i-1}}))|$. Let $\Delta(\alpha, T, r) = (\delta_1, \ldots, \delta_t)$. As information on the greedy algorithm work we will use the tuple $\Delta(\alpha, T, r)$, and numbers $|U(T, r)|$ and α. Note that $\delta_1 = \max\{|U(T, r, f_i)| : i = 1, \ldots, m\}$ and $t = L_{\text{greedy}}(\alpha, T, r)$.

Let us denote by P_{DR} the set of decision rule problems (T, r) with $U(T, r) \neq \emptyset$, and $D_{DR} = \{(\alpha, |U(T, r)|, \Delta(\alpha, T, r)) : \alpha \in \mathbb{R}, 0 \leq \alpha < 1, (T, r) \in P_{DR}\}$.

Lemma 11. $D_{DR} = D_{SC}$.

The proof of this lemma is similar to the proof of Lemma 9. Note that the set D_{SC} was described in Lemma 4.

Best Upper Bound for $L_{\min}(\alpha)$. We define a function $\mathcal{U}_{DR} : D_{DR} \to \mathbb{N}$. Let $(\alpha, n, (\delta_1, \ldots, \delta_t)) \in D_{DR}$. Then $\mathcal{U}_{DR}(\alpha, n, (\delta_1, \ldots, \delta_t)) = \max\{L_{\min}(\alpha, T, r) : (T, r) \in P_{DR}, |U(T, r)| = n, \Delta(\alpha, T, r) = (\delta_1, \ldots, \delta_t)\}$. It is clear that

$$L_{\min}(\alpha, T, r) \leq \mathcal{U}_{DR}(\alpha, |U(T, r)|, \Delta(\alpha, T, r))$$

is the best upper bound for $L_{\min}(\alpha)$ depending on α, $|U(T, r)|$ and $\Delta(\alpha, T, r)$.

Theorem 42. *Let $(\alpha, n, (\delta_1, \ldots, \delta_t)) \in D_{DR}$. Then $\mathcal{U}_{DR}(\alpha, n, (\delta_1, \ldots, \delta_t)) = t$.*

The proof of this theorem is similar to the proof of Theorem 29.

Thus, $L_{\min}(\alpha, T, r) \leq L_{\text{greedy}}(\alpha, T, r)$ is the best upper bound for $L_{\min}(\alpha)$ depending on α, $|U(T, r)|$ and $\Delta(\alpha, T, r)$.

Best Lower Bound for $L_{\min}(\alpha)$. We define a function $\mathcal{L}_{DR} : D_{DR} \to \mathbb{N}$. Let $(\alpha, n, (\delta_1, \ldots, \delta_t)) \in D_{DR}$. Then $\mathcal{L}_{DR}(\alpha, n, (\delta_1, \ldots, \delta_t)) = \min\{L_{\min}(\alpha, T, r) : (T, r) \in P_{DR}, |U(T, r)| = n, \Delta(\alpha, T, r) = (\delta_1, \ldots, \delta_t)\}$. It is clear that

$$L_{\min}(\alpha, T, r) \geq \mathcal{L}_{DR}(\alpha, |U(T, r)|, \Delta(\alpha, T, r))$$

is the best lower bound for $L_{\min}(\alpha)$ depending on α, $|U(T, r)|$ and $\Delta(\alpha, T, r)$.

Let $(\alpha, n, (\delta_1, \ldots, \delta_t)) \in D_{DR}$. We now remind the definition of parameter $l(\alpha, n, (\delta_1, \ldots, \delta_t))$. Set $\delta_0 = 0$. Then

$$l(\alpha, n, (\delta_1, \ldots, \delta_t)) = \max \left\{ \left\lceil \frac{\lceil (1-\alpha)n \rceil - (\delta_0 + \ldots + \delta_i)}{\delta_{i+1}} \right\rceil : i = 0, \ldots, t-1 \right\} .$$

Theorem 43. *Let* $(\alpha, n, (\delta_1, \ldots, \delta_t)) \in D_{DR}$. *Then* $\mathcal{L}_{DR}(\alpha, n, (\delta_1, \ldots, \delta_t)) = l(\alpha, n, (\delta_1, \ldots, \delta_t))$.

The proof of this theorem is similar to the proof of Theorem 30.

Thus, $L_{\min}(\alpha, T, r) \geq l(\alpha, |U(T, r)|, \Delta(\alpha, T, r))$ is the best lower bound for $L_{\min}(\alpha)$ depending on α, $|U(T, r)|$ and $\Delta(\alpha, T, r)$.

Properties of Best Lower Bound for $L_{\min}(\alpha)$. Assume that (T, r) is a decision rule problem from P_{DR}, and $\alpha \in \mathbb{R}$, $0 \leq \alpha < 1$. Let

$$l_{DR}(\alpha) = l_{DR}(\alpha, T, r) = l(\alpha, |U(T, r)|, \Delta(\alpha, T, r)) .$$

Lemma 12. *Let* $\alpha_1, \alpha_2 \in \mathbb{R}$ *and* $0 \leq \alpha_1 < \alpha_2 < 1$. *Then* $l_{DR}(\alpha_1) \geq l_{DR}(\alpha_2)$.

The proof of this lemma is similar to the proof of Lemma 10.

Corollary 6. $l_{DR}(0) = \max\{l_{DR}(\alpha) : 0 \leq \alpha < 1\}$.

The value $l_{DR}(\alpha)$ can be used for obtaining of upper bounds on the length of partial decision rules constructed by the greedy algorithm.

Theorem 44. *Let* α *and* β *be real numbers such that* $0 < \beta \leq \alpha < 1$. *Then* $L_{\text{greedy}}(\alpha) < l_{DR}(\alpha - \beta) \ln \left(\frac{1-\alpha+\beta}{\beta} \right) + 1$.

The proof of this theorem is similar to the proof of Theorem 31.

Corollary 7. *Let* $\alpha \in \mathbb{R}$ *and* $0 < \alpha < 1$. *Then* $L_{\text{greedy}}(\alpha) < l_{DR}(0) \ln \left(\frac{1}{\alpha} \right) + 1$.

If $l_{DR}(0)$ is a small number, then we have a good upper bound on $L_{\text{greedy}}(\alpha)$. If $l_{DR}(0)$ is a big number, then we have a big lower bound on $L_{\min}(0)$ and on $L_{\min}(\alpha)$ for some α.

4.6 Upper Bound on $L_{\text{greedy}}(\alpha)$

Assume that (T, r) is a decision rule problem from P_{DR}. In this subsection, we obtain an upper bound on $L_{\text{greedy}}(\alpha) = L_{\text{greedy}}(\alpha, T, r)$, which does not depend on $|U(T, r)|$, and show that, in some sense, this bound is unimprovable.

Theorem 45. *Let* α *and* β *be real numbers such that* $0 < \beta \leq \alpha < 1$. *Then* $L_{\text{greedy}}(\alpha) < L_{\min}(\alpha - \beta) \ln \left(\frac{1-\alpha+\beta}{\beta} \right) + 1$.

The proof of this theorem is similar to the proof of Theorem 32.

Let us show that obtained bound is, in some sense, unimprovable.

Theorem 46. *There is no real* $\delta < 1$ *such that for any decision rule problem* $(T, r) \in P_{DR}$ *and for any real* α *and* β, $0 < \beta \leq \alpha < 0$, *the following inequality holds:* $L_{\text{greedy}}(\alpha) \leq \delta \left(L_{\min}(\alpha - \beta) \ln \left(\frac{1-\alpha+\beta}{\beta} \right) + 1 \right)$.

The proof of this theorem is similar to the proof of Theorem 33.

4.7 On Decision Rules for the Most Part of Binary Decision Tables

On Tests and Local Tests for the Most Part of Binary Information Systems. A binary information system I is a table with n rows (corresponding to objects) and m columns labeled with attributes $f_1, \ldots, f_m$. This table is filled by numbers from $\{0,1\}$ (values of attributes). For $j = 1, \ldots, n$ we denote by r_j the j-th row of table I.

A subset $\{f_{i_1}, \ldots, f_{i_k}\}$ of attributes is a test for the information system I if these attributes separate any two rows r_j and r_l, where $j, l \in \{1, \ldots, n\}$ and $j \neq l$.

Adding an arbitrary decision attribute to the considered information system I we obtain a decision table T. For $j = 1, \ldots, n$ let $r_j = (b_1^j, \ldots, b_m^j)$ and d_j be the decision attached to r_j. If $\{f_{i_1}, \ldots, f_{i_k}\}$ is a test for the information system I, then for any $j \in \{1, \ldots, n\}$ the rule $(f_{i_1} = b_{i_1}^j) \wedge \ldots \wedge (f_{i_k} = b_{i_k}^j) \rightarrow d_j$ is a 0-decision rule for (T, r_j).

Let $m \geq \lceil 2\log_2 n \rceil + t$, where t is a natural number. Let $i_1, \ldots, i_{\lceil 2\log_2 n \rceil + t}$ be pairwise different numbers from $\{1, \ldots, m\}$. From Theorem 34 it follows that the fraction of information systems, for which $\{f_{i_1}, \ldots, f_{i_{\lceil 2\log_2 n \rceil + t}}\}$ is a test, is at least $1 - \frac{1}{2^{t+1}}$.

We will now fix a set D of decision attributes. From the considered result it follows, for example, that for 99% of binary decision tables with n rows, $m \geq \lceil 2\log_2 n \rceil + 6$ conditional attributes and decision attribute from D for each row there exists an exact decision rule which length is equal to $\lceil 2\log_2 n \rceil + 6$.

It is possible to improve this bound if we consider decision rules not for all rows but for one fixed row only.

Let $j \in \{1, \ldots, n\}$. A subset $\{f_{i_1}, \ldots, f_{i_k}\}$ of attributes will be called a *j-th local test* for the information system I if these attributes separate from the row r_j any row r_l, where $l \in \{1, \ldots, n\}$ and $l \neq j$.

Adding an arbitrary decision attribute to the considered information system I we obtain a decision table T. Let $r_j = (b_1, \ldots, b_m)$ and d be the decision attached to r_j. If $\{f_{i_1}, \ldots, f_{i_k}\}$ is a j-th local test for the information system I, then $(f_{i_1} = b_{i_1}) \wedge \ldots \wedge (f_{i_k} = b_{i_k}) \rightarrow d$ is a 0-decision rule for (T, r_j).

We will now fix a set D of decision attributes. If we prove the existence of good j-th local tests for the most part of binary information systems with n rows and m columns, then it means the existence of good decision rules for j-th row for the most part of binary decision tables with n rows, m conditional attributes and decision attributes from D.

Theorem 47. *Let us consider binary information systems with n rows and m columns labeled with attributes $f_1, \ldots, f_m$. Let $m \geq \lceil \log_2 n \rceil + t$, where t is a natural number, $j \in \{1, \ldots, n\}$ and $i_1, \ldots, i_{\lceil \log_2 n \rceil + t}$ be pairwise different numbers from $\{1, \ldots, m\}$. Then the fraction of information systems, for which $\{f_{i_1}, \ldots, f_{i_{\lceil \log_2 n \rceil + t}}\}$ is a j-th local test, is at least $1 - \frac{1}{2^t}$.*

Proof. Let $k = \lceil \log_2 n \rceil + t$, $l \in \{1, \ldots, n\}$ and $l \neq j$. The number of information systems, for which j-th and l-th rows are equal at the intersection with columns $i_1, \ldots, i_k$, is 2^{mn-k}. Therefore, the number of information systems, for which

$\{f_{i_1}, \ldots, f_{i_k}\}$ is not a j-th local test, is at most $n2^{mn-k} = 2^{mn-k+\log_2 n} \le 2^{mn-t}$. Thus, the fraction of information systems, for which $\{f_{i_1}, \ldots, f_{i_k}\}$ is a j-th local test, is at least $\frac{2^{mn}-2^{mn-t}}{2^{mn}} = 1 - \frac{1}{2^t}$. □

Let us fix a set D of decision attributes and a number $j \in \{1, \ldots, n\}$. From obtained result it follows that for 99% of binary decision tables with n rows, $m \ge \lceil \log_2 n \rceil + 7$ conditional attributes and the decision attribute from D for j-th row there exists an exact decision rule which length is equal to $\lceil \log_2 n \rceil + 7$.

On Partial Decision Rules Constructed by Greedy Algorithm for the Most Part of Binary Decision Tables. Now we study the behavior of greedy algorithm for the most part of binary decision tables, under some assumption on relationships between the number of rows and the number of columns in tables.

Let I be a binary information system with n rows and m columns labeled with attributes $f_1, \ldots, f_m$. For $j = 1, \ldots, n$, we denote by r_j the j-th row of I. The information system I will be called *strongly saturated* if, for any row $r_j = (b_1, \ldots, b_m)$ of I, for any $k \in \{1, \ldots, n-1\}$ and for any k rows with numbers different from j, there exists a column f_i which has at least $\frac{k}{2}$ numbers $\neg b_i$ at the intersection with considered k rows.

First, we evaluate the number of strongly saturated binary information systems. After that, we study the work of greedy algorithm on a decision table obtained from a strongly saturated binary information system by adding a decision attribute.

Theorem 48. *Let us consider binary information systems with n rows and $m \ge n + \log_2 n$ columns labeled with attributes $f_1, \ldots, f_m$. Then the fraction of strongly saturated information systems is at least $1 - \frac{1}{2^{m-n-\log_2 n+1}}$.*

Proof. Let us fix a number $j \in \{1, \ldots, n\}$, a tuple $\bar{b} = (b_1, \ldots, b_m) \in \{0,1\}^m$, a number $k \in \{1, \ldots, n-1\}$ and k rows with numbers different from j. Let us evaluate the number of information systems in which $r_j = \bar{b}$ and, for $i = 1, \ldots, m$, the column f_i has less than $\frac{k}{2}$ numbers $\neg b_i$ at the intersection with considered k rows. Such information systems will be called $(j, \bar{b})$-unsaturated in the considered k rows.

From Lemma 8 it follows that the number of tuples from $\{0,1\}^k$, which have less than $\frac{k}{2}$ numbers $\neg b_i$, is at most 2^{k-1}. Therefore, the number of information systems, which are $(j, \bar{b})$-unsaturated in the considered k rows, is at most 2^{mn-2m}.

There are n variants for the choice of j, at most 2^{n-1} variants for the choice of $k \in \{1, \ldots, n-1\}$ and k rows with numbers different from j, and 2^m variants for the choice of tuple $\bar{b}$. Therefore, the number of information systems, which are not strongly saturated, is at most $n2^{n-1}2^m 2^{mn-2m} = 2^{mn-2m+\log_2 n+n-1+m} = 2^{mn+\log_2 n+n-m-1}$, and the fraction of strongly saturated information systems is at least $\frac{2^{mn}-2^{mn+\log_2 n+n-m-1}}{2^{mn}} = 1 - \frac{1}{2^{m-n-\log_2 n+1}}$. □

For example, if $m \ge n + \log_2 n + 6$, then at least 99% of binary information systems are strongly saturated.

Let us consider the work of greedy algorithm on an arbitrary decision table T obtained from a strongly saturated binary information system. Let r be an arbitrary row of table T. For $i = 1, 2, \ldots$, after the step number i at most $\frac{|U(T,r)|}{2^i}$ rows from $U(T, R)$ are unseparated from r. It is not difficult to show that $L_{\text{greedy}}(\alpha) \leq \lceil \log_2 \frac{1}{\alpha} \rceil$ for any real α, $0 < \alpha < 1$. One can prove that $L_{\text{greedy}}(0) \leq \log_2 |U(T,r)| + 1$. It is easy to check that $l_{DR}(0) \leq 2$.

Results of Experiments. We made some experiments with binary decision tables T containing $n \in \{10, 50, 100, 1000, 3000, 5000\}$ rows, $m \in \{10, 40, 100\}$ conditional attributes and one decision attribute with values from the set $\{1, \ldots, c\}$, $c \in \{2, 10, 100\}$. For each triple of values (n, m, c) we generated randomly a decision table such that each element of this table is equal to b, $b \in \{0, 1\}$, with probability $\frac{1}{2}$, and each decision is equal to d, $d \in \{1, \ldots, c\}$, with probability $\frac{1}{c}$. For this table we choose randomly 10 rows r. The results of experiments are represented in Tables 8–11.

In Table 8 the average percentage of rows from $U(T, r)$ separated from r at i-th step of greedy algorithm, $i = 1, \ldots, 10$, is presented for the case, when $m = 40$ and $c = 10$. For example, 53.10 means that, on the average, 53.10% of rows remaining unseparated before i-th step are separated at i-th step.

Table 8. Average percentage of rows separated at i-th step of greedy algorithm ($m = 40$ and $c = 10$)

Number	Number i of step									
of rows n	1	2	3	4	5	6	7	8	9	10
10	85.79	100.00								
50	65.99	74.71	94.67	100.00						
100	61.90	67.42	79.38	100.00						
1000	54.05	55.05	56.54	56.56	64.01	76.50	100.00			
3000	52.04	52.50	53.77	55.52	57.06	61.51	71.01	82.94	100.00	
5000	51.57	52.09	53.10	54.31	56.28	59.01	64.85	74.46	92.07	100.00

Table 9. Average length of α-decision rules for decision tables with 10 conditional attributes

	α											
	0.1			0.01			0.001			0.0		
Number	Number of different decisions c											
of rows n	2	10	100	2	10	100	2	10	100	2	10	100
10	1.4	2.0	2.2	1.4	2.0	2.2	1.4	2.0	2.2	1.4	2.0	2.2
50	2.5	2.8	3.0	3.3	4.2	4.1	3.3	4.2	4.1	3.3	4.2	4.1
100	2.8	3.0	3.0	4.4	5.1	5.0	4.4	5.1	5.0	4.4	5.1	5.0
1000	3.2	3.5	3.9	5.8	6.1	6.2	7.8	8.4	8.7	7.8	8.4	8.7
3000	3.9	4.0	4.0	6.2	6.4	6.5	8.2	8.6	8.7	8.8	9.3	9.5
5000	4.0	4.0	4.0	6.4	6.8	6.8	8.6	8.9	9.1	9.0	9.9	9.9

Table 10. Average length of α-decision rules for decision tables with 40 conditional attributes

Number	α											
	0.1			0.01			0.001			0.0		
	Number of different decisions c											
of rows n	2	10	100	2	10	100	2	10	100	2	10	100
10	1.3	2.0	2.0	1.3	2.0	2.0	1.3	2.0	2.0	1.3	2.0	2.0
50	2.0	2.1	2.5	2.6	3.0	3.3	2.6	3.0	3.3	2.6	3.0	3.3
100	2.1	2.9	2.9	3.3	4.2	4.0	3.3	4.2	4.0	3.3	4.2	4.0
1000	3.0	3.0	3.1	5.0	5.8	5.8	6.1	7.0	7.0	6.1	7.0	7.0
3000	3.1	4.0	3.9	6.0	6.0	6.0	7.4	8.0	7.9	7.7	8.5	8.7
5000	3.9	4.0	4.0	6.0	6.2	6.1	8.0	8.1	8.7	8.5	9.1	9.3

Table 11. Average length of α-decision rules for decision tables with 100 conditional attributes

Number	α											
	0.1			0.01			0.001			0.0		
	Number of different decisions c											
of rows n	2	10	100	2	10	100	2	10	100	2	10	100
10	1.1	2.0	2.0	1.1	2.3	2.0	1.1	2.3	2.0	1.1	2.3	2.0
50	2.0	2.0	2.1	2.5	3.0	3.0	2.5	3.0	3.0	2.5	3.0	3.0
100	2.0	2.5	2.9	3.0	3.9	4.0	3.0	3.9	4.0	3.0	3.9	4.0
1000	3.0	3.0	3.0	5.0	5.1	5.3	6.0	6.4	6.8	6.0	6.4	6.8
3000	3.0	3.5	3.7	6.0	6.0	6.0	7.0	7.8	7.8	7.1	8.2	7.9
5000	3.4	4.0	4.0	6.0	6.0	6.0	7.6	8.0	8.0	8.0	8.9	8.7

In Table 9 for each $\alpha \in \{0.1, 0.01, 0.001, 0.0\}$ the average length of α-decision rules constructed by greedy algorithm is presented for decision tables with 10 conditional attributes.

In Table 10 for each $\alpha \in \{0.1, 0.01, 0.001, 0.0\}$ the average length of α-decision rules constructed by the greedy algorithm is presented for decision tables with 40 conditional attributes.

In Table 11 for each $\alpha \in \{0.1, 0.01, 0.001, 0.0\}$ the average length of α-decision rules constructed by the greedy algorithm is presented for decision tables with 100 conditional attributes.

The obtained results show that for the most part of the considered decision rule problems (not only for the case, when $m \geq n + \log_2 n$) during each step the greedy algorithm chooses an attribute which separates at least one half of unseparated rows.

It must be also noted that with increase of step number the percentage of rows, separated at this step, grows for the most part of the considered decision rule problems.

5 Conclusions

The paper is devoted (mainly) to the theoretical and experimental analysis of greedy algorithms for partial cover, reduct and decision rule construction.

The obtained results show that, under some natural assumptions on the class NP, these algorithms are close to the best polynomial approximate algorithms for the minimization of partial covers, reducts and rules. Based on an information received during greedy algorithm work it is possible to obtain lower and upper bounds on the minimal complexity of partial covers, reducts and rules. Experimental and some theoretical results show that, for the most part of randomly generated set cover problems and binary decision tables, greedy algorithms construct simple partial covers, reducts and rules with relatively high accuracy.

Acknowledgements

The authors are greatly indebted to Andrzej Skowron for stimulating discussions and to anonymous reviewers for helpful remarks and suggestions.

References

1. Cheriyan, J., Ravi, R.: Lecture Notes on Approximation Algorithms for Network Problems (1998), http://www.math.uwaterloo.ca/~jcheriya/lecnotes.html
2. Feige, U.: A threshold of $\ln n$ for approximating set cover (Preliminary version). In: Proceedings of 28th Annual ACM Symposium on the Theory of Computing, pp. 314–318 (1996)
3. Johnson, D.S.: Approximation algorithms for combinatorial problems. J. Comput. System Sci. 9, 256–278 (1974)
4. Karp, R.M.: Reducibility among combinatorial problems. In: Miller, R., Thatcher, J. (eds.) Complexity of Computer Computations, pp. 85–103. Plenum Press, New York (1972)
5. Kuzjurin, N.N.: Asymptotical investigation of set covering problem. In: Problems of Cybernetics, vol. 37, pp. 19–57. Nauka Publishers, Moscow (1980) (in Russian)
6. Lovász, L.: On the ratio of optimal integral and fractional covers. Discrete Math. 13, 383–390 (1975)
7. Moshkov, M.J.: Greedy algorithm for set cover in context of knowledge discovery problems. In: Proceedings of the International Workshop on Rough Sets in Knowledge Discovery and Soft Computing (ETAPS 2003 Satellite Event), Warsaw, Poland. Electronic Notes in Theoretical Computer Science, vol. 82 (2003)
8. Moshkov, M.J.: On greedy algorithm for partial cover construction. In: Proceedings of the Fourteenth International Workshop Design and Complexity of Control Systems, Nizhny Novgorod, Russia, p. 57 (2003) (in Russian)
9. Moshkov, M.J., Piliszczuk, M., Zielosko, B.: On partial covers, reducts and decision rules with weights. In: Peters, J.F., Skowron, A., Düntsch, I., Grzymała-Busse, J.W., Orłowska, E., Polkowski, L. (eds.) Transactions on Rough Sets VI. LNCS, vol. 4374, pp. 211–246. Springer, Heidelberg (2007)
10. Nguyen, H.S., Ślęzak, D.: Approximate reducts and association rules – correspondence and complexity results. In: Zhong, N., Skowron, A., Ohsuga, S. (eds.) RSFD-GrC 1999. LNCS (LNAI), vol. 1711, pp. 137–145. Springer, Heidelberg (1999)

11. Nigmatullin, R.G.: Method of steepest descent in problems on cover. In: Memoirs of Symposium Problems of Precision and Efficiency of Computing Algorithms, Kiev, USSR, vol. 5, pp. 116–126 (1969) (in Russian)
12. Pawlak, Z.: Rough Sets – Theoretical Aspects of Reasoning about Data. Kluwer Academic Publishers, Dordrecht (1991)
13. Pawlak, Z.: Rough set elements. In: Polkowski, L., Skowron, A. (eds.) Rough Sets in Knowledge Discovery 1. Methodology and Applications. Studies in Fuzziness and Soft Computing, vol. 18, pp. 10–30. Physica-Verlag, Heidelberg (1998)
14. Quafafou, M.: α-RST: a generalization of rough set theory. Information Sciences 124, 301–316 (2000)
15. Raz, R., Safra, S.: A sub-constant error-probability low-degree test, and sub-constant error-probability PCP characterization of NP. In: Proceedings of 29th Annual ACM Symposium on the Theory of Computing, pp. 475–484 (1997)
16. Skowron, A.: Rough sets in KDD. In: Proceedings of the 16-th World Computer Congress (IFIP 2000), Beijing, China, pp. 1–14 (2000)
17. Skowron, A., Rauszer, C.: The discernibility matrices and functions in information systems. In: Slowinski, R. (ed.) Intelligent Decision Support. Handbook of Applications and Advances of the Rough Set Theory, pp. 331–362. Kluwer Academic Publishers, Dordrecht (1992)
18. Slavík, P.: A tight analysis of the greedy algorithm for set cover (extended abstract). In: Proceedings of 28th Annual ACM Symposium on the Theory of Computing, pp. 435–441 (1996)
19. Slavík, P.: Approximation algorithms for set cover and related problems. Ph.D. thesis. University of New York, Buffalo (1998)
20. Ślęzak, D.: Approximate reducts in decision tables. In: Proceedings of the Congress Information Processing and Management of Uncertainty in Knowledge-based Systems, Granada, Spain, vol. 3, pp. 1159–1164 (1996)
21. Ślęzak, D.: Normalized decision functions and measures for inconsistent decision tables analysis. Fundamenta Informaticae 44, 291–319 (2000)
22. Ślęzak, D.: Approximate decision reducts. Ph.D. thesis. Warsaw University (2001) (in Polish)
23. Ślęzak, D.: Approximate entropy reducts. Fundamenta Informaticae 53, 365–390 (2002)
24. Vercellis, C.: A probabilistic analysis of the set covering problem. Annals of Oper. Res. 1, 255–271 (1984)
25. Wróblewski, J.: Ensembles of classifiers based on approximate reducts. Fundamenta Informaticae 47, 351–360 (2001)
26. Yablonskii, S.V.: Some problems of reliability and diagnosis in control systems. In: Mathematical Problems of Cybernetics, vol. 1, pp. 5–25. Nauka Publishers, Moscow (1988) (in Russian)
27. Ziarko, W.: Analysis of uncertain information in the framework of variable precision rough sets. Foundations of Computing and Decision Sciences 18, 381–396 (1993)

Evolutionary Rough k-Medoid Clustering

Georg Peters[1], Martin Lampart[2], and Richard Weber[3]

[1] University of Applied Sciences - München
Department of Computer Science and Mathematics
80335 Munich, Germany
georg.peters@cs.hm.edu
[2] munita e.V.
c/o University of Applied Sciences - München
Department of Computer Science and Mathematics
80335 Munich, Germany
martin_lampart@web.de
[3] University of Chile
Department of Industrial Engineering
República 701, Santiago, Chile
rweber@dii.uchile.cl

Abstract. Recently, clustering algorithms based on rough set theory have gained increasing attention. For example, Lingras et al. introduced a rough k-means that assigns objects to lower and upper approximations of clusters. The objects in the lower approximation surely belong to a cluster while the membership of the objects in an upper approximation is uncertain. Therefore, the core cluster, defined by the objects in the lower approximation is surrounded by a buffer or boundary set with objects with unclear membership status. In this paper, we introduce an evolutionary rough k-medoid clustering algorithm. Evolutionary rough k-medoid clustering belongs to the families of Lingras' rough k-means and classic k-medoids algorithms. We apply the evolutionary rough k-medoids to synthetic as well as to real data sets and compare the results to Lingras' rough k-means. We also introduce a rough version of the Davies-Bouldin-Index as a cluster validity index for the family of rough clustering algorithms.

Keywords: Rough Sets, Rough k-Medoids, Cluster Analysis, Evolutionary Clustering, Davies-Bouldin-Index.

1 Introduction

Since its introduction by Pawlak during the early 1980s [19,20,21] and elaborated in [24] rough set theory has gained increasing attention among researchers and practitioner and established itself as a central concept of soft computing [29] besides fuzzy sets [32], neural nets [5] and others.

The core idea of rough set theory is to separate between objects with clear membership to a set on the one hand and those with ambiguous memberships on the other hand.

J.F. Peters and A. Skowron (Eds.): Transactions on Rough Sets VIII, LNCS 5084, pp. 289–306, 2008.

An object with a clear membership is assigned to one and only one set. In rough set terminology it is a member of the lower approximation of the set. If the membership of an object is unclear it is assigned to the upper approximations of all possible sets. Obviously, such an object must be assigned to two or more upper approximations since, otherwise, its membership would be clearly defined.

Originally, rough set theory was a purely and strictly set-based approach (see [23]), where it is unambiguously possible to determine the objects of the upper and the lower approximations. In this context further powerful constructs - like reducts, global and local coverings besides others - have been introduced. However, in the meantime a new branch of rough set theory has been established which is basically an interval estimation using distances [31, 30]. In contrast to original rough sets the assignment of objects to the approximations is arguable and depends on the experimental set-up.

The main contribution of this paper is that we extend rough clustering by a rough k-medoids algorithm and its evolutionary version. The evolutionary rough k-medoids belongs to the family of rough clustering algorithms as introduced by Lingras et al. [12] and its refined version by Peters [26] on the one hand and Kaufmann's classic k-medoids [8] on the other hand. The evolutionary extension of the rough k-medoids goes along the lines with Mitra's work on evolutionary rough partitive clustering [16]. In experiments with synthetic and real data we compare the new rough k-medoids to the rough k-means algorithms proposed by Lingras and Peters. As cluster validity index we introduce a rough version of the Davies-Bouldin-Index [4].

The paper is structured as follows. In the next Section we give a short introduction to Lingras' rough k-means cluster algorithm and important refinements and extensions. In Section 3 we discuss the classic k-medoids and its relationship to crisp k-means. Then, in Section 4, we introduce the new rough k-medoids and its evolutionary extension. In the subsequent Section we present the results of our experiments. The paper concludes with a summary in Section 6.

2 Rough k-Means Algorithms

In this Section we describe rough k-means algorithms [12, 26] and their evolutionary extensions [16].

2.1 Basic Properties of Rough Sets

The family of rough clustering algorithms that are based on Lingras' rough k-means belongs to the interval based branch of rough set theory. In contrast to original rough set theory this branch has a reduced set of properties as defined next:

1. An object is a member of one lower approximation at most.
2. A member of a lower approximation is also a member of the corresponding upper approximation. So, the lower approximation is a subset of the upper approximation.

3. If an object does not belong to any lower approximation it is a member of at least two upper approximations.

The part of an upper approximation that is not covered by a lower approximation is often called boundary set. The actual memberships to a single cluster of the objects in the boundary set cannot be determined due to missing information.

Consider the following, set-based example. The tail-sides of 1-EURO coins are identical independently of the country that issued them. The head-side shows national symbols of the issuing country. Therefore, if one only sees the tail-side of an 1-EURO coin one can determine its value. However, due to missing information one cannot determine where is was issued.

2.2 Lingras' Rough k-Means Cluster Algorithm

The classic k-means was introduced by Hartigan and Wong [7]. In the meantime several variations have been suggested. They include fuzzy k-means by Bezdek [1], fast k-means by Darken and Moody [18,3] and indiscernibility based k-means by J.F. Peters and Borkowski [28]. In the context of our paper Lingras' [14] rough k-means is of special interest. It proceeds as follows:

1. *Initialization.* Assign each object to one and only one lower approximation. Due to the second property (Section 2.1) it is also a member of the corresponding upper approximation.
2. *Calculation of the means.* The means are calculated as the weighted sums of the cluster members. The objects in the lower approximation are weighted by the factor w_l and the objects in the boundary set by w_b.
3. *Assignment of the objects to the approximations.* First, each object is assigned to the upper approximation of the cluster of its closest mean. Second, the distances to the remaining means are checked. If a distance is not significantly larger than to the closest mean (defined by the threshold ϵ) then the object is assigned to the corresponding upper approximation of this mean. If no other mean is reasonably close then the object is assigned to the lower approximation of the cluster of its closest mean.
4. *Convergence of the algorithm.* If the algorithm has converged *Stop*, otherwise continue with *Step (2) Calculation of the means.*

Peters [25, 26] suggested some refinements of Lingras' rough k-means which improve the algorithm with respect to its numerical stability and in the presence of outliers, besides others. Applications of the rough k-means can be found for example in [11].

2.3 Mitra's Evolutionary Extension of the Rough k-Means

Mitra [16] argued that one of the main challenges of Lingras' rough k-means is to set the initial parameters. These are the number of clusters K, the weights of

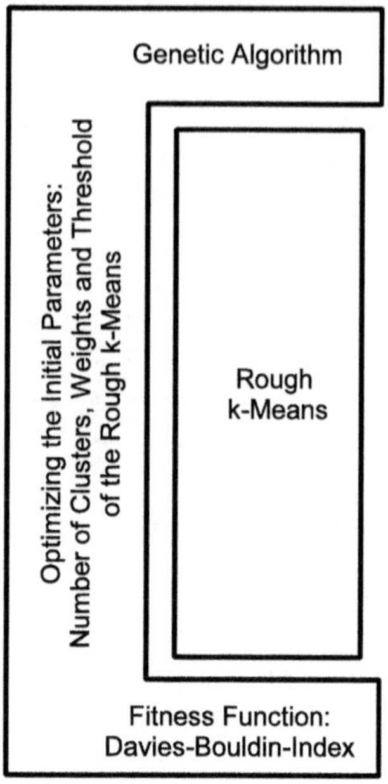

Fig. 1. Evolutionay Rough k-Means

the lower approximation (w_l) and the boundary set (w_b), and the threshold ϵ. She suggested an evolutionary extension of Lingras' rough k-means to optimize the initial parameters.

The principles of the evolutionary rough k-means are as follows: Lingras' algorithm is framed by an iteration that optimizes the weights w_l and w_b and the threshold ϵ for a given number of clusters K (Fig. 1). As optimization method Mitra chose a classic genetic algorithm [6] and as its fitness function the Davies-Bouldin-Index [4].

The Davies-Bouldin-Index (DBI) was also chosen as cluster validity index. It is considered as a general cluster validation criterion for partitive cluster algorithms since its results are independent from the number of clusters. Basically the Davies-Bouldin-Index is the ratio of the sum of the within-cluster scatter to the between cluster separation. Well separated clusters are obtained when the within-cluster scatter is small and the separation between different clusters is large [2]:

$$DBI = \frac{1}{K} \sum_{k=1}^{K} \max_{k \neq l} \left\{ \frac{S(U_k) + S(U_l)}{d(U_k, U_l)} \right\} \tag{1}$$

with $S(U_k)$ the average distance of the objects of cluster k to the cluster center (accordingly for $S(U_l)$) as a measure for the within-cluster distances:

$$S(U_k) = \left(\frac{1}{|C_k|} \sum_{\boldsymbol{X_n} \in C_k} \|\boldsymbol{X_n} - \boldsymbol{m_k}\|_2^q \right)^{1/q} \tag{2}$$

and $d(U_k, U_l)$ the distance between the cluster centers of the clusters k and l as an indicator for the between-cluster separation:

$$d(U_k, U_l) = \left\{ \sum_{f=1}^{F} |m_{kf} - m_{lf}|^p \right\}^{1/p} = \|\boldsymbol{m_k} - \boldsymbol{m_l}\|_p \tag{3}$$

The parameters p, q can be chosen separately. The Davies-Bouldin-Index has to be minimized for optimal cluster separation.

In a comparative study Mitra showed that the evolutionary rough k-means produces good results with respect to the Davies-Bouldin-Index in comparison to the classic k-means and k-medoids as well as to the fuzzy k-means and Lingras' rough k means. However, Mitra's study is limited by the following constraints:

1. The Davies-Bouldin-Index was designed for classic cluster algorithms where an object fully belongs to a cluster or it is not a member of a cluster at all. Since rough cluster algorithms have upper approximations with objects of ambiguous memberships to at least two clusters Mitra only considered the objects in the lower approximations of the cluster when calculating the Davies-Bouldin-Index. Obviously, the cluster separation of these data constellation is much better in comparison to the whole data set since the objects in the boundary sets are neglected. Therefore the significance of a comparison of the Davies-Bouldin-Indexes obtained by classic crisp algorithms, like the k-means and the k-medoids, and the rough k-means is limited.
2. The initial parameters pertain to the rough k-means only. Other algorithms like k-means, k-medoids do not have such tunable parameters.

So, although overall the evolutionary rough k-means performed well, the comparison with crisp clustering algorithms has limited expressiveness.

3 Classic k-Medoid Clustering

3.1 The Algorithm

The k-medoids was introduced by Kaufmann and Rousseeuw [8]. Instead of artificial cluster centers as in the k-means in k-medoid clustering each cluster is represented by a real object. The algorithm proceeds as follows:

1. *Initialization.* Select an objective function and define the number of clusters K. Randomly choose K objects as medoids and assign the remaining objects (non-medoids) to the clusters of their closest medoids.
2. *Determination of the medoids.* Swap each medoid with every non-medoid as long as the objective function improves.
3. *Convergence check.* If the algorithm has converged *Stop*, otherwise continue with *Step (2) Determination of the medoids.*

The objective function of the k-medoids is well separated from any other step of the clustering algorithm. Therefore it can be liberally defined by the data analyst. However, often the *compactness of clustering* (CPC) is chosen:

$$CPC = \sum_{k=1}^{K} CPC(C_k) \quad \text{with } CPC(C_k) = \sum_{\boldsymbol{X_n} \in C_k} d(\boldsymbol{X_n}, \boldsymbol{m_k}) \quad \text{and } CPC(C_k) \text{ the compactness of cluster } C_k. \tag{4}$$

A fuzzy version of the k-medoids was introduced by Krishnapuram et al. [9] and applied to web document and snippet clustering.

3.2 Comparison of k-Medoids and k-Means

The k-medoids has a real object as representative of a cluster while in k-means a cluster is represented by an artificial object. This leads to the following advantages of k-medoids in comparison to k-means:

1. Each cluster has a real, touchable object as its representative instead of an elusive artificial one.
2. The k-medoids delivers better results in presence of (extreme) outliers since the cluster center is always within the core cluster while applying the k-means could result in cluster centers that are "drawn" out of the core clusters. For example see Fig. 2 where the outlier draws a cluster center of the k-means out of the core cluster.
3. In a wide range the k-medoids is less noise sensitive in comparison to the k-means. However, this can also be interpreted as disadvantage (see below the last item of the drawbacks of the k-medoids).
4. The objective function of the k-medoids can be freely defined by the user.

The main drawbacks of k-medoids compared to k-means are:

1. The k-medoids algorithm is of combinatoric nature. This makes it less efficient in comparison to the k-means.
2. The need for a real object as representative of a cluster compromises on the quality of its cluster centers in comparison to the artificial cluster centers in the k-means.

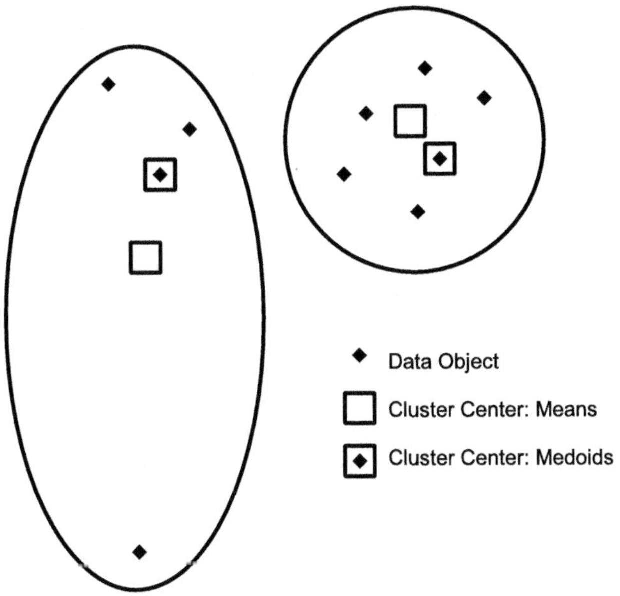

Fig. 2. Cluster Representatives and Outliers

3. For small changes in the distribution of the objects the cluster centers change discontinuously in certain circumstances: they jump from one object to another.

There is no general rule which cluster algorithms should be applied in certain circumstances. So it is up to the expert to choose an adequate one for an experiment[1].

4 Evolutionary Rough k-Medoids

In this Section we present two versions of the new rough k-medoids and their evolutionary extensions.

4.1 Rough k-Medoids Algorithms

We suggest rough k-medoids algorithms based on Lingras' as well as on Peters' versions of rough k-means. They are as follows.

Notations:

[1] It should be noted that, besides k-means and k-medoids, certainly a three digit number of variations and further cluster algorithms have been proposed up to date [15].

Abbreviation	Meaning
$\boldsymbol{X_n}, \mathbf{X}$	Data set: $\boldsymbol{X_n}$ for the nth data point and $\mathbf{X} = (\boldsymbol{X_1}, ..., \boldsymbol{X_N})^T$ with $n = 1, ..., N$ (N the number of objects).
$C_k, \underline{C_k}, \overline{C_k}, C_k^B$	Clusters: C_k the kth cluster, $\underline{C_k}$ its lower approximation, $\overline{C_k}$ the upper approximation and $C_k^B = \overline{C_k} - \underline{C_k}$ the boundary set.
$\boldsymbol{m_k}, \mathbf{M}$	Medoids: $\boldsymbol{m_k}$ the medoid of cluster C_k. The set of all medoids $\mathbf{M} = (\boldsymbol{m_1}, ..., \boldsymbol{m_K})^T$ with $k = 1, ..., K$ (K the number of clusters).
$d(\boldsymbol{X'_n}, \boldsymbol{m_k})$	Distance between object $\boldsymbol{X'_n}$ and medoid $\boldsymbol{m_k}$: $d(\boldsymbol{X'_n}, \boldsymbol{m_k}) = \Vert \boldsymbol{X'_n} - \boldsymbol{m_k} \Vert$.
w_l, w_u, w_b	Weights w_l, w_u and w_b for the lower and upper approximations and the boundary set.
ϵ, ζ	Thresholds in rough clustering.
T, T'	Set T respectively T': clusters that are considered to be close to an object.
ROF	Rough objective function ROF as defined in 4.2.

Algorithm (Model A):

First, we introduce a new rough cluster algorithm that has its roots in the classic k-medoids [8] and Lingras' rough k-means [13]. It proceeds as follows:

1. Randomly define K objects of the set $\boldsymbol{X}$ as medoids: $\boldsymbol{m_k}$, $k = 1, ..., K$. Assign them to the lower approximation of the set they are medoids of[2]: $\boldsymbol{m_k} \in \underline{C_k}$. The remaining objects are denoted as $\boldsymbol{X'_m}, m = 1, ..., (N - K)$.
2. Assign the remaining $(N - K)$ objects $\boldsymbol{X'_m}$ to the K clusters in a two step process. In the first step an object is assigned to the upper approximation of the cluster to which it is closest to. In the second step the object is assigned to the upper approximation of further reasonably close clusters *or* it is assigned to the lower approximation of the closest cluster. The details are as follows:
   (a) For a given object $\boldsymbol{X'_m}$ determine its closest medoid $\boldsymbol{m_k}$:

$$d(\boldsymbol{X'_m}, \boldsymbol{m_k}) = \min_{h=1,...K} d(\boldsymbol{X'_m}, \boldsymbol{m_h}) \tag{5}$$

   Assign $\boldsymbol{X'_m}$ to the upper approximation of the cluster k: $\boldsymbol{X'_m} \in \overline{C_k}$.
   (b) Determine the clusters C_h that are also close to $\boldsymbol{X'_m}$ - they are not farther away from $\boldsymbol{X'_m}$ than $d(\boldsymbol{X'_m}, \boldsymbol{m_k}) + \epsilon$ where ϵ is a given threshold:

$$T = \{h : d(\boldsymbol{X'_m}, \boldsymbol{m_h}) - d(\boldsymbol{X'_m}, \boldsymbol{m_k}) \leq \epsilon \wedge h \neq k\} \tag{6}$$

[2] Note, every cluster should have at least one sure member. Therefore we assign the medoids to the lower approximations.

- **If** $T \neq \emptyset$ ($\boldsymbol{X'_m}$ is also close to at least one other medoid besides $\boldsymbol{m_k}$)
- **Then** $\boldsymbol{X'_m} \in \overline{C_h}, \forall h \in T$.
- **Else** $\boldsymbol{X'_m} \in \underline{C_k}$.

3. Calculate $ROF_{current}$.
4. Swap every medoid $\boldsymbol{m_k}$ with every object $\boldsymbol{X'_m}$ and calculate $ROF_{k \leftrightarrow m}$[3]. Let $ROF_{k_0 \leftrightarrow m_0} = \min_{\forall k, \forall m} ROF_{k \leftrightarrow m}$ for $k = 1, ..., K$, $m = 1, ..., (N-K)$.
   - **If** $ROF_{k_0 \leftrightarrow m_0} < ROF_{current}$
   - **Then** swap the medoid $\boldsymbol{m_{k_0}}$ and object $\boldsymbol{X_{m_0}}$ and set $ROF_{current} = ROF_{k_0 \leftrightarrow m_0}$. Go back to *Step (2)*.
   - **Else** *Stop.*

Algorithm (Model B):

Second, we present a new rough cluster algorithm that has its foundations in the classic k-medoids [8] and Peters' refined rough k-means [26]. It differers from rough k-medoids *Model A* as presented above in Eq (6) of *Step (2) (b)*. Instead of the absolute distance as in Lingras' rough k-means Peters suggested to take a relative distance to improve its performance in the presence of outliers. Applying this to the rough k-medoids we get as new *Step (2) (b)*:

$$T' = \{h : \frac{d(\boldsymbol{X'_m}, \boldsymbol{m_h})}{d(\boldsymbol{X'_m}, \boldsymbol{m_k})} \leq \zeta \wedge h \neq k\} \quad \text{with } \zeta \text{ a threshold.} \tag{7}$$

The remaining parts of the algorithm are unchanged in comparison to *Model A*.

4.2 Objective Functions

In this Section we introduce two rough objective functions: (1) a Rough Compactness of Clustering and (2) a Rough Davies-Bouldin-Index.

Rough Compactness of Clustering. The classic objective function of k-medoids, the compactness of clustering CPC (see Eq (4)), can not be applied to the rough k-medoids since the objects are assigned to the two approximations. Therefore we have to adapt the compactness of clustering CPC to rough sets by splitting up the compactness of cluster $CPC(C_k)$ into a weighted sum. We obtain a rough compactness of clustering $RCPC$.

For the rough k-medoids that is based on Lingras' rough k-means the rough compactness of cluster $RCPC(C_k)$ consists of the weighted sum of the objects in the lower approximation and in the boundary region:

$$RCPC(C_k) = w_l \sum_{\boldsymbol{X_n} \in \underline{C_k}} d(\boldsymbol{X_n}, \boldsymbol{m_k}) + w_b \sum_{\boldsymbol{X_n} \in (\overline{C_k} - \underline{C_k})} d(\boldsymbol{X_n}, \boldsymbol{m_k}) \quad \text{with } w_l + w_b = 1. \tag{8}$$

[3] The properties of the ROF will be discussed in Section 4.2.

For the rough k-medoids that is based on Peters' refined rough k-means we get the weighted sum for the objects in the lower and upper approximations, respectively:

$$RCPC(C_k) = w_l \sum_{\boldsymbol{X_n} \in \underline{C_k}} d(\boldsymbol{X_n}, \boldsymbol{m_k}) + w_u \sum_{\boldsymbol{X_n} \in \overline{C_k}} d(\boldsymbol{X_n}, \boldsymbol{m_k}) \tag{9}$$

with $w_l + w_u = 1$.

Along the lines with the classic compactness of clustering CPC we get a rough objective function ROF:

$$ROF = RCPC = \sum_{k=1}^{K} RCPC(C_k) \tag{10}$$

Rough Davies-Bouldin-Index. Mitra et al. [17] suggested a rough Davies-Bouldin-Index using the weights of the rough approximations.

Alternatively one could follow Pawlak's [22] suggestion to generally weight the objects in an upper approximation by the factor 0.5. Following his arguments we could weight the distances of an object in the lower approximation to the cluster center by 1 and the distance of an object in an upper approximation by 0.5 when calculating the Davies-Bouldin-Index.

However, we adapt Laplace's famous *Principle of Indifference* [10] to rough clustering. Laplace argued that - if there are no reasons given that probabilities are not equally distributed - one has to assume that the probabilities are equally distributed.

Obviously we can apply Laplace's principle to the objects in the boundary sets since their memberships to a certain cluster is undefined. If an object n belongs to b_n boundary sets we can assume a degree of membership of $1/b_n$ to each of the corresponding clusters.

Utilizing Laplace's argument for the rough Davies-Bouldin-Index we weight the distance of the object n in a boundary set by $1/b_n$ and get:

$$S(U_k) = \left(\frac{1}{|\underline{C_k}|} \sum_{\boldsymbol{X_n} \in \underline{C_k}} \|\boldsymbol{X_n} - \boldsymbol{m_k}\|_2 + \frac{1}{|\overline{C_k} - \underline{C_k}|} \sum_{\boldsymbol{X_n} \in (\overline{C_k} - \underline{C_k})} \left(\frac{\|\boldsymbol{X_n} - \boldsymbol{m_k}\|_2}{b_n} \right)^q \right)^{1/q} \tag{11}$$

Each object n in a boundary set now has a total weight of 1 in the rough Davies-Bouldin-Index since it belongs with a fraction of $1/b_n$ to each of the b_n clusters.

4.3 An Evolutionary Extension of the Rough k-Medoids

Along the lines with Mitra's evolutionary extension of rough k-means [16] we implemented an evolutionary rough k-medoids and applied the rough Davies-Bouldin-Index as fitness function in the genetic optimization of the initial parameters.

5 Experiments

In this Section the non-evolutionary and the evolutionary versions of the rough k-medoid clustering algorithm are validated and compared to rough k-means in experiments with four different data sets: synthetic, colon cancer, forest and synthetically generated control data.

- *Non-evolutionary rough k-medoids.* In the non-evolutionary experiments a two step approach is chosen: (1) Up to 400 experiments are conducted with randomly selected initial parameters (weights and threshold). (2) The initial parameters that lead to the best Davies-Bouldin-Indexes (with $p = q = 2$) are analyzed in more detail. In the neighborhood of these initial parameters between 25 and 100 additional experiments are conducted to further optimize the Davies-Bouldin-Index. Since the algorithms converge towards local minima each experiment is repeated at least five times with identical initial settings.
- *Evolutionary rough k-medoids.* In the evolutionary experiments the initial parameters (weights and threshold) are optimized. To better compare the results to the non-evolutionary experiments the number of clusters K is excluded from the optimization and remains unchanged from the non-evolutionary test setup.

In our study we exclude comparisons to crisp cluster algorithms because of a missing well accepted cluster validity index for crisp and rough clustering.

5.1 Synthetic Data

Data Set. The two-dimensional, synthetic data set consists of 10 objects as depicted in Fig. 3 [27]. Basically they are arranged in two clusters.

Experiments. The number of clusters is set to $K = 2$. First, in rough k-medoid clustering the data are optimized according to the rough compactness of clustering. The results that are obtained for the rough k-means and the rough k-medoids are shown in the Tables 1 and 2[4].

Discussion. The Davies-Bouldin-Indexes for the rough k-means are better in comparison to the rough k-medoids which are optimized according to the $RCPC$ function. This shows that the $RCPC$ and the DBI are of different nature. Furthermore, in rough k-means the cluster centers are not limited to real objects as in rough k-medoids.

Optimizing rough k-medoids according to the Davies-Bouldin-Index provides much better results (see Table 1). Both versions of the rough k-medoids algorithm select the same objects as their medoids and have the same approximations (see

[4] Please note, that we use the following notations: DBI and $RCPC$ define the objective function applied in the rough k-medoids.

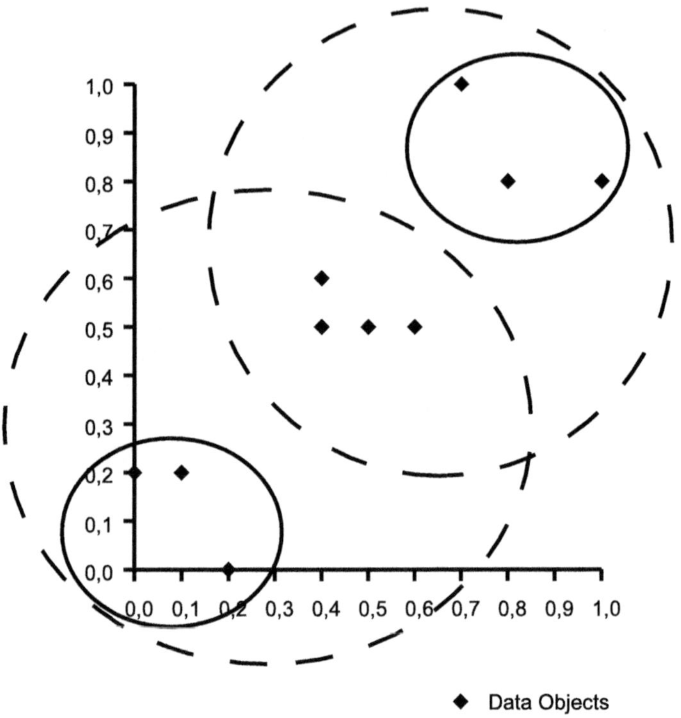

Fig. 3. Synthetic Data

Table 1. Synthetic Data: Results

Non-Evolutionary Algorithms	Davies-Bouldin-Index
Lingras' rough k-means	0.403
Peters' rough k-means	0.401
Rough k-medoids Model A ($RCPC$)	0.755
Rough k-medoids Model B ($RCPC$)	0.649
Rough k-medoids Model A (DBI)	0.404
Rough k-medoids Model B (DBI)	0.404
Evolutionary Algorithms	**Davies-Bouldin-Index**
Lingras' rough k-means	0.378
Peters' rough k-means	0.399
Rough k-medoids Model A ($RCPC$)	0.686
Rough k-medoids Model B ($RCPC$)	0.649
Rough k-medoids Model A (DBI)	0.404
Rough k-medoids Model B (DBI)	0.404

Table 2. Synthetic Data: Cluster Centers

Non-Evolutionary Algorithm	m_1	m_2
Lingras' rough k-means	(0.119, 0.153)	(0.815, 0.850)
Peters' rough k-means	(0.111, 0.145)	(0.823, 0.857)
Rough k-medoids Model A (*RCPC*)	(0.100, 0.200)	(0.500, 0.500)
Rough k-medoids Model B (*RCPC*)	(0.100, 0.200)	(0.600, 0.500)
Rough k-medoids Model A (*DBI*)	(0.000, 0.200)	(1.000, 0.800)
Rough k-medoids Model B (*DBI*)	(0.000, 0.200)	(1.000, 0.800)

Fig. 3 and Table 2). The results are similar to the results obtained by rough k-means.

The evolutionary rough k-medoids shows improved results in comparison to the non-evolutionary version when the *RCPC* is chosen as objective function.

When the Davies-Bouldin-Index is already implemented as objective function in the rough cluster algorithm the evolutionary optimization of the initial parameters did not improve the results any further. So the trial and error optimization of the initial parameters were found to be enough for this small data set.

5.2 Colon Cancer Data

Data Set. The colon cancer data (*http://www.molbio.princteon.edu/colondata*) consists of 62 gene expression records. Forty records stem from patients who suffer from cancer. The remaining 22 records belong to healthy people. Each data record has more than 2000 features. Due to performance requirements the experiments are conducted with a reduced number of 21 features [26]. The number of clusters is set to $K = 2$ to represent the two classes of patients.

Experiments. First, we apply the rough compactness of clustering in rough k-medoids. Second, we use the Davies-Bouldin-Index as objective function in the rough k-medoids algorithms. The results are depicted in Table 3.

Discussion. In the case of the colon cancer data the two objective functions, rough compactness of clustering and Davies-Bouldin-Index, lead to the same results. Lingras' rough k-means only performs a little bit inferior and, at the first sight, Peters' rough k-means produces significantly worse results than all other rough clustering algorithms.

Taking a closer look at the distribution of the objects to the cluster, we see that all algorithms expect Peters' tend to prefer unequally distributed solutions with no objects in the boundary sets (see Table 4). In contrast to that, Peters' algorithm provides a more intuitive solution where the lower approximations of both clusters have significant numbers of members.

The evolutionary rough k-medoids showed the same result ($DBI = 0.530$) for all tested versions. It did not improve the results compared to the

Table 3. Colon Cancer Data: Results

Non-Evolutionary Algorithms	Davies-Bouldin-Index
Lingras' rough k-means	0.602
Peters' rough k-means	1.178
Rough k-medoids Model A ($RCPC$)	0.530
Rough k-medoids Model B ($RCPC$)	0.530
Rough k-medoids Model A (DBI)	0.530
Rough k-medoids Model B (DBI)	0.530
Evolutionary Algorithms	**Davies-Bouldin-Index**
Lingras' rough k-means	0.542
Peters' rough k-means	1.116
Rough k-medoids Model A ($RCPC$)	0.530
Rough k-medoids Model B ($RCPC$)	0.530
Rough k-medoids Model A (DBI)	0.530
Rough k-medoids Model B (DBI)	0.530

Table 4. Colon Cancer Data: Assignment of the Objects

	Number of objects in		
Non-Evolutionary Algorithms	Cluster 1	Cluster 2	Boundary
Lingras' rough k-means	61	1	0
Peters' rough k-means	29	4	29
Rough k-medoids Model A ($RCPC$)	61	1	0
Rough k-medoids Model B ($RCPC$)	61	1	0
Rough k-medoids Model A (DBI)	61	1	0
Rough k-medoids Model B (DBI)	61	1	0
Evolutionary Algorithm			
Lingras' rough k-means	45	15	2

non-evolutionary version. However, for Lingras' rough k-means the evolutionary component significantly improved the results by two means. First, the Davies-Bouldin-Index was reduced from $DBI_{non-evolutionary} = 0.602$ to $DBI_{evolutionary} = 0.542$. Second, the distribution of the objects to the approximations is more intuitive in comparison to the non-evolutionary approach (see Table 4). The lower approximations of the clusters have 15 and 45 objects instead of the assignment of 61 objects to one cluster and only 1 to the other cluster in the non-evolutionary version (see Table 4).

Table 5. Forest Data: Results

Non-Evolutionary Algorithms	Davies-Bouldin-Index
Lingras' rough k-means	0.908
Peters' rough k-means	0.933
Rough k-medoids Model A (*RCPC*)	0.925
Rough k-medoids Model B (*RCPC*)	0.807
Rough k-medoids Model A (*DBI*)	0.484
Rough k-medoids Model B (*DBI*)	0.502
Evolutionary Algorithms	**Davies-Bouldin-Index**
Lingras' rough k-means	0.681
Peters' rough k-means	0.899
Rough k-medoids Model A (*RCPC*)	0.779
Rough k-medoids Model B (*RCPC*)	0.807
Rough k-medoids Model A (*DBI*)	0.482
Rough k-medoids Model B (*DBI*)	0.498

5.3 Forest Data

Data Set. The forest database (*http://kdd.ics.uci.edu*) consists of more than 500,000 data records. The features describe characteristics of a forest like slope, soil type and others. Ten of the features are quantitative. For our evaluation we use a randomly reduced data set of 241 objects and all 10 quantitative features. The number of clusters is arbitrarily set to $K = 3$ in all experiments.

Experiments. For the forest data we obtained Davies-Bouldin-Indexes as depicted in Table 5.

Discussion. Like in the experiments before, rough k-medoids using the Davies-Bouldin-Index as objective function performed better than rough k-medoids optimized according to the $RCPC$ and rough k-means.

This confirms the results of the previous experiments. Choosing the DBI as objective function is a distinct advantage when the DBI is also used as cluster validity index.

Again, the improvements for rough k-medoids with the Davies-Bouldin-Index as objective function are only small when the initial parameters are optimized evolutionarily. In contrast to that the improvements obtained for Lingras' rough k-means are significant.

5.4 Control Chart Data

Data Set. The control data set (*http://kdd.ics.uci.edu*) is synthetically generated and consists of 600 data records and 6 different classes of control charts.

Table 6. Control Chart Data: Results

Non-Evolutionary Algorithms	Davies-Bouldin-Index
Lingras' rough k-means	0.539
Peters' rough k-means	0.534
Rough k-medoids Model A (*RCPC*)	0.586
Rough k-medoids Model B (*RCPC*)	0.508
Rough k-medoids Model A (*DBI*)	0.462
Rough k-medoids Model B (*DBI*)	0.398
Evolutionary Algorithms	**Davies-Bouldin-Index**
Lingras' rough k-means	0.521
Peters' rough k-means	0.463
Rough k-medoids Model A (*RCPC*)	0.586
Rough k-medoids Model B (*RCPC*)	0.493
Rough k-medoids Model A (*DBI*)	0.453
Rough k-medoids Model B (*DBI*)	0.395

Experiments. In our experiments we used a reduced set of 21 objects and set the number of clusters to $K = 6$ to represent the 6 different classes of control charts. We obtained Davies-Bouldin Indexes as shown in Table 6.

Discussion. As observed before, those algorithms that are already optimized according to the Davies-Bouldin-Index outperformed both rough k-medoids with the *RCPC* as objective function as well as the rough k-means algorithms.

The results of the evolutionary algorithms confirm the observations of the previous experiments. If the objective function is identical to the cluster validity index the improvements of evolutionary optimization are limited. However, in contrast to some of the previous experiments, the results did not significantly improve by evolutionary optimizing the initial parameters in the other cases.

6 Conclusion

In this article we presented two versions of a new rough k-medoid clustering algorithm and tested them in four experiments. A comparison to rough k-means approaches shows that the special strength of rough k-medoids is its isolated objective function which can be freely defined by the user.

A comparison to the evolutionary versions shows that, in most cases, the improvements obtained by the evolutionary rough k-medoids are limited in comparison to the non-evolutionary versions. Therefore, in rough k-medoid clustering it is not as necessary to implement an evolutionary version as in rough k-means. The main reason for that is that in rough k-medoids the objective function can be liberally chosen, ideally identical to the selected cluster validity index.

Although several cluster validity indexes have been suggested already (see e.g. [2]) there is no criterion to compare the quality of crisp and rough cluster algorithms. Therefore we abstained from a comparison of the new rough k-medoids to the classic k-medoids. However, there is a need for a comprehensive class of cluster validity indexes that can be applied to crisp and rough as well as to fuzzy clustering approaches.

Acknowledgment

Support by the Scientific Millenium Institute "Complex Engineering Systems" is greatly acknowledged; see www.sistemasdeingenieria.cl.

References

1. Bezdek, J.: Pattern Recognition with Fuzzy Objective Algorithms. Plenum Press, New York (1981)
2. Bezdek, J., Pal, N.: Some new indexes of cluster validity. IEEE Transactions on Systems, Man, and Cybernetics 28, 301–315 (1998)
3. Darken, C., Moody, J.: Fast adaptive k-means clustering: Some empirical results. In: Proceeding 2nd International Joint Conference on Neural Networks, pp. 233–238 (1990)
4. Davies, D., Bouldin, D.: A cluster separation measure. IEEE Transactions on Pattern Analysis and Machine Intelligence 1, 224–227 (1979)
5. de Wilde, P.: Neural Network Models. Theory and Projects, 2nd edn. Springer, Berlin (1997)
6. Goldberg, D.: Genetic Algorithms in Search, Optimization, and Machine Learning. Addison-Wesley Professional, Boston (1989)
7. Hartigan, J., Wong, M.: A k-means clustering algorithm. Applied Statistics 28, 100–108 (1979)
8. Kaufman, J., Rousseeuw, P.: Finding Groups in Data: An Introduction to Cluster Analysis. John Wiley, New York (1990)
9. Krishnapuram, R., Joshi, A., Yi, L.: A fuzzy relative of the k-medoids algorithm with application to web document and snippet clustering. In: IEEE International Fuzzy Systems Conference, Seoul, Korea, pp. 1281–1286 (1999)
10. Laplace, P.: Philosophical Essay on Probabilities (Translated from the fifth French edition of 1825). Springer, Berlin (1998)
11. Lingras, P.: Applications of rough set based k-means, Kohonen SOM, GA clustering. Transactions on Rough Sets VII, 120–139 (2007)
12. Lingras, P., West, C.: Interval set clustering of web users with rough k-means. Technical Report 2002-002, Department of Mathematics and Computer Science, St. Mary's University, Halifax, Canada (2002)
13. Lingras, P., West, C.: Interval set clustering of web users with rough k-means. Journal of Intelligent Information Systems 23, 5–16 (2004)
14. Lingras, P., Yan, R., West, C.: Comparison of conventional and rough k-means clustering. In: Wang, G., Liu, Q., Yao, Y., Skowron, A. (eds.) RSFDGrC 2003. LNCS (LNAI), vol. 2639, pp. 130–137. Springer, Heidelberg (2003)
15. Mirkin, B.: Mathematical Classification and Clustering. Kluwer Academic Publishers, Boston (1996)

16. Mitra, S.: An evolutionary rough partitive clustering. Pattern Recognition Letters 25, 1439–1449 (2004)
17. Mitra, S., Banka, H., Pedrycz, W.: Rough-fuzzy collaborative clustering. IEEE Transactions on Systems, Man, and Cybernetics, Part B: Cybernetics 36, 795–805 (2006)
18. Moody, J., Darken, C.: Fast learning in networks of locally-tuned processing units. Neural Computation 1, 281–294 (1989)
19. Pawlak, Z.: Classification of objects by means of attributes. Report 429, Institute for Computer Science, Polish Academy of Sciences (1981)
20. Pawlak, Z.: Rough sets. Report 431, Institute for Computer Science, Polish Academy of Sciences (1981)
21. Pawlak, Z.: Rough sets. International Journal of Computer and Information Science 11, 341–356 (1982)
22. Pawlak, Z.: Rough sets and fuzzy sets. short communication. Fuzzy Sets and Systems 17, 99–102 (1985)
23. Pawlak, Z.: Rough Sets: Theoretical Aspects of Reasoning about Data. Kluwer Academic Publishers, Dordrecht (1992)
24. Pawlak, Z., Skowron, A.: Rudiments of rough sets. Information Sciences 177, 3–27 (2007)
25. Peters, G.: Outliers in rough k-means clustering. In: Pal, S.K., Bandyopadhyay, S., Biswas, S. (eds.) PReMI 2005. LNCS, vol. 3776, pp. 702–707. Springer, Berlin (2005)
26. Peters, G.: Some refinements of rough k-means. Pattern Recognition 39, 1481–1491 (2006)
27. Peters, G., Lampart, M.: A partitive rough clustering algorithm. In: Greco, S., Hata, Y., Hirano, S., Inuiguchi, M., Miyamoto, S., Nguyen, H.S., Słowiński, R. (eds.) RSCTC 2006. LNCS (LNAI), vol. 4259, pp. 657–666. Springer, Heidelberg (2006)
28. Peters, J., Borkowski, M.: K-means indiscernibility relation over pixels. In: Tsumoto, S., Słowiński, R., Komorowski, J., Grzymała-Busse, J.W. (eds.) RSCTC 2004. LNCS (LNAI), vol. 3066, pp. 580–585. Springer, Heidelberg (2004)
29. Polkowski, L.: Rough Sets. Physica-Verlag, Heidelberg (2003)
30. Yao, Y.: Two views of the theory of rough sets in finite universes. International Journal of Approximate Reasoning 15, 291–317 (1996)
31. Yao, Y., Li, X., Lin, T., Liu, Q.: Representation and classification of rough set models. In: Proceedings Third International Workshop on Rough Sets and Soft Computing, San Jose, CA, USA, pp. 630–637 (1994)
32. Zimmermann, H.: Fuzzy Set Theory and its Applications, 4th edn. Kluwer Academic Publishers, Boston (2001)

The Rough Set Database System

Zbigniew Suraj[1,2] and Piotr Grochowalski[2]

[1] Chair of Computer Science, Rzeszów University, Poland
{zsuraj,piotrg}@univ.rzeszow.pl
[2] Institute of Computer Science, State School of Higher Vocational Education in Jarosław, Poland

Abstract. This paper gives an overview of the current version of the Rough Sets Database System (in short the RSDS system). The current version of the RSDS system includes a number of modifications, extensions and functional improvements. The RSDS system is a freely available database system, developed to facilitate the creation of the rough sets bibliography for various types of publications as well as their fast popularization and application. This database is the most comprehensive online rough sets bibliography currently available at the Internet address http://rsds.univ.rzeszow.pl. The main functionalities of this system are presented along with a brief explanation of its exploitation methods.

Keywords: rough sets, data mining, knowledge discovery, pattern recognition, machine learning, database systems.

1 Introduction

The rough set theory [1] is a rapidly developing discipline of theoretical and applied computer science. It has become apparent during recent years that a bibliography on this subject is urgently needed as a tool for both the efficient research, and the use of the rough set theory.

This paper presents the latest edition of the Rough Set Database System (in short the RSDS system) for the creation of the bibliography on the rough sets and their applications. The first version of the RSDS system became available almost five years ago. The current version of this system includes a number of modifications, extensions and functional improvements, in particular:

- new versions of an advanced searching,
- information about co-authors for every author in the system,
- a module of a graph – statistical analysis of the content of the system,
- a module of a classification of scientific publications according to a projected classifier,
- an interactive map of the world showing who and where in the world works on the development of the rough set theory and its applications,
- some auxiliary sections, e.g. FAQ, Opinions.

J.F. Peters and A. Skowron (Eds.): Transactions on Rough Sets VIII, LNCS 5084, pp. 307–331, 2008.

The present version of the RSDS system includes new bibliographical descriptions of publications. Moreover, the existing contents of the database has been verified with respect to the accuracy and excessiveness of the stored data. As a result of conducted operations the data stored in the RSDS system is reliable and in most cases they have abstracts and keywords added. At the moment there are over 3400 publications in the database that have been written by over 1900 authors. Descriptions of publications in the system are classed in accordance with 12 publication types (specified in the specification *BibTeX*), i.e., article, book, booklet, inbook, incollection, inproceedings, manual, mastersthesis, phdthesis, proceedings, techreport, unpublished. Functionality of the created system is based on possibilities: adding, modifying, searching and data transforming (descriptions of publications) in the system. To simplify the maintenance of the system by users there were created separate sections (group of menu) that make it possible to move around the whole system.

A system user is able to login to it through the *Login* section. For registered users there are given possibilities such as: inserting new data in the system using a special form, modifying inserted data, classifying publications in accordance with a designed classifier. All system users are able to gain information (bibliographical descriptions publications).

In the section called *Search* there are two kinds of searching possible: alphabetical and advanced. The advanced search allows to find bibliographic data according to criteria. After finding the searched data, the system is able to generate and download the text file including these data (in the format of *BibTeX*).

In the section called *Statistics* there are dynamically generated statistics, that describe the system usage, data and their analysis.

In the system there is information concerning the software connected with the rough sets and biographies of outstanding people working actively into the rough set domain.

The actual version of the system has been given new extensions which make it more functional. The earlier versions of the RSDS system have been described in publications [2,3,4].

The rest of the paper is organized as follows. Section 2 presents an overview of the current possibilities of the RSDS system. The future plans for the RSDS system are discussed in section 3. Conclusions are given in section 4.

2 Capabilities of the System

2.1 Home Page

Having the system activated, the English version of the home page appears on a display. The service menu comprises several options making it possible to move around the whole system. The menu includes the following: *Home page*, *Login*, *Append*, *Search*, *Download*, *Send*, *Write to us*, *Statistics*, *Help*, *FAQ*, *Software*, *People*, *Opinions*, *Map* (see Figure 1).

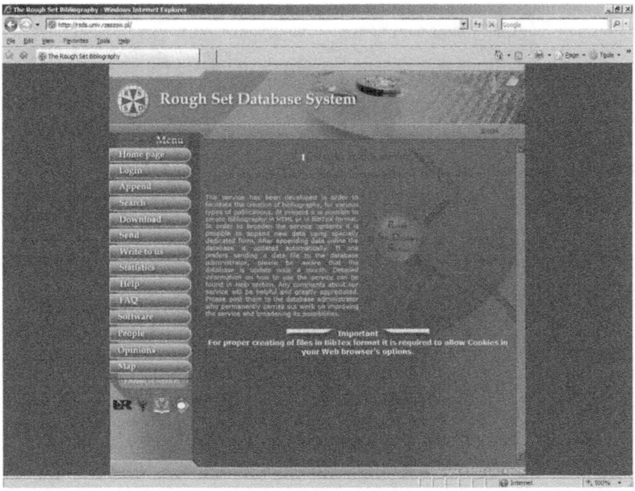

Fig. 1. The starting page of the RSDS system with a marked user menu

2.2 Adding Data – Online

The section *Append* is used for adding new data (publications) into the system. One must not use the national marks when introducing data into the system.

This section is available only for the users registered in the system (see Figure 2). In order to register into the system you have to fill in a form available in the section *Login*.

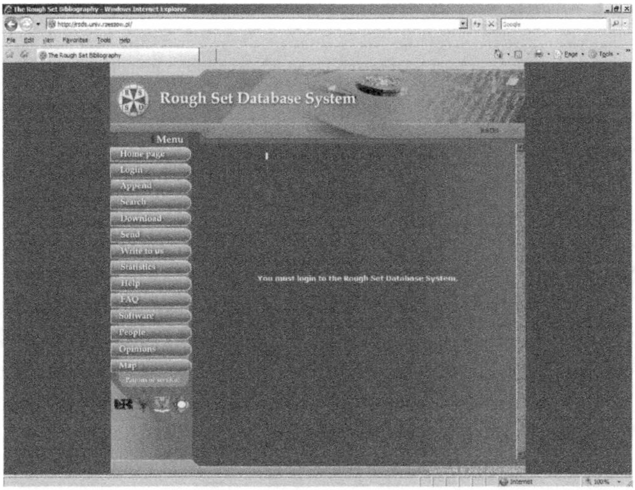

Fig. 2. A message displayed when trying to add new publications without logging into the system

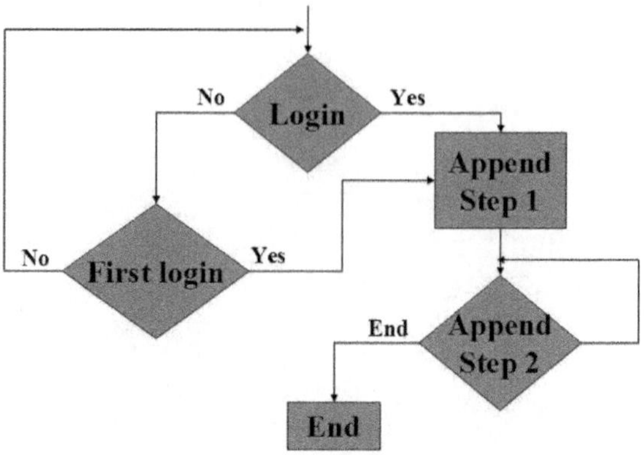

Fig. 3. Scheme for operation of adding data online

If a user logs into the system for the first time, he/she has to fill in a registration form available in the section *First Login* (after pressing the key *First Login*). In this form the user gives personal data. When the form has been filled in correctly and the *id* and the *password* has been defined, the user is automatically logged into the system. On the other hand, when the user is already registered in the system and wants to log in, then, in the section *Login* he fills in the form containing the id and the password, and when they are written correctly he is logged into the system. Next, the section *Append* becomes activated for the user. For the sake of safety the system automatically remembers which publications have been added by a given user. This information is also used when the data is edited. Adding new bibliographical descriptions has been divided into two phases (see Figure 3):

- During the first phase the user defines information describing a given publication, which is demanded by the system *BibTeX* specification, and corresponding to a particular type of a publication.
- During the second phase one defines information connected with authors or editors of a given publication.

At the beginning of introducing the data describing a publication the user defines the type of a publication (see Figure 4). There are twelve types of publications available. Depending on a chosen type, a form is generated which contains the data used for describing a given publication, i.e., a title, an editor, a year of publishing, a publishing etc. The data necessary for describing a given type are marked with an asterisk (*) (see Figure 5).

The list concerning publication types together with the fields describing them is as follows.

Publication	Description
article	An article from a journal. *Fields required:* author, title, journal, year. *Optional fields:* volume, number, pages, month, note.
book	A book with the known, given publisher. *Fields required:* author or editor, title, publisher, year. *Optional fields:* volume, series, address, edition, month, note.
booklet	Printed and bound matter, whilst the publisher is unknown. *Fields required:* title. *Optional fields:* author, address, month, year, note.
inbook	A part of a book, could be a chapter or given pages. *Fields required:* author or editor, title, chapter or pages, publisher, year. *Optional fields:* volume, series, address, edition, month, note.
incollection	A part of a book with its own title. *Fields required:* author, title, book title, publisher, year. *Optional fields:* editor, chapter, pages, address, month, note.
inproceedings	An article published in the conference proceedings. *Fields required:* author, title, book title, year. *Optional fields:* author, organization, publisher, address, month, note.
manual	Manual or documentation. *Fields required:* title. *Optional fields:* author, organization, address, edition, month, year, note.
mastersthesis	M.Sc. thesis. *Fields required:* author, title, school, year. *Optional fields:* address, month, note.
phdthesis	Ph.D. thesis. *Fields required:* author, title, school, year. *Optional fields:* address, month, note.
proceedings	Proceedings. *Fields required:* title, year. *Optional fields:* editor, publisher, organization, address, month, note.
techreport	Report, usually with a given number, being periodically issued. *Fields required:* author, title, institution, year. *Optional fields:* number, address, month, note.
unpublished	A document with a given author and title data, unpublished. *Fields required:* author, title, note. *Optional fields:* month, year.

Explanation on existing fields.

abstract	An abstract of a publication.
address	Publisher's address.
author	Forename and surname of an author (or authors).
booktitle	Title of a quoted in part book.
chapter	The chapter number.
edition	Issue, edition.
editor	Forenames and surnames of editors. If there also exists the field "author", the "editor" denotes the editor of a larger entity, of which the quoted work is a part.
institution	Institution publishing the printed matter.
ISBN	The International Standard Book Number.
ISSN	The International Standard Serial Number. Used to identify a journal.
journal	Journal's name.
keywords	Key words attached to a publication. This can be used for searching a publication.
month	Month of issue or completion of the manuscript.
note	Additional information useful to a reader.
number	The journal or the report number. Usually journals are being identified by providing their year and a number within the year of issue. A report, in general, has only a number.
organization	Organization supporting a conference.
pages	One or more page numbers; for example 42-11, 7,41,73-97.
publisher	Publisher's name.
school	University, college, where the thesis is submitted.
series	The name of a book series. If one quotes the book from a given series, then the "title" field denotes the title of a book, whilst the "series" field should contain the entire series' name.
title	The title of the work.
URL	The WWW Universal Resource Locator that points to the item being referenced. This often is used for technical reports to point to the http or ftp site where the postscript source of the report is located.
volume	The periodical's or the book's volume.
year	Year of issue. In case of unpublished work, the year of completing writing. Year only in number format e.g. 1984.

Note: All data must be appended in the Latin alphabet - without national marks.

When all required data is introduced, the system displays the introduced data again in order to verify its correctness. When the data is accepted the user gets on to another stage of adding a new publication - introduction of data related

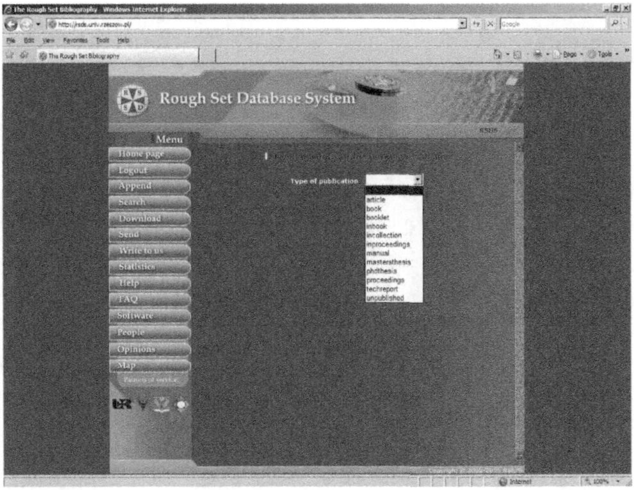

Fig. 4. Choosing the type of a publication in order to introduce it into the system

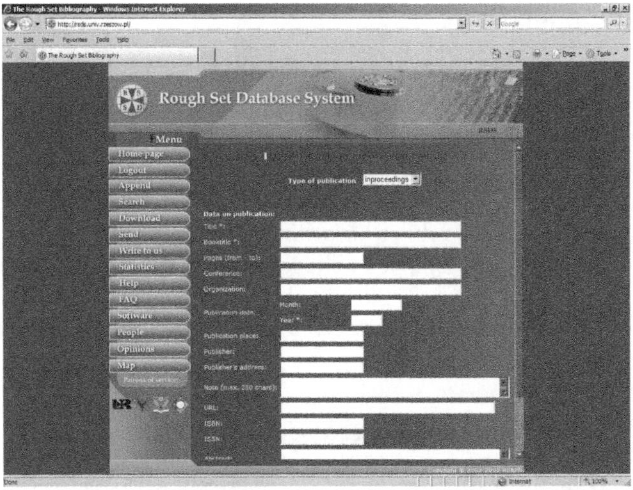

Fig. 5. A form of introducing data about a new publication displayed after stating the type of a publication

to the authors or editors of publications (see Figure 6). The data of the authors or editors are to be introduced singly (each author or editor separately) in order to add them correctly to the system. When the data of an author is introduced the system displays it again to verify it.This step will be repeated as long as the user decides he has introduced all data and accepts the whole process by pressing the key *End*. After being accepted, the data is sent to the database of the system.

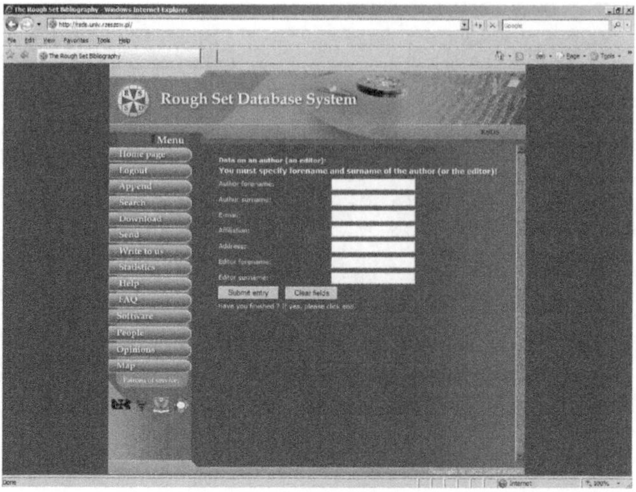

Fig. 6. A form used for introducing data about the authors and editors of publications

2.3 Searching for Data

In order to search for information one has to use the section *Search* (see Figure 7). In this section the following ways of searching has been detached: *alphabetical, advanced ver. 1, advanced ver. 2.*

In the alphabetical way of searching we can distinguish searching according to: *titles, authors, publishers, conferences, journals, year of issue* (Figure 8).

When we choose e.g. searching according to the authors of publications, the letters of the alphabet are displayed. The letters denote certain groups of authors segregated alphabetically (see Figure 8). When one clicks on any letter, the list of authors whose names start with a given letter appears. Next to each surname

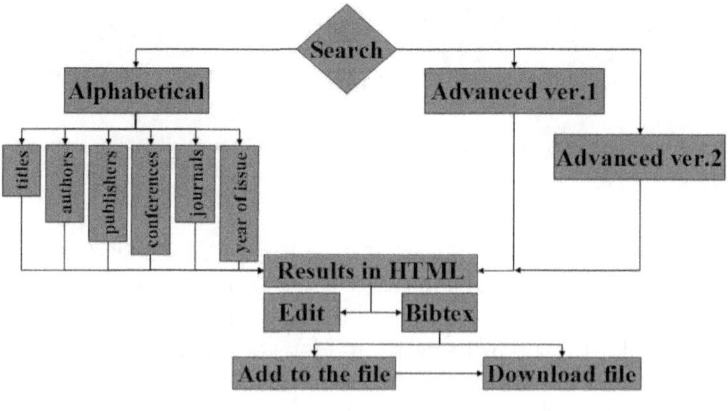

Fig. 7. Scheme for operation of searching for data

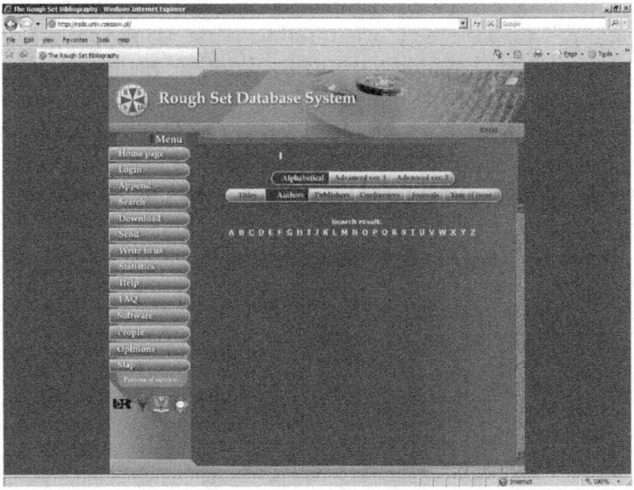

Fig. 8. An alphabetical search (according to authors - starting page)

there is information about the number of publications of this author available in the system. There are some icons that may appear next to a surname and they denote (see Figure 9):

– *an icon of a lupe* - information about an author,
– *an icon of an envelope* - an email address of an author,
– *an icon of a house* - www site of an author.

Fig. 9. The results of the alphabetical searching according to the authors starting with the letter S

When a user clicks on a chosen name, the list of all publications of this author appears. The publications which are found are first displayed in a *HTML* format (Figure 10), and after clicking the link to *BibTeX*, a description in a *BibTeX* format is generated (Figure 11).

In an *alphabetical search* according to the *titles of publications* (*publishers, conferences, journals, year of issue*), when we indicate a particular letter we receive the list of the titles of publications (publishers, conferences, journals,

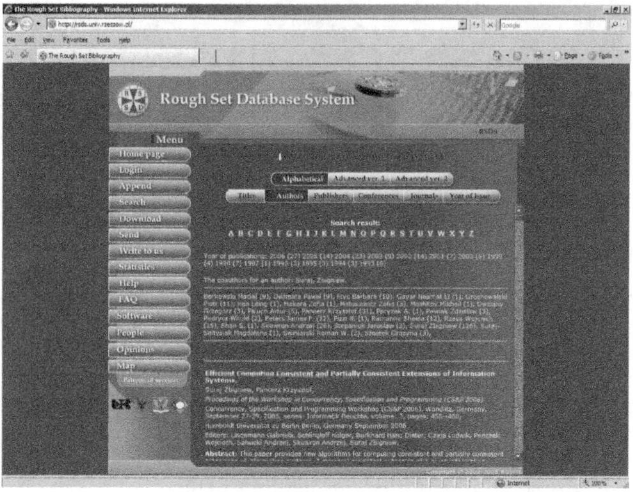

Fig. 10. A list of publications found thanks to the alphabetical searching for the author Suraj Zbigniew

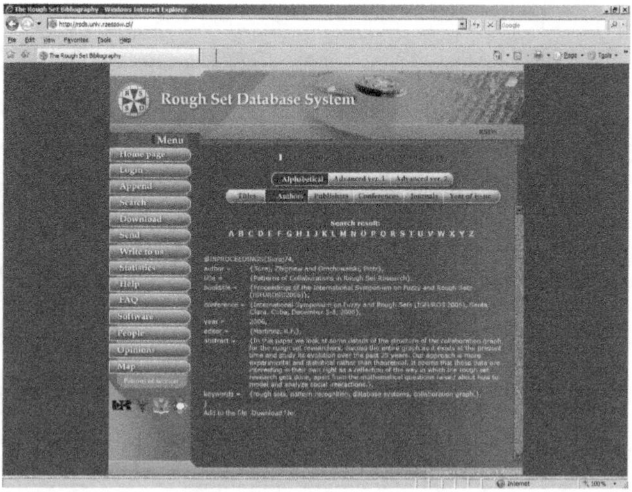

Fig. 11. A detailed description of a chosen publication

year of issue) in the system, starting with a given letter (year). Displaying the found publications for subcategories has been adequately prepared in order to shorten the time of searching for a proper publication. In searching according to:

- *Titles*, an alphabetical list of titles has been divided into successive years of publishing.
- *Authors*, the list of publications for particular authors has also been divided according to the years of publishing. In addition, for every author there is a dynamic list of co-authors writing publications with a given author built. Each surname of the co-authors allows for connection with his publications (see Figure 10).
- *Publishers*, the list of publications for particular editors has also been divided into the years of publishing.
- *Conferences*, in this subcategory there have been distinguished the main names of conferences and put in an alphabetical order. After choosing a particular name, the list according to the years is displayed and after choosing one year one gets the access to the publication connected with a given conference taking place in a chosen year.
- For *journals* the list of publications has been prepared in such a way, that each of the magazines has been divided according to years, which include successive numbers of magazines with adequate publications assigned.
- *Year of issue*, in this subsection there is a division of publications according to particular years of publishing.

In the subcategories every list is being built in a dynamic way, i.e., every change in the system causes the change in the list.

If we choose an *advanced ver. 1* search path we have a possibility to define accurately the conditions of searching for the publication (see Figure 12). The

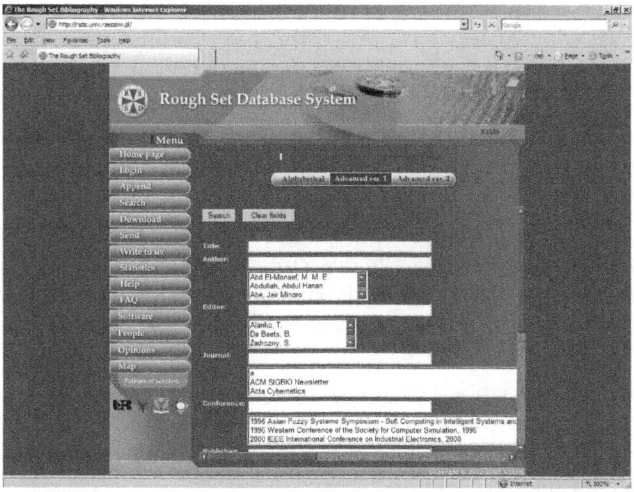

Fig. 12. A form used for defining the conditions of an advanced searching ver. 1

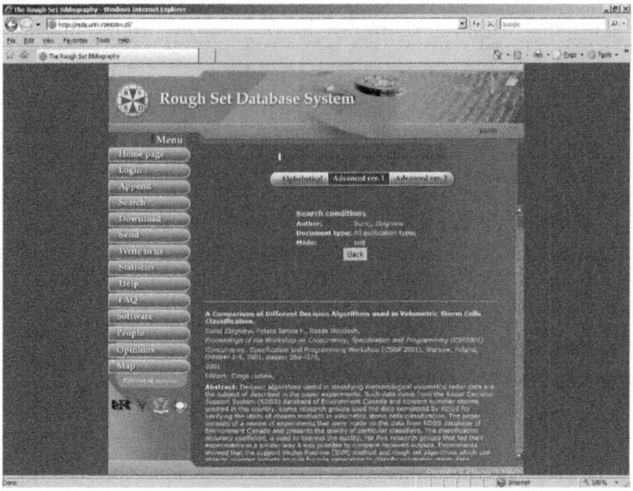

Fig. 13. An exemplary result of an advanced searching ver. 1

fields of the form used for developing the conditions of searching have been equipped with dynamic "self-organizing" lists, i.e., writing letters into the edition fields causes limiting the content of the list to data matching an introduced pattern. Between particular conditions of searching we can define the connections in forms of logical operators AND and OR. After filling a form in and clicking the button *Search*, we receive the list of publications which fulfill the criteria (see Figure 13).

When we choose an *advanced ver. 2* of the option of searching we gain the possibility of searching for publications with automatic grouping the results of searching defined by a classificator. After writing the query into the edition field and pressing the key *Search* we receive the list of found publications grouped according to proper categories. This list is presented in form of a developable "tree" (see Figure 14).

When the user finds an adequate publication, he has a possibility of getting its description in two formats:

- *HTML* - this is the format of displaying publications in the system (without the possibility of generating the description files),
- *BibTeX* - this is the format of displaying publications generated after clicking the link *BibTeX*. It has the possibility of generating the description files.

After generating a description of a publication in the *BibTeX* format, the possibility of adding the received description to the file created by the user - clicking the link *Add to the file* or downloading the created file is activated - clicking the link *Download the file* (see Figure 11).

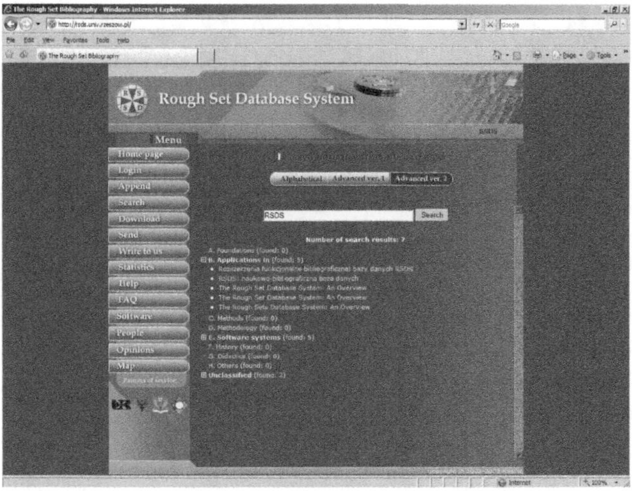

Fig. 14. An exemplary result of an advanced searching ver. 2

2.4 Editing the Existing Data

In the system there is also a possibility of edition of introduced data. Each user has a possibility to edit only the data which is introduced through one's own account into the system, i.e., during the session in which one was logged in. To go further to the edition of a publication introduced by the user, one has to log into the system. Then, one has to use the section *Search*, and when the publication is displayed in a *HTML* format, one has to click on the link *Edit*, and then, by means of a special form, make changes in the fields describing the publication. When the user wishes to stop the process of edition, he must click on the button *Submit entry* in order to send the data to the administrator of the system. The user logged into the system as an administrator has a possibility of removing repeated publications.

2.5 The Classification of Publications with the Use of a Defined Classificator

In the system there is a possibility of attributing to every publication a classificator describing in a detailed way the theory and application of the rough sets. The classifier had been divided into 8 main groups which include subgroups describing the parent groups.

Generally, the structure of the classifier looks as follows: in the classifier the main groups were marked with the letters A,..., H, while the subgroups with successive numbers 1,..., 48. The main groups have been prepared in order to describe all directions of the research over the rough sets and these are:

A. Foundations
B. Applications in

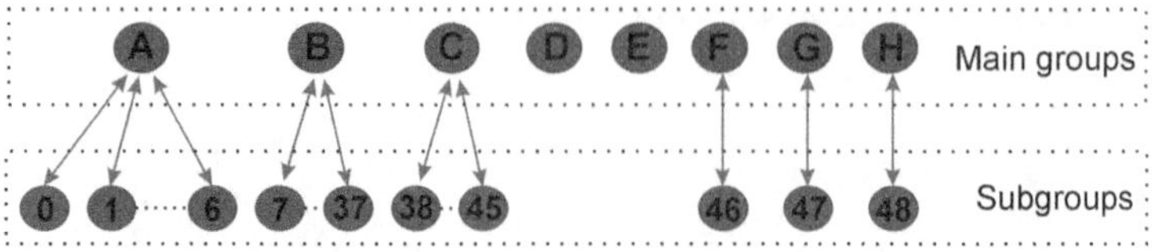

Fig. 15. A structure of a classifier

C. Methods
D. Methodology
E. Software systems
F. History
G. Didactics
H. Others

It is used for grouping publications during the advanced searching (ver. 2). Each user of the system has a possibility of classifying publications introduced through one's own account in the system, i.e., during the session in which one was logged in on that account. The option of classification becomes available only when the user logs into the system. After logging in and finding a given publication (by means of available ways of searching) during displaying a description of a publication in the *HTML* format, next to the link *Edit* appears the link *Classify* which causes entering the section of classificating a given publication. After indicating which subject presented in the classificator we can classify the publication to - there is a possibility of attributing several descriptions to one publication (see Figure 16). Then, a chosen classification is displayed again in

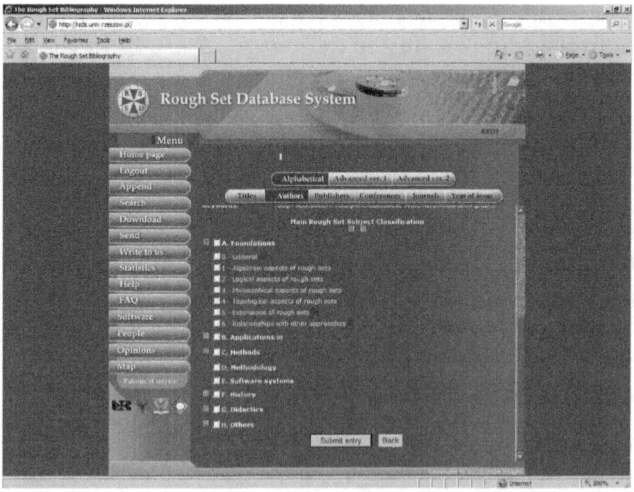

Fig. 16. The screenshot displaying the classification of a given publication

order to verify the correctness in the format e.g. B, A.1, A.2–5 (the format used by a publisher Springer-Verlag). After it is accepted with the key *Submit entry* it is sent to the system.

2.6 Registration of Users into the System

In order to make it possible to add data into the system, one has to log into the system as a user by going to a section *Login* and filling a form in with following data: *user name* and *user password* given during the first logging into the system.

In case of a lack of a name and password, one has to press the button *First login*. Then, one has to fill in a form with data such as: first name, surname, email, username and password. When the form is filled in correctly and the button *Login* is pressed, the user account will be set up and the user will be automatically logged into the system. If the form is not filled in correctly an adequate notice about an error will be displayed.

2.7 Saving Data in a File

When we display data in *BibTeX* format there is a possibility to form a file which would include bibliographical descriptions interesting for a user. One can do it by clicking on the link *Add to the file*. Then, a file *.tex will be formed and we can add the descriptions to this file during one session in the system. There is also a possibility to download the generated file by means of a *Download* section at any moment (see Figure 17). When the session is completed and the file is not downloaded, it is automatically deleted. In the system, two methods of

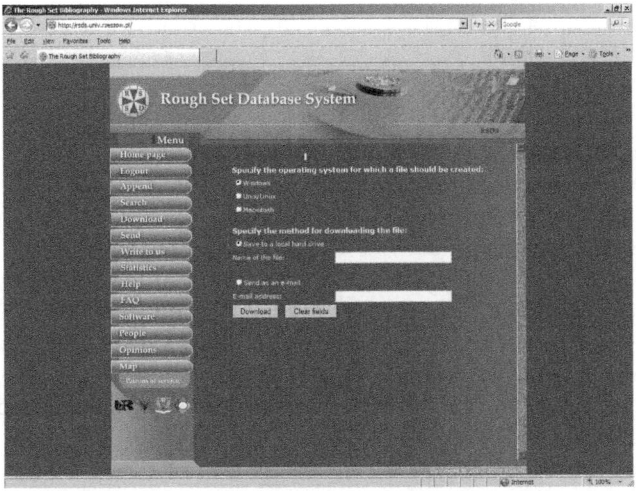

Fig. 17. A form used for stating the method of recording a generated file of a bibliography

downloading files are implemented for the comfort of the users: saving on a local disc of a computer, or sending the file as an attachment to an email message.

2.8 Sending Files with Data to an Administrator

It is possible to submit a file with the bibliographic data to the database administrator, who has the software allowing for appending automatically large data to the database (the section *Send*). In order to do it, one can use a special dedicated form. Submissions in the form of *BibTeX* files are preferred. Please note that submissions are not immediately available, as the database is updated in batches once a month.

2.9 Handling the Users Comments

The section *Write to us* allows to write and send the comments concerning the service to us by using a special dedicated form. Any comments about the service will be helpful and greatly appreciated. Please post them to the database administrator who permanently carries out work on improving the service and broadening its possibilities.

2.10 Statistics

The section *Statistics* includes statistical information concerning the system. This section has been divided into seven pages:

- *Page 1* contains information describing: the number of users' visits to the site, number of the authors in a database, as well as the dynamic diagrams related to: the number and types of publications in a database, number and years of published works.
- *Page 2* contains the statistics depicting the countries from which the users of the service come.
- *Page 3* contains the monthly and yearly statistics of visits to the service.
- *Page 4* includes information about what percent of all authors have written particular number of publications. Table 1 contains the listing, showing that the greatest number of authors 51% has written one publication related to the rough sets per person, two publications have been written by 11% of authors, three publications by 6% of authors. Only few authors have written a larger number of publications e.g. 6-10 publications have been written by about 4% of authors while 11-20 by about 2.4% of authors. At this point we can come to the conclusion that most authors learn about the rough sets because it is a fairly "young" research field.
- *Page 5* contains the analysis of data in the system with a division into defined time periods.
  The task of our analysis was to prepare data in the system i.e. we did not consider the publications without an author - reviewed works, collective publications - e.g. proceedings, publications without the year of publishing. Such works make up a small percentage of the works included in the system and

Table 1. The percentage share of authors with a given number of publications

Number of publications	Percentage of authors
1	51.32%
2	11.09%
3	6.08%
4	1.98%
5	1.67%
6-10	3.65%
11-20	2.38%
21-50	1.32%
51-100	0.61%
101-200	0.25%
> 200	0.05%

we have rejected them because during the analysis we interpret the relations between authors and publications as well as the publishing years and publications. Without these data we are not able to define such relations. After rejecting particular publications we have chosen 3071 publications written by 1578 authors for our analysis. The publications in the system can be divided into adequate groups: article - 869 items, book - 138 items, inbook - 19 items, incollection - 284, inproceedings - 1860, manual - 2 items, masterthesis - 12, phdthesis - 16, proceedings - 68, techreport - 148 items, unpublished - 2 item. The first analysis was done by dividing the period during which they had been created into the following intervals: 1981-85, 1986-90, ... Thanks to dividing the period into such intervals we can observe the development of the rough sets theory and use in numbers. By comparison of the five-year periods we can state during which period the development of the works on the rough sets was the greatest. Table 2 presents the results of the carried analysis.

When analyzing the results we can come to the following conclusions:

- The period during which the greatest number of publications were written and the greatest number of authors were creating the publications was the period between 2001-2005. This period is also the best when considering other parameters.
- As for the data concerning the number of publications written by one author the period between 1996-2000 was the best, while considering the number of authors falling on one publication the best period was between 2001-2005.
- Taking into consideration co-operation between the authors when creating publications we can observe that most authors write their works on their own or with one co-author. This means that until 1995 co-operation during writing common publications had hardly existed. After that year the situation has changed - the number of publications created by groups of co-authors increased.

Table 2. The data record for a given five-year period

In years	1981-1985	1986-1990	1991-1995	1996-2000	2001-2005
Number of publications	35	86	489	1007	1454
Number of authors	12	45	219	539	1102
Mean publications/author	4	2.98	3.71	5.32	4.09
Standard deviation publication/author	4.93	3.39	7.01	43.53	44.17
Mean authors/publication	0.37	0.56	0.66	0.88	1.11
Standard deviation authors/publication	0.76	0.84	1	1.14	1.28
Percentage share of publications with n coauthors					
Luck of authors (a publication under edition only)	2.86%	2.33%	6.13%	5.56%	7.08%
$n = 0$	68.57%	54.65%	44.58%	37.84%	27.79%
$n = 1$	20%	32.56%	34.36%	32.27%	30.61%
$n = 2$	5.71%	6.98%	9%	15.39%	21.94%
$n > 2$	2.86%	3.49%	5.93%	8.94%	12.59%
Number of authors sharing common publications	12	33	180	480	1033
Their percentage share	100%	73.33%	82.19%	89.05%	93.74%
Mean coauthors/author	2.17	1.96	2.27	2.77	3.06
Mean coauthors/author sharing a common publication	2.17	2.67	2.77	3.11	3.26

– *Page 6* contains the analysis of the data prepared before, according to the periods defined as follows: 1981-85, 1981-90, 1981-95, ... When analyzing the data according to time periods defined in this way our analysis has a global character showing the relation of a given five-year period to other periods. Table 3 presents the results of the carried analysis.
  On the basis of the results we have come to the following conclusions:
  - The greatest development of publications created in the field of the rough sets and in respect of people (authors) interested in this subject took place after 1995.
  - After 1995 the number of publications written by one author suddenly increased.
  - The year 1995 is a fateful year for the development of the rough sets theory and its use.
– *Page 7* contains the analysis of the so-called collaboration graph. (The vertices of the graph are the authors in our database, and two vertices are joined by an edge if the two authors have published a joint paper.) An exemplary structure of the collaboration graph is presented in the Figure 18, where the vertices at the graph were marked with the letters A,..., G.

  The co-operation graph on the basis of the data in the system contains 1904 vertices joined with 2954 edges. The average degree of the vertex in the obtained graph equals 3.10. Among all nodes of the graph there are 139 isolated vertices. These are the authors of publications who have not co-operated with anyone. We will not take these authors into consideration

Table 3. The cumulative data record to the end of a given five-year period

Year of completion	1981-1985	1981-1990	1981-1995	1981-2000	1981-2005
Number of publications	35	121	610	1617	3071
Number of authors	12	47	238	670	1578
Mean publications/author	4	3.87	4.18	6.67	5.69
Standard deviation publication/author	4.93	5.25	8.64	63.04	77.94
Mean authors/publication	0.37	0.5	0.63	0.79	0.94
Standard deviation authors/publication	0.76	0.83	0.97	1.08	1.19
Percentage share of publications with n co-authors					
Luck of authors (a publication under edition only)	2.86%	2.48%	5.41%	5.5%	6.25%
$n = 0$	68.57%	58.68%	47.38%	41.43%	34.97%
$n = 1$	20%	28.93%	33.28%	32.65%	31.68%
$n = 2$	5.71%	6.61%	8.52%	12.8%	17.13%
$n > 2$	2.86%	3.31%	5.41%	7.61%	9.96%
Number of authors sharing common publications	12	36	195	590	1465
Their percentage share	100%	76.6%	81.93%	88.06%	92.84%
Mean co-authors/author	2.17	2.13	2.38	2.83	3.15
Mean co-authors/author sharing a common publication	2.17	2.78	2.9	3.21	3.39

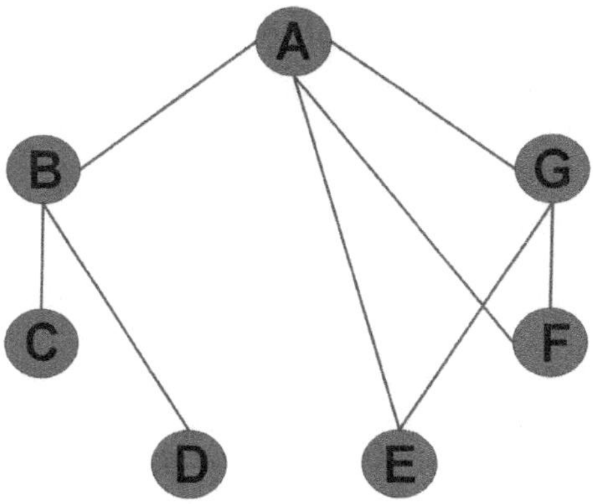

Fig. 18. An exemplary structure of a collaboration graph

in our analysis. For further analysis we will take the co-operation graph without the rejected authors (isolated vertices). Such graph we will call the reduced graph. It consists of 1765 vertices. The average degree of the vertex in the reduced graph equals 3.35. It means that every author in the system co-operated with three authors on average.

Further analysis of the co-operation graph included determining the components of the graph and analyzing the largest of them. We have determined 284 components containing 2 to 713 non-isolated vertices. Among these components there is one which consists of the largest number of vertices i.e. 713 and it is called the largest component. Each of the components represents the groups of authors co-operating with each other. These groups may include people closely co-operating with each other as well as people who are included thanks to indirect co-operators. The components can be also used to define the person one needs to get in touch with in order to find a particular author.

In order to describe the components, particularly the largest component, we we have accepted the following parameters: the average distance between two vertices, the diameter and the radius.

The distance between two vertices (authors) in a given group denotes that if we made a sphere from a given vertex (author), with the radius equal to the average distance between two vertices, we would obtain information about people closely co-operating with that particular author.

The diameter denotes how far from a particular author there is a person the least related to him, i.e., how far are the people working the least in the group.

The radius denotes the leaders, i.e., the most closely co-operating people.

These parameters we have determined fór the largest component and their values are: The average distance between the vertices is equal to 5.26 with a standard deviation 2.20, the diameter equals 17 whereas the radius equals 9. This means that the groups of co-operating authors are quite large and they have many distant branches.

On the other hand, if we made a sphere from every vertex of the component, with the radius equal to the radius of the component, all spheres would have a common part on some vertex (vertices). The vertex (vertices) from the common part denotes the leader (leaders) of a given group. Finding in a group authors outlying from the leader as far as a diameter denotes finding the "satellites" of a given group, and finding the authors outlying from the leader as far as a radius denotes finding the very leaders of a given group.

The statistics Page 1-page 3 are generated dynamically, i.e. every change in the data in the system is reflected in the statistics. Page 4-page 7 are generated by means of the software created by us and determined factors are updated once a month. A detailed analysis of the obtained results in generated statistics has been presented in the paper [6].

2.11 Help

This section provides information in what way one can use the RSDS system.

2.12 FAQ

This section provides answers concerning frequently asked questions about the RSDS system.

2.13 Software

In this section there is an opportunity to search for information concerning the software connected with the rough sets. There are two ways of searching demanded information:

- A search through submitting an application's name.
- An alphabetic search.

Apart from a description of the searched software, the RSDS system allows to download a searched application.

2.14 People

This section allows to find the biographies of outstanding people concerned with the rough sets methodology. After having found a person, this person's biography, e-mail, the name and address of the academy the person works at, is available.

2.15 Opinions

This section allows to present the represenative comments and remarks given by the users about the RSDS system.

2.16 Interactive Map of the World

The possibilities of the system have been augmented about an interactive map of the world illustrating where in the world the rough set theory is being developed and used - section *Map*, as well as allowing for different kind of searching for information in the system.

The realized map has been divided into 4 main parts:

- the map of the world,
- maps of the continents,
- maps of the countries,
- information about chosen rough set research groups (people).

After the map has been started (a map section in the main menu of the system) the map of the world is displayed with a division into continents. In this part one can obtain information about: how many rough set research groups are there on a given continent and how many authors come from a given continent.

After choosing the continent we go to a detailed map of the continent with the countries marked on it. In this part we can also obtain information about the number of rough set research groups and the number of authors depending on the country. In the top right-hand corner of the window there is a list of the countries where we can find people who deal with the rough sets. This is to facilitate navigation.

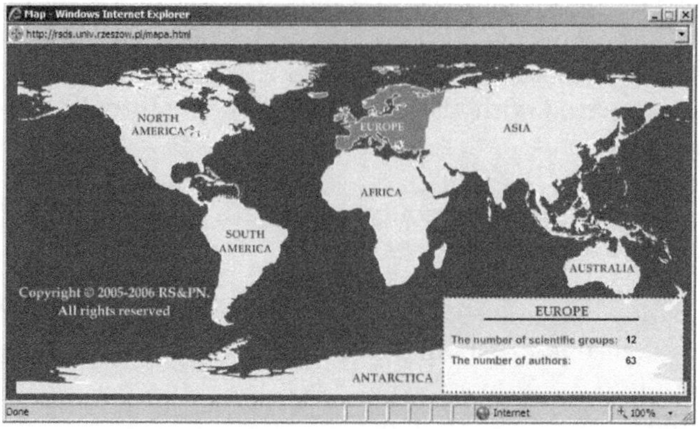

Fig. 19. An interactive map of the world - a map of the world (part 1)

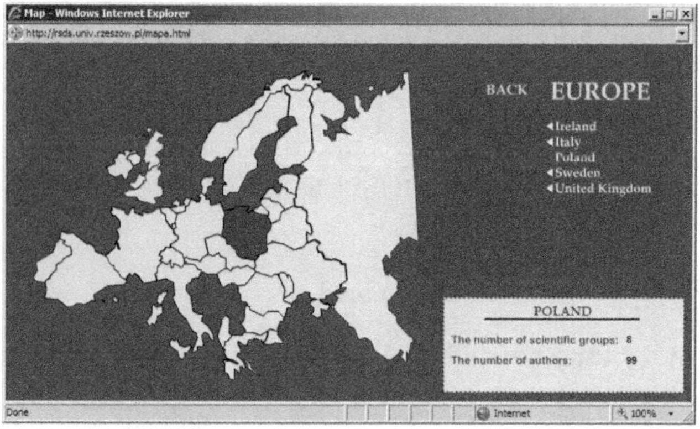

Fig. 20. An interactive map of the world - a map of particular continent (part 2)

When we choose a particular country we can move to the map with the cities (research centers), where we can find research groups (people) working on the rough set theory and its applications.

After choosing the city we obtain information about the research groups in a given city and information who is the leader of the group. When we choose a particular group we will move to the part with information about this group:

- The name of the research group (if the group has WWW web site the name is a reference to this site).
- The leader and members of the group (each name is a reference to the publication of a particular person in the system).
- E-mail address of a given person - an icon of an envelop (from this level, if we have the mail program configured in the system we can send an e-mail message to a given person).

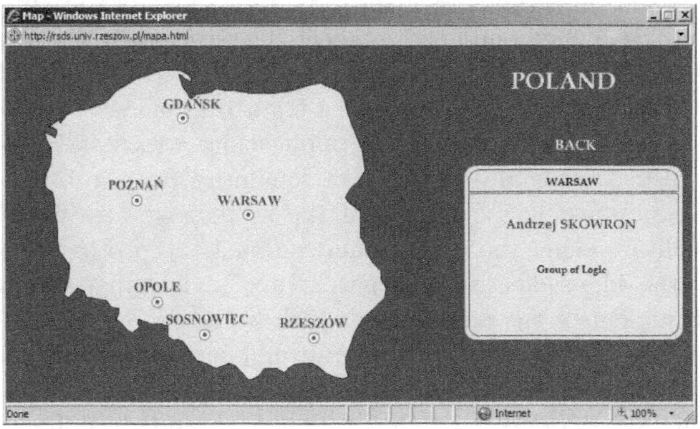

Fig. 21. An interactive map of the world - a map of a chosen country (part 3)

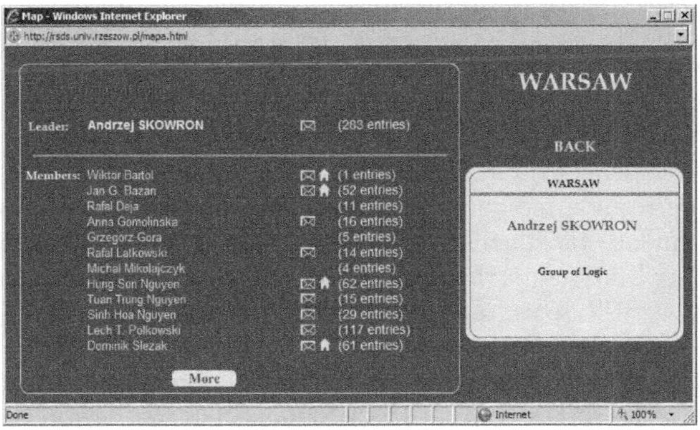

Fig. 22. An interactive map of the world - detailed information for a chosen scientific group (part 4)

– WWW web site of a person - an icon of a house (this icon symbolizes the WWW web site of a person and is a reference to this site).

The created map of the world can be also used to find descriptions of publications when we only have information about the origin of the author.

3 Plans for the Future

The system being created by us mainly gives users the access to data related to publications concerning the rough sets. As far as the conducted research is concerned it makes up the experimental environment for developing and testing the methods connected with intelligent administration and searching for information in the database (especially bibliographical database). The foundation of developing the above methods is preparing the ontology of concepts for the system. We want the defined ontology to reflect the structure of the system and the dependencies between the data which can be inferred from them. The next step towards the "intelligent system" will be a trial to implement new information-searching methods, i.e. descriptions of publications which will allow to obtain well chosen results especially in terms of semantics (on the basis of a defined ontology of concepts). For the users of the system we also want to create a helper to facilitate using the system, and making the process of searching for publications the most effective as regards time and quality. We also want to develop and implement the mechanism which would search for the data for the system in the Internet, so that the system would contain current information on publications being created. The data will be descriptions of publications, information about people working on the "rough sets" and the software related to this theory.

4 Conclusions

We have presented the current possibilities of the RSDS system. Using this system is an opportunity for an information exchange between the scientists and practitioners who are interested in the foundations and applications of the rough sets. The developers of the RSDS system hope that the increase in the dissemination of the results, methods, theories and applications based on the rough sets, will stimulate the further development of the foundations and methods for real-life applications in the intelligent systems. For future updating of the bibliography we will appreciate receiving all forms of help and advice. In particular, we would like to become aware of any relevant contributions which are not referred to in this bibliography database. All submitted material will also be included in the RSDS system. The RSDS system has been designed, implemented, and installed currently at Rzeszów University. The RSDS system runs on any computer with any operating system connected to the Internet. The service is based on the Internet Explorer 6.0, Mozilla 2.0 as well as Opera 7.03 (correct operation requires the web browser with the accepting cookie option enabled).

Acknowledgment. We are grateful to Professor Andrzej Skowron from Warsaw University for stimulating discussions about this work and providing the bibliographic data for the RSDS system. We wish to thank our colleagues from the Logic Group of Warsaw University as well as the Research Group on Rough Sets and Petri Nets that resides at the Chair of Computer Science of Rzeszów University for their help in searching for data. Our deepest thanks go to Katarzyna Garwol from Rzeszów University and Iwona Paluch from University of Information Technology and Management in Rzeszów for their support in the creation of the RSDS system.

References

1. Pawlak, Z.: Rough Sets - Theoretical Aspects of Reasoning About Data. Kluwer Academic Publishers, Dordrecht (1991)
2. Suraj, Z., Grochowalski, P.: The Rough Sets Database System: An Overview. In: Tsumoto, S., Miyamoto, S. (eds.) Bulletin of International Rough Set Society, 7(1/2) Shimane, Japan (2003)
3. Suraj, Z., Grochowalski, P.: The Rough Sets Database System: An Overview. In: Tsumoto, S., Słowiński, R., Komorowski, J., Grzymała-Busse, J.W. (eds.) RSCTC 2004. LNCS (LNAI), vol. 3066, pp. 841–849. Springer, Heidelberg (2004)
4. Suraj, Z., Grochowalski, P.: The Rough Set Database System: An Overview. In: Peters, J.F., Skowron, A. (eds.) Transactions on Rough Sets III. LNCS, vol. 3400, pp. 190–201. Springer, Heidelberg (2005)
5. Suraj, Z., Grochowalski, P.: Functional Extension of the RSDS System. In: Greco, S., Hata, Y., Hirano, S., Inuiguchi, M., Miyamoto, S., Nguyen, H.S., Słowiński, R. (eds.) RSCTC 2006. LNCS (LNAI), vol. 4259, pp. 786–795. Springer, Heidelberg (2006)
6. Suraj, Z., Grochowalski, P.: Patterns of Collaborations in Rough Set Research. In: Bello, R., Falcon, R., Gomez, Y. (eds.) Proceedings of the International Symposium on Fuzzy and Rough Sets, ISFUROS 2006, Santa Clara, Cuba, December 5-8 (2006)
7. Suraj, Z., Grochowalski, P.: About the Basic Possibilities of the RSDS System. In: Wakulicz-Deja, A. (ed.) Systemy Wspomagania Decyzji, Wyd. Uniwersytetu Śląskiego, Katowice (2007)

A Model of User-Oriented Reduct Construction for Machine Learning

Yiyu Yao[1], Yan Zhao[1], Jue Wang[2], and Suqing Han[2]

[1] Department of Computer Science, University of Regina, Regina, Saskatchewan, Canada S4S 0A2
{yyao,yanzhao}@cs.uregina.ca
[2] Laboratory of Complex Systems and Intelligence Science, Institute of Automation, Chinese Academy of Science, Beijing, China 100080
{jue.wang,suqing.han}@mail.ia.ac.cn

Abstract. An implicit assumption of many machine learning algorithms is that all attributes are of the same importance. An algorithm typically selects attributes based solely on their statistical characteristics, without considering their semantic interpretations. In order to resolve difficulties associated with this unrealistic assumption, many researchers attempted to introduce user judgements of the importance of attributes into machine learning. However, there is still a lack of formal framework. Based on decision theory and measurement theory, a model of user-oriented reduct construction is proposed for machine learning by considering the user preference of attributes. It seamlessly combines internal information and external information. User preferences of attributes are extended to user preferences of attribute sets. Accordingly, user preferred reducts can be constructed.

Keywords: user preference, attribute reduction, conditional user preference.

1 Introduction

One of the basic tasks of machine learning is to derive knowledge from data in terms of rules. The discovered rules should be concise, precise, general, easy to understand and practically useful. It is a crucial issue to select the most suitable attributes, features or properties of the objects in a dataset in a machine learning process.

The problem of *attribute selection* is studied in many different areas, such as machine learning, data mining and pattern recognition [3,12,13,16,22,28]. In the theory of rough sets, the attribute selection process is understood as *reduct construction* [16]. The difference between reduct construction and feature selection is their termination criteria. For feature selection, one might stop adding or deleting features when a predetermined condition is satisfied, such as a threshold of the importance of an attribute, a performance measure, or a computational cost measure. For reduct construction, the algorithm does not stop until the *minimum*

J.F. Peters and A. Skowron (Eds.): Transactions on Rough Sets VIII, LNCS 5084, pp. 332–351, 2008.

set of features that possesses some particular property is obtained. Reduct construction thus is a special case of feature selection. In fact, many feature selection algorithms can be viewed as the construction of approximate reducts.

Many proposals have been made regarding the importance of individual attributes or subsets of attributes. They can be broadly divided into two classes, the approaches based on internal information and the approaches based on external information. Internal information based approaches typically depend on the syntactic or statistical information of the dataset. For example, an attribute weighting function is designed by using attributes' distribution information or prediction power. External information based approaches assign weights to attributes, or rank attributes based on semantics or constraints. These two classes are complementary to each other. Once an ordering of attributes is obtained, independent of a particular approach, more important attributes are used first in a learning process. Therefore, it is possible to consider both syntactic and semantic information in a unified framework.

A review of the existing research in machine learning shows that the major research efforts have been focused on the internal information based approaches, although the external information based approaches may be more meaningful and effective. This may stem from the fact that external information covers a very diverse range, is highly subjective, and usually is not well-defined. Consequently, it may be difficult to build a well-accepted model. The concept of user preference has been studied by many researchers [8,9,10,14,21,24,27,31,35,36]. In this paper, we extend the results from paper [31]. We provide a formal model of reduct construction for machine learning by considering the user preference of attributes. The model seamlessly combines internal information and external information.

The rest of the paper is organized as follows. Section 2 discusses the user preference of attributes. Both quantitative and qualitative representations are discussed. Section 3 extends the user preference of attributes to attributes sets. Both quantitative and qualitative user preferences of attribute sets are discussed. Section 4 illustrates the usefulness of the proposed model by applying it to reduct construction. We extend the results to conditional and dynamic preferences in Section 5.

2 User Preference of Attributes

In many machine learning algorithms, it is implicitly assumed that all attributes are of the same importance. Consequently, attributes are selected solely based on their characteristics revealed in an information table. This results in a model that is simple and easy to analyze. At the same time, without considering the semantic information of attributes, the model is perhaps unrealistic. A more realistic model can be built by considering attributes with non-equal importance. This type of external information is normally provided by users, and is referred to as user judgement or user preference.

User judgement can be expressed in various forms. Quantitative judgement involves the assignment of different weights to different attributes. Qualitative judgement is expressed as an ordering of attributes. In many situations, user judgement is determined by semantic considerations. For example, it may be interpreted in terms of more intuitive notions, such as the cost of testing, the easiness of understanding, or the actionability of an attribute. It is virtually impossible to list all interpretations of user judgement. The meaning of a user judgement becomes clear only in a particular context of application. To simplify our discussion, we treat user judgement as a primitive notion. In other words, we only investigate the desirable properties of a user judgement, as well as how to incorporate it into a machine learning process.

2.1 Quantitative Judgement of Attributes

A simple and straightforward way to represent user judgement of attributes is to assign them with numerical weights. Formally, it can be described by a mapping:

$$w : At \longrightarrow \Re, \tag{1}$$

where At is a finite non-empty set of attributes, and $\Re$ is the set of real numbers. For an attribute $a \in At$, $w(a)$ is the weight of a. The numerical weight $w(a)$ may be interpreted as the degree of importance of a, the cost of testing a in a rule, or the number of occurrences of a in a set (which is also called the frequency of a). The weights of attributes naturally induce an ordering of attributes. For example, if the weights are interpreted as costs, a machine learning algorithm should use, if possible, attributes with lower costs first.

The use of numerical weights for attribute importance has been extensively considered in machine learning. In many learning algorithms, a numerical function is used to compute weights of individual attributes based on their distribution characteristics. According to the computed weights, attributes are selected. For example, entropy-theoretic measures have been studied and used for attribute selection [29].

2.2 Qualitative Judgement of Attributes

A difficulty with the quantitative method is the acquisition of the precise and accurate weights of all attributes. To resolve this difficulty, we consider a qualitative judgement that only relies on pairwise comparisons of attributes. For any two attributes, we assume that a user is able to state whether one is more important than, or more preferred to, the other. This qualitative user judgement can be formally defined by a binary relation $\succ$ on At. For any two $a, b \in At$:

$$a \succ b \Longleftrightarrow \text{the user prefers } a \text{ to } b. \tag{2}$$

The relation $\succ$ is called a preference relation. If $a \succ b$ holds, we say that the user strictly prefers a to b. In contrast to the quantitative representation, the preference does not say anything regarding the degree of preference, namely, how much a is preferred to b.

In the absence of preference, i.e., if both $\neg(a \succ b)$ and $\neg(b \succ a)$ hold, we say that a and b are indifferent. An indifference relation $\sim$ on At is defined as:

$$a \sim b \Longleftrightarrow \neg(a \succ b) \wedge \neg(b \succ a). \tag{3}$$

The indifference of attributes may be interpreted in several ways. A user may consider the two attributes are of the same importance. The indifference may also occur when the comparison of two attributes are not meaningful, as they are incompatible. When both a and b are unimportant, it may not make too much sense to compare them. The indifference represents such an absence of preference. In fact, in many practical situations, one is only interested in expressing the preference of a subset of crucial attributes, and considers all unimportant attributes to be the same.

Based on the strict preference and indifference relations, one can define a preference-indifference relation $\succeq$ on At:

$$a \succeq b \Longleftrightarrow (a \succ b) \vee (a \sim b). \tag{4}$$

If $a \succeq b$ holds, we say that b is not preferred to a, or a is at least as good as b. The strict preference can be re-expressed as $a \succ b \Longleftrightarrow (a \succeq b) \wedge \neg(b \succeq a)$.

A Weak Order Interpretation. A user preference relation must satisfy certain axioms in order to capture our intuitive understanding of preference. The following two axioms, studied extensively in decision theory [6] and measurement theory [18], seem to be reasonable, for any $a, b, c \in At$:

(1). $a \succ b \Longrightarrow \neg(b \succ a)$ (asymmetry);
(2). $(\neg(a \succ b) \wedge \neg(b \succ c)) \Longrightarrow \neg(a \succ c)$ (negative transitivity).

The asymmetry axiom states that a user cannot prefer a to b, and at the same time prefer b to a. The negative transitivity axiom states that if a user does not prefer a to b, nor b to c, then the user should not prefer a to c. If a preference relation $\succ$ on At is asymmetric and negatively transitive, it is called a *weak order* [6,18].

A weak order imposes structures on the set of attributes. Additional properties of a weak order are summarized in the following lemma [6].

Lemma 1. *Suppose a preference relation $\succ$ on a finite set of attributes At is a weak order. Then,*

- *Exactly one of $a \succ b$, $b \succ a$ and $a \sim b$ relations holds for any two $a, b \in At$;*
- *$\succ$ is transitive;*
- *$\sim$ is an equivalence relation that induces a partition $At/\sim$ of At;*
- *$\succeq$ is transitive and connected;*
- *The relation $\succ'$ on the partition $At/\sim$, defined by $[a]_\sim \succ' [b]_\sim \Longleftrightarrow a \succ b$, is a linear order, where $[a]_\sim$ is the equivalence class containing a.*

A Strict Partial Order Interpretation. Another class of user preference relations is defined by the following two axioms. For any $a, b, c \in At$:

$$\begin{aligned} &(1). \quad \neg(a \succ a) \ \text{(irreflexivity)};\\ &(2). \quad (a \succ b) \wedge (b \succ c) \Longrightarrow a \succ c \ \text{(transitivity)}. \end{aligned}$$

The irreflexivity axiom states that a user cannot prefer a to a itself. The transitivity axiom states that if a user prefers a to b, and b to c, then the user should prefer a to c. If a preference relation $\succ$ on At is irreflexive and transitive, then it is asymmetry and is called a *strict partial order* [6,18].

For a strict partial order, the indifference relation is no longer an equivalence relation. In terms of $\sim$, we can define a new binary relation $\approx$ as:

$$a \approx b \Longleftrightarrow \forall c \in At \ (a \sim c \Longleftrightarrow c \sim b). \tag{5}$$

A strict partial order also imposes structures on the set of attributes. Additional properties of a strict partial order are summarized in the following lemma [6].

Lemma 2. *Suppose a preference relation $\succ$ on a finite set of attributes At is a strict partial order. Then,*

- *Exactly one of $a \succ b$, $b \succ a$, $a \approx b$ and $(a \sim b \wedge \neg(a \approx b))$ relations holds for any two $a, b \in At$;*
- *$\approx$ is an equivalence relation that induces a partition $At/\approx$ of At;*
- *The relation $\succ''$ on the partition $At/\approx$, defined by $[a]_\approx \succ'' [b]_\approx \Longleftrightarrow a \succ b$, is a strict partial order, where $[a]_\approx$ is the equivalence class containing a.*

Linear Order Extensions. A linear order is a weakly connected weak order. That is, for $a, b \in At$, $a \succ b$ or $b \succ a$ whenever $a \neq b$. A linear order is a weak order in which any two distinct elements are comparable [6,7,18].

For the three orders, we have ($\succ$ is a linear order) $\Longrightarrow$ ($\succ$ is a weak order) $\Longrightarrow$ ($\succ$ is a strict partial order). In general, the converse implications do not hold. For a weak order or a strict partial order, it is possible to obtain its linear extensions. Given a weak order or a strict partial order, $\succ$, a linear order $\succ_l$ is called an extension of $\succ$ if for any $a, b \in At$,

$$a \succ b \Longrightarrow a \succ_l b.$$

That is, $\succ_l$ preserves the strict preference of $\succ$, but may introduce extra preference. Formally, this may be expressed equivalently as $\succ \ \subseteq \ \succ_l$. In the subsequent discussion, we also express a linear order in terms of the preference-indifference relation $\succeq_l$.

Example 1. Suppose a user preference relation $\succ$ is on a set of attributes $At = \{a, b, c, d\}$ by the following weak order:

$$c \succ a, \quad c \succ b, \quad d \succ a, \quad d \succ b, \quad d \succ c.$$

This relation $\succ$ satisfies the asymmetry and negative transitivity conditions. For attributes a and b, we have $a \sim b$. Thus, three equivalence classes $\{d\}, \{c\}, \{a, b\}$ can be formed. They can also be written as $[d]_\sim, [c]_\sim, [a]_\sim$ (or $[b]_\sim$), respectively. In turn, they can be arranged as three levels in a linear order: $[d]_\sim \succ' [c]_\sim \succ' [a]_\sim$. If one does not care the order of attributes in an equivalence class, we can extend the given weak order to either of the two distinct linear orders:

$$d \succeq_1 c \succeq_1 b \succeq_1 a,$$
$$d \succeq_2 c \succeq_2 a \succeq_2 b.$$

Suppose a user preference relation $\succ$ on a set of attributes $At = \{a, b, c, d, e\}$ is given by the following strict partial order:

$$a \succ b,\ a \succ c,\ a \succ d,\ a \succ e,\ b \succ d,\ b \succ e,\ c \succ e,\ d \succ e.$$

There is no preference relation between b and c, neither between c and d. According to the definition of $\approx$, we have $b \approx d$. Therefore, we obtain four equivalence classes $\{a\}, \{b, d\}, \{c\}, \{e\}$. They can also be written as $[a]_\approx, [b]_\approx$ (or $[d]_\approx$), $[c]_\approx$ and $[e]_\approx$, respectively. The equivalence classes $[b]_\approx$ and $[c]_\approx$ cannot be ordered. We can extend this order into the following three different linear orders:

$$a \succeq_1 b \succeq_1 c \succeq_1 d \succeq_1 e,$$
$$a \succeq_2 b \succeq_2 d \succeq_2 c \succeq_2 e,$$
$$a \succeq_3 c \succeq_3 b \succeq_3 d \succeq_3 e.$$

2.3 Connections between Quantitative and Qualitative Judgements of Attributes

A quantitative judgement can be easily translated into a qualitative judgement. The translation to a preference relation only preserves the ordering of attributes implied by the quantitative weights. Additional information given by the numerical values of weights is lost.

Given the weights of attributes, we can uniquely determine a preference relation. Suppose $w(a)$ and $w(b)$ represent the importance of $a, b \in At$, a preference relation is defined by:

$$a \succ b \Longleftrightarrow w(a) > w(b). \tag{6}$$

When $w(a)$ and $w(b)$ are the costs of testing attributes $a, b \in At$ in a rule, the following preference relation should be used instead,

$$a \succ b \Longleftrightarrow w(a) < w(b). \tag{7}$$

In general, two attributes may have the same weights.

The measurement of a user preference by a quantitative function depends on the properties of the preference relation. By imposing different sets of axioms on a user preference, it is possible to derive quantitative measurements using different scales. Different scales allow different arithmetic operations [6,7,18].

The following theorem states that a weak order is both necessary and sufficient for a numerical measurement [6].

Theorem 1. *Suppose $\succ$ is a preference relation on a finite non-empty set At of attributes. There exists a real-valued function $u : At \rightarrow \Re$ satisfying the condition:*

$$a \succ b \Longleftrightarrow u(a) > u(b), a, b \in At, \tag{8}$$

if and only if $\succ$ is a weak order. Moreover, u is uniquely defined up to a strictly monotonic increasing transformation.

The function u is referred to as an order-preserving utility function in ordinal scale. It provides a quantitative representation of a user preference. That is, the numbers of $u(a), u(b), \ldots$ as ordered by $>$ reflect the order of $a, b, \ldots$ under the preference relation $\succ$.

The utility function also trustfully represents the indifference relation, i.e.,

$$a \sim b \Longleftrightarrow u(a) = u(b), a, b \in At. \tag{9}$$

According to Theorem 1, for a given preference relation, there exist many utility functions. Under the ordinal scale, it is only meaningful to examine the order induced by a utility function. Although numerical values are used, it is not necessarily meaningful to apply arithmetic operations on them.

The next theorem states that a strict partial order is necessary and sufficient for a weaker numerical measurement [6].

Theorem 2. *Suppose $\succ$ is a preference relation on a finite non-empty set At of attributes. There exists a real-valued function $u : At \rightarrow \Re$ satisfying the condition:*

$$a \succ b \Longrightarrow u(a) > u(b), a, b \in At, \tag{10}$$

if and only if $\succ$ is a strict partial order.

Compared with Theorem 1, for a strict partial order we can only obtain a single implication. That is, we can conclude that a preferred attribute has a larger weight value. However, we cannot make a reverse inference. Equation (10) can be re-expressed as

$$u(b) \leq u(a) \Longrightarrow \neg(a \succ b).$$

That is, we can only infer that an attribute with a smaller weight value is not preferred to an attribute with a larger weight value.

3 User Preference of Attribute Sets

Conceptually, rule learning in an information system can be viewed as two tasks, the selection of a subset of attributes, and the construction of rules using such attributes. The two tasks can in fact be integrated in one algorithm without a clear separation. Ideally, the subset should contain more preferred attributes and avoid including unfavoured attributes. In this case, users should be able to express the preference over subsets of attributes. This requires a user preference relation on the power set 2^{At}. In this section, we present the way to derive a preference relation $\succ$ on 2^{At} based on a preference relation $\succ$ on At.

3.1 Basic Properties

For simplicity, we use the same symbol to denote the preference relation on At and the preference relation on 2^{At}. Obviously, the relation $\succ$ on 2^{At} needs to satisfy certain conditions. By definition, $\succ$ on 2^{At} must be an extension of $\succ$ on At. That is, for two singleton attribute sets $\{a\}$ and $\{b\}$:

$$\begin{array}{ll} \text{(E1).} & \{a\} \succ \{b\} \Longleftrightarrow a \succ b; \\ \text{(E2).} & \{a\} \sim \{b\} \Longleftrightarrow a \sim b; \\ \text{(E3).} & \{a\} \succeq \{b\} \Longleftrightarrow a \succeq b. \end{array}$$

According to Theorem 1, the axiom

$$\text{(T).} \qquad \succ \text{ on } 2^{At} \text{ is a weak order,}$$

is required. One may impose additional conditions depending on particular applications.

3.2 Quantitative Judgement of Attribute Sets

When the user judgement is given as weights of attributes, we can extend the weighting function w on At to a weighting function on 2^{At}. For simplicity, we use the same symbol to denote these two functions. Similarly, the extensions are not unique. For example, for $A \subseteq At$, we consider the following possible extensions:

Additive extension: $w(A) = \sum_{a \in A} w(a)$,
Average extension: $w(A) = \frac{\sum_{a \in A} w(a)}{|A|}$,
Maximal extension: $w(A) = \max_{a \in A} w(a)$,
Minimal extension: $w(A) = \min_{a \in A} w(a)$.

One can interpret the meaningful extensions based on the physical meaning of the weighting function on At. It is important to note that only some extensions are meaningful in a particular application. For example, if $w(a)$ is a cost measurement function, the above extensions are interpreted as the total cost, average cost, maximal cost, and minimal cost, respectively. An attribute set with the minimum cost is normally more useful. If $w(a)$ is an information measurement function, $w(A)$ is the joint information of all attributes in the set. Normally, an attribute set with the maximal information gain is preferred.

Based on the computed weights, we can order subsets of attributes in a similar way to the ones given by Equations (6) and (7).

3.3 Qualitative Judgement of Attribute Sets

For a set of attributes, we can arrange them in a linear order based on the preference-indifference relation $\succeq$. This enables us to derive a possible ordering of subsets by consecutively examining attributes one by one. Based on the directions in which attributes are examined, we define two lexical orders. In the left-to-right lexical order, we compare two sets of attributes from left to right, in order to determine which set of attributes is preferred. In the right-to-left lexical order, we compare attributes in the reverse order.

Definition 1. Left-to-right lexical order: *Given two attribute sets* $A : a_1 \succeq a_2 \succeq \ldots \succeq a_m$ *and* $B : b_1 \succeq b_2 \succeq \ldots \succeq b_n$, *let* $t = \min\{m, n\}$. *We say that* A *precedes* B *in the left-to-right lexical order, written* $A \succ B$, *if and only if*

(a) *there exists an* i: $1 \leq i \leq t$ *such that* $a_j \sim b_j$ *for* $1 \leq j < i$ *and* $a_i \succ b_i$, *or*
(b) $a_i \sim b_i$ *for* $1 \leq i \leq t$ *and* $m < n$.

Definition 2. Right-to-left lexical order: *Given two attribute sets* $A : a_1 \succeq a_2 \succeq \ldots \succeq a_m$ *and* $B : b_1 \succeq b_2 \succeq \ldots \succeq b_n$, *let* $t = \min\{m, n\}$. *We say that* A *precedes* B *in the right-to-left lexical order, written* $A \succ B$, *if and only if*

(a) *there exists an* i: $0 \leq i < t$ *such that* $a_{m-j} \sim b_{n-j}$ *for* $0 \leq j < i$ *and* $a_{m-i} \succ b_{n-i}$, *or*
(b) $a_{m-i} \sim b_{n-i}$ *for* $0 \leq i < t$ *and* $m < n$.

These two lexical orders represent two extreme views and define two different criteria for comparing attribute sets. The meaning of them can be interpreted as follows. The left-to-right method focuses on the preferred attributes of the two sets. That is, the winner is determined by comparing the strongest attributes in the sets. By the left-to-right lexical order, an attribute set A is preferred to another attribute set B if and only if the most preferred attribute of A is preferred to the most preferred attribute of B, or A is a proper subset consisting of the most preferred attributes of B.

On the other hand, the right-to-left method emphasizes the less preferred attributes of the two sets. The winner is determined by comparing the weakest attributes in the sets. By the right-to-left lexical order, an attribute set A is preferred to another attribute set B if and only if the least preferred attribute of A is preferred to the least preferred attribute of B, or A is a proper subset consisting of the least preferred attributes of B.

The left-to-right lexical order encourages an optimistic comparison, and the right-to-left lexical order promotes a pessimistic comparison. The left-to-right lexical order is also understood as the dictionary order.

Example 2. The running example can be used to illustrate the differences between two lexical orders. Recall that attributes in Example 1 can be arranged as $\{d\} \succ' \{c\} \succ' \{a, b\}$. For two attribute subsets $A : d \succeq c \succeq a$ and $B : d \succeq a$, since $d \sim d$ and $c \succ a$, then A is the winner according to the left-to-right lexical order. At the same time, since $a \sim a$ and $d \succ c$, thus B is the winner according to the right-to-left lexical order.

For two attribute subsets $C : d \succeq c \succeq a$ and $D : c \succeq b$, since $d \succ c$, then C is the winner according to the left-to-right lexical order. On the other hand, since $a \sim b, c \sim c$ and $|D| < |C|$, then D is the winner according to the right-to-left lexical order.

Example 3. Suppose A and B are two subsets of an attribute set At, which is under the following preference relations: $a \succ b$, $a \succ c$ and $b \succ c$. Seven distinct subsets of A and B can be identified. They are $\{a\}$, $\{b\}$, $\{c\}$, $\{a, b\}$, $\{a, c\}$, $\{b, c\}$

and $\{a,b,c\}$. The preference of attribute sets can be determined by the heuristic lexical orders. According to the left-to-right lexical order, we obtain:

$$\{a\} \succ \{a,b\} \succ \{a,b,c\} \succ \{a,c\} \succ \{b\} \succ \{b,c\} \succ \{c\}.$$

According to the right-to-left lexical order, we obtain:

$$\{a\} \succ \{b\} \succ \{a,b\} \succ \{c\} \succ \{a,c\} \succ \{b,c\} \succ \{a,b,c\}.$$

Both lexical orders satisfy Axioms (E1)-(E3) and (T), and should be considered as examples of potential extensions of the preference order from At to 2^{At}. They may provide different preference orders as shown in the example. It may be difficult to argue which one is better based solely on theoretical basis. In real applications, we might also need to consider other extensions.

A tree representation of all subsets of At has been studied by Zhang *et al.* [32,33]. We can explain and extend it to a graph representation. Under such a graph structure, a traversal method of all the subsets according to the left-to-right order or the right-to-left order can be determined.

4 User Preference of Reducts

The usefulness of the proposed model can be illustrated by attribute reduct construction.

4.1 Preliminaries

In many data analysis applications, information and knowledge is stored and represented in an *information table.* An information table provides a convenient way to describe a finite set of objects by a finite set of attributes [16]. It represents all available information and knowledge. That is, objects are only perceived, observed, or measured by using a finite number of attributes.

Definition 3. Information tables: *An information table is the following tuple:*

$$S = (U, At, \{V_a \mid a \in At\}, \{I_a \mid a \in At\}),$$

where U is a finite nonempty set of objects, At is a finite nonempty set of attributes, V_a is a nonempty set of values of $a \in At$, and $I_a : U \to V_a$ is an information function that maps an object of U to exactly one value in V_a.

Two objects are discernible if their values are different in at least one attribute. Skowron and Rauszer suggested a matrix form for storing the sets of attributes that discern pairs of objects [19].

Definition 4. Discernibility matrices: *Given an information table S, its discernibility matrix M is a $|U| \times |U|$ matrix with each element defined by*

$$m_{x,y} = \{a \in At \mid I_a(x) \neq I_a(y)\},$$

where $x, y \in U$.

Each matrix element maps an object pair into a set of attributes. Any attribute of the set can distinguish the two objects. A discernibility matrix M is symmetric, i.e., $m_{x,y} = m_{y,x}$, and $m_{x,x} = \emptyset$. The discernibility of objects can be also expressed as a set M by collecting only the distinct nonempty elements as:

$$M = \{m_{x,y} \mid m_{x,y} \neq \emptyset\}.$$

In the following discussion, we will use the set representation.

The theory of rough sets has been applied to data analysis, data mining and knowledge discovery. A fundamental notion supporting such applications is the concept of reducts [16]. The basic idea of reduct construction is to find the minimum subset of attributes that has the same property, or performs the same, as the entire set of attributes. Different algorithms, approaches and methodologies have been extensively studied [1,2,10,11,15,17,19,20,23,24,26,30,34,35,32]. The objective of reduct construction is to reduce the number of attributes, and at the same time, to preserve a desired property.

Definition 5. Reducts: *Given an information table S and a property p, an attribute set $R \subseteq At$ is called a p-reduct of At, or simply a reduct of At, if R satisfies the two conditions:*

(i). *R and At express the same property p of S;*
(ii). *for any $R' \subseteq R$, R' does not exprss the property p.*

The first condition indicates the joint sufficiency of the attribute set R, and the second condition indicates that each attribute in R is individually necessary.

Normally, an information table contains more than one reduct. If an attribute appears in at least one reduct, it is called a *reduct attribute*; if it appears in all reducts, it is called a *core attribute*; if it does not appear in any reduct, it is called a *non-reduct attribute.* A core attribute is a reduct attribute. Wei *et al.* [25] referred to the core attributes as the absolutely necessary attributes, the reduct attributes as the relatively necessary attributes, and the non-reduct attributes as the absolutely unnecessary attributes. For all attributes in At, we can classify them into three categories: the set REDUCT of all reduct attributes, the set CORE of all core attributes, and the set NREDUCT of all non-reduct attributes.

An attribute set $R' \subseteq At$ is called a *super-reduct* of a reduct R, if $R' \supseteq R$; an attribute set $R' \subseteq At$ is called a *partial reduct* of a reduct R, if $R' \subseteq R$. Given a reduct, there exist many super-reducts and many partial reducts. The set CORE is a partial reduct of any reduct, and the set REDUCT is a super-reduct of any reduct. For any reduct R, we have:

$$\emptyset \subseteq \text{CORE} \subseteq R \subseteq \text{REDUCT} \subseteq At.$$

The two sets CORE and REDUCT can be used as initial sets for reduct construction.

Given a user preference relation on the attribute set At, we can rank reducts according to the two lexical orders. Conceptually, internal information determines a set of reducts, and a user preference determines an ordering of reducts.

The two lexical orders immediately suggest two strategies for reduct construction, i.e., the deletion strategy and the addition strategy. While the deletion strategy can deterministically generate the winning reduct under the right-to-left order, the addition strategy cannot guarantee the generation of the winning reduct under the left-to-right order. In fact, constructing the winning reduct under the left-to-right order has been proved to be NP-hard [14]. More information about these two algorithms can be found in [30].

Before introducing the two algorithms, we need to introduce an important matrix operation. Given a discernibility matrix M, the *Absorb* operation absorbs any matrix element $m = m_{x,y}$ if there exists another matrix element $m' = m_{x',y'}$ such that $m' \subseteq m$. By the definition of a discernibility matrix, any attribute $a \in m'$ is sufficient to distinguish both object pairs (x, y) and (x', y'). Thus, attributes in $m - m'$ are superfluous. The *Absorb* operation of a discernibility matrix M is defined as:

$$\begin{aligned} \mathit{Absorb}(M): \ & \text{For all } m', m \in M, \\ & \text{if } m' \subseteq m \text{ and } m' \neq \emptyset, \text{ then } M = M - \{m\}. \end{aligned}$$

That is, all supersets of a set are deleted from M.

4.2 The Deletion Algorithm

A deletion algorithm starts from the entire attribute set At, or from the reduct attribute set REDUCT, deletes the superfluous attributes one-by-one until a reduct is constructed. The order for attribute deletion can be based on the user preference of attributes.

Input: The discernibility matrix M of an information table.
Output: A reduct R.

(1) Construct an attribute ordering of attribute set At.
(2) $R = At$.
(3) $M = \mathit{Absorb}(M)$.
(4) While M is not a family of singleton attribute sets:
  (4.1) Select the right-most attribute a in the ordering that satisfies the condition $\{a\} \notin M$;
  (4.2) $R = R - \{a\}$ and $M = \mathit{Absorb}(\{m - \{a\} \mid m \in M\})$.
(5) Update R by deleting the attributes not existing in M.

Algorithm 1. A deletion algorithm for reduct construction

A deletion algorithm is described as Algorithm 1, where the ordering of attributes can be constructed from an evaluation function or given by a user. According to the algorithm, given the attribute set At, an attribute a can be deleted from M if $\{a\}$ is not a singleton attribute set that uniquely distinguish any object pair. Thus, the attribute set $R = R - \{a\}$ is jointly sufficient to keep the discernibility of all object pairs, and is a super-reduct. The iteration process is stopped under the condition that M is a family of singleton attribute sets. All the remaining attributes in the discernibility matrix are necessary and sufficient to keep the discernibility of all object pairs, and hence a reduct is obtained. We update R by deleting the attributes not existing in M. In fact, the constructed reduct R is the union of all singleton attribute sets in M.

This deletion algorithm deletes the most unfavoured attributes first. That is, we check from the right, the most unfavourable attribute in At, to the left, the most preferred attribute in At. This results in the winning reduct under the right-to-left lexical order.

Example 4. Consider an information table S with an ordering $At : a \succeq b \succeq c \succeq d \succeq e \succeq f \succeq g$. Suppose after absorption, the set representation of the matrix is read as: $M = \{\{a, g\}, \{a, f\}, \{b, d, e\}, \{b, e, f\}, \{c, d, f\}, \{e, f, g\}\}$. According to the algorithm, suppose we select the least preferred attribute g for deletion. Consequently, M is updated as $\{\{a\}, \{b, d, e\}, \{c, d, f\}, \{e, f\}\}$ after the absorption. In the second round, we further select the least preferred attribute f for deletion. As a result, $M = \{\{a\}, \{c, d\}, \{e\}\}$. In the third round, we can repeat the same procedure by deleting the least preferred attribute d, and update $M = \{\{a\}, \{c\}, \{e\}\}$. At this stage, M contains only singleton attribute sets, and cannot be further simplified. Thus, $R = \{a, c, e\}$, which is the winning reduct under the right-to-left lexical order.

4.3 The Addition Algorithm

An addition algorithm starts from the empty set $\emptyset$, or from the core attribute set CORE, adds the necessary attributes one-by-one until a reduct is constructed. The order for attribute addition can be based on the user preference of attributes.

Input: The discernibility matrix M of an information table.
Output: A reduct R.

(1) Construct an attribute ordering of attribute set At.
(2) $R = \emptyset$.
(3) $M = Absorb(M)$.
(4) While M is not a family of singleton attribute sets:
  (4.1) Select the left-most attribute a in the ordering that satisfies the condition that there exists an element $m_a \in \{m \in M \mid a \in m\}$ such that $\forall m \in M \ (m - (m_a - \{a\})) \neq \emptyset$);
  (4.2) Select an element m_a that satisfies the condition of (4.1);
  (4.3) $R = R \cup \{a\}$ and $M = Absorb(\{m - (m_a - \{a\})) \mid m \in M\})$.

Algorithm 2. An addition algorithm for reduct construction

An addition algorithm is described as Algorithm 2 where the ordering of attributes can be constructed from an evaluation function or given by a user.

According to the algorithm, given the attribute set At, an attribute a can be added to R if it is necessary to distinguish at least one pair of objects. This is made possible by not considering attributes in $m_a - \{a\}$, also denoted as $Tail(a)$, in the construction of the partial reduct of R. To keep the discernibility property, we must ensure that for all $m \in M$, $m - Tail(a) \neq \emptyset$. It means that by deleting $Tail(a)$ from any matrix element m, the discernibility of all objects does not change. Otherwise, the element m_a cannot be selected. If all the tails of a cannot be deleted, then a is a non-reduct attribute, and cannot be added to R. The attributes in the set $R = R \cup \{a\}$ are individually necessary to keep the discernibility of some object pairs, and thus is a partial reduct. The iteration process is stopped when M is a family of singleton attribute sets. That is, all the remaining attributes are necessary and sufficient to keep the discernibility of all object pairs, and hence a reduct is obtained. In fact, the constructed reduct R is the union of all singleton attribute sets in M.

The addition adds the most preferred attributes first. That is, we check from the left, the most preferred attribute in At, to the right, the most unfavoured attribute in At. Comparing to the deletion algorithm which can deterministically generate the winning reduct under the right-to-left order, the addition algorithm cannot deterministically generate the reduct of the left-to-right order. The essential, but non-deterministic, part of the algorithm is in Step (4.2). For a chosen attribute a, there may exist more than one tail. The criterion for selecting one tail of a for deleting is to keep the best candidate attribute for further processing. Deleting a less preferred tail cannot guarantee the result of a more promising attribute in the future round. In other words, keeping the less preferred tail might induce a more preferred reduct attribute set, or, keeping the more preferred tail might induce a less preferred reduct attribute set in the future round. Liang *et al.* [14] have proved that the problem of constructing the winning reduct under the left-to-right lexical order is NP-hard.

Example 5. We can use the same setting in Example 4 to illustrate the complexity of the addition algorithm. Given an information table S with an ordering $At : a \succeq b \succeq c \succeq d \succeq e \succeq f \succeq g$, and the set representation of the discernibility matrix after absorption is read as: $M = \{\{a, g\}, \{a, f\}, \{b, d, e\}, \{b, e, f\}, \{c, d, f\}, \{e, f, g\}\}$. According to the algorithm, suppose we select the most preferred attribute a for addition. We obtain $R = \{a\}$, and we can either delete the tail $\{g\}$, or the tail $\{f\}$ of a.

- If the tail $\{g\}$ of a is selected, then accordingly, the discernibility matrix M is updated as $\{\{a\}, \{b, d, e\}, \{c, d, f\}, \{e, f\}\}$. In the second round, we select the second-most preferred attribute b for addition and obtain $R = \{a, b\}$. The tail $\{d, e\}$ of b needs to be deleted. This turns out $M = \{\{a\}, \{b\}, \{f\}\}$, containing only singleton attribute sets. Consequently, we obtain the reduct $R = \{a, b, f\}$.
- On the other hand, if the tail $\{f\}$ of a is selected, then accordingly, M is updated as $\{\{a\}, \{b, e\}, \{c, d\}, \{e, g\}\}$. In the second round, we still select the second-most preferred attribute b for addition and obtain $R = \{a, b\}$. The tail $\{e\}$ of b needs to be deleted. This turns out $M = \{\{a\}, \{b\}, \{c, d\}, \{g\}\}$. In the

third round, we select the third-most preferred attribute c for addition and obtain $R = \{a, b, c\}$. The tail $\{d\}$ of c needs to be deleted. This yields $M = \{\{a\}, \{b\}, \{c\}, \{g\}\}$ contains only singleton attribute sets. Consequently, we obtain the reduct $R = \{a, b, c, g\}$.

According to the left-to-right lexical order, we have $\{a, b, c, g\} \succ \{a, b, f\}$.

5 Conditional User Preferences

In the previous discussion, we simply assume that a user can precisely and completely express the preference over the entire attribute set. This enables us to investigate fundamental issues in a simple model. One may argue that a user might not be able to provide such information in practices. A practical issue is how to acquire the user preference of attributes. In this section, we discuss conditional user preferences of attributes and reducts.

5.1 Conditional User Preference of Attributes

User preferences under practical constraints are also understood as *conditional* user preferences, *constrained* user preferences, or *dynamic* user preferences. A simple conditional user preference can be interpreted as "given the condition x, a is preferred to b." The measurement of a simple conditional user preference relation states that "given the condition x, the weight of a is higher than the weight of b." A related notion of the conditional user preference of attribute values has been considered by some researchers [4,5,9].

Formally, a conditional user preference relation can be defined in two ways. Suppose there is a constraint attribute set A and a constraining attribute set B. A conditional user preference is the preference relation of B depending on the preference relation of A, i.e.,

$$(B, \succ) \mid (A, \succ).$$

Based on this constraint-based interpretation, given two constraint attributes x and y, we can express the conditional user preference of attributes a and b as $(a \succ b) \mid (x \succ y)$. It means that the preference of attributes a and b is constrained by the preference of attributes x and y. Suppose a set A is a partial reduct. A conditional user preference is the preference relation of the set A^c depending on the attributes in A, i.e.,

$$(A^c, \succ) \mid A.$$

Based on this dynamic-based interpretation, given a constraint attribute x, we can express the conditional user preference a and b as $(a \succ b) \mid \{x\}$.

For both interpretations, we can introduce preference on attribute sets. This leads to the following implication for a partial reduct A:

$$a \succ b \mid A \Rightarrow A \cup \{a\} \succ A \cup \{b\}.$$

We also can obtain the implication in another form for a super-reduct A:

$$a \succ b \mid A \Rightarrow A - \{b\} \succ A - \{a\}.$$

They can be immediately applied to the two reduct construction algorithms.

5.2 Reduct Construction Based on Conditional Preferences

Based on the conditional user preference of attributes, we can simply modify the above deletion Algorithm 1 and the addition Algorithm 2 as Algorithms 3 and 4 in order to cope with the dynamically changed attribute order. More specifically, by moving the Step (1) of these algorithms to be the first operation in Step (3) which results the attribute selection, we allow the user to specify the preference in the on-going construction procedure. The fitness value of attributes can be computed, modified and ordered in each iteration. This approach requires an interaction between the user and the system during the computation, and let the user guide the processing.

Input: The discernibility matrix M of an information table.
Output: A reduct R.

(1) $R = At$.
(2) $M = Absorb(M)$.
(3) While M is not a family of singleton attribute sets:
  (3.1) Construct an attribute ordering of all the attributes exist in M;
  (3.2) Select the right-most attribute a in the ordering that satisfies the condition $\{a\} \notin M$;
  (3.3) $R = R - \{a\}$ and $M = Absorb(\{m - \{a\} \mid m \in M\})$.
(4) Update R by deleting the attributes not existing in M.

Algorithm 3. A modified deletion algorithm for reduct construction

Input: The discernibility matrix M of an information table.
Output: A reduct R.

(1) $R = \emptyset$.
(2) $M = Absorb(M)$.
(3) While M is not a family of singleton attribute sets:
  (3.1) Construct an attribute ordering of all the attributes exist in M;
  (3.2) Select the left-most attribute a in the ordering that satisfies the condition that there exists an element $m_a \in \{m \in M \mid a \in m\}$ such that $\forall m \in M \ (m - (m_a - \{a\})) \neq \emptyset)$;
  (3.3) Select an element m_a that satisfies the condition of (3.1);
  (3.4) $R = R \cup \{a\}$ and $M = Absorb(\{m - (m_a - \{a\})) \mid m \in M\})$.

Algorithm 4. A modified addition algorithm for reduct construction

The construction of an attribute ordering in Step (3) of the two algorithms are based on the notion of the conditional user preference. In fact, we do not need a complete specification of all preferences of attributes. For the deletion algorithm, we only need to ask for the most unfavoured attribute, i.e., the right-most attribute under the right-to-left order. For the addition algorithm, we only ask for the most preferred attribute, i.e., the left-most attribute under the left-to-right order.

Example 6. Consider the same information table S with $At = \{a, b, c, d, e, f, g\}$. The set representation of the discernibility matrix after absorption is read as: $M = \{\{a, g\}, \{a, f\}, \{b, d, e\}, \{b, e, f\}, \{c, d, f\}, \{e, f, g\}\}$. Suppose the user wants to keep the most preferred attribute a in the partial reduct, i.e., $R = \{a\}$. According to the modified addition algorithm, we can delete either $\{g\}$ or $\{f\}$.

- If the tail $\{g\}$ of a is selected, then the discernibility matrix M is updated as $\{\{a\}, \{b, d, e\}, \{c, d, f\}, \{e, f\}\}$. In the second round, we re-order the remaining attributes and obtain the left-most attribute c for addition. The tail $\{d, f\}$ of c needs to be deleted. This turns out the partial reduct $R = \{a, c\}$ and $M = \{\{a\}, \{c\}, \{e\}\}$ containing only singleton attribute sets. Consequently, we obtain the reduct $R = \{a, c, e\}$ after update.
- On the other hand, if the tail $\{f\}$ of a is selected, then M is updated as $\{\{a\}, \{b, e\}, \{c, d\}, \{e, g\}\}$. In the second round, we re-order the remaining attributes and obtain the left-most attribute c for addition. The tail $\{d\}$ of c needs to be deleted. This turns out the partial reduct $R = \{a, c\}$ and $M = \{\{a\}, \{b, e\}, \{c\}, \{e, g\}\}$. In the third round, we order the remaining attributes again, and obtain the left-most attribute g for addition. The tail $\{e\}$ of $\{g\}$ need to be deleted. This yields the partial reduct $R = \{a, c, g\}$ and $M = \{\{a\}, \{b\}, \{c\}, \{g\}\}$ containing only singleton attribute sets. Consequently, we obtain the reduct $R = \{a, b, c, g\}$ after update.

6 Conclusion

In order to avoid an over-simplified assumption that all the attributes are equally important, we propose a model for machine learning based on the user preference of attributes. Based on utility theory and measurement theory, both the qualitative and quantitative representations of user preference of attributes, and the connection between the qualitative and quantitative representations are explored.

User preferences of attributes can be extended to the user preference of attribute sets. We explore several quantitative extensions and qualitative extensions. For the qualitative user preference of attribute sets, we need to consider two different forms: the left-to-right lexical order and the right-to-left lexical order, respectively. While the left-to-right order encourages an optimistic comparison between attribute sets, the right-to-left order promotes a pessimistic comparison.

With respect to the user preference of attribute sets, the computation of the most preferred reducts can be studied. Regarding the two lexical orders, we have the winning reduct under the left-to-right lexical order and the potential winner of the right-to-left lexical order, respectively.

It is also noted that a user preference can be either unconditional or conditional on some constraints. For conditional situations, a dynamic mechanism is required for the reduct construction. The paper proposed different algorithms to cope with both unconditional and conditional user preferences. They focus on user-oriented attribute selection and reduction.

Acknowledgement

The authors would like to thank Professors James Peters and Andrzej Skowron for the encouragement and anonymous reviewers for their constructive comments. This study is particularly supported by a NSERC discovery grant.

References

1. Bazan, J.G., Nguyen, H.S., Nguyen, S.H., Synak, P., Wroblewski, J.: Rough set algorithms in classification problem. In: Polkowski, L., Tsumoto, S., Lin, T.Y. (eds.) Rough Set Methods and Applications, pp. 49–88 (2000)
2. Beynon, M.: Reducts within the variable precision rough sets model: a further investigation. European Journal of Operational Research 134, 592–605 (2001)
3. Blum, A.L., Langley, P.: Selection of relevant features and examples in machine learning. Artificial Intelligence 97, 245–271 (1997)
4. Boutilier, C., Brafman, R.I., Domshlak, C., Hoos, H.H., Poole, D.: CP-nets: a tool for representing and reasoning with conditional ceteris paribus preference statements. Journal of Artificial Intelligence Research 21, 135–191 (2004)
5. Chen, S., Buffett, S., Fleming, M.W.: Reasoning with conditional preference across attributes. In: Proceedings of Canadian Artificial Intelligence, pp. 369–380 (2007)
6. Fishburn, P.C.: Utility Theory for Decision-Making. John Wiley & Sons, New York (1970)
7. Fishburn, P.C.: Mathematics for Decision Making. Unesco, Mouton (1972)
8. Greco, S., Matarazzo, B., Slowinski, R.: Rough approximation of a preference relation by dominance relations. European Journal of Operational Research 117, 63–83 (1999)
9. Grzymala-Busse, J.W.: LERS - A system for learning from examples based on rough sets. In: Slowinski, R. (ed.) Intelligent Decision Support, pp. 3–18. Kluwer Academic Publishers, Boston (1992)
10. Han, S.Q., Wang, J.: Reduct and attribute order. Journal of Computer Science and Technology archive 19(4), 429–449 (2004)
11. Inuiguchi, M.: Several approaches to attribute reduction in variable precision rough set model. Modeling Decisions for Artificial Intelligence, 215–226 (2005)
12. Jain, A., Duin, P., Mao, J.: Statistical pattern recognition: A review. IEEE Transactions on Pattern Analysis and Machine Intelligence 22(1), 4–37 (2000)
13. Kohavi, R., John, G.: Wrappers for feature subset selection. Artificial Intelligence 97(1), 273–324 (1997)
14. Liang, H., Wang, J., Yao, Y.Y.: User-oriented feature selection for machine learning. Computer Journal 50(4), 421–434 (2007)
15. Nguyen, H.S.: Approximate boolean reasoning: foundations and applications in data mining. Transactions on Rough Sets 5, 334–506 (2006)
16. Pawlak, Z.: Rough sets. International Journal of Computer Information and Science 11, 341–356 (1982)

17. Polkowski, L., Tsumoto, S., Lin, T.Y.: Rough Set Methods and Applications: New Developments in Knowledge Discovery in Information Systems. Physica-Verlag, Heidelberg (2000)
18. Roberts, F.S.: Measurement Theory. Academic Press, New York (1976)
19. Skowron, A., Rauszer, C.: The discernibility matrices and functions in information systems. In: Slowiński, R. (ed.) Intelligent Decision Support, Handbook of Applications and Advances of the Rough Sets Theory, Kluwer, Dordrecht (1992)
20. Slezak, D.: Various approaches to reasoning with frequency based decision reducts: a survey. In: Polkowski, L., Tsumoto, S., Lin, T.Y. (eds.) Rough set methods and applications, pp. 235–285. Physica-Verlag, Heidelberg (2000)
21. Slowinski, R., Greco, S., Matarazzo, B.: Mining decision-rule preference model from rough approximation of preference relation. In: Proceedings of IEEE Annual International Conference on Computer Software and Applications, pp. 1129–1134 (2002)
22. Swiniarski, R.W., Skowron, A.: Rough set methods in feature selection and recognition. Pattern Recognition Letters 24(6), 833–849 (2003)
23. Wang, J., Miao, D.: Analysis on attribute reduction strategies of rough set. Chinese Journal of Computer Science and Technology 13(2), 189–192 (1998)
24. Wang, J., Wang, J.: Reduction algorithms based on discernibility matrix: the ordered attributes method. Journal of Computer Science and Technology 16(6), 489–504 (2001)
25. Wei, L., Li, H., Zhang, W.: Knowledge reduction based on the equivalence relations defined on attribute set and its power set. Information Sciences, 177–15, 3178–3185 (2007)
26. Wu, W., Mi, J.: Knowledge reduction in incomplete information systems based on Dempster-Shafer theory of evidence. In: Wang, G.-Y., Peters, J.F., Skowron, A., Yao, Y. (eds.) RSKT 2006. LNCS (LNAI), vol. 4062, pp. 254–261. Springer, Heidelberg (2006)
27. Xu, F., Miao, D., Wei, L.: An approach for fuzzy-rough attribute reduction via mutual information. In: Proceedings of International Conference on Fuzzy Systems and Knowledge Discovery, pp. 107–112 (2007)
28. Yao, J.T., Zhang, M.: Feature selection with adjustable criteria. In: Ślezak, D., Wang, G., Szczuka, M.S., Düntsch, I., Yao, Y. (eds.) RSFDGrC 2005. LNCS (LNAI), vol. 3641, pp. 204–213. Springer, Heidelberg (2005)
29. Yao, Y.Y., Chen, Y.H., Yang, X.D.: A measurement-theoretic foundation for rule interestingness evaluation. In: Proceedings of Workshop on Foundations and New Directions in Data Mining in the IEEE International Conference on Data Mining, pp. 221–227 (2003)
30. Yao, Y.Y., Zhao, Y., Wang, J.: On reduct construction algorithms. In: Wang, G.-Y., Peters, J.F., Skowron, A., Yao, Y. (eds.) RSKT 2006. LNCS (LNAI), vol. 4062, pp. 297–304. Springer, Heidelberg (2006)
31. Yao, Y.Y., Zhao, Y., Wang, J., Han, S.: A model of machine learning based on user preference of attributes. In: Greco, S., Hata, Y., Hirano, S., Inuiguchi, M., Miyamoto, S., Nguyen, H.S., Słowiński, R. (eds.) RSCTC 2006. LNCS (LNAI), vol. 4259, pp. 587–596. Springer, Heidelberg (2006)
32. Zhang, X., Zhang, F., Li, M., Wang, N.: Research of optimal reduct under preference. Computer Engineering and Design 26(8), 2103–2106 (2005)
33. Zhang, X., Zhao, Y., Zou, H.: Finding the reduct subject to preference order of attributes. In: Kryszkiewicz, M., Peters, J.F., Rybinski, H., Skowron, A. (eds.) RSEISP 2007. LNCS (LNAI), vol. 4585, pp. 202–210. Springer, Heidelberg (2007)

34. Zhao, Y., Luo, F., Wong, S.K.M., Yao, Y.Y.: A general definition of an attribute reduction. In: Yao, J., Lingras, P., Wu, W.-Z., Szczuka, M.S., Cercone, N.J., Ślęzak, D. (eds.) RSKT 2007. LNCS (LNAI), vol. 4481, pp. 101–108. Springer, Heidelberg (2007)
35. Zhao, K., Wang, J.: A reduction algorithm meeting users' requirements. Journal of Computer Science and Technology 17, 578–593 (2002)
36. Ziarko, W.: Rough set approaches for discovering rules and attribute dependencies. In: Klösgen, W., Żytkow, J.M. (eds.) Handbook of Data Mining and Knowledge Discovery, Oxford, pp. 328–339 (2002)

Research on Rough Set Theory and Applications in China

Guoyin Wang[1,2], Qinghua Zhang[1,2], Houkuan Huang[3], Dongyi Ye[4], Qinghua Hu[5], Xuegang Hu[6], Zhongzhi Shi[7], Yongli Li[8], Lin Shang[9], Liping An[10], Ying Sai[11], Shanben Chen[12], Jiye Liang[13], Keyun Qin[14], Huanglin Zeng[15], Keming Xie[16], Duoqian Miao[17], Fan Min[18], Zhaocong Wu[19], Weizhi Wu[20,21], and Jianhua Dai[22]

[1] Institute of Computer Science & Technology, Chongqing University of Posts and Telecommunications, Chongqing, 400065, China
[2] School of Information Science & Technology, Southwest Jiaotong University, Chengdu, 610031, China
[3] School of Computer & Information Technology, Beijing Jiaotong University, Beijing, 100044, China
[4] College of Math. and Computer, Fuzhou University, Fuzhou, 350002, China
[5] Harbin Institute of Technology, PO Box 458, Harbin, 150006, China
[6] School of Computer & Information, Hefei University of Technology, Hefei, 230009, China
[7] Institute of Computing Technology, Chinese Academy of Sciences, Beijing, 100080, China
[8] Lanzhou University, No.222, Tianshui South Rd. Lanzhou, 730000, China
[9] State Key Laboratory for Novel Software Technology, Nanjing University, Nanjing, 210093, China
[10] Management Science & Engineering Department, Business School, Nankai University, Tianjin, 300071, China
[11] Shandong University of Finance, No.40 Shun-geng Road, Jinan, 250014, China
[12] Welding Engineering Institute, Shanghai Jiaotong University, Shanghai, 200030, China
[13] School of Computer & Information Technology, Shanxi University, Taiyuan, 030006, China
[14] Intelligent Control Development Center, Southwest Jiaotong University, Chengdu, 610031, China
[15] Artificial Intelligence of Key Laboratory, Sichuan University of Science and Engineering, Zigong, 643000, China
[16] Department of Automation, College of Information Engineering, Taiyuan University of Technology, Taiyuan, 030024, China
[17] Department of Computer Science and Technology, Tongji Universtiy (Jiading Campus), 4800 CaoAn Highway, Shanghai, 201804, China
[18] School of Computer Science and Technology, University of Electronic Science and Technology of China, Chengdu, 610054, China
[19] School of Remote Sensing Information Engineering, Wuhan University, Wuhan, 430079, China
[20] Institute for Information and System Sciences, Faculty of Science, Xi'an Jiaotong University, Xi'an, 710049, China
[21] Information College, Zhejiang Ocean University, Zhejiang, 316004, China
[22] Institute of Artificial Intelligence, Zhejiang University, Hangzhou, 310027, China

J.F. Peters and A. Skowron (Eds.): Transactions on Rough Sets VIII, LNCS 5084, pp. 352–395, 2008.

Abstract. This article gives a capsule view of research on rough set theory and applications ongoing at universities and laboratories in China. Included in this capsule view of rough set research is a brief description of the following things: Chinese research groups on rough set with their URLs for web pages, names of principal researchers (supervisors), numbers of graduate students, and topics being investigated. Statistical summaries showing the growth in the research on rough set theory and application in China are included. In addition, an introduction summarizing the research interests of Chinese researchers is included in this article. The contribution of this article is a complete overview of the principal research directions in rough set theory and its applications in China.

1 Introduction

Rough set theory is a valid mathematical tool to deal with imprecise, uncertain, and vague information proposed by Professor Z. Pawlak in 1982[151]. It has been developed and applied in many fields such as decision analysis, machine learning, dada mining, pattern recognition, and knowledge discovery successfully[152, 209, 336]. This theory has been studied for over 25 years now. The community of rough set is growing in the whole world. There are 3 series of international conferences on rough sets, that is, RSCTC (international conference on Rough Sets and Current Trends in Computing)[1, 15, 42, 153, 154, 186, 200, 212], RSFDGrC (international conference on Rough Sets, Fuzzy Sets, Data Mining, and Granular Computing)[2, 100, 155, 156, 201, 208, 213, 331, 332], RSKT (international conference on Rough Sets and Knowledge Technology)[157, 169, 202, 214]. There are many rough set references of books, journal special issues, proceedings, book chapters, etc[16, 101–107, 147–150, 158–168, 170–172, 187, 188, 203–207, 211, 273–276, 333–335]. A rough set database (`http://rsds.univ.rzeszow.pl`) has been set up with almost 4000 references.

The primary notions of rough set theory are the approximation space, lower and upper approximations of an object set. An approximation space is a classification of a domain of interest into disjoint categories. Human knowledge about a domain is expressed by the classification formally. Rough set theory treats knowledge as an ability to classify perceived objects relative to classes in partitions defined by the indiscernibility relation. Objects with matching feature values are considered to be indistinguishable to each other. The lower approximation of an object set (S) is a set of classes that are proper subsets of S, while its upper approximation is a set of classes having non-empty intersection with S. When the set difference between the lower and upper approximations (i.e., the boundary) is non-empty, the set S is called a rough set.

Research on rough set theory and applications in China began in the middle 1990s. Professor Qing Liu is one of the initiators at that time[115]. Subsequently, more and more researchers are interested in rough set theory. Some Chinese researchers like Huanglin Zeng, Guoyin Wang, Wenxiu Zhang, Duoqian Miao, etc, started their researches on rough set theory in the middle 1990s. The

Chinese rough set community is growing very quickly. The Chinese Rough Set and Soft Computing Society (CRSSC, `http://cs.cqupt.edu.cn/crssc`) was set up in 2003, which is a branch of the Chinese Association of Artificial Intelligence (CAAI). Several international conferences related to rough set theory have been organized in China. Chinese conference on rough sets and soft computing is held every year since 2001.

In this paper, we give a capsule view of research on rough set theory and applications ongoing in China, introduce key research groups on rough set theory and applications in China, summarize the research interests of Chinese rough set researchers.

In section 1, the Chinese Rough Set and Soft Computing Society (CRSSC) is introduced. Some key research groups of CRSSC are also introduced. A general introduction of the current developing status of the research on rough set theory and applications in China is also made in this section. In section 2, a brief overview of the principle research directions in rough set theory and its applications in China is presented. In section 3, a summary about the research on rough set theory and applications in China is made.

2 Development of the Chinese Rough Set and Soft Computing Society

The Chinese Rough Set and Soft Computing Society (CRSSC) was set up in 2003. It is a national scientific society issued by the Ministry of Civil Affairs (MCA) of China. It is a branch of the Chinese Association of Artificial Intelligence (CAAI). CRSSC is the home of Chinese researchers on rough set theory and applications.

In this section, we will introduce the organization and works of CRSSC, and some generation information of the research on rough set theory and applications in China in recent years.

2.1 Organization of CRSSC

The Chinese Rough Set and Soft Computing Society (CRSSC) was officially set up during the 2003 Chinese Conference on Artificial Intelligence (CAAI-10) in Guangzhou, November 21, 2003. Profesor Guoyin Wang at the Chongqing University of Posts and Telecommunications serves as its Chair, Profesors Duoqian Miao at Tongji University, Weizhi Wu at Zhejiang Ocean University, and Zhihua Zhou at Nanjing University serve as its Vice-Chairs, and Professor Qing Liu at Nanchang University serves as its honorary Chair. At present, there are over 70 CRSSC society members from different universities and institutes of the whole country. The Chinese conference on rough set and soft computing is its official conference. It is held every year since 2001. The following is a list of CRSSC conferences held until now.

- CRSSC2001 (Chongqing, May 2001)
- CRSSC2002 (Shuzhou, October 2002)
- CRSSC2003 (Chongqing, October 2003)

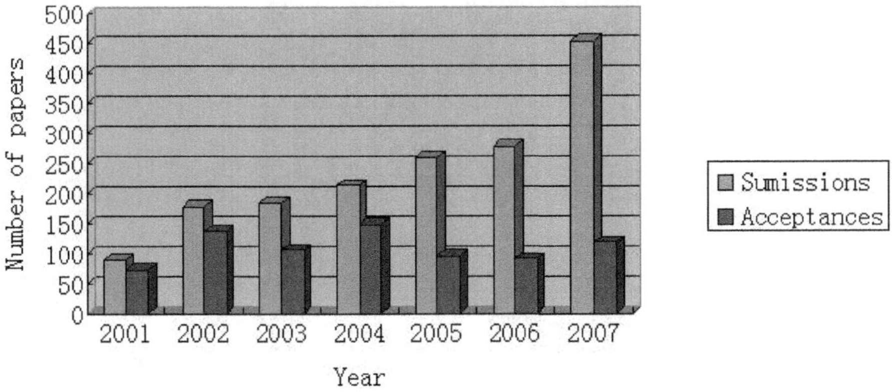

Fig. 1. Statistical information of CRSSC conferences

- CRSSC2004 (Zhoushan, October 2004)
- CRSSC2005 (Anshan, August 2005)
- CRSSC2006 (Jinhua, October 2006)
- CRSSC2007 (Taiyuan, August 2007)

Professor Zdzislaw Pawlak attended the CRSSC2001 in Chongqing and gave a talk. Many other oversea researchers on rough set theory and applications such as Professors Andrzej Skowron, Tsau Young Lin, Yiyu Yao, Ning Zhong, Shusaku Tsumoto, Leung Lee, Xiaohua (Tony) Hu, Jingtao Yao, also attended the conference to give talks and communicate with Chinese researchers. Some statistical information about the CRSSC conferences is shown in Figure 1.

Except for the CRSSC conferences, the society starts to sponsor other two series of workshops, Chinese Workshop on Granular Computing (CGrC) and Chinese Workshop on Web Intelligence (CWI) from 2007. The CGrC2007 and CWI2007 were jointly held together with CRSSC2007 in Taiyuan in August 2007. CRSSCs, CGrCs, and CWIs are good forums for Chinese researchers on rough set theory and applications to introduce, exchange their research results.

In order to communicate with oversea researchers, RSFDGrC2003 and RSKT 2006 were held in China. The RSKT2008 will also be held in Chengdu, China, in May 2008. In fact, CRSSC is one of the major sponsors of the RSKT conferences.

2.2 Key Research Groups of CRSSC

The Chinese rough set community is growing very quickly. Many outstanding research groups are emerging in Chinese universities and institutes. Here, we make a brief introduction to these research groups.

(1) Institute of Computational Intelligence, Department of Computer Science, Beijing Jiaotong University

Homepage: `Http://cit.njtu.edu.cn/`

Principal researchers: Houkuan Huang(hkhuang@center.njtu.edu.cn)
Jian Yu (jianyu@bjtu.edu.cn)
Postal Address: School of Computer & Information Technology, Beijing Jiaotong University

(2) Institute of Computer Science & Technology, Chongqing University of Posts and Telecommunications

Homepage: Http://cs.cqupt.edu.cn/wanggy
Principal researchers: Guoyin Wang(wanggy@ieee.org)
Yu Wu(wuyu@cqupt.edu.cn)
Jun Zhao(zhaojun@cqupt.edu.cn)
Hong Yu(yuhong@cqupt.edu.cn)
Xinghua Fan(fanxh@cqupt.edu.cn)
Postal Address: Institute of Computer Science & Technology,
Chongqing University of Posts and Telecommunications,
Chongqing, P. R. China, 400065
Graduated students: 18 PhD students and 40 Master students.

(3) College of Math. and Computer, Fuzhou University
Homepage: Http://youth.fzu.edu.cn/cetc/channel/
Principal researchers: Dongyi Ye(yiedy@fzu.edu.cn)
Postal Address: College of Math. and Computer,
Fuzhou University,

Fuzhou, 350002, P. R. China
Graduated students: 9 Master students.

(4) Institute of advanced power and propulsion, Harbin Institute of Technology

Homepage: Http://www.turbo.hit.edu.cn
Principal researchers: RenYu Da(yudaren@hcms.hit.edu.cn)
Qinghua Hu(huqinghua@hcms.hit.edu.cn)
Wen Bao(baowen@hit.edu.cn)
Zhiqiang Xu(xuzhiqiang@hcms.hit.edu.cn)
Postal Address: PO Box 458,
Harbin Institute of Technology,
Harbin, 150006, Heilongjiang, P. R. China
Graduated students: 25 PhD students and 20 Master students.

(5) Institute of Artificial Intelligence and Data Mining, Hefei University of Technology

Homepage: Http://www1.hfut.edu.cn/organ/kddweb/
Principal researchers: Xuegang Hu (jsjxhuxg@hfut.edu.cn)
Hao Wang(jsjxwangh@hfut.edu.cn)
Xindong Wu(xwu@cs.uvm.edu)
Dexing Wang(wangdexing198706@yahoo.com.cn)
Hongliang Yao(lhy_y@sohu.com)

Postal Address: School of Computer & Information,
Hefei University of Technology,
Hefei, 230009, Anhui, P.R.China
Graduated students: 8 PhD students and 45 Master students.

(6) Intelligence Science Laboratory, Institute of Computing Technology, Chinese Academy of Sciences

Homepage: Http://www.intsci.ac.cn/shizz/
Principal researchers: Zhongzhi Shi (shizz@ics.ict.ac.cn)
Qing He(heq@ics.ict.ac.cn)
Hong Hu(huhong@ics.ict.ac.cn)
Zhiping Shi(shizp@ics.ict.ac.cn)
Zuqiang Meng(Mengzq@ics.ict.ac.cn)
Postal Address: Institute of Computing Technology,
Chinese Academy of Sciences,
Beijing, 100080, P. R. China
Graduated students: 2 Post Doctors, 16 PhD students and 8 Master students.

(7) School of Information Science & Engineering, Lanzhou University
Homepage: Http://xxxy.lzu.edu.cn/
Principal researchers: Yongli Li(ylli@lzu.edu.cn)
Postal Address: Lanzhou University,
No.222, Tianshui South Rd.

Lanzhou, 730000, Gansu, P. R. China
Graduated students: 40 Master students.

(8) Dept. of Computer Science and Technology, State Key Laboratory for Novel Software Technology, Nanjing University

Homepage: `Http://cs.nju.edu.cn/shanglin`
Principal researchers: Lin Shang(shanglin@nju.edu.cn)
Xuri Yin(yinxuri@ai.nju.edu.cn)
Postal Address: State Key Laboratory for Novel Software Technology,
Nanjing University,
Mailbox 419, Hankou Road 22, Nanjing, 210093, P. R. China
Graduated students: 3 PhD students and 4 Master students.

(9) Management Science & Engineering Department, Business School, Nankai University

Homepage: `Http://www.nankai.edu.cn`
Principal researchers: Liping An(anliping@nankai.edu.cn)
Postal Address: Management Science & Engineering Department,
Business School, Nankai University,
Tianjin, 300071, P. R. China
Graduated students: 3 Master students.

(10) School of Computer & Information Engineering, Shandong University of Finance

Homepage: `Http://www2.sdfi.edu.cn/jsj/`
Principal researchers: Ying Sai(saiying@sdfi.edu.cn),
Jiqin Liu(sdfiljq@126.com)
Postal Address: Shandong University of Finance
No. 40 Shun-geng Road, Jinan City, 250014
Shandong Province, P.R. China
Graduated students: 5 Master students

(11) Intelligent Robotic Welding Laboratory, Shanghai Jiaotong University

Homepage: Http://www.sjtu.edu.cn/staff/teachers/182.xml
Principal researchers: Shanben Chen(sbchen@sjtu.edu.cn)
Postal Address: Welding Engineering Institute,
Shanghai Jiaotong University
No. 1954, Huashan Road, Shanghai, 200030, P. R. China
Graduated students: 2 PhD students and 4 Master students.

(12) Key Laboratory of the Ministry of Education for Computation Intelligence & Chinese Information Processing, Shanxi University

Homepage: Http://www.sxu.edu.cn/yuanxi/jikex/
Principal researchers: Jiye Liang(ljy@sxu.edu.cn)
Deyu Li(lidy@sxu.edu.cn)
Wenjian Wang(wjwang@sxu.edu.cn)
Yuhua Qian(jinchengqyh@126.com)
Kaishe Qu(quks@sxu.edu.cn)
Ru Li(liru@sxu.edu.cn)
Xia Zhang(zhangxia@sxu.edu.cn)
Postal Address: School of Computer & Information Technology,
Shanxi University

Taiyuan, 030006, Shanxi, P. R. China
Graduated students: 6 PhD students and 32 Master students.

(13) Intelligent Control Development Center, Southwest Jiaotong University

Principal researchers: Keyun Qin(qinkeyun@swjtu.edu.cn)
Tianrui Li(trli@swjtu.edu.cn)
Chaozhe Jiang(jiangchaozhe@163.com)
Postal Address: Intelligent Control Development Center,
Southwest Jiaotong University,
Chengdu, 610031, P. R. China
Graduated students: 6 PhD students and 14 Master students.

(14) Artificial Intelligence Key Laboratory, Sichuan University of Science and Engineering
Homepage: `Http://rgzn.suse.edu.cn/`
Principal researchers: Huanglin Zeng(zhl@suse.edu.cn)
Yi Yao(jwc@suse.edu.cn),
Chenghua Fu(dxx@suse.edu.cn)
Postal Address: Artificial Intelligence of Key Laboratory,
Sichuan University of Science and Engineering,
Zigong, Si Chuan, 643000, P.R. China
Graduated students: 30 Master students.

(15) Department of Automation, College of Information Engineering, Taiyuan University of Technology.
Homepage: `Http://auto.tyut.edu.cn/ (in building)`
Principal researchers: Keming Xie(kmxie@tyut.edu.cn),
Gang Xie(xiegang@tyut.edu.cn),
Gaowei Yan(yangaowei@tyut.edu.cn),
Zehua Chen(chenzehua@tyut.edu.cn),
Jun Xie(xiejun@tyut.edu.cn),
Hongbo Guo(ghb666@sohu.com),
Qiuyu Xia(qyxljl@yahoo.com.cn)

Postal Address: Department of Automation,
College of Information Engineering,
Taiyuan University of Technology,
Taiyuan, Shan Xi, P.R.C 030024

Graduated students: 7 PhD students and 65 Master students.

(16) Laboratory of Intelligent Information Processing, Tongji University

Homepage: Http://hpcc.tongji.edu.cn/~dqmiao

Principal researchers: Duoqian Miao(miaoduoqian@163.com)
Yan Wu(yanwu@mail.tongji.edu.cn)
Hongyun Zhang(zhangjiaye1972@hotmail.com)
Chunmei Liu(chunmei.liu@mail.tongji.edu.cn)
Postal Address: Department of Computer Science and Technology,
Tongji Universtiy(Jiading Campus),
4800 CaoAn Highway,
Shanghai, 201804, P. R. China

Graduated students: 15 PhD students and 19 Master students.

(17) Digital Multimedia Laboratory, University of Electronic Science and Technology of China

Homepage: Http://www.dm.uestc.edu.cn

Principal researchers: Leiting Chen(richardchen@uestc.edu.cn)
Fan Min(minfan@uestc.edu.cn)
Mingtian Zhou(mtzhou@uestc.edu.cn)
Qingxin Zhu(qxzhu@uestc.edu.cn)
Qihe Liu(qiheliu@uestc.edu.cn)
Hongbin Cai(caihb@uestc.edu.cn)
Jianzhong Zhang(jianzhang@uestc.edu.cn)
Hao Tan(tanhao@uestc.edu.cn)
Jinzhong Cui(jzcui@uestc.edu.cn)
Mingyun He(hmyun@uestc.edu.cn)
Guanghui Lu(lugh@uestc.edu.cn)
Postal Address: School of Computer Science and Technology,
University of Electronic Science and Technology of China,
Chengdu, 610054, P. R. China
Graduated students: 13 PhD students and 107 Master students.

(18) School of Remote Sensing Information Engineering, Wuhan University

Homepage: Http://rsgis.wtusm.edu.cn/
Principal researchers:Zhaocong Wu(zcwoo@whu.edu.cn)
Postal Address: School of Remote Sensing Information Engineering,
Wuhan University,

129 Luoyu Road, Wuhan, 430079, P. R. China
Graduated students: 3 Master students.

(19) Institute for Information and System Sciences, Faculty of Science, Xi'an Jiaotong University

Homepage: `Http://www.rcsci.xjtu.edu.cn/fsci/graduateSchool/index.aspx`

Principal researchers: Wenxiu Zhang(wxzhang@mail.xjtu.edu.cn)
Yee Leung(yeeleung@cuhk.edu.hk)
Weizhi Wu(wuwz@zjou.edu.cn)
Jusheng Mi(mijsh@263.net)
Degang Chen(chengdegang@263.net)
Lin Wei(qjjwv@nwu.edu.cn)
Mingwen Shao(mwshao@mail.xjtu.edu.cn)
Xiujiu Yuan(yuanxiujiu@sohu.com)
Lili Wei(liliwei@nxu.edu.cn)
Postal Address: Institute for Information and System Sciences,
Faculty of Science, Xi'an Jiaotong University,
Xi'an, 710049, Shanxi, P. R. China
Graduate students: 18 PhD students and 40 Master degree students.

(20) Institute of Artificial Intelligence, Zhejiang University

Homepage: `Http://www.cs.zju.edu.cn/`

Principal researchers: Jianhua Dai(jhdai@126.com)
Weidong Chen(chenwd@zju.edu.cn)
Postal Address: Institute of Artificial Intelligence,
Zhejiang University,
Hangzhou, 310027, P. R. China
Graduated students: 3 PhD students and 8 Master students.

2.3 General Status of Research on Rough Set Theory and Applications in China

Research on rough set theory and applications in China began in the middle 1990s. The Chinese rough sets and soft computing society has become one of

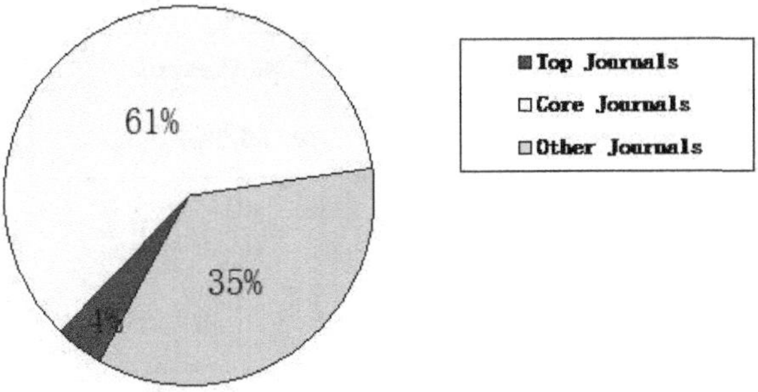

Fig. 2. Distribution of Chinese journal papers on rough set

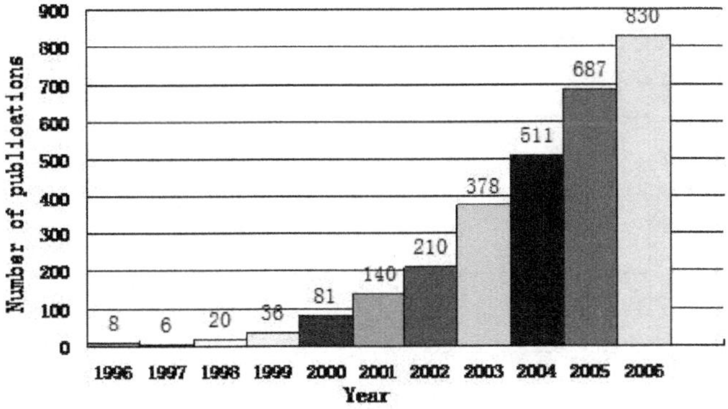

Fig. 3. Number of Chinese journal papers on rough set

the biggest national rough set societies in the whole world. Chinese researchers achieved many significant results on rough set theory and applications. There are 15 Chinese monographs related to rough set published in the past 20 years[69, 73, 92, 116, 136, 195, 215, 291, 292, 298, 299, 304–308]. Over 3000 Chinese papers about rough set theory and applications have been published. Some of them are published on international journals such as Fundamenta Informaticae, Information Sciences, Transactions on Rough Sets, etc. In China, Chinese journals are categorized into three levels, that is, top (leading) journals, core journals, and normal journals. Over 150 rough set related papers are published on top Chinese journals, and over 2000 papers on core Chinese journals. The distribution and number of publications of Chinese papers related to rough set each year are shown in Figure 2 and Figure 3. Figure 4 and Figure 5 show the status about Chinese PhD and MS degree dissertations in recent years. Table 1 shows the number of papers published in international conferences on rough sets by Chinese researchers. From Figures 2, 3, 4, and 5, one can find that both the quality and quantity of Chinese research papers are growing very quickly.

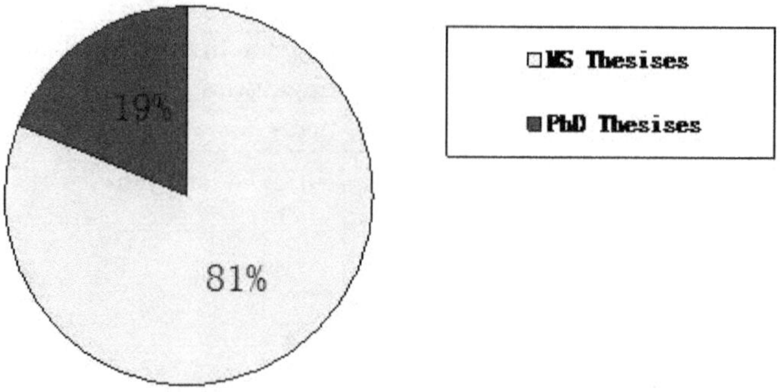

Fig. 4. Number of Chinese PhD and MS degree dissertations

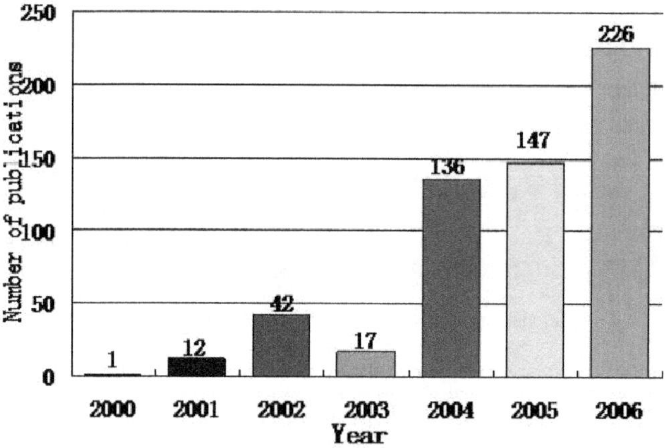

Fig. 5. Distribution of Chinese PhD and MS dissertations

Table 1. Rough set papers from China for international conferences

Confere-nce	RSCTC 2000	RSTGC 2001	RSCTC 2002	RSFDGrC 2003	RSCTC 2004	RSFDGrC 2005	RSKT 2006	RSCTC 2006	JRS 2007
Number of paper	7	3	10	64	30	44	83	19	44
Confere-nce site	Canada	Japan	USA	China	Sweden	Canada	China	Japan	Canada

The development of the research on rough set theory and applications in China gets many supports. NSFC (National Natural Science Foundation of China), one of the most important national research foundations in China, supported researches on rough set theory greatly. By far, there are about 20 NSFC projects founded by NSFC. The number of NSFC supported research projects is increasing year after year. A partial list of these projects is shown in Table 2. Some other national and provincial science & technology research programs in China also supported the researches on rough set theory and applications in the past 20 years.

Table 2. NSFC projects on rough set in China

Project No.	Project Title	Chief Scientist	University/ Institute	Project Dates
60773113	Basic theory and method for data-driven knowledge acquisition	G. Y. Wang	Chongqing University of Posts and Telecommunications	2008.01 ~2010.12
60775036	Study on principal curve theory and its application in character recognition	D. Q. Miao	Tongji University	2008.01 ~2010.12
60773174	Approaches to attribute reduction and knowledge acquisition based on concept lattices	J. S. Mi	Hebei Normal University	2008.01 ~2010.12
10771056	The theory of fuzzy concept lattice and its application to information science	Q. G. Li	Hunan University	2008.01 ~2010.12
60703013	Rough set models and algorithms for heterogeneous feature analysis	Q. H. Hu	Harbin Institute of Technology	2008.01 ~2010.12
60703038	Rough 3-valued Lukasiewicz algebra and application in granular computing in algebraic settings	J. H. Dai	Zhejiang University	2008.01 ~2010.12
60773133	Research on granulation structure and its knowledge acquisition in complex information systems	J. Y. Liang	Shanxi University	2008.01 ~2010.12
60673096	Mathematical structure of information granularity and its applications in data mining	W. Z. Wu	Zhejiang Ocean University	2007.01 ~2009.12
60663003	On the statistical evidence for rough set data analysis	L. L. Wei	Ningxia University	2007.01 ~2009.12
Continued ...				

Table 2. *(continued)*

60672178	The research of theoretical system of prevention rules of Aircrew's human errors based on rough set data mining method	Z. Tao	Civil Aviation University of China	2007.01 ~2007.12
10671173	Mapping theory and its applications in uncertain decision making	J. J. Li	Zhangzhou Normal College	2007.01 ~2009.12
60675031	Dynamic information analyze based on quotient space topology structural transformation	L. Zhang	Anhui University	2007.01 ~2009.12
70601013	Knowledge modeling and control methods by rough set software computation for complex industry process	L. P. An	Nankai University	2007.01 ~2009.12
60672173	Small spatial scale weather forecast for civil airport based on rough set and association rules	W. Fan	Civil Aviation University of China	2007.01 ~2007.12
70571032	Study on theory, method and application of knowledge acquisition from incomplete information system based on rough set	X. Z. Zhou	Nanjing University	2006.01 ~2008.06
60573068	Basic theory and effective algorithm of granular computing for huge dada processing	G. Y. Wang	Chongqing University of Posts and Telecommunications	2006.01 ~2008.12
10571151	Covering methods and its applications on rough set theory	S. Lin	Zhangzhou Normal College	2006.01 ~2008.12
60503022	Incremental learning based on rough sets theory	L. Shang	Nanjing University	2006.01 ~2008.12
60573074	Research on measurements of knowledge roughness and data analysis method based on rough set	D. Y. Li	Shanxi University	2006.01 ~2008.12
60475017	Theory of quotient space based granular computing and its applications	L. Zhang	Anhui University	2005.01 ~2007.12
Continued ...				

Table 2. *(continued)*

60475019	Study on methods of granular representation, model and inference	D. Q. Miao	Tongji University	2005.01 ~2007.12
60474036	Knowledge modeling and control methods by rough set software computation for complexindustry process	S. B. Cheng	Shanghai Jiaotong University	2005.01 ~2007.12
70471003	Research on uncertainty decision methods based on soft computations	J. Y. Liang	Shanxi University	2005.01 ~2007.12
60374029	Research on MEA-based rough-fuzzy intelligent control	K. M. Xie	Taiyuan University of Technology	2004.01 ~2006.12
60373078	Rough set theory under fuzzy and random environment and knowledge acquisition	W. Z. Wu	Zhejiang Ocean University	2004.01 ~2006.12
60373111	Problem of uncertain information processing based on rough set theory	G. Y. Wang	Chongqing University of Posts and Telecommunications	2004.01 ~2006.12
30271637	Rough set theory-based analysis of the effects of detoxification by acupuncture	P. Xu	Shanghai University of T.C.M	2003.01 ~2005.12
40201039	Rough sets approach to automatic classification of remote sensing image	Z. C. Wu	Wuhan University	2003.01 ~2005.12
60275019	Research on uncertainty fuzziness and knowledge acquisition in rough set theory	J. Y. Liang	Shanxi University	2003.01 ~2005.12
70271048	Study on the theory of dynamic rough managing and managing method	K. Q. Shi	Shandong University	2003.01 ~2005.12
60175016	Combination of rough sets & fuzzy sets and its application in data mining	D. Q. Miao	Shanxi University	2002.01 ~2002.12
60075013	Study on uncertain information processing	K. D. Liu	Hebei University of Engineering	2001.01 ~2003.12
Continued ...				

Table 2. *(continued)*

69875012	Study on intelligence pattern recognition on rough sets	H. L. Zeng	Sichuan University of Science and Engineering	1999.01 ~2001.12
69803014	Automatic knowledge acquisition technology based on rough setand its application	G. Y. Wang	Chongqing University of Posts and Telecommunications	1999.01 ~2001.12

3 Research on Rough Set Theory and Its Applications in China

There are many topics being investigated by Chinese researchers concerning rough sets such as the fundamentals of rough sets, knowledge acquisition, granular computing based on rough set, extended rough set models, rough logic, and applications of rough set. The work and achievements of Chinese researchers are briefly introduced in this section.

3.1 Fundamentals of Rough Sets

Research on the fundamentals of rough sets are developing very quickly in China. There are many fundamental topics studied by Chinese researchers in recent years that include algebra and information systems, information entropy, rough entropy, rough logics and rough algebras.

Guoyin Wang and his group found that the concepts of rough set theory in the algebra viewpoint and information viewpoint are not equivalent to each other[3, 41, 216–226]. A comparative study was done on the quantitative relationship between some basic concepts of rough set theory like attribute reduction, attribute significance and attribute core defined from these two viewpoints. It was found that the relationship between these conceptions from the two viewpoints is rather an inclusion instead of an equivalence due to the fact that rough set theory discussed from an information point-of-view restricts attributes and decision tables more strictly than it does when considered from the algebra point of view. The identity of the two viewpoints will only hold in consistent information decision systems. Dongyi Ye[278] found an error in the approach of Hu and Cercone for attribute core computation from the algebra viewpoint and corrected by introducing an improved discernibility matrix.

Duoqian Miao proposed a new method of knowledge representation based on information entropy[137, 138]. The relationship between entropy and the roughness of knowledge was analyzed. He compared the rough set methods with representative inductive learning algorithms[139]. A new method for the discretization of continuous attributes was proposed[140].

Jiye Liang, Deyu Li and Yuhua Qian established a relationship between inclusion degree and rough set data analysis[93, 140], introduced the concepts of information entropy, rough entropy, granulation measure, and knowledge granulation into information systems[93–97, 269]. Based on a granulation order, they developed the technique of positive approximation and converse approximation for dynamic knowledge reduction and rule generation from a decision table[98, 173–175].To evaluate the whole performance of a decision-rule set, three new measurements be introduced[174]. Deyu Li, Yee Leung and Bo Zhang studied algebraic structure of rough set for double universes, knowledge acquisition in incomplete information systems and knowledge reduction and decision rule optimization in inconsistent systems[71, 77, 78]. Kaishe Qu established the relationship between formal concept analysis and rough set theory. The notions of decision table and decision rule in rough set theory are introduced naturally into formal concept analysis as decision context and decision implication[183–185].

Wenxiu Zhang's group defined rough approximations by a constructive approach in fuzzy environments and in complex information systems[18, 72, 126–128, 256], the constructive approach is suitable and useful for practical applications of rough sets. They developed rough approximation operators based on different mathematical structures such as neighborhood systems, Boolean lattices, complete distributive lattices, coverings, random sets [19, 81, 82, 257]. Relationships between rough set theory and formal concept analysis was studied, in especial, rough set approximations in formal concept analysis were investigated in detail[124, 193, 194, 249].

Jianhua Dai's group proposed a uniform structure based on 3-valued Lukasiewicz algebra for rough algebras[22, 23], and the concept of rough 3-valued Lukasiewicz algebra. Rough sets are interpreted within the framework of BZ-lattices[24]. The relations between some rough algebras and BZ-lattices are established[25]. The rough set model was generalized based on molecular lattices. Logics for rough sets with rough double stone algebraic semantics and rough stone algebraic semantics are constructed[26–29]. Additionally, a minimal axiom group for rough sets based on quasi-ordering[30]was constructed by this group.

3.2 Knowledge Acquisition

Knowledge acquisition is one of the most important topics in the study of rough set theory. Reduction is one of the most important contributions of rough set theory to data mining. There are many Chinese groups working on knowledge acquisition based on rough set theory. Guoyin Wang's group developed several knowledge reduction algorithms and methods[4, 5, 17, 99, 181, 182, 227–230, 286, 287, 314, 315] using information entropy, discernibility matrix, etc. Based on Wang's new finding on the complexity analysis of quick sort for a two dimensions table, n*(m+logn) while not m*n*logn, some new efficient attribute reduction methods were developed[45–47]. Some other related issues of knowledge reduction based on rough set theory, such as attribution core computation[316, 321, 322], data pre-procession like data discretization[44, 317, 318], were also studied. Several efficient knowledge reduction algorithms based on entropy were developed by Duoqian Miao's group[141–143, 300]. Wenxiu Zhang's Group proposed many new concepts of attribute reduction and presented useful methods of knowledge reduction in various complex information tables and formal contexts. They introduced general approaches of attribute reduction in rough set theory and in concept lattice. At the same time, attribute features in information systems and in formal contexts were characterized[79, 80, 83, 129, 130, 210, 254, 258, 268, 309]. Houkuan Huang's group proved that the relative reduction in the view of information theory is equal to the μ-decision reduction and μ-reduction, the results based on Hu's improved discernibility matrix and function are general decision reductions, and the results based on Ye's improved discernibility matrix and function are relative reductions (also called Pawlak's reductions)[31–33]. Dongyi Ye's group[279–282] studied some interesting issues related to data inconsistency degree and the computation of all attribute reductions based on a modified matrix. A concise recursive relation between positive regions was proposed for developing faster reduction algorithms. A novel greedy maximum distribution reduction algorithm for inconsistent decision tables was proposed. Moreover, through establishing an equivalence relationship between the computation of the minimum attribute reduction and an unconstrained binary optimization problem, a PSO-based minimum reduction algorithm was developed. Some improved algorithms for attribute reduction and attribute value reduction were proposed[62] by Xuegang Hu's group. Keyun Qin and his research group discussed knowledge reduction approaches from logical point of view[177, 178]. They proved that the d-reduction (or distribute reduction, or approximate reduction) preserves the definite decision rules equivalent and under the distribution reduction or entropy reduction, all the decision rules derived from an information system and the corresponding reduced information system are logical equivalent. Keing Xie's group proposed Bit Granular Matrx(BGrM) to realize knowledge reduction including attribute reduction and attribute value reduction by BGrM operation[264].

Since human experts may play an important role in a machine learning process, their knowledge should be deliberately considered. Fan Min's group[117, 145, 146, 265] proposed M-reductions and M-relative reductions ensuring that an attribute subset M selected by human experts are always included in the reduction. Weighted reduction enables human experts to express their preferences in the process of reduction selection. The relationship between the new definitions and the traditional definitions are closely investigated. Instead of designing different reduction algorithms for inconsistent decision tables, they proposed a scheme converting the latter into consistent decision tables. Jue Wang and his group presented the concept of the information quantity of decision attributes. So a new reduction algorithm based on information quantity is developed, and they proved that the entropy of knowledge decreases monotonously as the granularity of information becomes smaller[123, 250]. Qinghua Hu's group proposed a new approach to attribute reduction of consistent and inconsistent covering decision systems with covering rough sets, and an information-preserving hybrid data reduction method based on fuzzy-rough techniques. And other some attribute reduction methods are presented by his group[20, 54–57].

In order to mine huge data sets, incremental knowledge acquisition is a topic in the field of knowledge discovery in database. Incremental knowledge reduction technologies were studied based on the simulation of human learning process[48, 64, 252, 323–325] in Guoyin Wang's group. They proposed incremental algorithms for attribute reduction based on modified discernibility matrix, developed a distributed model of incremental attribute reduction by decomposing values of decision attribute of positive region and boundary region in a non-tolerant decision table, a rough set and rule tree based incremental knowledge acquisition algorithm, and a new incremental learning algorithm based on variable precision rough set model, etc. An Incremental Updating Algorithm for Core Computing in Dominance-based Rough Set Model is proposed by Lin Shang's group[68].

Guoyin Wang and his group proposed a new understanding for data mining, that is, data mining is a process of knowledge transformation[231–233]. Based on this understanding, they designed a model of domain-oriented data-driven data mining (3DM)[231]. Some features of knowledge in data format, like uncertainty, were used to control a data mining process. Some valid data driven knowledge acquisition methods were developed[234–236, 253, 283–285, 319].

Ying Sai's group proposed a data analysis and data mining model for ordered information tables[189, 190]. Ordering of objects is a fundamental issue in human decision making and may play a significant role in the design of intelligent information systems. They generalized the notion of information tables to ordered information tables by adding order relations on attribute values. A data analysis method was thus proposed to describe the properties of ordered information tables. They defined the concepts of reduction and core etc. by analyzing the attribute dependency in ordered information tables. They formalized the problem of mining ordering rules, designed an ordered decision logic language (ODL-language), and presented a solution for mining ordering rules. Mining ordering rules based on ordered relations is a concrete example of application of generalizations of rough set model with non-equivalence relations[191, 277].

The classical rough set theory is based on complete information systems. That is, all values of each object are known. Unfortunately, there are always some missing values in real life information tables due to some reason. Almost all databases used for data mining contain imperfection, such as noise, unknown values or errors due to inaccurate measuring equipment, or some other limitations. Guoyin Wang and his group proposed an extended rough set model based on a new limited tolerance relation[237, 238, 288], and developed some knowledge reduction algorithms for extracting rules from incomplete information tables directly[21, 49, 65–67, 303]. Rule generation from incomplete information systems by GDT approach and a default rule extracting method were proposed by Lin Shang's group[192]. Jiye Liang defined combination entropy and combination granulation in incomplete information systems[175], and proposed an algorithm of knowledge reduction in incomplete information systems based on information entropy[96, 97]. Deyu Li introduced the concept of a maximal consistent block into rough set theory. It was used to construct an optimal depiction for elementary knowledge granulations and discernibility functions of knowledge reduction in incomplete information systems[71]. Deyu Li reviewed the relationship among the five kinds of reducts in an inconsistent decision information system and proposed the optimal methods for the five kinds of decision rules[78].

Uncertain information processing is a key issue of intelligent information processing. There are several different types of uncertainties in rough set based information processing[239–241]. Some methods for measuring the uncertainty of information systems were developed[51, 242, 267, 320]. New uncertain reasoning methods were developed also by Guoyin Wang's group. The positive region in rough set framework and Shannon conditional entropy are two traditional uncertainty measures, used usually as heuristic metrics in attribute reduction[243–245]. For the study of uncertainty measures related to rough set theory, Wenxiu Zhang and his group investigated the fuzziness, roughness, and entropy of crisp rough sets and fuzzy rough sets[43, 131, 132, 313]. They established the relationships between rough set theory and the Dempster-Shafer theory of evidence in various situations[259]. Jiye Liang and his group defined a new information entropy and compared it with Shannon entropy systematically[95, 96].

3.3 Granular Computing Based on Rough Sets

Rough sets are part of the foundations of granular computing. In the field of computational intelligence, granular computing is a new method for simulating human thinking and solving complicated problems. By effectively using levels of granularity, granular computing provides a systematic and natural way for analyzing, understanding, representing, and solving real world problems. With granular computing, one aims at structured thinking at the philosophical level, and structured problem solving at the practical level[34]. Guoyin Wang and his group developed some granular computing models and operators based on rough set theory[50, 52, 246]. Some data mining and knowledge acquisition algorithms were developed based on granular computing[6, 7, 11, 35, 36, 40, 53]. A tolerance granular space based information classification method was discussed by Zhongzhi Shi and his group[326–330]. Based on the theory of tolerance granular space and related knowledge of information classification, a tolerance granular space modeling algorithm TGM and a tolerance granular space based classification algorithm TGLC were developed. Simulation results show that the algorithms have higher classification rates and are more robust than other algorithms. The technique of granulation was also applied to image texture recognition. Duoqian Miao and his group studied granular representation, model and inference[74–76, 255]. Wenxiu Zhang's group developed some granular computing models and operators based on rough set theory and formal concept analysis[12–14, 125]. Some basic problems and research issues on granular computing and applications were discussed by Keming Xie[264, 270]. Yongli Li's group[88] found that a granular information world could be generated by different measures used to view the world. Qing Liu defined the granules of decision rules based on rough set[108, 118, 119]. Granular language, syntax, semantics, and operation rules of granular statements were also defined based on rough set. A reasoning model based on these concepts was proposed and used in logic reasoning.

3.4 Extended Rough Set Models

To process some problems more efficiently, many extended rough set models have been proposed, *e.g.*, fuzzy rough set model, rough fuzzy set model, covering rough set model, and variable precision rough set model. The relationship between fuzzy rough set models and fuzzy topologies on a finite universe is discussed by Keyun Qin's group[179]. They proved that there exist a one-to-one correspondence between the set of all reflexive and transitive fuzzy relations and the set of all fuzzy topologies which satisfy axiom. The extended covering rough set models were discussed[180], and based on the notion of neighborhood, five pairs of dual covering approximation operators were defined with their properties.

Rough set, fuzzy set, vague set, and Bayes theory are all useful technologies for uncertain information processing. Guoyin Wang's group studied their relationship and tried to integrate them. A mapping relationship between vague set and fuzzy set is set up[122, 266]. Knowledge acquisition from vague spaces[38, 39] and weighted naive Bayes classification based on rough set[37] were studied.

Wenxiu Zhang, Weizhi Wu and their group[133, 260, 261, 310] proposed a generalized fuzzy rough set model and an axiomatic method to extend rough set theory. In this method, they characterized relation-based rough approximation operators using the axiomatic approach and solved the problem of finding assumptions permitting a given set-theoretic operator to represent upper and lower approximations derived from a special binary relation. They presented different sets of axioms to characterize various classes of crisp and fuzzy approximation operators[84, 85, 134, 262]. Guilong Liu[110] defined a pair of dual approximation operators for rough fuzzy set and proved that an axiomatic system must be satisfied by the operators. The rough fuzzy set, fuzzy rough set, and covering rough set were discussed by Duoqian Miao's group also[144, 301]. Some new concepts of knowledge reduction based on variable precision rough set theory such as upper distribution reduction and lower distribution reduction are introduced by Jusheng Mi[135]. Bing Huang and Xianzhong Zhou[63] proposed a new rough set model based on tolerance degree in incomplete information systems. Yongli Li defined a uniform expression of rough set and fuzzy rough set using the body-shadow relation model of Pan-systems theory[89]. Kaiquan Shi and his group proposed a singular rough set(S-rough set) model, and studied its two kinds of mathematics structures: single-direction S-Rough set and two-direction S-Rough set[196–199]. Fuzzy-rough set model has been researched by Qinghua Hu's group[58–61]

3.5 Rough Logic

The rough logic defined on neighbor valued decision tables and its truth values of the formulas were discussed by Qing Liu's group[109, 116, 120]. Using the rough equality of lower and upper approximations with respect to the domain of formulas, they defined the rough equality between two logical formulas, established a rough logical reasoning system with rough equality relation and developed a temporal rough logic system. A neighborhood logic for neighborhood valued information tables, which is the logic used as operators with neighborhood topological interior and neighborhood topological closure, was discussed. The model of the proposed logic language was studied also[121].

3.6 Applications of Rough Sets

As a valid mathematical tool to deal with imprecise, uncertain, and vague information, rough set theory has been applied in such fields as decision analysis, machine learning, dada mining, and pattern recognition successfully. Guoyin Wang's group used rough set theory in many real life applications, like emotion recognition[271, 272], video retrieval[289, 290], network intrusion detection[90, 247, 248], junk email filtering[91], etc. Huanglin Zeng's group[293–297] proposed a fast learning algorithm for solving linear equations and related problems and a new selection and reduction method of system features in pattern recognition based on rough set. They developed a model of intelligent information processing. Houkuan Huang's group investigated conflict solving with rough sets and applied rough set theory into diagnosis systems[87, 311, 312]. Ying Sai's group[111–114] proposed two static rough communication models using rough set theory, and discussed their properties and applications. Combining the fuzzy set theory, they proposed two fuzzy rough communication models. A dynamic rough communication model was proposed using the two-direction s-rough sets theory. Zhaocong Wu developed a rough logical learning mechanism of RBFNN (Radial Basis Function Neural Network) based on rough set and used it in classification of remotely-sensed multi-spectral images[263]. Keyun Qin and his group did a lot of work on decision making, traffic and transportation controlling, selection of chemical solvents, evaluation of project management based on rough set[70, 86]. Duoqian Miao's group developed an automatic recognition system of bills based on principal curves, and applied the fuzzy rough set in bioinformatics[251, 302]. Liping An's group[8–10] proposed several methods for some real-world problems in banking, such as CRM in banking, risk and profit forecasting and controlling, and credit evaluating and decision making based on rough set theory. Qing Liu developed a medicine decision information system based on rough set[116].

The rough set theory has also been used in a lot of data mining areas, such as information retrieval, medical data analysis, aircraft pilot performance evaluation, image processing, and voice recognition, etc.

4 Summary

The rough set theory seems to be of fundamental importance in the fields of artificial intelligence and cognitive science, especially in the areas of machine learning, knowledge acquisition, decision analysis, and pattern recognition. It has been applied in many real-life applications like medical data analysis, finance, banking, voice recognition, image processing, etc. Developed independently from relational databases, rough set is another theory for processing relational databases (Information tables could be taken as relational databases). It could be used to analyze imprecise, uncertain or incomplete information in data. It is a new set theory complimentary to fuzzy set theory.

Many different problems can be addressed by rough set theory. Since it was proposed by Professor Z. Pawlak, it has attracted attentions of many researchers from different fields. In China, there is a rough set community of about 100 professors and hundreds of students. Although there are already many important theoretical contributions including fundamental of rough set, knowledge acquisition, granular computing, extended rough set model, and some applications about rough set in China, many essential research problems are still to be further studied. Some of them are listed as follows.

(1)General rough set model.
(2)Granular computing based on extended rough set model.
(3)Efficient reduction methods for a huge database.
(4)Decomposition of large information tables.
(5)Efficient and widely assessable software.
(6)Rough logic and rough control.
(7)Comparison with many other approaches to data analysis, like neural networks, genetic algorithms and others.

In this paper, we introduced the research on rough set theory and applications ongoing at universities and laboratories in China. We hope it would be helpful to establish better understanding, communication, and cooperation between Chinese researchers and overseas researchers.

Acknowledgements: This paper is supported by National Natural Science Foundation of P. R. China under Grants No. 60573068 and 60773113, Program for New Century Excellent Talents in University (NCET), Natural Science Foundation of Chongqing under Grants No. 2005BA2003 and 2008BA2017. Thanks go to Professors Andrzej Skowron and James F. Peters for their valuable suggestions and discussions.

References

1. Alpigini, J.J., Peters, J.F., Skowron, A., Zhong, N. (eds.): RSCTC 2002. LNCS (LNAI), vol. 2475. Springer, Heidelberg (2002)
2. An, A.J., Stefanowski, J., Ramanna, S., Butz, C.J., Pedrycz, W., Wang, G.Y. (eds.): RSFDGrC 2007. LNCS (LNAI), vol. 4482. Springer, Heidelberg (2007)
3. An, J.J., Chen, L., Wang, G.Y., Wu, Y.: Quantitative analysis of the difference between the algebra view and information view of rough set theory. In: Proceedings of WI/IAT 2003 Workshop on Applications, Products and Services of Web-based Support Systems, Halifax, Canada, pp. 159–164 (2003)
4. An, J.J., Wang, G.Y., Wu, Y.: A learning algorithm for multiple rule trees. In: The Second International Workshop on Web-based Support Systems, WSS 2004, Beijing, pp. 13–19 (2004)

5. An, J.J., Wang, G.Y., Wu, Y.: A learning algorithm based on multiple rule trees. Computer Science 31, 175–177 (2004)
6. An, J.J., Wang, G.Y., Wu, Y., Gan, Q.: A rule generation algorithm based on granular computing. In: 2005 IEEE International Conference on Granular Computing, GrC, pp. 102–107 (2005)
7. An, J.J., Gan, Q., Wang, G.Y., Wu, Y.: An algorithm for rule and knowledge acquisition based on granular computing. Computer Science 32, 137–140 (2005)
8. An, L.P.: Conflict analysis and negotiation model based on rough set theory. Journal of University of Science and Technology Beijing 24, 91–95 (2002)
9. An, L.P.: Determination of coalitions and strategy selection in conflict analysis. Journal of Tianjin University Science and Technology 35, 15–18 (2002)
10. An, L.P.: A rough neural expert system for medical diagnosis. In: Proceedings of 2005 International Conference on Services Systems and Services Management, Chongqing, China, vol. 2, pp. 1130–1135 (2005)
11. An, Q.S., Shen, J.Y., Wang, G.Y.: A clustering method based on information granularity and rough sets. Pattern Recognition and Artificial Intelligence 16, 412–417 (2003)
12. An, Q.S., Shen, J.Y.: Granular computing on functional dependencies for information system. In: Tsumoto, S., Słowiński, R., Komorowski, J., Grzymała-Busse, J.W. (eds.) RSCTC 2004. LNCS (LNAI), vol. 3066, pp. 186–191. Springer, Heidelberg (2004)
13. An, Q.S., Zhang, W.X.: The Measures relationships study of three soft rules based on granular computing. In: Wang, G.-Y., Peters, J.F., Skowron, A., Yao, Y. (eds.) RSKT 2006. LNCS (LNAI), vol. 4062, pp. 371–376. Springer, Heidelberg (2006)
14. An, Q.S., Zhang, Y.S., Zhang, W.X.: The study of rough relational database based on granular computing. In: IEEE International Conference on Granular Computing (GrC 2005), pp. 108–111 (2005)
15. Bazan, J.G., Kruczek, P., Bazan, S.S., Skowron, A., Pietrzyk, J.J.: Automatic Planning of Treatment of Infants with Respiratory Failure Through Rough Set Modeling. In: Greco, S., Hata, Y., Hirano, S., Inuiguchi, M., Miyamoto, S., Nguyen, H.S., Słowiński, R. (eds.) RSCTC 2006. LNCS (LNAI), vol. 4259, pp. 418–427. Springer, Heidelberg (2006)
16. Bazan, J.G., Kruczek, P., Bazan, S.S., Skowron, A., Pietrzyk, J.J.: Rough Set Approach to Behavioral Pattern Identification. Fundamenta Informaticae 75, 27–47 (2007)
17. Chang, L.Y., Wang, G.Y., Wu, Y.: An approach for attribute reduction and rule generation based on rough set theory. Journal of Software 10, 1206–1211 (1999)
18. Chen, D.G., Zhang, W.X.: The lower and upper approximations of fuzzy sets in fuzzy group. In: Pal, N.R., Sugeno, M. (eds.) AFSS 2002. LNCS (LNAI), vol. 2275, pp. 502–508. Springer, Heidelberg (2002)
19. Chen, D.G., Zhang, W.X., Daniel, D., Tsang, E.C.C.: Rough approximations on a complete completely distributive lattice with applications to generalized rough sets. Information Sciences 176, 1829–1848 (2006)
20. Chen, D.G., Wang, C.Z., Hu, Q.H.: A new approach to attribute reduction of consistent and inconsistent covering decision systems with covering rough sets. Information Sciences 177, 3500–3518 (2007)
21. Dai, J., Hu, F., Wang, G.Y.: Research and application of text classification based on incomplete information system. Journal of Chongqing University of Posts and Telecommunications 18, 397–401 (2006)

22. Dai, J.H.: Rough algebras and 3-valued Lukasiewicz algebras. Journal of Computers 30, 161–167 (2007)
23. Dai, J.H., Pan, Y.H.: On rough algebras. Journal of Software 16, 1197–1204 (2005)
24. Dai, J.H.: Rough sets and Brouwer-Zadeh lattices. In: Wang, G.-Y., Peters, J.F., Skowron, A., Yao, Y. (eds.) RSKT 2006. LNCS (LNAI), vol. 4062, pp. 200–207. Springer, Heidelberg (2006)
25. Dai, J.H., Lv, H.F., Chen, W.D.: Two kinds of rough algebras and Brouwer-Zadeh lattices. In: Greco, S., Hata, Y., Hirano, S., Inuiguchi, M., Miyamoto, S., Nguyen, H.S., Słowiński, R. (eds.) RSCTC 2006. LNCS (LNAI), vol. 4259, pp. 99–106. Springer, Heidelberg (2006)
26. Dai, J.H.: Logic for rough sets with rough double stone algebraic semantics. In: Ślęzak, D., Wang, G., Szczuka, M., Düntsch, I., Yao, Y. (eds.) RSFDGrC 2005. LNCS (LNAI), vol. 3641, pp. 141–148. Springer, Heidelberg (2005)
27. Dai, J.H., Chen, W.D., Pan, Y.H.: Sequent caculus system for rough sets based on rough stone algebras. In: Proc. of IEEE International Conference on Granular Computing, pp. 423–426. IEEE Press, New Jersy (2005)
28. Dai, J.H.: Generalization of rough set theory using molecular lattices. Journal of Computers 27, 1436–1440 (2004)
29. Dai, J.H.: Structure of rough approximations based on molecular lattices. In: Tsumoto, S., Słowiński, R., Komorowski, J., Grzymała-Busse, J.W. (eds.) RSCTC 2004. LNCS (LNAI), vol. 3066, pp. 69–77. Springer, Heidelberg (2004)
30. Dai, J.H., Pan, Y.H.: Acqusition algorithm of decision rules based on classification consistency ratio. Control and Decision 19, 1086–1090 (2004)
31. Deng, D.Y., Huang, H.K., Li, X.J.: Comparison of various types of reductions in inconsistent systems. Acta Electronica Sinica 35, 252–255 (2007)
32. Deng, D.Y., Huang, H.K.: A new discernibility matrix and function. In: Wang, G.-Y., Peters, J.F., Skowron, A., Yao, Y. (eds.) RSKT 2006. LNCS (LNAI), vol. 4062, pp. 114–121. Springer, Heidelberg (2006)
33. Deng, D.Y.: Research on data reduction based on rough sets and extension of rough set models. A Dissertation of Beijing Jiaotong University, pp. 162–165 (2007)
34. Deng, W., Wang, G.Y., Wu, Y.: A survey of granular computing. Computer Science 31, 178–181 (2004)
35. Deng, W., Wang, G.Y., Wu, Y.: A rule acquisition algorithm of granular computing based on approximate partition. In: Proceedings of 2005 National Conference on Artificial Intelligence (CAAI-11), Wuhan, pp. 690–695 (2005)
36. Deng, W., Wang, G.Y., Wu, Y.: An improved algorithm of evolutionary computation based on granular computing. Computer Science 31, 209–211, 214 (2004)
37. Deng, W.B., Wang, G.Y., Wang, Y.: Weighted naive bayes classification algorithm based on rough set. Computer Science 34, 204–206, 219 (2007)
38. Feng, L., Wang, G.Y.: Knowledge acquisition in vague information systems. In: Proc. of 2006 Int. Conf. on Artificial Intelligence, Beijing, pp. 533–537 (2006)
39. Feng, L., Wang, G.Y., Liu, Y., Zhu, Z.G.: Knowledge acquisition in vague objective information systems. In: Wang, L., Jiao, L., Shi, G., Li, X., Liu, J. (eds.) FSKD 2006. LNCS (LNAI), vol. 4223, pp. 344–354. Springer, Heidelberg (2006)
40. Gan, Q., Wang, G.Y., Hu, J.: A self-learning model based on granular computing. In: 2006 IEEE International Conference on Granular Computing, IEEE GrC 2006, Atlanta, Georgia, USA, pp. 530–533 (2006)
41. Gong, X., Wang, G.Y.: Illustrating the algebra view of rough ret theory with information view. Computer Science 33, 151–154 (2006)

42. Greco, S., Hata, Y., Hirano, S., Inuiguchi, M., Miyamoto, S., Nguyen, H.S., Slowinski, R. (eds.): Rough Sets and Current Trends in Computing. In: Greco, S., Hata, Y., Hirano, S., Inuiguchi, M., Miyamoto, S., Nguyen, H.S., Słowiński, R. (eds.) RSCTC 2006. LNCS (LNAI), vol. 4259. Springer, Heidelberg (2006)
43. Guo, Z.X., Mi, J.S.: Uncertainty measures in rough fuzzy sets. Fuzzy Systems and Mathematics 19, 135–140 (2005)
44. Hou, L.J., Wang, G.Y., Wu, Y., Nie, N.: Discretization in rough set theory. Computer Science 27, 89–94 (2000)
45. Hu, F., Wang, G.Y.: Analysis of the complexity of quick sort for two dimension table. Chinese Journal of Computers 30, 693–698 (2007)
46. Hu, F., Wang, G.Y.: Quick reduction algorithm based on attribute order. Chinese Journal of Computers 30, 1429–1435 (2007)
47. Hu, F., Wang, G.Y., Xia, Y.: Attribute core computation based on divide and conquer method. In: Kryszkiewiczet, M., et al. (eds.) RSEISP 2007. LNCS (LNAI), vol. 4585, pp. 310–319. Springer, Heidelberg (2007)
48. Hu, F., Wang, G.Y., Huang, H., Wu, Y.: Incremental attribute reduction based on elementary sets. In: Ślęzak, D., Wang, G., Szczuka, M., Düntsch, I., Yao, Y. (eds.) RSFDGrC 2005. LNCS (LNAI), vol. 3641, pp. 185–193. Springer, Heidelberg (2005)
49. Hu, F., Huang, H., Wang, G.Y., Wu, Y.: Granular computing in incomplete information systems. Mini-micro Systems 26, 1335–1339 (2005)
50. Hu, F., Huang, H., Wang, G.Y., Wu, Y.: A granular computing model based on tolerance relation. Computer Science 31, 48–50 (2004)
51. Hu, J., Wang, G.Y., Zhang, Q.H.: Uncertainty measure of covering generated rough set. In: 2006 IEEE/WIC/ACM International Conference on Web Intelligence and Intelligent Agent Technology (WI-IAT 2006 Workshops) (WI-IATW 2006), pp. 498–504 (2006)
52. Hu, J., Wang, G.Y., Zhang, Q.H., Liu, X.Q.: Approximation operators based on general relation. Computer Science 33, 60–62, 69 (2006)
53. Hu, J., Wang, G.Y., Zhang, Q.Z., Liu, X.Q.: Attribute reduction based on granular computing. In: Greco, S., et al. (eds.) RSCTC 2006. LNCS (LNAI), vol. 4259, pp. 458–466. Springer, Heidelberg (2006)
54. Hu, Q.H., Yu, D.R., Xie, Z.X.: Information-preserving hybrid data reduction based on fuzzy-rough techniques. Pattern recognition letters 27, 414–423 (2006)
55. Hu, Q.H., Zhao, H., Xie, Z.X., Yu, D.R.: Consistency based attribute reduction. In: Zhou, Z.-H., Li, H., Yang, Q. (eds.) PAKDD 2007. LNCS (LNAI), vol. 4426, pp. 96–107. Springer, Heidelberg (2007)
56. Hu, Q.H., Wang, M.Y., Yu, D.R.: Construct rough decision forests based on sequentially data reduction. In: 2006 IEEE Conference on machine learning and cybernetics, Dalian, China, August 2006, pp. 13–16 (2006)
57. Hu, Q.H., Li, X.D., Yu, D.R.: Analysis on classification performance of rough set based reducts. In: Yang, Q., Webb, G. (eds.) PRICAI 2006. LNCS (LNAI), vol. 4099, pp. 423–433. Springer, Heidelberg (2006)
58. Hu, Q.H., Xie, Z.X., Yu, D.R.: Hybrid attribute reduction based on a novel fuzzy-rough model and information granulation. Pattern Recognition, 3509–3521 (2007)
59. Hu, Q.H., Xie, Z.X., Yu, D.R.: Weighting algorithm for text classification based on rough set approach. Journal of the China society for scientific and technical information 24, 59–63 (2005)

60. Hu, Q.H., Yu, D.R., Xie, Z.X.: Fuzzy probabilistic approximation spaces and their information measures. IEEE transactions on fuzzy systems 14, 191–201 (2006)
61. Hu, Q.H., Yu, D.R., Wang, M.Y.: Constructing rough decision forests. In: Ślęzak, D., Yao, J., Peters, J.F., Ziarko, W., Hu, X. (eds.) RSFDGrC 2005. LNCS (LNAI), vol. 3642, pp. 147–156. Springer, Heidelberg (2005)
62. Hu, X.G., Shi, J.H., Wu, X.D.: A new algorithm for attribute reduction in decision tables. In: An, A., Stefanowski, J., Ramanna, S., Butz, C.J., Pedrycz, W., Wang, G. (eds.) RSFDGrC 2007. LNCS (LNAI), vol. 4482, pp. 37–46. Springer, Heidelberg (2007)
63. Huang, B., Zhou, X.Z., Hu, Z.J.: A variable precision rough set model in incomplete information systems. Journal of North University of China 27, 283–287 (2006)
64. Huang, H., Wang, G.Y., Hu, F., Wu, Y.: An incremental attribute reduction algorithm. Computer Science 32, 13–15 (2005)
65. Huang, H., Wang, G.Y., Wu, Y.: Direct reduction method for incomplete information system. Mini-micro Systems 26, 1761–1765 (2005)
66. Huang, H., Wang, G.Y., Wu, Y.: A direct approach for incomplete information systems, in Data Mining and Knowledge Discovery: Theory, Tools, and Technology VI. In: Dasarathy, B.V. (ed.) Proceedings of SPIE, vol. 5433, pp. 114–121 (2004)
67. Huang, H., Wang, G.Y., Wu, Y.: A direct approach for incomplete information systems. In: The 13th National Conference on Neural Networks, pp. 349–354 (2003)
68. Jia, X.Y., Shang, L.: An Incremental Updating Algorithm for Core Computing in Dominance-based Rough Set Model. In: An, A., Stefanowski, J., Ramanna, S., Butz, C.J., Pedrycz, W., Wang, G. (eds.) RSFDGrC 2007. LNCS (LNAI), vol. 4482, pp. 403–410. Springer, Heidelberg (2007)
69. Jiang, C.Z.: Rough set multiple attribute decision making theories and methods. Southwest Jiaotong University Press, Sichuan (2007)
70. Jiang, C.Z., et al.: The research of rough evaluation system to the project manager's quality. Chinese Journal Of Management Science 14, 207–210 (2006)
71. Leung, Y., Li, D.Y.: Maximal consistent block technique for rule acquisition in incomplete information systems. Information Sciences: An International Journal 153, 85–106 (2003)
72. Leung, Y., Wu, W.Z., Zhang, W.X.: Knowledge acquisition in incomplete information systems: a rough set approach. European Journal of Operational Research 168, 164–180 (2006)
73. Li, D.G.: Study of Information granule-Computing theory-Model and Application. Shanxi Science & Technology Press, Shanxi (2006)
74. Li, D.G.: Research on granular computing models and their applications in intelligence information Processing, [Ph.D Dissertations], Tongji University (2006)
75. Li, D.G., Miao, D.Q., Zhang, D.X., Zhang, H.Y.: An overview of granular computing. Computer Science 32, 1–12 (2005)
76. Li, D.G., Miao, D.Q., Du, W.L.: Application of granular computing to artificial neural network. Journal of Tongji University (Natural Science) 34, 960–964 (2006)
77. Li, D.Y., Ma, Y.C.: Invariant characters of information systems under some homomorphisms. Information Sciences: An International Journal 129, 211–220 (2000)

78. Li, D.Y., Zhang, B., Leung, Y.: On knowledge reduction in inconsistent decision information systems. International Journal of Uncertainty Fuzziness and Knowledge-Based Systems 12, 651–672 (2004)
79. Li, H.R., Zhang, W.X.: Applying indiscernibility attribute sets to knowledge reduction. In: Zhang, S., Jarvis, R. (eds.) AI 2005. LNCS (LNAI), vol. 3809, pp. 816–821. Springer, Heidelberg (2005)
80. Li, H.R., Zhang, W.X., Xu, P., Wang, H.: Rough set attribute reduction of decision systems. In: Wang, G.-Y., Peters, J.F., Skowron, A., Yao, Y. (eds.) RSKT 2006. LNCS (LNAI), vol. 4062, pp. 135–140. Springer, Heidelberg (2006)
81. Li, T.J.: Rough approximation operators in covering approximation space. In: Greco, S., Hata, Y., Hirano, S., Inuiguchi, M., Miyamoto, S., Nguyen, H.S., Słowiński, R. (eds.) RSCTC 2006. LNCS (LNAI), vol. 4259, pp. 174–182. Springer, Heidelberg (2006)
82. Li, T.J., Ma, J.M.: Fuzzy approximation operators based on coverings. In: An, A., Stefanowski, J., Ramanna, S., Butz, C.J., Pedrycz, W., Wang, G. (eds.) RSFDGrC 2007. LNCS (LNAI), vol. 4482, pp. 55–62. Springer, Heidelberg (2007)
83. Li, T.J., Zhang, W.X., Ma, J.M.: Attribute reductions and attribute features of formal contexts based on a type of rough sets. Computer Science 33, 178–180 (2006)
84. Li, T.J.: On axiomatic characterization of approximation operators based on atomic Boolean algebras. In: Wang, G.-Y., Peters, J.F., Skowron, A., Yao, Y. (eds.) RSKT 2006. LNCS (LNAI), vol. 4062, pp. 129–134. Springer, Heidelberg (2006)
85. Li, T.J., Zhang, W.X., Yang, X.P.: Axiomatic characterization of generalized fuzzy approximation operators. Fuzzy Systems and Mathematics 20, 122–128 (2006)
86. Li, T.R., Ruan, D., wets, Greet, Song, J., Xu, Y.: A rough sets based characteristic relation approach for dynamic attribute generalization in data mining. Knowledge-based Systems 20, 485–494 (2007)
87. Li, X.J., Tian, S.F., Deng, D.Y., Chen, J.N.: A method of multi-agent system conflict analysis based on rough set theory. In: International Conference on Granular Computing (IEEE), vol. 2, pp. 180–184 (2005)
88. Li, Y.L.: Boundary of sets in pansystems information system Kybernetes. The International Journal of Systems and Cybernetics 26, 596–601 (1997)
89. Li, Y.L., Liu, Y., Wang, X.Y., Lin, H.: An initial comparison of fuzzy sets and rough sets from the view of pan-systems theory. In: Proceeding of 2005 IEEE International Conference on Granular Computing, vol. 2, pp. 520–525 (2005)
90. Li, Z., Wu, Y., Wang, G., Hai, Y., He, Y.: A new framework for intrusion detection based on rough set theory, in Data Mining and Knowledge Discovery: Theory, Tools, and Technology VI. In: Dasarathy, B.V. (ed.) Proceedings of SPIE, vol. 5433, pp. 122–130 (2004)
91. Li, Z.J., Wang, G.Y., Wu, Y.: An e-mail classification system based on rough set. Computer Science 31, 58–60, 66 (2004)
92. Liang, J.Y., Li, D.Y.: Uncertainty and knowledge acquisition in information systems. Science Press, Beijing, China (2005)
93. Liang, J.Y., Shi, Z.Z., Li, D.Y.: Applications of inclusion degree in rough set theory. International Journal of Computational Cognition 1, 67–78 (2003)
94. Liang, J.Y., Shi, Z.Z., Li, D.Y., Wierman, W.J.: The information entropy, rough entropy and knowledge granulation in incomplete information systems. International Journal of General Systems 35, 641–654 (2006)

95. Liang, J.Y., Shi, Z.Z.: The information entropy, rough entropy and knowledge granulation in rough set theory. International Journal of Uncertainty, Fuzziness and Knowledge-Based Systems 12, 37–46 (2004)
96. Liang, J.Y., Chin, K.S., Dang, C.Y., Yam, Richard, C.M.: A new method for measuring uncertainty and fuzziness in rough set theory. International Journal of General Systems 31, 331–342 (2002)
97. Liang, J.Y., Xu, Z.B.: The algorithm on knowledge reduction in incomplete information systems. International Journal of Uncertainty, Fuzziness and Knowledge-Based Systems 24, 95–103 (2002)
98. Liang, J.Y., Qian, Y.H., Chu, C.Y., Li, D.Y., Wang, J.H.: Rough set approximation based on dynamic granulation. In: Ślęzak, D., Wang, G., Szczuka, M., Düntsch, I., Yao, Y. (eds.) RSFDGrC 2005. LNCS (LNAI), vol. 3641, pp. 701–708. Springer, Heidelberg (2005)
99. Liao, L., Wang, G.Y., Nie, N., Wu, Y.: A knowledge discovery model and its quantification and implementation based on rough set. Computer Science 28, 27–30 (2001)
100. Lin, T.Y.: Granular computing: structures, representations, applications and future directions. In: Wang, G., Liu, Q., Yao, Y., Skowron, A. (eds.) RSFDGrC 2003. LNCS (LNAI), vol. 2639, pp. 16–24. Springer, Heidelberg (2003)
101. Lin, T.Y.: From Rough Sets to Soft Computing: Introduction. Information Sciences 104, 1–2 (1998)
102. Lin, T.Y.: Neighborhood systems and relational database. In: Proceedings of CSC 1988, pp. 725–726 (1988)
103. Lin, T.Y.: Granular Computing on binary relations I: Data Mining and Neighborhood Systems. In: Skowron, A., Polkowski, L. (eds.) II: Rough Set Representations and Belief Functions, Rough Sets In Knowledge Discovery, pp. 107–140. Physica-Verlag (1998)
104. Lin, T.Y.: Data mining: granular computing approach. In: Zhong, N., Zhou, L. (eds.) PAKDD 1999. LNCS (LNAI), vol. 1574, pp. 24–33. Springer, Heidelberg (1999)
105. Lin, T.Y.: Granular computing: Fuzzy logic and rough sets. In: Zadeh, Kacprzyk (eds.) Computing with Words in Information/Intelligent Systems, pp. 183–200. Physica-Verlag (1999)
106. Lin, T.Y.: Granular Computing Rough Set Perspective. IEEE connections, The newsletter of the IEEE Computational Intelligence Society 2, 1543–4281 (2005)
107. Lin, T.Y.: Granular Computing: A Problem Solving Paradigm. In: The Proceedings of the 2005 IEEE International Conference on Fuzzy Systems, Atlantis Casino Resort Reno, Nevada, USA, pp. 2132–2137 (2005)
108. Lin, T.Y., Liu, Q.: First Order Rough Logic 1: Approximate Reasoning Via Rough Sets. Fundamenta Informaticae 27, 137–154 (1996)
109. Lin, T.Y., Liu, Q.: First order rough logic I: Approximate reasoning via rough sets. Fundamenta Informatica 27, 137–154 (1996)
110. Liu, G.L.: The axiom systems of rough fuzzy sets on fuzzy approximation spaces. Chinese Journal of Computers 27, 1189–1191 (2004)
111. Liu, J.Q., Shi, K.Q.: Information rough communication based on rough sets. Journal of Systems engineering and Electronics 29, 437–442 (2007)
112. Liu, J.Q.: A new fuzzy rough communication model and its application. In: The 3rd International Conference on Natural Computational and the 4th International Conference on Fuzzy Systems and Knowledge Discovery, vol. 1, pp. 62–66 (2007)

113. Liu, J.Q.: A fuzzy rough communication model and its properties. In: IEEE proceedings of the Sixth International Conference on Machine Learning and Cybernetic, vol. 7, pp. 3699–3703 (2007)
114. Liu, J.Q., Shi, K.Q.: Dynamic rough communication. An International Journal Advances in Systems Sciences and Applications (accepted, 2007)
115. Liu, Q., Huang, Z.H.: Rough theory: present state and perspective. Computer Science 24, 1–5 (1997)
116. Liu, Q.: Rough Set and Rough Reasoning. Science Press, Beijing (2003)
117. Liu, Q., Min, F., Cai, H., Yang, G.: Converting decision tables. In: 11th World Congress of international Fuzzy Systems Association, pp. 1391–1396 (2005)
118. Liu, Q.: Operator rough logic and its resolution principle. Chinese Journal of Computers 21, 435–476 (1998)
119. Liu, Q., Liu, S.H., Zheng, F.: Rough logic and its applications in data mining. Journal of Software 12, 415–419 (2001)
120. Liu, Q., Liu, S.H., Zheng, F.: Rough logic and its applications in data reduction. Journal of Software 12, 415–419 (2001)
121. Liu, Q.: Neighborhood logic and its data reasoning on neighborhood-valued information table. Chinese Journal of Computers 24, 405–410 (2001)
122. Liu, Y., Wang, G.Y., Feng, L.: A general model for transforming vague sets into fuzzy sets. In: Wang, G.-Y., Peters, J.F., Skowron, A., Yao, Y. (eds.) RSKT 2006. LNCS (LNAI), vol. 4062, pp. 341–348. Springer, Heidelberg (2006)
123. Liu, Z.H., Liu, S.Y., Wang, J.: An attribute reduction algorithm based on the information quantity. Journal of Xi'an University 30, 835–838 (2003)
124. Ma, J.M., Zhang, W.X., Wang, X.: Dependence space of concept lattices based on rough set. In: Proceedings of the 2006 IEEE International Conference on Granular Computing, GrC 2006, pp. 200–204 (2006)
125. Ma, J.M., Zhang, W.X., Li, T.J.: A covering model of granular computing. In: The Fourth International Conference on Machine Learning and Cybernetics, Guangzhou, pp. 1625–1630 (2005)
126. Mi, J.S., Zhang, W.X.: Indirect learning based on rough set theory. Computer Science 29, 96–97, 104 (2002)
127. Mi, J.S., Zhang, W.X.: Composition of general fuzzy approximation spaces. In: Pal, N.R., Sugeno, M. (eds.) AFSS 2002. LNCS (LNAI), vol. 2275, pp. 497–501. Springer, Heidelberg (2002)
128. Mi, J.S., Zhang, W.X.: An axiomatic characterization of a fuzzy generalization of rough sets. Information Sciences 160, 235–249 (2004)
129. Mi, J.S., Wu, W.Z., Zhang, W.X.: Knowledge reductions within the variable precision rough sets model. Information Sciences 159, 255–272 (2004)
130. Mi, J.S., Wu, W.Z., Zhang, W.X.: Approaches to approximation reducts in inconsistent decision table. In: Wang, G., Liu, Q., Yao, Y., Skowron, A. (eds.) RSFDGrC 2003. LNCS (LNAI), vol. 2639, pp. 283–286. Springer, Heidelberg (2003)
131. Mi, J.S., Guo, Z.X., Feng, T., Zhao, H.Y.: Uncertainty in generalized fuzzy rough sets. In: Proceedings of IEEE International Conference on Granular Computing, July 2005, pp. 213–216 (2005)
132. Mi, J.S., Leung, Y., Wu, W.Z.: An uncertainty measure in partition-based fuzzy rough sets. International Journal of General Systems 34, 77–90 (2005)
133. Mi, J.S., Wu, W.Z., Zhang, W.X.: Constructive and axiomatic approaches for the study of the theory of rough sets. Pattern Recognition and Artificial Intelligence 15, 280–284 (2002)

134. Mi, J.S., Zhang, W.X.: An axiomatic characterization of a fuzzy generalization of rough sets. Information Sciences 160, 235–249 (2004)
135. Mi, J.S., Wu, W.Z., Zhang, W.X.: Knowledge reducts based on variable precision rough set theory. Systems Engineering Theory & Practice 24, 76–82 (2004)
136. Miao, D.Q., Wang, G.Y., Liu, Q., Lin, T.Y., Yao, Y.Y.: Granular Computing: Past, Present and Future Prospects. Science Press, Beijing (2007)
137. Miao, D.Q., Wang, J.: On the relationships between information entropy and roughness of knowledge in rough set theory. Pattern Recognition and Artificial Intelligence 11, 34–40 (1998)
138. Miao, D.Q., Wang, J.: An information representation of the concepts and operations in rough set theory. Journal of Software 10, 113–116 (1999)
139. Miao, D.Q., Hou, L.S.: A comparison of rough set methods and representative inductive learning algorithms. Fundamenta informaticae 59, 203–219 (2004)
140. Miao, D.Q., Wang, J.: A new method of discretization of continuous attributes in rough sets. In: Proceedings of Chinese Intelligence Automatization Academic Conference, Shanghai, pp. 838–843 (1998)
141. Miao, D.Q., Hou, L.S.: A heuristic algorithm for reduction of knowledge based on discernibility matrix. In: International Conference on Intelligent Information Technology (ICIIT 2002), Beijing, China, September 22-25, pp. 276–279 (2002)
142. Miao, D.Q., Wang, J.: Information-based algorithm for reduction of knowledge. In: Proceedings of the 1997 IEEE International Conference on Intelligent Processing Systems, ICIPS 1997, Beijing, China, October 28-31, vol. 2, pp. 1155–1158 (1997)
143. Miao, D.Q., Wang, J.: Rough sets based approach form multivariate decision tree construction. Journal of Software 8, 425–431 (1997)
144. Miao, D.Q., Li, D.G., Fan, S.D.: Fuzzy rough set and its improvement. In: Proceedings of GrC 2005 IEEE International Conference on Granular Computing, Beijing, China, vol. 1, pp. 17–20. IEEE press, Los Alamitos (2005)
145. Min, F., Bai, Z., He, M., Liu, Q.: The reduct problem with specified attributes. In: Rough Sets and Soft Computing in Intelligent Agent and Web Technology, International Workshop at WI-IAT 2005, pp. 36–42 (2005)
146. Min, F., Liu, Q., Tan, H., Chen, L.: The M-relative reduct problem. In: Wang, G.-Y., Peters, J.F., Skowron, A., Yao, Y. (eds.) RSKT 2006. LNCS (LNAI), vol. 4062, pp. 170–175. Springer, Heidelberg (2006)
147. Pal, S.K., Shankar, B.: Granular computing, rough entropy and object extraction. Pattern Recognition Letters 26, 2509–2517 (2005)
148. Pal, S.K.: Soft data mining, computational theory of perceptions, and rough-fuzzy approach. Information Sciences 163, 5–12 (2004)
149. Pal, S.K., Mitra, P.: Rough Fuzzy MLP: Modular Evolution, Rule Generation and Evaluation. IEEE Transactions on Knowledge and Data Engineering 15, 14–25 (2003)
150. Pal, S.K., Mitra, P.: Case Generation Using Rough Sets with Fuzzy Representation. IEEE Trans. Knowledge and Data Engineer 16, 292–300 (2004)
151. Pawlak, Z.: Rough set. International Journal of Computer and Information Sciences 11, 341–356 (1982)
152. Pawlak, Z.: Rough Sets: Theoretical Aspects of Reasoning about Data. Kluwer Academic Publishers, Dordrecht (1991)
153. Pawlak, Z.: Decision Networks. In: Tsumoto, S., Słowiński, R., Komorowski, J., Grzymała-Busse, J.W. (eds.) RSCTC 2004. LNCS (LNAI), vol. 3066, pp. 1–7. Springer, Heidelberg (2004)

154. Pawlak, Z.: Decision Trees and Flow Graphs. In: Greco, S., Hata, Y., Hirano, S., Inuiguchi, M., Miyamoto, S., Nguyen, H.S., Słowiński, R. (eds.) RSCTC 2006. LNCS (LNAI), vol. 4259, pp. 1–11. Springer, Heidelberg (2006)
155. Pawlak, Z.: Flow Graphs and Decision Algorithms. In: International Workshop on Rough Sets, Fuzzy Sets, Data Mining, and Granular Computing (RSFDGrC 2003), Chongqing, China, vol. 2639, pp. 1–10 (2003)
156. Pawlak, Z.: Rough Sets and Flow Graphs. In: Ślęzak, D., Wang, G., Szczuka, M., Düntsch, I., Yao, Y. (eds.) RSFDGrC 2005. LNCS (LNAI), vol. 3641, pp. 1–11. Springer, Heidelberg (2005)
157. Pawlak, Z.: Probability, Truth and Flow Graph. In: Proceedings of The International Workshop on Rough Sets in Knowledge Discovery and Soft Computing (RSKD 2003), Warsaw, Poland, April 2003, pp. 1–9 (2003)
158. Pawlak, Z., Skowron, A.: Rudiments of rough sets. Information Sciences 177, 3–27 (2007)
159. Pawlak, Z., Skowron, A.: Rough sets: Some extensions. Information Sciences 177, 28–40 (2007)
160. Pawlak, Z., Skowron, A.: Rough sets and Boolean reasoning. Information Sciences 177, 41–73 (2007)
161. Pawlak, Z.: Some remarks on conflict analysis. European Journal of Operational Research 166, 649–654 (2005)
162. Pawlak, Z.: Flow Graphs and Data Mining. Transactions on Rough Sets III 3, 1–36 (2005)
163. Pawlak, Z., Peters, James, F., Skowron, A.: Approximating functions using rough sets. In: North American Fuzzy Information Processing Society (NAFI 2004), Banff, Alberta, pp. 358–371 (2004)
164. Pawlak, Z.: Decision Rules and Dependencies. Fundamenta Informaticae 60, 33–39 (2004)
165. Pawlak, Z.: Rough sets, decision algorithms and Bayes' theorem. European Journal of Operational Research 136, 181–189 (2002)
166. Pawlak, Z.: Rough sets and intelligent data analysis. Information Sciences: An International Journal 147, 1–12 (2002)
167. Pawlak, Z.: Theorize with Data Using Rough Sets. In: Proceedings of the 26th International Computer Software and Applications Conference on Prolonging Software Life: Development and Redevelopment, pp. 1125–1128 (2002)
168. Pedrycz, W., Bargiela, A.: Fuzzy Fractal Dimensions and Fuzzy Modeling. Information Sciences 153, 199–216 (2003)
169. Peters, J.F., Skowron, A.: Some Contributions by Zdzislaw Pawlak. In: Wang, G.-Y., Peters, J.F., Skowron, A., Yao, Y. (eds.) RSKT 2006. LNCS (LNAI), vol. 4062, pp. 1–11. Springer, Heidelberg (2006)
170. Peters, J.F., Skowron, A., Stepaniuk, J.: Nearness in Approximation Spaces. In: Proceedings of the Workshop in Concurrency, Specification and Programming (CS&P 2006), pp. 434–445 (2006)
171. Pomykala, J.: The Stone Algebra of Rough Sets. Bull. Polish Academy of Sciences Math 36, 495–508 (1998)
172. Pomykala, J.: Approximation, Similarity and Rough Constructions, ILLC Prepublication Series CT-93-07, University of Amsterdam (1993)
173. Qian, Y.H., Liang, J.Y., Zhang, X., Dang, C.Y.: Rough set approach under dynamic granulation in incomplete information systems. Lecture Note in Artificial Intelligence, vol. 4827, pp. 1–8. Springer, Heidelberg (2007)

174. Qian, Y.H., Liang, J.Y.: Evaluation method for decision rule sets. In: An, A., Stefanowski, J., Ramanna, S., Butz, C.J., Pedrycz, W., Wang, G. (eds.) RSFDGrC 2007. LNCS (LNAI), vol. 4482, pp. 272–279. Springer, Heidelberg (2007)
175. Qian, Y.H., Liang, J.Y., Deyu, L., Haiyun, Z., Chuangyin, D.: Measures for evaluating the decision performance of a decision table in rough set theory. Information Sciences 178, 181–202 (2008)
176. Qian, Y.H., Liang, J.Y.: Combination entropy and combination granulation in incomplete information system. In: Wang, G.-Y., Peters, J.F., Skowron, A., Yao, Y. (eds.) RSKT 2006. LNCS (LNAI), vol. 4062, pp. 184–190. Springer, Heidelberg (2006)
177. Qin, K.Y., Zheng, P.: The relationship among several knowledge reduction approaches. In: Wang, L., Jin, Y. (eds.) FSKD 2005. LNCS (LNAI), vol. 3613, pp. 1232–1241. Springer, Heidelberg (2005)
178. Qin, K.Y., Du, W.F.: The logical characteristics of knowledge reduction approaches. Computer Engineering and Applications 42, 7–9 (2006)
179. Qin, K.Y., Zheng, P.: On the topological properties of fuzzy rough sets. Fuzzy Sets and Systems 151, 601–613 (2005)
180. Qin, K.Y., Gao, Y., Zheng, P.: On covering rough sets. In: Yao, J., Lingras, P., Wu, W.-Z., Szczuka, M., Cercone, N.J., Ślęzak, D. (eds.) RSKT 2007. LNCS (LNAI), vol. 4481, pp. 34–41. Springer, Heidelberg (2007)
181. Qin, Z.R., Wu, Y., Wang, G.Y.: A partition algorithm for huge data sets based on rough set. Pattern Recognition and Artificial Intelligence 19, 249–256 (2006)
182. Qin, Z.R., Wang, G.Y., Wu, Y., Xue, X.R.: A scalable rough set knowledge reduction algorithm. In: Tsumoto, S., Słowiński, R., Komorowski, J., Grzymała-Busse, J.W. (eds.) RSCTC 2004. LNCS (LNAI), vol. 3066, pp. 445–454. Springer, Heidelberg (2004)
183. Qu, K.S., Zhai, Y.H., Liang, J.Y., Li, D.Y.: Representation and Extension of Rough Set Theory Based on Formal Concept Analysis. Journal of Software 18, 2174–2182 (2007)
184. Qu, K.S., Zhai, Y.H., Liang, J.Y., Chen, M.: Study of decision implications based on formal concept analysis. International Journal of General Systems 36, 147–156 (2007)
185. Qu, K.S., Zhai, Y.H.: Posets, inclusion degree theory and FCA 29, 219–226 (2006)
186. Ramanna, S., Peters, J.F., Skowron, A.: Generalized Conflict and Resolution Model with Approximation Spaces. In: Greco, S., Hata, Y., Hirano, S., Inuiguchi, M., Miyamoto, S., Nguyen, H.S., Słowiński, R. (eds.) RSCTC 2006. LNCS (LNAI), vol. 4259, pp. 274–283. Springer, Heidelberg (2006)
187. Ramanna, S., Peters, J.F., Skowron, A.: Approaches to Conflict Dynamics Based on Rough Sets. Fundamenta Informaticae 75, 453–468 (2007)
188. Saha, S., Murthy, C.A., Pal, S.K.: Rough Set Based Ensemble Classifier for Web Page Classification. Fundamenta Informaticae 76, 171–187 (2007)
189. Sai, Y., Yao, Y.Y.: Analyzing and mining ordered information tables. Journal of Computer Science & Technology 18, 771–779 (2003)
190. Sai, Y., Yao, Y.Y.: Data analysis and mining in ordered information tables. In: Proceedings of the 2001 IEEE International Conference on Data Mining, ICDM 2001, San Jose, California, USA, vol. 2, pp. 497–504. Springer, Heidelberg (2001)
191. Sai, Y., Chen, W.W.: Design and implementation of a rough set data mining approach MIF-RS. Mini-Micro Systems 22, 565–567 (2001)
192. Shang, L., Wang, J.G., Yao, W.S., Chen, S.F.: A classification approach bBased on evolutionary neural networks. Journal of Software 16, 1577–1583 (2005)

193. Shao, M.W., Liu, M., Zhang, W.X.: Rough fuzzy set approximations in fuzzy formal contexts. In: Greco, S., Hata, Y., Hirano, S., Inuiguchi, M., Miyamoto, S., Nguyen, H.S., Słowiński, R. (eds.) RSCTC 2006. LNCS (LNAI), vol. 4259, pp. 137–146. Springer, Heidelberg (2006)
194. Shao, M.W., Zhang, W.X.: Approximation in formal concept analysis. In: Ślęzak, D., Wang, G., Szczuka, M., Düntsch, I., Yao, Y. (eds.) RSFDGrC 2005. LNCS (LNAI), vol. 3641, pp. 43–53. Springer, Heidelberg (2005)
195. Shi, K.Q., Cui, Y.Q.: S-rough set and rough decision making. Science Press, Beijing (2006)
196. Shi, K.Q., Cui, Y.Q.: S-Rough Set and Its General Structures. Journal of Shandong University 37, 471–474 (2002)
197. Shi, K.Q.: S-rough sets and knowledge separation. Journal of Systems Engineering and Electronics, 403–410 (2005)
198. Shi, K.Q., Chang, T.C.: One direction S-rough sets. International Journal of Fuzzy Mathematics, 319–334 (2005)
199. Shi, K.Q.: Function S-rough sets and function transfer. International Journal Advances in Systems Sciences and Applications 5, 1–8 (2005)
200. Skowron, A., Peters, J.F., Pawlak, Z.: Commemorating His Life and Work. In: Greco, S., Hata, Y., Hirano, S., Inuiguchi, M., Miyamoto, S., Nguyen, H.S., Słowiński, R. (eds.) RSCTC 2006. LNCS (LNAI), vol. 4259, pp. 49–52. Springer, Heidelberg (2006)
201. Skowron, A., Peters, J.F.: Rough Sets: Trends and Challenges. In: Wang, G., Liu, Q., Yao, Y., Skowron, A. (eds.) RSFDGrC 2003. LNCS (LNAI), vol. 2639, pp. 25–34. Springer, Heidelberg (2003)
202. Skowron, A., Ramanna, S., Peters, J.F.: Conflict Analysis and Information Systems: A Rough Set Approach. In: Wang, G.-Y., Peters, J.F., Skowron, A., Yao, Y. (eds.) RSKT 2006. LNCS (LNAI), vol. 4062, pp. 233–240. Springer, Heidelberg (2006)
203. Skowron, A., Komorowski, H.J., Pawlak, Z., Polkowski, L.T.: A rough set perspective on data and knowledge. In: Handbook of Data Mining and Knowledge Discovery, pp. 134–149. Oxford University Press, Oxford (2002)
204. Skowron, A., Stepaniuk, J., Peters, J.F.: Calculi of Approximation Spaces. Fundamenta Informaticae 72, 363–378 (2006)
205. Skowron, A.: Rough Sets and Vague Concepts. Fundamenta Informaticae 64, 417–431 (2005)
206. Skowron, A., Pal, S.K.: Special Issue: Rough Sets, Pattern Recognition and Data Mining. Pattern Recognition Letters 24, 829–831 (2003)
207. Skowron, A.: Tolerance Approximation Spaces. Fundamenta Informaticae 27, 245–253 (1996)
208. Slezak, D., Wang, G.Y., Szczuka, M.S., Dntsch, I., Yao, Y.Y. (eds.): RSFDGrC 2005. LNCS (LNAI), vol. 3641. Springer, Heidelberg (2005)
209. Słowiński, R.: Intelligent Decision Support: Handbook of Applications and Advances of the Rough Sets Theory. Kluwer Academic Publishers, Dordrecht (1992)
210. Song, X.X., Zhang, W.X.: Knowledge reduction and attributes characteristics in set-valued information system. Computer Engineering 32, 26–28 (2006)
211. Synak, P., Bazan, J.G., Skowron, A., Peters, J.F.: Spatio-Temporal Approximate Reasoning over Complex Objects. Fundamenta Informaticae 67, 249–269 (2005)
212. Tsumot, S., Słowiński, R., Komorowski, H.J., Grzymala-Busse, J.W.: RSCTC 2004. LNCS (LNAI), vol. 3066. Springer, Heidelberg (2004)
213. Wang, G.Y., Liu, Q., Yao, Y.Y., Skowron, A. (eds.): RSFDGrC 2003. LNCS (LNAI), vol. 2639. Springer, Heidelberg (2003)

214. Wang, G.Y., Peters, J.F., Skowron, A., Yao, Y.Y. (eds.): RSKT 2006. LNCS (LNAI), vol. 4062. Springer, Heidelberg (2006)
215. Wang, G.Y.: Rough Set Theory and Knowledge Discovery. Xi'an Jiaotong University Press, Xi'an (2001)
216. Wang, G.Y., Zhao, J., An, J.J., Wu, Y.: A comparative study of algebra viewpoint and information viewpoint in attribute reduction. Fundamenta Informaticae 68, 289–301 (2005)
217. Wang, G.Y.: Rough reduction in algebra view and information view. International Journal of Intelligent systems 18, 679–688 (2003)
218. Wang, G.Y., Zhao, J., An, J.J., Wu, Y.: Theoretical study on attribute reduction of rough set theory: in algebra view and information view. In: 3rd International Conference on Cognitive Informatics, Canada, pp. 148–155 (2004)
219. Wang, G.Y., An, J.J., Wu, Y.: Quantitative analysis of the difference between the algebra view and information view of rough set theory. Mini-micro Systems 26, 1187–1190 (2005)
220. Wang, G.Y.: Relationship between the algebra view and information view of rough set, in Data Mining and Knowledge Discovery: Theory, Tools, and Technology V. In: Dasarathy, B.V. (ed.) Proceedings of SPIE, vol. 5098, pp. 103–113 (2003)
221. Wang, G.Y.: On the relationship between the algebra view and information view of rough set theory. World Science-Technology R&D 24, 20–26 (2002)
222. Wang, G.Y.: Algebra view and information view of rough sets theory, in Data Mining and Knowledge Discovery: Theory, Tools, and Technology III. In: Dasarathy, B.V. (ed.) Proceedings of SPIE, vol. 4384, pp. 200–207 (2001)
223. Wang, G.Y.: Attribute core of decision table. In: Alpigini, J.J., Peters, J.F., Skowron, A., Zhong, N. (eds.) RSCTC 2002. LNCS (LNAI), vol. 2475, pp. 213–217. Springer, Heidelberg (2002)
224. Wang, G.Y.: Attribute core of inconsistent decision information systems. Journal of Shanghai Jiaotong University 38, 2094–2098 (2004)
225. Wang, G.Y., Chen, L.: Attribute core of decision table in the algebra view and information view. WSEAS Transactions on Mathematics 2(1 and 2), 1–6
226. Wang, G.Y.: On the difference of the core attributes in the algebra view and information view of rough set theory. In: Proceedings of 2003 National Conference on Artificial Intelligence (CAAI-10), pp. 1336–1340 (2003)
227. Wang, G.Y.: On some problems of rough set in knowledge acquisition. In: Proceedings of 2001 National Conference on Artificial Intelligence, pp. 443–447 (2001)
228. Wang, G.Y., Yu, H., Yang, D.C.: Decision table reduction based on conditional information entropy. Chinese Journal of Computers 25, 759–766 (2002)
229. Wang, G.Y., Yu, H., Yang, D.C., Wu, Z.F.: Knowledge reduction based on rough set and information entropy. In: The 5th World Multi-Conference on systemics, sybernetics and informatics (SCI 2001), pp. 555–560 (2001)
230. Wang, G.Y., Wu, Y., Fisher, P.S.: Rule generation based on rough set theory, in Data Mining and Knowledge Discovery: Theory, Tools, and Technology II. In: Dasarathy, B.V. (ed.) Proceedings of SPIE, vol. 4057, pp. 181–189 (2000)
231. Wang, G.Y.: Domain-oriented data-driven data mining (3DM): Simulation of human knowledge understanding. In: WICI International Workshop on. Web Intelligence (WI) meets Brain Informatics (BI) (2006)
232. Wang, G.Y.: Domain-oriented data-driven data mining based on rough set. In: Proc. Int. Forum on Theory of GrC from Rough Set Perspective (IFTGrCRSP), p. 46 (2006)

233. Wang, G.Y.: Initiative machine learning based on rough Set. Computer Science 31, 18–20 (2004)
234. Wang, G.Y., Xiao, H.: A self-learning model under uncertain condition. Journal of Software 14, 1096–1102 (2003)
235. Wang, G.Y., He, X.: Initiative learning algorithm based on rough set, in Data Mining and Knowledge Discovery: Theory, Tools, and Technology V. In: Dasarathy, B.V. (ed.) Proceedings of SPIE, vol. 5098, pp. 94–102 (2003)
236. Wang, G.Y., He, X.: A self-learning model based on rough set. Computer Science 29, 24–26 (2002)
237. Wang, G.Y.: Extension of rough set under incomplete information systems. Journal of Computer Research and Development 39, 1238–1243 (2002)
238. Wang, G.Y.: Extension of rough set under incomplete information systems. In: 2002 IEEE International Conference on Fuzzy Systems (FUZZ-IEEE), pp. 1098–1103 (2002)
239. Wang, G.Y., Yang, S.X.: Rough set based uncertain information processing. Dynamics of Continuous Discrete and Impulsive Systems-Series B-Applications & Algorithms, 410–416 (2005)
240. Wang, G.Y.: Rough set based uncertain information processing. In: Proceedings of 2005 National Conference on Artificial Intelligence (CAAI-11), Wuhan, vol. 11, p. 53 (2005)
241. Wang, G.Y.: On some topics of uncertain information processing based on rough set. In: The 12th National Conference on Neural Networks, pp. 56–64 (2002)
242. Wang, G.Y.: Uncertainty measurement of decision table information systems. Computer Science 28, 23–26 (2001)
243. Wang, G.Y., Wu, Y., Liu., F.: Generating rules and reasoning under inconsistencies. In: 2000 IEEE International Conference on Industrial Electronics, Control and Instrumentation, Japan, pp. 2536–2541 (2000)
244. Wang, G.Y., Liu, F.: The inconsistency in rough set based rule generation. In: Ziarko, W., Yao, Y. (eds.) RSCTC 2000. LNCS (LNAI), vol. 2005, pp. 370–377. Springer, Heidelberg (2001)
245. Wang, G.Y., Liu, F., Wu, Y., Nie, N.: The inconsistency in rough set based rule generation. Journal of Chongqing University of Posts and Telecommunications 12, 16–21 (2000)
246. Wang, G.Y., Hu, F., Huang, H., Wu, Y.: A granular computing model based on tolerance relation. The Journal of China Universities of Posts and Telecommunications 12, 86–90 (2005)
247. Wang, G.Y., Chen, L., Wu, Y.: Rough set based solutions for network security. In: Dunin-Keplicz, B., Jankowski, A., Skowron, A., Szczuka, M. (eds.) Monitoring, Security, and Rescue Techniques in Multiagent Systems, pp. 455–465. Springer, Heidelberg (2005)
248. Wang, G.Y., Chen, L., Wu, Y.: Rough set based solutions for network security (extended abstract). In: MSRAS 2004, 1st International Workshop on Monitoring, Security and Rescue Techniques in Multiagent Systems, Poland, pp. 114–118 (2004)
249. Wang, H., Zhang, W.X.: Relationships between concept lattice and rough set. In: Rutkowski, L., Tadeusiewicz, R., Zadeh, L.A., Żurada, J.M. (eds.) ICAISC 2006. LNCS (LNAI), vol. 4029, pp. 538–547. Springer, Heidelberg (2006)
250. Wang, J., Liu, S.Y., Zhang, J., Liu, Z.H.: Multivalued rough set based attribute reduction algorithm. Journal of Xi'an University 30, 277–280 (2003)
251. Wang, J., Miao, D.Q.: Rough set theory and its application: A Survey. Pattern Recognition and Artificial Intelligence 9, 337–344 (1996)

252. Wang, L., Wang, G.Y., Wu, Y.: An incremental rule acquisition algorithm based on variable precision rough set modes. Journal of Chongqing University of Posts and Telecommunications (Natural Sciences Edition) 17, 709–713 (2005)
253. Wang, Y., Wang, G.Y., Deng, W.B.: Concept lattice based data-driven uncertain knowledge acquisition. Pattern Recognition and Artificial Intelligence (to appear, 2007)
254. Wei, L., Li, H.R., Zhang, W.X.: Knowledge reduction based on the equivalence relations defined on attribute set and its power set. Information Sciences 177, 3178–3185 (2007)
255. Wei, L., Miao, D.Q.: Application of granular computing in knowledge reduction. In: Wang, G.-Y., Peters, J.F., Skowron, A., Yao, Y. (eds.) RSKT 2006. LNCS (LNAI), vol. 4062, pp. 357–362. Springer, Heidelberg (2006)
256. Wu, W.Z., Zhang, W.X., Xu, Z.B.: On Characterization of rough fuzzy sets in constructive and axiomatic approaches. Chinese Journal of Computers 27, 197–203 (2004)
257. Wu, W.Z., Zhang, W.X.: Neighborhood operator systems and approximations. Information Sciences 144, 201–217 (2002)
258. Wu, W.Z., Zhang, M., Li, H.Z., Mi, J.S.: Knowledge reduction in random information systems via Dempster-Shafer theory of evidence. Information Sciences 174, 143–164 (2005)
259. Wu, W.Z., Leung, Y., Zhang, W.X.: Connections between rough set theory and Dempster-Shafer theory of evidence. International Journal of General Systems 31, 405–430 (2002)
260. Wu, W.Z., Zhang, W.X.: Neighborhood operator systems and approximations. Information Sciences 144, 201–217 (2002)
261. Wu, W.Z., Mi, J.S., Zhang, W.X.: Generalized fuzzy rough sets. Information Sciences 151, 263–282 (2003)
262. Wu, W.Z., Mi, J.S., Zhang, W.X.: Generalized fuzzy rough sets. Information Sciences 151, 263–282 (2003)
263. Wu, Z.C.: RBFNN representation based on rough sets and its application to remote Sensing image classification. Acta Geodaetica et Cartographica Sinica 32, 53–57 (2003)
264. Xie, K.M., Chen, Z.H., Xie, G., Lin, T.Y.: BGrC for Superheated Steam Temperature System Modeling in Power Plant. In: Proc. of 2006 IEEE International Conference on Granular Computing, Atlanta, USA, pp. 708–711 (2006)
265. Xu, C., Min, F.: Weighted reduction for decision tables. In: Wang, L., Jiao, L., Shi, G., Li, X., Liu, J. (eds.) FSKD 2006. LNCS (LNAI), vol. 4223, pp. 246–255. Springer, Heidelberg (2006)
266. Xu, J.C., An, Q.S., Wang, G.Y.: Disposal of information with uncertain borderline-fuzzy sets and vague sets. Computer Engineering and Applications 38, 24–26 (2002)
267. Xu, J.C., Shen, J.Y., Wang, G.Y.: Measure of similarity between rough sets. Computer Science 30, 55–57, 60 (2003)
268. Xu, W., Xu, P., Zhang, W.X.: Information concept lattice and its reduction. In: Yao, J., Lingras, P., Wu, W.-Z., Szczuka, M., Cercone, N.J., Ślęzak, D. (eds.) RSKT 2007. LNCS (LNAI), vol. 4481, pp. 60–67. Springer, Heidelberg (2007)
269. Xu, Z.B., Liang, J.Y., Dang, C.Y., Chin, K.S.: Inclusion degree: a perspective on measures for rough set data analysis. Information Sciences 141, 229–238 (2002)

270. Yan, G.W., Xie, G., Qiu, Y.X., Chen, Z.H.: MEA based nonlinearity correction algorithm for the VCO of LFMCW radar level gauge. In: Ślęzak, D., Wang, G., Szczuka, M., Düntsch, I., Yao, Y. (eds.) RSFDGrC 2005. LNCS (LNAI), vol. 3641, pp. 461–470. Springer, Heidelberg (2005)
271. Yang, Y., Wang, G.Y., Chen, P.J., Zhou, J., He, K.: Feature selection in audio-visual emotion recognition based on rough set theory. In: Peters, J.F., Skowron, A., Marek, V.W., Orłowska, E., Słowiński, R., Ziarko, W. (eds.) Transactions on Rough Sets VII. LNCS, vol. 4400, pp. 283–294. Springer, Heidelberg (2007)
272. Yang, Y., Wang, G.Y., He, K.: An approach for selective ensemble feature selection based on rough set theory. In: Yao, J., Lingras, P., Wu, W.-Z., Szczuka, M., Cercone, N.J., Ślęzak, D. (eds.) RSKT 2007. LNCS (LNAI), vol. 4481, pp. 518–525. Springer, Heidelberg (2007)
273. Yao, Y.Y.: Constructive and Algebraic Methods of Theory of Rough Sets. Information Sciences 109, 21–47 (1998)
274. Yao, Y.Y.: A Comparative Study of Fuzzy Sets and Rough Sets. Information Sciences 109, 227–242 (1998)
275. Yao, Y.Y.: Relational Interpretations of Neighborhood Operators and Rough Set Approximation Operators. Information Sciences 101, 239–259 (1998)
276. Yao, Y.Y.: Neighborhood Systems and Approximate Retrieval. Information Sciences 176, 3431–3452 (2007)
277. Yao, Y.Y., Sai, Y.: On mining ordering rules. In: Terano, T., Nishida, T., Namatame, A., Tsumoto, S., Ohsawa, Y., Washio, T. (eds.) JSAI-WS 2001. LNCS (LNAI), vol. 2253, pp. 316–321. Springer, Heidelberg (2001)
278. Ye, D.Y., Chen, Z.J.: A new discernibility matrix and the computation of a core. Acta Eletronica Sinica 30, 1086–1088 (2002)
279. Ye, D.Y., Chen, Z.J.: Inconsistency classification and Discernibility-Matrix-Based approaches for computing an attribute core. In: Wang, G., Liu, Q., Yao, Y., Skowron, A. (eds.) RSFDGrC 2003. LNCS (LNAI), vol. 2639, pp. 269–273. Springer, Heidelberg (2003)
280. Ye, D.Y., Chen, Z.J.: An improved discernibility matrix for computing all attribute reducts. In: Proceedings of the 5th IEEE International Conference on Cognitive Informatics, pp. 305–308. IEEE Computer Society Press, Los Alamitos (2006)
281. Ye, D.Y.: Recursive computation of positive regions and approximation quality in rough sets and its application. Pattern Recognition and Artificial Intelligence 15, 474–477 (2002)
282. Ye, D.Y., Chen, Z.J., Liao, J.K.: A new algorithm for minimum attribute reduction based on binary particle swarm optimization with vaccination. In: Zhou, Z.-H., Li, H., Yang, Q. (eds.) PAKDD 2007. LNCS (LNAI), vol. 4426, pp. 1029–1036. Springer, Heidelberg (2007)
283. Yin, D.S., Wang, G.Y., Wu, Y.: Initiative learning algorithm for decision tree pre-pruning. Journal of Tongji University 32, 194–196 (2004)
284. Yin, D.S., Wang, G.Y., Wu, Y.: Data-driven decision tree learning algorithm based on rough set theory. In: Tarumi, H., Li, Y., Yoshida, T. (eds.) Proc. of the 2005 International Conference on Active Media Technology (AMT), Takamatsu, Kagawa, Japan, pp. 579–584 (2005)
285. Yin, D.S., Wang, G.Y., Wu, Y.: A self-learning algorithm for decision tree pre-pruning. In: Proceedings of the Third International Conference on Machine Learning and Cybernetics, Shanghai, pp. 2140–2145 (2004)
286. Yu, H., Yang, D.C., Wang, G.Y., Wu, Z.F.: Rough set based knowledge reduction algorithms. Computer Science 28, 31–34 (2001)

287. Yu, H., Wang, G.Y., Yang, D.C., Wu, Z.F.: Knowledge reduction algorithms based on rough set and conditional information entropy, in Data Mining and Knowledge Discovery: Theory, Tools, and Technology IV. In: Dasarathy, B.V. (ed.) Proceedings of SPIE, vol. 4730, pp. 422–431 (2002)
288. Yu, Y.Z., Wang, G.Y., Wu, Y.: A rough set based analysis approach for incomplete information systems. Computer Science 28, 35–38 (2001)
289. Yuan, Z., Wu, Y., Wang, G.Y., Li, J.B.: Motion-information-based video retrieval system using rough pre-classification. In: Peters, J.F., Skowron, A. (eds.) Transactions on Rough Sets V. LNCS, vol. 4100, pp. 306–333. Springer, Heidelberg (2006)
290. Yuan, Z., Wu, Y., Wang, G.Y., Li, J.B.: Video retrieval via rough-set-based video pre-classification. Computer Science 32, 20–23 (2005)
291. Zeng, H.L.: Theoretical Aspects and Applications of Rough Sets. Chongqing University Press, Chongqing (1996)
292. Zeng, H.L.: Intelligent Computation. Chongqing University Press, Chongqing (2003)
293. Zeng, H.L.: A New Method of Selection and Reduction of System Feature in Pattern Recognition based on Rough Sets (in English). J. of Advance in Modeling & Analysis (French) 29 (2002)
294. Zeng, H.L.: A fast learning algorithm for solving systems of linear equations and related problems. Journal of Advance in Modelling & Analysis (French) 29 (1995)
295. Zeng, H.L.: On dynamics of a learning associative neural network. In: Proc of ICCAS/IEEE, USA (1995)
296. Zeng, H.L., Yuan, Z.R., Zeng, X.H.: A new method of selection and reduction of system feature in pattern recognition based on rough sets. Journal of Communication and Computer 3, 25–28 (2006)
297. Zeng, H.L.: Image recognition using adaptive fuzzy neural network and wavelet transform. In: Rough Sets, Fuzzy Sets, Data Mining, and Granular Computing, pp. 635–639. Springer, Heidelberg (2003)
298. Zhang, B., Zhang, L.: The Theory and Applications of Problem Solving-Quotient Space Based Granular Computing (The Second Version). Tsinghua University Press, Beijing (2007)
299. Zhang, B., Zhang, L.: Theory and Applications of Problem Solving. Tsinghua University Press, Beijing (1990)
300. Zhang, D.X., Miao, D.Q., Du, W.L., Li, D.G.: Contrast and analysis of attribute reduction algorithms. Computer Science 32, 89–92 (2005)
301. Zhang, D.X., Miao, D.Q., Li, D.G.: A variable precision rough set model based on database system. Computer Science 32, 179–184 (2005)
302. Zhang, H.Y., Miao, D.Q.: Automatic recognition system of bills based on principal curves. Journal of Information and Computational Science 1, 199–204 (2004)
303. Zhang, Q.H., Wang, G.Y., Hu, J., Liu, X.Q.: Incomplete information systems processing based on fuzzy- clustering. In: 2006 IEEE/WIC/ACM International Conference on Web Intelligence and Intelligent Agent Technology (WI-IAT 2006 Workshops) (WI-IATW 2006), pp. 486–489 (2006)
304. Zhang, W.X., Wu, W.Z., Liang, J.Y., Li, D.Y.: Theory and Methods of Rough Sets. Science Press, Beijing (2001)
305. Zhang, W.X., Leung, Y., Wu, W.Z.: Information Systems and Knowledge Discovery. Science Press, Beijing (2003)
306. Zhang, W.X., Qiu, G.F.: Uncertain Decision Making Based on Rough Sets. Tsinghua University Press, Beijing (2005)

307. Zhang, W.X., Leung, Y., Xu, P.: Uncertainty Reasoning Based on Inclusion Degree. Tsinghua University Press, Beijing (2007)
308. Zhang, W.X., Yao, Y.Y., Leung, Y.: Rough set and concept lattice. Xi'an Jiaotong University Press, Xi'an (2006)
309. Zhang, W.X., Wei, L., Qi, J.J.: Attribute reduction in concept lattice based on discernibility matrix. In: Ślęzak, D., Yao, J., Peters, J.F., Ziarko, W., Hu, X. (eds.) RSFDGrC 2005. LNCS (LNAI), vol. 3642, pp. 157–165. Springer, Heidelberg (2005)
310. Zhang, W.X., Wei, L., Xu, P.: Rough sets theory on generalized information systems. Fuzzy System and Mathematics 18, 29–33 (2004)
311. Zhang, X.P., Zhang, J.S., Wang, J., Li, J.: Study on diagnosis in sintering process based on rough set theory and neural network. Sintering and Pelletizing 30, 24–26 (2005)
312. Zhang, X.P.: Study on fault diagnosis of mining conveyance machine based on rough set. China Mining Magazine 14, 74–76 (2005)
313. Zhang, X.Y., Xu, W.H.: A novel approach to roughness measure in fuzzy rough sets. Advances in Soft Computing, 775–780 (2007)
314. Zhao, J., Wang, G.Y., Wu, Z.F., Tang, H., Li, H.: The study on technologies for feature selection. In: The First Int. Conference on Machine Learning and Cybernetics, pp. 689–693 (2002)
315. Zhao, J., Wang, G.Y., Wu, Z.F., Tang, H., Li, H., Liao, X.F.: Algorithms based on rough set theory for feature subset selection. Computer Science 29, 83–86 (2002)
316. Zhao, J., Wang, G.Y., Wu, Z.F., Tang, H., Li, H., Liao, X.F.: An efficient approach to compute the feature core. Mini-micro Systems 24, 1950–1963 (2003)
317. Zhao, J., Wang, G.Y., Wu, Z.F., Tang, H., Li, H.: Method of data discretization based on rough set theory. Mini-micro Systems 25, 60–64 (2004)
318. Zhao, J., Wang, G.Y., Wu, Z.F., Li, H.: New algorithms for data discretization based on rough set theory. Journal of Chongqing University 25, 18–21 (2002)
319. Zhao, J., Wang, G.Y.: A data-driven knowledge acquisition method based on system uncertainty. In: 4th IEEE International Conference on Cognitive Informatics, pp. 267–275 (2005)
320. Zhao, J., Wang, G.Y.: Research on system uncertainty measures based on rough set theory. In: Wang, G., et al. (eds.) RSKT 2006. LNCS (LNAI), vol. 4062, pp. 227–232. Springer, Heidelberg (2006)
321. Zheng, Z., Wang, G.Y., Wu, Y.: Objects' combination based simple computation of attribute core. In: The 17th IEEE International Symposium on Intelligent Control, ISIC 2002, Vancouver, Canada, pp. 514–519 (2002)
322. Zheng, Z., Wang, G.Y., Wu, Y.: Objects' combination based computation of core attributes. Computer Science 29, 204–207 (2002)
323. Zheng, Z., Wang, G.Y.: RRIA.: A rough set and rule tree based incremental knowledge acquisition algorithm. Fundamenta Informaticae 59, 299–313 (2004)
324. Zheng, Z., Wang, G.Y., Wu, Y.: A rough set and rule tree based incremental knowledge acquisition algorithm. In: Wang, G., Liu, Q., Yao, Y., Skowron, A. (eds.) RSFDGrC 2003. LNCS (LNAI), vol. 2639, pp. 122–129. Springer, Heidelberg (2003)
325. Zheng, Z., Wang, G.Y., Wu, Y.: An incremental acquisition knowledge algorithm. Computer Science 30, 13–15 (2003)
326. Zheng, Z.: Tolerance granular space and its applications. PhD thesis, Institute of Computing Technology, Chinese Academy of Sciences (2006)

327. Zheng, Z., Hu, H., Shi, Z.Z.: Tolerance relation based information granular space. In: Ślęzak, D., Wang, G., Szczuka, M., Düntsch, I., Yao, Y. (eds.) RSFDGrC 2005. LNCS (LNAI), vol. 3641, pp. 682–691. Springer, Heidelberg (2005)
328. Zheng, Z., Hu, H., Shi, Z.Z.: Tolerance granular space and its applications. In: Proceedings of IEEE International Conference on Granular Computing, pp. 367–372 (2005)
329. Zheng, Z., Hu, H., Shi, Z.Z.: Granulation based image texture recognition. In: Tsumoto, S., Słowiński, R., Komorowski, J., Grzymała-Busse, J.W. (eds.) RSCTC 2004. LNCS (LNAI), vol. 3066, pp. 659–664. Springer, Heidelberg (2004)
330. Zheng, Z., Hu, H., Shi, Z.Z.: Rough set based image texture recognition algorithm. In: Negoita, M.G., Howlett, R.J., Jain, L.C. (eds.) KES 2004. LNCS (LNAI), vol. 3213, pp. 772–778. Springer, Heidelberg (2004)
331. Zhong, N., Skowron, A., Ohsuga, S. (eds.): RSFDGrC 1999. LNCS (LNAI), vol. 1711. Springer, Heidelberg (1999)
332. Zhong, N., Skowron, A., Ohsuga, S. (eds.): RSFDGrC 1999. LNCS (LNAI), vol. 1711. Springer, Heidelberg (1999)
333. Zhong, N., Yao, Y., Ohshima, M.: Peculiarity Oriented Multidatabase Mining. IEEE Trans. Knowledge and Data Engineer 15, 952–960 (2003)
334. Zhu, F., Wang, F.Y.: Some Results on Covering Generalized Rough Sets. Pattern Recognition and Artificial Intelligence 15, 6–13 (2002)
335. Zhu, W.: Topological Approaches to Covering Rough Sets. Information Sciences 177, 1499–1508 (2007)
336. Ziarko, W.: The discovery, analysis and representation of data dependencies in databases. In: Piatesky-Shapiro, G., Frawley, W.J. (eds.) Knowledge discovery in databases, pp. 177–195. AAAI Press/MIT Press (1991)

Rough Neural Fault Classification of Power System Signals

Liting Han and James F. Peters

Department of Electrical and Computer Engineering, University of Manitoba
Winnipeg, Manitoba R3T 5V6, Canada
liting@ee.umanitoba.ca
jfpeters@ee.umanitoba.ca

Abstract. This paper proposes an approach to classify faults that commonly occur in a High Voltage Direct Current (HVDC) power system. These faults are distributed throughout the entire HVDC system. The most recently published techniques for power system fault classification are the wavelet analysis, two-dimensional time-frequency representation for feature extraction and conventional artificial neural networks for fault type identification. The main limitation of these systems is that they are commonly designed to focus on a group of faults involved in a specific area of a power system. This paper introduces a framework for fault classification that covers a wider range of faults. The proposed fault classification framework has been initiated and developed in the context of the HVDC power system at Manitoba Hydro, which uses what is known as the Transcan^{TM} system to record and archive fault events in files. Each fault file includes the most active signals (there are 23 of them) in the power system. Testing the proposed framework for fault classification is based on fault files collected and classified manually over a period of two years. The fault classification framework presented in this paper introduces the use of the rough membership function in the design of a neural fault classification system. A rough membership function makes it possible to distinguish similar feature values and measures the degree of overlap between a set of experimental values and a set of values representing a standard (*e.g.*, set of values typically associated with a known fault). In addition to fault classification using rough neural networks, the proposed framework includes what is known as a linear mean and standard deviation classifier. The proposed framework also includes a classifier fusion technique as a means of increasing the fault classification accuracy.

Keywords: Power system faults, knowledge-based fault recognition, rough membership, rough neuron, rough neural network, classification, classify fusion.

1 Introduction

With the rapid increase of electrical power consumption by utilities and industries, more stability and efficiency in power delivery is needed. A report by CEIDS (Consortium for Electric Infrastructure to Support a Digital Society) shows that the U.S. economy is losing between \$104 billion and \$164 billion a

J.F. Peters and A. Skowron (Eds.): Transactions on Rough Sets VIII, LNCS 5084, pp. 396–519, 2008.

year due to power outages [10]. The analysis and classification of power system disturbances are becoming mandatory in working towards minimizing and even eliminating power outages. Typically, an effort is made to identify the most significant patterns of system faults that provide input to a region-based analysis system for decision support. Operators or engineers make use of the summary reports to operate and maintain a power system.

In the power fault classification research area, the existing literature and methods [8,11,12,13,16,17,20,21,26,29,37,38,55,69,70,71,72,76,77,82,83,85] explained in section 4 are focused on the wavelet analysis, two-dimensional time-frequency representation for signal pre-processing, feature extraction and conventional artificial neural networks for fault type identification. In this research project for Manitoba Hydro power fault classification system, autocorrelation, cross-correlation, Wavelets, FFT, IFFT, low pass filter, phase shifting, derivatives and coding have been used to analyze and extract feature information of the 23 most active signals recently recorded at the Dorsey Station. With the feature values as input, the conventional artificial neural network has been applied to determine the fault type in the beginning stage of this research and the results addressed in section 8 are undesirable due to the complexity and uncertainty of the feature information. The instinct is to introduce the rough neuron for rough membership computation to distinguish similar feature values by assigning each of them with the degree of each type of fault. It greatly improves the quality of the feature information and consequently the classification performance. However, the rmNNs successfully classify 10 types of faults with 100% accuracy while for fault 7 and 10 with only 83% and 75% accuracy respectively. The bringing in of the second classifier LMD and classifier fusion techniques [1,6,22,30,34,35,67,74,75] is to profit from the complementary information that different classifiers provide and to improve the classification performance for some types of faults.

This paper is organized as follows: Section 1 (this section) is an introduction for this paper. Section 2 briefly introduces power system fundamentals and a brief overview of power system faults. Section 3 reviews the rough set and classifier fusion theories. Section 4 gives an overview of fault identification techniques commonly used in the electrical power industry. The main parts of the research completed for this Manitoba Hydro power system fault classification are presented in section 5 to section 9. Appendices A(Correlation Theory), B(Conventional Fast Fourier Transform), C (Wavelet Transform), and D (Time-Frequency Representation Theory) summarize the basic theory used in this article.

2 Power System Fundamentals

This section briefly introduces the power system fundamentals [84,23] required for an understanding of power system faults.

2.1 Power Systems

Electric power transmission was originally developed with direct current (DC). The availability of transformers and the development and improvement of induction motors at the beginning of the 20th century, led to the use of alternating current (AC) transmission. Even so, d.c transmission is generally used for the following reasons:

1. An overhead DC transmission line with its towers can be designed to be less costly per unit of length than an equivalent AC line designed to transmit the same level of electric power. However the DC converter stations at each end are more costly than the terminating stations of an AC line and so there is a break-even distance above which the total cost of DC transmission is less than the cost of AC transmission. In addition, DC transmission line can have a lower visual profile than an equivalent AC line, which contributes to a lower, perceived environmental impact. An environmental advantage to a DC transmission line over an AC line is the presence of lower electromagnetic fields.
2. If transmission is by underground cable, the break-even distance is less than overhead transmission. It is not practical to consider AC cable systems exceeding 50 kilometers but hundreds of kilometers of underground DC cable transmission systems are feasible.
3. Some AC electric power systems are not synchronized with neighboring networks even though the physical distance between them is quite small. This occurs in Japan where half the country has a 60Hz network and the other has a 50Hz system. It is physically impossible to connect the two by direct AC methods for the purpose of exchanging electric power between them. However, if a DC converter station is located in each system with an interconnecting DC link, it is possible to transfer power flow from one system to the other.

The integral part of an HVDC power converter is the valve or valve arm. It may be non-controllable if constructed from one or more power diodes in series or controllable if constructed from one or more thyristors in series. Figure 1 depicts the International Electrotechnical Commission (IEC) graphical symbols for valves and bridges (valve groups). The standard bridge or converter connection is defined as a 2-way connection consisting of six valves or valve arms, which are shown in Figure 2. Electric power flowing between an HVDC valve group and an AC system is three phase. When electric power flows into a DC valve group from an AC system, it flows through a rectifier. If power flows from the DC valve group into the AC system, it flows through an inverter.

The most common building block for HVDC valves is the thyristor (see Figure 3 for characteristics of a thyristor). In the 'off' state, a thyristor blocks the flow of current as long as the reverse or forward breakdown voltages (V_{br} or V_{bo}) are not exceeded. A thyristor can be made to attain an 'on' state if it is forward biased ($V_{ak} > 0$) and a small positive 'gate' voltage is applied between the gate and the cathode. This 'firing pulse' need not be present once the thyristor is ignited, although in practice, a train of pulses in rapid succession is often maintained over an entire conduction period. Once turned on, a thyristor follows its 'on' characteristic as shown in Figure 3. Note that the forward voltage drop in the on condition is relatively small and an actual thyristor characteristic

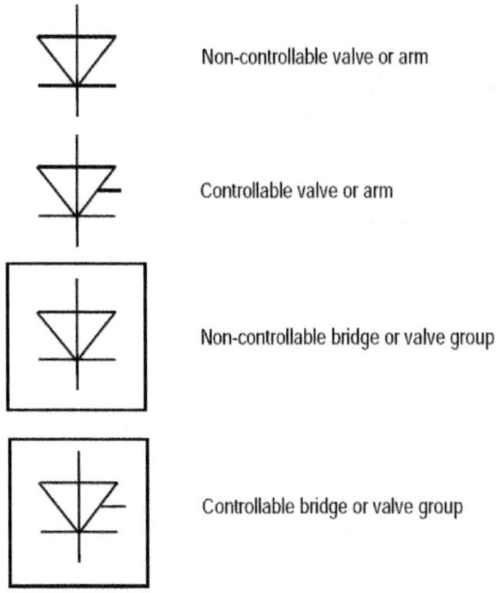

Fig. 1. Standard graphical symbols for valves and bridges [84]

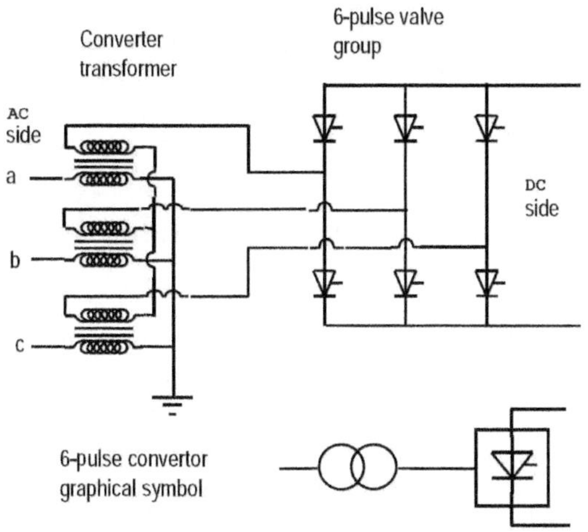

Fig. 2. Electric circuit configuration of the basic 6-pulse valve group with its converter transformer in star-star connection [84]

closely follows that of an ideal switch (horizontal line for the 'off' state, vertical y axis for the 'on' state). The thyristor can also turn on if the voltage across it exceeds the forward break-over voltage V_{bo}. This mechanism is often used to protect a thyristor against excessive voltage.

The normal state transition diagram for a thyristor is shown in Figure 4. The thyristor attains its 'off' state when the current through it attempts to reverse. One other factor that is necessary for a successful turn-off is that a thyristor must not be subject to a forward biasing voltage too soon after the current has

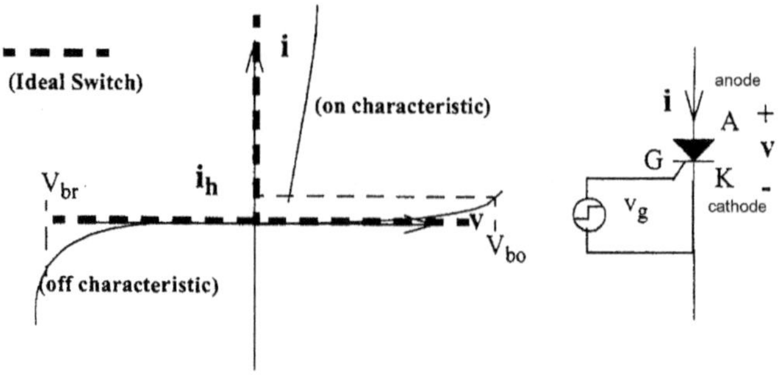

Fig. 3. Thyristor characteristic [23]

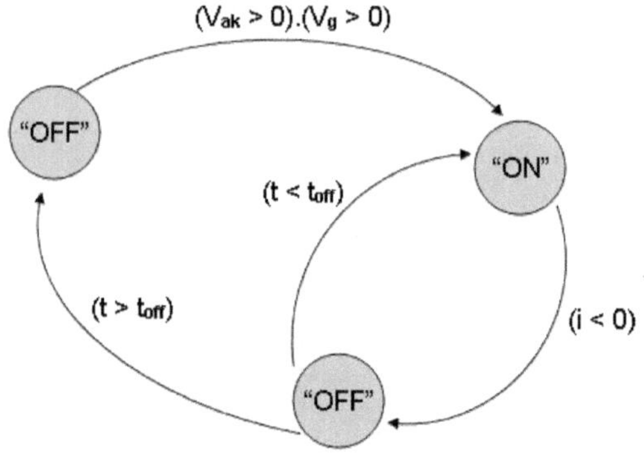

Fig. 4. State transition diagram for thyristor switching [23]

extinguished. Otherwise, there is a possibility of re-ignition even in the absence of a pulse. Re-ignition occurs when the charge carriers in the semi-conductor have not had sufficient time to be re-absorbed. This critical time is referred to as the turn-off time t_{off} and often expressed in terms of a so called "extinction angle" $\gamma = \omega t_{off}$, if AC waveforms of angular frequency ω are involved. This phenomenon in which a thyristor fails to attain its forward blocking state, 'off' state, is referred to as commutation failure.

The 6-pulse bridge is the most widely used HVDC converter configuration. Figure 5 shows a typical 6-pulse thyristor bridge with the AC supply, the converter transformer X_c and the DC-side smoothing reactance. A 6-pulse bridge consists of an upper and a lower half as seen in Figure 6(a). It is assumed initially that the converter transformer is ideal so that there is no leakage inductance. It is also assumed that ideal thyristors behave like diodes, *i.e.*, zero voltage drop when the device is on and an ideal open circuit when off. The device is in a conducting state as soon as the forward biased voltage ($V_{ak} > 0$) causes current to flow in the forward (anode to cathode) direction and no 'firing pulse' is required.

The upper bridge half is a standard maximum select circuit that selects the largest of the three voltages V_a, V_b and V_c at the common cathode terminal. This

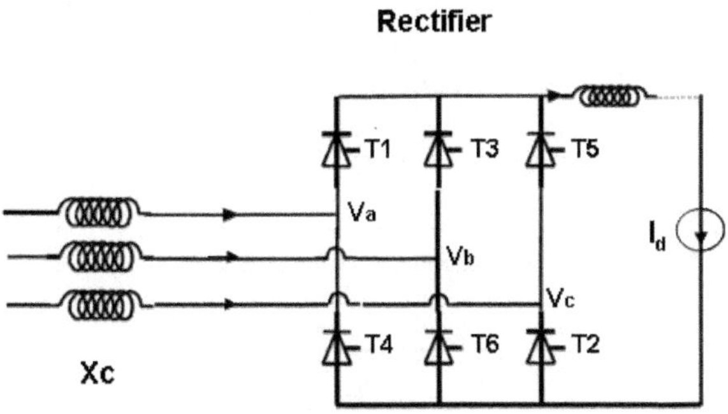

Fig. 5. Three phase (6-pulse) bridge

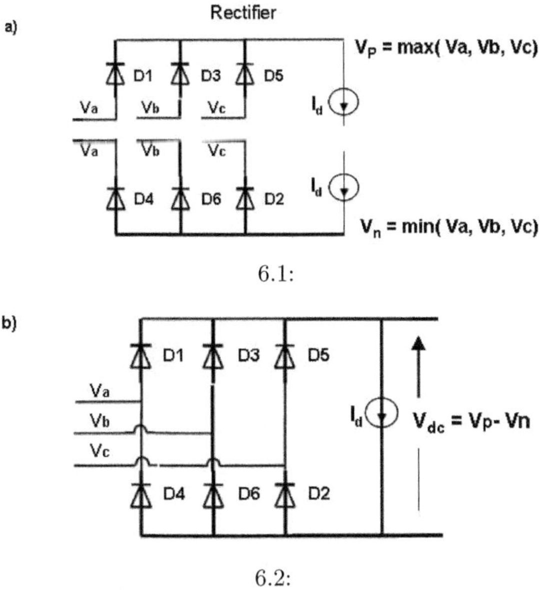

Fig. 6. Analysis of three phase (6-pulse) bridge

can be proved by contradiction. To see this , assume $V_a < V_b$ but that $V_p = V_a$ because D1 is assumed to be conducting. Then D3 should also conduct since it is forward biased because $V_b > V_a$, hence, $V_a = V_b$, which is a contradiction. The only possibility that does not lead to a contradiction is for V_p to be equal to the largest of the three voltages.

Similarly the lower bridge half causes a voltage $V_n = min(V_a, V_b, V_c)$ to appear at the common anode terminal of devices D2, D4 and D6. Thus the total DC side voltage as can be seen from Figure 6(2) must be the difference $V_{dc} = V_p - V_n$. The waveforms for the bridges are shown in Figure 7. The current on the AC side in phase a is I_d when D1 conducts and $-I_d$ when D4 conducts. The conduction period for D1 can be determined from the waveforms as the period in which the

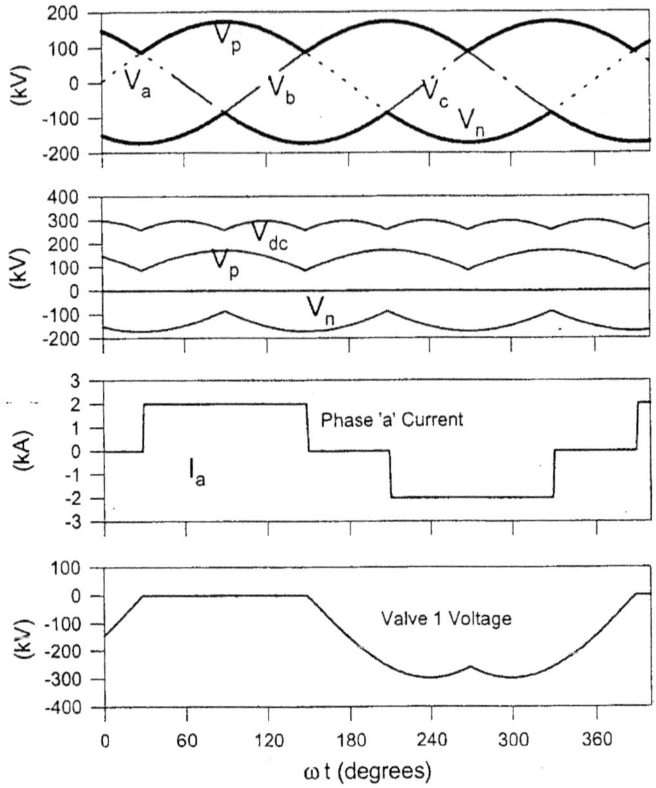

Fig. 7. Three phase diode bridge waveforms (no overlap) [23]

voltage V_a of phase a is the largest of the three phase voltages. Similarly, D4 is on when V_a is at its smallest in magnitude.

The sequence of conduction for the valves in the upper bridges is D1, D3, D5, D1, D3, D5, and so on, since each successive phase dominates over a 120° interval. In the bottom bridge, the sequence is D2, D4, D6, D2, D4, D6, and so on. Considering the two halves together, each valve enters conduction 60° after its predecessor in the sequence D1, D2, D3, D4, D5, D6, D1, D2, D3, D4, D5, D6, and so on.

Without any series inductance in the circuit, the current instantaneously rises to the value $\pm I_d$ on turn-on and makes an instantaneous transition to zero on turn-off when the current transfers to the next phase. The valve voltage is an important parameter in determining the valve rating. The voltage in the forward direction across valve 1 is determined to be $V_a - V_p$, and while the valve is conducting this voltage is zero.

In practice, transformer leakage inductance must be considered. With the inclusion of transformer reactance X_c shown in Figure 8, the current can no longer make an instantaneous transition from one phase to another because that would require a discontinuous change in inductor current as is evidenced from

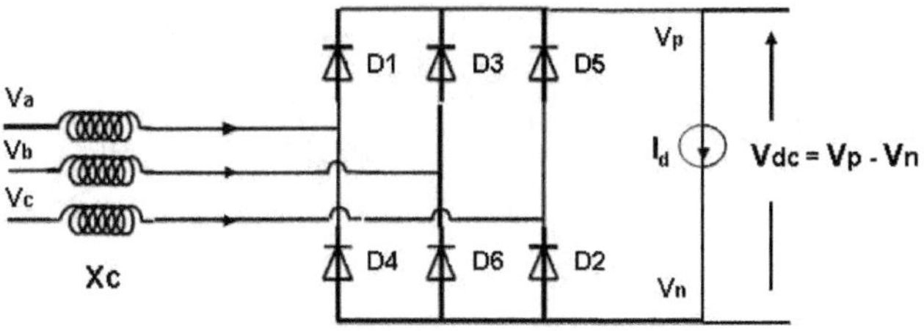

Fig. 8. Three phase (6-pulse) bridge: transformer inductance included

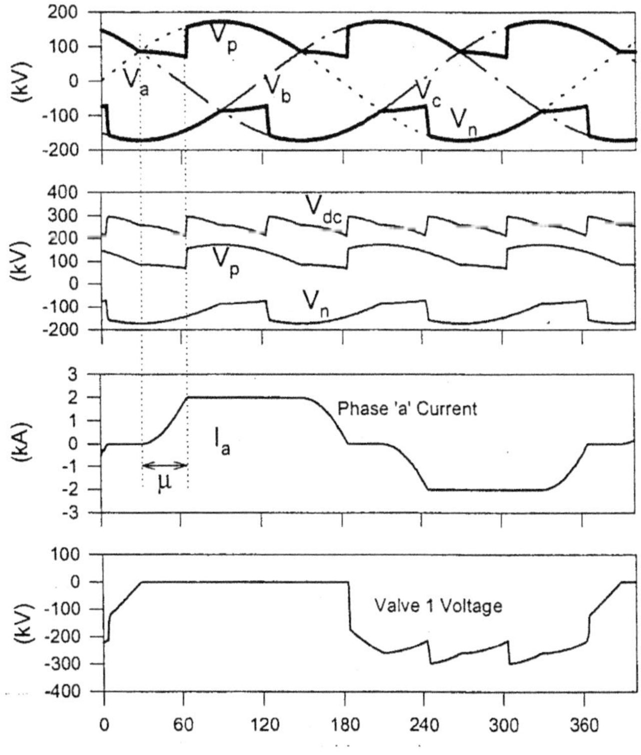

Fig. 9. Three phase diode bridge waveforms [23]

the waveforms shown in Figure 9. In this case, when valve 1 is turned on, there is an "overlap" between valve 1 and valve 5, *i.e.*, valve 1 is turned on while valve 5 starts to be turned off. The overlap interval is represented by the angle μ. During this interval, the DC-side voltage V_p (similarly V_n) is the average of the two conducting phase voltages, *i.e.*, V_a and V_c. Also note from Figure 9 that the valve voltage waveform now has additional commutation "spikes".

The thyristors in a controlled bridge are idealized, *i.e.*, a thyristor behaves like a diode, except that mere forward bias (positive anode-cathode voltage) is not

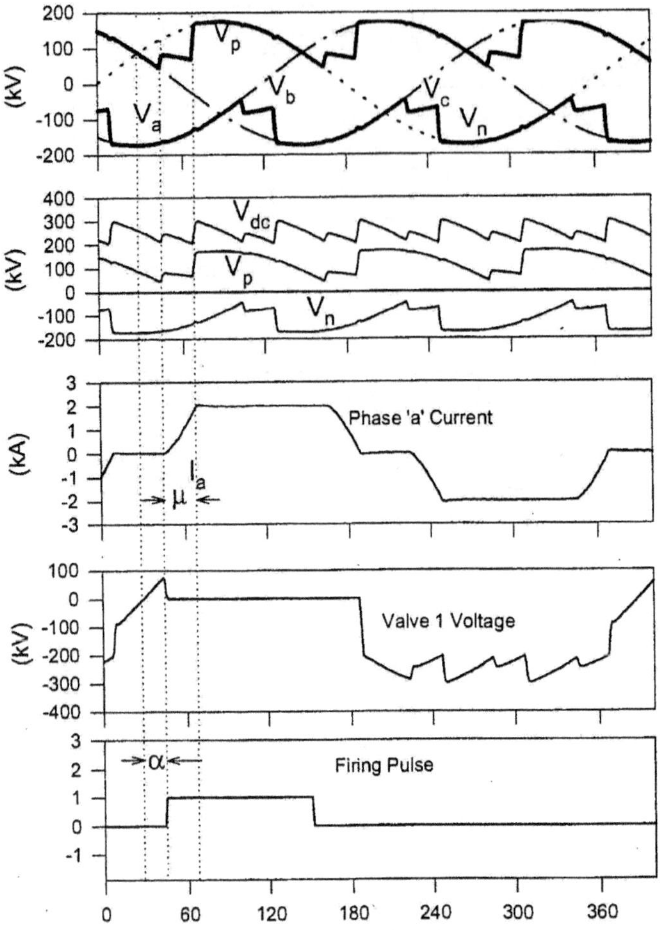

Fig. 10. Controlled thyristor bridge waveforms: $\alpha = 15°$ [23]

sufficient to ensure conduction. The additional condition to attain the conducting state is a required gate, 'firing pulse' that must be present in addition to a forward bias. Hence, the main difference in analyzing the operation of a thyristor bridge is that the maximum (or minimum) select action only commences on the issue of a firing pulse. The thyristor valves are fired in the sequence T1, T2, T3, T4, T5, and T6. The elapsed angle from the earliest instant at which a thyristor may conduct (*i.e.*, the point at which forward bias first appears) to the instant at which the firing pulse is issued and the valve commences conduction is called the "firing" or "delay" angle and is denoted by the Greek letter α.

In the waveforms shown in Figure 10, $\alpha = 15°$ has been used. Also note that in Figure 10, the pulse duration is a full 120°. This is not strictly necessary, since a thyristor valve that has been triggered on continues conducting until the current through it attempts to flow in the reverse direction. However, in HVDC systems, it is common practice to keep pulsing continuous over a valve's

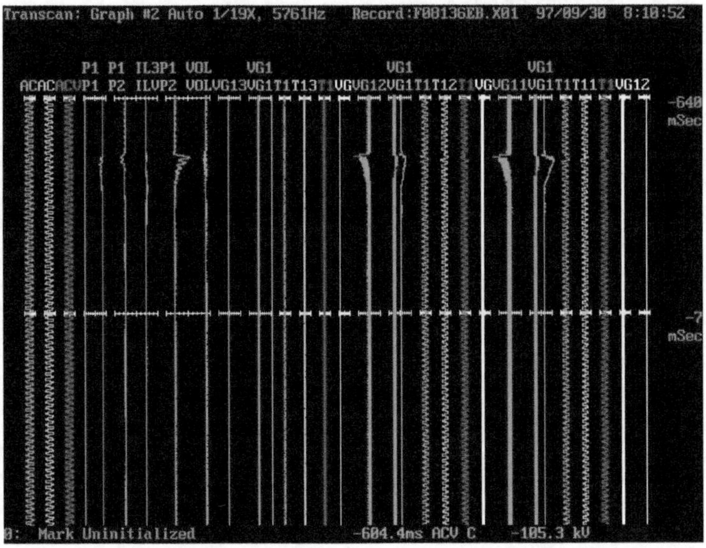

Fig. 11. 31 signals in the "Valve Current Commutation Failure" fault

nominal conduction interval of 120° (in the form of a train of high-frequency pulses) in case a premature current zero occurs because of waveform distortions. Note that for this value of the firing angle ($\alpha = 15°$), the DC voltage is positive and the power flows from the AC to the DC side. This is the "rectifier" mode of operation. Note that if continuous current is maintained in the circuit by some external device, the firing angle α can be made to have a value in excess of 90°. In this situation, the voltage V_p turns out to be negative and V_n is positive, which causes the DC voltage to be negative. Thus, power transfer is from the DC side to the AC side, although the direction of the DC current remains the same. This is the "inverter" mode of operation.

2.2 Power System Faults

A power system fault is the result of an electrical disturbance. At the Manitoba Hydro Dorsey Station, the Transcan^{TM} recording system is deployed as a power system monitoring tool. It archives 31 power signals in a fault file whenever a power system fault occurs. A typical screen snapshot of 31 signals recorded by Transcan^{TM} is shown in Figure 11. Transcan^{TM} is capable of recording power system faults in a real-time manner. However, this system cannot identify the type and cause of a recorded fault. Engineers at the Dorsey Station must visually assess all the 31 signals then manually log the cause of the fault into the database of the Transcan^{TM} system and consolidate this information into an archived fault file. The graphical user interface (GUI) of the Transcan^{TM} system is shown in Figure 12. The 23 most active and informative signals referenced in the proposed fault classification system are listed in Table 1.

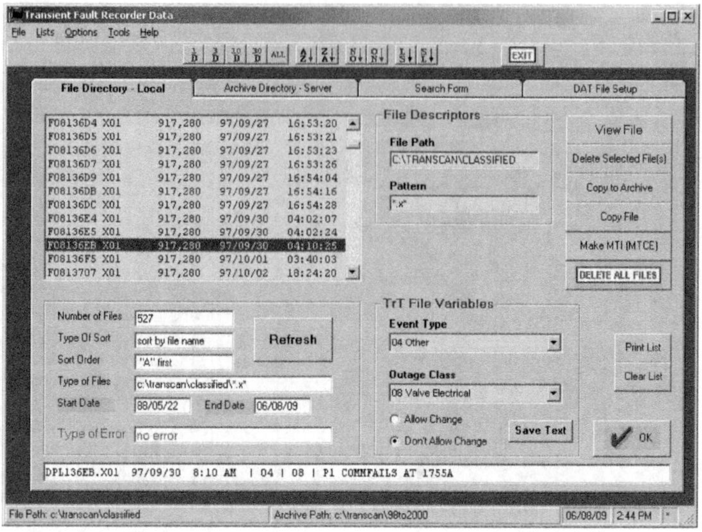

Fig. 12. The Transcan^{TM} system GUI

Table 1. Most active power system signals

Bus signals	AC Phase A, B, C Pole voltages and currents Pole current order	Sinusoidal Constant Constant
Valve signals (total 3 valve groups)	6-pulse Current A phase, B phase, C phase Start pulse	Periodic Sinusoidal Periodic

Table 2. Common power system faults

Fault index	Fault name	Number of fault files
Fault 1	Minor AC Disturbance	240
Fault 2	Severe AC Disturbance	148
Fault 3	Valve Current Closed/Blocked/Deblocked	114
Fault 4	Line Fault	81
Fault 5	Valve Current Commutation Failure	95
Fault 6	Pole Voltages/Currents Closed/Blocked/Deblocked	64
Fault 7	Phase Current Arc Back	26
Fault 8	Parallel Operation	29
Fault 9	Pole Current Oscillation	31
Fault 10	Normal Affected by Another Pole	18
Fault 11	Asymmetric Protection	25
Fault 12	Disturbance on DC Voltage	25

The twelve most common power system faults are listed in Table 2. An information table for fault classification cannot be established without a good understanding of the mechanism underlying each fault and the behavior of the signal associated with each fault.

- AC Voltage Disturbance. This is a bus error that will induce some other faults such as valve current commutation failure, line fault and valve current blocked. Normally, three AC phase voltages are sinusoidal signals that have a fixed 120° phase delay relative to each other. The AC voltage line will be impacted by different disturbances such as a falling tree hitting a transmission line, heavy snowfall or severe wind, and sometimes radiation or magnetic field interference.

- Valve Current Closed/Blocked/Deblocked. This fault happens in one or two valve groups. There are three valve groups in poles 1 and 2, and two valve groups in poles 3 and 4. Vg11, Vg12, Vg13 designate pole 1; Vg21, Vg22, Vg23, pole 2; Vg31, Vg32, pole 3; Vg41, Vg42, pole 4. A failure of a 6-pulse signal in a valve group will shut down or block the valve currents. An AC voltage disturbance also has the same effect. The restart of the 6-pulse signal will unblock the valve currents.
- Line Fault. This fault is due to the AC voltage disturbance, the pole line short to the ground or the energy of a DC line decreases (line force retard) causing a pole line voltage flashover or shutdown. The power system will restart in a short time if the control system responds quickly.
- Valve Current Commutation Failure. This happens when a valve is not turned off successfully because the valve is subject to a forward biasing voltage too soon after the current has been extinguished. This causes a minor valve current distortion for a very short period of time.
- Pole Voltages/Currents Closed/Blocked/Deblocked. This happens when all the valve groups in one pole are closed, blocked or deblocked.
- Phase Current Arc Back. This happens only in one valve group. The valve current increases sharply for a short period of time and then shuts down. This type of power system fault is caused when valve lines short together or short to ground.
- Parallel Operation. This is not a fault but an indicator that the line maintenance is in progress. When a pole current line needs to be tested, the current will be switched to another pole line. Inside the power station, the current of this pole line goes down to 0; outside the station, the current provided does not decrease, and the pole voltage remains normal.
- Pole Current Oscillation. This fault is caused by oscillation of the pole current order. Usually with this fault, the pole voltage remains relatively constant.
- Normal Affected by Another Pole. This fault happens occasionally. There is a bi-pole power system at the Dorsey Station. Pole 1 and pole 2 compose one active station. Pole 3 and pole 4 are usually for a back-up station. If a fault, especially a line fault, occurs in pole 1, Transcan^{TM} will generate 2 fault files: one for pole 1, and one for pole 2 even in the case where pole 2 is absolutely normal. This occurs because the bus signals, pole voltages and currents are shared and reordered in both pole 1 and pole 2 fault files.
- Asymmetric Protection. If the pulse to open the valve arrives in an abnormal sequence, this fault will cause more than two valves to open at the same time. The circuit control system will then force this valve group to close. The most noticeable event associated with this fault is that the 6-pulse signal will have 7 cycles of severe oscillation and will be closed until the control system restarts the valve group.

– Disturbance on DC Voltage. At the Dorsey Station, the AC voltage is converted from the DC voltage. The long distance transmission of DC voltage is easier and the interference problem is greatly decreased. However, sometimes snow on DC transmission lines or windy weather will cause changes in the DC voltages and a DC voltage disturbance is recorded.

3 Mathematics Underlying Fault Classification and Recognition Techniques

This section gives an overview of the mathematics underlying fault classification and fault recognition techniques.

3.1 Rough Set Theory

This section briefly presents the basic rough set approach to the approximation of sets [57] that provides a foundation for classifying power system fault signals. The rough set approach introduced by Zdzisław Pawlak [40,48,49,50,51,52,53] and elaborated by others [32,43,44,45,54,57,68,73,58,78,42,41,65,3,59,60,61] provides the grounds for approximating a set X. Let B denote a set of functions that represent object features (traditionally, also called attributes in rough set theory [51]) of objects in a set U. The basic approach in rough set theory is to use an equivalence relation $\sim_B$ [62]

$$\sim_B = \{(x, x') \in U \times U \mid \forall f \in B,\ f(x) = f(x')\},$$

to define the partition of a set U into non-empty, pairwise disjoint subsets (equivalence classes). An equivalence class in a partition is denoted by $[x]_B$, where

$$[x]_B = \{x' \in U \mid \forall f \in B, f(x) = f(x')\}.$$

The equivalence classes in a partition form a new set, denoted by $U/\sim_B$, where

$$U/\sim_B = \{[x]_B \mid x \in U\},$$

for a given set of objects U. Let $X \subseteq U$ be a set of objects of interest. After the partition of the set U has been defined, the lower and upper approximations of the set X are defined relative to the equivalence classes in the partition.

Preliminaries

The notation and terminology in Table 3 is important for an understanding of basic rough set theory. Let $U, \mathcal{F}$ denote a set of sample objects and a set of functions, respectively. The functions in $\mathcal{F}$ represent the features (attributes) of the objects in U. Assume that $B \subseteq \mathcal{F}$, the notation (U, B) denotes an information system, which is usually represented in table form.

In keeping with current notation for equivalence relations, $\sim$ denotes an equivalence relation on a set U [19]. The $\sim$ symbol is used extensively to express equivalence [9,18,19].

Table 3. Rough Set Theory Symbols

Symbol	Interpretation
U	Set of sample objects,
$\mathcal{F}$	Set of functions representing object features,
B	$B \subseteq \mathcal{F}$,
X	$X \subseteq U$,
x	$x \in X$,
$\sim_B$	$\sim_B = \{(x, x') \in U \times U \mid \forall x \in U, f(x) = f(x')\}$,
$[x]_B$	$[x]_B = \{x' \in U \mid x' \sim_B x\}$,
$U/\sim_B$	$U/\sim_B = \{[x]_B \mid x \in U\}$, a partition of U,
B_*X	$\bigcup_{x:[x]_B \subseteq X} [x]_B$, B-lower approximation of X,
B^*X	$\bigcup_{x:[x]_B \cap X \neq \emptyset} [x]_B$, B-upper approximation of X,
$Bnd_B X$	$Bnd_B X = B^*X \setminus B_*X = \{x \mid x \in B^*X \text{ and } x \notin B_*X\}$.

The notation $U/\sim$ denotes a partition of U. Let $[x]$ denote a class belonging to $U/\sim$, where

$$[x] = \{x' \in U \mid x \sim x'\}.$$

The classes of a partition are disjoint, *i.e.*, if $[x], [y] \in U/\sim$, then $[x] \cap [y] = \emptyset$. In addition, every object in U is in only one class in $U/\sim$.

The use of $\sim_B$ drew attention to the role of the set B in partitioning a set U. The basic idea here is that the relation $\sim_B$ provides a classification of objects according to knowledge contained in the system (U, B) [33].

The class $[x]_B$ is called a B-elementary set [48,51]. If $(x, x') \in \sim_B$ (also written $x \sim_B x'$), then x and x' are said to be *indiscernible* with respect to all functions in B, or simply, B-*indiscernible.* In the case where $B = \{f\}$, $\sim_{\{f\}}$ denotes an equivalence relation defined relative to a set of feature f and $[x]_{\{f\}}$ denotes the equivalence class $x/\sim_{\{f\}}$ represented by x and defined by $\sim_f$. For simplicity, write $\sim_f$ to denote $\sim_{\{f\}}$.

A sample $X \subseteq U$ can be approximated from information contained in B by constructing a B-lower approximation

$$B_*X = \bigcup_{x:[x]_B \subseteq X} [x]_B,$$

and a B-upper approximation

$$B^*X = \bigcup_{x:[x]_B \cap X \neq \emptyset} [x]_B.$$

The B-lower approximation B_*X is a collection of classes of sample elements that can be classified with full certainty as members of X. By contrast, the B-upper approximation B^*X is a collection of classes representing both certain and possibly uncertain knowledge about X because it is possible for B^*X to have one or more classes that are not subsets of X but still have a non-empty intersection with X. An approximation boundary $Bnd_B X$ is defined by

$$Bnd_B X = B^*X \setminus B_*X = \{x \mid x \in B^*X \text{ and } x \notin B_*X\}.$$

Table 4. Decision system notation

Symbol	Interpretation
d	Decision function,
U	Set of sample objects,
$\mathcal{F}$	Set of functions representing features,
$(U, \mathcal{F}, d)$	Decision system.

The set $Bnd_B X$ contains all objects in the upper approximation B^*X that are not in the lower approximation B_*X. Whenever $B_*X \subsetneq B^*X$, the sample X has been classified imperfectly, and is considered a rough set. In other words, a set X is a rough set, if and only if, the boundary $Bnd_B X$ is not empty.

Information Tables
For computational reasons, a syntactic representation of information systems is usually given in the form of tables. Discovering objects in the composition of a class $[x]_B \subseteq U/\sim_B, x \in U$ in the partition $U/\sim_B$ in the system $(U, \mathcal{F})$ is accomplished by gathering together inside the class all of those objects that have matching function values. Identifying the classes in $U/\sim_B$ is greatly aided by a table representation of $(U, \mathcal{F})$.

Decision Systems
Of particular interest is the extension of information systems made possible by including a function d representing what is known as a decision attribute in rough set theory. A decision is defined by a function $d : X \longrightarrow V_d$, where V_d is the range of d. In addition, $(U, \mathcal{F}, d)$ denotes a decision system. It is typical in rough set theory to start with an information system $(U, \mathcal{F})$ and introduce a decision function d as a means of separating sample objects in U into decision classes, *i.e.*, sets of objects representing a particular value of d. Decision systems are also represented by tables.

Rough Membership Function
Because it is important to determine the extent to which a set of sample signals match a class of signals representing a particular power system fault, the rough membership function defined by (1) has been used in this research. The degree of overlap between X and $[x]_B$ containing x can be quantified with the rough membership function (rmf),

$$\mu_X^B : U \rightarrow [0,1] \;\; defined\, by \;\; \mu_X^B(x) = \frac{|[x]_B \cap X|}{[x]_B}. \tag{1}$$

The rough membership function has proven to be very useful in measuring the extent that classes of signals for known faults overlap with sets of signals representing power system faults to be classified. This is explained in detail in Section 7, where the rmf is used in the design of a neural network useful in classifying power system faults.

3.2 Classifier Fusion Theory

Classifier combination has received considerable attention in the past decade and is now an established pattern recognition offspring. It has been recognized for some time that the classical approach to designing a pattern recognition system, which focuses on finding the best classifier has a serious drawback. Any complementary discriminatory information that other classifiers may encapsulate is not tapped. Multiple expert fusion aims to make use of many different designs to improve classification performance. Over the last few years, a myriad of methods for fusing the output of multiple classifiers have been proposed.

Let $\mathcal{D} = \{D_1, D_2, ..., D_L\}$ be a set of classifiers and $\Re^n$ be the feature space. All classifiers produce soft class labels. We assume that $d_{j,i}(x) \in [0,1]$ is an estimate of the degree of set c_i offered by classifier D_j for an input $x \in \Re^n, i = 1,2$; $j = 1, ..., L$. There are two possible classes $C = \{c_1, c_2\}$ and L classifiers $\mathcal{D} = \{D_1, D_2, ..., D_L\}$ [34]. Simple fusion methods are the most obvious choice when constructing a multiple classifier system [30,35,74,75,6], *i.e.*, the support for class c_i, $d_i(x)$, yielded by the set of classifiers is [34]

$$d_i(x) = \mathcal{F}(d_{1,i}(x), ..., d_{L,i}(x)), i = 1,2, \tag{2}$$

where $\mathcal{F}$ is the chosen fusion method. Here, it is necessary to study the fusion methods compared in [1]:

- minimum
- maximum
- average
- median
- majority vote
- oracle.

For the majority vote, the first step is to harden the individual decisions by assigning class labels $D_j(x) = c_1$ if $d_{j,1}(x) > 0.5$, and $D_j(x) = c_2$ if $d_{j,1}(x) \leq 0.5$, $j = 1, ..., L$. Next, the class label most represented among the L (label) output is chosen.

The oracle model is an abstract fusion model. In this model, if at least one of the classifiers produces the correct class label, then the team produces the correct class label too. Usually, Oracle is used in comparative experiments.

In order to achieve a high overall performance of the classification function, the performance of each individual classifier has to be optimized prior to using it within any fusion schemes. That is, the fusion scheme will be able to improve the overall classification result relative to the performance of the individual, optimized classifiers. If several classifiers with only marginal performance are being used, the results cannot necessarily be expected to reach the high performance sought. On the other hand, if several classifiers are used that work exceptionally well, any further gains will be exceedingly hard to accomplish because the opportunity for diversity will be diminished. Individual classifier optimization can be performed by selecting object features, appropriate parameters, and classifier structure that governs the performance.

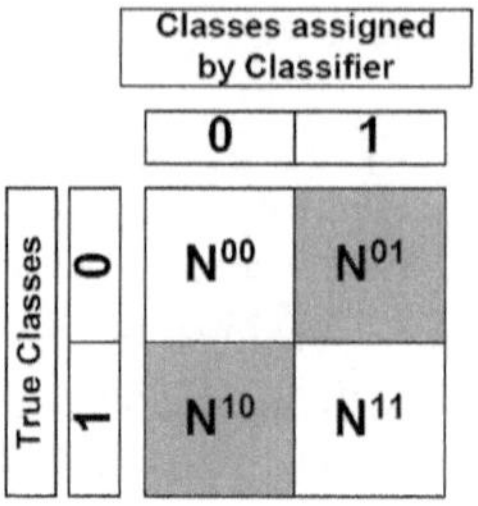

Fig. 13. Typical 2-class confusion matrix [22]

After designing a classifier fusion scheme, a confusion matrix $\boldsymbol{M}$ can be generated for each classifier using labeled training data [22]. The confusion matrix lists the true classes c versus the estimated classes $\widehat{c}$. Because all classes are enumerated, it is possible to obtain information not only about correctly classified states (N^{00} and N^{11}), but also about false positives (N^{01}) and false negatives (N^{10}). A typical two-class confusion matrix $\boldsymbol{M}$ is shown in Figure 13.

From the confusion matrix of each classifier, the false positive (FP) error, the false negative (FN) error, the total error rate (TER), and the total success rate (TSR) can be calculated for the classifier. These error rates are defined as in (3) to (6). The total error rate (TER) or the total success rate (TSR) is typically used as a simple measure for overall performance of a classifier.

$$FP = \frac{N^{01}}{N^{00} + N^{01}}. \tag{3}$$

$$FN = \frac{N^{10}}{N^{10} + N^{11}}. \tag{4}$$

$$TER = \frac{N^{01} + N^{10}}{N^{00} + N^{11} + N^{01} + N^{10}}. \tag{5}$$

$$TSR = 1 - TER. \tag{6}$$

Although each individual classifier's performance is very important to the performance of a classifier fusion, the dependency between the classifiers to be fusioned also affects the fusion results. Some studies [67] have shown that the degree of correlation between the classifiers adversely affects the performance of the subsequent classifier fusion. If two classifiers agree everywhere, the fusion of the two classifiers will not achieve any accuracy improvement no matter what fusion method is used. For classifier fusion design, classifier correlation analysis is, therefore, equally as important as the classifier performance analysis. Based on the classifier output on the labeled training data, a 2x2 matrix $\boldsymbol{N}$ as shown in Figure 14 can be generated for each classifier pair. The off-diagonal numbers directly indicate the correlation degree of the two classifiers. The smaller the two off-diagonal numbers are, the higher the correlation between the two classifiers will be. The proportion of specific agreement, which here is called the correlation, ρ_2, is defined in [67] as

$$\rho_2 = \frac{2 \times N^{FF}}{N^{TF} + N^{FT} + 2 \times N^{FF}}, \qquad (7)$$

where, as further shown in Figure 14, N^{TT} implies that both classifiers classified correctly; N^{FF} means both classifiers classified incorrectly; N^{TF} represents the case of the 1st classifier classified correctly and the 2nd classifier classified incorrectly; and N^{FT} stands for the 2nd classifier classified correctly and the 1st classifier classified incorrectly. In order for classifier fusion to be effective in performance improvement, the correlation, ρ_2, has to be small (low correlation).

Consider the output of two classifiers as enumerated in Table 5. The calculation of ρ_2 yields $\rho_2 = 0.36$. Had classifier 2 been completely redundant to classifier 1, the correlation would have been $\rho_2 = 1$.

The 2-class correlation coefficient can be extended to n different classifiers [22]. The notion that redundancy is described by the individual true and false answers of the classifiers is retained from the 2 class correlation analysis. The larger the ρ-correlation, the larger the redundancy. In particular, the ρ-correlation goes to zero if the individual incorrect answers are disjoint for all answers. That implies that there is always at least one correct answer from some classifier for any case available. The ρ-correlation coefficient gets larger as the number of wrong

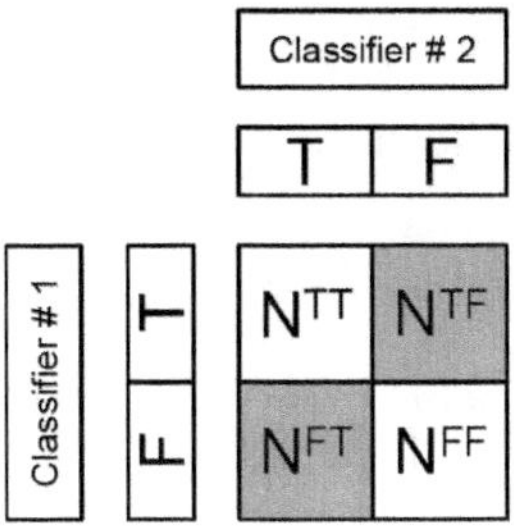

Fig. 14. Correlation analysis matrix [22]

Table 5. Results from experiment for 2 classifiers [22]

Answer classifier 1	Answer classifier 2
T	T
T	F
F	T
T	F
F	F
F	F
T	F
F	T
T	T
T	T
T	T
T	T
T	F
T	T
T	T
F	T

answers are the same for many answers. Let N^f be the number of experiments where all classifiers give a wrong answer; N_i^c be the number of experiments with combinations of correct and incorrect answers; c is the combination of correct and incorrect answers (for 2 classifiers: $c \in \{wr, rw\}$; for 3 classifiers: $c \in \{wwr, wrw, rww, wrr, rwr, rrw\}$ *etc.*); n is the number of classifiers. The ρ-correlation coefficient is then [22]

$$\rho_n = \frac{nN^f}{\sum_{i=1}^{2^n-2} N_i^c + nN^f}. \tag{8}$$

If N is the number of experiments and N^t is the number of experiments for which all classifiers had a right answer, (8) can more conveniently be rewritten as [22]

$$\rho_n = \frac{nN^f}{N - N^f - N^t + nN^f}. \tag{9}$$

Consider a 3-classifier example, which is the same as the previous 2-classifier example except that a third classifier was added that will get answer wrong in 50% of the cases. The calculation of ρ_n yields: $\rho_n = 0.21$.

Although the newly added classifier has poor performance, its addition reduces the overall redundancy of the classifier assembly.

Note that the ρ-correlation does not record redundancy for any particular classifier (for $n > 2$) but for a set of classifiers only. For illustrative purposes, consider two simplistic cases shown in Table 6 and Table 7 [22].

Table 6. Output for 3 classifiers (case 1) [22]

Answer classifier 1	Answer classifier 2	Answer classifier 3
T	F	F
F	T	F
F	T	T
T	T	T
F	F	F

The ρ-correlation is $\rho_n = 0.5$.

Table 7. Output for 3 classifiers (case 2) [22]

Answer classifier 1	Answer classifier 2	Answer classifier 3
T	F	T
F	T	T
F	T	F
T	T	T
F	F	F

The ρ-correlation is $\rho_n = 0.5$.

Obviously the third classifier is different in the two example cases above. However, the degree of correlation is the same because it does not matter whether it is correlated to the first or to the second classifier. Rather, it is only relevant that it is correlated to the combination of the first two classifiers. Note that the calculation of the ρ-correlation factor can be performed on multi-class scenarios as well because the factor is only concerned with the correctness of the outcome.

4 Technology Review of Power System Fault Classification (PSFC)

4.1 Wavelet Applications in Power Systems

The main difficulty in dealing with power engineering phenomena is the extreme variability of the signals and the necessity to operate on a case–by–case basis. Another aspect of power disturbance signals is often localized temporally or spatially (*e.g.*, transients in power systems). This requires the efficient use of analysis methods, which are versatile enough to handle signals in terms of their time-frequency localization. Wavelets localize the information in a time frequency plane. In particular, wavelets are capable of trading one type of resolution for another, which makes them especially suitable for the analysis of non-stationary signals. The fundamentals of wavelets are explained in Appendix C. Considerable work has been done in applying the wavelet transform to power systems in analyzing and processing the voltage-current signals to make a real-time identification of transients in a fast and accurate way [20].

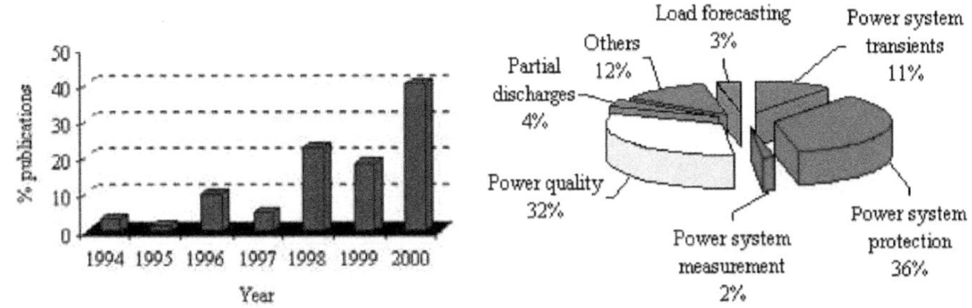

15.1: Evolution of wavelet publications in power systems

15.2: Percentage of wavelet publications in different power system areas

Fig. 15. Overview of wavelet applications in power systems [20]

The wavelet transform was first applied to power systems in 1994 by Robertson [70] and Rebeiro [69]. Since then, the number of publications in this area has rapidly increased as Figure 15.1 shows. Figure 15.2 illustrates the most popular wavelet transform applications in power systems:

- Power system protection
- Power quality
- Power system transients
- Partial discharges
- Load forecasting
- Power system measurement

The field of power system transients is the area in which wavelets were first applied to power system applications by Robertson [70]. In this paper, the authors presented a methodology for the development of software for classifying power system disturbances by type from the transient waveform signature. Transients are signals with a finite life, *i.e.*, a transient reduces to zero in a finite time. Electromagnetic transients are caused by sudden changes in system topology or parameters. For instance, short circuit faults are one of the most common causes of transients in a power system. Power system switching causes transients as well. Robertson [71] distinguished single-phase faults from capacitor switching using waveform signatures.

An example of transient analysis using wavelets was given by Ramaswamy [72]. Using the Electromagnetic Transient Package provided in the Power System Simulation Software, MIPOWER, and the wavelet transform toolbox provided in MATLAB *Ver. 5.3*, the authors analyzed a group of simulated transients namely the phase BC-Ground fault, three phase-Ground fault and phase C-Ground fault, in a simple power system network (Figure 16) consisting of a generator, a load, two buses and a transmission line. Figure 17 shows a typical waveform of a certain type of transient disturbance in power systems.

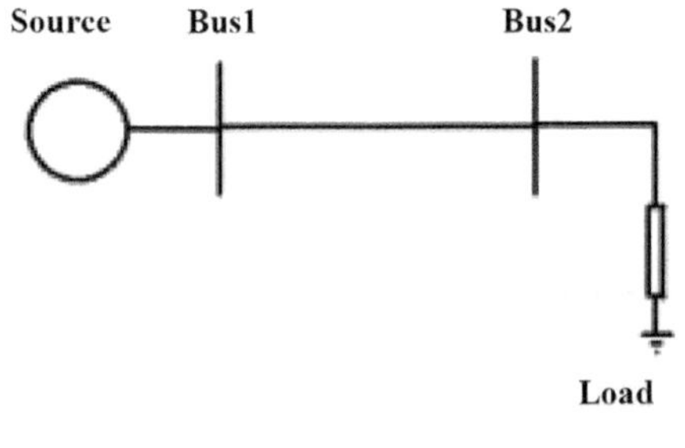

Fig. 16. A typical power system network [72]

The authors applied different types of wavelets to the transient disturbance signal to perform Multiple Level Decomposition. The Meyer wavelet (Figure 18) was found to work better as the fundamental source signal was restored at the 4th approximation. Other wavelets such as a 'Haar' wavelet, added noise to the fundamental wave. The transients were analyzed by the 'Meyer' mother wavelet and Figure 19 shows Multiple Level Decomposition of the transient disturbance, where s is the source signal, a_4 is the 4th level approximation, d_4 is the 4th level

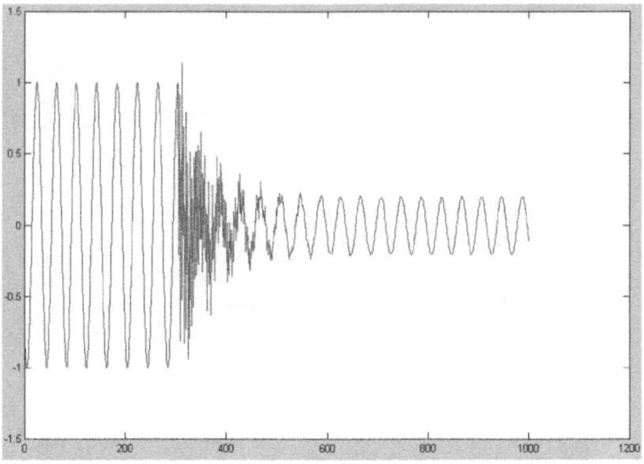

Fig. 17. Example of transient disturbance for certain types of faults indistinguishable by the naked eye [72]

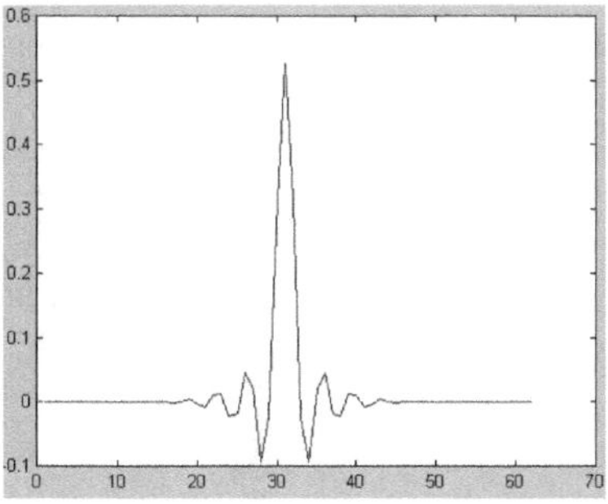

Fig. 18. A typical Meyer wavelet [72]

detail coefficient, d_3 is the 3rd level detail coefficient, d_2 is the 2nd level detail coefficient, and d_1 is the 1st level detail coefficient.

The detail coefficients of faults are given in Figure 20 for the phase BC-Ground fault, three phase-Ground fault and phase C-Ground fault.

In power quality applications, several studies have been carried out to detect and locate disturbances using the wavelet transform to analyze interference, impulses, notches, glitches, interruptions, harmonics, flicker, etc. of non-stationary signals. Drisen [16] analyzed power system harmonics while Santoso [76] analyzed power system interference.

In power system protection applications, the potential benefits of applying the wavelet transform to improve the performance of protection relays and fault classification have been recognized in recent years. Charri [11] analyzed the transient information of a resonant grounded distribution system using the wavelet transform. Imriš [26] presented the analysis of ground fault transients in high voltage networks for earth fault location purposes using the Gaussian mother wavelet method and discussed the main sources of error affecting the accuracy

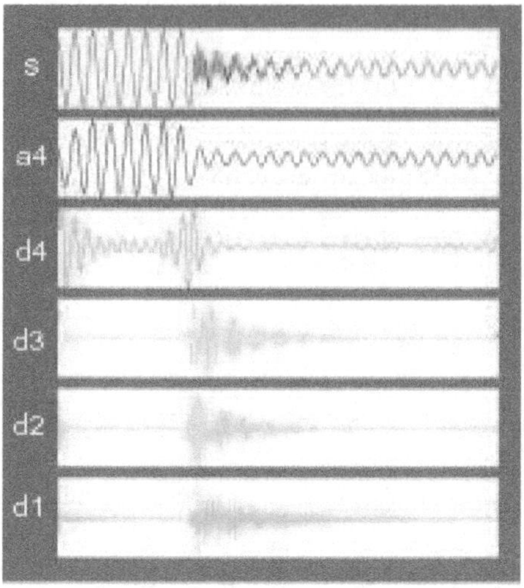

Fig. 19. Multiple level decomposition of a transient disturbance [72]

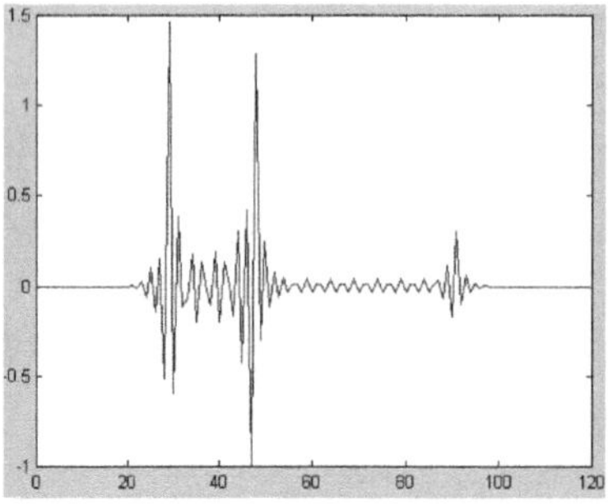

20.1: Phase BC-ground fault

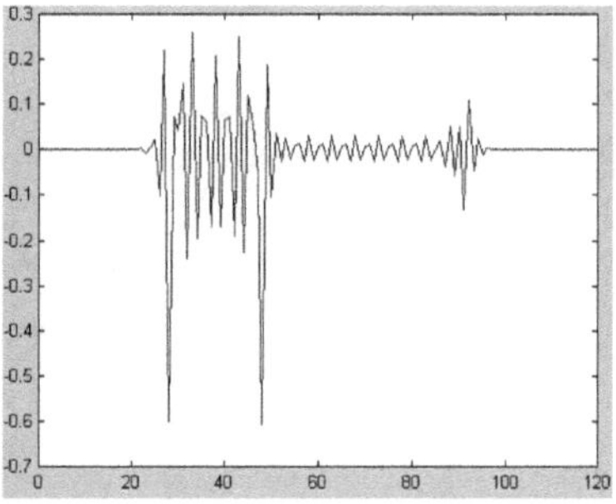

20.2: Three phase-ground fault

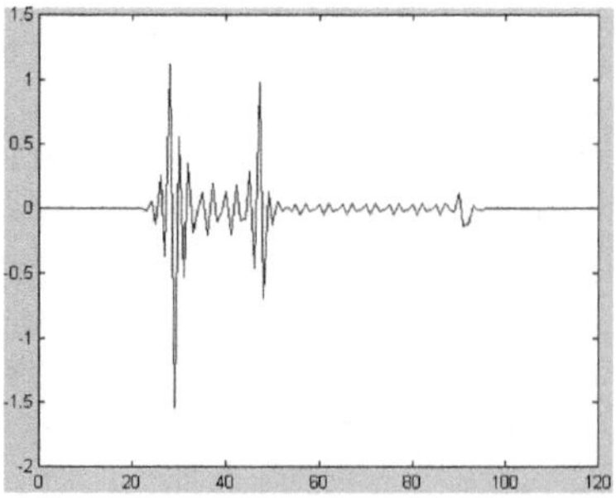

20.3: Phase C-ground fault

Fig. 20. The detail coefficient of faults [72]

of the method. Liang [37] proposed an algorithm for fault classification based on Wavelet Multiresolution Analysis (MRA) with Daubechies four (D-4) wavelet measuring and comparing sharp variation in the values of the currents for the three phases in the first stage MRA detail signals extracted from the original signal. Cheng [13] used a B-Spline wavelet transform for fault classification purposes based on threshold values as in [37]. Zhao [85] proposed an algorithm with Daubechies eight (D-8) wavelet for fault detection and classification in an underground cable system using two different levels of MRA detail signals. Chanda [12] presented an algorithm for classification of faults based on MRA with Daubechies eight (D-8) wavelet transforms of the three phase currents on a transmission line fed from both ends.

Imriš [26] and Chanda [12] were both using wavelets for data preprocessing before applying the fault location and classification algorithms to the recorded transients on transmission lines. Imriš analyzed ground fault transients in 110kV networks using low frequency records for fault location purposes. As shown in Figure 21, ground fault signals consist of different frequency components, which result from charging or discharging of the network capacitances. The charge transient is generated by the voltage rise in sound phases during a single-phase to ground fault. This means that a charge transient is always a side effect of the ground fault. Moreover, it is typically of strong amplitude and, therefore, is reasonable to use for single-phase to ground fault location. The fault transients are mixed with the other signals as noise and fundamental frequency components. Sometimes the transient can be short in duration and also small in amplitude. Moreover, the transient can be very close to the fundamental frequency signal in the frequency domain. Therefore, the 50Hz component can negatively affect the fault transient frequency estimation. To enable a more precise analysis of the fault transient, preprocessing is performed with a wavelet filter [26].

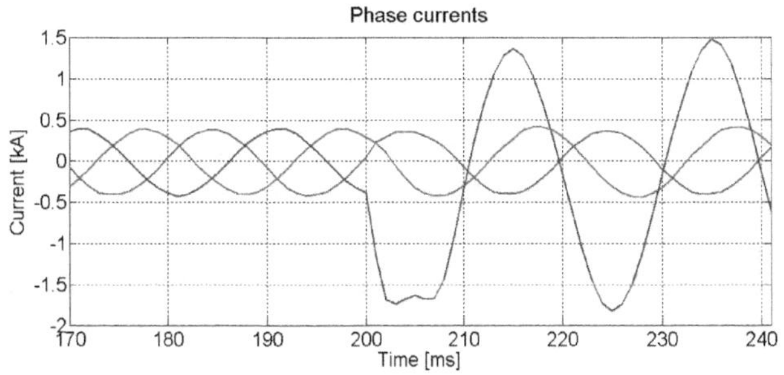

Fig. 21. The recorded single phase to ground fault: Phase currents [26]

The filtering of the signal is performed using a wavelet filter to get the fault transient precisely out of the measured signal. The wavelet filter is set exactly on the frequency of the measured (charge) fault transient estimated by the Fourier transform. The filter's coefficient and its frequency response with an example fault current are shown in Figure 22. The filter coefficients are represented by a Gaussian mother wavelet. After removing the 50Hz component, the charge transient frequency is detected. In the case of the phase currents shown in

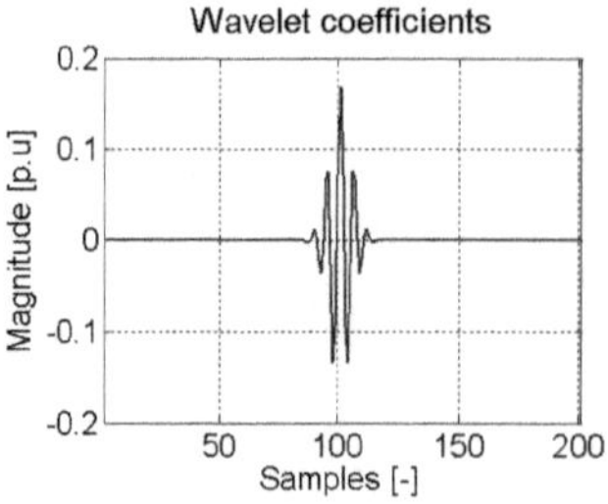

22.1: Wavelet coefficients

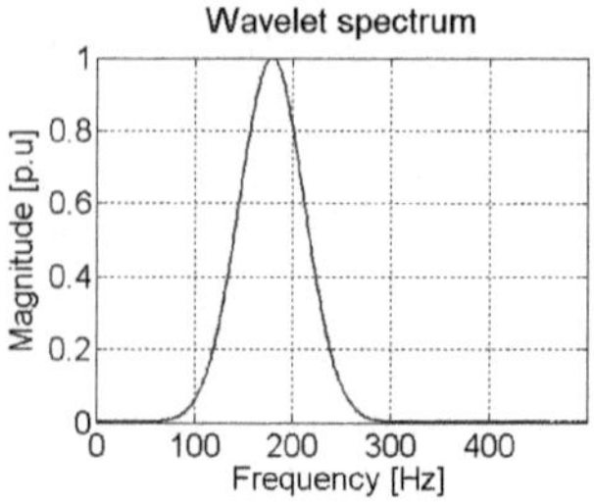

22.2: Wavelet spectrum

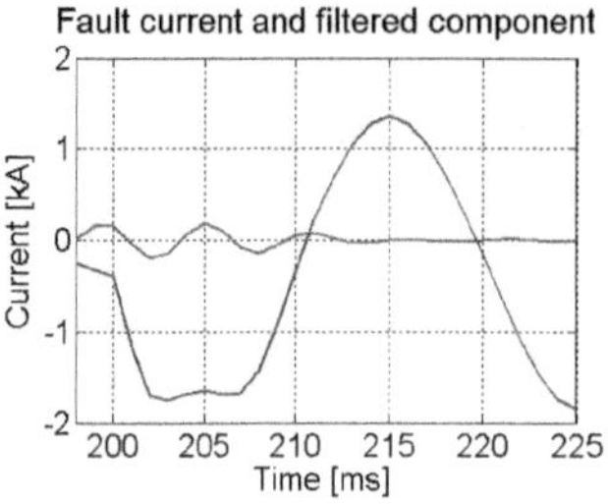

22.3: Fault current and filtered component

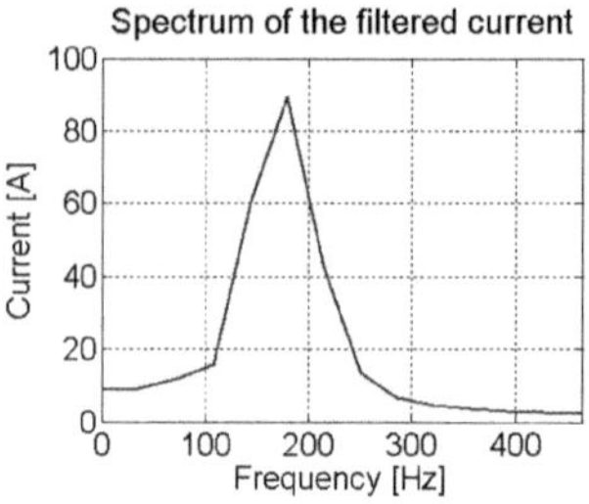

22.4: Spectrum of the filtered current

Fig. 22. Pre-processing of the fault signals using wavelet [26]

Figure 21, the charge transient frequency is detected at 178.57Hz. These transients can then be used for fault location if they are detected. Transient fault location is based on the estimation of the fault path inductance L_f from the detected fault transients. The fault path inductance can be calculated directly from the filtered signal (the charge transient) [26],

$$L_f = \frac{1}{\omega_c} Im \left[\frac{v_c(t,f)}{i_c(t,f)} \right] = \frac{1}{3}(L_0 + L_1 + L_2) \cdot l_f, \tag{10}$$

where ω_c, v_c and i_c are the angular frequency, voltage and current of the charge transient. The fault distance is l_f. The constants L_0, L_1 and L_2 are the zero-, positive- and negative-sequence inductances of the faulty line per km. In (10), t represents time and f the frequency.

Chanda, on the other hand, simulated the application of Wavelet MRA theory for the classification of faults on a power transmission line as shown in Figure 23. The base values of the voltage and power in the system are taken as 230kV and 100MVA. The frequency of the system is taken to be 50Hz. The phase current signals are recorded at the two ends (P, Q). The generated time domain signals are sampled every 80μs and then used for the analysis using wavelet transform. The data considered in the analysis is assumed to be of finite duration and of length 2^N, where N is an integer. If N is chosen to be 9, the total duration of the analysis comes to 2^9 (=512) $\times$ 80μs = 40.96ms, which is about two cycles and is sufficient for the fault analysis. With $N = 9$, there are $(N + 1) = 9 + 1 = 10$ wavelet levels. If these 10 levels are added together, then the original signal is faithfully reproduced at each of the sample points.

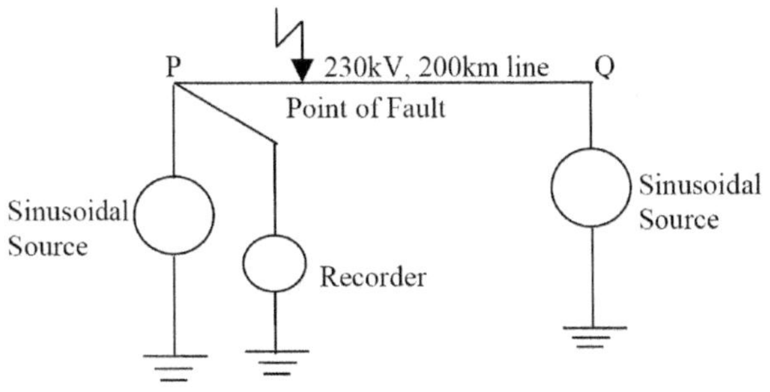

Fig. 23. 230kV, 200km transmission line system used for simulation studies [12]

Daubechies Eight (D-8) wavelet is used in this work for the analysis, since it closely matches the signal to be processed (this is of the utmost importance in wavelet applications). Due to the unique feature of providing multiple resolution in both time and frequency by wavelets, the sub-band information can be extracted from the original signal. When applied to faults, this sub-band information of a faulted power system is seen to provide useful signatures for faults. By randomly shifting the point of fault on the transmission line, a number of simulations can be carried out. The generated time domain signal for each case is analyzed using the wavelet transform. From the different decomposed levels, only 3rd level output is considered for the analysis.

The types of faults considered in the analysis are L-G, L-L-G, L-L, L-L-L. The simulations show that the fault inception angle (α_F) has a considerable effect on the phase current samples and, therefore, also on the wavelet transform output of post-fault signals. Through exhaustive experimentation, the authors have concluded that the parameter identified for classification is the summation of 3rd level output for the three phase currents. The results are shown in Figure 24 and Figure 25, where, S_a = Summation of 3rd level values for current in phase 'a', S_b = Summation of 3rd level values for current in phase 'b', and S_c = Summation of 3rd level values for current in phase 'c'.

If $S_a + S_b + S_c \cong 0$, then the fault is classified as an L-L-L fault, in which the magnitude of all the summation values, S_a, S_b and S_c are comparable to each other. This can be verified from the simulation results shown in Figure 24.1 (an L-L-L fault at 5km) and Figure 24.2 (an L-L-L fault at 195km).

If $S_a + S_b + S_c \cong 0$ and also if the sum of two of the summations S_a, S_b and S_c is equal to zero, *i.e.*, the magnitude of one of the summations is very small and almost negligible in comparison to the equal magnitudes of other two summations, then the fault is classified as an L-L fault, *i.e.*, if $S_a + S_b = 0$, it is a fault involving the a and b phase; $S_a + S_c = 0$, it is a fault involving the a and

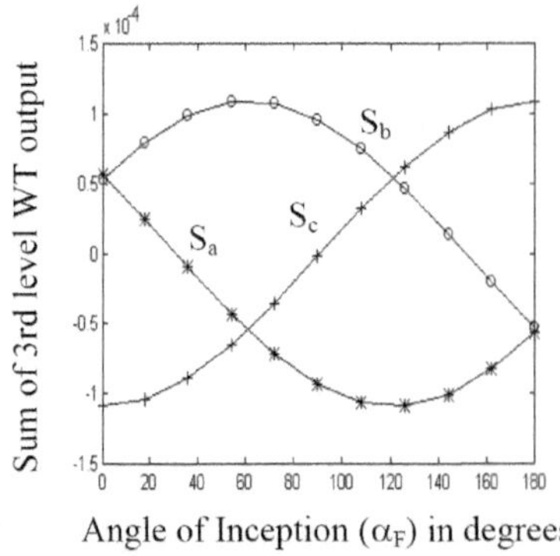

24.1: Effect of Inception angle (α_F) for L-L-L Fault at 5km

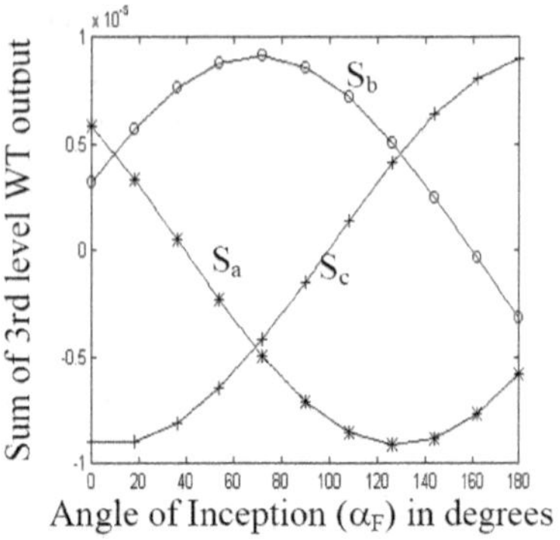

24.2: Effect of Inception angle (α_F) for L-L-L Fault at 195km

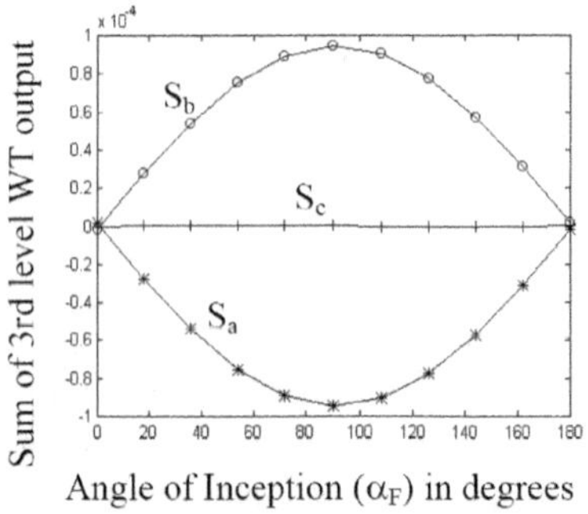

24.3: Effect of inception angle (α_F) for L-L fault involving phases 'a', 'b' at 5km

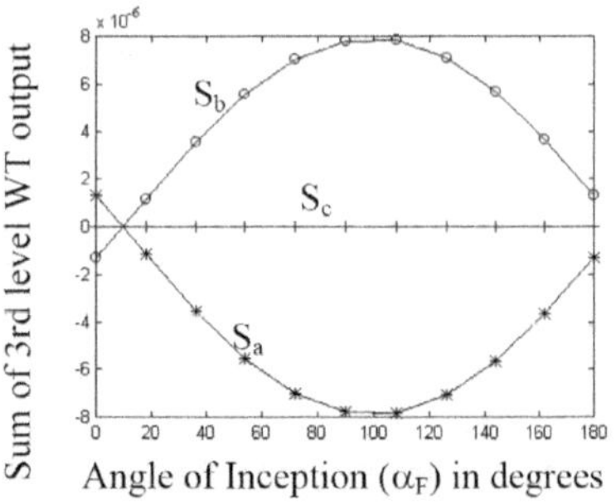

24.4: Effect of inception angle (α_F) for L-L fault involving phases 'a', 'b' at 195km

Fig. 24. Preprocessing of the L-L and L-L-L fault signals using wavelet [12]

c phase; and $S_b + S_c = 0$, it is a fault involving the b and c phase. The results of classifying an L-L fault involving the a and b phase are shown in Figure 24.3 (an L-L fault at 5km) and Figure 24.4 (an L-L fault at 195km).

If $S_a + S_b + S_c \neq 0$, then it is either an L-G or L-L-G fault. If the absolute value of any two summations (S_a, S_b, S_c) is equal and is always much smaller than the absolute value of the 3rd summation, then it is an L-G fault. If $|S_b| = |S_c|$ & $<< |S_a|$, it is an L-G fault involving phase a; if $|S_a| = |S_c|$ & $<< |S_b|$, it is an L-G fault involving phase b; and if $|S_a| = |S_b|$ & $<< |S_c|$, it is an L-G fault involving phase c. The results of classifying an L-G fault involving the a phase are shown in Figure 25.1 (an L-G fault involving the a phase at 5km) and Figure 25.2 (an L-G fault involving the a phase at 195km).

If the absolute value of any two summations (S_a, S_b, S_c) is not equal and is always much higher than the absolute value of the 3rd summation, then it is an L-L-G fault as shown in Figure 25.3 and 33.4. Furthermore, provided that $S_{min} = min(|S_a|, |S_b|, |S_c|)$, if $S_{min} = |S_c|$ and $<< |S_a|$ or $|S_b|$, then it is an L-L-G fault involving phases a, b and ground; if $S_{min} = |S_b|$ and $<< |S_a|$ or $|S_c|$,

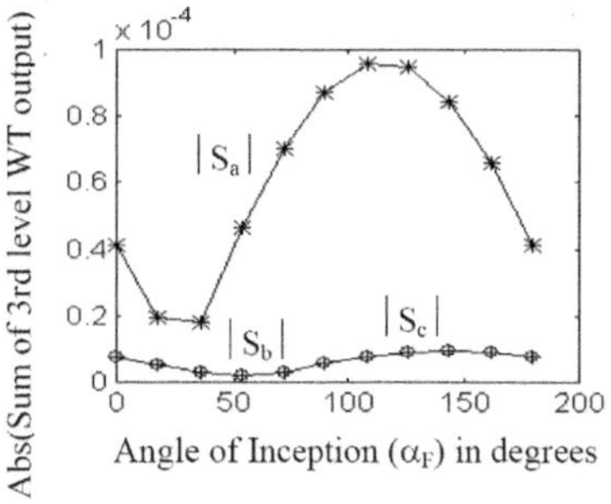

25.1: Effect of inception angle (α_F) for L-G fault involving phase 'a' and ground at 5km

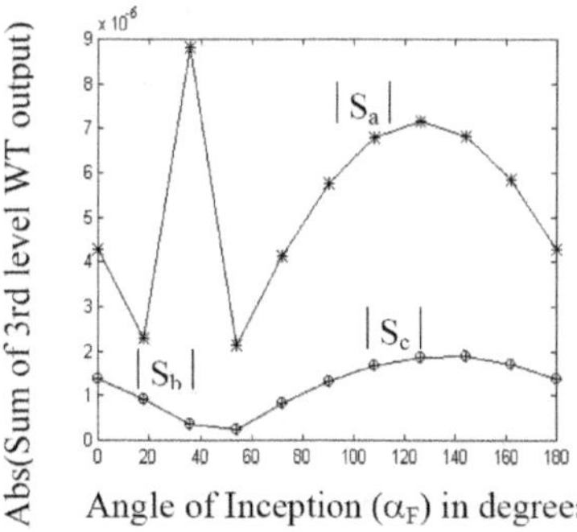

25.2: Effect of inception angle (α_F) for L-G fault involving phase 'a' and ground at 195km

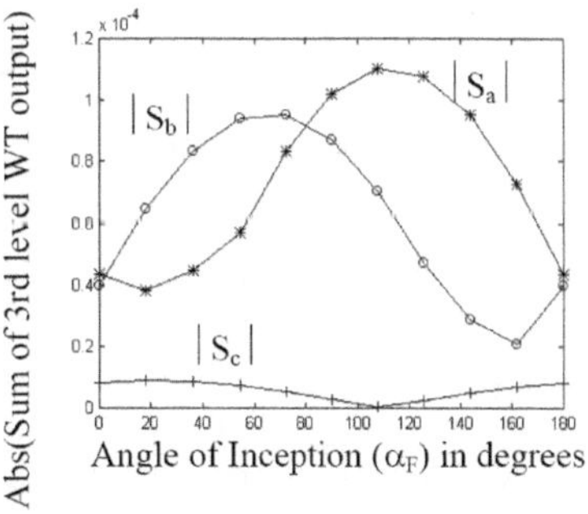

25.3: Effect of inception angle (α_F) for L-L-G fault involving phases 'a', 'b' and ground at 5km

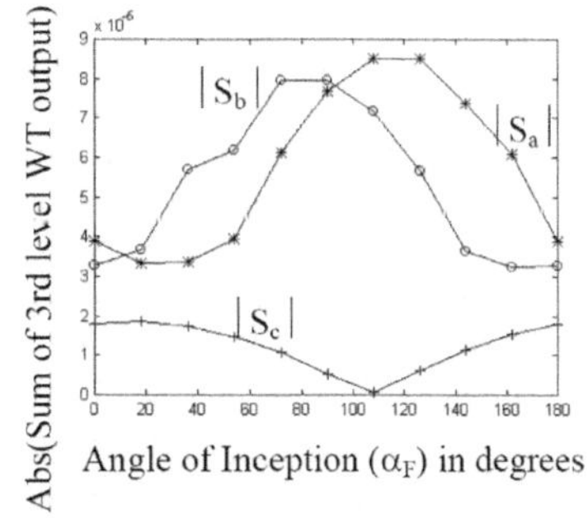

25.4: Effect of inception angle (α_F) for L-L-G fault involving phases 'a', 'b' and ground at 195km

Fig. 25. Preprocessing of the L-G and L-L-G fault signals using wavelet [12]

then it is an L-L-G fault involving phases a, c and ground; and if $S_{min} = |S_a|$ and $<< |S_b|$ or $|S_c|$, then it is an L-L-G fault involving phases b, c and ground.

4.2 Combination of the Wavelet and Neural Network Techniques for Fault Detection

Recently, research has been focused more on combining the wavelet and neural network algorithms for fault identification in power systems. Wavelet analysis is applied to analyze transient signals, then a neural network algorithm is utilized for the identification of problems. The basic neural network structures and design algorithms are given in an Appendix (see [15]).

Ramaswamy [72] and Kashyap [29] proposed a method that incorporates a Probabilistic Neural Network (PNN) for detecting the type of power system fault. The PNN has preference over other Artificial Neural Network (ANN) algorithms in the application of power system fault classification. It combines the merits of

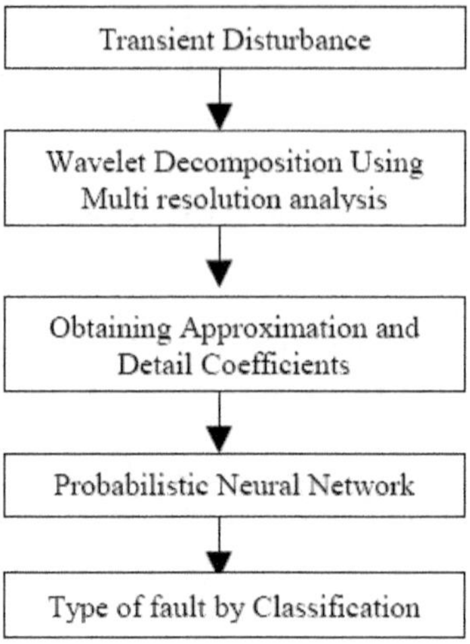

Fig. 26. Procedure for fault detection and classification [72]

statistical theory with that of ANN. Figure 26 shows the entire procedure for fault recognition.

Three power system faults, *i.e.*, phase A-Ground fault, double phase AB-Ground fault and 3-phase symmetrical fault are simulated and investigated. Transients are analyzed by the Meyer mother wavelet, and Multiple Level Decomposition of the transient disturbance was generated. The final level detail coefficient is considered for the feature detection and used in the Probabilistic Neural Network.

Figure 27 shows the model of a Probabilistic Neural Network, which classifies these three power faults [29]. The PNN Architecture consists of four layers, *i.e.*, the Input Layer: consisting 119 Neurons, number of samples of the detail coefficient; the Exemplar Layer: consisting of 9 Neurons, 3 faults $\times$ 3 sets of data for each fault; the Summation Layer: consisting of 3 Neurons, equal to the number of faults; and the Decision Layer: follows the "Winner take all" mechanism.

Researchers also proposed solutions for digital relays for transmission line protection. Martin has simulated a system with two generators and three lines (distributed parameters model) [38]. Simulations include 3 different faults at different distances from the beginning of each line, several fault resistances, inception angles, and steady states. The process consists of a preprocessing module based on Discrete Wavelet Transform (DWT) combined with an ANN for detecting and classifying fault events.

Wavelets of length six (N=6) are used for the relay to operate in real time. These wavelets can be expressed as functions of two parameters α and β [8]. By varying parameters α and β, a family of length-6 wavelets can be generated. For a certain range of variation of these parameters, the generated wavelets are classified according to their performance for this particular application. The

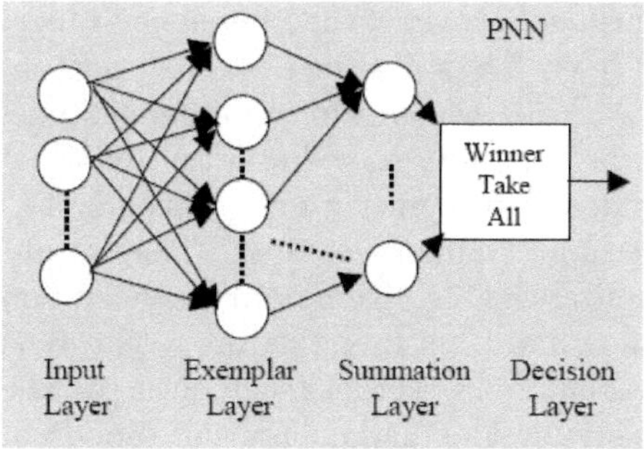

Fig. 27. Model of a Probabilistic Neural Network. Detail coefficient is fed to the input layer and the type of fault is obtained at the output [29].

parameters for the length-6 wavelet with quasi-optimal performance are $\alpha = 0.48\pi$ and $\beta = -0.35\pi$.

Three independent multilayer (two hidden-layers), feed-forward neural networks have been used for detection, classification and location of fault transients. The ANNs are fed with the six detail signals (three currents and three voltages). The input data of the ANN is organized in a sliding-window of a quarter of a cycle, thus a faster response is obtained since only a quarter of a cycle from the occurrence of the fault is required. The input vector has 24 elements. The detection ANN has one output neuron, which indicates the existence of a fault. The location net has one neuron that indicates if the fault has occurred in the protected zone. The classification ANN output layer has four neurons indicating which phases (A, B, C) or ground are involved in the fault event. An error back-propagation algorithm has been used for training the ANN.

4.3 Time-Frequency Representation Technique for Classifying Power Quality Disturbances

Voltage disturbances are the most frequent cause of a broad range of disruption in power supply systems. Power quality (PQ) disturbances cover a broad frequency range and significantly different magnitude variations. Typically, there are five major PQ related waveform events: harmonics, voltage sags, capacitor high frequency switching, capacitor low frequency switching, and normal voltage variations. Harmonics distortion is the most common power quality problem [17].

Approaches for automated detection and classification of PQ disturbances proposed recently are based on wavelet analysis and artificial neural networks [21,55,77]. To enhance the sufficiency for supporting a robust PQ monitoring system is one of the most interesting research areas for scientists.

A wavelet transform on a PQ signal produces a multiresolution decomposition (MRD) matrix, which contains time domain information for the signal at different scales. This property has made wavelets a promising tool for detecting and extracting disturbance features for various types of PQ events [21,55,77]. However, there are still some issues to be resolved in wavelet-based methods.

First, while PQ disturbances cover a wide frequency range, a very small subset of the MRD matrix (*e.g.*, five scales in [77]) may not be a sufficient or optimized selection for capturing features for all different types of PQ events. This feature selection scheme may filter some important information for classification and potentially degrade the recognition rates. Second, the wavelet-based methods relatively require more training examples. They result in greater efforts or difficulties when adapting the algorithm onto a new system.

Wang and Mamishev had been investigating a feature extraction tool, time-frequency ambiguity plane with kernel techniques [15,82,83], which is new to the power engineering field. The fundamentals of time-frequency representation (TFR) is presented in Appendix D. The essence of the feature extraction is to project a PQ signal onto a low-dimension time-frequency representation (TFR), which is deliberately designed for maximizing the separability between classes. A distinct TFR is designed for each class. The classifiers include a Heaviside-function linear classifier and neural networks with feedforward structures.

A set of 860 real world voltage signals from five event classes were collected from industrial databases for the training and testing of the algorithm. Each voltage signal to be identified consists of five cycles of a voltage waveform sampled 128 times per cycle, and has a length of 640 sampling points. In the training stage, four classification-optimal kernels are designed for separating five classes sequentially. The kernel design process selects nine locations from the time-frequency ambiguity plane.

Classification kernels are designed for training according to Fisher's discriminant function. Fisher's discriminant function (FDF), which was developed by R. A. Fisher in the 1930s, is a method that projects high dimensional data onto low-dimensional space for classification. The projection maximizes the distances between the means of the different classes while minimizing the variances within each class.

The kernel $\varphi_i[\eta,\tau]$ is defined as a binary matrix (each matrix element is either 0 or 1). Feature points are ambiguity plane points of locations (η,τ) where $\varphi_i[\eta,\tau]=1$. Therefore, the process of feature extraction is to select points that are optimal for the classification task from the ambiguity plane.

A total number of $N-1$ kernels need to be designed for an N-class PQ classification system. A kernel K_s works as either a single-class separator or a group-class separator. In the single-class separator case, kernel K_i is dedicated to discriminate class i from all remaining classes $\{i+1,...,N\}$. In the group-class separator case, kernel K_i is dedicated to discriminate a class group $\{i,i+1,...,i+m\}$ from all remaining classes $\{i+m+1,i+m+2,...,i+m+N\}$. In the second case, additional kernels are needed in order to uniquely identify class i from the class group $\{i,i+1,...,i+m\}$, and the total number of kernels required for an N-class classification is still N-1.

Ambiguity planes for all training signals are calculated before the Fisher's discriminant function is applied for the kernel design. Assume there are n classes and totally N_i training examples for class i. The notation $A_{ij}[\eta,\tau]$ represents the ambiguity plane of the jth training example in the ith class.

With the Fisher's criterion, locations on the ambiguity plane are ranked according to their importance for classification. A certain amount of training data from each class is needed for feature ranking in this statistical method. For example, when designing kernel i, a Fisher's discriminant score is calculated for each location (η, τ) on the ambiguity plane,

$$J_{Fi}(\eta, \tau) = \frac{(m_i[\eta, \tau] - m_{i-remain}[\eta, \tau])^2}{D_i^2[\eta, \tau] - D_{i-remain}^2[\eta, \tau]}, \tag{11}$$

where $m_i[\eta, \tau]$ and $m_{i-remain}[\eta, \tau]$ represent two means of location (η, τ),

$$m_i[\eta, \tau] = \frac{1}{N_i}\sum_{j=1}^{N_i} A_{ij}[\eta, \tau], \tag{12}$$

$$m_{i-remain}[\eta, \tau] = \frac{\sum_{k=i+1}^{5}\sum_{j=1}^{N_k} A_{kj}[\eta, \tau]}{\sum_{k=i+1}^{5} N_k}, \tag{13}$$

and $D_i^2[\eta, \tau]$ and $D_{i-remain}^2[\eta, \tau]$ represent two variances of location (η, τ),

$$D_i^2[\eta, \tau] = \frac{1}{N_i}\sum_{j=1}^{N_i}(A_{ij}[\eta, \tau] - m_i[\eta, \tau])^2, \tag{14}$$

$$D_{i-remain}^2[\eta, \tau] = \frac{\sum_{k=i+1}^{5}\sum_{j=1}^{N_k}(A_{kj}[\eta, \tau] - m_{i-remain}[\eta, \tau])^2}{\sum_{k=i+1}^{5} N_k}. \tag{15}$$

Locations (η, τ) that receive the highest discriminant score $J_{Fi}(\eta, \tau)$ are selected as feature locations.

By examining Fisher's discriminant score $J_{Fi}(\eta, \tau)$, the optimal numbers of feature points for each individual kernel have been found: one for the harmonics kernel; two for the voltage sag kernel; three for the capacitor switching kernel; and three for the capacitor high-frequency switching kernel. Therefore, nine feature locations are selected for these four kernels.

Each classification node consists of a kernel function and a classifier. A Heaviside-function linear classifier is used for the task of separating harmonics that is a great distance apart from other fault cases and is relatively easy to discriminate. Neural networks with small numbers of input nodes are used for all other classification tasks. The structure of the ANN for discriminating sags is 2-12-2 (input layer node number - hidden layer node number - output layer node number); the one for capacitor switching is 3-10-2; and the one for capacitor high-frequency switching is 3-10-2. The transfer and training functions adopted for the ANN include: the hyperbolic tangent sigmoid transfer function as the transfer function for the hidden layer, the linear transfer function as the transfer function for the output layer, backpropagation as the network training function, the gradient descent learning function as the weight learning function, and the mean squared error function as the performance evaluation function.

5 Data Preparation for Manitoba Hydro HVDC PSFC

Prior to feature extraction, data preparation and signal preprocessing are required to define the characteristics of power system signals. The fault data from the Transcan^{TM} is in binary format and non-editable. Data preparation consists of two steps. The first step is to convert the data from binary format to ASCII (American Standard Code for Information Interchange) format. The second step is to separate the signals into different groups according to their physical nature (*i.e.*, Pole voltages/currents, 3 AC phase voltages, valve control signals, valve currents).

5.1 Data Conversion

The data recorded by Transcan^{TM} is in binary format and compressed as *.x01 files [80], which are unreadable by humans. Together with the .x01 files, Transcan^{TM} provides *.scf files. The *.scf file is a configuration file and contains the information for data arrangement. It tells how many channels have been scanned. At the Manitoba Hydro Dorsey Station, a fault file has 48 analog and 4 digital channels, with some of them being spares. The *.scf file indicates the scanning order and the physical name for each channel. The first 52 bytes in a .x01 file are used for recording the file name and date. Every 16 bits that follow are allocated for storing one channel data. In each 16-bit data field, the first 12 bits store one digitalized data for a channel and the last 4 bits indicate the channel number.

For this research, a *C++* program has been designed to convert the data to ASCII format (*.dat) from binary format (.x01). Each .x01 file can be converted into 48 *.dat files. Among these files, 23 files are selected to represent the most active and informative signals in the power system for fault classification.

5.2 Signal Grouping

Among the 23 converted signals, some are constant signals and the others are periodic signals. Bus signals, *i.e.*, the 3 AC phase voltages and the pole voltages and currents should be grouped separately from the valve signals. Bus signals will induce more than one fault and usually cause significant problems. Valve signals will affect only one valve group and cause a certain level of decrease or increase of either the pole voltage or current. Table 8 lists the signal groups for a pole 1 file. The number of signal groups will guide the number of the feature sets to be extracted.

6 Signal Preprocessing and Feature Extraction for PSFC

To set up the information table for fault classification, the normal behavior of each signal needs to be clarified and the abnormality of each signal related to each type of fault can then be identified. Signal preprocessing and feature extraction is presented in this section.

Table 8. Signal groups

group 1 (3 signals)	AC voltage A phase, B phase, C phase
group 2 (5 signals)	Pole 1 and 2 voltage, Pole 1 and 2 current, pole current order
group 3 (3 signals)	Vg11 current A phase, B phase, C phase (first valve group in pole 1)
group 4 (3 signals)	Vg12 current A phase, B phase, C phase (second valve group in pole 1)
group 5 (3 signals)	Vg13 current A phase, B phase, C phase (third valve group in pole 1)
group 6 (3 signals)	6-pulse in 3 valve groups
group 7 (3 signals)	Start pulse in 3 valve groups

6.1 Signal Characteristics in Normal Condition

Standard value or waveform of each signal in normal condition is described in the following two tables.

Constant Signals

In the 23 signals converted from .x01 file, the constant signals are pole-current order, alpha order, pole current, pole voltage. Under normal conditions, their standard values are given in Table 9:

Table 9. Constant signals in the 23 converted signals

Pole Current Order	Alpha Order	Pole Current	Pole Line Voltage
±1400 amps	150 degrees	±1400 amps	±450KV, ±300KV, ±150KV

Periodic Signals

The periodic signals are AC Phase Voltages, Phase Currents and 6-pulse Voltages. Their normal waveform and standard peak values are shown in Table 10.

Table 10. Periodic signals in the 23 converted signals

	AC Phase Voltage	Phase Current	6-Pulse Voltage
Amplitude (peak to peak)	200 KV	1400 Amps	27 KV
Waveform	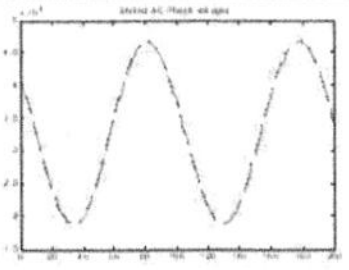	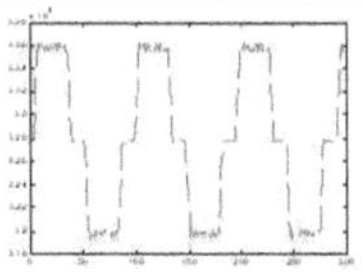	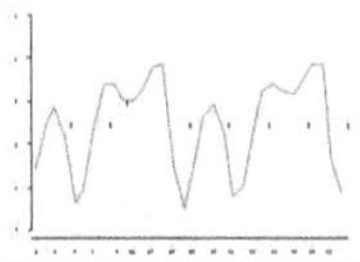

6.2 Feature Extraction of 12 Types of Faults

Extensive time has been spent in studying 676 fault files provided by the Manitoba Hydro Dorsey Station. The 676 .x01 fault files recorded all the events that happened in two recent years and covered 12 types of faults. Together with .x01 files, 676 .trt files are also provided. A .trt file contains the fault information, *i.e.*, the fault cause and type. This fault information is created manually by power system engineers and provides a reliable basis for the target for PSFC training and testing. Various signal processing techniques are applied to analyze the fault signals. They are auto-correlation, cross-correlation, the FFT and inverse FFT, low pass filter, Wavelet MRD, phase shifting, derivatives and coding techniques. The mathematics underlying these techniques can be found in appendices A-D. A total of 17 features or attributes in Table 11 are generated for power system fault classification. The 17 functions that represent these 17 features (attributes) are further described.

A portion of the information table for power system fault classification training is shown in Table 12. This information table is derived from 508 fault files and consists of 508 lines in total, with each line containing 17 features. This table is further processed to prepare for the training sets to calibrate the rough membership Neural Network (rmNN) for fault classification. Also, a portion of the information table for testing is illustrated in Table 13. The testing table consists of 168 rows generated from an additional 168 fault files. The complete training and testing information tables are attached in an appendix available at [15].

Table 11. 17 features/attributes for power system fault classification

A1	Pole voltage sharp dropping
A2	AC voltage disturbance severity
A3	Pole index
A4	Pole 1 or 3 voltage trend
A5	Pole 2 or 4 voltage trend
A6	Pole 1 or 3 current trend
A7	Pole 2 or 4 current trend
A8	Valve current trend - valve group 1, vg*1
A9	Valve current trend - valve group 2, vg *2
A10	Valve current trend - valve group 3, vg *3
A11	Valve current minor disturbance
A12	Pole current closed with normal pole voltage
A13	3 valve groups all closed (True = 1, False = 0)
A14	Same current trend in 3 valve groups (True = 1, False = 0)
A15	Voltage flashover in 6-pulse signal
A16	Valve currents flashover
A17	Valve currents flashover happens only in one valve group (True = 1, False = 0)
Notation: * represents the pole index, i.e. for pole 1, the valve groups are Vg11, Vg12, Vg13	

The following sections cover the details of signal processing for feature extraction and the 17 functions that represent these features (attributes).

Feature 1 – Pole Voltage Sharp Dropping

The pole line voltage is a constant signal and the standard values are ±450KV, ±300KV, ±150KV. In fault 12 *i.e.*, "Disturbance on DC Line", the pole voltage is affected by high frequency interference and causes a sharp drop at the tripping edge. The pole voltage sharp dropping sometimes happens in fault 4 as well. As described in Section 2.2, there are two cases in fault 4, "Pole Line Fault" and "Force Retard". The pole voltage in "Force Retard" decreases slowly while the pole voltage in "Pole Line Fault" drops as sharp and quick as in fault 12. Figure 28 shows the pole voltages in fault 4 and 12. F1121E8D.x01 is a "Force Retard Fault", and F2213569.x01 is a "Pole Line Fault".

Table 12. Partial information table for power system fault classification training

File index	Fault types	Fault file names	A1	A2	A3	A4	A5	A6	A7	A8	A9	A10	A11	A12	A13	A14	A15	A16	A17
1	Fault 1	F08101FE.x01	0	1	1	4	43	23	23	2	2	2	0	0	0	1	0	0	0
31	Fault 1	F224045A.x01	0	1	4	414	4	313	3	2	2	2	0	0	0	1	0	0	0
69	Fault 1, 3	F2913FDD.x01	0	1	1	3	4	2	2	2	2	1	0	0	0	0	0	0	0
70	Fault 1, 3, 5	F04102CE.x01	0	1	1	43	4	3	3	21	2	2	11	0	0	0	1	0	0
91	Fault 1, 3, 5, 7	F082016A.x01	0	1	2	4	43	2	2	2	21	212	100	0	0	0	2	1	1
93	Fault 1, 3, 5, 11	F0822697.x01	0	1	2	4	43	3	34	2121	2	2	10	0	0	0	10	0	0
99	**Fault 1, 3, 11**	**F0822405.x01**	**0**	**1**	**2**	**3**	**43**	**34**	**3234**	**2**	**2**	**21**	**0**	**0**	**0**	**0**	**7**	**0**	**0**
103	Fault 1, 3, 11	F0922884.x01	0	1	2	4	42	34	34	2	2	2121	0	0	0	0	28	0	0
104	Fault 1, 3, 11	F0810140.x01	0	1	1	4	43	34	34	2	2	2121	0	0	0	0	7	0	0
128	**Fault 1, 4**	**F1112E8D.x01**	**0**	**1**	**1**	**414**	**4**	**313**	**3**	**212**	**212**	**212**	**0**	**0**	**0**	**1**	**1**	**0**	**0**
131	Fault 1, 4	F2212CD7.x01	1	1	1	431	4	31	34	21	21	21	0	0	1	1	1	0	0
150	**Fault 1, 4**	**F2213569.x01**	**1**	**1**	**1**	**414**	**4**	**3123**	**3**	**212**	**212**	**212**	**0**	**0**	**0**	**1**	**1**	**0**	**0**
202	**Fault 1, 5**	**F08101CA.x01**	**0**	**1**	**1**	**4**	**4**	**3**	**3**	**2**	**2**	**2**	**101**	**0**	**0**	**1**	**1**	**0**	**0**
209	Fault 1, 5	F0820165.x01	0	1	2	4	434	3	3	2	2	2	10	0	0	1	2	0	0
219	Fault 1, 6	F112267F.x01	0	1	2	4	41	23	21	21	21	21	0	0	1	1	0	0	0
222	Fault 1, 6	F22225C4.x01	0	1	2	41	1	41	31	1	1	1	0	0	1	1	0	0	0
229	Fault 1, 6, 7	F08226DB.x01	0	1	2	4	1	3	1	1	1	1	0	0	1	1	1	1	1
239	Fault 1, 6, 8	F11226A5.x01	1	1	2	4	14	32	1	12	12	12	0	1	0	1	1	0	0
252	Fault 2	F20404C1.x01	0	2	4	41	4	341	34	2	2	2	0	0	0	1	0	0	0
265	**Fault 2**	**F2713113.x01**	**0**	**2**	**1**	**4**	**3**	**2**	**2**	**2**	**2**	**2**	**111**	**0**	**0**	**1**	**1**	**0**	**0**
267	Fault 2	F2713116.x01	0	2	1	434	323	343	3	2	2	2	111	0	0	1	1	0	0
305	Fault 2, 3	F041075C.x01	0	2	1	3	32	3213	3	1	2	2	0	0	0	0	0	0	0
349	Fault 2, 3, 7	F0820715.x01	0	2	2	3	313	2	2	212	121	2	0	0	0	0	2	6	1
374	**Fault 2, 6**	**F08226BF.x01**	**0**	**2**	**2**	**4**	**421**	**3**	**1**	**21**	**21**	**21**	**0**	**0**	**1**	**1**	**1**	**32**	**0**
392	Fault 3	F0121F8B.x01	0	0	2	3	23	21	21	12	1	2	0	0	0	0	0	0	0
403	**Fault 3**	**F0121F8D.x01**	**0**	**0**	**2**	**3**	**212**	**2**	**212**	**21**	**1**	**12**	**0**	**0**	**0**	**0**	**1**	**0**	**0**
415	Fault 3, 5	F1140866.x01	0	0	4	4	3	3	313	2	1	1	100	0	0	0	0	0	0
420	Fault 4	F2410189.x01	1	0	1	41	41	31	31	21	21	21	0	0	1	1	5	0	0
421	Fault 5	F1122499.x01	0	0	2	4	434	2	2	2	2	2	110	0	0	1	1	0	0
422	Fault 5	F27200E4.x01	0	0	2	4	4	3	3	2	2	2	10	0	0	1	0	0	0
440	Fault 8	F2212F95.x01	0	0	1	4	4	1	2	2	2	2	0	1	0	1	0	0	0
441	Fault 8	F222260C.x01	0	0	2	4	4	1	2	2	2	2	0	1	0	1	0	0	0
448	Fault 9	F1121E5D.x01	0	0	2	13	4	12	32	2	2	2	0	0	0	1	0	0	0
465	Fault 9	F2410185.x01	0	0	1	4	14	32	12	2	2	2	0	0	0	1	3	0	0
466	Fault 10	F0813030.x01	0	0	1	4	1	3	1	2	2	2	0	0	0	1	0	0	0
475	**Fault 10**	**F272015F.x01**	**0**	**0**	**2**	**41**	**4**	**21**	**2**	**2**	**2**	**2**	**0**	**0**	**0**	**1**	**0**	**0**	**0**
490	**Fault 12**	**F224070E.x01**	**1**	**0**	**4**	**3**	**3**	**2**	**2**	**2**	**2**	**2**	**0**	**0**	**0**	**1**	**0**	**0**	**0**
504	Fault 12	F2510DC5.x01	1	0	1	4	4	3	3	2	2	2	0	0	0	1	0	0	0
508	Fault 12	F2522498.x01	1	0	2	4	4	3	3	2	2	2	0	0	0	1	0	0	0

Table 13. Partial information table for power system fault classification testing

File index	Fault types	Fault file names	A1	A2	A3	A4	A5	A6	A7	A8	A9	A10	A11	A12	A13	A14	A15	A16	A17
1	Fault 1	**F0812D94.x01**	0	1	1	4	43	3	3	2	2	2	0	0	0	1	0	0	0
30	Fault 1	**F2240598.x01**	0	1	4	414	4	313	3	2	2	2	0	0	0	1	0	0	0
46	Fault 1, 3	**F2513620.x01**	0	1	1	3	4	3	3	2	2	1	0	0	0	0	0	0	0
53	Fault 1, 3, 5, 7	F2720944.x01	0	1	2	3	34	2	2	2	2	1212	10	0	0	0	2	1	1
54	Fault 1, 3, 5	**F08136DC.x01**	0	1	1	3	4	3	3	2	2	1	110	0	0	0	1	0	0
56	Fault 1, 3, 8	F08228BE.x01	0	1	2	4	434	3	1	2	2	212	0	1	0	0	2	0	0
57	Fault 1, 3, 11	**F08224D9.x01**	0	1	2	4	43	3	3	21	2	2	0	0	0	0	7	0	0
82	Fault 1, 4	**F2512DE2.x01**	0	1	1	414	4	2312	2	212	212	212	0	0	0	1	1	0	0
86	Fault 1, 5	**F2710116.x01**	0	1	1	4	4	3	3	2	2	2	1	0	0	1	1	0	0
87	Fault 1, 5	**F27101E6.x01**	0	1	1	4	4	3	3	2	2	2	1	0	0	1	1	0	0
90	Fault 1, 6	F2420314.x01	0	1	2	41	41	431	431	21	21	21	0	0	1	1	7	0	0
93	Fault 1, 6, 8	F1122405.x01	0	1	2	4	41	1	21	21	21	21	0	1	1	1	1	0	0
94	Fault 1, 8	F1213907.x01	0	1	1	4	43	1	343	2	2	2	0	1	0	1	0	0	0
111	Fault 2	**F22124E5.x01**	1	2	1	4	4	3	3	2	2	2	0	0	0	1	0	0	0
116	Fault 2, 3	**F22135CA.x01**	0	2	1	3	4	3	3	2	2	1	0	0	0	0	0	0	0
121	Fault 2, 3, 5	F08163D5.x01	0	2	1	3	4	3	3	2	2	1	110	0	0	0	1	0	0
128	Fault 2, 3, 5, 7	F27208F3.x01	0	2	2	3	3	2	2	12	21	2	1	0	0	0	3	1	1
129	Fault 2, 4	F22124FD.x01	1	2	1	41	1	321	31	21	21	21	0	0	1	1	1	0	0
130	Fault 2, 5	F0814371.x01	0	2	1	4	1	343	1	2	2	2	1	0	0	1	1	0	0
132	Fault 2, 6	F113044A.x01	0	2	3	431	431	31	321	21	21	21	0	0	1	1	8	0	0
134	Fault 2, 6	F242042B.x01	0	2	2	4	41	34	321	21	21	21	0	0	1	1	1	0	0
135	Fault 2, 6, 7	F0827040.x01	0	2	2	3	1	3	1	1	1	1	0	0	1	1	0	2	1
136	Fault 2, 4, 7	F2923565.x01	0	2	2	3	1	23	1	1	1	1	0	0	1	1	2	4	1
137	Fault 2, 8	F1213908.x01	0	2	1	4	31	1	31	2	2	2	0	0	0	1	0	0	0
138	Fault 2, 9	F1123FF6.x01	0	2	2	3	4	3	34	2	2	2	0	0	0	1	0	0	0
139	Fault 3	**F0121F8B.x01**	0	0	2	3	23	21	21	12	1	2	0	0	0	0	0	0	0
144	Fault 3, 5	**F0121F8E.x01**	0	0	2	3	3	2	2	12	1	2	1	0	0	0	1	0	0
145	Fault 3, 5, 9	F1112BAE.x01	0	0	1	4	3	3	323	2	1	1	100	0	0	0	0	0	0
146	Fault 4	F2410819.x01	1	0	1	41	41	31	31	21	21	21	0	0	1	1	5	0	0
147	Fault 6	F1130FA4.x01	0	0	3	1	1	1	1	1	1	1	0	0	1	1	0	0	0
148	Fault 6, 8	F082B62B.x01	1	0	2	4	14	32	1	12	12	12	0	1	0	1	2	0	0
151	Fault 8	F22129F5.x01	0	0	1	4	4	1	2	2	2	2	0	1	0	1	0	0	0
152	Fault 8	F2210A1C.x01	0	0	1	4	4	1	2	2	2	2	0	1	0	1	0	0	0
153	Fault 9	F1122E5D.x01	0	0	2	13	4	12	32	2	2	2	0	0	0	1	0	0	0
156	Fault 9	F2410810.x01	0	0	1	4	4	32	32	2	2	2	0	0	0	1	0	0	0
158	Fault 10	F1121E95.x01	0	0	2	31	4	21	2	2	2	2	0	0	0	1	0	0	0
167	Fault 12	F2510CD5.x01	1	0	1	4	4	3	3	2	2	2	0	0	0	1	0	0	0
168	Fault 12	F2522894.x01	1	0	2	4	4	3	3	2	2	2	0	0	0	1	0	0	0

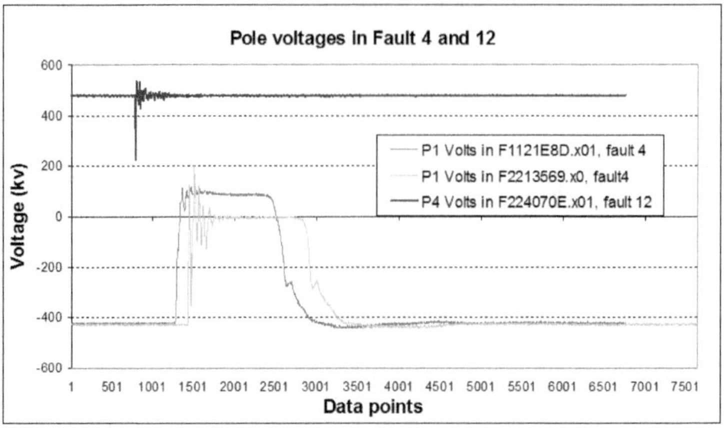

Fig. 28. Pole line voltages with sharp dropping

The derivative of pole line voltages is an efficient method to detect the sharp dropping of the pole line voltage. A 4-point averaging for noise compression is applied before the derivative. The derivative result is shown in Figure 29. It is noticeable that the derivative of P1 in F2213569.x01 and F224070E.x01 both have sharp peaks, while the P1 in F1121E8D.x01 has fairly small output. The threshold to determine a sharp pole voltage drop is 100. The value of feature 1 is 1 for F2213569.x01 and F224070E.x01, and 0 for F1121E8D.x01.

The function f_1 representing this feature is defined by (16)

$$f_1(x) = \begin{cases} 1 & if \ max(derivitive(average(x))) > 100, \\ 0 & otherwise, \end{cases} \tag{16}$$

where x is the discrete pole voltage signal in a fault file.

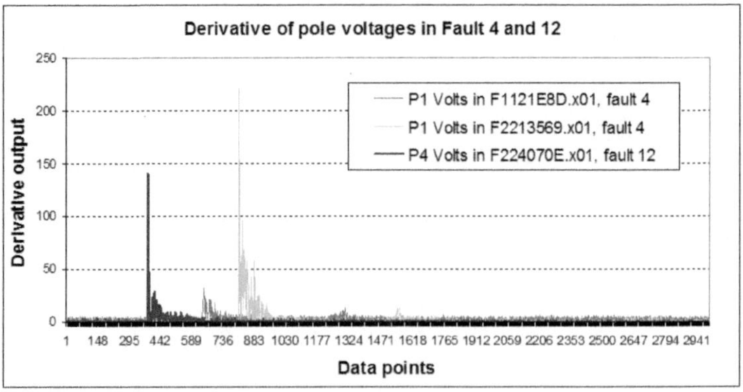

Fig. 29. Derivative of pole line voltages

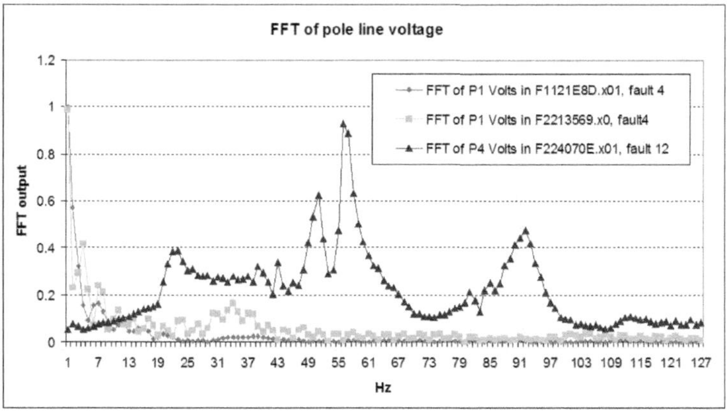

Fig. 30. FFT analysis of pole line voltages

Figure 28 shows that the pole voltage oscillates at the tripping edge in both "Pole Line Fault" (F2213569.x01) and "Disturbance on DC Line" fault (F224070E.x01). The FFT analysis in Figure 30 shows that "DC Disturbance on DC Line" contains higher frequency components. It has a FFT peak at 60Hz, which indicates that the interference is possibly from the AC line. The FFT peaks for both cases of fault 4 are located lower than 6Hz. This feature is very useful and will be added to improve the accuracy of the fault classification system.

Feature 2 – AC Disturbance

The three AC phase voltages, namely the A-phase, B-phase and C-phase, have a fixed 120° phase difference from each other. It is found by studying the data file that one period of AC phase voltage is represented by 96 data points. So if B-phase is shifted 32 points and C-phase is shifted 64 points, the shifted B-phase and C-phase will be exactly the same as the A-phase in normal condition. If the AC voltages have distortion, it can be detected by an error signal, which is calculated by

$$err = \frac{|(A{-}phase) - (shifted\,B{-}phase)|}{3} + \frac{|(shifted\,B{-}phase) - (shifted\,C{-}phase)|}{3} + \frac{|(shifted\,C{-}phase) - (A{-}phase)|}{3}. \quad (17)$$

Taking file F2713113.x01 as an example, the AC phase analysis results are shown in Figure 31. The first graph shows the original three AC phase voltages; the second graph shows the shifted AC phase voltages; and the last one displays the error output of AC voltage signals.

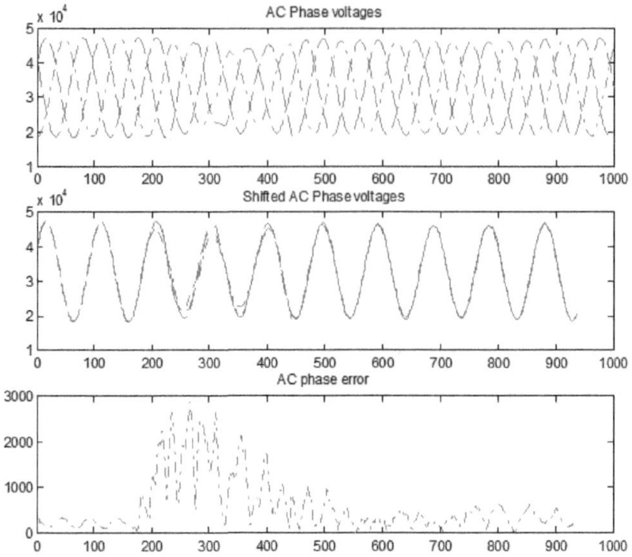

Fig. 31. Analysis of AC phase voltages by phase shift method

The AC disturbance error can be discretized by granule algorithm. The granule functions are designed based on the Gaussian function and can be written as

$$\phi_j(x) = exp\left(-\frac{|x-\mu_j|^2}{2\sigma_j^2}\right), j = 1, 2, 3. \tag{18}$$

Three granule functions need to be designed to discretize the AC disturbance error into three intervals: *low*, *medium* and *high*. The center μ_j and σ_j is estimated based on the 676 files provided by the Manitoba Hydro Dorsey Station. Among those 676 fault files, 240 files are indicated as Minor AC disturbance and 148 files as AC disturbance. The averaged AC phase voltage error calculated from the first 240 files is close to 700 while the averaged AC phase voltage error from the other 148 files is approximately 2100. This leads to a supervised procedure for optimizing the granule function parameters. The Gaussian granule functions for the AC disturbance error discretization are defined as (19) and plotted in Figure 32.

$$\begin{aligned}
\phi_1(x) &= exp\left(-\frac{|x-200|^2}{2\times 200^2}\right), \\
\phi_2(x) &= exp\left(-\frac{|x-700|^2}{2\times 500^2}\right), \\
\phi_3(x) &= exp\left(-\frac{|x-2100|^2}{2\times 900^2}\right).
\end{aligned} \tag{19}$$

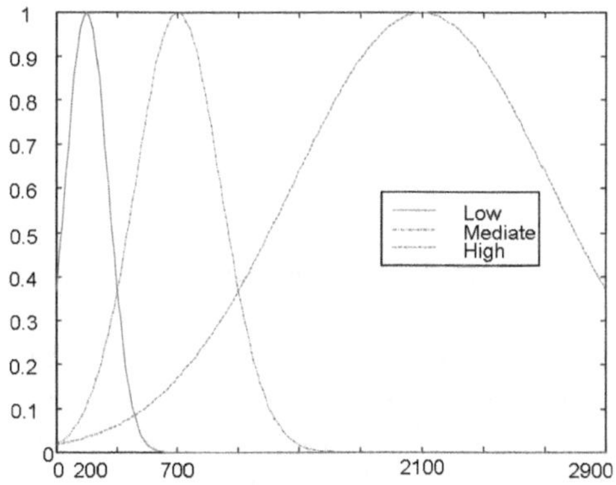

Fig. 32. The granule formula

For an input x, the peak value of the AC voltage error, three granule output $\phi_1(x)$, $\phi_2(x)$ and $\phi_3(x)$ are calculated respectively. If $\phi_1(x)$ is the biggest, the AC error is small enough to be considered as normal and 0 will be assigned. If $\phi_2(x)$ is the biggest, the AC error is moderate implying a minor disturbance and 1 will be assigned. If $\phi_3(x)$ is the biggest, it is a severe AC disturbance and 2 will be assigned.

The function f_2 representing feature 2 can be defined as

$$f_2(x) = \begin{cases} 2 & if \;\; max(\phi_1(x), \phi_2(x), \phi_3(x)) = \phi_3(x), \\ 1 & if \;\; max(\phi_1(x), \phi_2(x), \phi_3(x)) = \phi_2(x), \\ 0 & if \;\; max(\phi_1(x), \phi_2(x), \phi_3(x)) = \phi_1(x), \end{cases} \tag{20}$$

where $x = max(err(V_a, V_b, V_c))$, V_a, V_b and V_c are discrete A-phase, B-phase and C-phase voltages in a fault file.

Feature 3 – Pole Index

The information about the pole index is very easy to retrieve but helpful to identify the fault, "Normal Affected by Another Pole". According to the *.scf file, it is known that the 4th character of the file name indicates the pole index, *i.e.*, F272015F.x01 file is a pole 2 file. Seventeen features of this fault file are listed in Table 14. It is observed that attributes 4 and 6 notify the pole 1 voltage and the current was blocked. All the other features are for pole 2, and they are normal. The fault file of pole 2 was created due to the effect from pole 1. It is reasonable to classify this file as the fault, "Normal Affected by Another Pole".

Table 14. Features for F272015F.x01

File index	Fault types	Fault file names	A1	A2	A3	A4	A5	A6	A7	A8	A9	A10	A11	A12	A13	A14	A15	A16	A17
475	Fault 10	F272015F.x01	0	0	2	41	4	21	2	2	2	2	0	0	0	1	0	0	0

The function f_3 representing this feature (pole index) is defined as

$$f_3(x) = pole\ index, \tag{21}$$

where x is the file name of a fault file.

Features 4, 5, 6 and 7 – Pole Voltage and Current Trend

To derive the trend of the pole line voltages and currents, it is necessary to smooth the waveform by applying a low pass filter. A high order FFT followed by a low order inverse FFT is an alternative to a digital low pass filter. The sampling rate of the Transcan^{TM} system is 6000 points per second. Transcan^{TM} itself is a low pass filter with a cutoff frequency of 3 kHz. An FFT of 8192 points followed by a 32 point inverse FFT is a low pass filter with cutoff frequency around 11.7 Hz. Most interference on the pole line voltage has a frequency of 16 – 90 Hz and is removed by the low pass filter. An example of pole line voltages and currents and their simplified waveforms are shown in Figures 33 and 34, respectively.

The simplified waveforms are represented by a sequence of numbers (codes) based on (22) and (23).

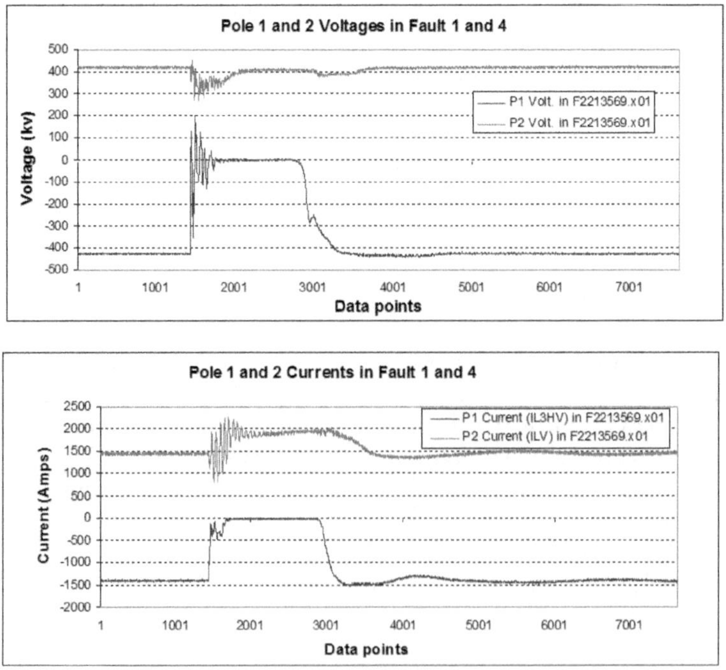

Fig. 33. Pole voltages and currents in fault 1 and 4

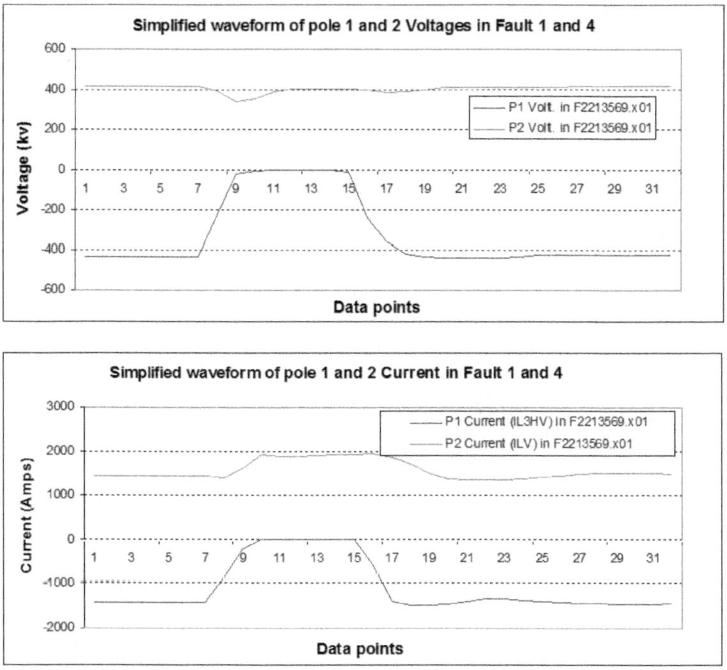

Fig. 34. Simplified waveform of pole voltages and currents in faults 1 and 4

$$Code_V = \begin{cases} 1 & if \ \ |Pole\ voltage| \leq 100, \\ 2 & if \ \ 100 < |Pole\ voltage| \leq 150, \\ 3 & if \ \ 150 < |Pole\ voltage| \leq 300, \\ 4 & if \ \ 300 < |Pole\ voltage|. \end{cases} \tag{22}$$

$$Code_I = \begin{cases} 1 & if \ \ |Pole\ current| \leq 400, \\ 2 & if \ \ 400 < |Pole\ current| \leq 1000, \\ 3 & if \ \ 1000 < |Pole\ current| \leq 2000. \end{cases} \tag{23}$$

The codes for pole voltages and currents in fault F2213569.x01 are listed in Table 15. The original codes contain 32 numbers; the simplified codes remove all the duplicated numbers and for some special cases, *i.e.*, "43134", "42124" and "32123", they are further condensed to "414", "414" and "313", respectively.

Table 15. Codes for pole voltage and current trend

Signal names	Original codes	Simplified codes
Pole 1 volt.	44444443134444444444444444444444	414
Pole 2 volt.	44444444444444444444444444444444	4
Pole 1 current	3333333123333333333333333333333	3123
Pole 2 current	33333333333333333333333333333333	3

Functions f_4, f_5, f_6 and f_7 represent features 4, 5, 6 and 7 respectively and are defined as follows:

$$f_4(x) = Code_V(LF(x)), \tag{24}$$

where x is the discrete pole 1 voltage in a fault file;

$$f_5(x) = Code_V(LF(x)), \tag{25}$$

where x is the discrete pole 2 voltage in a fault file;

$$f_6(x) = Code_I(LF(x)), \tag{26}$$

where x is the discrete pole 1 current in a fault file; and

$$f_7(x) = Code_I(LF(x)), \tag{27}$$

where x is the discrete pole 2 current in a fault file.

In (24) to (27) $Code_V(\cdot)$ and $Code_I(\cdot)$ represent the coding processes and $LF(\cdot)$ represents a lowpass filter.

Features 8, 9 and 10 – Valve Current Trend Vg*1, Vg*2, Vg*3
A normal valve current is a periodic signal with 96 samples per cycle. Reference to A-phase, B-phase and C-phase are 32 and 64 points delayed respectively. The amplitude of the phase current should match the current order in normal condition. A normalized phase current is calculated by (28) and illustrated in Figure 35

$$Normalized\ phase\ current = \frac{phase\ current}{phase\ current\ order}. \tag{28}$$

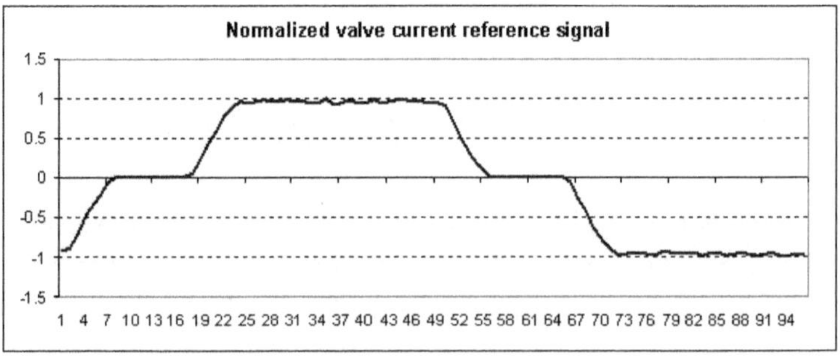

Fig. 35. Normalized valve current reference signal

When a "Valve Current Closed/Blocked/Deblocked" happens, A, B and C phase currents in this group are all closed and/or blocked and/or deblocked. An example of this fault, F0121F8D.x01 is illustrated in Figure 36. In this file, valve group 1 is blocked and valve group 3 is deblocked at a different time. To describe

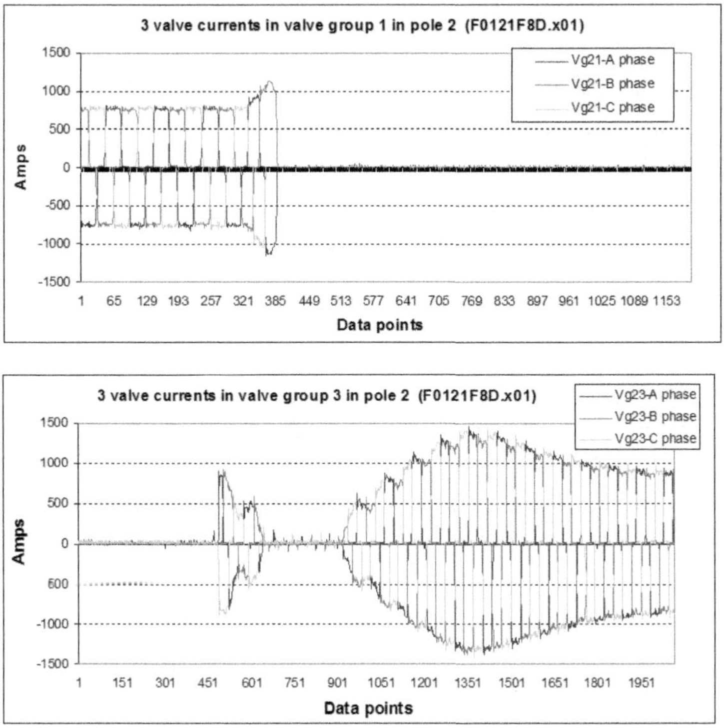

Fig. 36. Valve Current Closed/Blocked/Deblocked error

the trend of a valve current, the correlation algorithm plus a coding method is applied. The correlation theory was described in Appendix A.

The autocorrelation of the normalized valve reference signal $r(j)$ at origin point, denoted ρ_0 is first evaluated by (29).

$$\rho_0 = \sum_{j=0}^{95} r(j)r(j). \tag{29}$$

The Maximum of the cross-correlation of the normalized valve current reference signal and every 96-point segment of the normalized input valve current signal $x(j)$, denoted ρ_0' is calculated by (30).

$$\rho_0' = Max(\rho_0'(i)) = Max(\sum_{j=0}^{95} r(j) \cdot x(j+i)). \tag{30}$$

If ρ_0'/ρ_0 is less than 30%, valve current is considered to be closed and a code 1 will be assigned to this input segment. The valve is considered to be normal or deblocked back to normal if the ratio is bigger than 80% and a code 2 will be assigned. Ratio ρ_0'/ρ_0 for valve group 1 and 3 in fault file F0121F8D.x01

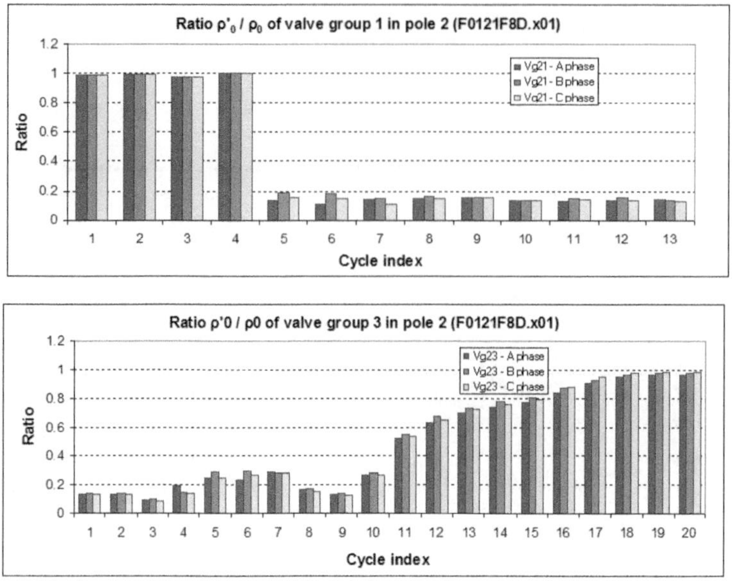

Fig. 37. Ratio ρ_0'/ρ_0 for attribute Valve Current Trend

is illustrated in Figure 37. Usually during the interim from normal to valve closed status and vice versa, various ratio of ρ_0'/ρ_0 will be observed. No code will be assigned to the ratio between 30% and 80%. It is because only the closing and normal stages need to be extracted to describe the trend of valve current. After all the input segments are processed, codes, "2222111111111" for valve group 1 and "1111111111111111111111122222" for valve group 3 are derived. To simplify the codes, only the turn points are maintained. The simplified codes for valve group 1 and 3 are "21" and "12" respectively (see attributes 8 and 10 of fault file 403 in Table 12).

Functions f_8, f_9 and f_{10} representing features 8, 9 and 10 are defined as follows:

$$f_8(x) = Code(max(r \star r)/max(r \star x)), \tag{31}$$

where x is the discrete normalized A, B and C phase valve currents in valve group *1* in a fault file;

$$f_9(x) = Code(max(r \star r)/max(r \star x)), \tag{32}$$

where x is the discrete normalized A, B and C phase valve currents in valve group *2* in a fault file; and

$$f_{10}(x) = Code(max(r \star r)/max(r \star x)), \tag{33}$$

where x is the discrete normalized A, B and C phase valve currents in valve group *3* in a fault file.

In (31) to (33), r denotes the discrete normalized valve current reference signals, $\star$ is the correlation operator, and $Code(\cdot)$ represents the coding processes.

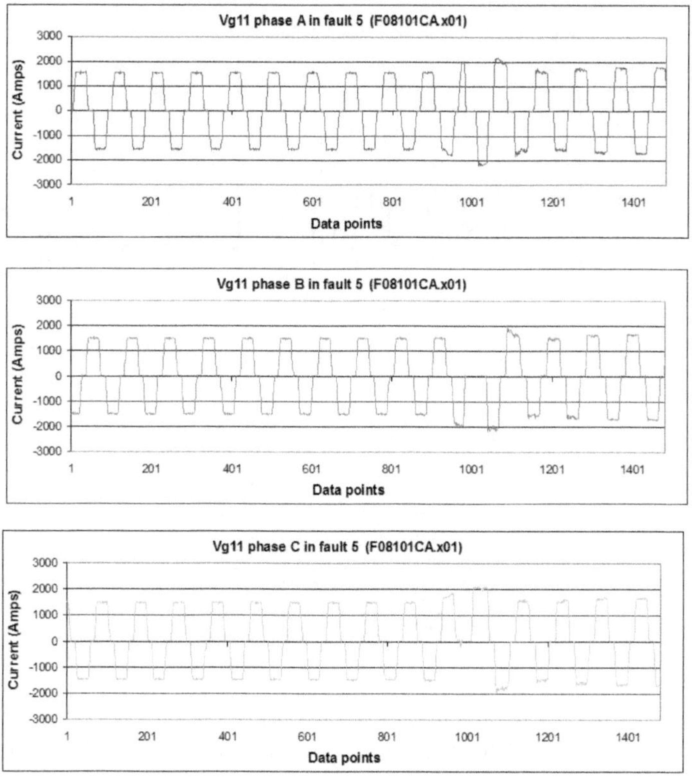

Fig. 38. Valve currents in fault 5

The feature "Valve Current Trend" is very useful for classification of particular faults, *e.g.*, fault 4 ("Line Fault") and fault 6 ("Pole Voltages/Currents Closed/Blocked/Deblocked"). With the occurrence of "Line Fault", all 3 valve groups will have the same trend (the most common pattern is "212"). With this type of fault, pattern "12" or "21" also happens occasionally. With the occurrence "Pole Voltages/Currents Closed/Blocked/Deblocked", all three valve groups behave in the same way. The most commonly observed pattern for this type of fault is "1" (sometimes "12" or "21" can also occur).

Feature 11 – Valve Currents Minor Disturbance

"Valve Currents Minor Disturbance" happens very frequently and usually associated with a fault of "AC Disturbance" or "Valve Current Commutation Failure". The typical waveforms of valve currents minor disturbance is shown in Figure 38, which includes A, B, C, three phases of valve group Vg11 in fault F08101CA.x01. A few cycles present distortions and happen in all three phases. It is considered a valve minor disturbance as long as any one of three phases shows a disturbance.

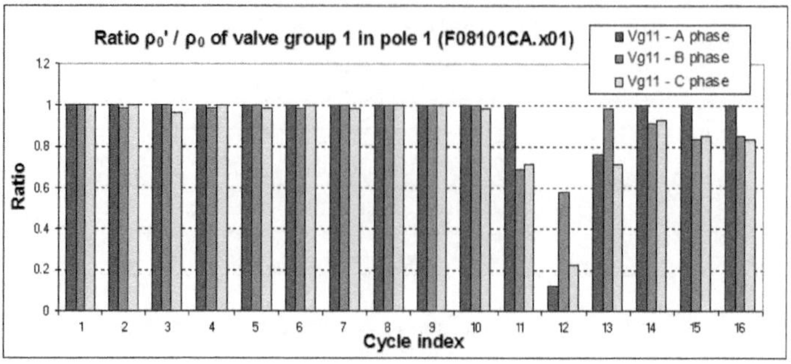

Fig. 39. Ratio ρ_0'/ρ_0 for attribute Valve Currents Minor Disturbance

Table 16. Feature (attribute) codes

Feature code 1	2,2,2,1,2,2,2
Feature code 2	2,2,2,0,2,2,2
Feature code 3	2,2,2,1,1,2,2,2
Feature code 4	2,2,2,0,0,2,2,2
Feature code 5	2,2,2,1,0,2,2,2
Feature code 6	2,2,2,0,1,2,2,2
Feature code 7	2,2,2,1,0,1,2,2,2

To detect "Valve Current Minor Disturbance", the method applied in the feature extraction for "Valve Current Trend" is adopted here. The ratio ρ_0'/ρ_0 is estimated and displayed in Figure 39. The same threshold is used to assign the code. If the ratio $> 80\%$, code "2" is assigned; ratio $< 30\%$, code "0" is assigned; in addition, between 30% and 80%, code "1" is assigned. The feature patterns indicating a valve current minor distortion are listed in Table 16 and the codes derived for phase A, B and C currents of Vg11 in F08101CA.x01 are listed in Table 17. A minor disturbance is detected in all three phases of Vg11 and a final code "1" is assigned. Only when all three phases are normal, should a final code "0" be assigned. For each fault file, the same procedure is applied to all three valve groups, *i.e.*, Vg11, Vg12 and Vg13 for pole 1. Three final codes are simply combed together as the value for feature 11. The value of attribute 11 for fault F08101CA.x01 is 101.

Function f_{11} representing feature 11 is similar to the functions for features 8, 9 and 10, except that there is a different coding process.

$$f_{11}(x) = Code(max(r \star r)/max(r \star x)), \tag{34}$$

Table 17. Codes for valve currents minor disturbance in Vg11 in F08101CA.x01

Signal names	Original codes
A phase valve current	2222222222201222
B phase valve current	2222222222112222
C phase valve current	2222222222101222

where x accepts discrete normalized A, B and C phase valve currents from all valve groups.

Feature 12 – Pole Current Closed with Normal Pole Voltage (True = 1, False = 0)
Section 6.2 explains how to extract the pole voltage and current trend. In one specific case, the pole current recorded in Transcan^{TM} gives an output of zero while the pole voltage is perfectly normal. This event happens during a parallel operation. For instance, a "pole 1 to pole 3 parallel operation" is to switch the pole 1 current to pole 3 to unload the pole 1 current line for maintenance. In Table 12, the information table for training, file F2212F95.x01 is an example of "pole 1 to pole 3 parallel operation". The 17 attributes of this file are listed in Table 18. Attribute 4 indicates the pole 1 voltage is normal, while attribute 6 shows the pole 1 current is closed, therefore the value of attribute 12 is 1.

Function f_{12} representing feature 12 is defined by (35).

$$f_{12}(x,y) = \begin{cases} 1 & if\ \ ((f_4(x) == 4) or (f_4(x) == 3)) AND (f_6(y) == 1), \\ 1 & if\ \ ((f_5(x) == 4) or (f_5(x) == 3)) AND (f_7(y) == 1), \\ 0 & otherwise, \end{cases} \tag{35}$$

where x is the discrete pole voltage and y is the discrete pole current.

Table 18. Features for F2212F95.x01

File index	Fault types	Fault file names	A1	A2	A3	A4	A5	A6	A7	A8	A9	A10	A11	A12	A13	A14	A15	A16	A17
440	Fault 8	F2212F95.x01	0	0	1	4	4	1	2	2	2	2	0	1	0	1	0	0	0

Feature 13 – 3 Valve Groups All Closed (True = 1, False = 0)
Features 8, 9 and 10 of valve current trends have been discussed in Section 6.2. Based on features 8, 9, 10, feature 13 can be derived. If the codes of valve current trends for three valve groups all end in 1, meaning the three valve groups are all closed in the end, feature 13 yields an output of 1. This usually implies that the whole pole line is closed.

Function f_{13} representing feature 13 is defined as

$$f_{13}(x,y,z) = \begin{cases} 1 & if\ codes\ f_8(x), f_9(y)\ and\ f_{10}(z)\ are\ all\ ended\ in\ 1, \\ 0 & otherwise, \end{cases} \tag{36}$$

where x, y and z are the discrete normalized A, B and C phase currents in valve groups 1, 2 and 3, respectively.

Feature 14 – Same Current Trend in 3 Valve Groups (True = 1, False = 0)
Based on features 8, 9 and 10, feature 14 can also be derived. If the valve current trends of three valve groups are all the same, feature 14 gives an output of 1, which produces a high possibility of the following three faults, fault 4, "Pole Line Fault"; fault 10, "Normal Affected by Another Pole"; and fault 12, "Disturbance on DC Voltage".

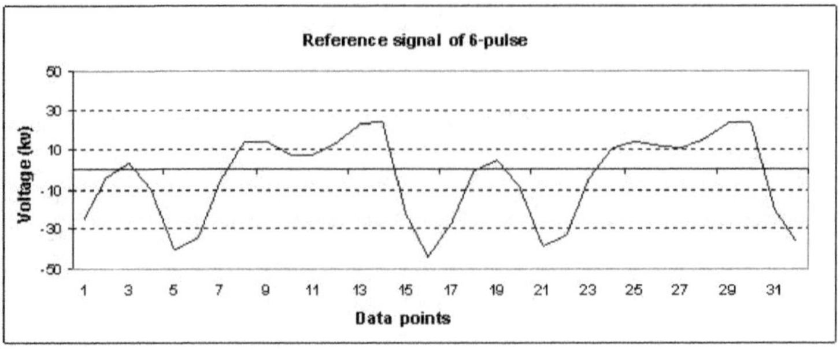

Fig. 40. 6-pulse reference signal (2 cycles)

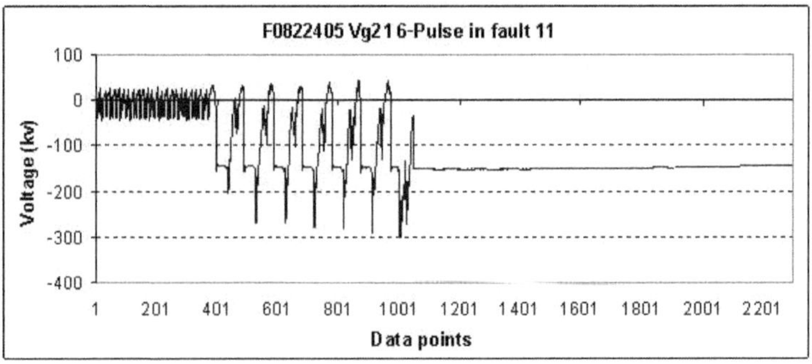

Fig. 41. 6-pulse signal in Asymmetric Protection fault

Function f_{14} representing feature 14 is defined by (37).

$$f_{14}(x, y, z) = \begin{cases} 1 & if\ \ f_8(x) == f_9(y) == f_{10}(z), \\ 0 & otherwise, \end{cases} \tag{37}$$

where x, y and z are the discrete normalized A, B and C phase currents in valve groups 1, 2 and 3, respectively.

Feature 15 – Voltage Flashover in 6-Pulse

Feature 15 records the number of cycles of voltage flashover that happened in a 6-pulse signal. A normal 6-pulse signal shown in Figure 40 is a periodic signal. When fault 11, "Asymmetric Protection", happens, the 6-pulse does 7 cycles of voltage flashover and closes the valves for protection. The typical waveform of a 6-pulse signal in fault 11, F0822405.x01, is illustrated in Figure 41.

To detect those 7 cycles of flashover, the Wavelet Multi-resolution Decomposition (MRD) method has been applied to extract different levels of details for the recorded signals. A number of experiments have been done to evaluate the performances of different wavelet functions such as Daubechies wavelets and the Meyer wavelet. The decomposition can be carried out in Matlab^{TM} using functions `wavedec` and `wrcoef`. The MRD with the Daubechies 2 wavelet (DB2)

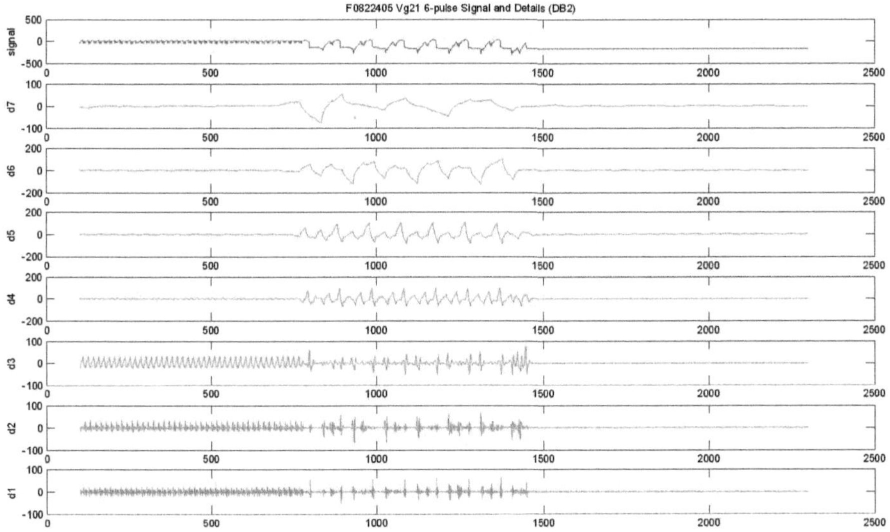

Fig. 42. The multi-level details of 'DB2' MRD applied to a 6-pulse signal in Asymmetric Protection fault

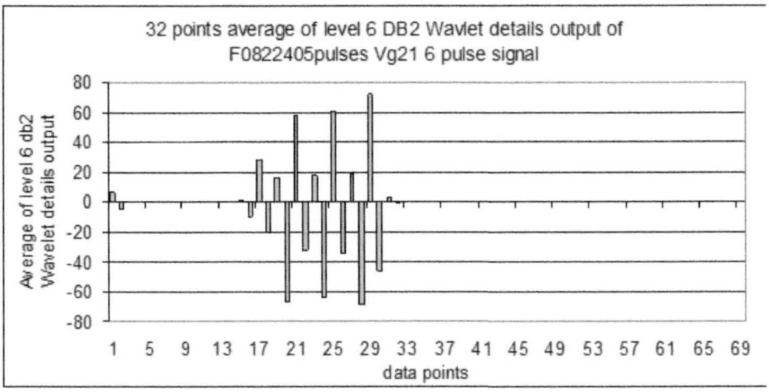

Fig. 43. Seven peaks detected in a 6-pulse signal in Asymmetric Protection fault

function extracts the 7 cycles of flashover at the 6th level detail coefficient output. Figure 42 shows the transient signal and 7-level 'DB2' MRD details. The experiments with the 'DB3', 'DB4' and 'Meyer' wavelets extract 8 cycles of flashover, which does not agree with the 6-pulse transient signal. The 6th level detail coefficient output from 'DB2' MRD is further processed by 32-point averaging. In addition, 7 positive peaks with values greater than 18 are detected and shown in Figure 43. Occasionally, the first 7 cycles of voltage flashover failed to close the valves and the control system continues with another 7 cycles until the valves are closed. An example of this phenomenon is F0922884.x01.

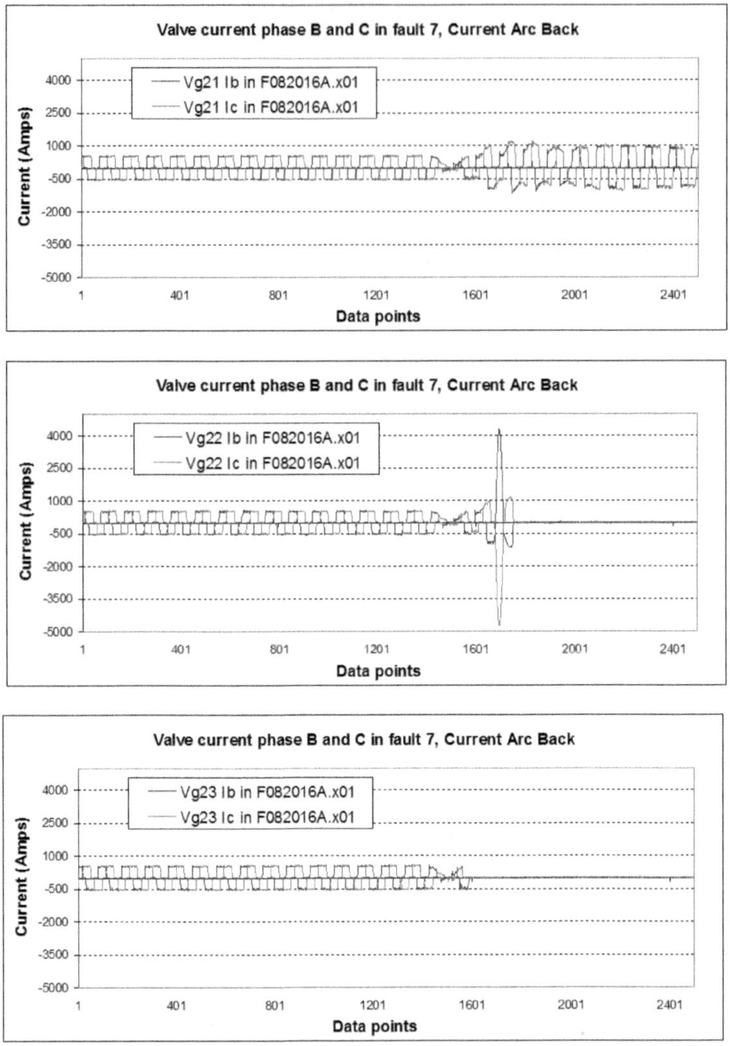

Fig. 44. B and C phase currents in F082016A.x01, phase currents flashover in Vg22 valve group

Function f_{15} representing feature 15 is given by (38).

$$f_{15}(x, y, z) = max(g_{15}(average(MRD(x))), g_{15}(average(MRD(y))), \\ g_{15}(average(MRD(z)))), \quad (38)$$

where $g_{15}(\cdot)$ picks up the points with values of $average(MRD(\cdot))$ greater than 18. The discrete 6-pulse signals in valve groups 1, 2 and 3 are x, y and z respectively.

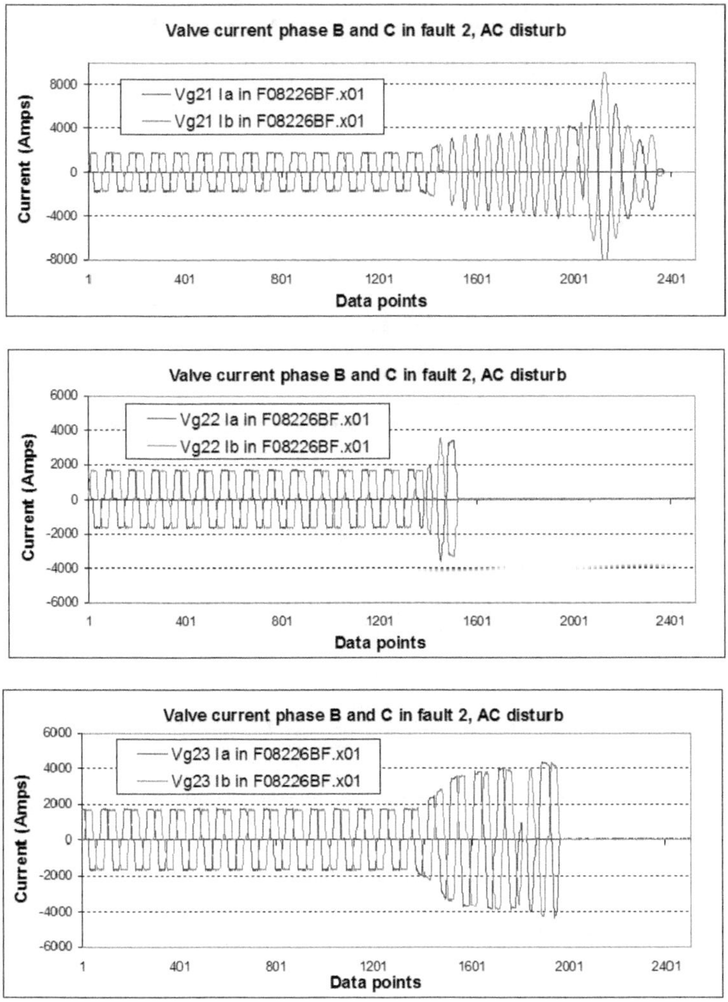

Fig. 45. B and C phase currents in F08226BF.x01, phase currents flashover in Vg21 and Vg23 valve group

Feature 16 – Valve Currents Flashover

Feature 11, the valve current minor disturbance, has been discussed in Section 6.2. In this section, a severe fault is addressed that is involved with valve current flashover. A standard peak value for a valve current is 1400 Amps. Occasionally with a severe AC bus error or the valve line shorted together or shorted to ground, valve currents increase dramatically to an excess of 4000 Amps. Usually this happens within a pair of valves in a valve group. Two valve currents increase in opposite directions to prevent the pole current from overshooting. Examples are illustrated in Figures 44 and 45. The first example is fault 7, "Current Arc Back", valve current flashover happens only in one valve

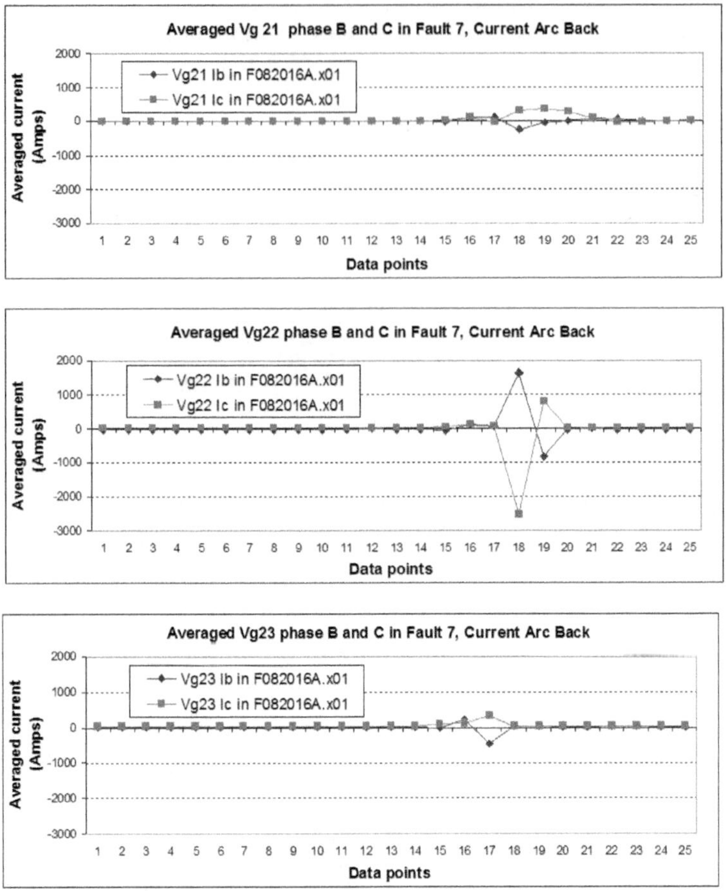

Fig. 46. Averaged waveforms for B and C phase currents in F082016A.x01

group. The second one is fault 2, “AC Disturbance”, valve current flashover happens in two valve groups.

To detect the valve current flashover, 96-point averaging is applied to derive a mean value for each cycle. The mean value of a normal cycle is 0 and a flashover cycle is over 1800, which is the threshold used to detect the event of current flashover. The averaged waveforms of B and C phase currents for 3 valve groups in F082016A.x01 and F08226BF.x01 are displayed in Figures 46 and 47, respectively. For F082016A.x01, there is only one point over 1800 in Vg22; for F08226BF.x01, there are three points over 1800 in Vg21 and two points over 1800 in Vg23. The value for feature 16 is 1 for F082016A.x01 and 32 for F08226BF.x01.

Function f_{16} representing feature 16 is defined as

$$f_{16}(x, y, z) = Code(g_{16}(average(x)), g_{16}(average(y)), g_{16}(average(z))), \quad (39)$$

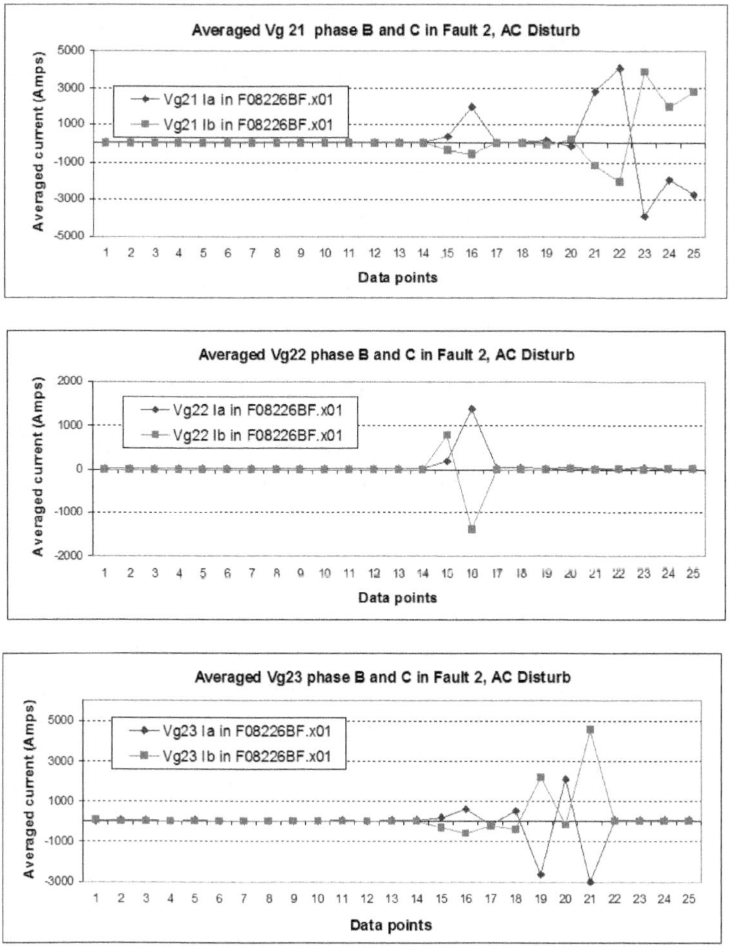

Fig. 47. Averaged waveforms for B and C phase currents in F08226BF.x01

where $g_{16}(\cdot)$ picks up the points with values of $average(\cdot)$ greater than 1800. $Code(\cdot)$ is the coding process used to concatenate the number of points from 3 valve groups and x, y and z are the discrete normalized A, B and C phase currents in valve groups 1, 2 and 3, respectively.

Feature 17 – Valve Current Flashover Happens Only in One Valve Group (True = 1, False = 0)

As seen in the discussion of feature 16, valve current flashover happens in both faults 2 and 7. "Current Arc Back" usually comes with an "AC Disturbance". However, the current flashover in "Current Arc Back" is only due to the electronic faults in the valve group itself. A severe AC disturbance is a bus error and affects all the valve groups. If the severe AC disturbance induces current flashover, it will affect almost all valve groups. To further separate these two faults, feature 17 is added.

It is simple to obtain feature 17 based on the results of feature 16. Feature 17 will yield an output of 1 if only one valve group gives an output of a non-zero number in feature 16.

Function f_{17} representing feature 17 is defined as

$$f_{17}(x, y, z) = \begin{cases} 1 & if\ only\ g_{16}(average(x)) \neq 0, \\ 1 & if\ only\ g_{16}(average(y)) \neq 0, \\ 1 & if\ only\ g_{16}(average(z)) \neq 0, \\ 0 & otherwise. \end{cases} \tag{40}$$

7 Rough Membership Neural Network (rmNN) for PSFC

A form of rough neural computing based on rough sets and rough membership functions [46,63,64,66] is introduced in this section. A rough membership function neural network (rmNN) has been designed and applied to classify power system faults [24,25,56].

7.1 Sample Information System For PSFC

The fault files recorded by Transcan^{TM} form the universe of events U. Table 12 and Table 13 in Section 6.2 represent the information system and have 17 features, which form the knowledge domain for the PSFC system. Further research will be addressed to examine the possibility of reducing the feature dimension by applying the discernibility-based reduction algorithm [50,4,5]. The 17 features in the information table are sub-grouped into 11 feature sets, $B = \{B_1, B_2, ..., B_{11}\}$ (Table 19). B_3 contains 5 features, feature 3, 4, 5, 6 and 7. B_4 has 3 features, feature 8, 9 and 10. A feature set is a collection of the attributes which represent the signals of same nature. Features 3, 4, 5, 6 and 7 represent pole signals, which are constant signals in the normal condition. Features 8, 9 and 10 characterize the valve currents, the periodic signals in 3 valve groups. By grouping 17 features into 11 feature sets, the dimension of the knowledge domain is reduced, whereas the classification generalization is decreased. The information system is then represented by (U, B).

A simple information system containing sample fault events and feature set B_4 (Table 20) is discussed in this section to illustrate the rough set basic theory in the application of the power system. In this example, assume that U is a set of sample fault events. By way of approximation of a set of objects, consider $X \subseteq U$ defined as

$$\begin{aligned} X &= \{x | x \text{ is a fault event in the power system}\} \\ &= \{F08101FE.x01, F1113009.x01, F0420695.x01, F2913FDD.x01, F1112E8D.x01, \\ &\quad F2212CD7.x01, F223079B.x01, F0820165.x01, F112267F.x01, ...\}, \end{aligned}$$

$F = \{f_8, f_9, f_{10}\}$, defined in Section 6.2, is a set of functions representing the feature set $B_4 = \{A_8, A_9, A_{10}\}$.

Table 19. 11 Feature Sets

B1	Feature/attribute 1
B2	Feature/attribute 2
B3	Feature/attribute 3, 4, 5, 6, 7
B4	Feature/attribute 8, 9, 10
B5	Feature/attribute 11
B6	Feature/attribute 12
B7	Feature/attribute 13
B8	Feature/attribute 14
B9	Feature/attribute 15
B10	Feature/attribute 16
B11	Feature/attribute 17

Table 20. Sample information system

Events	B_4		
Fault file names	f_8	f_9	f_{10}
F08101FE.x01	2	2	2
F1113009.x01	2	2	2
F0420695.x01	2	2	21
F2913FDD.x01	2	2	1
F1112E8D.x01	212	212	212
F2212CD7.x01	21	21	21
F223079B.x01	212	212	212
F0820165.x01	2	2	2
F112267F.x01	21	21	21
F22225C4.x01	1	1	1
F20406CC.x01	2	2	2
F2713113.x01	2	2	2
F2713116.x01	2	2	2
F041075C.x01	1	2	2
F0820715.x01	1	1	1
F08226BF.x01	21	21	21
F0112939.x01	2	2	1
F1140866.x01	2	1	1
F1112BAE.x01	2	1	1
F2410189.x01	21	21	21
F1122499.x01	2	2	2
F2212F95.x01	2	2	2
F1121E5D.x01	2	2	2
F272015F.x01	2	2	2
F111302A.x01	2	2	2

The fault events and their associated fault types are listed in Table 21. Notice that each of the events in class

$$[\text{F08101FE.x01}]_{B_4} = \{\text{F08101FE.x01, F1113009.x01, F0820165.x01, F20406CC.x01, F2713113.x01, F2713116.x01, F1122499.x01, F2212F95.x01, F1121E5D.x01, F272015F.x01, F111302A.x01}\},$$

has exactly the same B_4 output, namely, {2, 2, 2} (Table 20).

The partition of U defined by the relation $\sim_{B_4}$ is as follows:

$$[\text{F08101FE.x01}]_{B_4} = \{\text{F08101FE.x01, F1113009.x01, F0820165.x01, F20406CC.x01, F2713113.x01, F2713116.x01, F1122499.x01, F2212F95.x01, F1121E5D.x01, F272015F.x01, F111302A.x01}\},$$

$$[\text{F0420695.x01}]_{B_4} = \{\text{F0420695.x01}\},$$

$$[\text{F2913FDD.x01}]_{B_4} = \{\text{F2913FDD.x01, F0112939.x01}\},$$
$$[\text{F1112E8D.x01}]_{B_4} = \{\text{F1112E8D.x01, F223079B.x01}\},$$
$$[\text{F2212CD7.x01}]_{B_4} = \{\text{F2212CD7.x01, F112267F.x01, F08226BF.x01, F2410189.x01}\},$$
$$[\text{F22225C4.x01}]_{B_4} = \{\text{F22225C4.x01, F0820715.x01}\},$$
$$[\text{F041075C.x01}]_{B_4} = \{\text{F041075C.x01}\},$$
$$[\text{F1140866.x01}]_{B_4} = \{\text{F1140866.x01, F1112BAE.x01}\}.$$

Table 21. Fault events and associated fault types

Events	Decision
F08101FE.x01	Fault 1
F1113009.x01	Fault 1
F0420695.x01	Fault 1 and 3
F2913FDD.x01	Fault 1 and 3
F1112E8D.x01	Fault 1 and 4
F2212CD7.x01	Fault 1 and 4
F223079B.x01	Fault 1 and 4
F0820165.x01	Fault 1 and 5
F112267F.x01	Fault 1 and 6
F22225C4.x01	Fault 1 and 6
F20406CC.x01	Fault 2
F2713113.x01	Fault 2
F2713116.x01	Fault 2
F041075C.x01	Fault 2 and 3
F0820715.x01	Fault 1 and 6 and 7
F08226BF.x01	Fault 2 and 6
F0112939.x01	Fault 3
F1140866.x01	Fault 3 and 5
F1112BAE.x01	Fault 3 and 5 and 9
F2410189.x01	Fault 4
F1122499.x01	Fault 5
F2212F95.x01	Fault 8
F1121E5D.x01	Fault 9
F272015F.x01	Fault 10
F111302A.x01	Fault 12

Now select a particular set X, which contains all the events of fault 1: (Table 21), *i.e.*,

$$X = \{\text{F08101FE.x01, F1113009.x01, F0420695.x01, F2913FDD.x01, F1112E8D.x01, F2212CD7.x01, F223079B.x01, F0820165.x01, F112267F.x01, F22225C4.x01, F0820715.x01}\}.$$

This choice leads to the following lower and upper approximations of the set X.

$$\begin{aligned} B_{4*}X &= [\text{F0420695.x01}]_{B_4} \cup [\text{F1112E8D.x01}]_{B_4} \cup [\text{F22225C4.x01}]_{B_4} \\ &= \{\text{F0420695.x01, F1112E8D.x01, F223079B.x01, F22225C4.x01,} \\ &\quad \text{F0820715.x01}\}, \end{aligned}$$

$$\begin{aligned} B_4{}^*X &= [\text{F0420695.x01}]_{B_4} \cup [\text{F1112E8D.x01}]_{B_4} \cup [\text{F22225C4.x01}]_{B_4} \\ &\quad \cup [\text{F08101FE.x01}]_{B_4} \cup [\text{F2913FDD.x01}]_{B_4} \cup [\text{F2212CD7.x01}]_{B_4} \\ &= \{\text{F0420695.x01, F1112E8D.x01, F223079B.x01, F22225C4.x01,} \\ &\quad \text{F0820715.x01, F08101FE.x01, F1113009.x01, F0820165.x01,} \\ &\quad \text{F20406CC.x01, F2713113.x01, F2713116.x01, F1122499.x01,} \end{aligned}$$

F2212F95.x01, F1121E5D.x01, F272015F.x01, F111302A.x01, F2913FDD.x01, F0112939.x01, F2212CD7.x01, F112267F.x01, F08226BF.x01, F2410189.x01},

$Bnd_{B4}X = B_4{}^*X$ - $B_{4*}X$
= $[\text{F08101FE.x01}]_{B_4} \cup [\text{F2913FDD.x01}]_{B_4} \cup [\text{F2212CD7.x01}]_{B_4}$
= {F08101FE.x01, F1113009.x01, F0820165.x01, F20406CC.x01, F2713113.x01, F2713116.x01, F1122499.x01, F2212F95.x01, F1121E5D.x01, F272015F.x01, F111302A.x01, F2913FDD.x01, F0112939.x01, F2212CD7.x01, F112267F.x01, F08226BF.x01, F2410189.x01}.

In effect, the lower approximation $B_{4*}X$ indicates that the events in $[\text{F0420695.x01}]_{B_4} \cup [\text{F1112E8D.x01}]_{B_4} \cup [\text{F22225C4.x01}]_{B_4}$ certainly are the members of set X. $B_{4*}X$ is called the "Yes" set in Section 7.2. Meanwhile, the non-empty boundary $Bnd_{B4}X$ indicates that set X is a rough set and the events in $Bnd_{B4}X$ might belong to set X. $Bnd_{B4}X$ is called the "YesOrNo" set in Section 7.2.

Next, consider the degree of overlap of class $[\text{F08101FE.x01}]_{B_4}$ with the set X, *i.e.*,

X = {F08101FE.x01, F1113009.x01, F0420695.x01, F2913FDD.x01, F1112E8D.x01, F2212CD7.x01, F223079B.x01, F0820165.x01, F112267F.x01, F22225C4.x01, F0820715.x01},

and

$[\text{F08101FE.x01}]_{B_4}$ = {F08101FE.x01, F1113009.x01, F0820165.x01, F20406CC.x01, F2713113.x01, F2713116.x01, F1122499.x01, F2212F95.x01, F1121E5D.x01, F272015F.x01, F111302A.x01},

where the degree of overlap is calculated using (41)

$$\mu_X^{B_4}(x) = \frac{|[F08101FE.x01]_{B_4} \cap X|}{|[F08101FE.x01]_{B_4}|} = \frac{3}{11} = 0.273. \tag{41}$$

This demonstrates that the degree to which the events in class $[\text{F08101FE.x01}]_{B_4}$ belong to X is 27.3%. This shows that fault events in class $[\text{F08101FE.x01}]_{B_4}$ and the faults in the set X are partially related.

7.2 Rough Membership Functions

A rough membership function (rm function) makes it possible to measure the degree to which any specified object belongs to a given set X. In the power fault classification system, there are 11 feature sets and 12 types of faults; the universe is divided into 132 Rough Sets. A mapping of these 132 Rough Sets and the information table is illustrated in Table 22.

RS_{ij} is a set of the ith $(1 \le i \le 12)$ faults represented by function values for functions in B_j $(1 \le j \le 11)$. Eleven sets, RS_{i1}, RS_{i2}, RS_{i3}, RS_{i4}, RS_{i5}, RS_{i6}, RS_{i7}, RS_{i8}, RS_{i9}, RS_{i10}, RS_{i11}, are derived to represent the 11 features

Table 22. The mapping of 132 Rough Sets and information table

	B1	B2	B3	B4	B5	B6	B7	B8	B9	B10	B11
Fault1	RS 11	RS 12	RS 13	RS 14	RS 15	RS 16	RS 17	RS 18	RS 19	RS 110	RS 111
Fault2	RS 21	RS 22	RS 23	RS 24	RS 25	RS 26	RS 27	RS 28	RS 29	RS 210	RS 211
Fault3	RS 31	RS 32	RS 33	RS 34	RS 35	RS 36	RS 37	RS 38	RS 39	RS 310	RS 311
Fault4	RS 41	RS 42	RS 43	RS 44	RS 45	RS 46	RS 47	RS 48	RS 49	RS 410	RS 411
Fault5	RS 51	RS 52	RS 53	RS 54	RS 55	RS 56	RS 57	RS 58	RS 59	RS 510	RS 511
Fault6	RS 61	RS 62	RS 63	RS 64	RS 65	RS 66	RS 67	RS 68	RS 69	RS 610	RS 611
Fault7	RS 71	RS 72	RS 73	RS 74	RS 75	RS 76	RS 77	RS 78	RS 79	RS 710	RS 711
Fault8	RS 81	RS 82	RS 83	RS 84	RS 85	RS 86	RS 87	RS 88	RS 89	RS 810	RS 811
Fault9	RS 91	RS 92	RS 93	RS 94	RS 95	RS 96	RS 97	RS 98	RS 99	RS 910	RS 911
Fault10	RS 101	RS 102	RS 103	RS 104	RS 105	RS 106	RS 107	RS 108	RS 109	RS 1010	RS 1011
Fault11	RS 111	RS 112	RS 113	RS 114	RS 115	RS 116	RS 117	RS 118	RS 119	RS 1110	RS 1111
Fault12	RS 121	RS 122	RS 123	RS 124	RS 125	RS 126	RS 127	RS 128	RS 129	RS 1210	RS 1211

Table 23. "YesOrNo" set of feature sets from B4 to B11 in faults 1, 2, 3 and 4

	YesOrNo SET							
	B4	**B5**	**B6**	**B7**	**B8**	**B9**	**B10**	**B11**
Fault 1	72 of {2 2 2 }; 10 of {2 2 21 }; 12 of {2 2 1 }; 1 of {212 2 2 }; 1 of {21 1 2 }; 7 of {1 2 2 }; 3 of {12 2 2 }; 6 of {21 2 2 }; 2 of {2 21 21 }; 4 of {2 21 1 }; 2 of {2 2 212 }; 1 of {212 21 2 }; 1 of {2 21 212 }; 1 of {2 2 1212 }; 1 of {2121 2 2 }; 1 of {2 212 12 }; 1 of {2 1 1 }; 5 of {2 2 2121 }; 4 of {2 21212 2 }; 4 of {2 21 2 }; 63 of {212 212 212 }; 21 of {21 21 21 }; 1 of {21 21 1 }; 3 of {21212 21212 21212 }; 9 of {1 1 1 }; 1 of {21 21 2121 }; 1 of {1 21 21 }; 1 of 21 1 21 ; 1 of 12 12 12 ;	199 of 0; 3 of 111; 5 of 11; 7 of 110; 9 of 10; 8 of 100; 7 of 1; 2 of 101;	232 of 0; 8 of 1;	206 of 0; 34 of 1;	170 of 1; 70 of 0;	75 of 0; 79 of 1; 27 of 2; 10 of 4; 1 of 10; 22 of 3; 8 of 7; 3 of 6; 6 of 8; 4 of 28; 4 of 29; 1 of 5;	227 of 0; 9 of 1; 3 of 2; 1 of 6;	227 of 0; 13 of 1;
Fault 2	77 of {2 2 2 }; 6 of {1 1 1 }; 10 of {1 2 2 }; 29 of {2 2 1 }; 1 of {21212 21212 1 }; 1 of {21212 1 1 }; 1 of {212 212 1 }; 1 of {21 1 2 }; 1 of {12 21 2 }; 1 of {212 121 2 }; 1 of {2 1 2 }; 1 of {2 21 2 }; 10 of {21 21 21 }; 4 of {21212 21212 21212 }; 1 of {1 21 21 }; 1 of {21 1 1 }; 2 of {21 21 1 };	51 of 111; 57 of 0; 22 of 110; 4 of 10; 4 of 1; 7 of 11; 2 of 100; 1 of 101;	148 of 0;	128 of 0; 20 of 1;	97 of 1; 51 of 0;	80 of 1; 35 of 0; 7 of 3; 14 of 2; 3 of 6; 4 of 4; 2 of 8; 1 of 10; 1 of 5; 1 of 7;	126 of 0; 5 of 11; 2 of 32; 2 of 22; 4 of 2; 3 of 1; 1 of 6; 1 of 7; 1 of 3; 2 of 5; 1 of 4;	127 of 0; 21 of 1;
Fault 3	10 of {2 2 21 }; 49 of {2 2 1 }; 1 of {212 2 2 }; 2 of {21 1 2 }; 20 of {1 2 2 }; 3 of {12 2 2 }; 6 of {21 2 2 }; 3 of {2 21 21 }; 4 of {2 21 1 }; 2 of {2 2 212 }; 1 of {212 21 2 }; 7 of {2 1 1 }; 1 of {2 21 212 }; 1 of {2 2 1212 }; 1 of {2121 2 2 }; 1 of {2 212 12 }; 5 of {2 2 2121 }; 4 of {2 21212 2 }; 5 of {2 21 2 }; 1 of {21212 21212 1 }; 1 of {21212 1 1 }; 1 of {212 212 1 }; 1 of {12 21 2}; 1 of {212 121 2 }; 5 of {2 1 2 }; 2 of {12 1 2 }; 2 of {212 1 2 }; 1 of {2 1 21 };	75 of 0; 8 of 11; 27 of 110; 6 of 10; 22 of 100; 7 of 1;	143 of 0; 2 of 1;	145 of 0;	145 of 0;	41 of 0; 17 of 2; 53 of 1; 1 of 4; 1 of 10; 6 of 3; 7 of 7; 4 of 6; 7 of 8; 4 of 28; 4 of 29;	134 of 0; 6 of 1; 3 of 2; 1 of 6; 1 of 7;	134 of 0; 11 of 1;
Fault 4	3 of {2 2 2 }; 63 of {212 212 212 }; 12 of {21 21 21 }; 3 of {21212 21212 21212 };	81 of 0;	81 of 0;	69 of 0; 12 of 1;	81 of 1;	3 of 0; 31 of 1; 21 of 3; 16 of 2; 9 of 4; 1 of 5;	81 of 0;	81 of 0;

for the faults of type i. For simplicity, fault file names will be replaced by sets of values of functions, *e.g.*, the equivalence class $[F08101FE.x01]_{B_4}$ is represented by $11of\,\{2,2,2\}$ as discussed in Section 7.1. 11 fault files have same output, which is $\{2,2,2\}$, from B_4.

Consider two sets, RS_{ij} and RS_{kj}, $(1 \leq k \leq 12, k \neq i)$, the intersection $RS_{ij} \cap RS_{kj}$ belongs to the "YesOrNo" set of RS_{ij} in the case where RS_{ij} and RS_{kj} have feature values in common. The elements that exist only in set RS_{ij} constitute what is known as the "Yes" set. In other words, each RS_{ij} set is divided into two sets, the "YesOrNo" set and the "Yes" set. The RS_{ij} set is a rough set if its "YesOrNo" set is non-empty. Examples of "YesOrNo" sets of B4

Table 24. "Yes" set of 11 feature sets in 12 faults

	Yes SET										
	B1	B2	B3	B4	B5	B6	B7	B8	B9	B10	B11
Fault 1			{1 4 43 23 23 } {1 4 41 3 3241 } {1 4 434 3 343 } {1 4 41 23 21 } {1 4 41 23 21 } {4 43 4 3 3 } {4 41 4 321 34 } {1 4 4134 3 313 } {2 41 4 31 34 } {2 4 4 3 23 } {2 4314 4 3123 3 } {2 414 4 323 3 } {2 323 4 3 3 } {2 3 4 2 2 } {3 4 414 3 313 } {3 4 414 232 }								
Fault 2			{2 3 434 3 3 } {2 323 434 2 2 } {1 434 43 23 23 } {1 434 4 2 2 } {4 4241 434 3141 34 } {4 41 4 341 34 } {4 3 4 34 34 } {2 434 434 3 343 } {2 434 4 32 32 } {3 4 4 3 3 } {1 414 434 232 12 } {1 4 434 323 3 } {1 4 3 2 2 } {1 4234 323 3 3 } {1 434 323 343 3 }								
Fault 3			{1 34 23 21 21 } {1 3 312 2 2 } {1 3 2 2 212 } {2 3 23 21 21 } {1 3 4 3 32 } {2 2 3 3 3 } {1 13 4 12 2 } {1 3 12 2 12 } {2 3 12 2 12 } {2 3 212 2 212 } {4 4 313 3 3213 }	{12 12 1 } {1 1 12 } {21 1 12 } {1 212 212}							
Fault 4			{1 41 41 31 31 }								
Fault 5											
Fault 6			{3 1 1 1 1 } {4 1 1 1 1 } {1 1 1 1 1 } {2 1 1 1 1 }								
Fault 7											
Fault 8			{1 4 4 1 2 } {2 4 4 1 2 } {2 4 4 1 2 } {1 4 4 1 2 } {1 4 4 1 2 } {4 431 4 321 34 }								
Fault 9			{3 4 13 43 13 } {1 4 14 32 12 } {2 13 4 12 32 } {2 143 4 12 32 } {1 4 434 23 23 } {1 4 4 321 321 } {1 4 4 23 23 } {2 4 4 32 32 } {2 3 4 3 43 } {1 123 4 12 32 } {1 4 4 321 321 } {1 13 4 12 32 } {1 4 4 23 23 } {1 3 4 3 34 } {1 3 4 3 43 }								
Fault 10			{2 31 4 21 2 } {2 143 4 12 2 } {4 3 4 3 3 } {2 41 4 21 2 } {1 4 143 2 12 } {1 4 31 2 21 } {1 4 3 3 3 } {1 4 123 2 12 } {1 4 4 3 2 } {1 4 42 2 32 } {1 4 31 2 21 }								
Fault 11											
Fault 12			{4 3 3 2 2 } {1 3 4 2 3 } {4 3 3 2 2 } {2 4 3 3 2 } {1 3 4 2 3 } {4 3 3 2 2 } {2 4 3 3 2 } {4 3 3 2 2 } {4 3 3 2 2 } {1 3 4 2 3 } {2 4 3 3 2 } {4 3 3 2 2 }								

to B11 in faults 1, 2, 3 and 4 are listed in Table 23. The complete tables of the "YesOrNo" set of all feature sets in all 12 faults are attached in an Appendix available at [15]. The "Yes" sets of all feature sets in all 12 faults are listed in Table 24. The equivalence classes of all feature sets are listed in Tables 25 and 26.

For a fault file, if its output of B_j falls in the "YesOrNo" set of RS_{ij}, it indicates that the file could represent an ith fault; and the degree of ith fault will be estimated based on the rough membership function in (42).

$$\mu_X^B(x) = \frac{|[x]_B \cap X|}{|[x]_B|}. \tag{42}$$

As an example, the rough membership calculation for the fault file F0822405.x01 is explained in (43). For fault file F0822405.x01, the feature set B_4 gives an output of {2, 2, 21}. Since set {2, 2, 21} is identified in the "YesorNo" set of RS_{34} 10 times, the fault is possibly a fault 3. In (42), the equivalence class $[x]_B$ is substituted by $[F0822405.x01]_{B_4}$, where x is represented by fault file F0822405.x01 and B is replaced with B_4. Meanwhile, X is

Table 25. The equivalence classes of feature sets B1, B2 and B3

Equivalence classes		
B1	**B2**	**B3**
700 of 0;	495 of 1;	10 of {1 4 43 3 3 }; 7 of {1 4 1 3 1 }; 8 of {2 3 4 3 3 }; 6 of {4 4 4 3 3 }; 16 of {1 4 4 3 3 };
227 of 1;	279 of 2;	3 of {1 4 41 3 31 }; 20 of {2 4 4 3 3 }; 3 of {4 41 4 31 34 }; 7 of {1 4 4 2 2 };
	153 of 0;	9 of {1 434 4 3 3 }; 2 of {2 3 43 34 34 }; 2 of {1 3 43 3 3143 }; 14 of {1 3 43 3 3 };
		2 of {2 3 434 3 343 }; 60 of {1 3 4 3 3 }; 2 of {2 3 312 23 23 }; 8 of {1 3 4 2 2 };
		2 of {1 3 34 3 3 }; 8 of {1 3 3 2 2 }; 2 of {1 3 413 2 2 }; 2 of {2 3 34 3 3 };
		2 of {3 43 43 3 3 }; 2 of {4 43 43 3 3 }; 3 of {1 43 4 3 3 }; 27 of {1 323 4 3 3 };
		6 of {1 34 4 3 3 }; 3 of {3 43 4 3 3 }; 7 of {3 3 4 3 3 }; 4 of {2 4 434 2 2 };
		3 of {1 43 4 3 23 }; 3 of {4 4 43 3 3 }; 3 of {1 312 43 23 23 }; 3 of {2 3 413 2 2 };
		4 of {2 4 43 2 2 }; 4 of {2 3 34 2 2 }; 4 of {2 4 43 3 34 }; 3 of {2 4 4 1 1 }; 3 of {2 4 2 3 1};
		3 of {2 4 434 3 1 }; 3 of {2 4 43 1 343 }; 3 of {2 3 43 34 3134 }; 3 of {2 3 43 34 3234 };
		3 of {2 4 43 3 3 }; 3 of {2 4 42 3 3 }; 3 of {2 4 43 34 34 }; 3 of {2 4 42 34 34 };
		6 of {1 3 42 3 3 }; 6 of {1 4 424 3 3 }; 9 of {1 4 434 3 3 }; 3 of {1 3 43 34 3134 };
		3 of {1 3 43 34 3234 }; 3 of {1 4 42 3 3 }; 3 of {1 4 43 34 34 }; 3 of {1 4 42 34 34 };
		4 of {2 431 4 321 34 }; 2 of {2 414 4 313 3 }; 20 of {1 414 4 313 3 }; 8 of {1 414 4 3123 3};
		2 of {1 414 4 2 2 }; 2 of {1 431 4 31 3 }; 2 of {1 41 4 31 3 }; 2 of {1 431 4 31 34 };
		4 of {1 4134 4 3123 3 }; 4 of {1 4134 4 323 3 }; 2 of {1 414 4 232 2 }; 2 of {1 414 4 212 2};
		2 of {1 414 4 2132 2}; 2 of {1 414 4 3213 3}; 2 of {1 4314 4 3213 3}; 2 of {2 4 414 2 212};
		2 of {2 4 4134 3 3143 }; 2 of {2 4 41 34 31 }; 2 of {3 431 4 321 34 }; 36 of {3 414 4 313 3};
		16 of {3 414 4 3123 3 }; 2 of {3 414 4 3143 3 }; 2 of {3 41 4 31 243 }; 4 of {3 41 4 31 34};
		2 of {3 41 4 31 3 }; 2 of {3 431 4 31 34 }; 2 of {3 424 4 3134 34 }; 2 of {3 424 4 313 3};
		2 of {3 414 4 212 2342 }; 2 of {3 414 4 212 232 }; 2 of {3 414 4 212 242 };
		2 of {3 414 4 212 2432}; 2 of {3 414 3 313 3}; 4 of {4 4 414 3 313}; 4 of {4 4 414 232 212};
		2 of {1 4314 4 313 3 }; 2 of {1 414 4 2312 2 }; 2 of {1 4 2 3 2 }; 9 of {2 4 4 2 2 };
		3 of {2 4 434 3 3 }; 2 of {1 31 4 231 2 }; 2 of {2 3 41 34 31 }; 2 of {2 4 431 3 1 };
		2 of {2 4 431 23 1 }; 4 of {2 4 41 23 21 };2 of {1 41 4 21 23 }; 2 of {2 41 1 41 31 };
		2 of {1 31 31 31 31 }; 2 of {2 41 41 431 431 }; 2 of {2 321 31 31 321 }; 25 of {2 4 1 3 1};
		3 of {4 4 41 343 31 }; 3 of {2 4 1 343 1 }; 4 of {2 4 41 1 1 }; 3 of {2 4 41 1 21 };
		3 of {2 4 41 1 31 }; 3 of {2 4 431 1 341 }; 9 of {2 4 14 32 1 }; 2 of {1 4 43 1 343 };
		3 of {2 434 4 3 3 }; 2 of {1 3 32 3213 3 }; 2 of {3 323 4 34 34 }; 2 of {2 313 313 323 323};
		18 of {2 3 3 2 2 }; 2 of {1 3 41 3 341 }; 2 of {3 313 414 3413 3413 }; 3 of {1 313 434 2 2};
		3 of {1 3 434 2 2 }; 3 of {1 3 434 1 1 }; 6 of {1 3 323 2 2 }; 3 of {1 313 3123 323 323 };
		6 of {1 3 434 3 3 }; 3 of {2 4 323 3 3 }; 3 of {1 323 4 32 32 }; 3 of {1 323 434 32 32 };
		3 of {1 3 31 3 31 }; 3 of {1 3 21 23 2341 }; 8 of {2 3 323 2 2 }; 4 of {2 323 32 23 23 };
		5 of {2 3 313 2 2 }; 3 of {2 4 43 23 23 }; 2 of {1 41 1 321 31 }; 2 of {1 434 434 3 3 };
		2 of {2 434 434 343 3 }; 2 of {1 434 1 343 1 }; 2 of {1 4 1 343 1 }; 5 of {1 4 4 32 32 };
		8 of {4 313 414 3413 3413 }; 2 of {2 3 41 34 341 }; 2 of {3 431 431 31 321 };
		2 of {4 431 431 321 321 }; 2 of {3 41 4 431 4 }; 2 of {3 41 4 341 34 }; 4 of {2 4 41 34 321};
		2 of {2 4 421 3 1 }; 6 of {2 3 1 3 1 }; 3 of {3 41 4 321 34 }; 3 of {2 3 31 4 431 };
		3 of {2 3 1 23 1 }; 3 of {2 4 21 3 321 }; 3 of {2 4 31 1 31 }; 7 of {1 4 31 1 31 };
		4 of {4 4 4 32 32 }; 3 of {2 3 4 3 34 }; 6 of {2 3 3 3 3 }; 3 of {4 4 3 3 3413 };
		3 of {4 4 3 3 313 }; 6 of {1 4 3 3 323 }; 3 of {1 4 3 3 3413 }; 3 of {1 4 3 3 3423 };
		6 of {2 4 1 1 1 }; 4 of {1 14 4 1 32 }; 4 of {1 4 1 1 1 }; 4 of {1 4 14 32 1 };

Table 26. The equivalence classes of feature sets B4 to B11

Equivalence classes							
B4	**B5**	**B6**	**B7**	**B8**	**B9**	**B10**	**B11**
253 of {2 2 2 }; 29 of {2 2 21 }; 122 of {2 2 1 };	614 of 0;	874 of 0;	783 of 0;	536 of 1;	268 of 0;	830 of 0;	832 of 0;
2 of {212 2 2 }; 6 of {21 1 2 }; 47 of {1 2 2 };	56 of 111;	53 of 1;	144 of 1;	391 of 0;	356 of 1;	41 of 1;	95 of 1;
8 of {12 2 2 }; 17 of {21 2 2 }; 6 of {2 21 21 };	32 of 11;				110 of 2;	23 of 2;	
12 of {2 21 1 }; 6 of {2 2 212 }; 3 of {212 21 2 };	88 of 110;				27 of 4;	6 of 6;	
4 of {2 21 212 }; 4 of {2 2 1212 }; 4 of {2121 2 2 };	36 of 10;				6 of 10;	6 of 11;	
3 of {2 212 12 }; 21 of {2 1 1 }; 15 of {2 2 2121 };	62 of 100;				68 of 3;	3 of 32;	
12 of {2 21212 2 }; 15 of {2 21 2 };	32 of 1;				25 of 7;	3 of 22;	
126 of {212 212 212 }; 70 of {21 21 21 };	7 of 101;				15 of 6;	3 of 7;	
14 of {21212 21212 21212 };8 of {21 21 1 };					22 of 8;	3 of 3;	
2 of {21 21 2121 }; 54 of {1 1 1 }; 5 of {1 21 21 };					12 of 28;	6 of 5;	
2 of {21 1 21 }; 17 of {12 12 12 };					12 of 29;	3 of 4;	
2 of {21212 21212 1 }; 2 of {21212 1 1 };					6 of 5;		
3 of {212 212 1 }; 4 of {12 21 2 };							
3 of {212 121 2 }; 11 of {2 1 2 }; 3 of {21 1 1 };							
3 of {12 1 2 }; 3 of {212 1 2 }; 2 of {2 1 21 };							

replaced by RS_{34}. $[x]_{B_4}$ has 29 elements. Then the degree of overlap between $[x]_{B_4} = [F0822405.x01]_{B_4}$ and RS_{34} can be estimated by (43).

$$\mu^{B_4}_{RS_{34}}(x) = \frac{|[F0822405.x01]_{B_4} \cap RS_{34}|}{|[F0822405.x01]_{B_4}|} = \frac{10}{29} = 34.4\%. \tag{43}$$

It demonstrates that the degree to which class $[F0822405.x01]_{B_4}$ belongs to fault 3 is 34.4%.

7.3 Rough Membership Tables for rmNN Training and Verification

By simply repeating the rough membership computation procedure described in Section 7.2, 11 degrees to which the file F0822405.x01 belongs to fault 3 are obtained based on RS_{3j} $(1 \leq j \leq 11)$ and $[F0822405.x01]_{B_j}$ $(1 \leq j \leq 11)$. The 11 degrees of membership are represented as a vector, *e.g.*, (0.19, 0.14, 0.33, 0.34, 0.12, 0.16, 0.19, 0.37, 0.28, 0.16, 0.16). In addition, for a fault file *.x01, the membership for each type of fault based on each feature set is derived to transform the training information table (Table 12) and testing information table (Table 13) into 12 rough membership training tables and 12 rough membership testing tables, respectively. One of these rough membership tables is shown in Table 27. The first column indicates the file index. The following 11 columns contain 11 rough memberships for fault 3. The last column is the target that indicates whether it is a fault 3 or not. The value of 1 is for "Yes" and 0.01 is for "No". Each row in Table 27 is employed as a training set to calibrate the rmNN for fault 3. Table 28 is the rough membership table for fault 3 rmNN verification. It is necessary to point out that both Table 27 and 28 are partial rough membership tables. Twelve complete rough membership training tables as well as 12 testing tables are listed in an Appendix available at [15].

7.4 Design of rmNNs for PSFC

Neural networks are collections of massively parallel computation units called neurons. A neuron is a processing element in a neural network. To design a rough membership neural network, the rough neurons should first be defined.

Architecture of the Rough Membership Neural Network (rmNN) for PSFC

The architecture of an rmNN for fault classification is dependent on the number of types of available faults. Each fault will have its own rmNN. In the research reported in this paper, 12 separate rmNNs are employed to classify 12 types of faults. Each rmNN will output an estimation of the degree of one type of fault for a given object. For instance, the output neuron of the kth rmNN will aggregate all contributions from the rough neurons in the first layer, process in the hidden layer and finally output an estimation of the degree of kth fault. The output from 12 rmNNs will be forwarded to a fault type decider neuron. The decider neuron simply picks up all the faults with degree values above 80%. Almost 50% of the power system fault events are multiple faults jointed. Compared to the k-Nearest Neighbor (k-NN) method, the neural network method was proved superior when the input deviated in shape from the pattern used in training [39]. Since k-NN classifiers are sensitive to outliers and noise contained in the training data [27]. The k-NN classifier may have advantageous in applications where there is less deviation in the sampled pattern.

Concepts of Rough Neurons

Typically, a neuron y maps its weighted input from R_n to [0, 1]. A selection of different types of neurons is given in Table 29: common neurons, rough neurons, fuzzy neurons.

Table 27. Partial rough membership table for fault 3 rmNN training

File Index	B1	B2	B3	B4	B5	B6	B7	B8	B9	B10	B11	Target
	11	1	508	1								
1	0.19	0.14	0	0	0.12	0.16	0.19	0	0.15	0.16	0.16	0.01
4	0.19	0.14	0	0	0.12	0.16	0.19	0	0.15	0.16	0.16	0.01
7	0.19	0.14	0	0	0.12	0.16	0.19	0	0.15	0.16	0.16	0.01
31	0.19	0.14	0	0	0.12	0.16	0.19	0	0.15	0.16	0.16	0.01
55	0.19	0.14	0.5	0.34	0.12	0.16	0.19	0.37	0.15	0.16	0.16	1
69	0.19	0.14	0.5	0.4	0.12	0.16	0.19	0.37	0.15	0.16	0.16	1
70	0.19	0.14	0.33	0.35	0.25	0.16	0.19	0.37	0.15	0.16	0.16	1
91	0.19	0.14	0.25	0.25	0.35	0.16	0.19	0.37	0.15	0.15	0.12	1
92	0.19	0.14	0.25	0.25	0.17	0.16	0.19	0.37	0.15	0.15	0.12	1
93	0.19	0.14	0.25	0.25	0.17	0.16	0.19	0.37	0.17	0.16	0.16	1
99	0.19	0.14	0.33	0.34	0.12	0.16	0.19	0.37	0.28	0.16	0.16	1
103	0.19	0.14	0.33	0.33	0.12	0.16	0.19	0.37	0.27	0.16	0.16	1
104	0.19	0.14	0.36	0.34	0.12	0.16	0.19	0.37	0.32	0.16	0.16	1
128	0.19	0.14	0	0	0.12	0.16	0.19	0	0.15	0.16	0.16	0.01
131	0.05	0.14	0	0	0.12	0.16	0	0	0.15	0.16	0.16	0.01
141	0.19	0.14	0	0	0.12	0.16	0.19	0	0.15	0.16	0.16	0.01
150	0.05	0.14	0	0	0.12	0.16	0.19	0	0.15	0.16	0.16	0.01
162	0.05	0.14	0	0	0.12	0.16	0.19	0	0.15	0.16	0.16	0.01
202	0.19	0.14	0	0	0	0.16	0.19	0	0.15	0.16	0.16	0.01
209	0.19	0.14	0	0	0.17	0.16	0.19	0	0.15	0.16	0.16	0.01
219	0.19	0.14	0	0	0.12	0.16	0	0	0.15	0.16	0.16	0.01
222	0.19	0.14	0	0	0.12	0.16	0	0	0.15	0.16	0.16	0.01
229	0.19	0.14	0	0	0.12	0.16	0	0	0.15	0.15	0.12	0.01
236	0.19	0.14	0	0	0.12	0.04	0	0	0.15	0.16	0.16	0.01
239	0.05	0.14	0	0	0.12	0.04	0.19	0	0.15	0.16	0.16	0.01
252	0.19	0.17	0	0	0.12	0.16	0.19	0	0.15	0.16	0.16	0.01
253	0.19	0.17	0	0	0	0.16	0.19	0	0.15	0	0.12	0.01
254	0.19	0.17	0	0	0.12	0.16	0.19	0	0.15	0.16	0.16	0.01
265	0.19	0.17	0	0	0	0.16	0.19	0	0.15	0.16	0.16	0.01
267	0.19	0.17	0	0	0	0.16	0.19	0	0.15	0.16	0.16	0.01
305	0.19	0.17	0.5	0.43	0.12	0.16	0.19	0.37	0.15	0.16	0.16	1
349	0.19	0.17	0.4	0.33	0.12	0.16	0.19	0.37	0.15	0.17	0.12	1
374	0.19	0.17	0	0	0.12	0.16	0	0	0.15	0	0.12	0.01
391	0.19	0.2	1	0.4	0.12	0.16	0.19	0.37	0.15	0.16	0.16	1
392	0.19	0.2	1	0.67	0.12	0.16	0.19	0.37	0.15	0.16	0.16	1
403	0.19	0.2	1	1	0.12	0.16	0.19	0.37	0.15	0.16	0.16	1
415	0.19	0.2	0.33	0.33	0.35	0.16	0.19	0.37	0.15	0.16	0.16	1
420	0.05	0.2	0	0	0.12	0.16	0	0	0	0.16	0.16	0.01
421	0.19	0.2	0.25	0	0.31	0.16	0.19	0	0.15	0.16	0.16	0.01
422	0.19	0.2	0	0	0.17	0.16	0.19	0	0.15	0.16	0.16	0.01
440	0.19	0.2	0	0	0.12	0.04	0.19	0	0.15	0.16	0.16	0.01
441	0.19	0.2	0	0	0.12	0.04	0.19	0	0.15	0.16	0.16	0.01
447	0.19	0.2	0	0	0.12	0.16	0.19	0	0.09	0.16	0.16	0.01
448	0.19	0.2	0	0	0.12	0.16	0.19	0	0.15	0.16	0.16	0.01
465	0.19	0.2	0	0	0.12	0.16	0.19	0	0.09	0.16	0.16	0.01
466	0.19	0.2	0	0	0.12	0.16	0.19	0	0.15	0.16	0.16	0.01
472	0.19	0.2	0	0	0.12	0.16	0.19	0	0.15	0.16	0.16	0.01
475	0.19	0.2	0	0	0.12	0.16	0.19	0	0.15	0.16	0.16	0.01
484	0.05	0.2	0	0	0.12	0.16	0.19	0	0.15	0.16	0.16	0.01
490	0.05	0.2	0	0	0.12	0.16	0.19	0	0.15	0.16	0.16	0.01
504	0.05	0.2	0	0	0.12	0.16	0.19	0	0.15	0.16	0.16	0.01
507	0.05	0.2	0	0	0.12	0.16	0.19	0	0.15	0.16	0.16	0.01
508	0.05	0.2	0	0	0.12	0.16	0.19	0	0.15	0.16	0.16	0.01

In the design of the rough membership function neural network (rmNN), the hidden layer consists of fuzzy neurons defined using the t-norm, s-norm and → (imply operators) from fuzzy set theory. The formal definition for a hidden neuron in an rmNN is given in (44) using the t-norm, s-norm and → operators. The reason that this form of hidden neuron is applied is because it provides a numerical representation of set intersection (t-norm), set union

Table 28. Partial rough membership table for fault 3 rmNN testing

File Index	B1	B2	B3	B4	B5	B6	B7	B8	B9	B10	B11	Target
	11	1	168	1								
1	0.19	0.14	0	0	0.12	0.16	0.19	0	0.15	0.16	0.16	0.01
19	0.19	0.14	0	0	0.12	0.16	0.19	0	0.15	0.16	0.16	0.01
30	0.19	0.14	0	0	0.12	0.16	0.19	0	0.15	0.16	0.16	0.01
46	0.19	0.14	0.4	0.4	0.12	0.16	0.19	0.37	0.15	0.16	0.16	1
53	0.19	0.14	0.25	0.25	0.17	0.16	0.19	0.37	0.15	0.15	0.12	1
54	0.19	0.14	0.4	0.4	0.31	0.16	0.19	0.37	0.15	0.16	0.16	1
56	0.19	0.14	0.33	0.33	0.12	0.04	0.19	0.37	0.15	0.16	0.16	1
57	0.19	0.14	0.33	0.35	0.12	0.16	0.19	0.37	0.28	0.16	0.16	1
63	0.19	0.14	0	0	0.12	0.16	0.19	0	0.15	0.16	0.16	0.01
82	0.19	0.14	0	0	0.12	0.16	0.19	0	0.15	0.16	0.16	0.01
86	0.19	0.14	0	0	0.22	0.16	0.19	0	0.15	0.16	0.16	0.01
87	0.19	0.14	0	0	0.22	0.16	0.19	0	0.15	0.16	0.16	0.01
90	0.19	0.14	0	0	0.12	0.16	0	0	0.28	0.16	0.16	0.01
91	0.19	0.14	0	0	0.12	0.16	0	0	0.15	0.15	0.12	0.01
93	0.19	0.14	0	0	0.12	0.04	0	0	0.15	0.16	0.16	0.01
94	0.19	0.14	0	0	0.12	0.04	0.19	0	0.15	0.16	0.16	0.01
111	0.05	0.17	0	0	0.12	0.16	0.19	0	0.15	0.16	0.16	0.01
112	0.19	0.17	0	0	0.12	0.16	0.19	0	0.15	0.16	0.16	0.01
116	0.19	0.17	0.4	0.4	0.12	0.16	0.19	0.37	0.15	0.16	0.16	1
120	0.19	0.17	0.39	0.43	0.12	0.16	0.19	0.37	0.15	0.16	0.16	1
121	0.19	0.17	0.4	0.4	0.31	0.16	0.19	0.37	0.15	0.16	0.16	1
127	0.19	0.17	0.4	0.4	0.35	0.16	0.19	0.37	0.15	0.16	0.16	1
128	0.19	0.17	0.39	0.25	0.22	0.16	0.19	0.37	0.09	0.15	0.12	1
129	0.05	0.17	0	0	0.12	0.16	0	0	0.15	0.16	0.16	0.01
130	0.19	0.17	0	0	0.22	0.16	0.19	0	0.15	0.16	0.16	0.01
131	0.19	0.17	0	0	0.17	0.16	0.19	0	0.15	0.16	0.16	0.01
132	0.19	0.17	0	0	0.12	0.16	0	0	0.32	0.16	0.16	0.01
134	0.19	0.17	0	0	0.12	0.16	0	0	0.15	0.16	0.16	0.01
135	0.19	0.17	0	0	0.12	0.16	0	0	0.15	0.13	0.12	0.01
136	0.19	0.17	0	0	0.12	0.16	0	0	0.15	0	0.12	0.01
137	0.19	0.17	0	0	0.12	0.16	0.19	0	0.15	0.16	0.16	0.01
138	0.19	0.17	0	0	0.12	0.16	0.19	0	0.15	0.16	0.16	0.01
139	0.19	0.2	1	0.67	0.12	0.16	0.19	0.37	0.15	0.16	0.16	1
141	0.19	0.2	1	1	0.12	0.16	0.19	0.37	0.15	0.16	0.16	1
144	0.19	0.2	0.39	0.67	0.22	0.16	0.19	0.37	0.15	0.16	0.16	1
145	0.19	0.2	0.33	0.33	0.35	0.16	0.19	0.37	0.15	0.16	0.16	1
146	0.05	0.2	0	0	0.12	0.16	0	0	0	0.16	0.16	0.01
147	0.19	0.2	0	0	0.12	0.16	0	0	0.15	0.16	0.16	0.01
148	0.05	0.2	0	0	0.12	0.04	0.19	0	0.15	0.16	0.16	0.01
150	0.19	0.2	0	0	0.12	0.04	0	0	0.15	0.16	0.16	0.01
151	0.19	0.2	0	0	0.12	0.04	0.19	0	0.15	0.16	0.16	0.01
152	0.19	0.2	0	0	0.12	0.04	0.19	0	0.15	0.16	0.16	0.01
153	0.19	0.2	0	0	0.12	0.16	0.19	0	0.15	0.16	0.16	0.01
156	0.19	0.2	0	0	0.12	0.16	0.19	0	0.15	0.16	0.16	0.01
158	0.19	0.2	0	0	0.12	0.16	0.19	0	0.15	0.16	0.16	0.01
161	0.19	0.2	0	0	0.12	0.16	0.19	0	0.15	0.16	0.16	0.01
167	0.05	0.2	0	0	0.12	0.16	0.19	0	0.15	0.16	0.16	0.01
168	0.05	0.2	0	0	0.12	0.16	0.19	0	0.15	0.16	0.16	0.01

(s-norm) and implication ($\rightarrow$) that works well as a means of aggregating the input from the rough membership functions in the input layer. Let $B, X, [x]_B$ denote a set of features, a set of files with matching fault type based on knowledge, and an equivalence class derived from known objects, respectively. The basic computation steps performed by a rough membership neuron are reflected in the flow chart in Figure 48.

In Figure 48 x is a newObj, an unclassified fault file. A rough neuron measures the degree of overlap of sets $[x]_B$ and X, and represents certain as well as uncertain classification of the input newObj, x.

Table 29. Different types of neurons

Common Neuron	Rough membership Neuron	Fuzzy Neurons
$y(x) = g(\sum_i w_i * x_i + v)$, where input x_i has connection (weight) w_i, which denotes a modifiable neural connection and bias v, $g(\bullet)$ is called activation function which can be a sigmoidal function such as logistic, tanh and Gaussian function	$y(x) = rmf(x)$ $= \mu_X^B(x) = \frac{\lvert [x]_B \cap X \rvert}{\lvert [x]_B \rvert}$	$y(x1, x2) =$ t-norm $(x1, x2)$ $= x1 * x2$ $y(x1, x2) =$ s-norm $(x1, x2)$ $= x1 + x2 - x1 * x2$ $y(r, x) = r \rightarrow x$ $= \begin{cases} \min(1, \frac{x}{r}) = \frac{x}{r} & \text{if } r \geq x \\ 1, & \text{otherwise} \end{cases}$

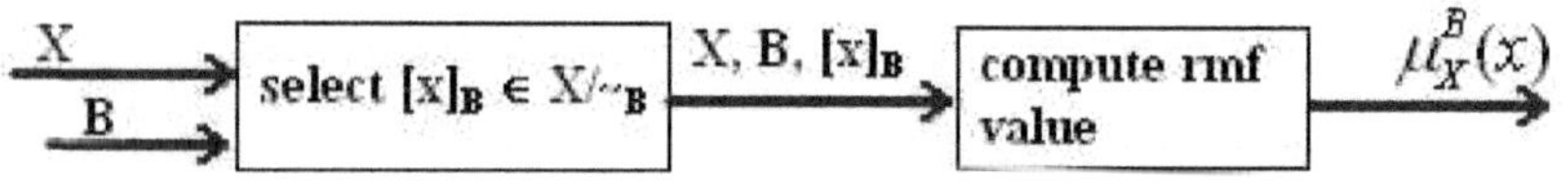

Fig. 48. Flow chart for basic Rough Neuron computation

Architecture of the Rough Membership Neural Network (rmNN) for PSFC

The architecture of a rmNN for fault classification is dependent on the number of types of available faults. Each fault will have its own rmNN. In the research reported in this paper, 12 separate rmNNs are employed to classify 12 types of faults. Each rmNN will output an estimation of the degree of one type of fault for a given object. For instance, the output neuron of the kth rmNN will aggregate all contributions from the rough neurons in the first layer, process in the hidden layer and finally output an estimation of the degree of kth fault. The output from 12 rmNNs will be forwarded to a fault type decider neuron. The decider neuron simply picks up all the faults with degree values above 80%. Almost 50% of the power system fault events are multiple faults jointed.

The whole computational framework that contains 12 rmNNs is shown in Figure 49. An example of the kth rmNN is shown in Figure 50, where $rmf^k(B_i(obj_n))$ is the rough membership for fault k based on the output of attribute set B_i for a given object n. The interconnections, *i.e.*, r_{ij}, w_{ij} and u_j inside the rmNN are shown in Figure 51. There are 11 rough neurons in the first layer and 11 fuzzy neurons in the hidden layer. Each neuron in the first layer is fully connected to the neuron in the hidden layer and each neuron in the hidden layer is fully connected to the output neuron. $rmf^k(B_i(obj_n))$ is simplified as rmf_i^k in Figure 51 as well as in the weights updating formulas.

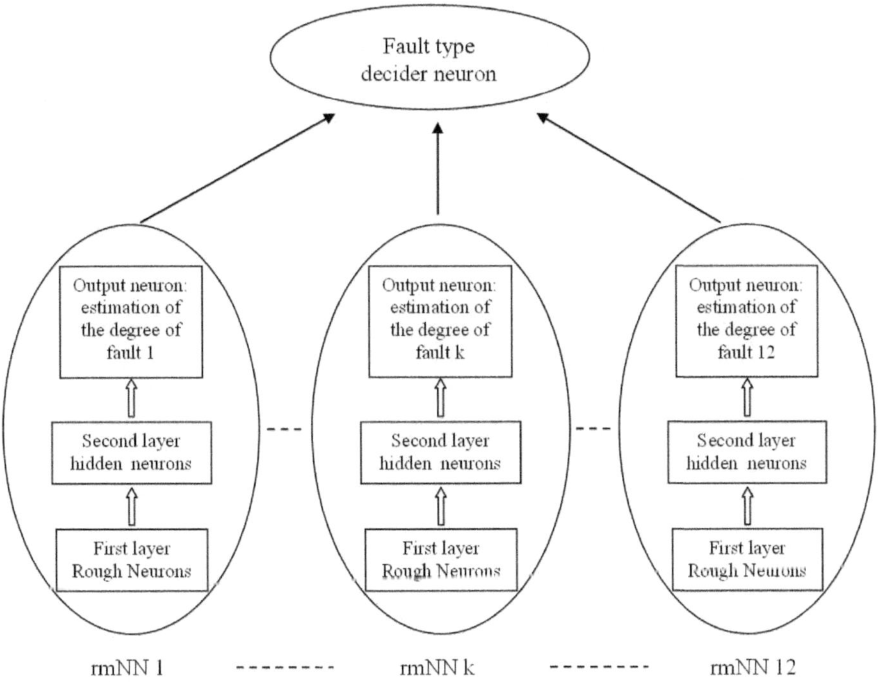

Fig. 49. Diagram of connection of 12 rmNNs

The formula for the activation function of the hidden neuron and the output neuron is described in (44) and (47) respectively. The formulas here are with respect to the kth rmNN. The *t-norm* is defined in (44). The imply operator $\rightarrow$ is defined in (45). The *s-norm*, namely probability sum, is defined in (46).

$$\begin{aligned} h_j^k &= t_{i=1}^{11}[r_{ij} \rightarrow rmf_i^k \; s \; w_{ij}] \\ &= [(r_{1j} \rightarrow rmf_1^k) \; s \; w_{1j}][(r_{2j} \rightarrow rmf_2^k) \; s \; w_{2j}]... \\ &\quad ...[(r_{11j} \rightarrow rmf_{11}^k) \; s \; w_{11j}], \end{aligned} \tag{44}$$

where

$$\begin{aligned} (r_{ij} \rightarrow rmf_i^k) &\equiv min\left(1, \frac{rmf_i^k}{r_{ij}}\right) \\ &= \begin{cases} \frac{rmf_i^k}{r_{ij}}, & if \;\; r_{ij} \geq rmf_i^k, \\ 1, & otherwise. \end{cases} \end{aligned} \tag{45}$$

$$(r_{ij} \rightarrow rmf_i^k) \; s \; w_{ij} = (r_{ij} \rightarrow rmf_i^k) + w_{ij} - (r_{ij} \rightarrow rmf_i^k)w_{ij}. \tag{46}$$

$$O^k = s_{j=1}^{11}[h_j^k u_j]. \tag{47}$$

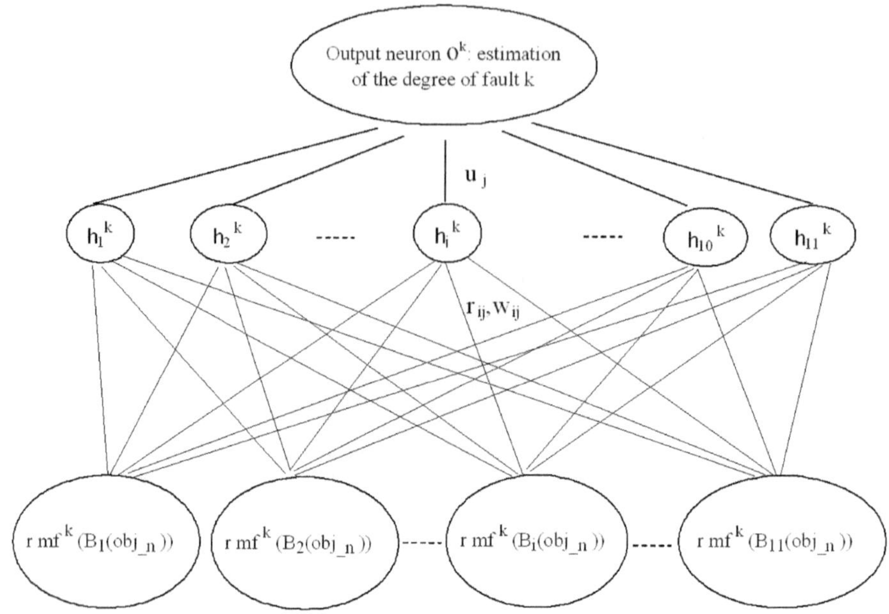

Fig. 50. Diagram of rmNN for k^{th} fault classification

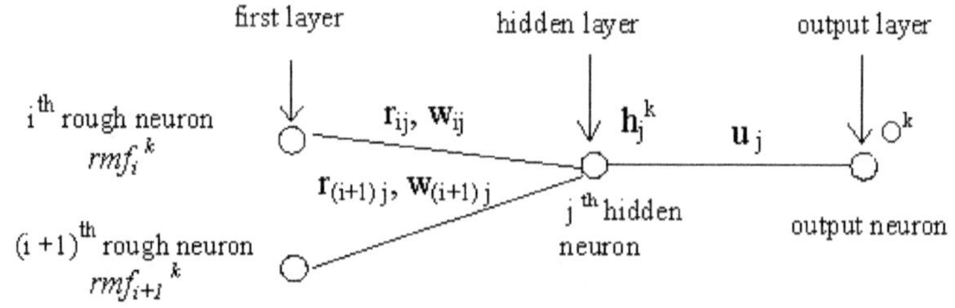

Fig. 51. Interconnection of the rmNN for fault k

Weights Adjustment with a Fixed Step Size Gradient Algorithm
In the process of tuning weights in the kth rmNN, r_{ij}, w_{ij} and u_j will be updated by the partial derivative of the error. The error is the square of the difference between the target and the output of the kth rmNN.

$$\begin{aligned} Error &= \left(target^k - O^k\right)^2 \\ &= \left(target^k - s_{j=1}^{11}\left(\left(t_{i=1}^{11}\left(\left(r_{ij} \to rmf_i^k\right)\ s\ w_{ij}\right)\right)u_j\right)\right)^2 \\ &= F\left(r_{ij}, w_{ij}, u_j\right), \end{aligned} \tag{48}$$

and

$$u_j^{(new)} = u_j^{(old)} - \alpha \frac{\partial Error}{\partial u_j},$$

$$w_{ij}^{(new)} = w_{ij}^{(old)} - \alpha \frac{\partial Error}{\partial w_{ij}},$$
$$r_{ij}^{(new)} = r_{ij}^{(old)} - \alpha \frac{\partial Error}{\partial r_{ij}}, \tag{49}$$

where α is the learning rate. The effects of the learning rate are discussed in Section 7.6. $u_j^{(new)}$ will be updated by the partial derivative of the error function with respect of $u_j^{(old)}$ in (49).

$$\frac{\partial Error}{\partial u_i} = -\left(target^k - O^k\right) \frac{\partial O^k}{\partial u_i}$$
$$= -\left(target^k - O^k\right) \frac{\partial}{\partial u_i} \left[s_{j=1}^{11}[h_j^k u_j]\right], \tag{50}$$

where $i = 1, 2, \ldots 11$. The overall expression can be rewritten by separating the ith component in the overall s-norm composition,

$$\frac{\partial O^k}{\partial u_i} = \frac{\partial}{\partial u_i} \left[A + u_i h_i^k - A u_i h_i^k\right]$$
$$= h_i^k (1 - A), \tag{51}$$

where factor A summarizes the remaining components of the s-norm composition, *i.e.*,

$$A = s_{j=1, j \neq i}^{11} \left[h_j^k u_j\right]. \tag{52}$$

The computation of the connections between the input layer and the hidden layer, *i.e.*, w_{ij} given by the second formula in (49) requires the use of the chaining rule of differentiation. This implies the following,

$$\frac{\partial Error}{\partial w_{ij}} = -\left(target^k - O^k\right) \frac{\partial O^k}{\partial h_j^k} \frac{\partial h_j^k}{\partial w_{ij}}, \tag{53}$$

where w_{ij} refers to the connection from the ith node in the first layer to the jth node in the hidden layer. The $\frac{\partial O^k}{\partial h_j^k}$ factor is expressed as,

$$\frac{\partial O^k}{\partial h_j^k} = u_j (1 - A). \tag{54}$$

For factor $\frac{\partial h_j^k}{\partial w_{ij}}$, the activation function will be applied governing the hidden neuron,

$$\frac{\partial h_j^k}{\partial w_{ij}} = \frac{\partial}{\partial w_{ij}} \left(t_{l=1}^{11} \left[\left(r_{lj} \rightarrow rmf_l^k\right) \ s \ w_{lj}\right]\right)$$
$$= \frac{\partial}{\partial w_{ij}} \left[\left(r_{ij} \rightarrow rmf_i^k\right) \ s \ w_{ij}\right] \left(t_{l=1, l \neq i}^{11} \left[\left(r_{lj} \rightarrow rmf_l^k\right) \ s \ w_{lj}\right]\right). \tag{55}$$

By introducing the notation

$$B = t_{l=1,l\neq i}^{11}\left[\left(r_{lj} \rightarrow rmf_l^k\right)\ s\ w_{lj}\right], \tag{56}$$

(55) can be rewritten as

$$\begin{aligned}\frac{\partial h_j^k}{\partial w_{ij}} &= \frac{\partial}{\partial w_{ij}}\left(B\left[\left(r_{ij} \rightarrow rmf_i^k\right)\ s\ w_{ij}\right]\right)\\ &= B\frac{\partial}{\partial w_{ij}}\left[\left(r_{ij} \rightarrow rmf_i^k\right)\ s\ w_{ij}\right]\\ &= B\frac{\partial}{\partial w_{ij}}\left(\left(r_{ij} \rightarrow rmf_i^k\right) + w_{ij} - \left(r_{ij} \rightarrow rmf_i^k\right) w_{ij}\right)\\ &= B\left(1 - \left(r_{ij} \rightarrow rmf_i^k\right)\right).\end{aligned} \tag{57}$$

The same procedure applies to update the r_{ij} parameter by computing

$$\frac{\partial Error}{\partial r_{ij}} = -\left(target^k - O^k\right)\frac{\partial O^k}{\partial h_j^k}\frac{\partial h_j^k}{\partial r_{ij}}, \tag{58}$$

where r_{ij} refers to the connection from the ith node in the first layer to the jth node in the hidden layer. The $\frac{\partial O^k}{\partial h_j^k}$ factor is expressed in (54), which is the same thing when updating w_{ij}.

$$\begin{aligned}\frac{\partial h_j^k}{\partial r_{ij}} &= \frac{\partial}{\partial r_{ij}}\left(B\left[\left(r_{ij} \rightarrow rmf_i^k\right)\ s\ w_{ij}\right]\right)\\ &= B\frac{\partial}{\partial r_{ij}}\left[\left(r_{ij} \rightarrow rmf_i^k\right)\ s\ w_{ij}\right]\\ &= B\frac{\partial}{\partial r_{ij}}\left(\left(r_{ij} \rightarrow rmf_i^k\right) + w_{ij} - \left(r_{ij} \rightarrow rmf_i^k\right) w_{ij}\right)\\ &= B\left(1 - w_{ij}\right)\frac{\partial\left(r_{ij} \rightarrow rmf_i^k\right)}{\partial r_{ij}},\end{aligned} \tag{59}$$

and

$$\begin{aligned}\frac{\partial\left(r_{ij} \rightarrow rmf_i^k\right)}{\partial r_{ij}} &= \frac{\partial}{\partial r_{ij}}\begin{cases}\frac{rmf_i^k}{r_{ij}}, & if\ r_{ij} \geq rmf_i^k,\\ 1, & otherwise.\end{cases}\\ &= \begin{cases}-\frac{rmf_i^k}{r_{ij}^2}, & if\ r_{ij} \geq rmf_i^k,\\ 0, & otherwise.\end{cases}\end{aligned} \tag{60}$$

The program flowchart for the rmNN calibration is illustrated in Figure 52.

Calibration Results of Selected rmNNs for PSFC

By way of illustration, only the calibration and verification results for fault 3 and 5 rmNNs are illustrated here. The results of all 12 rmNNs are attached in an Appendix available at [15].

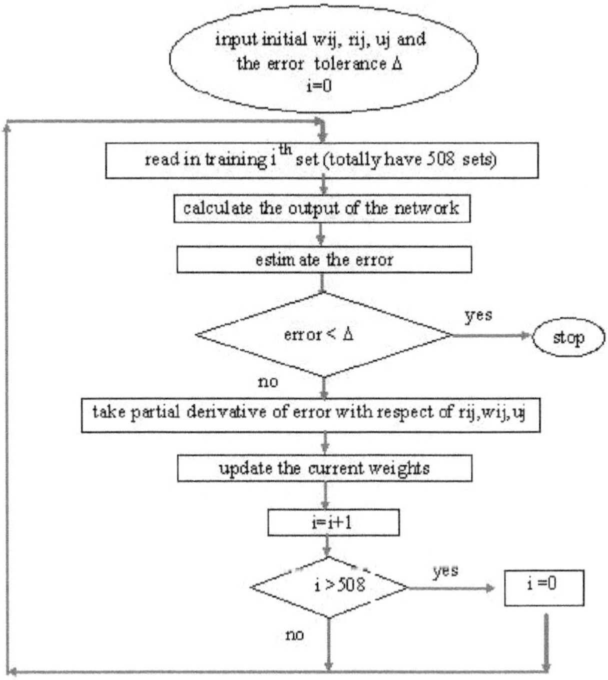

Fig. 52. Flowchart of weights updating for rmNN calibration

The error output during the training of the fault 3 and fault 5 rmNNs is shown in Figures 53 and 54 respectively. After 200 learning cycles, the error between the target and the output of fault 3 rmNN is less than 0.8 and of fault 5 rmNN is less than 0.2.

The output for fault 3 and 5 rmNNs compared to their targets after the first learning cycle is shown in Figures 55 and 56 respectively. Figures 57 and 58 illustrate that after 200 learning cycles, the output for fault 3 and 5 rmNNs matches their targets well.

It is observed, from Figures 57 and 58, that approximately 70 fault files, which are files from file #73 to file #96, from file #318 to file #348 and from file #405 to file #419, are indicating combinations of fault 3 and 5.

The trained r, w and u parameters for fault 3 and 5 rmNNs are shown in Tables 30 and 31 respectively.

Verification Results of Selected rmNNs for PSFC

In this section, 168 additional fault files have been used in a test set to verify the performance of the rmNN power fault classification system. The test results for fault 3 and 5 rmNNs are shown in Figures 59 and 60. With the fault detection threshold set to 80%, both rmNNs have 100% classification accuracy.

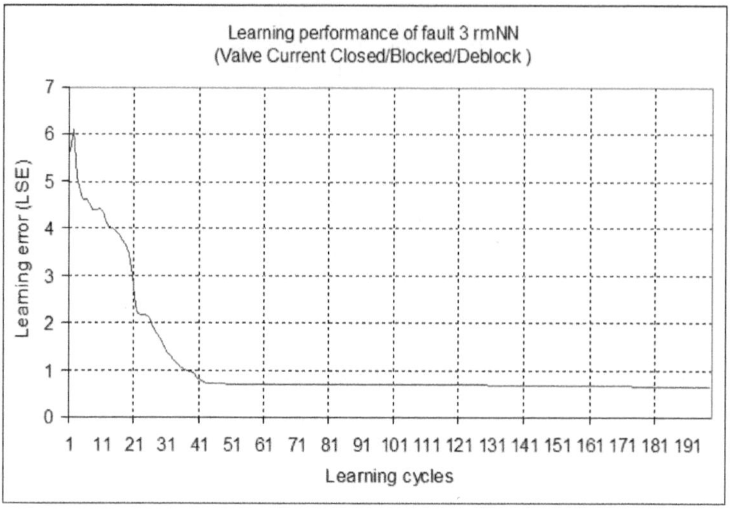

Fig. 53. Learning performance of fault 3 rmNN

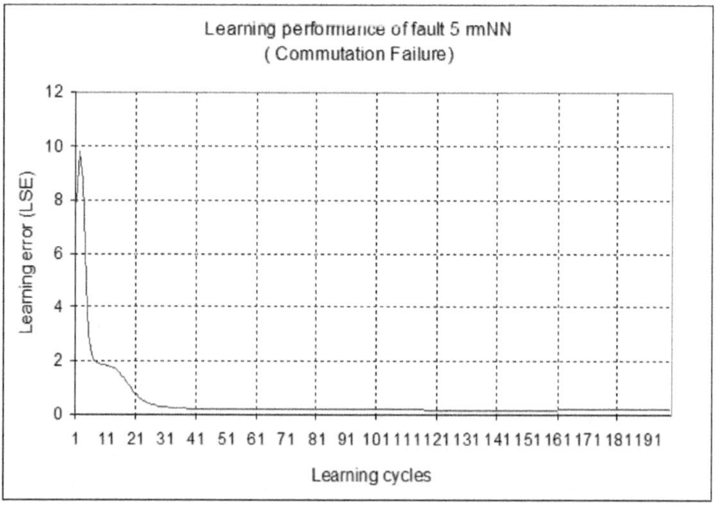

Fig. 54. Learning performance of fault 5 rmNN

Table 32 summarizes the accuracy of the rmNN power fault classification system. It has been found that for each type of fault, the more fault files used in training, the more accurate the test results will be. For instance, a greater number of fault files were used to calibrate the rmNNs for faults "Minor AC disturbance", "AC Disturbance", "Valve Current Closed/Blocked/Deblocked", "Line Fault", and "Commutation Failure". The test results for these types of

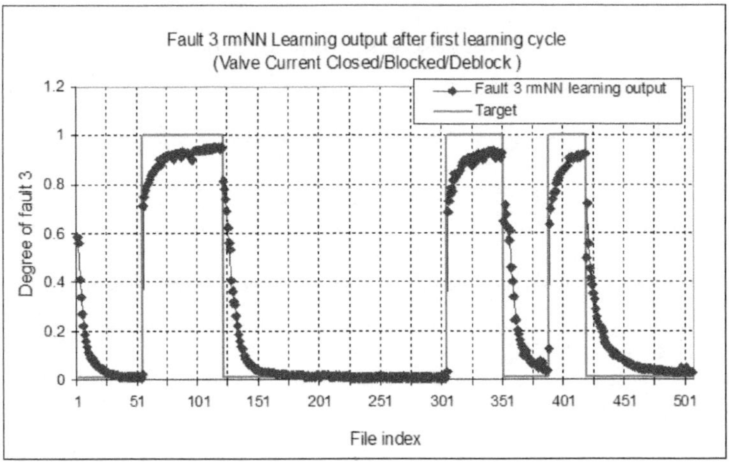

Fig. 55. The output for fault 3 rmNN after one learning cycle

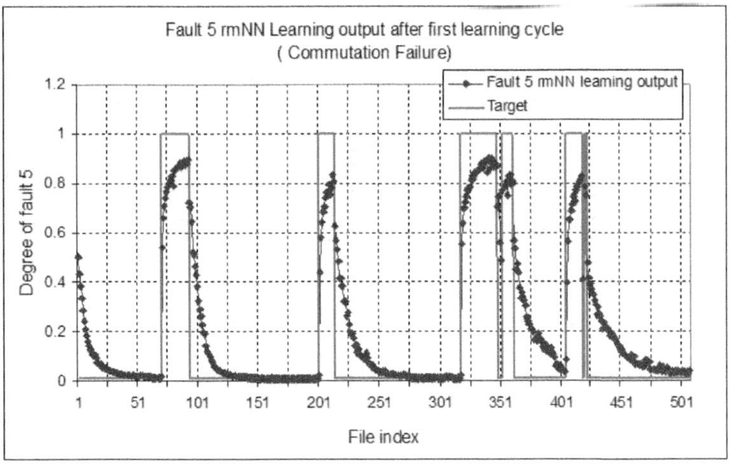

Fig. 56. The output for fault 5 rmNN after one learning cycle

faults showed 100% accuracy. By contrast, for faults "Current Arc Back" and "Normal affected by another pole", 26 and 18 fault files are employed for calibration respectively; and the corresponding accuracy of the test results were comparatively low (Table 32).

7.5 Effects of the Number of Neurons in the Hidden Layer

For verifying the rmNN performance, the rmNN sensitivity with respect to different numbers of neurons in the hidden layer was analyzed. The rmNNs with

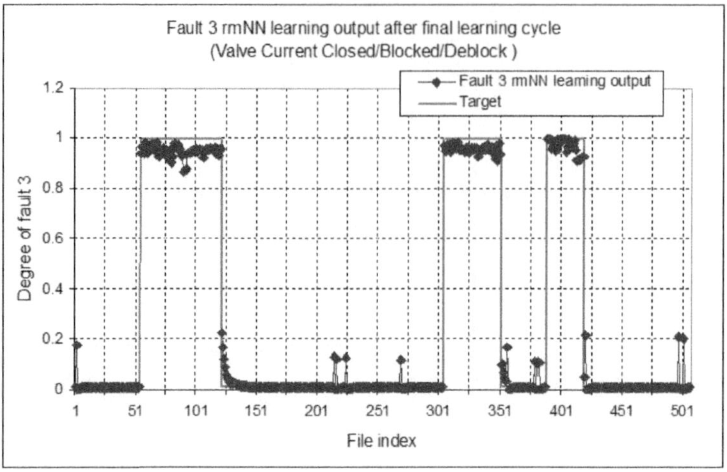

Fig. 57. The output for fault 3 rmNN after 200 learning cycles

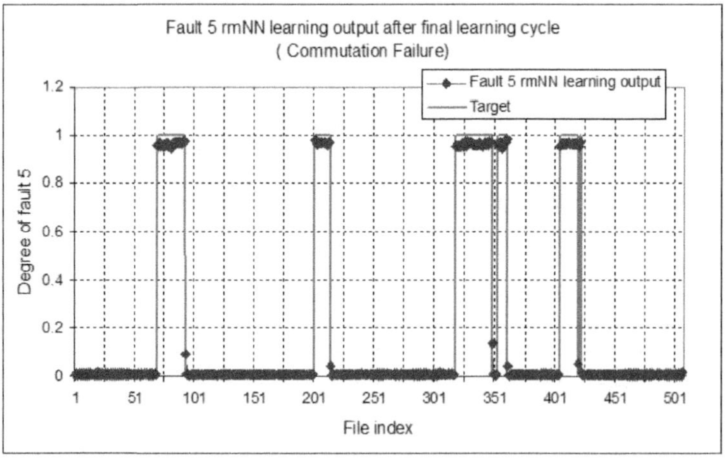

Fig. 58. The output for fault 5 rmNN after 200 learning cycles

11, 9, 7 or 3 hidden neurons are tested for the learning and verification set. The numerical results provide a very good performance index.

To decide on the number of hidden neurons, the square root of the product of the input and output is a good number with which to begin, in this case, $\sqrt{11 \times 1}$ is 3.3. The test results of the learning and verification performance with 3, 7, 9 and 11 neurons for 12 rmNNs are shown in an Appendix available at [15]. The test results indicate that the performance with 3 hidden neurons is always the worst case and unacceptable. The performance with 7 and 9 hidden neurons is very close. The case with 11 neurons has the best performance for fault 7 and fault 8 rmNNs.

Table 30. r, w and u parameters trained for fault 3 rmNN

r										
0.000	0.060	0.974	0.330	0.120	0.058	0.190	0.999	0.039	0.072	0.120
0.037	0.003	0.255	0.995	0.068	0.040	0.190	0.370	0.043	0.130	0.119
0.198	0.140	0.996	0.517	0.120	0.040	0.190	0.390	0.040	0.130	0.003
0.198	0.140	0.996	0.512	0.120	0.041	0.190	0.390	0.050	0.130	0.047
0.185	0.170	0.827	0.462	0.250	0.317	0.113	0.399	0.340	0.133	0.399
0.048	0.342	0.455	0.708	0.053	0.320	0.169	0.519	0.327	0.044	0.367
0.050	0.133	0.348	0.406	0.214	0.040	0.194	0.996	0.108	0.153	0.120
0.050	0.009	0.253	0.995	0.120	0.040	0.190	0.370	0.040	0.105	0.044
0.363	0.273	0.667	0.607	0.361	0.081	0.292	0.539	0.345	0.150	0.120
0.049	0.137	0.255	0.995	0.120	0.040	0.190	0.370	0.040	0.067	0.030
0.413	0.383	0.361	0.543	0.284	0.146	0.262	0.557	0.289	0.266	0.058
w										
0.176	0.253	0.003	0.000	0.218	0.268	0.116	0.001	0.114	0.050	0.158
0.002	0.118	0.000	0.004	0.206	0.114	0.156	0.000	0.095	0.233	0.307
0.000	0.074	0.003	0.000	0.302	0.106	0.000	0.000	0.023	0.003	0.287
0.000	0.216	0.003	0.000	0.123	0.302	0.000	0.000	0.195	0.111	0.070
0.243	0.308	0.000	0.055	0.011	0.145	0.289	0.162	0.074	0.022	0.192
0.274	0.168	0.000	0.000	0.157	0.113	0.165	0.037	0.131	0.115	0.000
0.207	0.033	0.000	0.000	0.082	0.110	0.000	0.003	0.184	0.148	0.133
0.279	0.311	0.000	0.004	0.200	0.067	0.000	0.000	0.295	0.103	0.229
0.192	0.227	0.000	0.000	0.238	0.036	0.000	0.000	0.191	0.022	0.063
0.088	0.073	0.000	0.004	0.298	0.297	0.000	0.000	0.312	0.193	0.137
0.000	0.071	0.154	0.000	0.177	0.145	0.181	0.041	0.213	0.111	0.231
u										
1.000	1.000	0.990	0.990	0.000	0.000	1.000	1.000	0.000	1.000	0.000

Table 31. r, w and u parameters trained for fault 5 rmNN

r										
0.000	0.060	0.122	0.100	0.998	0.110	0.120	0.050	0.040	0.103	0.100
0.037	0.003	0.512	0.397	0.476	0.238	0.144	0.342	0.293	0.368	0.384
0.050	0.080	0.187	0.100	0.998	0.110	0.120	0.050	0.040	0.100	0.003
0.048	0.080	0.123	0.100	0.998	0.110	0.120	0.050	0.040	0.101	0.047
0.050	0.080	0.126	0.100	0.998	0.110	0.120	0.044	0.040	0.103	0.101
0.048	0.329	0.419	0.420	0.421	0.309	0.172	0.286	0.324	0.071	0.357
0.209	0.191	0.457	0.353	0.493	0.332	0.212	0.389	0.305	0.087	0.325
0.039	0.009	0.123	0.100	0.998	0.110	0.120	0.034	0.040	0.101	0.044
0.050	0.080	0.217	0.100	0.998	0.110	0.120	0.036	0.040	0.100	0.093
0.050	0.079	0.170	0.100	0.998	0.110	0.120	0.050	0.040	0.101	0.030
0.376	0.358	0.306	0.334	0.482	0.206	0.259	0.335	0.270	0.247	0.058
w										
0.176	0.253	0.000	0.107	0.002	0.256	0.080	0.009	0.092	0.009	0.143
0.002	0.118	0.000	0.078	0.034	0.095	0.176	0.211	0.046	0.196	0.257
0.049	0.074	0.067	0.139	0.002	0.118	0.060	0.021	0.007	0.000	0.287
0.086	0.222	0.000	0.000	0.002	0.282	0.000	0.103	0.152	0.068	0.070
0.255	0.314	0.042	0.120	0.002	0.157	0.264	0.283	0.057	0.000	0.231
0.274	0.166	0.000	0.045	0.040	0.107	0.159	0.210	0.104	0.113	0.000
0.204	0.029	0.157	0.036	0.000	0.080	0.065	0.058	0.140	0.236	0.108
0.284	0.311	0.000	0.000	0.002	0.044	0.000	0.232	0.246	0.050	0.229
0.240	0.239	0.018	0.006	0.002	0.030	0.010	0.142	0.182	0.014	0.059
0.101	0.072	0.063	0.018	0.002	0.287	0.065	0.311	0.270	0.175	0.137
0.000	0.072	0.181	0.003	0.037	0.137	0.173	0.231	0.193	0.105	0.231
u										
1.000	0.050	1.000	1.000	1.000	0.010	0.030	1.000	1.000	1.000	0.000

The learning output comparison for fault 7 rmNN with 3, 7 and 11 hidden neurons is shown in Figure 61(a). The learning output for fault 7 rmNN with 9 hidden neurons is omitted since its performance is close to the rmNN with 7 hidden neurons (see Appendix available at [15]). Figures 61(b), 61(c), 61(d)

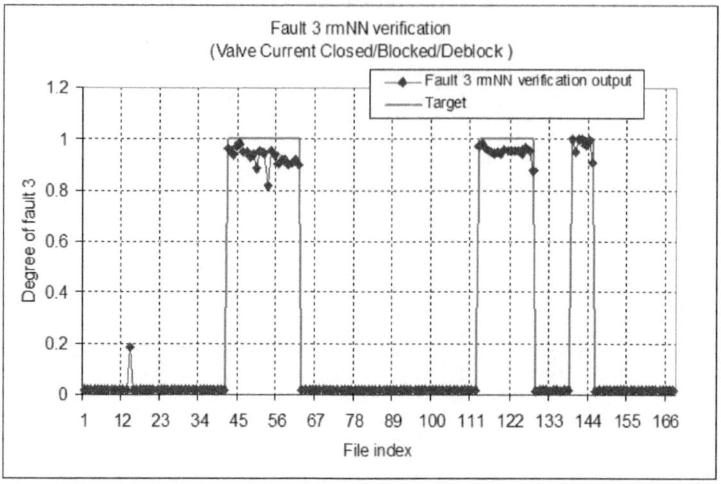

Fig. 59. Testing results for fault 3 rmNN

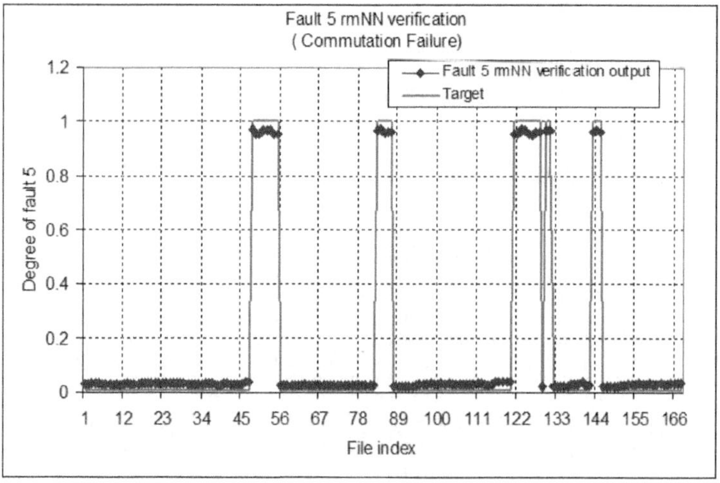

Fig. 60. Testing results for fault 5 rmNN

and 61(e) show the details of the A, B, C and D parts in Figure 61(a) respectively. They clearly indicate that, for all the true cases, the rmNN with 11 hidden neurons present the highest output, and the rmNN with 3 hidden neurons give the lowest output. The verification output comparison for fault 7 rmNN with 3, 7 and 11 hidden neurons is shown in Figure 62(a). The verification output for fault 7 rmNN with 9 hidden neurons is omitted for the same reason. Figures 62(b), 62(c), 62(d) and 62(e) show the details of the A, B, C and D parts of Figure 62(a) and clearly confirm the results from learning for all the true cases.

Table 32. Accuracy of the rmNN power fault classification system

Fault type	# of files for verification	# of files incorrect	Accuracy
Minor AC Disturbance	94	0	1.00
AC Disturbance	44	0	1.00
Valve Current Closed/Blocked/Deblocked	44	0	1.00
Line Fault	22	0	1.00
Commutation Failure	25	0	1.00
Pole Voltages/Current Closed/Blocked/Deblocked	15	0	1.00
Current Arc Back	6	1	0.83
Parallel Operation	9	0	1.00
Pole Current Oscillation	7	0	1.00
Normal Affected by Another Pole	4	1	0.75
Asymmetric Protection	6	0	1.00
Disturbance on DC Voltage	7	0	1.00

The rmNN with 11 hidden neurons gives the highest verification score, and the verification output from the one with 3 hidden neurons yields the lowest.

As another example, the learning and verification output comparison for fault 8 rmNN with 3, 7 and 11 hidden neurons is shown in Figures 63 and 64 respectively. It agrees with the learning and verification results for fault 7 rmNN, *i.e.*, different numbers of neurons in the hidden layer considerably affect the performance of the rmNN. The goal is to have not too many but enough hidden neurons to be able to learn correctly. There are no analytically shown facts about the necessary number of hidden neurons, instead more tests are required to find an appropriate number. In addition, some research shows that the redundancy on hidden-layer neurons is useful in the fault tolerance of neural networks, especially for the feedforward networks.

7.6 Effects of Learning Cycles, Learning Rate and Least Square Error

To obtain satisfactory performance for an rmNN, the selection of appropriate learning rates (α) for the rmNN is critical and challenging. Applying an improper learning rate to the rmNN may cause the learning curve of the rmNN to oscillate. For example, suppose a network produces an error of -0.5 and the error was adjusted at an improper learning rate by the network. The new error is +0.5, and the next error is -0.5 again..., so on and so forth. Apparently the learning period this system takes will be endless. On the other hand, if the learning rate is too small, the network parameters will improve toward the best solution, but at a very low speed. It might take hours, even days, to optimize such a network. To gain a good learning rate requires interactive processing to achieve an acceptable overall direction for the search.

It is sometimes seen that the learning error decreases for the learning set of data with more and more learning cycles (LCs), but still does not lead to better classification performance. This suggests that the network is "overfitting" due to some local minimum.

An example of an "overfitting" rmNN is observed when the rmNN is trained for classifying fault 10. Figure 65.1 shows the learning least square error (LSE)

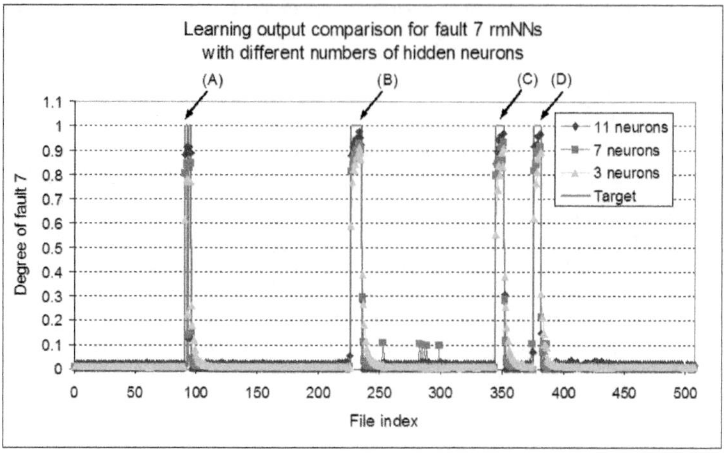

61.1: A broad view of the performance comparison for 508 rmNN training files

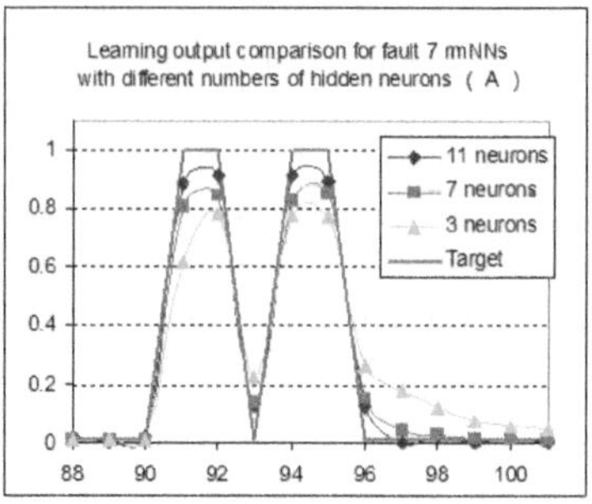

61.2: A zoom-in detail of part A

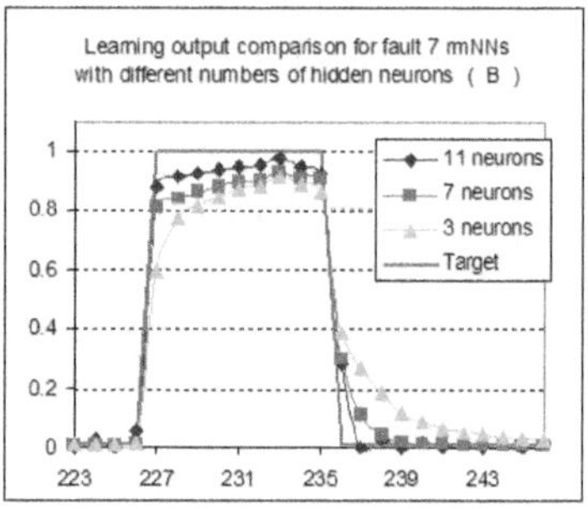

61.3: A zoom-in detail of part B

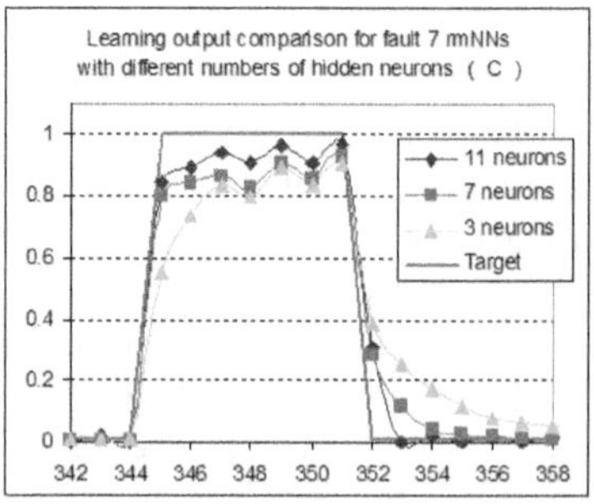

61.4: A zoom-in detail of part C

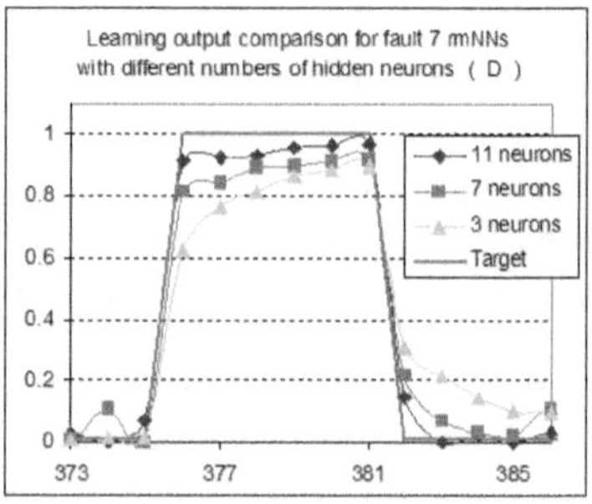

61.5: A zoom-in detail of part D

Fig. 61. The learning output comparison for fault 7 rmNNs with 3, 7 and 11 hidden neurons

comparison for fault 10 rmNN with different LCs and α. The details of A and B parts in Figure 65.1 are illustrated in Figure 65.2 and Figure 65.3, respectively. Learning case 1 has LCs = 100, $\alpha = 0.22$ and LSE = 0.233; learning case 2

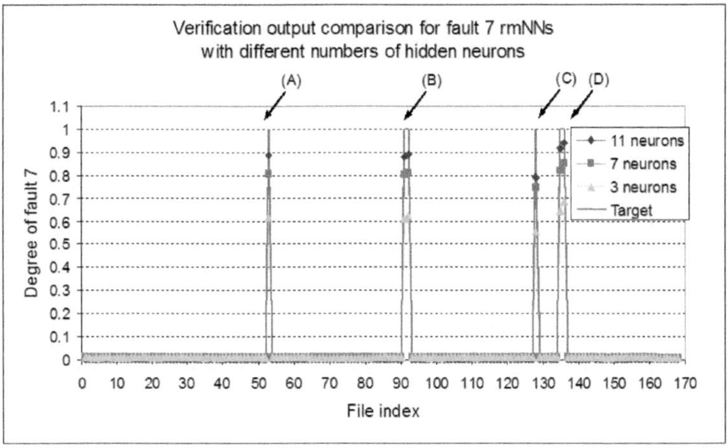

62.1: A broad view of the performance comparison for 168 rmNN testing files

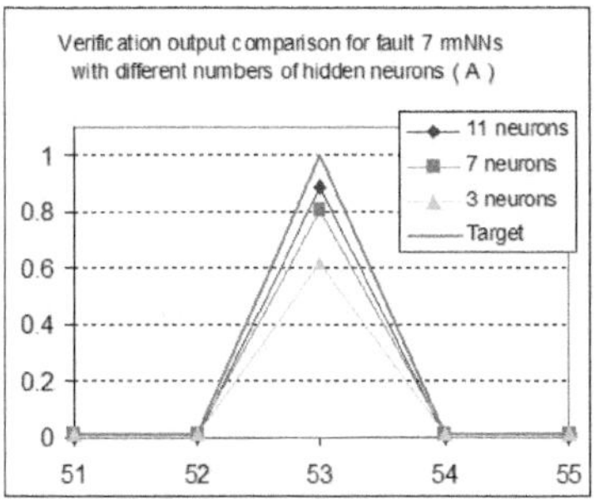

62.2: A zoom-in detail of part A

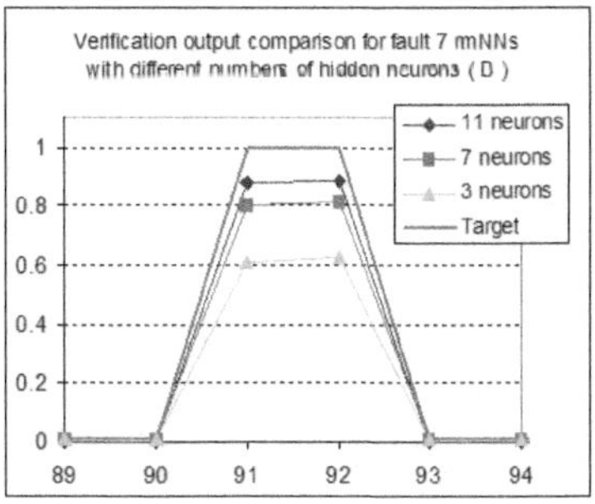

62.3: A zoom-in detail of part B

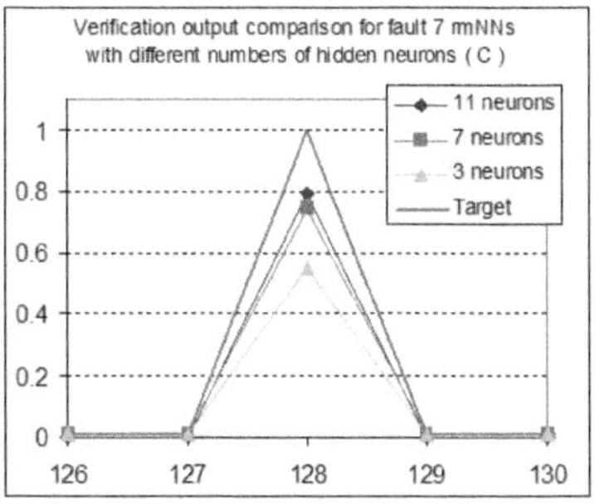

62.4: A zoom-in detail of part C

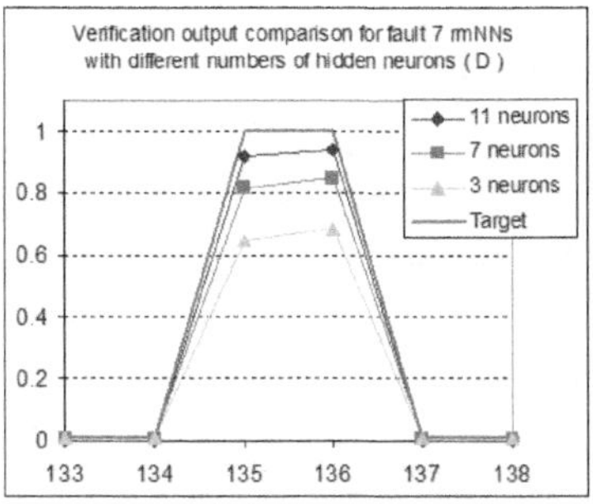

62.5: A zoom-in detail of part D

Fig. 62. The verification output comparison for fault 7 rmNNs with 3, 7 and 11 hidden neurons

has LCs = 800, $\alpha = 0.22$ and LSE = 0.21; and learning case 3 has LCs = 800, $\alpha = 0.3$ and LSE = 0.19. The learning cycles are increasing and the LSEs are decreasing.

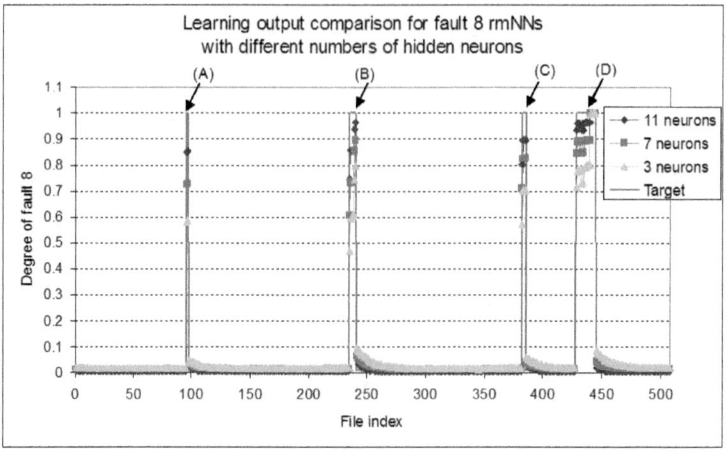

63.1: A broad view of the performance comparison for 508 rmNN training files

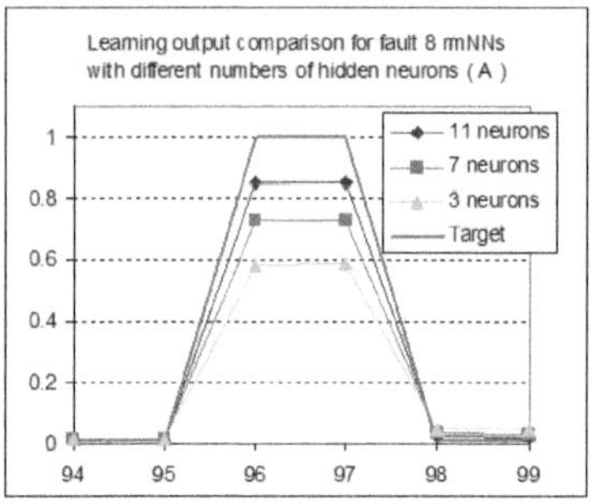

63.2: A zoom-in detail of part A

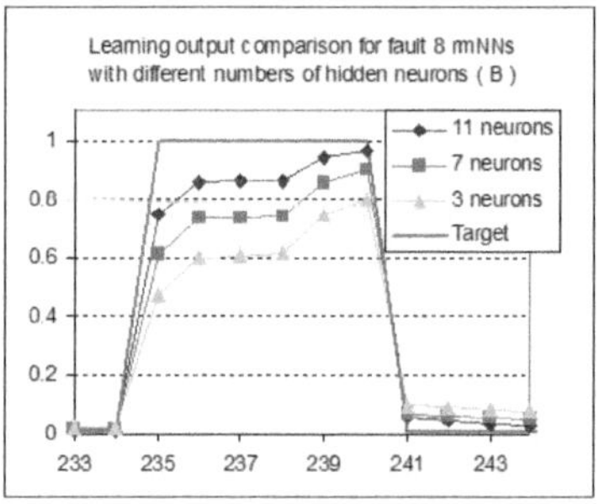

63.3: A zoom-in detail of part B

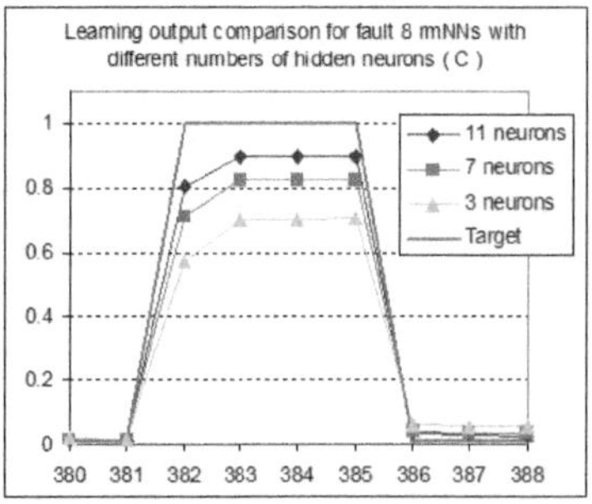

63.4: A zoom-in detail of part C

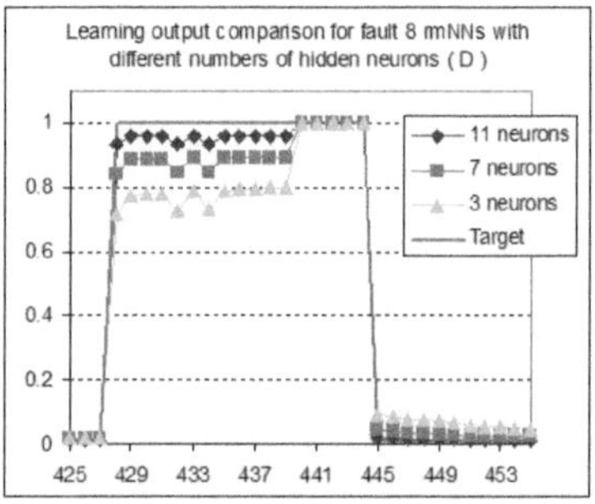

63.5: A zoom-in detail of part D

Fig. 63. The learning output comparison for fault 8 rmNNs with 3, 7 and 11 hidden neurons

Figure 66.1 illustrates the learning output comparison for fault 10 rmNN in the three cases shown in Figure 65. Figures 66.2 and 66.3 show the learning output at points A and B. Points A and B represent file 422 and file 471, which

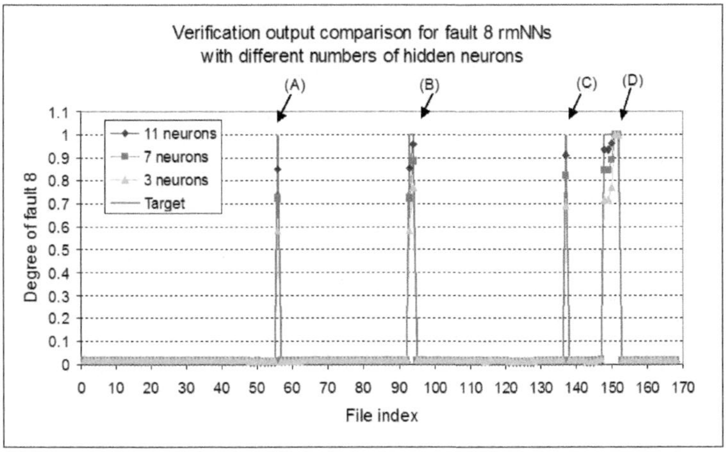

64.1: A broad view of the performance comparison for 168 rmNN testing files

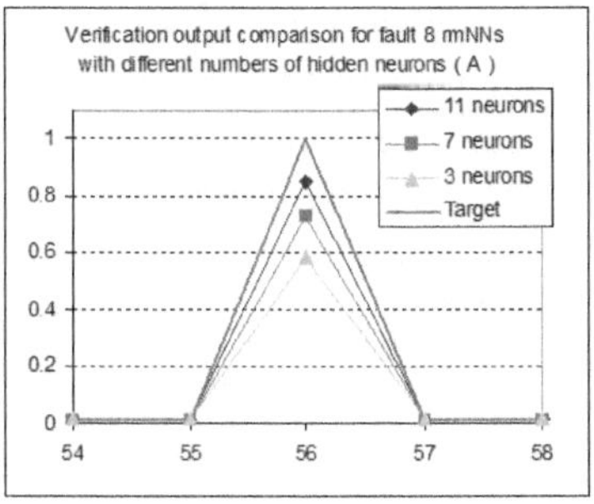

64.2: A zoom-in detail of part A

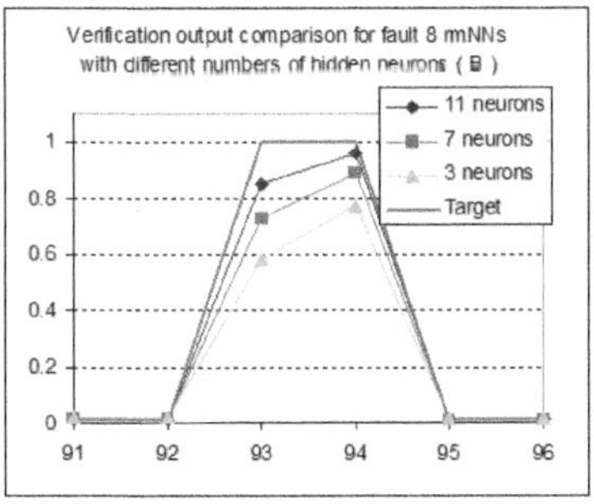

64.3: A zoom-in detail of part B

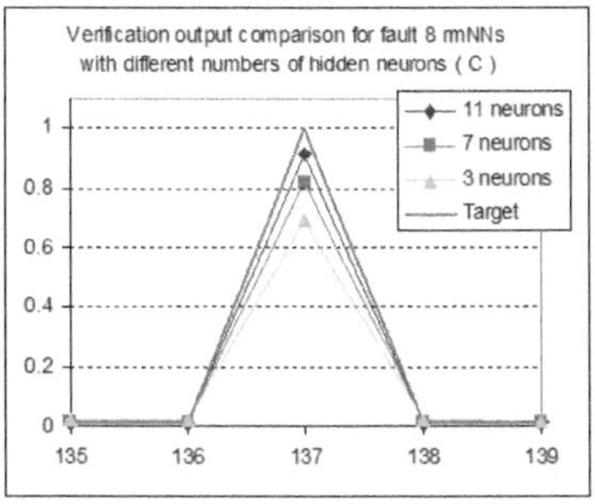

64.4: A zoom-in detail of part C

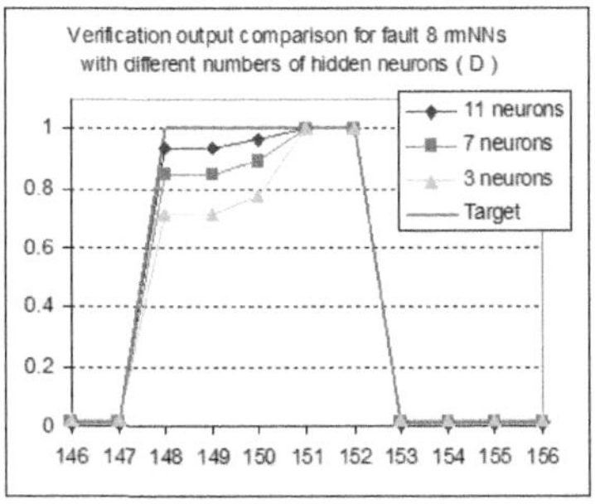

64.5: A zoom-in detail of part D

Fig. 64. The verification output comparison for fault 8 rmNNs with 3, 7 and 11 hidden neurons

belong to false case and true case respectively. From Figures 66.2 and 66.3, it is found that for case 1, points A and B are 0.09 apart from each other; for case 2, A and B are closer with a distance of 0.07; and for case 3, A and B locate almost

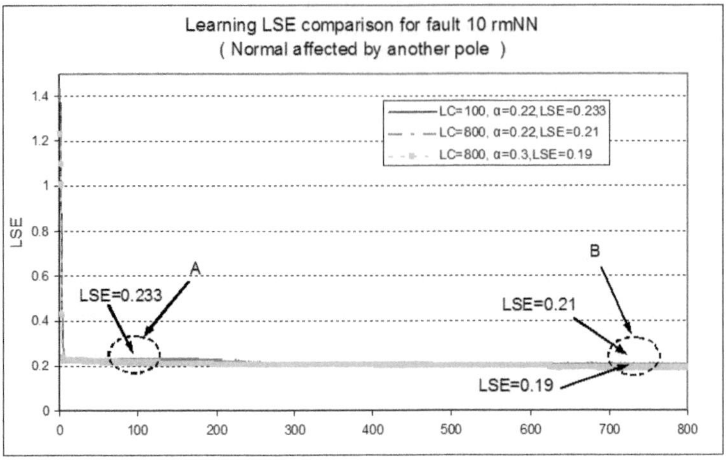

65.1: Learning LSE comparison for fault 10 rmNN

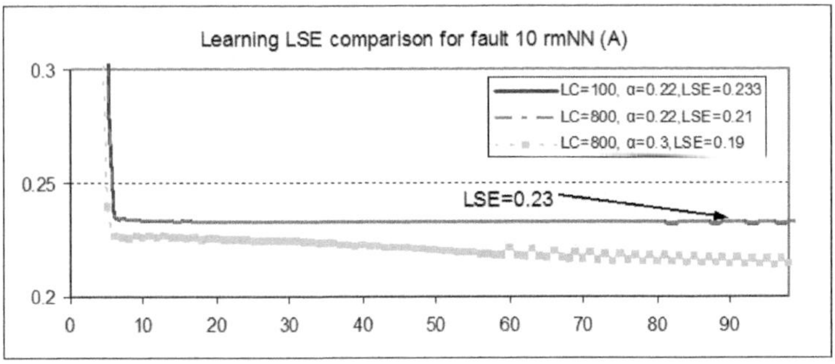

65.2: A zoom in detail of part A

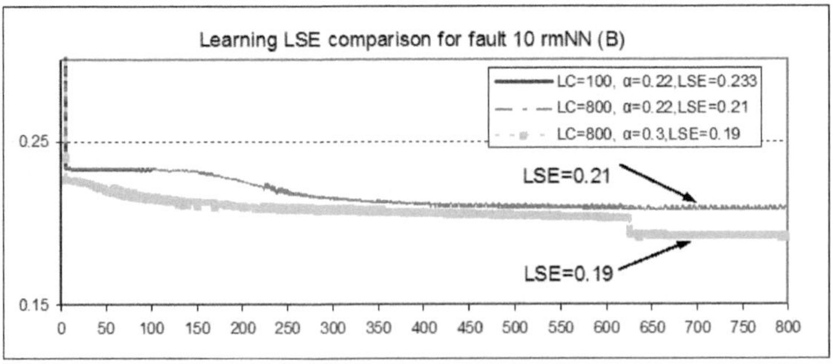

65.3: A zoom in detail of part B

Fig. 65. Learning LSE comparison for fault 10 rmNN with different LCs and α

at the same line with a distance of 0.01, and could barely be distinguished. It has been found that the smaller LSE does not lead to better classification performance. The conjecture is that the rmNN is "overfitting", which causes the vagueness (slight difference) between the true and false cases.

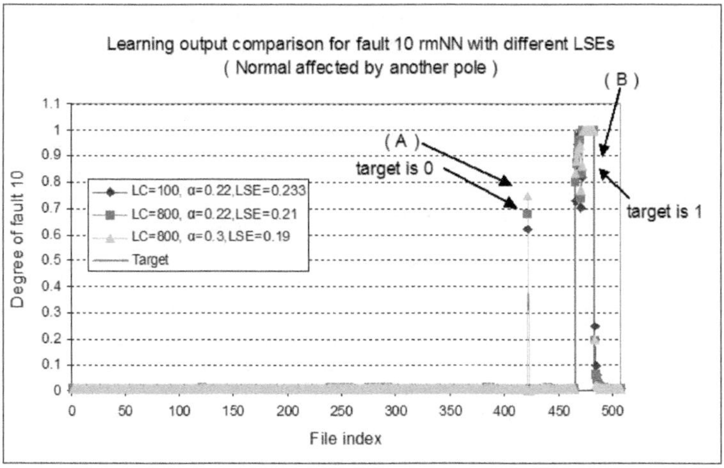

66.1: A broad view of the performance comparison for 508 rmNN training files

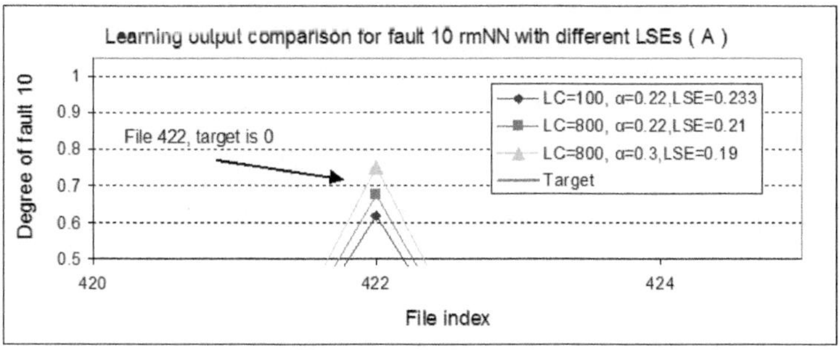

66.2: A zoom-in detail of point A

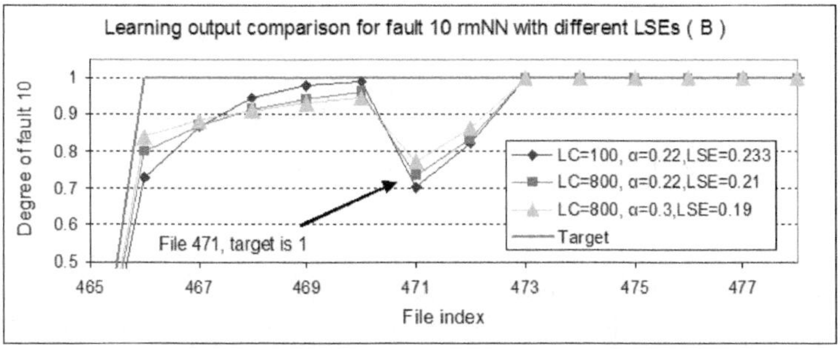

66.3: A zoom-in detail of point B

Fig. 66. The learning output comparison for fault 10 rmNN with different LCs and α

Another example shows the lower output for both learning and verification output for true case files with a smaller LSE. Figure 67 shows the learning least square error (LSE) comparison for fault 7 rmNN with different LCs and α. In

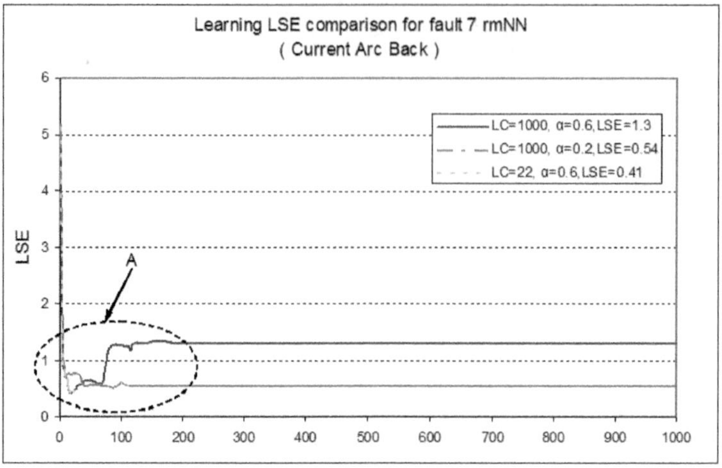

67.1: Learning LSE comparison for fault 7 rmNN

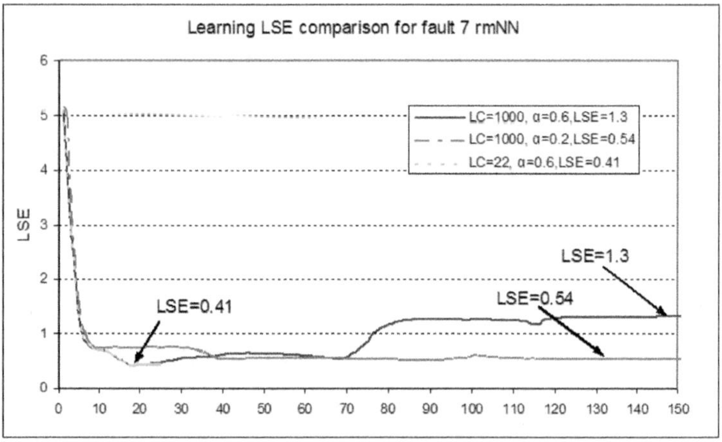

67.2: A zoom-in detail of part A

Fig. 67. Learning LSE comparison for fault 7 rmNN with different LCs and α

learning case 1, LSE = 1.3; learning case 2, LSE = 0.54; and learning case 3, LSE = 0.41. From case 1 to case 3, the LSE decreases.

Figure 68 shows the learning output for the 3 cases. Figures 68.2, 68.3, 68.4 and 68.5 show the details of the A, B, C and D parts in Figure 68.1.

Figure 69 shows the verification output for the 3 cases. Figures 69.2, 69.3, 69.4 and 69.5 show the details of the A, B, C and D parts in Figure 69.1.

It has been shown that, for all the true case files with a target value of 1, case 1 with the higher LSE has the highest output and case 3 with the lowest LSE has the lowest output.

The learning cycles, learning rates applied and the learning errors received for 12 rmNNs are listed in Table 33.

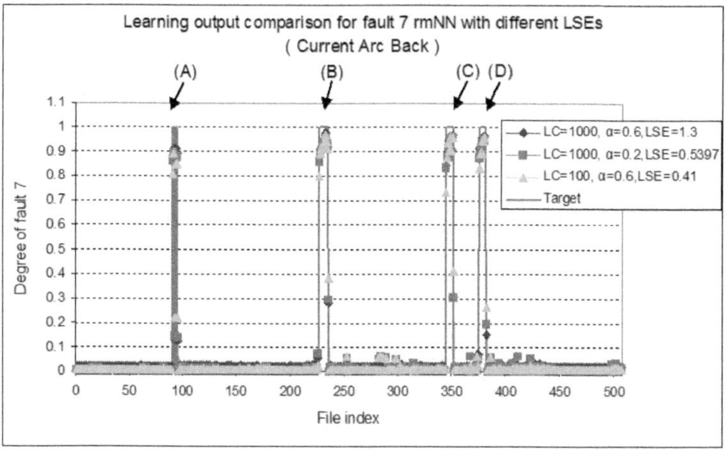

68.1: A broad view of the performance comparison for 508 rmNN training files

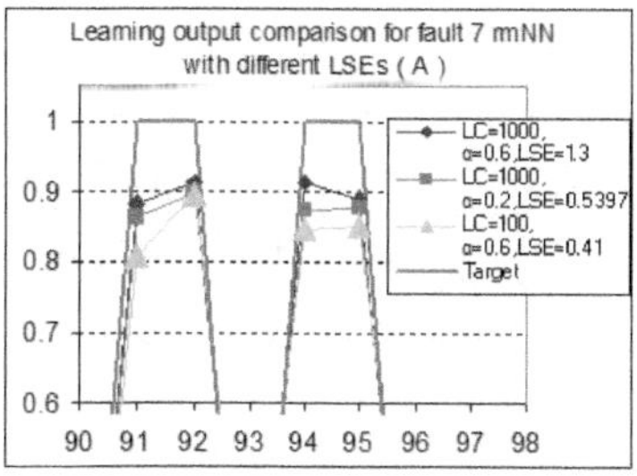

68.2: A zoom-in detail of part A

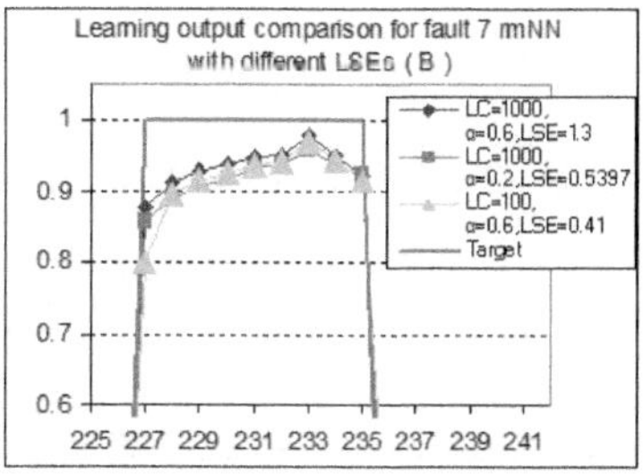

68.3: A zoom-in detail of part B

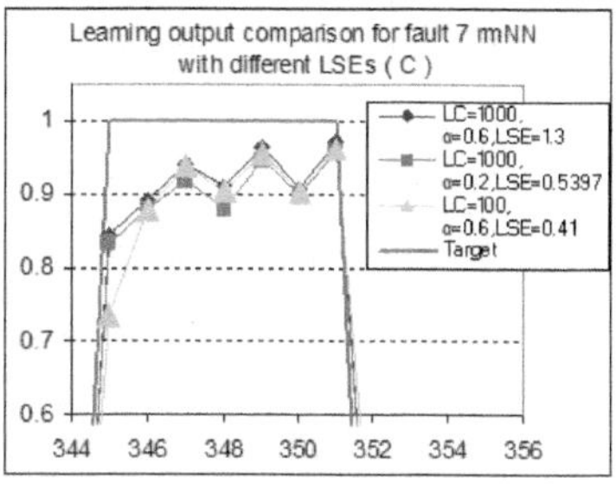

68.4: A zoom-in detail of part C

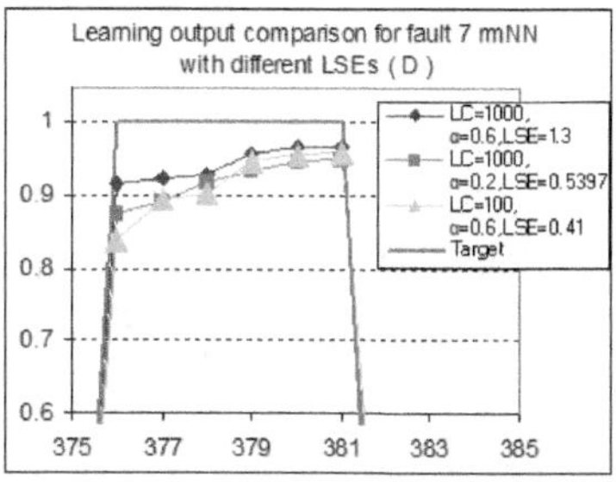

68.5: A zoom-in detail of part D

Fig. 68. The learning output comparison for fault 7 rmNN with different LCs and α

7.7 Implementation of rmNN for PSFC

After theoretical development and computer simulation, the next sought-after step is to build the software package for the implementation of rmNN power system fault classification with a user friendly interface. The software package provides the following functions:

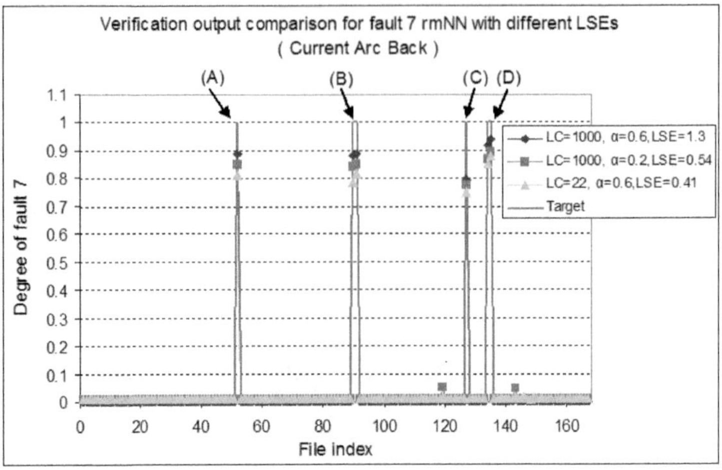

69.1: A broad view of the performance comparison for 168 rmNN testing files

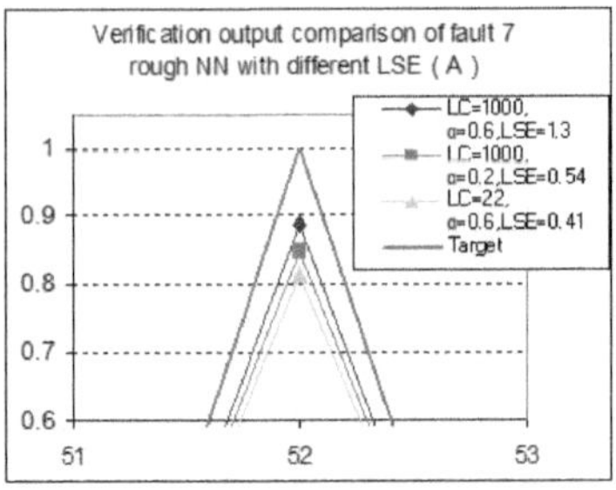

69.2: A zoom-in detail of part A

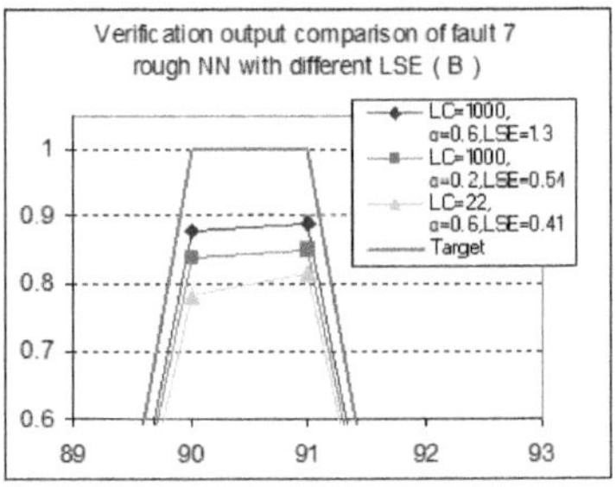

69.3: A zoom-in detail of part B

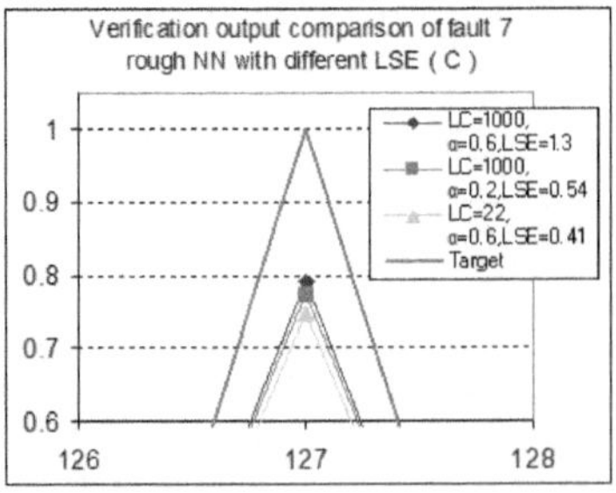

69.4: A zoom-in detail of part C

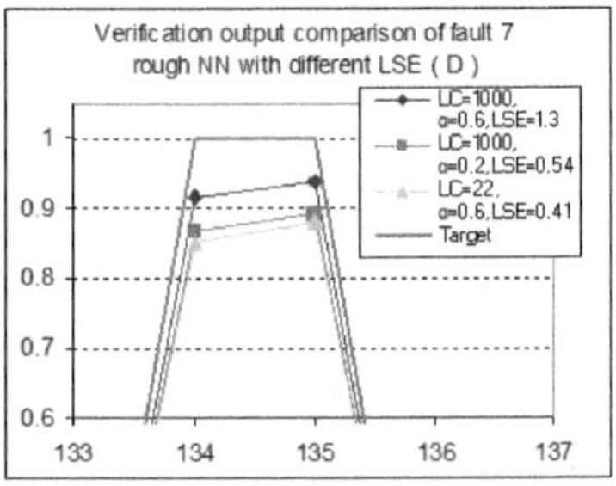

69.5: A zoom-in detail of part D

Fig. 69. The verification output comparison for fault 7 rmNN with different LCs and α

- Feature extraction
- Rough set construction
- Rough membership computation
- Rough membership neural network calibration and verification
- Power system fault type detection

Table 33. The learning cycles, learning rate and learning error for 12 rmNNs

	Training parameters		
	Learning cycles (LC)	Learning rate (α)	Least square error (LSE)
Fault 1 rmNN	200	0.01	0.2658
Fault 2 rmNN	200	0.01	0.8579
Fault 3 rmNN	200	0.1	0.6608
Fault 4 rmNN	800	0.1	0.1653
Fault 5 rmNN	200	0.1	0.2295
Fault 6 rmNN	1000	0.2	0.4061
Fault 7 rmNN	1000	0.6	1.3
Fault 8 rmNN	800	0.02	0.1439
Fault 9 rmNN	800	0.1	0.0892
Fault 10 rmNN	100	0.22	0.233
Fault 11 rmNN	800	0.1	0.86
Fault 12 rmNN	100	0.01	0.0634

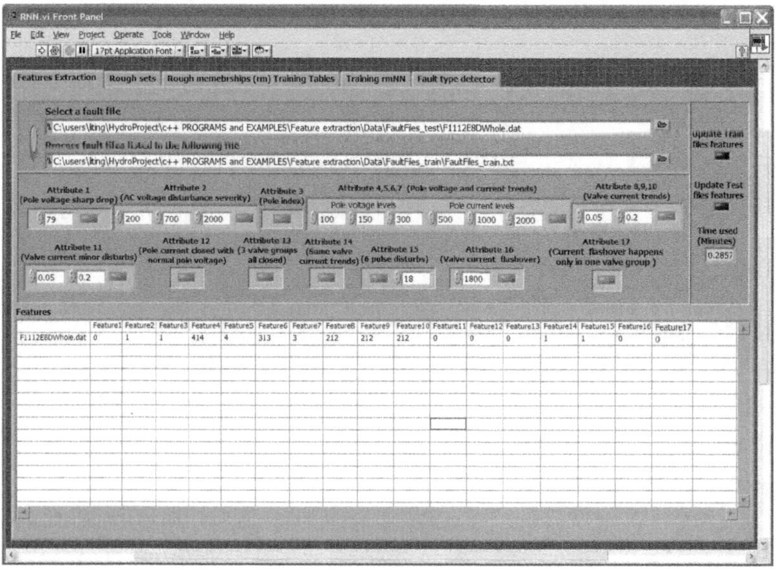

Fig. 70. The user interface for rmNN power system fault classification

C++ programs as well as the executable codes have been developed for each function. The executables are called and embedded in a LabVIEW program, which creates a flexible and scalable user interface. With LabVIEW, users can interface with real-world signals, analyze data for meaningful information, and share results through intuitive displays and reports. The screen snapshot of the user interface for rmNN power system fault classification is shown in Figure 70.

The main GUI (graphic user interface) window for the rmNN PSFC contains five tabs that are created for the five functions. The first tab is designed for the features extraction function seen in Figure 70. Users are allowed to select either a fault file list or a specific fault file to be processed. An example of a fault file list is FaultFiles_train.txt, which contains the file names of all fault files

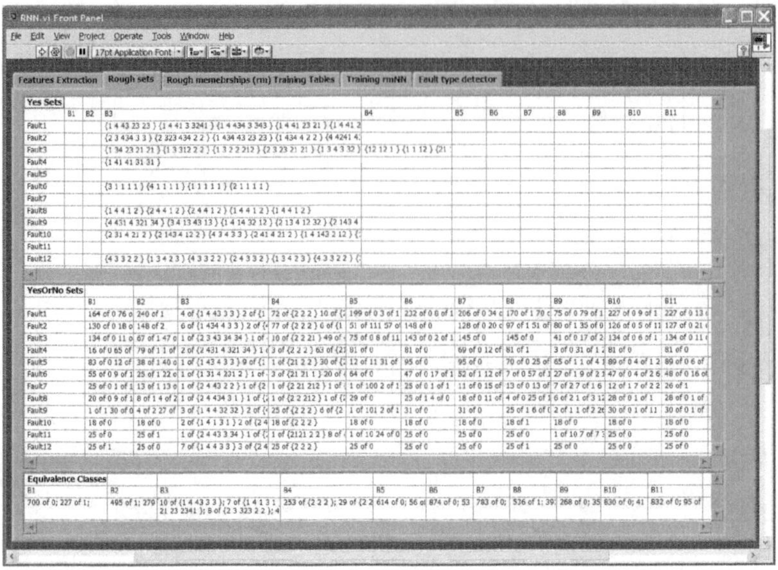

Fig. 71. Rough set construction for rmNN power system fault classification

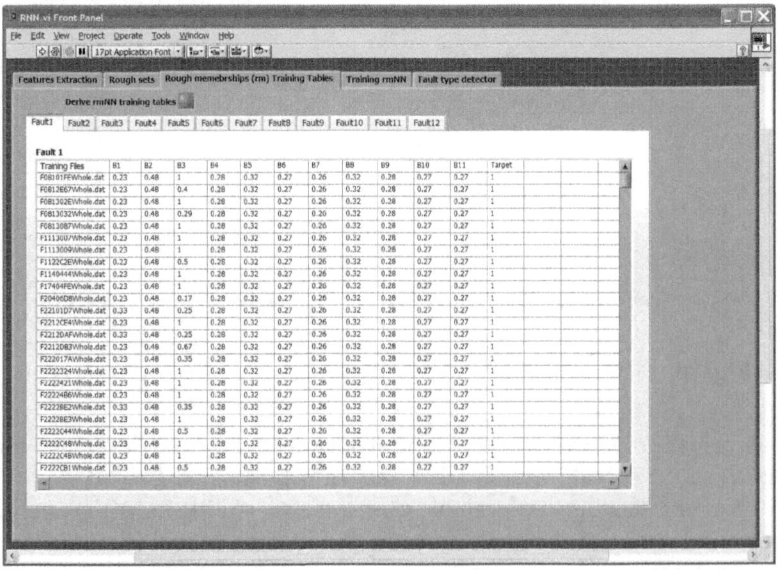

Fig. 72. Rough membership computation for rmNN power system fault classification

for training. The 23 signals are analyzed and 17 features are derived for each fault file. The thresholds applied in the features extraction can be adjusted to achieve optimized feature values for the best fault classification performance.

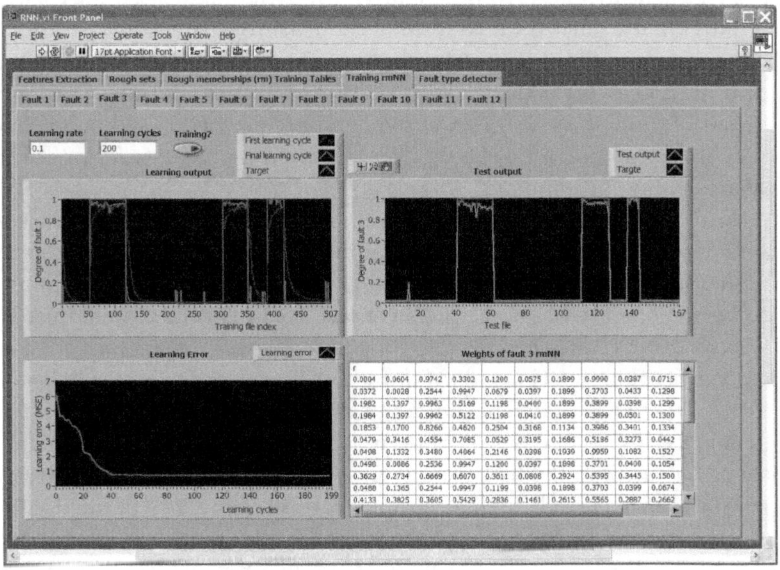

Fig. 73. rmNN calibration and verification function for rmNN power system fault classification

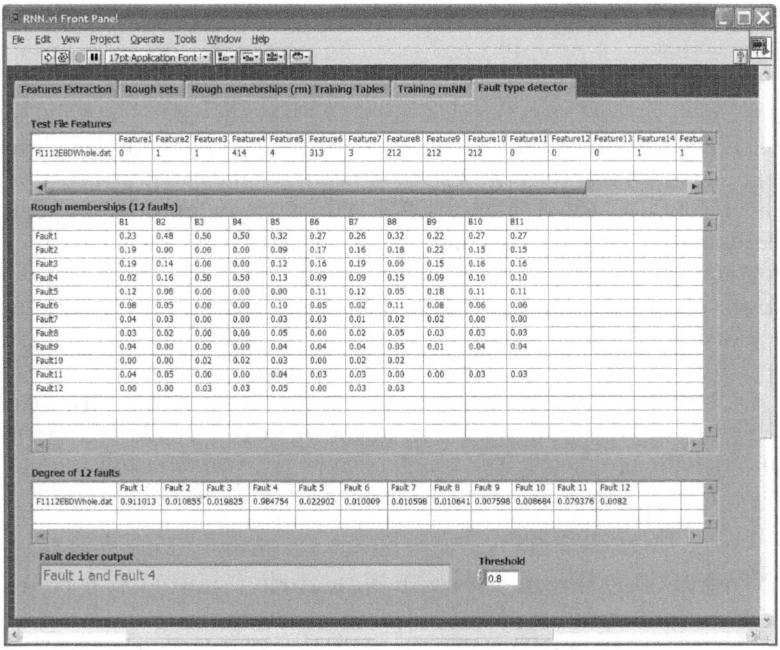

Fig. 74. Power system fault type detection

When partial features need to be updated, the switch allocated for each feature can be individually turned off to disable the feature extraction operation.

The second tab allows accessing the rough set construction function shown in Figure 71. The equivalence classes, the B-low approximations (Yes Sets) and approximation boundary sets (YesOrNo Sets) are created based on features values of all training fault files. The rough sets need to be restructured whenever the features values are updated.

The third tab is for the rough membership computation function illustrated in Figure 72. Twelve rough membership training tables, which contain the training sets for the calibration of twelve rmNNs, have been derived. These training tables need to be re-generated as well whenever the features values are adjusted.

The forth tab allows accessing the rough membership neural network calibration and verification function shown in Figure 73. Twelve pages have been developed inside this function for calibration of twelve rmNNs respectively. The learning rate and cycles are the parameters to be adjusted until the best classification performance is achieved. The learning error, learning output and test result for each rmNN are displayed in three graphs respectively and the final calibrated weights are reported in the table at the bottom-right quarter of the tab window.

The final function is for the power system fault type detection, which is illustrated in Figure 74. For a fault file to be classified, 17 features are derived by function 1 and loaded by this fault type detector. The 11 rough memberships, a rough membership set, based on 11 feature sets associated with a type of fault are estimated. Consequently a total of 12 rough membership sets are derived and input to 12 trained rmNNs respectively. Twelve degrees, one for each fault type, are estimated by the 12 rmNNs and forwarded to a decider neuron, which picks up the faults with degree output greater than the preset threshold. For example, in Figure 74, the threshold is set as 0.8; the faults detected for fault file F1112E80Whole.dat are a combination of fault 1 and fault 4 with a degree of 0.911 and 0.985 respectively. The fault types are indicated in a text box. The threshold can be adjusted lower to allow more fault types to be considered.

The developed graphic user interface makes the power system fault classification easy to operate for users. The software integrates features extraction, rough sets construction, rough membership computation, rmNNs training, verification and fault type detection into one system, which provides the possibility for further implementation of an adaptive learning real-time fault classification system. Currently, the most time consuming component of the software package is the features extraction, which takes two to three seconds and needs to be computationally optimized. The rough membership computation and fault type detection take less than one second, which meets the need of real-time.

8 Normal Artificial Neural Network (ANN) for PSFC

Before using the rmNN to classify the power system faults, two normal artificial neural networks were investigated. Each row in the information table (Table 12) is an input to the neurons in the first layer of both ANNs. No rough membership

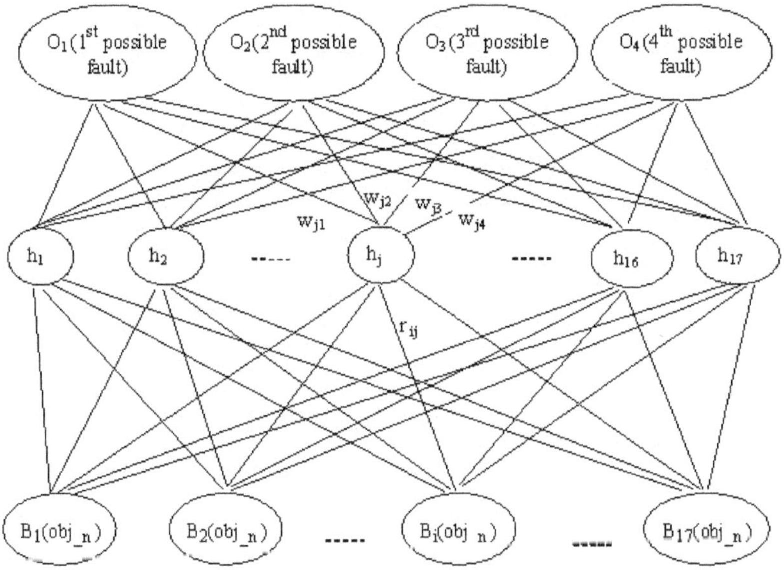

Fig. 75. The architecture of the single ANN power fault classification system

is computed. The performance of these two types of ANNs is addressed in detail in the following two sections.

8.1 A Single ANN for PSFC

First, a single ANN was considered to classify 12 types of faults. The single ANN is designed with 3 layers, containing 17 neurons in the input layer and hidden layer and 4 neurons in the output layer. Each neuron in the output layer indicates one possible type of fault. As seen in Table 12, some *.x01 files are involved with 4 types of faults, *i.e.*, F082016A.x01 is intervened with faults 1, 3, 5 and 7. The four output neurons are expected to output 1, 3, 5 and 7 respectively when processing fault F082016A.x01.

The architecture of the single ANN is illustrated in Figure 75, where

$$h_j = g\left(\sum_{i=1}^{17} r_{ij} B_i\left(obj\_n\right)\right), \tag{61}$$

$$O_k = \sum_{j=1}^{17} w_{jk} h_j, \tag{62}$$

and $g(\cdot)$ is the logistic sigmoid activation function given by

$$g(a) = \frac{1}{1+e^{-a}}. \tag{63}$$

The 17 neurons in the first layer receive values from functions representing 17 features. Unlike an rmNN, the features are not grouped. Back-propagation is

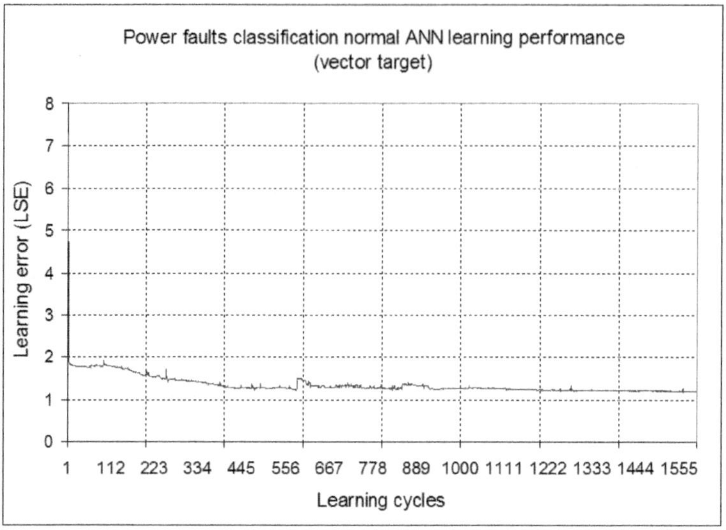

Fig. 76. Learning performance of the single ANN for power system fault classification

adopted as the network training function. A gradient descent learning function is used as the weights updating function, and the least squared error function is used as the learning performance evaluation function. The learning performance is shown in Figure 76. After 1600 learning cycles, the LSE converged to 1.23.

Once the ANN is calibrated, 168 test files are applied to evaluate the accuracy of this ANN fault classification system. Four neuron output compared with four targets are displayed in Figures 77, 78, 79 and 80, respectively. It is noticeable that over 60% of testing files failed the verification.

8.2 Twelve Sub-ANNs for PSFC

The basic architecture of the second ANN power fault classification system is similar to the rmNNs described earlier. Twelve ANNs are created and each of them classifies one type of fault. An example of ANN for k^{th} fault classification is illustrated in Figure 81. The output from the 12 ANNs are the estimations of the degrees of the 12 faults respectively. The output from the 12 ANNs are forwarded to a decider neuron, which simply picks up the faults with degrees above a preset threshold.

In Figure 81,

$$h_j^k = g\left(\sum_{i=1}^{17} r_{ij} B_i \left(obj\_n\right)\right), \tag{64}$$

$$O^k = \sum_{j=1}^{17} w_j h_j^k, \tag{65}$$

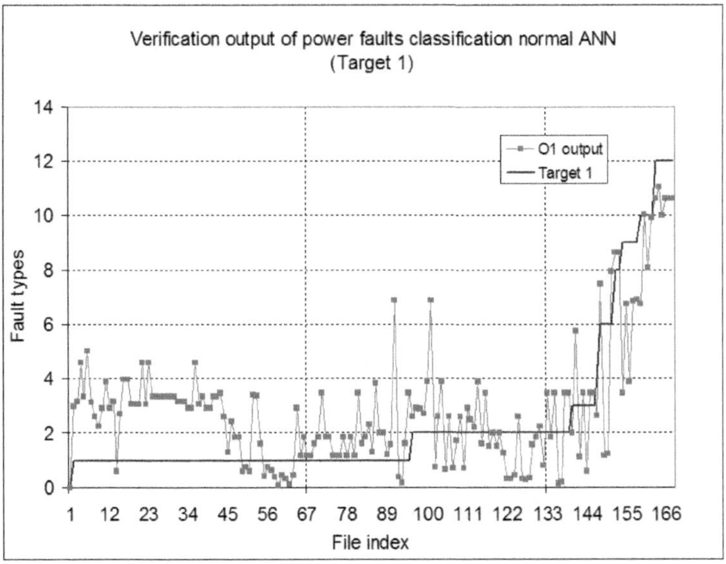

Fig. 77. Target 1 verification

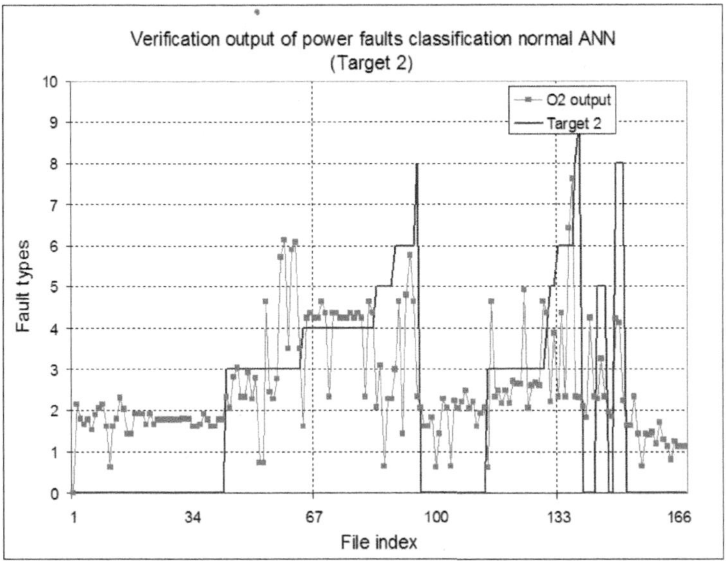

Fig. 78. Target 2 verification

and $g(\cdot)$ is the logistic sigmoid activation function same as the one applied in the first type of ANN.

The 17 neurons in the first layer receive values from functions representing 17 features. Again, the features are not grouped. Backpropagation is still used

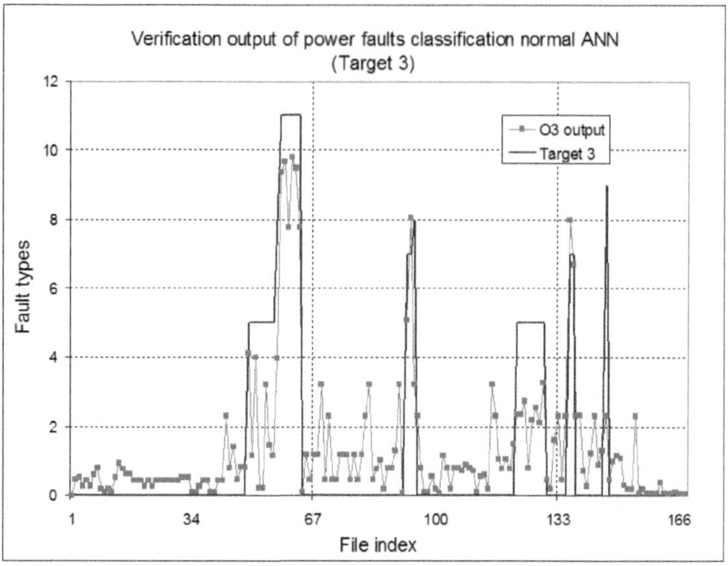

Fig. 79. Target 3 verification

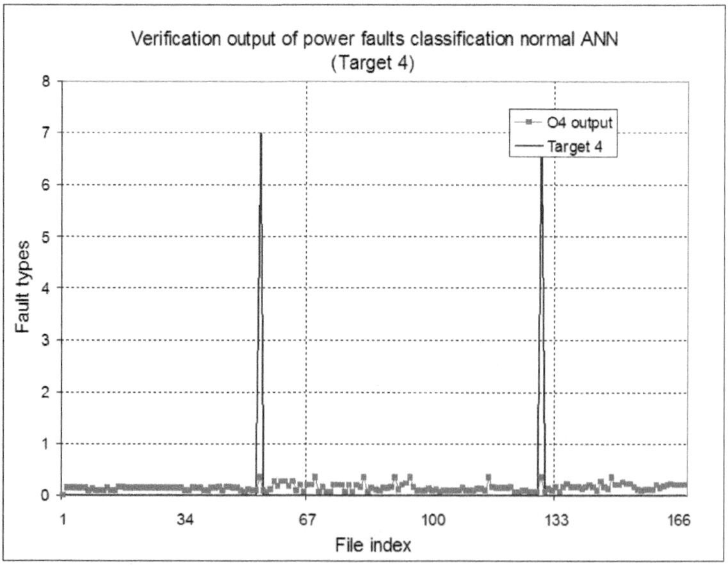

Fig. 80. Target 4 verification

as the network training function, the gradient descent learning function as the weights learning function, and the least squared error function as the learning performance evaluation function. The learning performance for fault 3 and 5

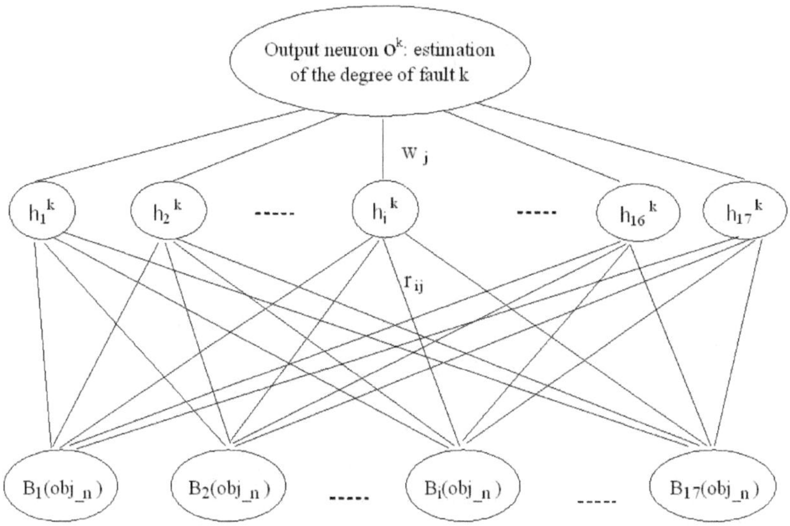

Fig. 81. The architecture of a sample ANN for k^{th} fault classification

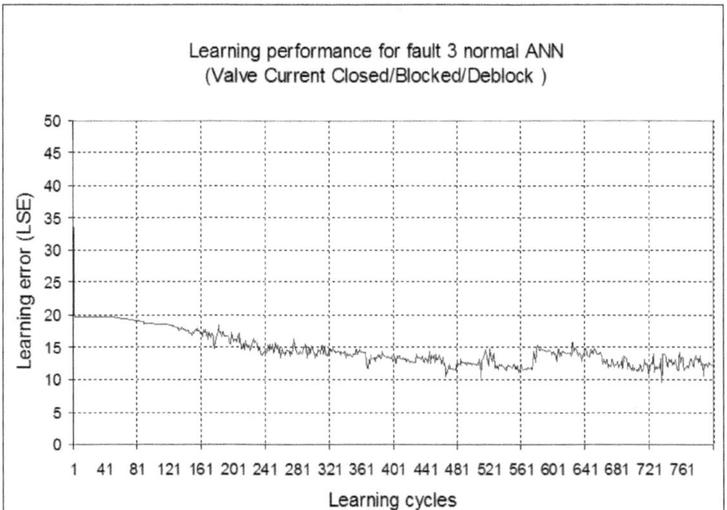

Fig. 82. The learning performance for fault 3 ANN

ANN is shown in Figures 82 and 83 respectively. After 800 learning cycles, both LSEs are approximately 10.

After 800 learning cycles, 168 test files are applied to evaluate the performance of the ANNs. The testing output for fault 3 and 5 ANN are displayed in

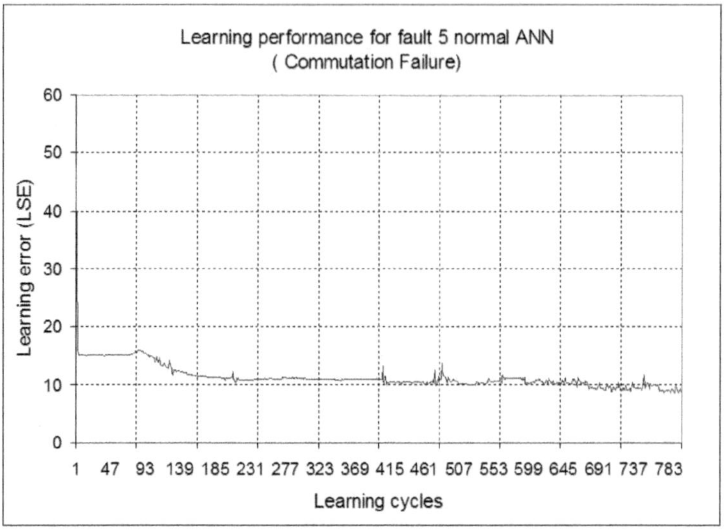

Fig. 83. The learning performance for fault 5 ANN

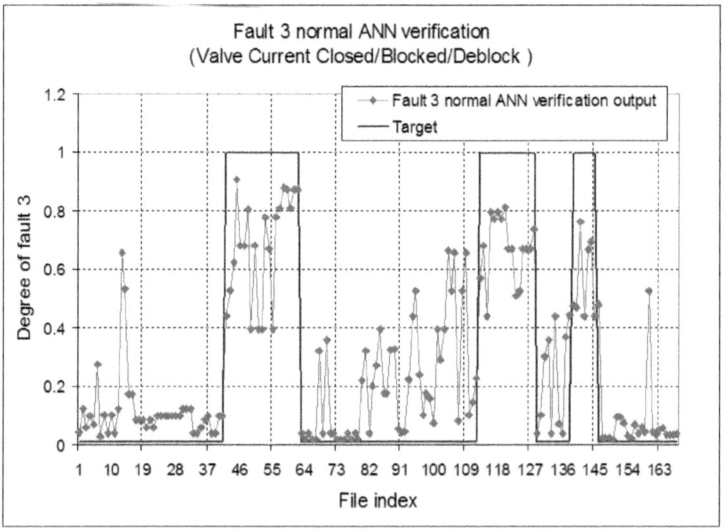

Fig. 84. Fault 3 ANN verification

Figures 84 and 85 respectively. It is obvious that the threshold to pick up the fault has to be reduced to 60% to generate better accuracy. The results for the calibrations and verifications of the 12 ANNs are attached in an Appendix available at [15]. The classification accuracy is listed in Table 34. Compared with the rmNN system, the accuracy of the ANN fault classification system is fairly poor.

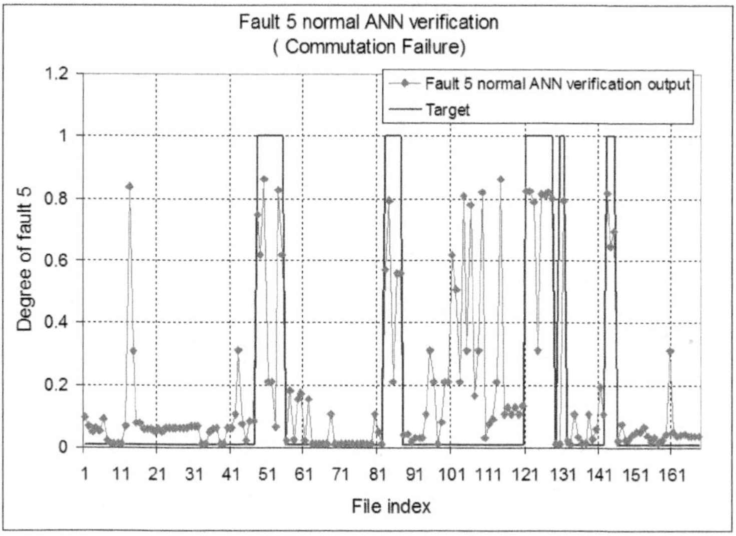

Fig. 85. Fault 5 ANN verification

Table 34. Accuracy of 12 ANNs for PSFC

Fault type	# of files for verification	# of files incorrect	Accuracy
Minor AC Disturbance	94	93	0.011
AC Disturbance	44	25	0.432
Valve Current Closed/Blocked/Deblocked	44	14	0.680
Line Fault	22	13	0.410
Commutation Failure	25	5	0.800
Pole Voltages/Current Closed/Blocked/Deblocked	15	7	0.530
Current Arc Back	6	1	0.830
Parallel Operation	9	5	0.440
Pole Current Oscillation	7	7	0.000
Normal Affected by Another Pole	4	4	0.000
Asymmetric Protection	6	5	0.170
Disturbance on DC Voltage	7	7	0.000

It either produces low detection accuracy for the desired faults or generates a great number of unexpected false alarms.

The failure of both ANN fault classification systems is possibly due to the input, which are the 17 feature values. Consider the feature 5 (Pole Current Trend), two possible codes are "313" and "343". They are very close in terms of the values of these two numbers when treated by the ANN system. But "313" usually happens in fault 4 (Line Faults), and "343" happens in fault 1 (Minor AC Disturbance). The rough membership computation distinguishes these two numbers by assigning each of them with the degree of each type of fault, which greatly improves the quality of the feature information and consequently the classification performance.

Table 35. Maximum and minimum rmNN output for false and true cases, respectively

	Learning		Verification	
	Lowest output for true cases	Highest output for false cases	Lowest output for true cases	Highest output for false cases
fault 1	**0.94**	**0.08**	**0.93**	**0.01**
fault 2	**0.9**	**0.16**	**0.9**	**0.04**
fault 3	0.87	0.22	0.82	0.19
fault 4	**0.98**	**0.06**	**0.98**	**0.01**
fault 5	**0.94**	**0.14**	**0.95**	**0.04**
fault 6	0.74	0.5	0.84	0.27
fault 7	0.85	0.31	0.79	0.01
fault 8	0.75	0.06	0.85	0.01
fault 9	0.87	0.15	0.87	0.01
fault 10	0.68	0.56	0.66	0.01
fault 11	0.61	0.34	0.81	0.39
fault 12	**0.99**	**0.01**	**0.99**	**0.01**

9 Classifier Fusion

9.1 Motivation in Using a Second Complementary Classifier

A number of classifier fusion methods have been recently developed and lead to potential improvement in classification performance [1,6,30,34,35,67,74,75]. In this section, a second successful classifier based on mean and standard deviation evaluation of the sum of 11 rough memberships is proposed. The goal is to take advantage of the diversity of two classifiers to improve the performance of PSFC.

To achieve high overall performance of the classification function, the performance of each individual classifier has to be optimized prior to using it within any fusion schemes. That is, the fusion scheme will be able to improve the overall classification result relative to the performance of the individual classifiers. If several classifiers with only marginal performance are being used, the results cannot necessarily be expected to reach high performance. On the other hand, if several classifiers are used that work exceptionally well, any further gains will be exceedingly hard to accomplish because the opportunity for diversity is diminished.

Recall the performance of the 12 rmNNs. Table 35 lists the minimum rmNN output for true cases and the maximum rmNN output for false cases in both learning and verification. The classification performance of the rmNNs for fault 1, fault 2, fault 4, fault 5 and fault 12 are excellent and both the learning and verification output for the true cases have high scores over 0.9, while for the false cases have low scores less than 0.16.

Faults 1, 2, 4, 5 and 12 do not need to be reinforced by a second complementary classifier. However, for faults 3, 6, 7, 8, 9, 10 and 11 classification, a second LMD classifier is introduced to fusion the output from rmNNs in order to increase the overall PSFC accuracy.

9.2 Linear Mean-Deviation (LMD) Based Classifier

The input for the linear mean and deviation based (LMD) classifier is the sum of 11 rough memberships (SORM) in the training and testing tables for rmNNs. Figure 86 shows the SORMs of 508 training files for fault 7.

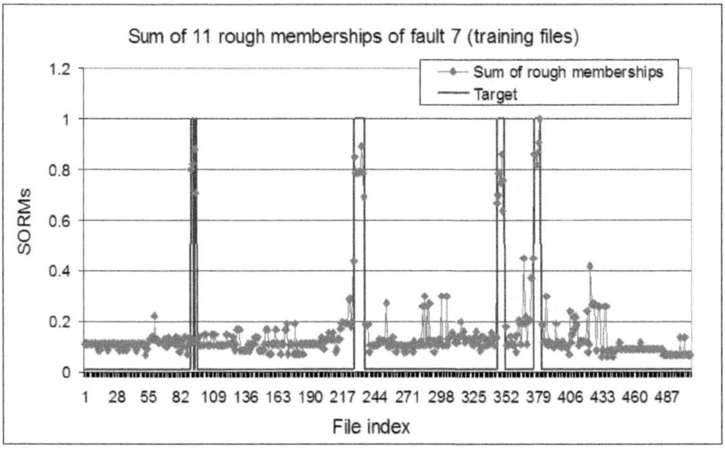

Fig. 86. The SORMs of 508 training files for fault 7

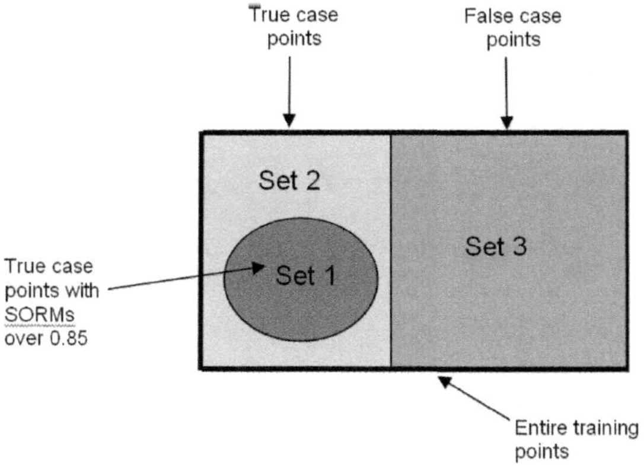

Fig. 87. Three sets of SORMs

From all the points of SORM output, three sets are constructed. Set 1 consists of all the points of true case with SORM values over 0.85. Set 2 contains all the points of true case with SORM values less than 0.85. Set 3 collects all the points of false case. These 3 sets are illustrated in Figure 87.

The points in sets 2 and 3 will be employed to estimate the mean and deviation values to establish the distribution functions of set 2 and set 3. Assuming that x_{ij} is the jth point in set i and there are N_i points in set i. The mean and absolute deviation for set i, *i.e.*, μ_i and dev_i are defined as follows.

$$\mu_i = \frac{1}{N_i} \sum_{j=1}^{N_i} x_{ij}. \tag{66}$$

$$dev_i = \frac{1}{N_i}\sum_{j=1}^{N_i} |x_{ij} - \mu_i|. \qquad (67)$$

The Gaussian distribution function of set i, $f_i(x)$, is defined as,

$$f_i(x) = \frac{1}{\sqrt{2\pi(dev_i)^2}} e^{\frac{-(x-\mu_i)^2}{2(dev_i)^2}}. \qquad (68)$$

Take fault 7 as an example, the mean and deviation of sets 2 and 3 are calculated, and listed in Table 36.

Table 36. The mean and deviation of sets 2 and 3 for fault 7 training files

Fault 7 training files			
set 2		set 3	
μ_2	dev_2	μ_3	dev_3
0.758	0.12	0.126	0.324

The degree of fault 7 will then be calculated as described in (69)

$$deg(x) = \begin{cases} x, & if\ x \geq 0.85, \\ \frac{f_2(x)+1-f_3(x)}{2}, & if\ x < 0.85. \end{cases} \qquad (69)$$

Keep the SORM as the degree of fault 7, if it is bigger or equal to 0.85. $\frac{f_2(x)+1-f_3(x)}{2}$ is only applied to the points with SORM values that are less than 0.85. In this way, the degree of fault 7 of the points in set 2 is raised. This method is applied to faults 3, 6, 7, 8, 9, 10 and 11. For example, the degree of fault 7 of 508 training files is shown in Figure 88.

Use the training files, the mean and deviation of sets 2 and 3 can be estimated to set up the distribution function for the points of true case with SORM values less than 0.85 and the distribution function for the points of false case. The trained distribution functions will be applied to the test points to estimate the degree of a type of fault. In this example, the degree of fault 7 is estimated.

The SORMs of fault 7 for 168 testing files are shown in Figure 89, and the fault 7 LMD classifier output for 168 testing files are shown in Figure 90. The degrees of fault 7 for all the true case points are above 0.87. One point, file 128, exists in the verification output for fault 7 rmNN. It has a low estimation of the degree of fault 7, which is only 0.79 (Table 35, Section 9.1).

The fault 10 LMD classifier results are illustrated in Figures 91, 92, 93 and 94. The SORMs of 508 training files are shown in Figure 91, and the LMD classifier training output is shown in Figure 92. The SORMs of 168 testing files are shown in Figure 93, and the LMD classifier testing output is shown in Figure 94. In Figure 91, it is very clear that only two points (file 471 and file 472) have low SORM output (*i.e.*, 0.745 and 0.746) and they are almost at the same level. The distribution function of set 2 is designed based on these two points and the

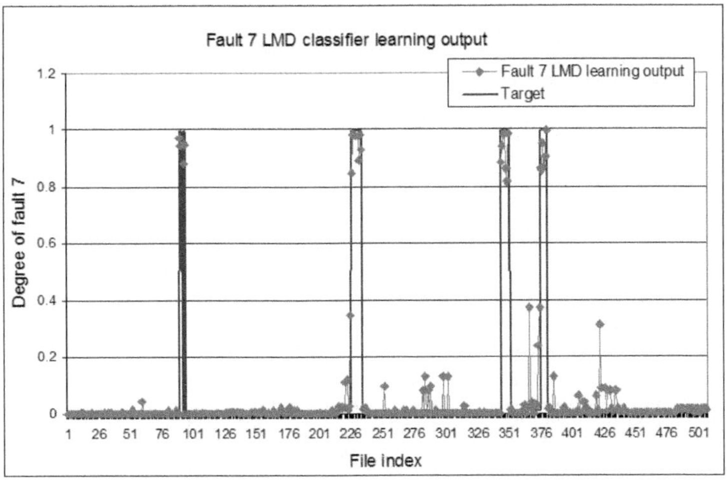

Fig. 88. Fault 7 LMD classifier output for 508 training files

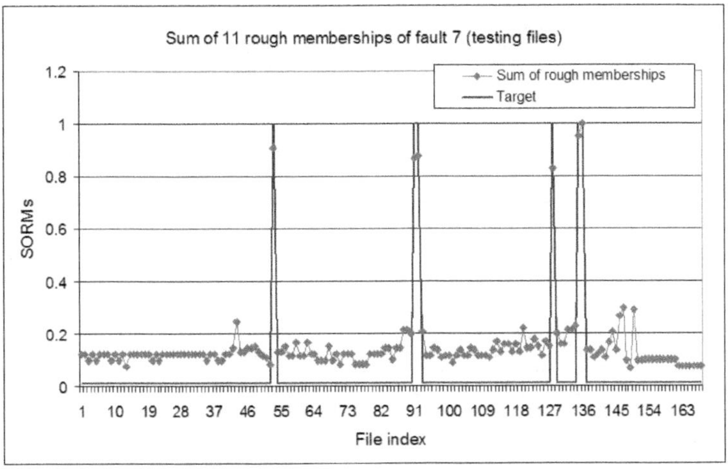

Fig. 89. The SORMs of 168 testing files for fault 7

degree of fault 10 for these two points from the LMD classifier is high and raised to 0.985. In the test, the degree of fault 10 for file 159 is boosted to 0.984 as well. The rmNN classifier testing output for this point, however, is as low as 0.66 (Table 35, Section 9.1).

The LMD classifier training and testing results for fault 6 are illustrated in Figures 95 to Figure 98. There is one point, file 90, in the fault 6 LMD classifier testing output, which gives a low estimation of the degree of fault 6. The degree of fault 6 is only 0.786.

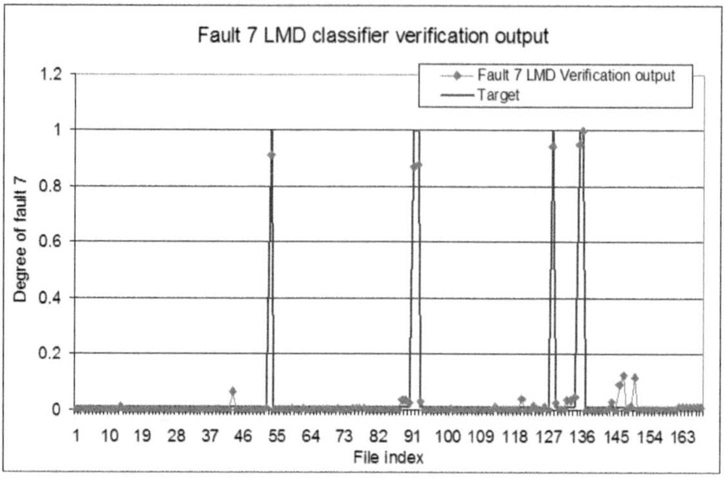

Fig. 90. Fault 7 LMD classifier output for 168 testing files

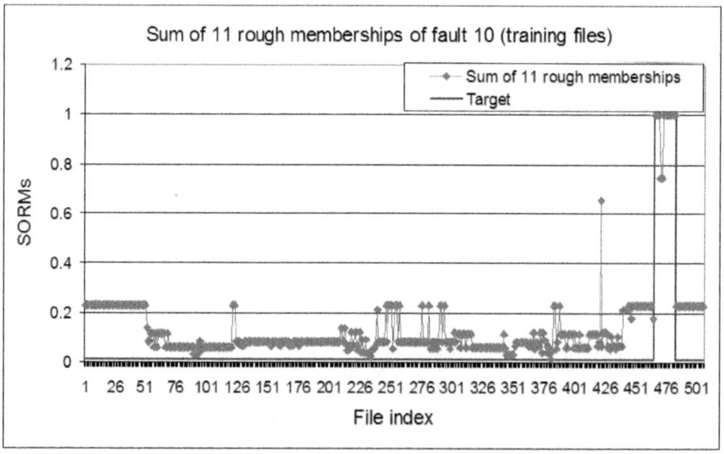

Fig. 91. The SORMs of 508 training files for fault 10

The SORMs for faults 3, 6, 7, 8, 9, 10 and 11 of both training and testing files are listed in an Appendix available at [15]. The fault 3, 6, 7, 8, 9, 10 and 11 LMD classifier training and verification output is included in an Appendix available at [15].

Table 37 summarizes the accuracy of the LMD classifiers. Except that the accuracy for fault 6 classification is 0.93, the accuracy for all of the other 6 faults is 100%. LMD classifier considers the isolated points with medium and low SORM values. For fault 7, "Current Arc Back", and fault 10, "Normal affected by another pole", only 26 and 18 fault files are employed for calibration

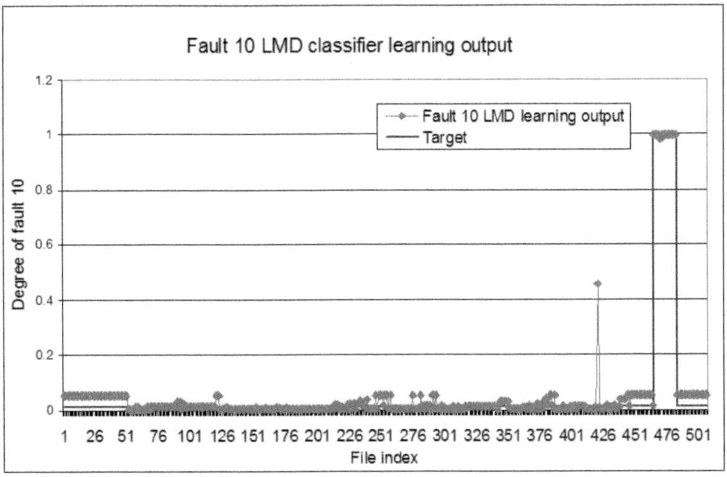

Fig. 92. Fault 10 LMD classifier output for 508 training files

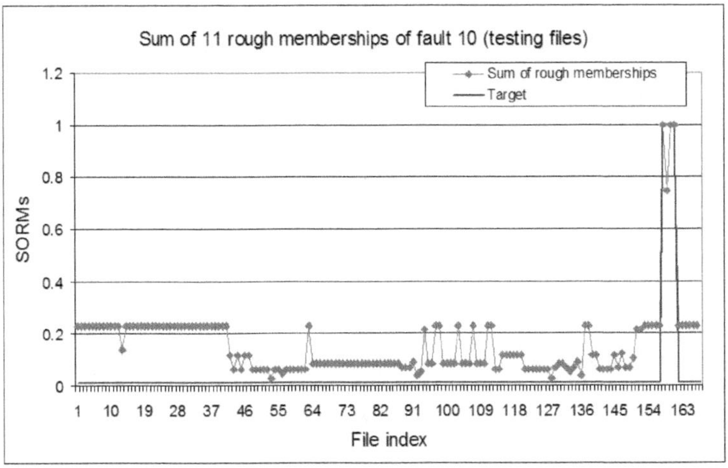

Fig. 93. The SORMs of 168 testing files for fault 10

respectively; but the LMD classifier test result is 100% accurate. On the other hand, the rmNN classifier gives poorer results when it deals with a fault with less files participate in learning. For fault 7 and 10, the rmNN classifier verification accuracy is only 0.83 and 0.75 respectively (Table 32, Section 7.4).

One point that needs mentioning is that the LMD classifier is not suitable for the classification for all 12 faults. Look at the SORMs for the 508 training files of fault 1 and fault 2 (Figures 99 and 100), where the SORMs of many false and true cases are comparable, which causes the failure of the LMD classifier.

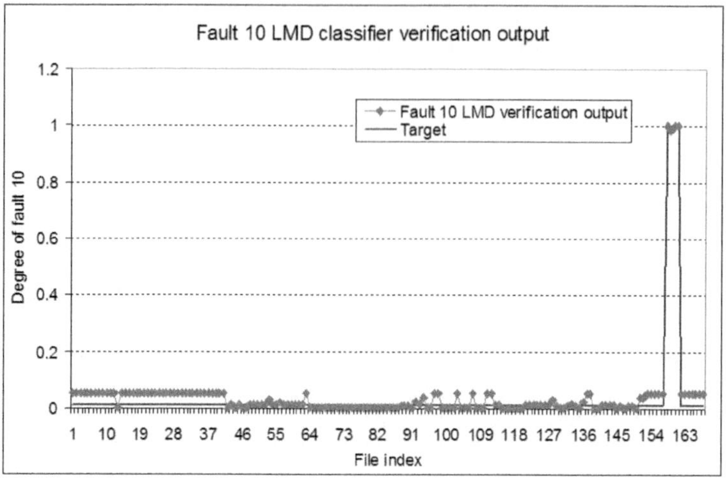

Fig. 94. Fault 10 LMD classifier output for 168 testing files

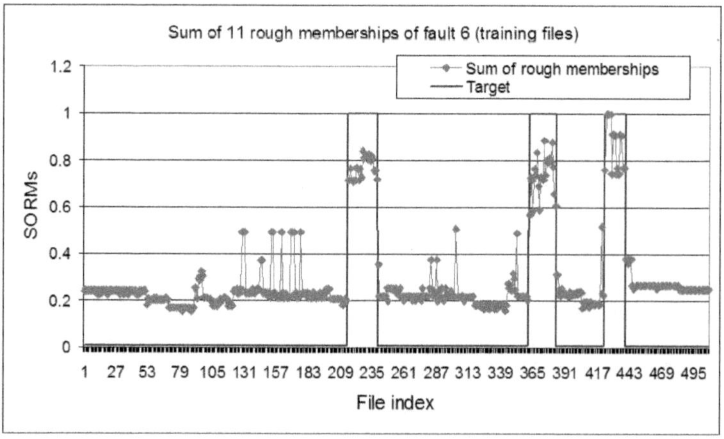

Fig. 95. The SORMs of 508 training files for fault 6

The good thing is that the accuracy of the rmNN classifier for these two faults is excellent and compensates the weakness of the LMD classifier.

9.3 Correlation of the rmNN and LMD Classifier

In classifier fusion, it is desirable to use classifiers that not only offer reasonable performance but also have a mutually low correlation. If two classifiers are completely redundant, many fusion schemes not only will not gain anything, but will actually exhibit poorer performance. Obviously, some degree of confirmatory

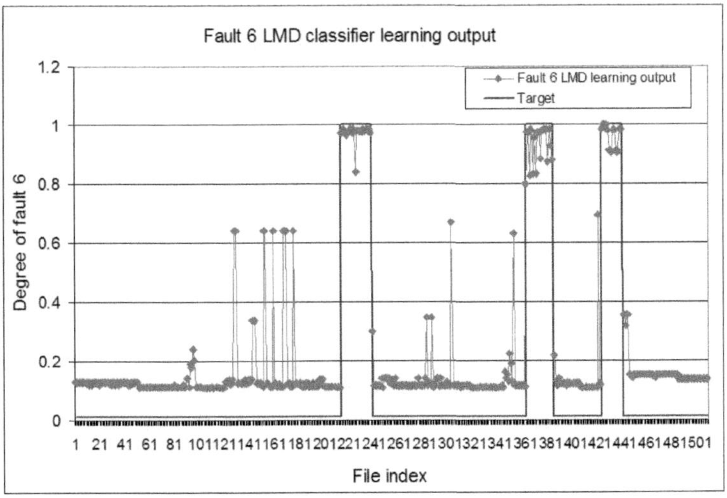

Fig. 96. Fault 6 LMD classifier output for 508 training files

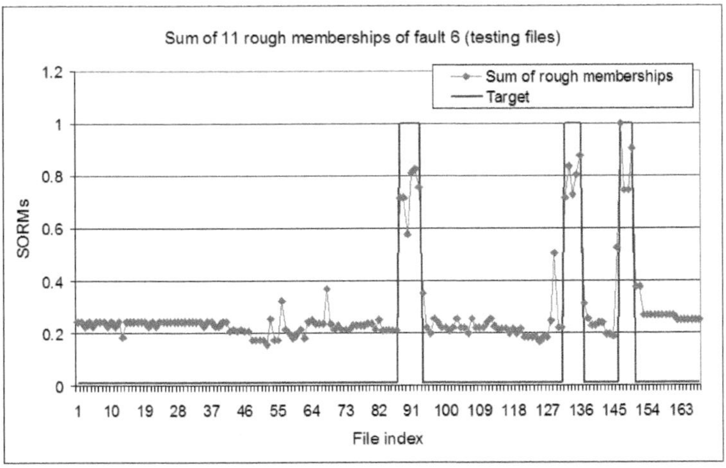

Fig. 97. The SORMs of 168 testing files for fault 6

information is desirable, but it is the complementary information that gives the multi-classifier fusion a chance for success.

In this section, the correlation of the rmNN and LMD classifier will be estimated to prove that they are good candidates for classifier fusion.

According to the 2-Classifier correlation analysis mentioned in Section 3.2, the formula for the correlation is:

$$\rho_2 = \frac{2 \times N^{FF}}{N^{TF} + N^{FT} + 2 \times N^{FF}}, \tag{70}$$

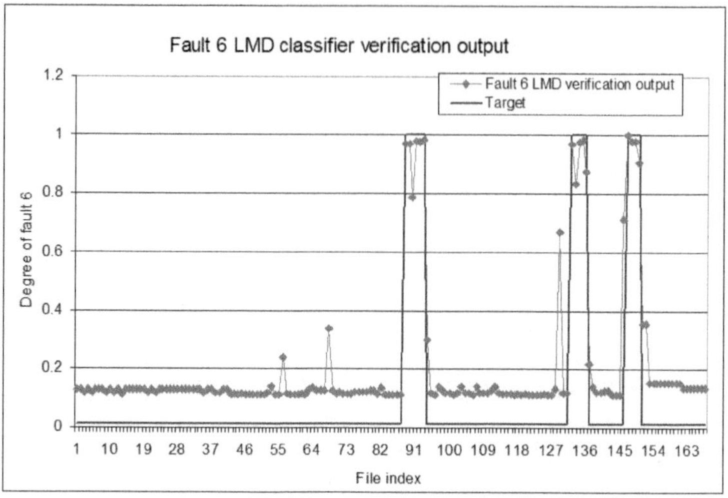

Fig. 98. Fault 6 LMD classifier output for 168 testing files

Table 37. Accuracy of the LMD power fault classification system

Fault type	# of files for verification	# of files incorrect	Accuracy
Fault 3: Valve Current Closed/Blocked/Deblocked	44	0	1
Fault 6: Pole Voltages/Current Closed/Blocked/Deblocked	15	1	0.93
Fault 7: Current Arc Back	6	0	1
Fault 8: Parallel Operation	9	0	1
Fault 9: Pole Current Oscillation	7	0	1
Fault 10: Normal Affected by Another Pole	4	0	1
Fault 11: Asymmetric Protection	6	0	1

where,

TT represents that the output of the rmm_NN is T and the output of the LMD is T;

TF represents that the output of the rmm_NN is T and the output of the LMD is F;

FT represents that the output of the rmm_NN is F and the output of the LMD is T; and

FF represents that the output of the rmm_NN is F and the output of the LMD is F;

and the following two methods are also applied for the correlation evaluation.

1. Try 3 thresholds for the 'true' decision making;
2. Consider both the training and testing files.

The correlation estimations of rmNN and LMD classifier for faults 3, 6, 7, 8, 9, 10 and 11 are listed in the following 3 tables for 3 thresholds respectively. Table 38 shows that the correlations are all 0 for 7 faults when the threshold

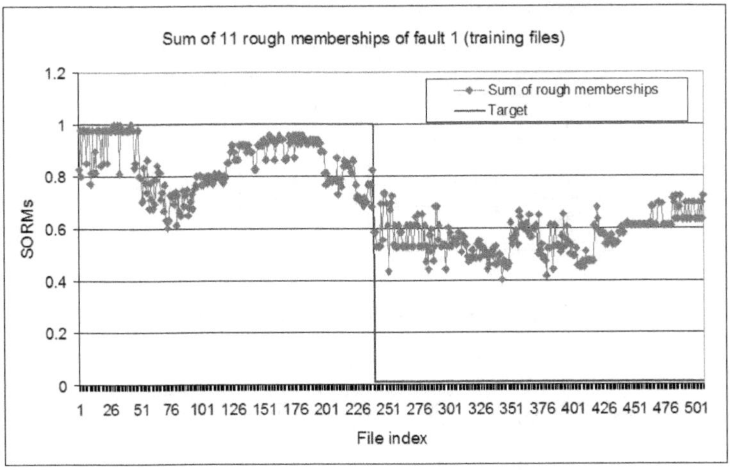

Fig. 99. The SORMs of 508 training files for fault 1

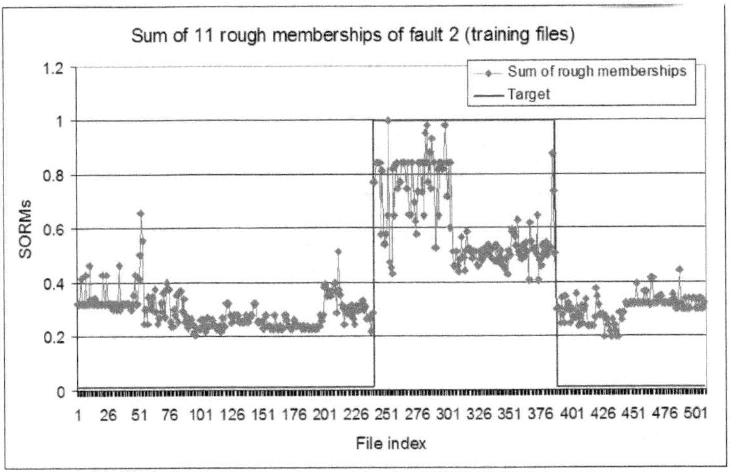

Fig. 100. The SORMs of 508 training files for fault 2

= 0.8 for 'true' decision making. Table 39 shows that the correlations are still 0 when the threshold is pushed to 0.85. Table 40 shows that the correlation for fault 6 and 11 are increased to 0.2 and 0.267 respectively when the threshold is pushed to 0.86. But 0.2 and 0.267 still have a reasonably low correlation level to ensure the success of the classifier fusion.

9.4 Results of the rmNN and LMD Classifier Fusion

The fusion methods are less important than the diversity of the classifier team, but still need to consider which method is more suitable for specific problem solving.

Table 38. 2-Classifier correlation estimation (Threshold for true case is 0.8)

Threshold 0.8						
	Total true cases	TT	TF	FT	FF	correlation
fault 3	189	189	0	0	0	0
fault 6	79	77	1	1	0	0
fault 7	32	31	0	1	0	0
fault 8	38	37	0	1	0	0
fault 9	38	38	0	0	0	0
fault 10	22	19	0	3	0	0
fault 11	31	30	0	1	0	0

Table 39. 2-Classifier correlation estimation (Threshold for true case is 0.85)

Threshold 0.85						
	Total true cases	TT	TF	FT	FF	correlation
fault 3	189	188	0	1	0	0
fault 6	79	70	6	3	0	0
fault 7	32	29	1	2	0	0
fault 8	38	33	3	2	0	0
fault 9	38	37	1	0	0	0
fault 10	22	18	0	4	0	0
fault 11	31	20	0	11	0	0

Table 40. 2-Classifier correlation estimation (Threshold for true case is 0.86)

Threshold 0.86						
	Total true cases	TT	TF	FT	FF	correlation
fault 3	189	188	0	1	0	0
fault 6	79	70	5	3	1	0.2
fault 7	32	27	3	2	0	0
fault 8	38	27	3	8	0	0
fault 9	38	37	1	0	0	0
fault 10	22	18	0	4	0	0
fault 11	31	18	0	11	2	0.266667

The classifier fusion function for two classifiers can be minimum, maximum, average, median and oracle. The majority vote usually applies when having more than two classifiers. The minimum will not help in this PSFC system. The maximum and oracle emphasize the possible true points and it is easy to generate a false alarm. The average and median methods are relatively soft and safe and their performances are approximately the same. The average method is tried in this PSFC system and tested out to gain excellent classification performance.

Once again, take fault 7 as an example. The training output for fault 7 LMD and the rmNN classifier are shown in Figures 101.1 and 101.2 respectively. The output is the degree of fault 7. The average of the two training output is shown in Figure 101.3. The lowest point in the true cases from the LMD classifier is at file 350 with a value of 0.817. On the other hand, the lowest point in the true cases from the rmNN classifier is at file 345 with a value of 0.845. After averaging, the degree of fault 7 for file 350 is increased to 0.862, which is the lowest point after classifier fusion. In the training, the rmNN classifier helps to lift the lowest point and improve the PSFC performance.

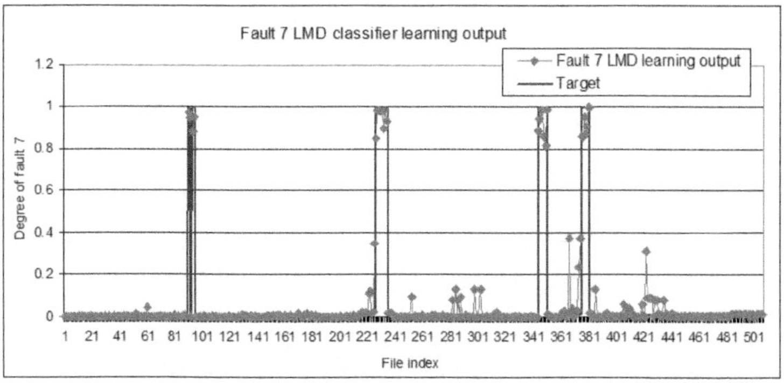

101.1: The learning output for fault 7 from the LMD classifier

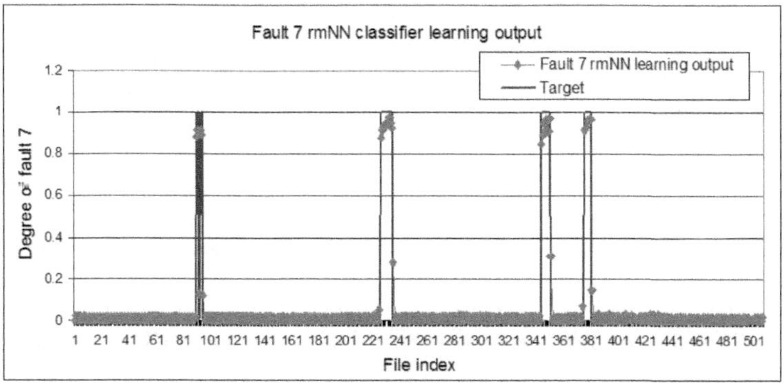

101.2: The learning output for fault 7 from the rmNN classifier

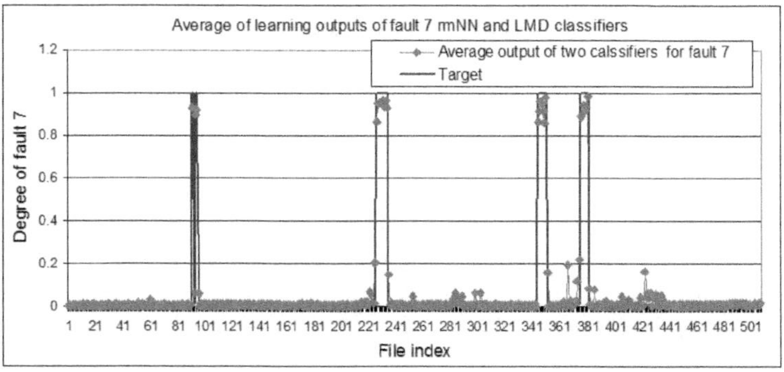

101.3: The average of two learning output for fault7 from the LMD and rmNN classifiers

Fig. 101. The learning output for fault 7 after the fusion of the LMD and rmNN classifiers

Now consider the verification results, which are illustrated in Figures 102.1, 102.2 and 102.3. The lowest point in the true cases from the LMD classifier is at file 91 with a value of 0.869. On the other hand, the lowest point in the true cases

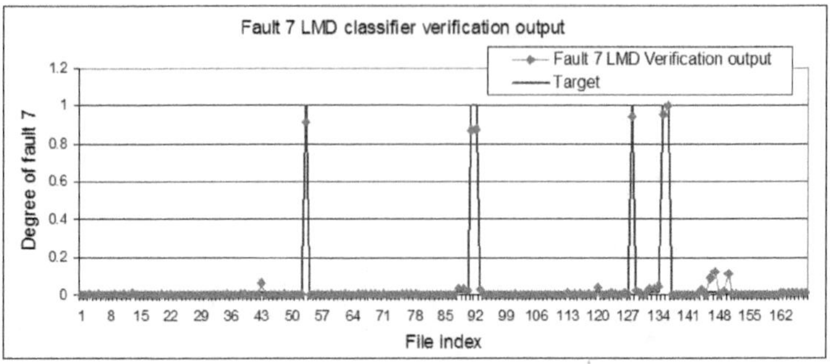

102.1: The verification output for fault 7 from the LMD classifier

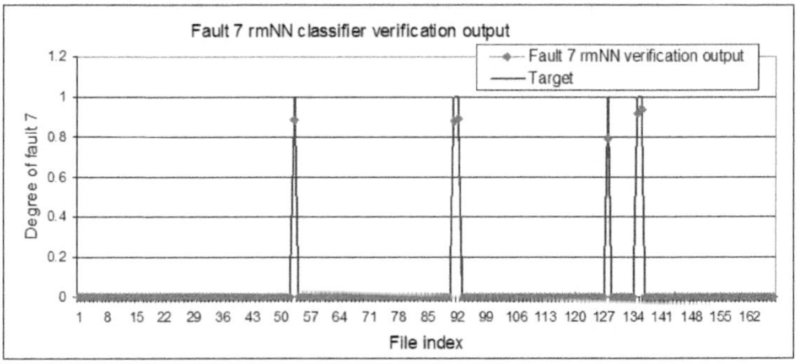

102.2: The verification output for fault 7 from the rmNN classifier

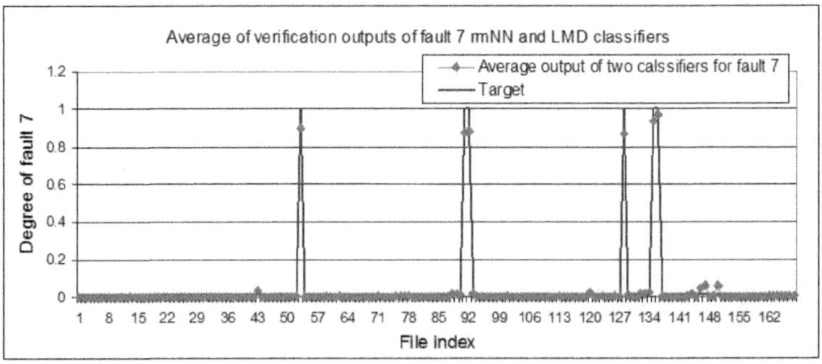

102.3: The average of two verification output for fault7 from the LMD and rmNN classifiers

Fig. 102. The verification output for fault 7 after the fusion of the LMD and rmNN classifiers

from the rmNN classifier is at file 128 with a value of 0.792. After averaging, the degree of fault 7 for file 128 is increased to 0.867, which is the lowest point after classifier fusion; and the overall performance of the PSFC is improved. In the testing process, the LMD classifier helps to lift the lowest point and improve the PSFC performance.

Table 41. Minimum learning output from the rmNN, LMD and fusioned classifier

Evaluation of classifier fusion performance (Training)						
	Minimum learning output of rmNN for true cases		Minimum learning output of LMD for true cases		Minimum learning output after classifier fusion for true cases	
	value	file index	value	file index	value	file index
Fault 3	0.871	92	0.899	401	0.901	92
Fault 6	**0.742**	383	**0.793**	362	**0.811**	383
Fault 7	0.845	345	0.817	350	0.862	350
Fault 8	**0.748**	235	0.835	384	**0.861**	382
Fault 9	0.87	414	0.84	387	0.896	387
Fault 10	**0.68**	471	0.984	471	**0.832**	471
Fault 11	**0.61**	93	0.854	103	**0.803**	93

Table 42. Minimum verification output from the rmNN, LMD and fusioned classifier

Evaluation of classifier fusion performance (Verification)						
	Minimum verification output of rmNN for true cases (true cases)		Minimum verification output of LMD for true cases		Minimum verification output after classifier fusion for true cases	
	value	file index	value	file index	value	file index
Fault 3	0.819	53	0.894	139	0.863	53
Fault 6	0.839	93	**0.786**	90	**0.833**	90
Fault 7	**0.792**	128	0.869	91	**0.867**	128
Fault 8	0.852	56	0.838	137	0.875	137
Fault 9	0.868	145	0.92	138	0.932	145
Fault 10	**0.656**	159	0.984	159	**0.82**	159
Fault 11	0.808	60	0.926	60	0.867	60

Table 43. The accuracy of the PSFC system

Fault type	# of files for verification	# of files incorrect	Accuracy
Minor AC Disturbance	94	0	1.00
AC Disturbance	44	0	1.00
Valve Current Closed/Blocked/Deblocked	44	0	1.00
Line Fault	22	0	1.00
Commutation Failure	25	0	1.00
Pole Voltages/Current Closed/Blocked/Deblocked	15	0	1.00
Current Arc Back	6	0	1.00
Parallel Operation	9	0	1.00
Pole Current Oscillation	7	0	1.00
Normal Affected by Another Pole	4	0	1.00
Asymmetric Protection	6	0	1.00
Disturbance on DC Voltage	7	0	1.00

The Learning and testing results for faults 3, 6, 7, 8, 9, 10 and 11 after the classifier fusion have been illustrated in an Appendix available at [15]. The overall improvement of the PSFC performance will be discussed via Tables 41 and 42.

Table 41 lists the minimum learning output from the rmNN, LMD and fusioned classifier for all the true cases in the training set. It is apparent that, for faults 6, 8, 10 and 11, the minimums from rmNNs are lower than 0.8. After classifier fusion the minimums are all above 0.8. On the other hand, for fault 6, the minimum from LMD is 0.793, lower than 0.8. After classifier fusion, the minimum output is raised to 0.811.

Table 42 lists the minimum verification output from the rmNN, LMD and fusioned classifier for all the true cases in the testing set. It is apparent that, for faults 7 and 10, the minimums from rmNNs are 0.792 and 0.656, both lower than 0.80. After classifier fusion, the minimums are all raised above 0.82. On the other hand, for fault 6, the minimum from LMD is 0.786, lower than 0.8. After classifier fusion, the minimum output is raised to 0.833.

The accuracy of the PSFC, which benefits from the fusion of the rmNN and LMD classifiers is listed in Table 43. The threshold for 'true' decision making is still 0.8.

It is obvious that the overall performance of the PSFC was improved via the fusion of the two classifiers, the rmNN and LMD. The two classifiers provide complementary information that gives the 2-classifier fusion method a chance to succeed. The accuracy of the PSFC is 100%, which provides confident information for fault decision making and enhances the quality of the power system protection functionality.

10 Conclusion

This paper introduces a rough set approach to power system fault classification. A form of rough neural computing based on the use of rough membership functions is introduced in the design of what is known as a rough membership function neural network (rmNN). A rough membership function makes it possible to measure the degree that any specified object with given feature values belongs to a given set X. The set X in this application is a set of fault files, which represent the same type of fault. Each rmNN has 3 layers: input, hidden, and output. The input layer contains what are known as rmf neurons, *i.e.*, neurons that compute the degree overlap between a specific class containing objects representing a fault type and a set of sample objects representing fault signals to be classified. The neurons in the hidden layer aggregate the output from the rmf neurons. The hidden layer neurons are designed using fuzzy set theory, which is ideally suited for numerical representation of aggregated rmf neuron output. The output neuron of an rmNN estimates the degree of a specific type of fault.

The most significant contribution of this research is a demonstration that the rough membership function successfully distinguishes objects with similar feature values. This makes rmNN a reasonable choice as a power system fault classifier.

A C++ and Labview based graphic user interface is implemented for the rmNN classifier, which makes the power system fault classification easy to operate.

To further improve the performance of the proposed approach to power system fault classification, a 2-Classifier fusion method has been introduced. This fusion method takes into account both the results of the rmNN classifier as well as a linear mean and standard deviation (LMD) based classifier. The correlation of the rmNN and LMD classifiers was estimated and has proved to be low enough to ensure that these two classifiers provide complementary information and are good candidates for classifier fusion. The 'average' method was selected as a fusion function.

Future work will include an extension of the Transcan^{TM} system used by Manitoba Hydro. In addition, it is possible to reduce the complexity of this classification system by searching for minimal subsets of attributes approximately preserving the decision information using rough set algorithms based on discernibility and Boolean reasoning. It is possible that the method of hierarchical learning with domain knowledge can be well adjusted. It is also possible to consider various forms of unsupervised, adaptive learning as a means of classifying power system faults.

Acknowledgements

We gratefully acknowledge the suggestions, comments, and insights concerning this monograph by Professors Gert Martens, Mirek Pawlak, David Gunderson, Zbigniew Ras and Andrzej Skowron. This research has been supported by Manitoba Hydro and Natural Sciences & Engineering Research Council (NSERC) grant 185986. The assistance provided by Mr. L. Crowe and the other engineers and staff at Manitoba Hydro Dorsey Station is also very much appreciated.

References

1. Alkoot, F., Kittler, J.: Experimental Evaluation of Expert Fusion Strategies. Pattern Recognition Letters 20(11), 1361–1369 (1999)
2. Alpagini, J., Peters, J.F., Skowron, A., Zhong, N. (eds.): RSCTC 2002. LNCS (LNAI), vol. 2475. Springer, Heidelberg (2002)
3. Bazan, J.G., Peters, J.F., Skowron, A.: Behavioral pattern identification through rough set modelling. In: Slezak, D., et al. (eds.). LNCS (LNAI), pp. 685–694. Springer, Berlin (2005)
4. Bazan, J., Osmolski, A., Skowron, A., Slezak, D., Szczuka, M., Wroblewski, J.: Rough set approach to the survival analysis. In: Alpigini, J.J., Peters, J.F., Skowron, A., Zhong, N. (eds.) RSCTC 2002. LNCS (LNAI), vol. 2475, pp. 522–529. Springer, Heidelberg (2002)
5. Bazan, J., Skowron, A., Slezak, D., Wroblewski, J.: Searching for the complex decision reducts: The case study of the survival analysis. In: Zhong, N., Raś, Z.W., Tsumoto, S., Suzuki, E. (eds.) ISMIS 2003. LNCS (LNAI), vol. 2871, pp. 160–168. Springer, Heidelberg (2003)
6. Van Breukelen, M., Duin, R., Tax, D., Den Hartog, J.: Combining Classifiers for the Recognition of Handwritten Digits. In: Proceedings of First IAPR TC1 Workshop Statistical Techniques in Pattern Recognition, pp. 13–18 (1997)
7. Brigham, E.O.: The Fast Fourier Transform and Its Applications. Prentice-Hall, Englewood Cliffs (1988)
8. Burrus, C.S., Gopinath, R.A., Guo, H.: Introduction to Wavelets and Wavelet Transforms. Prentice-Hall, New York (1998)
9. Cajori, F.: A History of Mathematical Notations, vol. 1-2. The Open Court Publishing Co., Illinois (1929)

10. CEIDS (Consortium for Electric Infrastructure to Support a Digital Society) The Cost of Power Disturbances to Industrial and Digital Economy Companies - Executive Summary (June 2001)
11. Chaari, O., Meunier, M., Brouaye, F.: Wavelets: A New tool for the Resonant Grounded Power Distribution system Relaying. IEEE Transactions on Power Delivery 11(3), 1301–1308 (1996)
12. Chanda, D., Kishore, N.K., Sinha, A.K.: Application of Wavelet Multiresolution Analysis for Classification of Faults on Transmission lines. In: Poster Session, IEEE TENCON 2003: Proceedings of the IEEE Region 10 Conference on Convergent Technologies for Asia-Pacific Region, Bangalore, India, vol. 4, pp. 1464–1469 (2003)
13. Cheng, H., Elangovan, S.: A B-Spline Wavelet Based Fault Classification Scheme for High Speed Protection Relaying. Electric Machines and Power Systems 28(4), 313–324 (2000)
14. Cohen, L.: Time-Frequency Analysis. Prentice-Hall, NJ, USA (1994)
15. Computational Intelligence Laboratory, Department of Electrical & Computer Engineering, University of Manitoba, `http://130.179.230.218/cilab/`
16. Drisen, J., Van Craenenbroeck, T., Reekmans, R., Van Dommelen, D.: Analyzing Time-Varying Power System Harmonics Using Wavelet Transform. In: Proceedings of IEEE Instrumentation and Measurement Technology Conference, Brussels, Belgium, vol. 1, pp. 474–479 (1996)
17. Dugan, R.C., McGranaghan, M.F., Beaty, H.W.: Electrical Power Systems Quality. McGraw-Hill, NY, USA (1996)
18. Eistenstein, G.: Note. Crelle's Journal XXVII, 91 (1844)
19. Equivalence class notation, `http://en.wikipedia.org/wiki/Equivalence_class`
20. Fernández, R., Rojas, H.N.: An Overview of Wavelet Transforms Application in Power Systems. In: 14th Power Systems Computation Conference, Sevilla, Spain, Session 1, Paper 6, pp. 1–8 (2002)
21. Gaouda, A.M., Salama, M.A., Sultan, M.R., Chikhani, A.Y.: Power Quality Detection and Classification Using Wavelet-Multiresolution Signal Decomposition. IEEE Transactions on Power Delivery 14(4), 1469–1476 (1999)
22. Goebel, K., Yan, W., Cheetham, W.: A method to calculate classifier correlation for decision fusion. In: Proceedings IDC 2002, Adelaide, pp. 135–140 (2002)
23. Gole, A.M.: Course 24.799 - HVDC Transmission Course Notes, Department of Electrical and Computer Engineering, University of Manitoba, Canada (1995)
24. Han, L., Peters, J.F., Ramanna, S., Suraj, Z.: Towards a power system fault classification system: A rough neurocomputing approach. In: Suraj, Z. (ed.) Sixth International Conference on Soft Computing and Distributed Processing, Rzeszow, Poland, pp. 89–92 (2002)
25. Han, L., Peters, J.F.: Heuristic method in searching for global minima in high-voltage AC error. In: Kinsner, W., Sebak, A., Ferens, K. (eds.) Canadian Conference on Electrical and Computer Engineering (CCECE 2002), Winnipeg, Manitoba, Canada, pp. 776–781 (2002)
26. Imriš, P., Lehtonen, M.: Transient Based Earth Fault Location In 110KV Subtransmission Networks. In: 15th Power Systems Computation Conference, Liége, Belgium, vol. 23, pp. 1–4 (2005)
27. Joanna, G., Wieslaw, L., Waldemar, T.: A comparison of the KNN method and neural networks in spirometric tests classification. Bio-algorithms and med-systems 1(1/2), 9–12 (2005)
28. Kaiser, G.: A Friendly Guide to Wavelets. Birkhäuser, Boston (1994)

29. Kashyap, K.H., Shenoy, U.J.: Classification of Power System Faults using Wavelet Transforms and Probabilistic Neural Networks. In: Proceedings of the 2003 International Symposium on Circuits and Systems, vol. 3, pp. 423–426 (2003)
30. Kittler, J., Hatef, M., Duin, R., Matas, J.: On Combining Classifiers. IEEE Transaction on Pattern Analysis and Machine Intelligence 20(3), 226–239 (1998)
31. Komorowski, J., Pawlak, Z., Polkowskis, L., Skowron, A.: Rough Fuzzy Hybridization - A New Trend in Decision Making. In: Pal, S.K., Skowron, A. (eds.). Springer Verlag Publisher, Singapore (1999)
32. Komorowski, J., Pawlak, Z., Polkowski, L., Skowron, A.: Rough sets: A tutorial. In: [31], pp. 3–98 (1999)
33. Konrad, E., Orłowska, E., Pawlak, Z.: Knowledge Representation Systems. Institute for Computer Science, Polish Academy of Sciences. Report 433 (1981)
34. Kuncheva, L.I.: A theoretical Study on six Classifier Fusion Strategies. IEEE Transactions on Pattern Analysis and Machine Intelligence 24(2), 281–286 (2002)
35. Kuncheva, L., Bezdek, J., Duin, R.: Decision Templates for Multiple Classifier Fusion: An Experimental Comparison. Pattern Recognition 34(2), 299–314 (2001)
36. Lee, C.H., Wang, Y.J., Huang, W.L.: A Literature Survey of Wavelets in Power Engineering Applications. Proceedings National Science Council. ROC(A) 24(4), 249–258 (2000)
37. Liang, J., Elangovan, S., Devotta, J.B.X.: A wavelet multiresolution analysis approach to fault detection and classification in transmission lines. International Journal of Electrical Power and Energy Systems 20(5), 327–332 (1998)
38. Martin, F., Aguado, J.A.: Wavelet-based ANN approach for transmission line protection. IEEE Transactions on Power Delivery 18(4), 1572–1574 (2003)
39. O'Farrell, M., Lewis, E., Flanagan, C., Lyons, W., Jackman, N.: Comparison of k-NN and neural network methods in the classification of spectral data from an optical fibre-based sensor system used for quality control in the food industry. In: The 18th European Conference on Solid-State Transducers. Sensors and Actuators B: Chemical, vol. 111-112, pp. 354–362 (2005)
40. Orłowska, E.: Semantics of Vague Concepts, Applications of Rough Sets, Institute for Computer Science, Polish Academy of Sciences. Report 469 (1982)
41. Orłowska, E., Peters, J.F., Rozenberg, G., Skowron, A.: New Frontiers in Scientific Discovery. In: Commemorating the Life and Work of Zdzisław Pawlak. IOS Press, Amsterdam (2007)
42. Orłowska, E. (ed.): Incomplete Information: Rough Set Analysis. Studies in Fuzziness and Soft Computing, vol. 13, pp. 1–22. Physica-Verlag, Heidelberg (1998)
43. Pal, S.K., Peters, J.F., Polkowski, L., Skowron, A.: Rough-neural computing: An introduction. In: [44], pp. 15–42 (2004)
44. Pal, S.K., Polkowski, L., Skowron, A. (eds.): Rough-Neural Computing. Techniques for Computing with Words. Springer, Berlin (2004)
45. Pal, S.K., Skowron, A. (eds.): Rough-Fuzzy Hybridization: A New Trend in Decision-Making. Springer, Singapore Pte. Ltd (1999)
46. Pal, S.K., Polkowski, L., Peters, J., Skowron, A.: 2003. In: Pal, S.K., Polkowski, L., Skowron, A. (eds.) Rough-Neuro Computing, pp. 16–43. Springer, Berlin (2003)
47. Papoulis, A.: The Fourier Integral and Its Applications, pp. 241–242. McGraw-Hill, NY (1962)
48. Pawlak, Z., Skowron, A.: Rudiments of rough sets. Information Sciences 177, 3–27 (2007)
49. Pawlak, Z., Skowron, A.: Rough sets: Some extensions. Information Sciences 177, 28–40 (2007)

50. Pawlak, Z., Skowron, A.: Rough sets and Boolean reasoning. Information Sciences 177, 41–73 (2007)
51. Pawlak, Z.: Classification of Objects by Means of Attributes. Institute for Computer Science, Polish Academy of Sciences. Report 429 (1981)
52. Pawlak, Z.: Rough Sets, Institute for Computer Science, Polish Academy of Sciences. Report 431 (1981)
53. Pawlak, Z.: Rough sets. International J. Comp. Inform. Science (11), 341–356 (1982)
54. Pawlak, Z., Skowron, A.: Rough membership functions. In: Yager, R., et al. (eds.) Advances in Dempster Shafer Theory of Evidence, pp. 251–271. Wiley, N.Y., U.S.A (1994)
55. Perunicic, B., Mallini, M., Wang, Z., Liu, Y., Heydt, G.T.: Power Quality Disturbance Detection and Classification Using Wavelets and Artificial Neural Networks. In: The 8th International Conference on Harmonics and Quality of Power, pp. 77–82 (1998)
56. Peters, J.F., Han, L., Ramanna, S.: Rough neural computing in signal analysis. Computational Intelligence 1(3), 493–513 (2001)
57. Peters, J.F., Skowron, A.: Transactions on Rough Sets, I-VII (2004-2007)
58. Peters, J.F.: Classification of Perceptual Objects by Means of Features. International Journal of Information Technology and Intelligent Computing (in press, 2007)
59. Peters, J.F.: Near sets. General theory about nearness of objects, Applied Mathematical Sciences 1(53), 2029–2609 (2007)
60. Peters, J.F.: Near sets. Special theory about nearness of objects. Fundamenta Informaticae 75(1-4), 407–433 (2007)
61. Peters, J.F., Skowron, A., Stepaniuk, J.: Nearness of Objects: Extension of Approximation Space Model. Fundamenta Informaticae 79, 1–16 (2007)
62. Peters, J.F., Henry, C., Gunderson, D.S.: Biologically-inspired approximate adaptive learning control strategies: A rough set approach. International Journal of Hybrid Intelligent Systems 4(4), 203–216 (2007)
63. Peters, J.F., Skowron, A., Han, L., Ramanna, S.: Towards rough neural computing based on rough membership functions: Theory and Application. In: Ziarko, W., Yao, Y. (eds.) RSCTC 2000. LNCS (LNAI), vol. 2005, pp. 611–618. Springer, Heidelberg (2001)
64. Peters, J.F., Ramanna, S., Suraj, Borkowski, Z., Rough, M.: neurons: Petri net models and applications. In: Pal, S.K., Polkowski, L., Skowron, A. (eds.) Rough-Neuro Computing, pp. 471–489. Springer, Berlin (2003)
65. Peters, J.F., Skowron, A., Synak, P., Ramanna, S.: Rough Sets and Information Granulation. In: De Baets, B., Kaynak, O., Bilgiç, T. (eds.) IFSA 2003. LNCS, vol. 2715, pp. 370–377. Springer, Heidelberg (2003)
66. Peters, J.F., Skowron, A., Suraj, Z., Han, S.: Design of rough neurons: Rough set foundation and Petri net model. In: Ohsuga, S., Raś, Z.W. (eds.) ISMIS 2000. LNCS (LNAI), vol. 1932, pp. 283–291. Springer, Berlin (2000)
67. Petrakos, M., Kannelopoulos, I., Benediktsson, J., Pesaresi, M.: The effect of correlation on the accuracy of the combined classifier in decision level fusion. In: Proceedings IEEE 2000 International. Geo-science and Remote Sensing Symposium, vol. 6, pp. 2623–2625 (2000)
68. Polkowski, L.: Rough Sets Mathematical Foundation. Physica-Verlag, Berlin (2002)

69. Ribeiro, P.F.: Wavelet transform: An advanced tool for analysing non-stationary harmonic distortion in power system. In: Proceedings of IEEE ICHPS, VI. Bologna, Italy, pp. 925–293 (1994)
70. Robertson, D.C., Camps, O., Mayer, J.: Wavelets and power system transients. In: SPIE international symposium on optical engineering in aerospace sensing, pp. 474–487 (1994)
71. Robertson, D.C., Camps, O.I., Mayer, J.S., Gish, W.: Wavelets and electromagnetic power system transients. IEEE Transactions on Power Delivery 11(2), 1050–1058 (1996)
72. Ramaswamy, S., Kiran, B.V., Kashyap, K.H., Shenoy, U.J.: Classification of Power System Transients using Wavelet Transforms and Probabilistic Neural Networks. In: IEEE Tencon 2003: Proceedings of the IEEE Region 10 Conference on Convergent Technologies for the Asia-Pacific, Taj Residency, Bangalore, India, vol. 4, pp. 1272–1276 (2003)
73. Rough Set Database System (RSDS), `http://rsds.wsiz.rzeszow.pl/rsds.php`
74. Tax, D., Duin, R., Van Breukelen, M.: Comparison between Product and Mean Classifier Combination Rules. In: Proceedings Workshop Statistical Pattern Recognition, pp. 165–170 (1997)
75. Tumer, K., Ghosh, J.: Linear and Order Statistics Combiners for Pattern Classification. In: Sharkey, A. (ed.) Combining Artificial Neural Nets, pp. 127–161 (1999)
76. Santoso, S., Powers, E., Hofmann, P.: Power quality assessment via wavelet transform analysis. IEEE Transactions on Power Delivery 11(2), 924–930 (1996)
77. Santoso, S., Powers, E.J., Grady, W.M., Parsons, A.C.: Power Quality Disturbance Waveform Recognition Using Wavelet-Based Neural Classifier. I. Theoretical Foundation. IEEE Transactions on Power Delivery 15(1), 222–228 (2000)
78. Skowron, A., Peters, J.F.: Rough-Granular Computing. In: Pedrycz, W., Skowron, A., Kreinovich, V. (eds.) Handbook on Granular Computing, Wiley, NY (in press, 2007)
79. Skowron, A, Stepaniuk, J., Peters, J.F.: Rough neurocomputing based on hierarchical classifiers. In: Alpigini, J.J., et al. (eds.) [2], pp. 308-315
80. Transcan^{TM} Recording System - Operation and Software. Mehta Tech. Inc. (1994)
81. Valens, C.: A Really Friendly Guide to Wavelets ©1999-2004 `http://perso.orange.fr/polyvalens/clemens/wavelets/wavelets.html`
82. Wang, M., Mamishev, A.V.: Classification of Power Quality Events Using Optimal Time-Frequency Representations – Part 1: Theory. IEEE Transactions on Power Delivery 19(3), 1488–1495 (2004)
83. Wang, W., Rowe, G.I., Mamishev, A.V.: Classification of Power Quality Events Using Optimal Time-Frequency Representations – Part 2: Application. IEEE Transactions on Power Delivery 19(3), 1496–1503 (2004)
84. Woodford, D.A.: HVDC Transmission. Technical Report, Manitoba HVDC Research Centre (1998)
85. Zhao, W., Song, Y.H., Min, Y.: Wavelet analysis based scheme for fault detection and classification in underground power cable systems. Electric Power Systems Research (53), 23–30 (2000)
86. Zwillinger, D. (ed.): CRC Standard Mathematical Tables and Formula, p. 223. CRC Press, Boca Raton (1995)

A Correlation Theory

Correlation is the degree to which two or more quantities are linearly associated. The cross-correlation of two complex functions $f(t)$ and $g(t)$ of a real variable t, denoted $f \star g$, is defined by (71) [47]

$$f \star g = \bar{f}(-t) * g(t), \tag{71}$$

where $*$ denotes convolution and $\bar{f}$ is the complex conjugate of $f(t)$. Since convolution is defined as (72)

$$f(t) * g(t) = \int_{-\infty}^{\infty} f(\tau)g(t-\tau)d\tau, \tag{72}$$

it follows that

$$f(t) \star g(t) = \int_{-\infty}^{\infty} \bar{f}(-\tau)g(t-\tau)d\tau. \tag{73}$$

Let $\tau' \equiv -\tau$, $d\tau' = -d\tau$, then (73) is equivalent to

$$\begin{aligned} f \star g &= \int_{\infty}^{-\infty} \bar{f}(\tau')g(t+\tau')(-d\tau') \\ &= \int_{-\infty}^{\infty} \bar{f}(\tau)g(t+\tau)d\tau. \end{aligned} \tag{74}$$

Similarly, for a complex function $f(t)$, the autocorrelation $\rho_f(t)$ is defined by (75) [47]

$$\begin{aligned} \rho_f(t) &\equiv f \star f \\ &= \bar{f}(-t) * f(t) \\ &= \int_{-\infty}^{\infty} \bar{f}(\tau)f(t+\tau)d\tau, \end{aligned} \tag{75}$$

Let series $\{a_i, i = 0, 1, ..., N-1\}$ be a periodic sequence, then the autocorrelation of the sequences, sometimes called the periodic autocorrelation, is written as (76) [86]

$$\rho_i = \sum_{j=0}^{N-1} a_j a_{j+i}, \tag{76}$$

where the final subscript is understood to be taken modulo N. The cross-correlation and autocorrelation discard phase information, returning only the power, and are therefore irreversible operations.

The most important property of correlation is that $f \star f$ is maximum at the origin ($x = 0$), in other words,

$$\int_{-\infty}^{\infty} f(u)f(u+x)du \leq \int_{-\infty}^{\infty} f^2(u)du. \tag{77}$$

It is efficient to classify the waveforms of fault signals for differentiating one fault from others by applying the cross-correlation and autocorrelation operations.

B Conventional Fast Fourier Transform (FFT)

Fourier methods such as the Fourier series and Fourier integral are used in analyzing continuous time signals. That is, Fourier methods are applicable in systems where there is a characteristic signal $s(t)$ defined for all values of t in the interval [-∞, ∞].

A Fourier transform decomposes a waveform into a sum of sinusoids of different frequencies [7]. The signal $s(t)$ in the time domain is decomposed into the sum of its sinusoids $S(f)$ in the frequency domain by,

$$S(f) = \int_{-\infty}^{\infty} s(t)e^{-j2\pi ft}dt, \tag{78}$$

where $j = \sqrt{-1}$.

In this paper, the focus is on the application of what is known as the Discrete Fourier Transform (DFT) that is applicable to discrete-time signals. A discrete time signal $s[n]$ is defined for values of n in the interval [-∞, ∞]. A discrete Fourier transform is used in studying finite collections of sampled data $\{s_0, ..., s_{N-1}\}$ relative to the sequence $\{S_0, ..., S_{N-1}\}$. The DFT is given by,

$$S_k = \sum_{n=0}^{N-1} s_n e^{-j\frac{2\pi}{N}nk}, k = 0, 1, ..., N-1. \tag{79}$$

A fast Fourier transform results from the application of a particular algorithm that can compute the DFT more rapidly than other available algorithms [7].

C Wavelet Transform

The big disadvantage of a Fourier expansion is that it has only frequency resolution and no time resolution. This means that although we might be able to determine all the frequencies present in a signal, we do not know when they are present [81]. The wavelet transform provides a means of overcoming the shortcomings of the Fourier transform. In wavelet analysis, the use of a fully scalable modulated window makes it possible to know the exact frequency and the exact time of occurrence of this frequency in a signal. In other words, a signal can simply be represented as a point in the time-frequency space. The window is shifted along the signal and for every position the spectrum is calculated. Then this process is repeated many times with a slightly shorter (or longer) window for every new cycle. In the end, the result will be a collection of time-frequency representations of a signal, all with different resolutions.

Wavelets provide a form of multiresolution analysis resulting from the collection of representations produced by applying a set of functions of different scales to a signal. Large scales are used to paint the big picture, while small scales expose the details. Thus, going from large scale to small scale is analogous to zooming in.

The Continuous Wavelet Transform (CWT) in general is formally defined by (80) [28]:

$$\gamma(s, \tau) = \int f(t)\Psi^*_{s,\tau}(t)dt, \tag{80}$$

where $*$ denotes complex conjugation. Equation (80) shows how a function $f(t)$ is decomposed into a set of basis functions called wavelets. The variables s and τ, scale and translation, are the new dimensions after the wavelet transform. The inverse wavelet transform can be written as shown in (81) [28]:

$$f(t) = \int\int \gamma(s,\tau)\Psi_{s,\tau}(t)dsd\tau. \tag{81}$$

The wavelets $\Psi_{s,\tau}(t)$, sometimes called child wavelets, are generated from a single basic wavelet $\Psi(t)$, the so-called mother wavelet, by scaling (parameter s) and translation (parameter τ) [28]. For a wavelet $\Psi_{s,\tau}(t)$, a family of curves with parameters s and τ can be formed as:

$$\Psi_{s,\tau}(t) = \frac{1}{\sqrt{s}}\Psi(\frac{t-\tau}{s}), \tag{82}$$

where s is the scale factor, τ is the translation factor and $\frac{1}{\sqrt{s}}$ is the factor for energy normalization across the different scales.

Unlike the Fourier transform or other transforms, the wavelet basis function, $\Psi(t)$ is not specified. The theory of wavelet transforms deals with the general properties of the wavelets and wavelet transforms only. It defines a framework for designing wavelets that satisfy different applications.

When discrete wavelets are used to transform a continuous signal, functions of the form shown in (83) are selected [8].

$$\Psi_{j,k}(t) = \frac{1}{\sqrt{s_0^j}}\Psi(\frac{t-k\tau_0 s_0^j}{s_0^j}), \tag{83}$$

which is normally a piecewise continuous function, where j and k are integers and $s_0 > 1$ is a fixed dilation step. The translation factor τ_0 depends on the dilation step. The effect of discretizing the wavelet is that the time-scale space is now sampled at discrete intervals. We usually choose $s_0 = 2$ so that the sampling of the frequency axis corresponds to dyadic sampling as shown in Figure 103. This is a very natural choice for computers, the human ear and music for instance. For the translation factor, it is usual to choose $\tau_0 = 1$ so that there is also a dyadic sampling of the time axis.

Practical applications require Discrete Wavelet Transforms (DWT). The discrete wavelets can be made orthogonal to their own dilations and translations by special choices of the mother wavelet. There is a large class of wavelet functions for which the set of child wavelets is an orthogonal basis. The simplest of these is the Haar wavelet. An arbitrary signal can be reconstructed by summing the orthogonal wavelet basis functions weighted by wavelet transform coefficients.

The DWT and Inverse DWT (IDWT) of a signal $f(t)$ are defined in (84) and (85), respectively.

$$\gamma_{j,k} = \int_{-\infty}^{\infty} f(t)\Psi_{j,k}(t)dt. \tag{84}$$

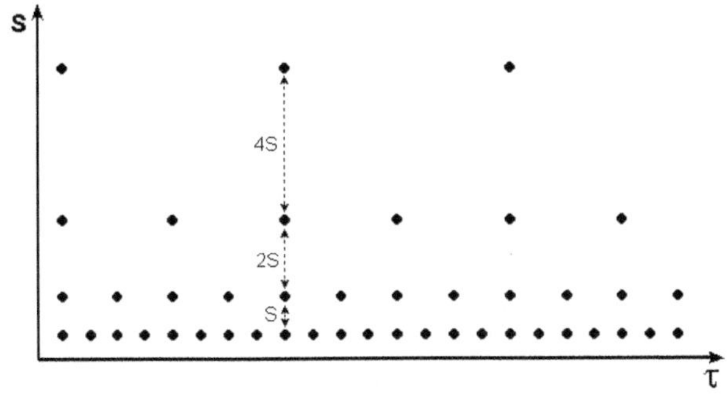

Fig. 103. Localization of discrete wavelets in the time-scale space on a dyadic grid [81]

$$f(t) = \sum_j \sum_k \gamma_{j,k} \Psi_{j,k}(t), \tag{85}$$

Such wavelets give rise to a Wavelet Multiresolution Analysis (MRA) derived as follows.

Define W_j to be a set of all signals $f(t)$ which can be synthesized from the child wavelets $\Psi_{j,k}(t)$, $-\infty < k < \infty$. These spaces are orthogonal to each other and we can synthesize any signal $f(t)$ using (86)

$$f(t) = \sum_{j=-\infty}^{\infty} f_j(t),$$

$$f_j(t) = \sum_{k=-\infty}^{\infty} \gamma_{j,k} \Psi_{j,k}(t), \tag{86}$$

where $f_j(t)$ is in the space W_j.

There is another way to express this idea. Define V_j to be the set of all signals, $f(t)$, which can be synthesized from the child wavelets $\Psi_{i,k}(t)$ where $i < j$ and $-\infty < k < \infty$ as in (87)

$$f(t) = \sum_{i=-\infty}^{j-1} \sum_k \gamma_{i,k} \Psi_{i,k}(t). \tag{87}$$

The spaces V_j are nested inside each other, as follows:

$$\{0\} \subset ... \subset V_{-2} \subset V_{-1} \subset V_0 \subset V_1 \subset V_2 \subset ... \subset L^2. \tag{88}$$

As j goes to ∞, V_j enlarges to become all energy signals (L^2). As j goes to $-\infty$, V_j shrinks down to only the zero signal. It is clear from the definitions that every signal in V_{j+1} is the sum of a signal in V_j and W_j because

$$f(t) = \sum_{i=-\infty}^{j} \sum_k \gamma_{i,k} \Psi_{i,k}(t) = \sum_{i=-\infty}^{j-1} \sum_k \gamma_{i,k} \Psi_{i,k}(t) + \sum_k \gamma_{j,k} \Psi_{j,k}(t). \tag{89}$$

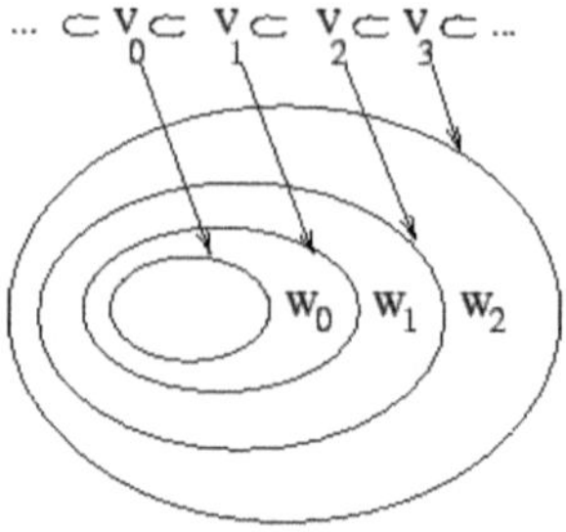

Fig. 104. MRA: nested subspaces

Hence, it can be written:

$$V_{j+1} = V_j + W_j. \tag{90}$$

This shows that the spaces W_j are the differences (in the subspace sense) between adjacent spaces V_{j+1} and V_j. The spaces V_j and W_j can be visualized as shown in Figure 104.

The term *Wavelet Multiresolution Analysis* (MRA) refers to the analysis of signals in relation to a nested sequence of subspaces like the one shown in Figure 104. For example, to decompose a signal, $f(t)$, in space V_0 a few times, use the following decomposition:

$$\begin{aligned} V_0 &= V_{-1} + W_{-1} \\ &= V_{-2} + W_{-2} + W_{-1} \\ &= V_{-3} + W_{-3} + W_{-2} + W_{-1} \\ &= V_{-4} + W_{-4} + W_{-3} + W_{-2} + W_{-1}. \end{aligned} \tag{91}$$

This leads to various decompositions:

$$\begin{aligned} f(t) &= A_1(t) + D_1(t) \\ &= A_2(t) + D_2(t) + D_1(t) \\ &= A_3(t) + D_3(t) + D_2(t) + D_1(t) \\ &= A_4(t) + D_4(t) + D_3(t) + D_2(t) + D_1(t), \end{aligned} \tag{92}$$

where $D_i(t)$, in W_{-i}, is called the detail at level i and $A_i(t)$, in V_{-i}, is called the approximation at level i.

Figure 105 gives an example of how the decomposition can be carried out in Matlab^{TM} using the `wavemenu` interface. There are a number of sample signals, which can be used for a demonstration analysis. The signal `sumsin` is the sum of two sine waves, and is decomposed four times in this example.

Notice that different aspects of the signal appear at different levels of the details and approximations in Figure 105.

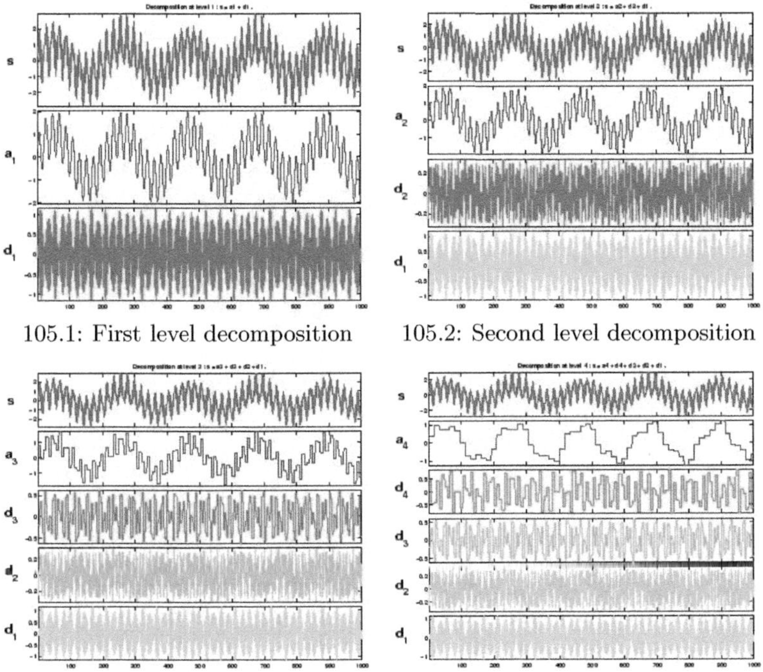

105.1: First level decomposition

105.2: Second level decomposition

105.3: Third level decomposition

105.4: Forth level decomposition

Fig. 105. An example of Wavelet Multiresolution Analysis (MRA) decomposition

The space V_j has a very important property related to time compression by factors of 2. The MRA Two Scale Property asserts that a signal $f(t)$ is in the space V_j if and only if, $f(2t)$ is in the next space V_{j+1}. Therefore, investigation of the multiresolution analysis leads to a scaling function, a pair of discrete time filters, and a perfect reconstruction filter bank, which can be used to calculate the DWT quickly. In other words, a wavelet has a band-pass like spectrum. Given that compression in time is equivalent to stretching the spectrum and shifting it upwards, a time compression of the wavelet by a factor of 2 will stretch the frequency spectrum of the wavelet by a factor of 2 and also shift all frequency components up by a factor of 2. Using this insight, the finite spectrum of a signal can then be covered with the spectra of dilated wavelets in the same way that the signal is covered in the time domain with translated wavelets. Alternatively, if one wavelet can be seen as a band-pass filter, then a series of dilated wavelets can be seen as a band-pass filter bank.

The filter bank can be built in several ways. One way is to build many band-pass filters to split the spectrum into frequency bands. Another way is to split the signal spectrum into two (equal) parts, a lowpass and a highpass part. The low-pass part can be split into a lowpass and a highpass part again. This splitting process continues until the details of a signal that has been exposed are satisfied. In this way, an iterated filter bank is created as shown in Figure 106.

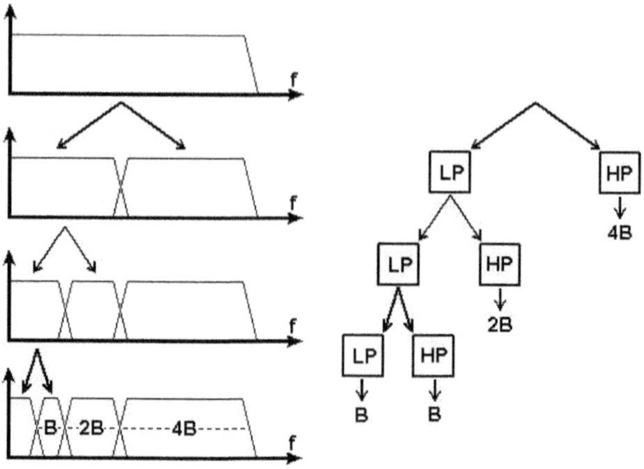

Fig. 106. Splitting the signal spectrum with an iterated filter bank [81]

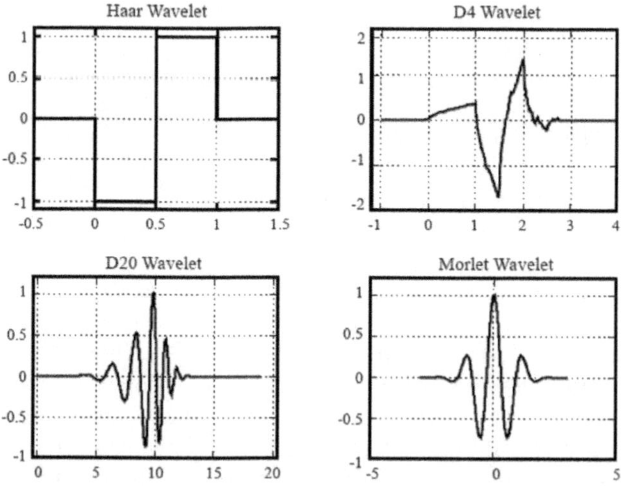

Fig. 107. Four mother wavelets often used in wavelet analysis [36]

Four mother wavelets often used in wavelet analysis are shown in Figure 107. The difference between these wavelets is mainly due to the different lengths of filters that define the wavelet and scaling functions [36].

The scaled (dilated) and translated (shifted) versions of the Daubechies mother wavelet are shown in Figure 108. Daubechies wavelets belong to a special class of mother wavelets and are actually used most often for detection, localization, identification and classification of power disturbances.

Transient signals in a power system are non-stationary, time-varying voltage and current signals. Wavelet transforms are feasible to provide efficient and localized analysis of non-stationary, fast transient fault signals for power systems.

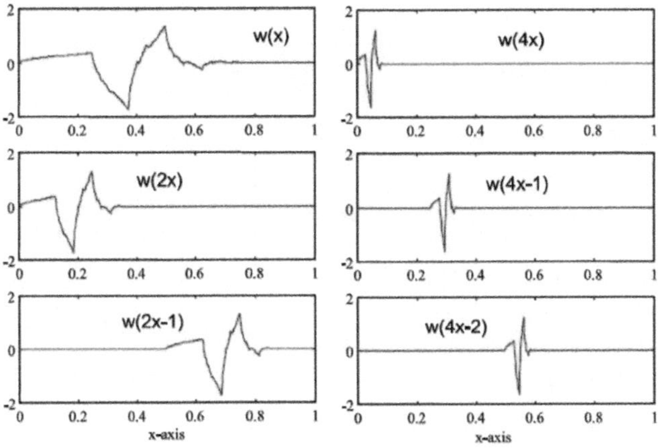

Fig. 108. Scaled and translated versions of the D4 wavelet [36]

More detailed discussion on the application of wavelets analysis for classification of fault signals for power systems will be addressed in Section 4.2.

D Time-Frequency Representation (TFR) Theory

In addition to applying wavelet theory to power system fault classification, the Time-Frequency Representation (TFR) algorithm is becoming attractive to scientists and engineers in the power industry. This section will introduce the basics of TFR theory, and Section 4.3 will present the TFR in classifying power system faults.

TFR $P(t, f)$ can be expressed as a two-dimensional Fourier transform of the product of the ambiguity plane $A(\eta, \tau)$ of the signal and a kernel function $\varphi(\eta, \tau)$ [14]:

$$P(t, f) = \int_{-\infty}^{\infty} \int_{-\infty}^{\infty} A(\eta, \tau)\varphi(\eta, \tau)e^{j2\pi\eta t}e^{-j2\pi f\tau}d\eta d\tau, \tag{93}$$

where t represents time, f represents frequency, η represents continuous frequency shift, and τ represents continuous time lag. The ambiguity plane $A(\eta, \tau)$ for a given signal $s(t)$ is defined as:

$$A(\eta, \tau) = \int_{-\infty}^{\infty} s(t)s^*(t+\tau)e^{j2\pi\eta t}dt, \tag{94}$$

where $s(t)$ represents the signal at time t, and $s(t+\tau)$ represents the signal at a future time $t+\tau$, and the $s^*(t+\tau)$ means the complex conjugate of $s(t+\tau)$.

The kernel $\varphi_i[\eta, \tau]$ is defined as a binary matrix (each matrix element is either 0 or 1). Feature points are ambiguity plane points of locations (η, τ) where $\varphi_i[\eta, \tau] = 1$.

Author Index

GPSR Compliance

The European Union's (EU) General Product Safety Regulation (GPSR) is a set of rules that requires consumer products to be safe and our obligations to ensure this.

If you have any concerns about our products, you can contact us on ProductSafety@springernature.com

In case Publisher is established outside the EU, the EU authorized representative is:

Springer Nature Customer Service Center GmbH
Europaplatz 3
69115 Heidelberg, Germany

Batch number: 10844619

Printed by Printforce, the Netherlands